APCTP-ASEAN Workshop on Advanced Materials Science and Nanotechnology 2008

(AMSN 08)

AA002669

Journal of Physics: Conference Series Volume 187

Nha Trang, Vietnam
15-21 September 2008

ISBN: 978-1-61738-289-5

Printed from e-media with permission by:

Curran Associates, Inc.
57 Morehouse Lane
Red Hook, NY 12571
www.proceedings.com

Some format issues inherent in the e-media version may also appear in this print version.

Copyright© (2008) by the Institute of Physics
All rights reserved.

Printed by Curran Associates, Inc. (2010)

For permission requests, please contact the Institute of Physics
at the address below.

Institute of Physics
Dirac House, Temple Back
Bristol BS1 6BE UK

Phone: 44 1 17 929 7481
Fax: 44 1 17 920 0979

techtracking@iop.org

TABLE OF CONTENTS

Dear friends

To contribute to the enhancement of the international scientific cooperation of the ASEAN countries and in reply to the proposal of the Vietnam Academy of Science and Technology (VAST), the Asia-Pacific Center for Theoretical Physics (APCTP) and the Sub Committee on Materials Science and Technology (SCMST) of the ASEAN Committee of Science and Technology (ASEAN COST) agreed to organize this APCTP-ASEAN Workshop on Advanced Materials Science and Nanotechnology with the participation of the Ministry of Science and Technology of Vietnam, the Vietnam Academy of Science and Technology, Rencontres du Vietnam, the Vietnam Physical Society, the Vietnam National University in Ho Chi Minh City and the Vietnam National University in Hanoi.

As well as the participants from 9 of the 10 ASEAN countries and many other countries/regions of APCTP (Australia, China, Chinese Taipei, Japan and Korea) we warmly welcome the guests from Europe, the United States, Canada and Israel.

Without the financial support of the Asia-Pacific Center for Theoretical Physics APCTP, Abdus Salam International Center for Theoretical Physics ICTP, the Asian Office of Aerospace Research and Development AOARD, the US Office of Naval Research Global-Asia ONRG, the Ministry of Science and Technology of Vietnam MOST, the Vietnam Academy of Science and Technology VAST, the Vietnam National University in Ho Chi Minh City VNU HCMC and other Sponsors, we would have been unable to hold this Workshop. On behalf of the International and Local Organizing Committees I would like to express our deep gratitude to the Sponsors.

We highly appreciate the support and advice of the members of the International Advisory Committee, the scientific contribution of the invited speakers and all participants. We acknowledge the warm reception of the Khanh Hoa province Administration and citizens, and the hard work of the VAST staff for the success of the Workshop.

We cordially wish all participants lively scientific discussions and enjoyable meetings at the Workshop and a pleasant stay in beautiful Nha Trang. We do hope that all foreign participants will take away good impressions of Vietnamese hospitality.

Nguyen Van Hieu
VAST and APCTP
Chairman of the Workshop

All papers published in this volume of *Journal of Physics: Conference Series* have been peer reviewed through processes administered by the proceedings Editors. Reviews were conducted by expert referees to the professional and scientific standards expected of a proceedings journal published by IOP Publishing.

Micro/nano electro mechanical systems for practical applications

Masayoshi Esashi

The World Premier International Research Center Advanced Institute for Materials Research, Tohoku University; 6-6-01 AzaAoba, Aramaki, Aobaku, Sendai, 980-8579, Japan

E-mail: esashi@cc.mech.tohoku.ac.jp

Abstract. Silicon MEMS as electrostatically levitated rotational gyroscope, 2D optical scanner and wafer level packaged devices as integrated capacitive pressure sensor and MEMS switch are described. MEMS which use non-silicon materials as diamond, PZT, conductive polymer, CNT (carbon nano tube), LTCC with electrical feedthrough, SiC (silicon carbide) and $LiNbO_3$ for multi-probe data storage, multi-column electron beam lithography system, probe card for wafer-level burn-in test, mould for glass press moulding and SAW wireless passive sensor respectively are also described.

Keywords: MEMS, gyroscope, packaging, switch, data storage.

1. Introduction

An extended IC fabrication technology based on deep etching, anodic bonding and other advanced process technologies is used to produce value added MEMS (Micro Electro Mechanical Systems) and NEMS (Nano Electro Mechanical Systems) which have multi-functions. Examples of application oriented MEMS developed by open collaboration with industry will be described below.

2. Silicon MEMS

2.1. Electrostatically levitated rotational gyroscope

Silicon rotational gyroscopes have been developed for the purpose of motion control and navigation [1]. The principle and the photograph are shown in figure 1. A 1.5 mm diameter silicon ring which is electrostatically levitated by digital control using capacitive position sensing and electrostatic actuation is rotated at 74,000 rpm. The rotation is based on the principle of a variable capacitance motor. A 5μm radial gap between the ring rotor and stator electrodes is formed using deep RIE (Reactive Ion Etching) of a silicon wafer. The silicon is anodically bonded on both sides to specimens which have electrodes. The chip is packaged in a vacuum cavity to prevent a viscous dumping. This inertia measurement system (Tokimec Inc. MESAG (Micro Electrostatically Suspended Accelerometer Gyro)) can measure two axes rotation and three axes acceleration simultaneously with high precision (sensitivity 0.01 deg/s and 0.2mG, respectively).

© 2009 IOP Publishing Ltd

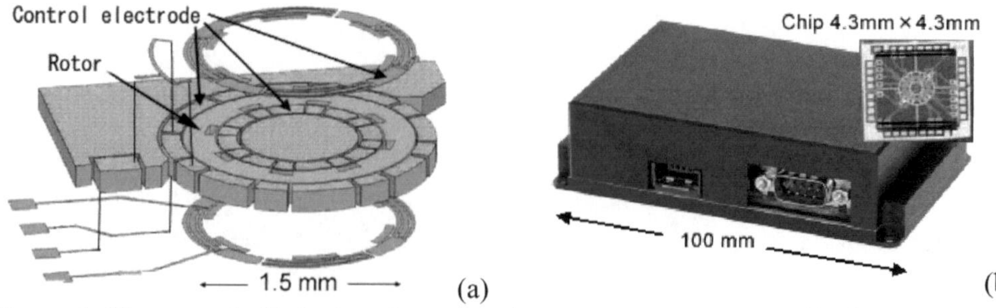

(a) (b)

Figure 1. Electrostatically levitated rotational gyroscope: a) structure, b) photographs of chip and inertia measurement system.

2.2. Capacitive silicon microphone

Capacitive silicon microphones ware developed in the NHK research center [2] and has been produced by Panasonic Co. Ltd. The structure and an example of its application in TV programmes are shown in figure 2. The microphone is composed of a silicon membrane and a back plate and the motion of the membrane by sound is capacitively detected. This silicon microphone can be used even in humid environments.

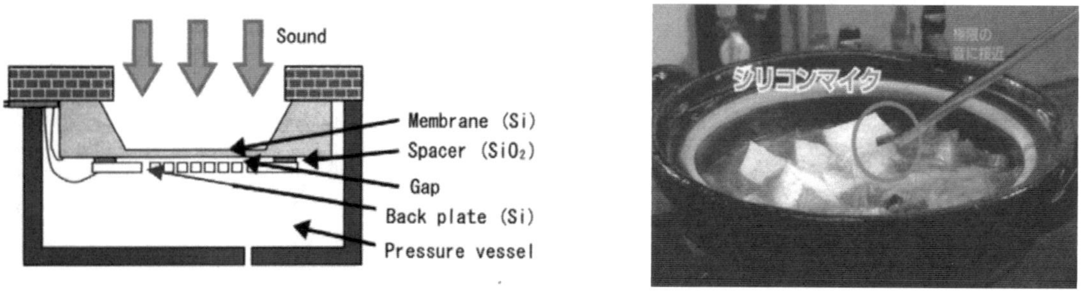

Figure 2. Capacitive silicon microphone and its application.

2.3. Two dimensional optical scanner

Electromagnetic 2D (two-dimensional) optical scanner as shown in figure 3a was developed [3]. Mirror is deflected electromagnetically using planer coils on a silicon gimbal structure and external permanent magnets. The scanner (Nippon Signal Co. Ltd., ECO Scan) has been applied to a 3D imaging system by measuring the distance to the object using time-of-flight of laser light as shown in figure 3b.

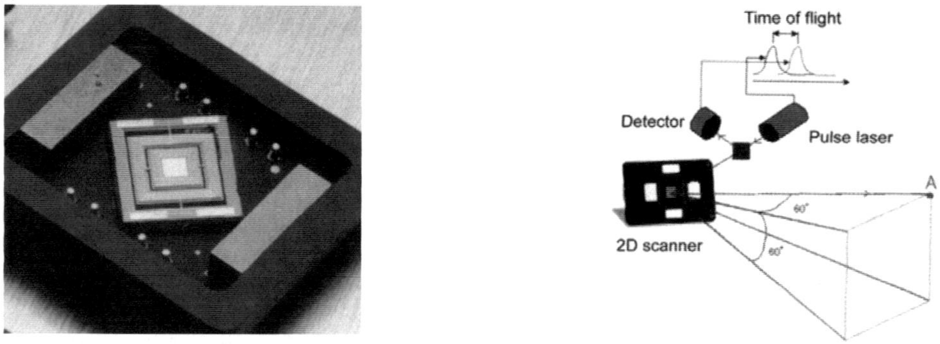

(a) (b)

Figure 3. 2D optical scanner and its application to a time-of-flight 3D imaging system: a) 2D optical scanner, b) Time-of-flight 3D imaging system.

2.4. Integrated capacitive pressure sensor

The fabrication process of the integrated capacitive pressure sensor [4] is shown in figure 4. The integrated capacitive pressure sensor has been produced in JTECT Inc. and used for low pressure measurement. The silicon is used not only for a capacitive diaphragm pressure sensor and a CMOS capacitance detection circuit, but also as a package. Packaged chip-size MEMS are fabricated by anodic bonding of a silicon wafer and a glass wafer and dicing the bonded wafer to chips. This process called wafer level packaging is effective for the MEMS because it makes batch assembly possible and hence assembly machines can be eliminated [5]. The method has advantages of reduced test cost, small chip-size packaging and high reliability. Electrical feedthrough from the glass hole plays an important role in the wafer level packaging.

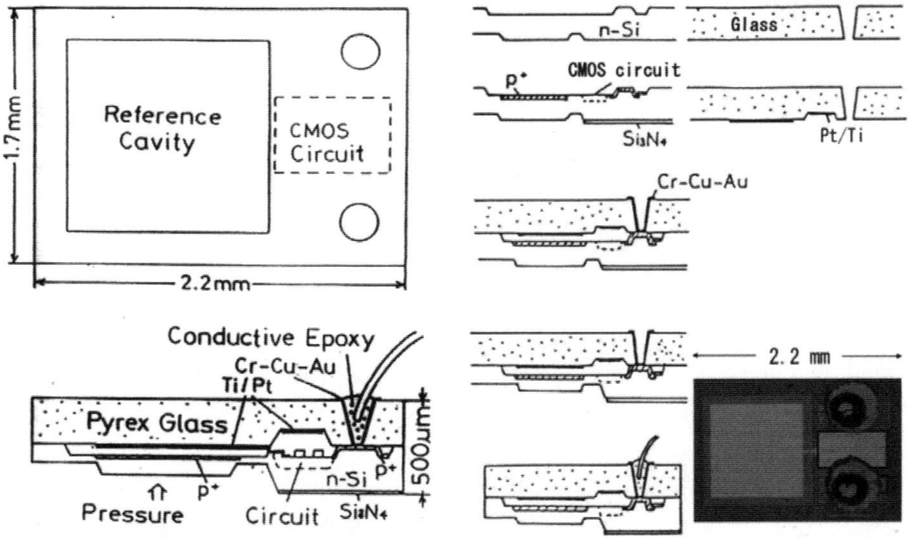

Figure 4. Structure and fabrication process of integrated capacitive pressure sensor by wafer level packaging.

2.5. Capacitive vacuum sensor

Vacuum cavity can be made by putting a non-volatile getter which absorbs oxygen gas generated during the anodic bonding [6]. Capacitive vacuum sensors which have thin diaphragms and reference vacuum cavity with the getter in it (figure 5) are wafer level packaged and have been produced by Canon Anelva Corp. [7].

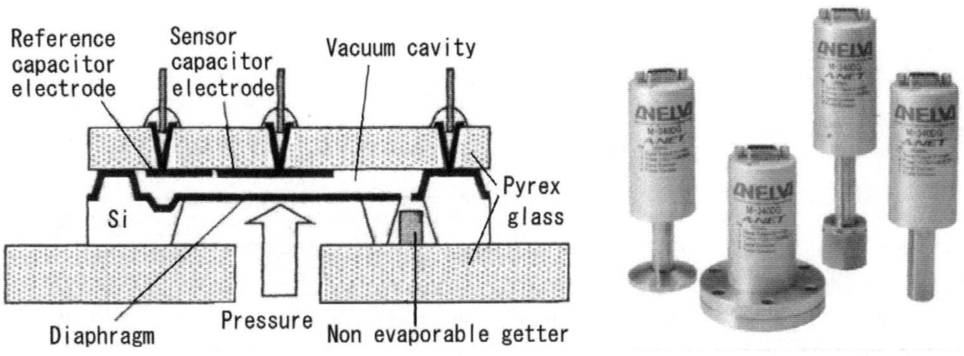

Figure 5. Capacitive vacuum sensor.

2.6. MEMS switch

MEMS switch which uses a thermal bimetal actuator for making electrical contact was developed [8] (figure 6). The electrical connections are made to the backside using the electrical feedthrough in a glass. The wafer level packaging results in not only low cost fabrication but also excellent reliability because the contact surface is kept contamination-free owing to the hermetic sealing. Operation more than 10^7 cycles and high frequency response up to 20 GHz were achieved. This switch has been produced and used for latest high speed LSI tester by Advantest Inc. [9].

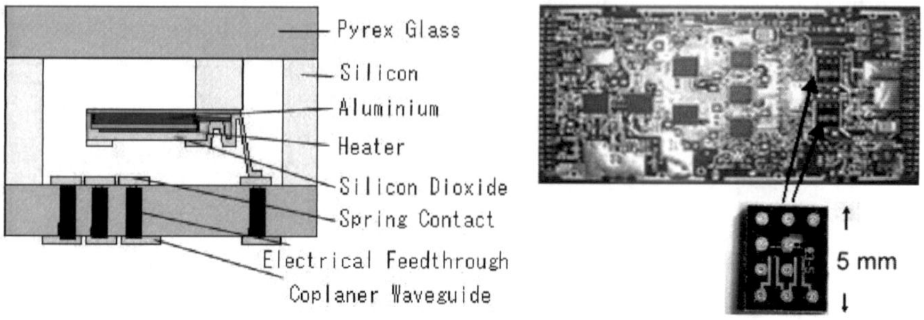

Figure 6. MEMS switch.

3. MEMS with non-silicon materials

3.1. Multiprobe data storage (PZT, diamond and polyaniline)

Parallel data recording and reading can be made using a multiprobe data storage system as shown in figure 7 (left) [10]. Since electrical interconnection from the probe array to the control IC chip is needed, high density electrical feedthrough in a glass (figure 7 (right)) was developed for the purpose [11].

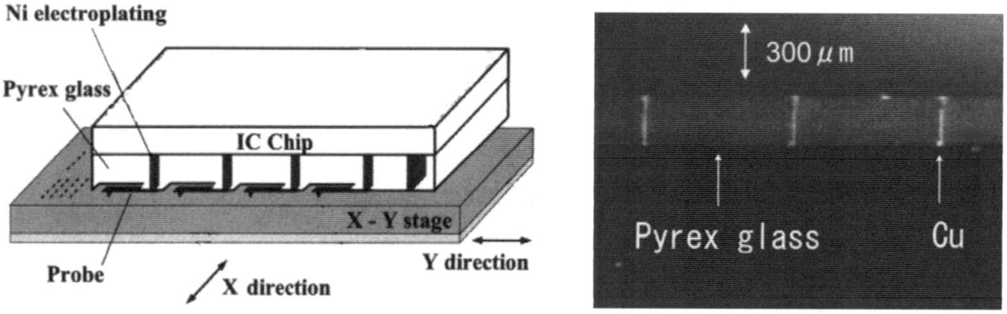

Figure 7. Concept of multiplobe data storage (left) and electrical feedthrough in Pyrex glass (right).

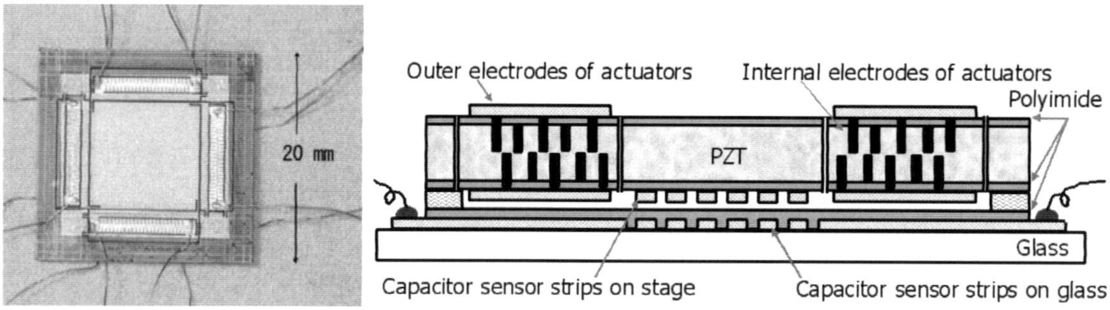

Figure 8. Monolithic PZT stage, photograph of monolithic PZT stage (left) and feedback control system (right).

The stage in figure 7 has to move in X-Y direction to cover the area determined by the probe pitch. A monolithic XYZ stage was fabricated on a PZT (lead zirconium titanate) ceramic plate (figure 8). Multi-layer piezoelectric actuators were fabricated on both sides of the PZT by dicing and metal filling. The stage motion could be controlled precisely with a feedback using capacitive position sensor as shown in figure 8 (right) [12].

Electrical recording to a ferroelectric recording media ($LiTaO_3$) using a diamond multiprobe system was developed. The fabrication process of the diamond probes, hot filament diamond CVD and diamond probes with an example of ferroelectric recording are shown in figure 9 [13].

Electrical recording using the microprobe array on a conductive polymer (doped polyaniline) film has been also studied for the multiprobe data storage as shown in figure 10 [14]. A conductive polymer deposited on a conductive substrate was used as a patterned media with 30nm pitch which corresponds to 0.7 Tbit/inch^2. A +3V substrate voltage changes the recording media from conductive to nonconductive, on the other hand a -3V changes it vice versa reversibly. The conductance of the recording media was decreased more than ten times. The conductance modification is made in the atmosphere and the mechanism of the rewritable recording is supposed that the polymer film becomes conductive and nonconductive by electrochemical reduction and oxidation, respectively, as shown in figure 11 [15].

The fabrication process of the patterned media is shown in figure 12 [14]. The microphase-separation of polystyrene (PS)-block-polymethylmethacrylate (PMMA) was used to make a perpendicular-oriented nano-structure. Nano-holes made by removing the PMMA were filled by polyaniline.

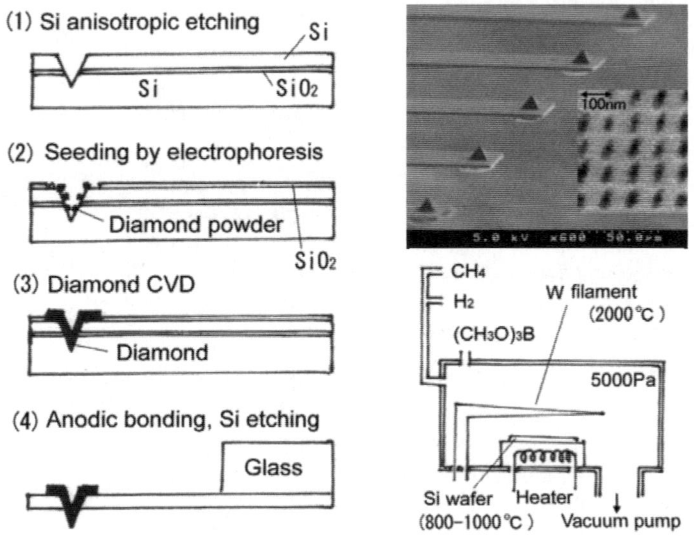

Figure 9. Fabrication process of diamond probe (left), hot filament diamond CVD system (right top) and diamond probes with an example of ferroelectric recording (right bottom).

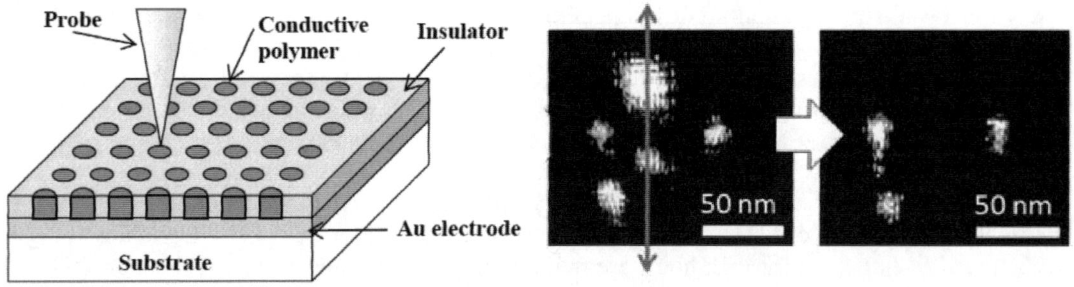

Figure 10. The principle of electrical recording on a conductive polymer (left) and an example of recording and erasing (right).

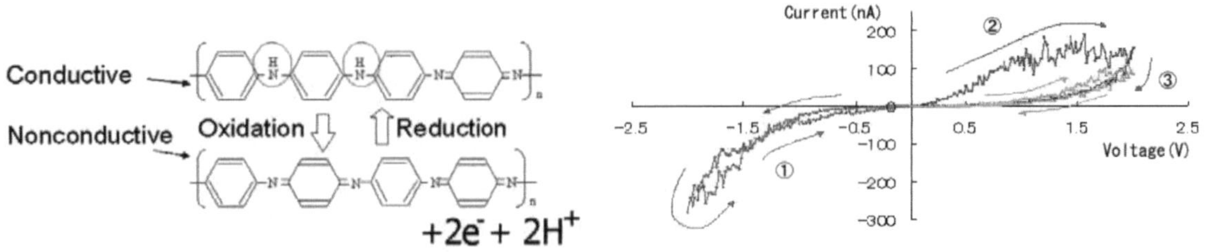

Figure 11. The mechanism of the rewritable recording by conductance modification.

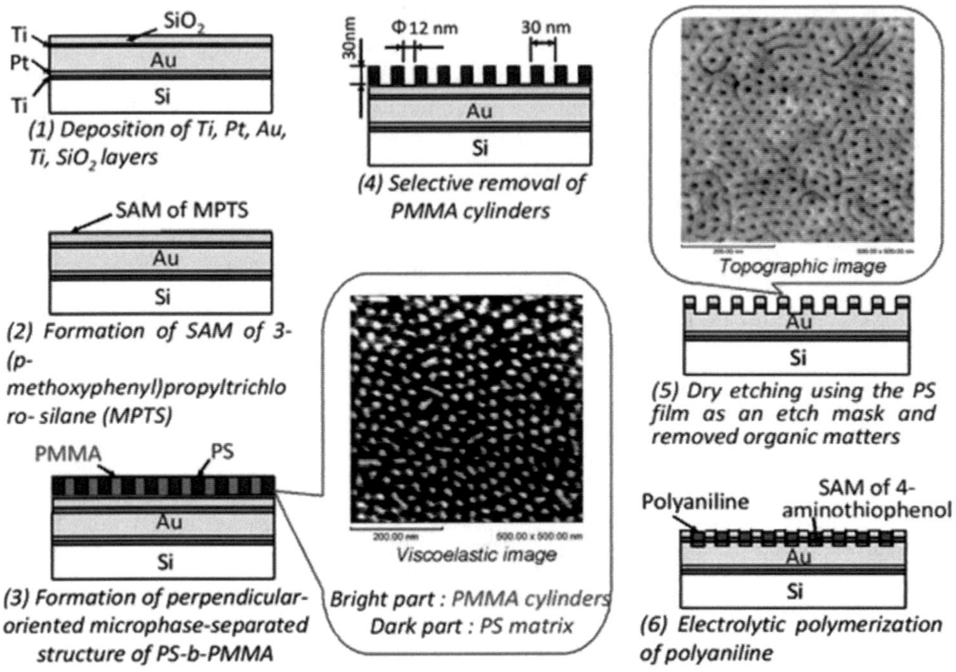

Figure 12. Fabrication process of patterned media of polyaniline using diblocked coplymer.

3.2. Massive parallel electron beam lithography (carbon nano tube)

Electron field emitter array with electrostatic focusing lens has been developed for the purpose of high throughput multi-column electron beam lithography system [16]. A CNT (Carbon Nano Tube) was applied to the electron field emitter. The CNT was grown selectively at an apex of catalytic nickel by negatively biasing the substrate in a hot filament CVD system [17]. The CNT was also grown at the apex by remaining the catalytic nickel only at the apex by annealing after nickel deposition as shown in figure 13 [18]. Low threshold voltage for electron field emission and small emitter size for fine focusing were achieved owing to the essential feature of the CNT. Hydrogen coverage on the CNT enhanced the emission and reduced the noise of the emission current [19].

3.3. Probe card for wafer-level burn-in test (low thermal expansion coefficient LTCC)

Electrical feedthroughs can be made in LTCC (Low Temperature Co-fired Ceramics) [19]. As shown in figure 14 (left), the through holes are made by puncturing the soft green sheet and the holes are filled with a silver or a gold paste. The sheet is sintered to make a ceramic and the lateral dimension can be controlled by sintering under pressure. This LTCC wafer can be anodically bonded to a silicon wafer. A multilayered ceramic wafer is fabricated by laminating the green sheets before sintering and the cross sectional photograph of the laminated LTCC with electrical feedthrough is shown in figure 14 (center). The thermal expansion of the LTCC is matched with silicon, as shown in figure 9 (right).

The LTCC with the electrical feedthrough was applied to a MEMS probe card [20]. The fabrication process and a photograph are shown in figure 15. Nickel probes are made by electroplating nickel into a silicon mold and then soldered to the LTCC wafer with AuSn. Finally the silicon mold is etched out. The probe card has similar thermal expansion with silicon, which makes wafer-level burn-in test (reliability test at elevated temperature on a wafer) possible.

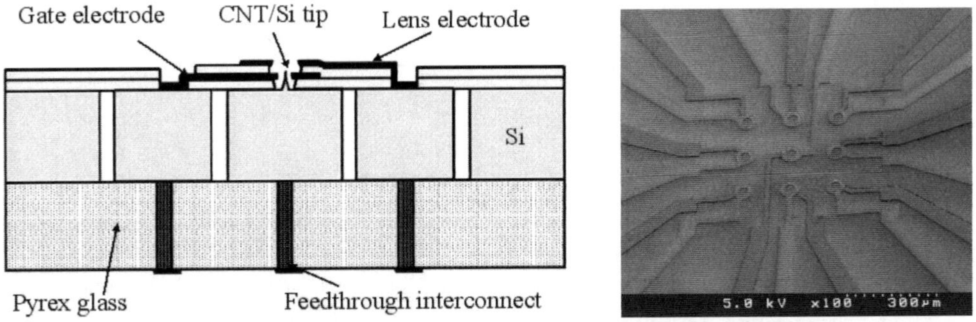

(a) Electron field emitter array with electrostatic focusing lens

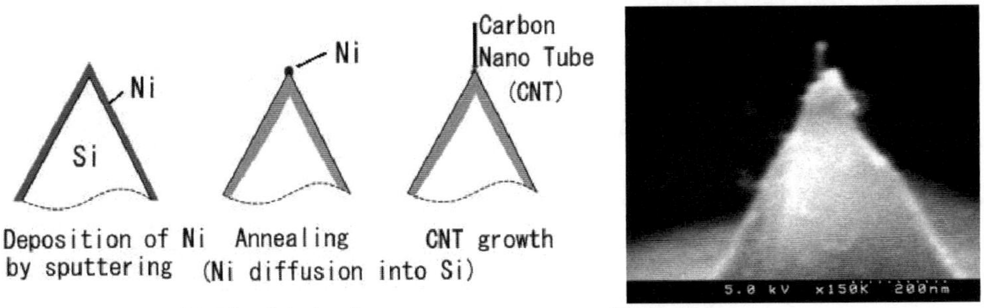

(b) The fabrication process to grow a CNT at the apex.

Figure 13. Cabon nano tube electron field emitter.

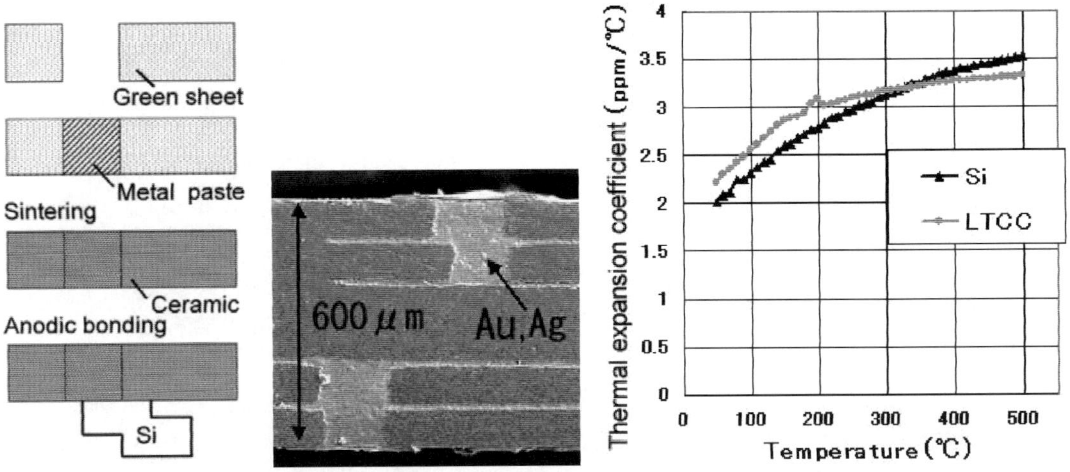

Figure 14. LTCC with electrical feedthrough. Fabrication process (left), cross-sectional photograph (center) and thermal expansion of silicon and LTCC (right).

3.4. SiC microstructure for glass press moulding (SiC)

SiC (silicon carbide) microstructure can be used as a mold to form glass parts by pressing. This takes advantages of the hardness of the SiC at high temperature. The fabrication process is shown in figure16a [21]. The SiC was deposited on a micromachined silicon wafer by CVD. The SiC surface is ground and bonded to a SiC ceramic plate by reaction bonding using an interfacial nickel film. Finally the silicon wafer is etched out. The micromachined silicon wafer which has a surface profile for non-spherical lens was made by transferring the resist profile using RIE. The resist profile was made by programmable mask less exposure system using DMD (Digital Micro mirror Device) [22]. The photographs of the SiC mould for a lens and the Pyrex glass press formed are shown in figures 16b and c, respectively.

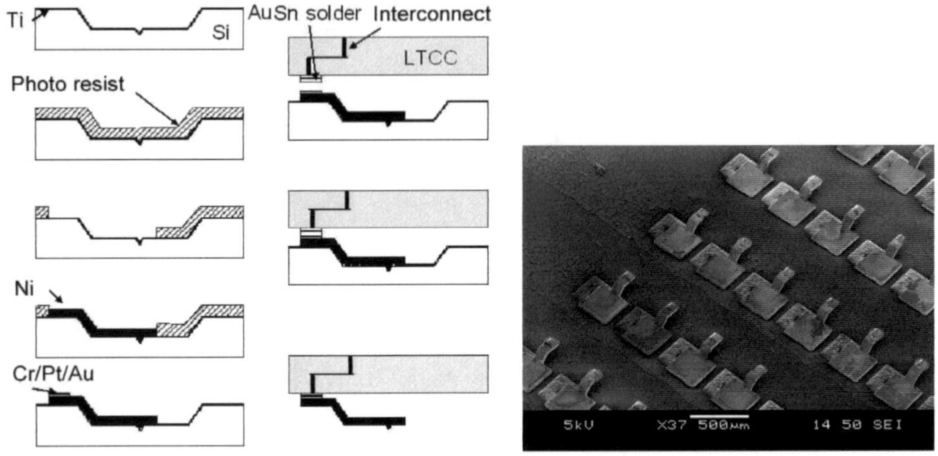

Figure 15. Fabrication process and photograph of probe card for wafer-level burn-in test.

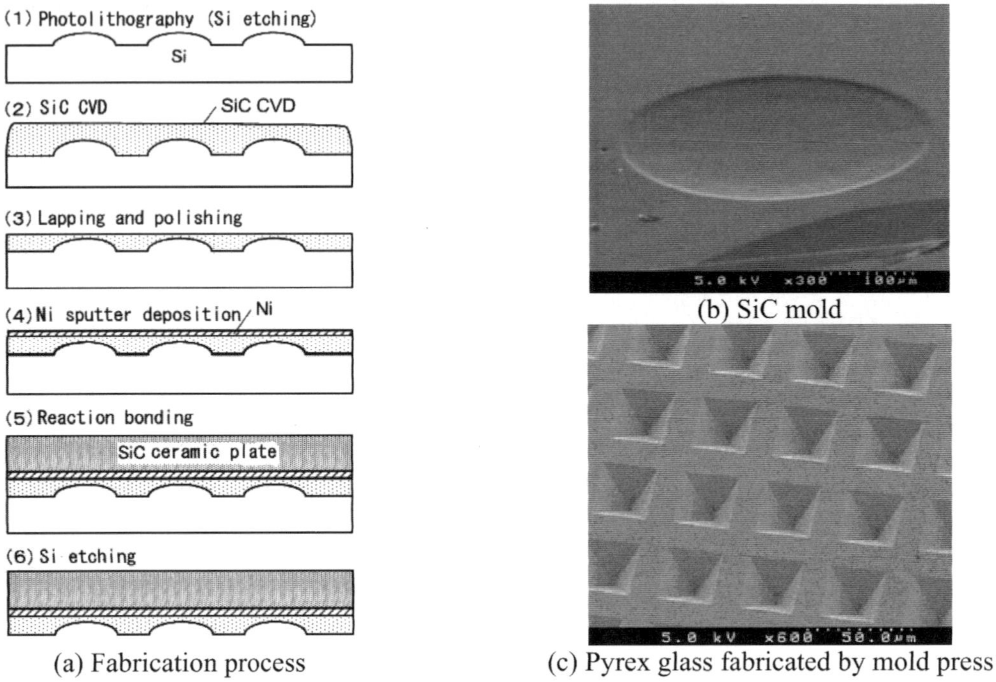

Figure 16. SiC microstructure for glass press molding.

3.5. SAW passive wireless sensor ($LiNbO_3$)

2.45GHz SAW (Surface Acoustic Wave) based passive transponders for wireless sensing has been developed. The principle, the photograph and the structure of the pressure sensor are shown in figure 17. Receiving the 2.45GHz electromagnetic wave, a surface acoustic wave generated by the IDE (Inter Digital Transducer) on a $LiNbO_3$ substrate propagates. It is reflected and a 2.45GHz electromagnetic wave is transmitted back and the sensing can be performed by measuring the delay time. The delay time is modulated by the deformation of the diaphragm by the pressure. Temperature can be also measured from the temperature dependency of the delay time and therefore multiple reflectors are formed for the temperature compensation in the pressure measurement. The SAW pressure sensor was developed to apply to TPMS (Tire Pressure Measurement System) [23].

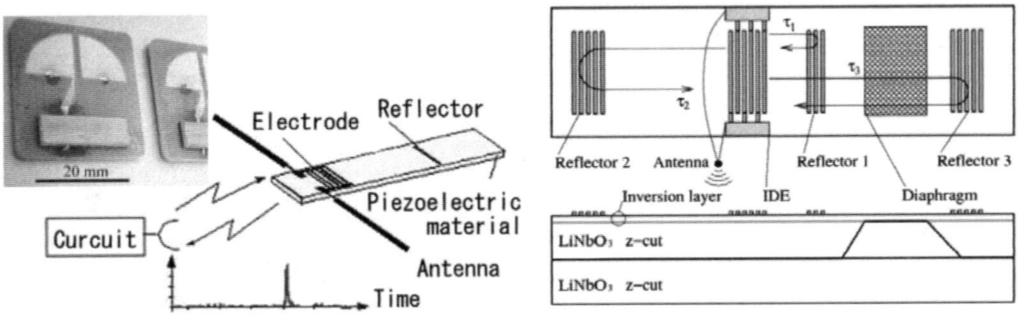

Figure 17. SAW passive wireless sensor for pressure measurement.

Fabrication process of diaphragm by selective etching of $LiNbO_3$ is shown in figure 18. Thermal inversion of polarization and polarization dependent selective etching are used for this process [24].

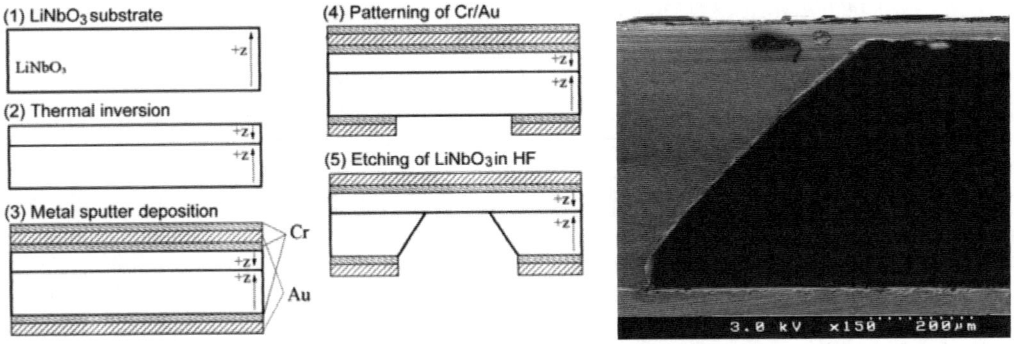

Figure 18. Fabrication process of diaphragm by selective etching of $LiNbO_3$.

4. Conclusions

MEMS play important roles as key devices in various systems. Packaging and electrical interconnection are needed for high reliability and arrayed MEMS respectively. Not only silicon but also other materials such as diamond, conductive polymer, carbon nanotube and silicon carbide could be used effectively for MEMS.

References

[1] Murakoshi T,Endo Y,Sigeru K, Nakamura S and Esashi M 2003 *Jpn. J. Appl. Phys.* **42** 2468
[2] Tajima T, Nishiguchi T, Chiba S, Morita A, Abe M, Tanioka K, Saito N and Esash M 2003 *Microelectronic Engineering* **67-68** 508
[3] Asada N, Matsuki H, Minami K and Esashi M 1994 *IEEE Trans. on Magnetics* **30** 4647

[4] Matsumoto Y and Esashi M 1992 *Electronics and Communications in Japan* **76** 93
[5] Esashi M 2008 *J. of Micromachining and Microengineering* **18** 073001
[6] Henmi H, Shoji S, Shoji Y, Yoshimi K, and Esashi M 1994 *Sensors and Actuators A* **43** 243
[7] Miyashita H and Kitamura Y 2005 *Anelva Technical Report* **11** 37 (in Japanese)
[8] Liu Y, Li X, Abe X, Haga Y and Esashi M 2001 *Proc. of MEMS'2001 (IEEE, Interlaken, Switzerland)* p 220
[9] Nakamura A, Takayanagi F, Moro Y, Sanpei H, Onozawa M and Esashi M 2004 *Advantest Technical Report* **22** 9 (in Japanese)
[10] Lee D W, Ono T, Abe E and Esashi M 2002 *J. of Micro-electromechanical Systems* **11** 215
[11] Abe T, Li X and Esashi M 2003 *Sensors and Actuators A* **108** 234
[12] Xu H G, Ono T and Esashi M 2006 *J. of Micromech. Microeng.* **16** 2747
[13] Takahashi H, Onoe A, Ono T, Cho Y and Esashi M 2006 *Jap. J. of App. Phys.* **45** 1530
[14] Yoshida S, Ono T and Esashi M 2008 *The 4th Asia Pacific Conf. on Transducers and Micro/Nano Technologies (APCOT 2008, Tainan, Taiwan)*
[15] Yoshida S, Ono T and Esashi M 2007 *Nanotechnology* **18** 505302
[16] Minh P N, Hong P N, Cuong T M, Ono T and Esashi M 2004 *Proc. of MEMS'2004 (IEEE, Maastricht, The Netherlands)* p 430
[17] Ono T, Miyashita H and Esashi M 2002 *Nanotechnology* **13** 62
[18] Tsai C H, Ho J Y, Ono T and Esashi M 2008 *MEMS 2008 Technical Digests (IEEE, Tueson, USA)* p 430
[19] Mori M, Kobayashi Y and Kidani N 2007 *SEMICON Japan STS seminar Microsystem / MEMS session (Makuhari, Japan)* p 15
[20] Choe S-H, Tanaka S and Esashi M 2007 *IEEE International Test Conference 2007 (IEEE, Santa Clara, USA)* p 20.2
[21] Min KO, Tanaka S and Esash M 2004 *Proc. of the 21th Sensor Symposium (IEEJ, Kyoto, Japan)* p 473
[22] Totsu K and Esashi M 2005 *J. Vac.Sci.Technol.* **B23** 1487
[23] Hashimoto S, Kuypers J H, Tanaka S and Esashi M 2007 *Proc. of the 24th Sensor Symposium (IEEJ, Tokyo, Japan)* p 1487
[24] Randles A B, Esashi M and Tanaka S 2007 *Jap. J. of Applied Physics* **46** L1099

Acknowledgments
This work is supported in part by Special Coordination Funds for Promoting Science and Technology, Formation of Innovation Center for Fusion of Advanced Technologies from Ministry of Education, Culture, Sports, Science and Technology.

Carbon nanotube: a novel material for applications

Phan Ngoc Minh and Phan Hong Khoi

Institute of Materials Science, Vietnam Academy of Science and Technology
18 Hoang Quoc Viet Road, Cau Giay Distr., Hanoi, Vietnam

E-mail: minhpn@ims.vast.ac.vn

Abstract. In this paper, current state of research and development on carbon nanotube material is overviewed. A brief description on the author's current results of synthesis and purification technologies, characterization and application potentials of multiwall carbon nanotubes in advanced composite materials, thermal dissipation media, electro-magnetic wave absorption, scanning probe and electron field emitters are presented. Future trends and potential breakthrough in applications using the materials will be discussed.

Keywords: Field emission, Schottky emission, electron sources, biomedical, nanocomposite

1. Introduction

Carbon is a key element of the nature and the life. It is recognized that carbon nanotubes (CNTs) discovered in 1991 by Iijima [1], is an ideal building block for nanotechnology because the material has many unique properties such as high aspect ratio structure (1-20 nm diameter, upto several cm long), extremely large surface area, chemical inertness, super hard materials (100 times stronger than steel), high melting temperature in vacuum, super high thermal conductivity, bio compatibility and so on. The number of published papers on finding unique properties of the CNTs has been exponentially increasing. Such properties of the CNTs lead to great application potentials, and the number of papers and patents related to the application of the CNTs material strongly increased. The number of conferences and workshops related to the CNTs material is also drastically increased.

A full description on the progress of the CNTs science and technology can be found in several text books [2]. In general, there are several main approaches on both singlewall and multiwall CNTs research and developments (i) synthesis technology of low cost, large quantity, high quality and high reproducibility CNTs; (ii) control of the CNTs structure such as diameter, length, orientation, uniformity, chirality structure; (iii) purification, separation and functionalization technologies; (iv) investigation of novel properties; (v) realization and discovery of novel applications; (vi) study on other carbon related materials such as graphene, diamond, etc.

Indeed, the materials have been commercialized with the price being drastically decreased. This is the consequence of the growing technology progress. For instance, super long and super high growing speed CNTs (6 cm long and above 16 μm/s growths speed) has been reported by Qian Wen (Tsinghua University-Beijing) recently [3]. In fact the super long CNTs has been synthesized since 2002 as shown in figure 1 [4-5]. Several companies in China, USA, Korea, etc. announced the CNTs production capability of several tons per year and worldwide production capability of several hundreds tons CNTs per year. The number of companies and enterprises related to the CNTs products increased worldwide.

© 2009 IOP Publishing Ltd

The CNTs have many unique properties that lead to many application potentials. The CNTs was confirmed as an excellent electron field emission and thermal field emission source for high brightness electron emitters, low power consumption flat panel display, cathode-ray-tube lighting elements, and as field-emission sources in modern electron microscopes, X-ray source, etc. [6].

The CNTs has been tried as a conduction channel in a CNT-Field Effect Transistor (FET) as well as ultra small wires for next generation of electronic devices. Individual CNT-FET has been already achieved, however to realize a complex circuit is still a challenge.

Figure 1. Images of super-long CNTs [4-5].

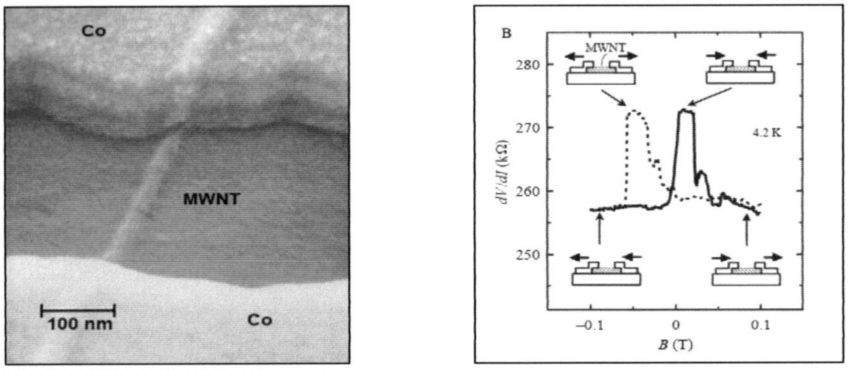

Figure 2. CNTs for spintronic application (MWCNT + Co electrodes) and the characteristic at 4.2 K [7].

Many works have been focused on the optical properties of the CNTs including absorption, emission, fluorescence, and Raman spectroscopy, phonon and electron-phonon interactions and the relation to their geometric, electronic, band structures, defects, etc. of the materials. A large interest was paid on the functionalization of the CNTs. The CNTs must be chemically functionalized in order to be well dispersed in the base materials (metal, polymer, ceramic, etc.) for advanced composite materials. The CNTs must also be functionalized for application in biomedicine or drug delivery system, where the motion and heating of the CNTs can be magnetically controlled. The CNTs materials were chosen because the CNTs itself are biological compatible high drug loading, easily penetrate cell membranes, etc. The functionalized CNTs (CNTs-COOH) can also be attached with fluorescent and anti-cancer drug or other bio components such as Peptides, DNA, surfactants, etc.

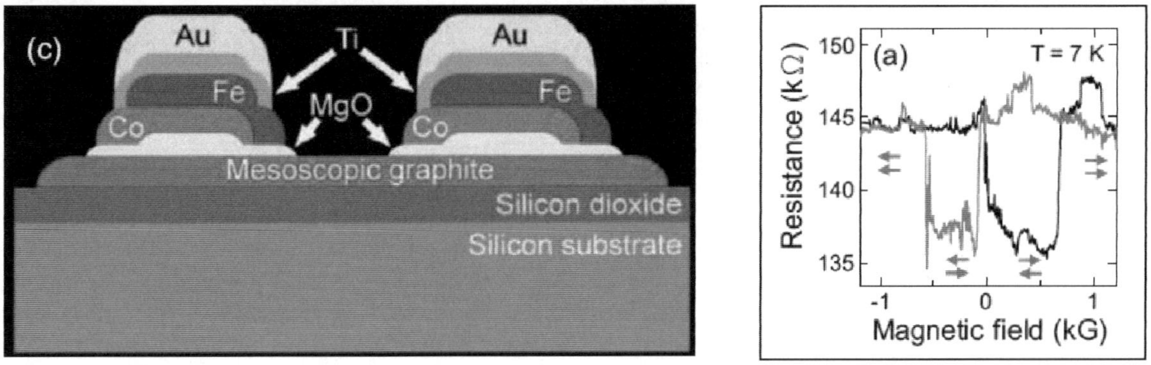

Figure 3. Graphen for spintronic application [8].

Table 1. Applications roadmap for Carbon Nanotubes by M. Endo, M. S. Strano, P. M. Ajayan (NT08 Conference, 2008).

	Large Volume Application	Limited Volume Application
Present	- Batery electrode additives (MWCNTs) - Composite (sporting goods, MWCNTs) Composites (Electrical shielding Devices applications; MWCNTs)	- Scanning probe tips (MWCNTs) - Specialized Medical Appliances (catheters) (MWCNTs)
Near term (less than 10 years)	- Battery and Super-capacitor Electrodes - Multifunctional Composites (3D, damping) - Fuel Cell Electrodes (catalyst support) - Transparent Conducting Films - Field Emission Displays / Lighting - CNT based Inks for Printing	- Single Tip Electron Guns - Multi-Tip Array X-ray Sources - Probe Array Test Systems - CNT Brush Contacts - CNT Sensor Devices - Electro-mechanical Memory Device - Thermal Management Systems
Long term (beyond 10 years)	- Power Transmission Cables - Structural Composites (aerospace and automobile etc.) - CNT in Photovoltaic Devices	- Nano-electronics (FET,Interconnects) - Opto-electronics (Photovoltaic, OLEDs, transparent electrode) - Flexible Electronics - CNT based Chemical, bio-sensors - CNT Filtration/Separation Membranes - Drug-delivery Systems

The CNTs are used or will be used soon in more and more commercial products (sport equipments, cars, paints and coatings, medical applications etc.). Table 1 listed the application roadmap of the CNTs material according to M. Endo, M. S. Strano, P. M. Ajayan, presented at the NT08 Conference, 2008.

2. Results of synthesis and application of CNTs

2.1. Synthesis and purification of the CNTs

We have started the work with CNTs materials since 2003 and the following is a brief description of the works which have been done in our laboratory. We developed a thermal chemical vapor deposition (CVD) system for synthesis of multiwall CNTs using a thermal furnace with C_2H_2 as hydrocarbon and H_2 + Ar as carrier gases. Carbon nanomaterials with several configurations such as nanotubes, nanorods, nanoballs were successfully synthesized. Using iron mesh as catalyst (content of Fe ingredient in the iron mesh catalyst was about 78 wt.% estimated by EDS analysis) and stainless steel tube as a thermal CVD reactor, the CNTs were successfully grown with high reproducibility. For each time of growth, an amount of the iron mesh was settled into the center part of a stainless steel tube. A nitrogen flow of 500 standard cubic centimeters per minute (sccm) was passed through the thermal furnace until the temperature reached 750°C. After 10 minutes for the deoxidization of the catalyst surface by 300 sccm of H_2, the decomposition of 200 sccm C_2H_2 flow was carried out for 60 minutes. A thick layer of MWCNTs was formed on the iron mesh. The CNTs product was then separated from the iron mesh by mechanical technique [13]. Vertically-aligned CNTs was also successfully synthesized by the thermal CVD in $C_2H_2 + N_2$ using Fe_3O_4 nanoparticles or thin Fe film on AlO_2/Si substrate. Acetylene gas was used as the carbon source and nitrogen gas was used as a carrier gas [14]. Figure 4 is a typical SEM image of the grown vertically aligned CNTs we have achieved recently. Typical TEM image of the CNTs is shown in figure 5. The multiwall structure of the grown tubes it is clearly seen. The grown CNTs were effectively *in-situ* purified by oxidation in pure oxygen at 400°C. Figure 6 shows the TGA curves measured in air of the CNTs before and after the purification treatment in oxygen for 60 min. In curve (a), the weight starts to reduce at near 350°C and completely evaporates at above 700°C. In curve (b), the weight of purified CNTs starts to burn at 510°C and completely burn-out at 710°C. The catalyst contents in the CNTs before and after the purification process are approximately 6 %. Figure 7 shows Raman spectra of the CNTs after oxidation in oxygen at 400°C for 0 min., 30 min., 60 min. and 210 min. All spectra show two Raman peaks at 1318 cm^{-1} ÷ 1323 cm^{-1} (D band) and at 1583 cm^{-1} ÷ 1591 cm^{-} (G band). The D band indicates the defective structure of graphite sheets.

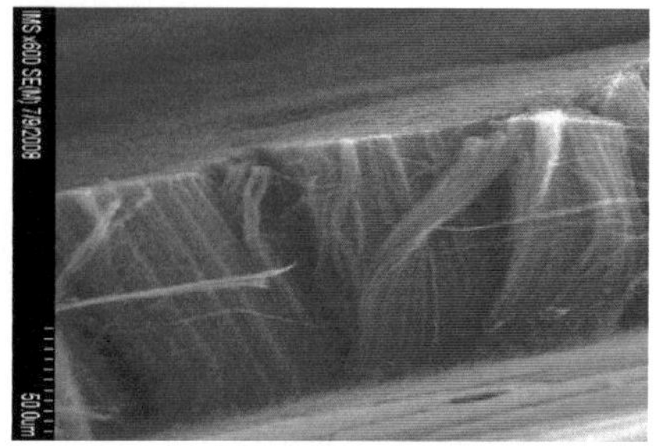

Figure 4. SEM image of the vertically aligned CNT.

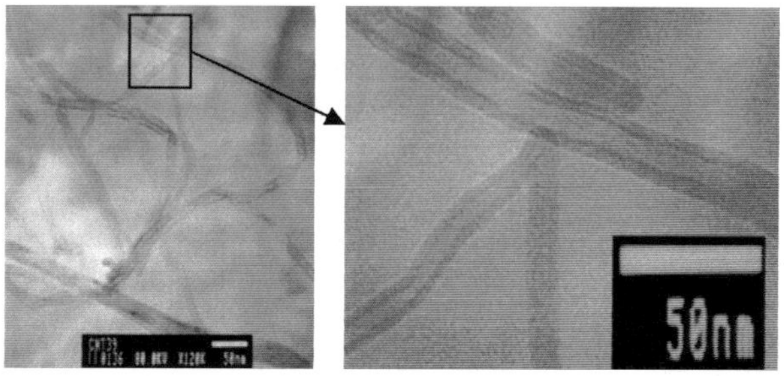

Figure 5. TEM images of the multiwall CNTs.

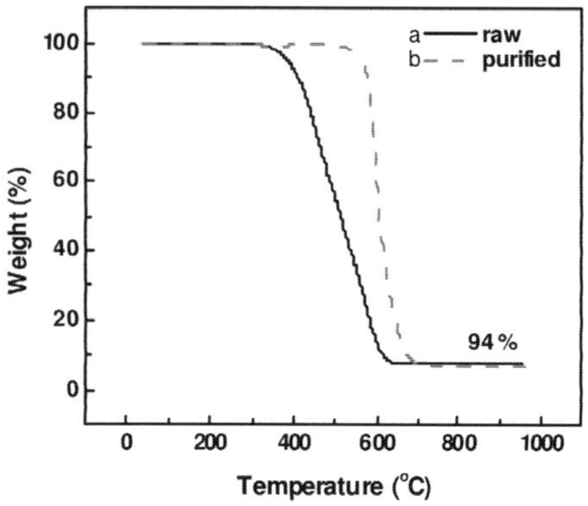

Figure 6. TGA curves of the CNTs before and after purification by in-situ treatment in O_2 at 400°C.

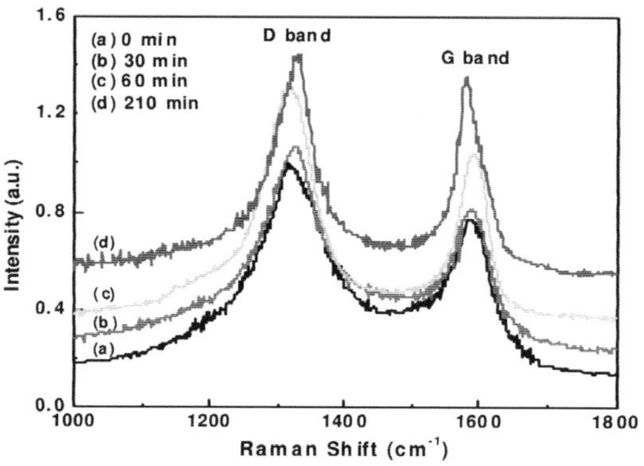

Figure 7. Raman spectra of the CNTs after oxidation in oxygen at 400°C for different oxidation time.

The peak intensity ratios of G and D bands increase when increasing the time of oxidation. It is seen that the I_G/I_D ratio would increase after purification because of the improvement in nanotubes percentage by eliminating amorphous carbon.

Beside the *in-situ* oxidation purification method, the material was annealed in 10^{-4} Torr vacuum at high temperature (1900°C at the filament). Detailed studies using SEM, XRD, Raman, TGA have confirmed that the contents of a-C, Si and related catalysts (Fe, Ni) were strongly reduced without damaging of the tubes' structure. The raw material MWCNTs sample (commercial product) used in this study contains impurities, such as a-C, Si, Fe, Ni of about 10%, 0.55%, 1.61%, 1.04% weight concentration, respectively. By annealing in 3 hours the Si, Fe and Ni impurities reduced to 0, 0.065% and 0.1%, respectively. This indicated that the impurities (Si, Fe Ni) were extracted from the MWCNTs by high temperature annealing in vacuum.

In summary, using a simple thermal CVD equipment with one thermal tube, it is possible to produce 95% purified CNTs with productivity of several hundreds gram/day. Obviously the equipment can be multiplied and the synthesis/purification techniques can be utilized for low cost, high purity and mass volume production [15].

2.2. Application of MWCNTs in advanced composite

The CNTs were used as an additive material for fabricating Ni-CNTs film by electrodeposition technique. The work aimed at creating an advanced nano-composite coating by taking mechanical properties advantages of the CNTs material. Nickel was plated onto steel substrates in Watts solution with and without CNTs additives. CNTs were ultrasonically mixed in the plating solution in a ratio of 3g/l. The temperature was kept at 30°C in a 3 liter bath, with a pH of 4.5. We used 470 Hz pulsed current for all plating process. High current density was 20A/dm^2 in 0.08 ms. Low current density was -30A/dm^2 in 0.02 ms, and zero current was kept in 2.03 ms. The measurement results finally showed that the Ni-MWCNTs electrodeposited films were 1.5 times harder than the conventional Ni electrodeposited films and the corrosion resistance of the Ni-MWCNTs electrodeposited films were greatly enhanced [16].

The Cr-CNTs composite coatings were also prepared on steel substrates by electro-deposition in a Cr plating bath (sulfate solution) containing carbon nanotubes (CNTs). The high dispersion of the CNTs in the plating bath has been achieved by using the surfactant (ammonium citrate) and the functionalized CNTs. The CNTs were functionalized by a diazo salt which is (p- ammino) phenylen diazonium chloride. The functionalized CNTs contained ammino group that protonlised in plating solution, so they are positive charged that supports the deposition of CNTs on the plated Cr coating.

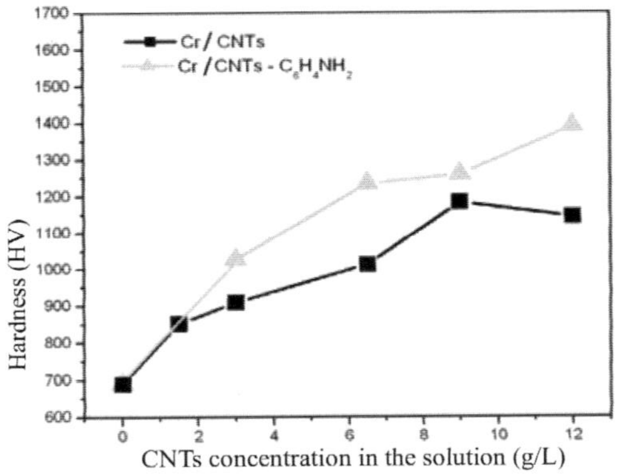

Figure 8. Hardness of the Cr/CNTs and Cr/functionalized CNTs plated films as a function of CNTs concentration in the plating solution.

Figure 9. Distribution of CNTs inside the plated Cr film (a part of Cr was etched out).

The microhardness of the electroplated coatings was determined by using a Vicker's microhardness indenter with a load of 200 g for 15 s. The abrasion tests were carried out by the abrader (APGI-G13.01). The morphology of the coatings was analyzed by using a SEM measurement. Figure 8 shows the dependence of the measured hardness of the Cr, Cr-CNTs, Cr-CNTs-$C_6H_4NH_2$ coatings on the concentration of CNTs in the plating solution. It is clear that the hardness of the Cr with CNTs additive is much higher than the Cr coating. Hardness of the Cr/CNTs-$C_6H_4NH_2$ coating is higher than that of the Cr/non-functionalized CNTs. This is probably related to the higher dispersion of the functionalized CNTs-$C_6H_4NH_2$ in the solution. The hardness of Cr/CNTs increases with increasing the CNT content in the solution and reaches to a value near 1400 HV at the CNTs concentration of 12 g/l that is twice compared to that of the Cr coating. The abrasion testing also confirmed that the Cr-CNTs coatings are almost two times better than those of the conventional Cr coating. By partly etching the plated Cr film, we can see the distribution of the CNTs in the plated film. This opened a promising way for application of the CNTs in creating super hard coating by electroplating technique [17].

2.3. Application of MWCNTs in advanced rubber

As mentioned above, the CNTs materials are very attractive for nanocomposite application because of the nano size and excellent mechanical and electrical properties. We have utilized the grown MWCNTs as an additive component to reinforce natural rubber (NR) vulcanizates. The preparation is the following: NR was dissolved in toluene and then MWCNTs was mixed with the sonication. The solvent in the resulting dispersions were then evaporated and the samples dried under a vacuum. The resulting mixtures were mixed with other additives in a Brabender mixer and vulcanizations were carried out in a standard hot press at 150°C for 10 mins. Table 2 shows several mechanical properties of the natural rubber, rubber doped with graphite powder, rubber doped with 7% MWCNTs and rubber doped with 10% MWCNTs. It is clear that the rubber doped with MWCNTs greatly enhanced the mechanical properties of the rubber. Also, in order to improve the solubility and dispersion of the MWCNTs into the natural rubber matrix, thereby improving properties of NR vucanizates, we are also dealing with the grafting of hydroxyl-terminated liquid natural rubber (HTNR) onto MWCNTs. The functionalized MWCNTs was carried out in three steps: acid treatment in H_2SO_4/HNO_3 (3:1) mixture, reaction with $SOCl_2$ and reaction with hydroxyl-terminated liquid natural rubber (HTNR). More detail research on the influence of the dispersion, alignment, interfacial properties and other factors, such as methods for the preparation of MWCNTs/NR composite materials are still under investigation. The advanced rubber (10% MWCNTs/NR) was used for producing essential components of water in Vietnam. The advanced rubber is not only utilized for components in the water pumping systems but also applicable for other field such as rubber wheel of airplane, car, and so on [18].

Table 2. Mechanical properties of the Natural rubber doped with Graphite MWCNTs of different contents.

Parameters	Natural Rubber	Rubber+doped with Graphite Powders	Rubber doped with 7% MWCNTs	Rubber doped with 10% MWCNT
Hardness (Shore A)	42	54	53	53
Specific Mass (g/cm^3)	0.995	1.019	1.016	1.015
Broken durability (N/cm^2)	21.03	47.6	64.15	79.62
Broken length (%)	293.6	226.4	298.4	341.8
Arkon Abrasion (g/1.61 km)	4.5	2.4	1.7	1.3
Heat resistant (°C)	200	350	350	350

2.4. Application of MWCNTs in electromagnetic absorption materials and paints

The shielding of radio frequency radiation is important for many military and commercial applications. CNTs composite material was announced as a good additive for making a light weight and effective

EMI shielding media due to their geometric and electrical properties. The conductivity and shielding effectiveness of these composites are strongly correlated to the structure of the CNTs. We have tried to make the CNTs/epoxy (EPIKOTE 828 resin-Epon 828, and EPIKURE 3072 curing agent-Epicure 3072) and measured the Rada absorption efficiency of the composite using a network vector analyzer (HP E8363B) in 8–12 GHz (X-band) range. Here we would like to again impress that the CNTs materials can be used for Rada shielding application. Figure 10, 11 are the SEM image of the functionalized MWCNTs/epoxy composite and the measured shielding efficiency, respectively. The measured sample size was (10x10 x0,5) cm^3 corresponding to the requirement of the measurement system. The target value of the shielding efficiency for commercial applications is around 20 dB. As shown in figure 11, the composites with CNTs of 15-20 % wt. can meet such requirements. The work is now continued to explore the shielding effect in other frequency range and also to reduce the thickness of the media for actual application.

The grown CNTs was used for preparing the CNTs/Polyaniline conducting polymer. The electrical conductivity of the CNTs/polyaniline composite (S/cm) was measured as a function of the CNTs concentration. The electrical conductivity of the polyaniline/CNTs composite reached to the value of 30 S/cm that is absolutely comparable with the conductivity range of metallic materials. With this high electrical conductivity, polyaniline/CNTs material opened a promising way for application in producing protective paint, especially protective paint for sea-ship by eliminating the electro-chemical etching effect.

Figure 10. Typical SEM image of the functionalized CNTs/ epoxy composite.

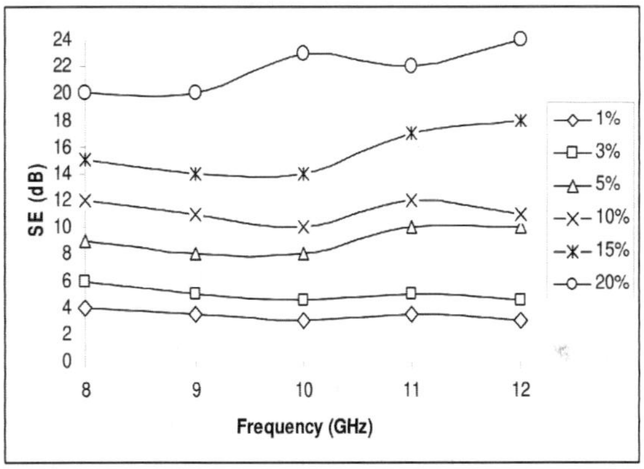

Figure 11. EM shielding efficiency of the CNTs/epoxy composite (size: 10x10x0.5 cm) measured in 8–12 GHz range with different CNTs concentration.

2.5. Application of MWCNTs for thermal dissipation

As a material with highest thermal conductivity (2000 W/m.K compared to thermal conductivity of Ag 419 W/m.K), CNTs therefore are an ideal candidate for thermal dissipation media to improve the performance of computer processor and other high power electronic devices. We have employed the grown CNTs as the heat dissipation media in a μ-processor of a Personal Computer with configuration: Intel Pentium IV 3.066 GHz, 512Mb of RAM and Windows XP Service Pack 2 Operating System. The temperature of the μ-processor was measured directly during the operation of the computer in two modes: 100% usage CPU and overclocking modes.

Figure 12 is the measured temperature of the chip of the computer during the operation of the computer using different type of thermal matching media: commercial thermal matching and 2% wt. CNTs added thermal glues. The results showed that when a mixture of a commercial thermal compound and 2% wt. MWCNTs is used, the temperature of the μ-processor decreased 5°C, and the time for increasing the temperature of the μ-processor was three times longer than that when using commercial thermal compound. In overclocking mode, the processor speed reached 3.8 GHz with 165 MHz of system bus clock speed that was 1.24 times higher than that in non-overclocking mode. This is a direct evidence that confirmed the advantage of the MWCNTs for the thermal dissipation media for μ-processor of the PC or high power electronic devices in general.

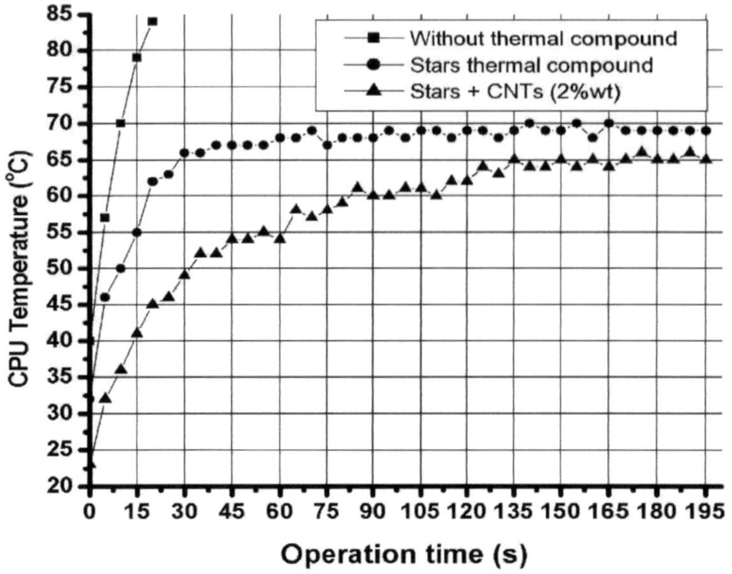

Figure 12. The measured temperature of the chip of a computer as a function of working time using different thermal matching media.

2.6. Application of limited volume CNTs for field emission and scanning tunneling probes

Besides the application in advanced nanocomposite mentioned above, the CNTs especially the individual CNT can be utilized for advanced scanning probe microscopy and electron field emitter. Individual CNT can emit electrons as a pointed source at a low voltage and the energy spread of the emitted electrons by the CNTs is very narrow (0.1-0.2 eV). This means that the emitted electron beam can be easily focused into a small spot that is very essential for application such as in electron beam based instrumentation, parallel electron beam nanolithography, field emission information storage technology. The individual CNT has been successfully grown on the Fe coated Si tip formed by a standard microfabrication process and the electron field emission of the individual CNT on the Si tip was measured in comparison with the Si tip itself. It was shown that the individual CNT emit electrons at a threshold electric field of 4 V/μm that is 5 times lower than that of the Si tip itself [22-23]. In another approach we tried to growth the CNT on W tip for an advanced electron emitter and scanning tunneling microscopy probes application.

Sharp tungsten (W) tips of 50 nm-diameter were formed by electrochemical etching of 0.3 mm diameter W wires in 10% wt. KOH solution. By electroplating in 0.01% wt. $FeCl_2$ solution with electroplated-pulse mode, Fe nanograins catalyst were created at the apex of the W tip. Carbon nanotube was then grown at the Fe nanograins of the sharp W tip using a Hot-filament CVD apparatus [21]. The electron field emission of the W bare tip and the CNTs/W tip was measured in 10^{-6} Torr vacuum. Figure 13 is the SEM images of the W and CNTs/W and the corresponding electron field emission. As shows in this figure, the CNTs/W tip started to emit electron at 0.7 V/μm and to 26 μA at the electric field of 1 V/μm. Whereas, the W tip could emit electron at 5.8 V/μm and the current rises to 13 μA at the electric field 6.3 V/μm. Thus, the CNTs/W tip could emit electron at threshold voltage nearly 10 times lower than that of the W tip itself. Individual CNT on W tip has been successfully tested as an advanced Scanning Tunneling Microscopy probe. Atomic resolution image in air was successfully recoded on the HOPG sample using the developed CNT/W tip. The result will be published elsewhere.

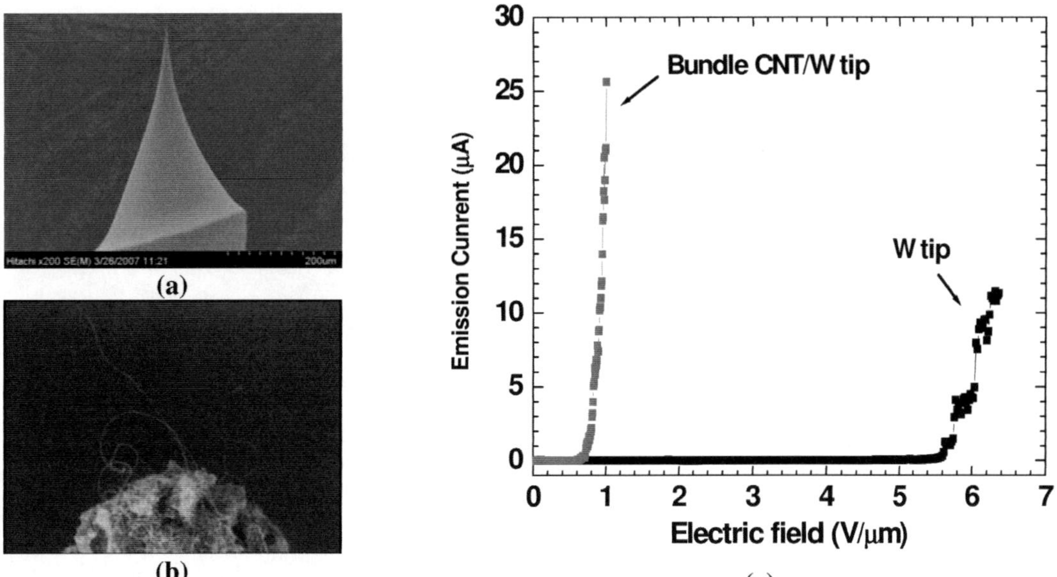

(c)

Figure 13. SEM images of electronchemical etched W tip (a); CNTs on W tip (b) and the corresponding field emission characteristics measured in 10^{-6} Torr vacuum.

3. Conclusion

Carbon nanotube is a key element of the nanotechnology-a leading technology in the century. Unique properties of the materials have been discovered that leads to great application potentials. The materials are already used and will be used soon in commercial products. Findings a good way for synthesis, characterization and application realization of the CNTs require a broad interdisciplinary of materials science, physics, chemistry, biology, etc.

References

[1] Iijima S 1991 *Nature* **354** 56
Iijima S and Ichihashi T 1993 *Nature* **363** 603
[2] Saito R, Dresselhaus G and Dresselhaus M S 1998 *Physical Properties of Carbon Nanotubes* (Imperial College Press London)
Meyyappan M 2005 *Carbon nanotubes science and applications* (CRC Press LLC)
[3] Wen Q 2008 *The 9th International Conference on the Science and Application of Nanotubes* (Montpellier- NT08 France)
[4] Ajayan P M 2002 *Science* 296
[5] Hata et al. 2004 *Science* **306** 1362
[6] Minh P N, Tuyen L T T, Takahito Ono, Miyashita H, Suzuki Y, Mimura H and Esashi M 2003 *Journal of Vacuum Science and Technology* B21 **4** 1705-1709
[7] Tsukagoshi K et al. 1999 *Nature* **401** 572
[8] Tombros N et al. 2007 *Nature* **448** 571
[9] Cho S et al. 2007 *Appl. Phys. Lett.* **91** 123105
[10] Nishioka M et al. 2007 *Appl. Phys. Lett.* **90** 252505
[11] Goto H et al. 2008 *Appl. Phys. Lett.* **93** 212110
[12] Wang W H et al. 2008 *Phys. Rev. B* **77** 020402
[13] Chuc N V, Dung N D, Hong P N, Quang L D, Khoi P H and Minh P N May 2008 *Journal of the Korean Physical Society* Vol. 52 No. 5 pp 1368-1371
[14] Tam N T T, Ha D T, Hieu N T, Dung N D, Minh P N and Khoi P H 2007 *Proc. of the IWWNA Conference* (Vung Tau Vietnam) pp 485-488

[15] Minh P N, Quang L D and Khoi P H *Patent pending*
[16] Thang B H et al. July 25-27 2007 *Proceedings of the BME Conference* (Hanoi-Vietnam) pp 308-311
[17] Khoai N N et al. July 25-27 2007 *Proceedings of the BME Conference* (Hanoi-Vietnam) pp 325-328
[18] Ly P H and Khoi P H 2005 Preparation and Characterization of Carbon nanotubes/polymers composites *Advances in Natural Science* Vol. 6 No. 4 pp 339-361
[19] Minh P N, Tuyen L T T, Ono T, Mimura H, Yokoo K and Esashi M 2002 Carbon nanotube on a Si tip for electron field emitter *Japanese Journal of Applied Physics* Vol. 41, Part 2, No. 2A L1409-L1411
[20] Minh P N et al. June 2005 *Technical Digest of the 13th International Conference on Solid-state Sensors Actuators and Microsystems (Transducers' 05)* (Seoul-Korea) pp 267-270
[21] Hong P N, Cuong V V, Thang B H, Khoi P H and Minh P N May 2008 *Journal of the Korean Physical Society* Vol. 52, No. 5 pp 1386-1389

Acknowledgements
We wish to thank the National Program on Basic Research, Vietnam-Korea and Vietnam-French Bilateral Research Projects. We wish to thank the AFOSR/AOARD organization for the financial supports related to CNTs works. We would like thank the colleagues and students for their contribution to the works. Without their effort, we can not have data to report here. A part of the work was done with the help of the Key Laboratory in Electronic Materials and Devices, Institute of Materials Science, Vietnam Academy of Science and Technology, Vietnam.

An Erbium -Ytterbium DFB laser with a simple and compact structure

S. W. Harun[1], P. Hofmann[2], A. Schülzgen[2], L. Li[2], D. Nguyen[3], Ch. Spiegelberg[4], R. Matei Rogojan[4], A. Laronche[4], J. Albert[4], H. Ahmad[1] and N. Peyghambarian[2]

[1]Department of Electrical Engineering, University of Malaya 50603 Kuala Lumpur, Malaysia
[2]College of Optical Sciences, University of Arizona, Tucson, Arizona 85721, USA
[3]NP Photonics Inc., UA Science and Technology Park, Tucson, Arizona 85747, USA
[4]Department of Electronics, Carleton University, Ottawa, Ontario K1S 5B6, Canada

E-mail: swharun@um.edu.my

Abstract. A distributed feedback fiber (DFB) laser operating at 1540 nm having more than 45 dB signal to amplified spontaneous emission ratio with a simple and compact structure is introduced. The DFB laser consists of 45 mm long highly Er-Yb co-doped phosphate glass fiber with a grating structure inside the gain medium. The laser emission is stable, but the power is still relatively lower due to un-optimized grating structure.

Keywords: DFB laser, Er-Yb doped fiber laser, phosphate fiber laser.

1. Introduction

A distributed-feedback laser (DFB) is a laser where the whole resonator consists of a grating structure, which acts as a distributed reflector in the wavelength range of laser action, and contains a gain medium. Typically, the grating structure is made with a phase shift in its middle. This structure is essentially the direct concatenation of two Bragg gratings with internal optical gain. Most distributed-feedback lasers are either semiconductor lasers or fiber lasers and are operating on a single frequency. Semiconductor DFB lasers can be built with an integrated grating structure, e.g. a corrugated waveguide [1]. In the case of a fiber laser, the distributed reflection occurs in a fiber Bragg grating, typically with a length of a few millimeters or centimeters. Efficient pump absorption can be achieved only with a high doping concentration of the fiber, and unfortunately it is often not easy to write Bragg gratings into fibers with a composition (e.g. phosphate glass) which allows for a high doping concentration. However, this kind of single-frequency fiber laser is very simple and compact.

Recently, a study of the photosensitive in different phosphate glass has been reported, which allows high-quality Bragg gratings to be constructed in both phosphate fibers and planar glass waveguides [2]. In an earlier work, a DFB laser has been demonstrated using phosphate Er/Yb co-doped fiber in conjunction with a multimode pump [3]. In this paper, an Er-Yb co-doped phosphate DFB laser is demonstrated using a single-mode diode laser as the pump source. The laser resonator is formed by an asymmetric grating structure that provides distributed feedback for a signal light propagating in the single-mode core of an active Er–Yb codoped phosphate fiber.

© 2009 IOP Publishing Ltd

2. Experimental

The schematic diagram of the linear-cavity Er-Yb DFB fiber laser is shown in figure 1. The active fiber is fabricated from phosphate glasses that allow for high levels of rare-earth-ion doping in the active core with 1.1 x 10^{26} Er^{3+} ions/m^3 and 8.6 x 10^{26} Yb^{3+} ions/m^3. The gain fiber is 45 mm long and has a core diameter of 6 μm and a numerical aperture of 0.12. The DFB grating structure is written directly into the active fiber by exposure to 193 nm UV light through a phase mask. The grating consists of a 20 mm long section close to the pump side and a 15 mm long section at the single-mode DFB emission side that are separated by a 0.05 mm wide gap that creates the defect state inside the grating's reflection band.

The overall ~35 mm long grating structure is located approximately at the center of the active fiber piece, and the asymmetric DFB grating design results in unidirectional DFB laser emission. The single mode pump from 976 nm laser diode is launched into the Er-Yb fiber through a single-mode fiber which is spliced into the phosphate active fiber. The total splice loss at both ends of phosphate fiber is measured to be around 6 dB at 1310 nm. The output of the DFB laser is tapped from 1550 nm port of WDM coupler and is characterized by an optical spectrum analyzer (OSA) with a resolution of 0.015 nm.

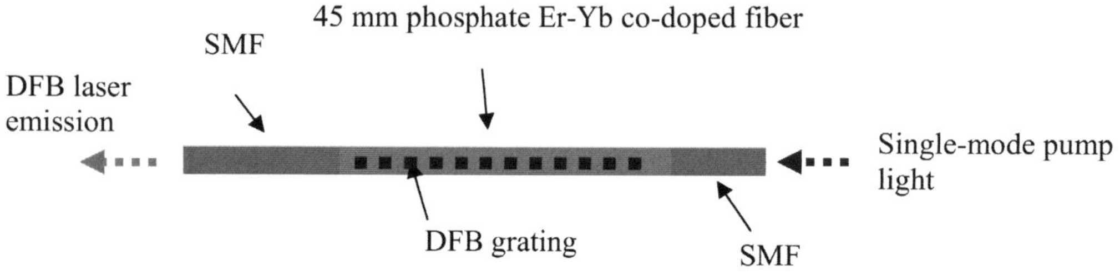

Figure 1. The configuration of DFB phosphate fiber laser with a 35 mm long DFB grating section inside the gain medium.

3. Results and discussion

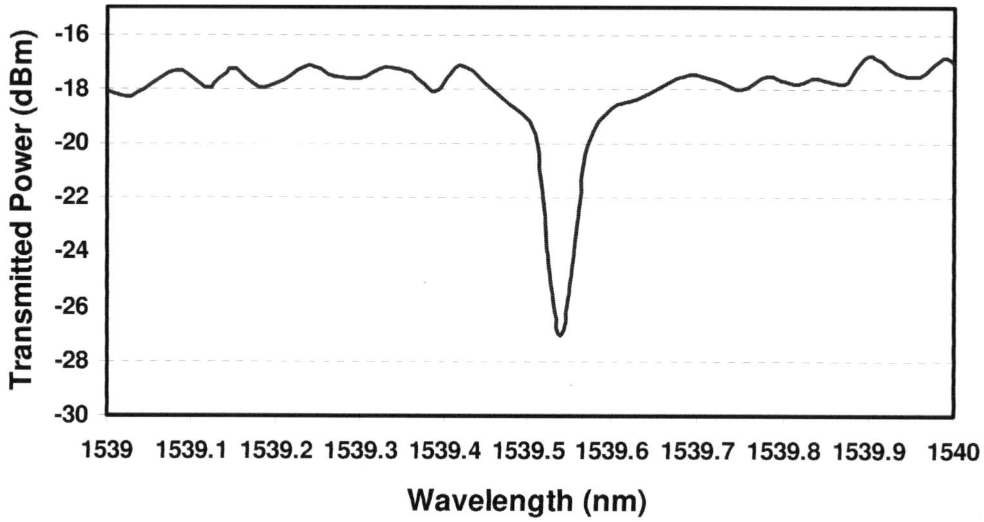

Figure 2. Transmission spectrum for the DFB grating (Maximum reflectivity = 89%).

Journal of Physics: Conference Series **187** (2009) 012003 doi:10.1088/1742-6596/187/1/012003

A transmittance spectrum of the gratings in the phosphate fiber is measured using an erbium amplified spontaneous emission (ASE) and the result is shown in figure 2. The spectrum shows a transmission dip of 9.6 dB, which translates to 89% reflectivity. The central wavelength of the grating is around 1539.5 nm at room temperature. The emission spectrum of the DFB fiber laser at different launching 976 nm pump power is shown in figure 3. The launching power is defined as the power measured right after the pump laser delivery fiber, which is fusion-spliced with the SMF. At pump power 468 mW, the DFB laser emits a narrow laser line located at the grating structure designed wavelength of 1540 nm. The width of the line is measured to be approximately 0.02 nm which is limited by the resolution of the OSA, while the true emission line-width is expected to be much narrower. A shown in figure 3, the side-mode suppression ratio (SMSR) is better than 45 dB, indicating stable single longitudinal mode operation of the DFB laser. Figure 3 also shows that no lasing is achieved at other pump powers. The laser is only achieved if both left and right grating wavelengths are matched.

The grating wavelength changes with the pump power, and therefore the DFB laser only achieved the maximum power -15 dBm after precise adjustment of pump power. The relatively low output power is a consequence of high splicing loss between the phosphate and SMF fiber on both sides of the 45 mm DFB laser. The pumping into active fiber sections that are on both sides of the 35 mm long DFB grating structure but do not contribute to the DFB cavity and the limited pump absorption within the DFB section are other factors that reduces the efficiency of the DFB laser. The efficiency can be improved by reducing the splicing loss, eliminating doped fiber segments that are not part of the DFB and also increasing the grating reflectivity.

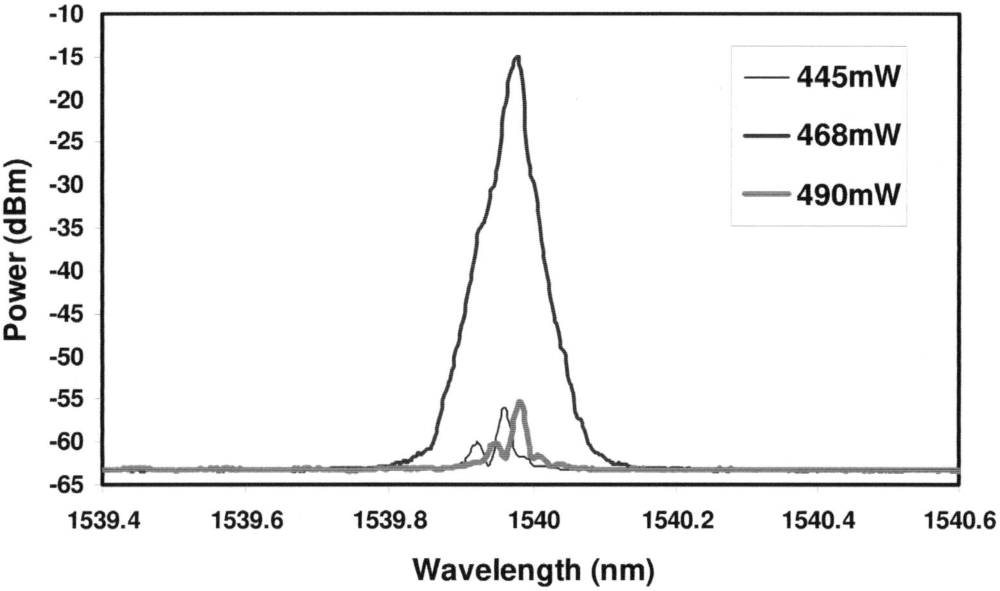

Figure 3. Emission spectrum of the core pumped DFB fiber laser measured by an optical spectrum analyzer with resolution of 0.015 nm.

In order to study the lasing wavelength stability at a fixed temperature, 14 successive scans of the system output were carried out within 10 min and the results are recorded in figure 4. The variations in output power and emission wavelength were below 2% and 0.05 nm, respectively. Even though the achieved power level is relatively low for fiber laser, any improvements in grating fabrication as well as the optimization of the laser set-up are expected to give a reasonably higher output power, indicating the potential of our laser scheme for the development of high-power DFB fiber lasers.

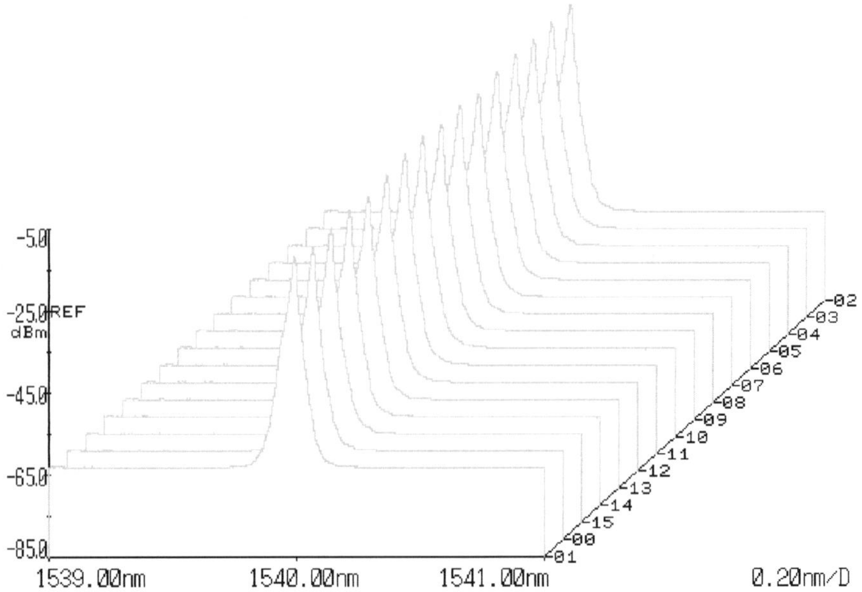

Figure 4. Laser output spectrum scanned every 10 minutes.

4. Conclusions

We have demonstrated a stable and compact DFB fiber laser scheme based on the Er-Yb co-doped phosphate fiber despite the large amount of splicing loss at both ends of the doped fiber and our low reflectivity grating. A stable and narrow line-width lasing is obtained at 1540 nm with SMSR of more than 45 dB by adjusting the amount of pump power that is injected. The achieved device performance makes our DFB laser a promising candidate for the fabrication of single-frequency fiber lasers that can be optimized for high-power, simple structure, compact and low noise.

References

[1] Han J-H and Park S-W 2007 *Microwave and Optic. Technol. Lett.* **49** 636
[2] Honkanen S, Schülzgen A, Albert J and Peyghambarian N 2008 *Proc. SPIE* **6897** p 3
[3] Schülzgen, Li L, Nguyen D, Spiegelberg Ch, Rogojan R M, Laronche A, Albert J and Peyghambarian N 2008 *Optics Letters* **33** 614

Acknowledgments

S. W. Harun would like to thank Academy Science of Malaysia for the post-doctoral Fellowship under the Brain-Gain Malaysia scheme.

Manipulating emission of CdSe/ZnS nanocrystals embedded in synthetic opals

Paul Benalloul[1], Celine Vion[1], Charles Barthou[1], Catherine Schwob[1], Jean-Marc Frigerio[1], Agnes Maître[1], Alex Gruzintsev[2], Gennadi Emelchenko[3], Wladimir Masalov[3] and Pham Thu Nga[4]

[1]Institut des NanoSciences de Paris, UPMC Uni. Paris 06, CNRS, UMR 7588 Campus Boucicaut, 140 rue de Lourmel, 75015 Paris, France.
[2]Institute of Microelectronics Technology and High Purity Materials, Russian Academy of Science, 142432, Chernogolovka, Moscow District, Russia
[3]Institute of Solid State Physics, Russian Academy of Science, 142432, Chernogolovka, Moscow District, Russia
[4]Institute of Materials Science, Vietnam Academy of Science and Technology, Hanoi, Vietnam

E-mail: paul.benalloul@insp.jussieu.fr

Abstract. Photonic crystals (PCs) are the object of great interest due to the possibility, for appropriate PCs, to modify and control light propagation and even to influence the emission properties of an emitter, such as its emission diagram and its life time. One of the most common approaches to prepare 3D PCs takes advantage of the spontaneous self-organisation of spherical colloidal particles. Various self-assembly techniques such as sedimentation, convective or Langmuir-Blodgett ones have been studied as they provide a low cost and relatively easy protocol to obtain artificial opals. SiO_2 opals exhibit a pseudo-band gap. Nevertheless the coupling of II-VI nanocrystal emitters in such PCs allows one to recognize and study some basic problems. Large opals have been prepared by the sedimentation method and the size of the balls has been adjusted so that the pseudo-band gap of those PCs lies in the same region as the emission band of CdSe/ZnS nanocrystals. Diagrams of radiation and the modification of the spontaneous life time of the embedded nanocrystals will be presented and discussed. Introducing well-defined defects in PCs which are necessary to guide the photons through the crystal remains a hard technological challenge. Several top-down methods have been investigated. We will present different bottom-up routes proposed by different groups to engineer planar defects into colloidal PCs.

Keywords: Photonic crystal, opal, quantum dots, decay rate, planar defect.

1. Introduction

Photonic crystals (PCs) in the visible and near infra-red ranges are characterized by a periodic dielectric constant at wavelength scale. As a consequence, the emission of light sources embedded inside PCs can be strongly affected by the band gap such as its propagation with minimal losses, its emission diagram and its life time [1]. Many approaches have been proposed to fabricate 3-D periodically modulated dielectric materials. Some of them come from more or less well established

© 2009 IOP Publishing Ltd

microelectronic industry processes such as semiconductor layer-by-layer nanomachining [2], layer-by-layer direct laser writing [3-5], interference lithography using a single diffraction mask [6], holographic lithography with at least 4 laser beams [7, 8] or two-beam only but with multi-exposure [9]. Alternatively to these methods which have known different sophisticated modifications proposed by several groups, bottom–up techniques taking advantage of the spontaneous self-organisation of spherical colloidal particles have been investigated. Various self-assembly techniques have been studied as they provide a low cost and relatively easy protocol to obtain artificial opals. Each one has received large attention and several groups have proposed for each technique modifications and improvements. Well orientated fcc crystals have been formed by gravity [10-12] or controlled sedimentation onto patterned substrate [13]. Different convective methods, vertical evaporation [14], induced by a temperature gradient [15] or by isothermal heating evaporation-induced self assembly (IHEISA) at a heating temperature of the solvent very close to its boiling point, method indicated for large balls [16] have provided large mono-domain opals. Langmuir-Blodgett (LB) technique is a layer-by-layer procedure for the preparation of large area synthetic opals. Solid films of particles are transferred from the water surface onto supports. The results suggest that a successful synthesis of ordered monolayers of monodisperse silica with the LB technique critically depends on the hydrophilic/hydrophobic balance [17, 18]. Silica spheres opals have been the object of a large number of papers even if these opals do not present a complete photonic bandgap due to the low index contrast between silica and air. Nevertheless the pseudo photonic band gap affects the propagation of light through the opal and causes transmission dip or reflectance peaks in the sample spectra depending with direction. Experimental studies on quantum dots [19-24] or molecules [25-28] embedded in such opals have been reported. However, the corresponding measured modification of the decay rate is limited since the refractive index contrast is low (below 5% for an index contrast of 1.2) [29]. A way to obtain larger effects is to use inverted opals as they present high refractive contrast and complete photonic bandgap [25, 30, 31]. As an example, lifetime reductions up to 30% have been measured for CdSe Qdots infiltrated in titania inverted opals [1]. But the production of good quality inverted opals over large scale is not yet totally mastered.

2. Opals characterization

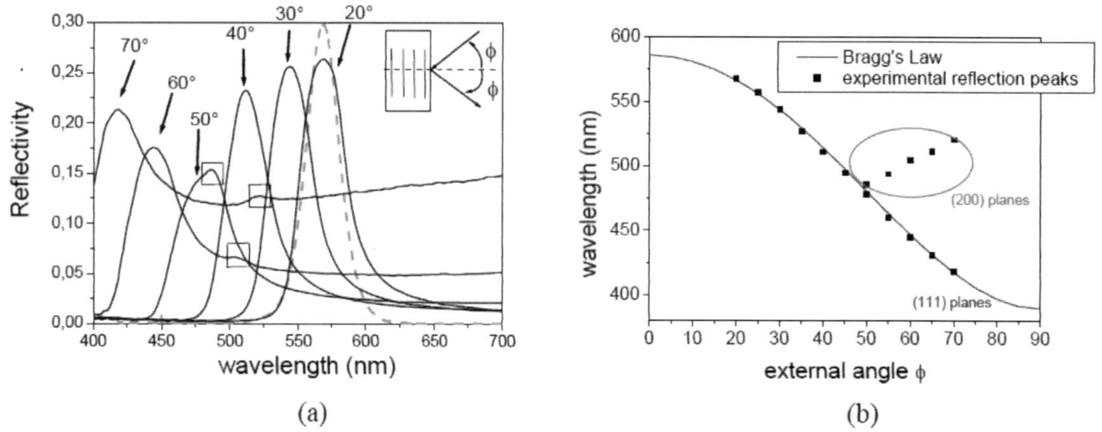

Figure 1. (a) Reflectivity spectra of an opal for various incidence angles. Dashed line: nanocrystals luminescence spectrum. (b) Comparison of the experimental results with the Bragg's law (D=269 nm, n_{eff}=1.34).

The opals characterized and then infiltrated with nanocrystals were prepared by sedimentation of 270 nm diameter silica balls in suspension in water. They were optically characterized by specular reflection. Details on the experimental set-up are given in ref. [32].

The figure 1a shows spectra corresponding to specular reflection for non polarized light at different incidence angles. The maxima are obtained for wavelengths linked to the incidence angles φ by the Bragg's law:

$$\sqrt{(n_{eff}^2 - \sin^2 \phi)}_{22}$$

$$\lambda_{max} = 2\sqrt{2/3}^{\,2}{}_{3}\, D\,(n_{eff} - \sin \varphi)$$

where D is the diameter of the balls, φ is the incidence angle relative to the normal to the (111) plane and n_{eff} the effective refractive index of the medium. The high measured reflectivity (about 20% in spite of the fact that we illuminate a large area of the sample because of our large spot size) demonstrates the good quality of the samples. In figure 1b the values of λmax deduced from the above spectra are plotted according to the incidence angle. An adjustment with the Bragg's law yields to a diameter of the balls D=269 nm with a precision of 5% and to an effective refractive index of the medium n_{eff}=1.34, corresponding to an index of silica n_{silica}=1.44 (deduced from the expression

$$n_{eff} = \sqrt{\alpha \varepsilon_{silica} + (1-\alpha)\varepsilon_0}$$

where α=0.74 represents the filling factor for closed packed structures). This diameter is in good agreement with the measurements performed with an atomic force microscope.

3. Modification of nanocrystals spontaneous emission diagram

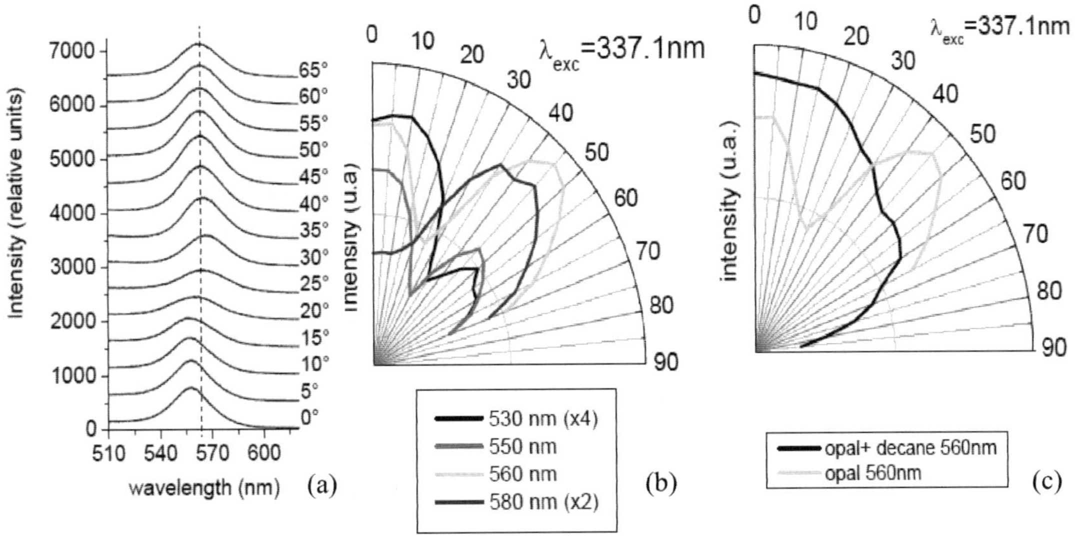

Figure 2. (a) Luminescence spectra of infiltrated nanocrystals in an opal sample at different collection angles. (b) Diagrams of emission of infiltrated nanocrystals at different wavelengths.
(c) Diagrams of emission at 560 nm in opal (green plain line) compared to nanocrystals in opal infiltrated with decane (dashed black line).

We first study the emission diagram modification of nanocrystals embedded in the opal. We infiltrate 1 μl of a nanomolar solution of CdSe/ZnS nanocrystals (Qdot Invitrogen 565 ITK) diluted in decane. Briefly, the nanocrystals are excited by a pulsed nitrogen laser at 337.1 nm, their luminescence is collected by an optical fiber mounted on a rotary stage and analyzed with a spectrometer coupled with a nitrogen-cooled Si CCD detector (see [32] for experimental details). Depending on the angle of collection, the central wavelength of the luminescence spectra is shifted from 556 nm to 566 nm and the linewidth is reduced from 30 to 25 nm (see figure 2a). In figure 2b, the radiation diagrams are plotted according to the angle of detection for different wavelengths. For each wavelength, a strong reduction of luminescence is observed for specific angles, demonstrating the modification of the nanocrystals emission diagram by the photonic bandgap of the opal. In order to confirm the role played by the photonic crystal, the opal is filled with decane (ndecane = 1.41). The decane infiltrated opal can be considered as a homogeneous medium. In this case, the dip in the emission diagram disappears (see figure 2c). In figure 3, the wavelengths for which the luminescence is reduced are plotted as a function of the external angle. The curse is well fitted by the Bragg's law, like the central wavelengths of the specular reflection spectra previously measured (see figure 1).

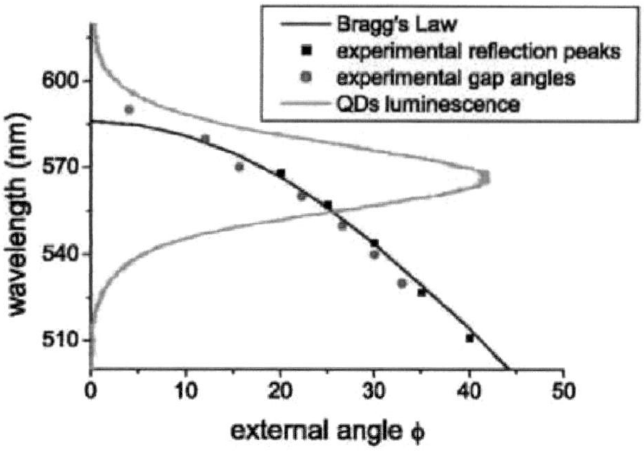

Figure 3. Comparison of the experimental results with Bragg's law.

4. Modification of the decay time

We now consider the influence of the opal on the nanocrystals decay time. We experimentally compare the lifetime of the nanocrystals embedded in an opal and in the same opal infiltrated with a solution of decane which provides a uniform medium used as a reference. The experimental set-up is the same as the one used for the luminescence study. The emission at 560 nm is selected by a spectrometer and the lifetime measurement is provided by a photomultiplier. The resolution of our setup is equal to one nanosecond. As shown on figure 4, the self-luminescence of the silica balls influence on the decay rate becomes negligible after 40 ns.

We experimentally demonstrate an increase of the lifetime of 16% (28 *ns* ±1 *ns* for quantum dots embedded in the opal, 24 *ns* ±1 *ns* for quantum dots embedded in the opal filled with decane). The same experiment performed with similar nanocrystals infiltrated in a 320 nm balls diameter opal gives the same result.

In order to explain these lifetime modifications, we have performed a calculation of the variation of the local density of states (LDOS) at the emitters position [32]. The 16% increase of the lifetime observed for the 320 nm balls diameter opals is in good agreement with the LDOS calculations, whereas, for the 269 nm balls diameter opals, no decay time modification is predicted. These results prove that the observed lifetime variation is due to an effect not fully taken into account in the LDOS calculation. In the previous calculations, we have assumed that the dipole moment is randomly oriented and that its value is kept constant whatever the environment is. Our experimental results suggest that this assumption is too restrictive and that the variations of the dipole moment due to the local electromagnetic field may play a key role, which can hide the effect of the photonic bandgap on the radiative lifetime.

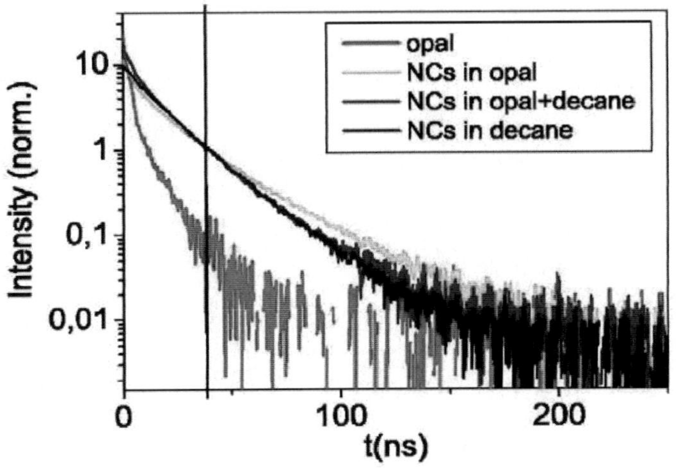

Figure 4. Luminescence decay of nanocrystals embedded in an opal (green line), dissolved in decane (black line), in the opal infiltrated with decane (blue line) Self-luminescence of the opal (red line).

5. Planar defects

The behavior of nano-emitters in PCs is somehow well understood. To go further in controlling the light emission, one has to introduce defects in the photonic structure. This leads to the occurrence of localized states in the forbidden region of the band gap [33]. The defect results in the observation of a pass band in the photonic band gap. The appearance of the defect mode changes with the thickness of the crystal, the dip decreasing as the thickness increases. The position of the pass band is a function of the size (thickness, diameter of the balls) of the defect. A point-like defect could serve as a microcavity to localize light: a high-Q single mode cavity could be created by the introduction of a local defect in a PC, leading to a thresholdless laser. Linear defects offer great technological appeal as they can serve as optical linear waveguides. As it adds an extra dimension, the introduction of planar defects in PCs is a way to strongly increase the electromagnetic field confinement and leads to planar waveguides [34]. Introducing well-defined defects in PCs which are necessary to guide the photons through the crystal remains a hard technological challenge. In synthetic opals, several top-down methods have been investigated like multi-photon polymerization and laser micro-annealing introducing defects like points and lines [35, 36] or photolithography allowing linear defect [37]. These methods are precise and allow design flexibility; but they are time consuming and seem to be limited to small area. We will present different bottom-up routes proposed by different groups to engineer planar defects into colloidal PCs. Planar defects have been fabricated by combining convective colloidal self-assembly and chemical vapour deposition to introduce a silica thin layer between two opal PCs [38, 39]. While these defects are passive, the Ozin group has also proposed active and tunable planar defects by incorporating polyelectrolyte multilayers having refractive index and thickness tuned by exposure to different solvent vapour pressures [40], UV-exposure as well as temperature. In this latter case, they obtain a full reversibility [41]. Another way to introduce a planar defect has been proposed by incorporating a monolayer of microspheres prepared by the Langmuir-Blodgett technique between two opal films with spheres of different diameters prepared by convective self-assembly. The LB technique supplies a well-defined single layer when it is difficult to realize such objective by using other self-assembly methods [42, 43]. Recently, the whole structure, the planar defect between two opals, has been prepared by the same LB technique, providing good quality samples [44]. The kind of defect introduced in the PC depends on the size of the spheres: when the size of the middle layers spheres is larger /lower than that of the lower and upper spheres it is a donor/acceptor like impurity peak which is relatively closer to the high-energy edge/low energy edge (low/high wavelength) of the optical stop band [45]. The technologies developed to engineer planar defects are now mastered. The next step consists in introducing emitters such as nanocrystals in such defects. As the LDOS scales like the square of the field, strong effects on the emission and particularly on the lifetime are expected.

References

[1] Lodahl P, Driel F van, Nikolaev I S, Irman A, Overgaag K, Vanmaekelbergh D and Vos W L 2004 *Nature* **430** 654
[2] Fleming J G and Lin Shawn-Yu 1999 *Optics letters* **24** 49
[3] H B Sun, S Matsuo and H Misawa 1999 *App. Phys. Lett.* **74** 786
[4] Deubel M, Freymann G Von, Wegener M, Pereira S, Bush K and Soukoulis C M 2004 *Nature Materials* **3** 444
[5] Mizeikis V, Seet K Khuen, Juodkazis S and Mizawa H 2004 *Optics Letters* **29** 2061
[6] Divliansky, Mayer T S, Holiday K Sand Crespi V 2003 *App. Phys. Lett.* **82** 1667
[7] Campbell M, Sharp D N, Harrison M T, Denning R G and Turberfield A 2000 *Nature* **404** 53
[8] Miklyaev Y V, Meisel D C, Blanco A, Freymann G von, Bush K, Koch W, Enkrich C, Deubel M and Wegener M 2003 *App. Phys. Lett.* **82** 1284
[9] Lai N D, Wen P L, Jian H L, Chia C H and Cheng H L 2005 *Optics Express* **13** 9605
[10] Astratov V N, Vlasov Yu A, Karimov O Z, Kaplyanskii A A, Musikhin Y U G, Bert N A, Bogomolov V N, Prokofiev A V 1996 *Physics Letters A* **222** 349
[11] Miguez H , Lopez C, Meseguer F, Blanco A , Vasquez L, Mayoral R, Ocana M, Fornes V and Mifsud A 1997 *App. Phys. Lett.* **71** 1148
[12] Miguez H, Meseguer F, Lopez C, Mifsud A, Moya J.S and Vasquez L 1997 *Langmuir* **13** 6009
[13] Braun P V, Zehner R W, White C A, Weldon M K, Kloc C, Patel S S and Wiltzius P 2001 *Adv. Mater.* **13** 721
[14] Jiang P, Bertone J F, Hwang K S and Colvin V L 1999 *Chemical Materials* **11** 2132
[15] Vlasov Y A, Xiang-Zheng Bo, Sturm J C and Norris D J 2001 Nature **414** 289
[16] Wong S, Kitaev V and Ozin G A 2003 *J. Am. Chem. Soc.* **125** 15589
[17] Szekeres M, Kamalin O, Schoonheydt R A, Wostyn K, Clays K, Persoons A and Dekany I 2002 *J. Mater. Chem.* **12** 3268
[18] Reculusa S and Ravaine S 2003 *Chem. Mater* **15** 598
[19] Barth M, Schuster R, Gruber A and Cichos F 2006 *Phys. Rev. Lett* **96** 243902
[20] Blanco A, Lopez C, Mayoral R, Miguez H, Meseguer F, Mifsud A and Herrero J 1998 *App. Phys. Lett.* **73** 1781
[21] Valsov Y A 1997 *App. Phys. Lett.* **71** 1616
[22] Lin Y, Zhang J, Sargent E H and Kumacheva E 2002 *App. Phys. Lett.* **81** 3134
[23] Chuang C M, Lu W B, Su W F, Lin C M and Chen Y F 2005 *J Applied Physics* **97** 096104
[24] Liu K, Schmedake T, Daneshvar K and Tsu R 2007 *Microelectronics Journal* **38** 700
[25] Megens M Wijnhoven, J E G, Lagendijk A and Vos W L 1999 *Phys. Rev. A* **59** 4727
[26] Gaponenko S 2000 *J. Lumni.* **152** 87
[27] Romavov S G, Maka T, Torres C M S, Müller M and Zentel R 2002 *J. Appl. Phys.* **91** 9426
[28] Vallée R A L, Baert K, Kolaric B, Auweraer M V der and Clays K 2007 *Phys. Rev. B* **76** 045113
[29] Li Z Y and Zhang Z Q 2001 *Phys. Rev. B* **63** 125106
[30] Solovyev V 2003 *J. Appl. Phys.* **94** 1205
[31] Nikolaev I S, Lodahl P, Driel A F van, Koenderink A F and Vos W L 2007 *Phys. Rev. B* **75** 115302
[32] Vion C, Barthou C, Bénalloul P, Schwob C, Coolen L, Gruzintev A, Emelchenko G, Masalov W, Frigerio J-M and Maître A *submitted to Journal of Applied Physics*
[33] Yablonovitch E, Gmitter T, Meade R D, Rappe A M, Brommer K D, Joannopoulos J D 1991 *Phys. Rev. Lett.* **67** 3380
[34] Joannopoulos J D, Meade R D and Winn J N 1995 *Photonic crystals* (New Jersey: Princeton Press)
[35] Lee W, Pruzinsky S A and Braun P V 2002 *Adv. Mater.* **14** 271
[36] Tétreault N, Miguez H, Yang S M, Kitaev V and Ozin G A 2003 *Adv. Mater.* **15** 1167
[37] Vekris E, Kitaev V, Freymann G von, Petrovic D D, Aitchison J Stewart and Ozin G A 2005

Adv. Mater. **17** 1269

[38] Palacios-Lidon E, Galisteo-Lopez J F, Juarez B H and Lopez C 2004 *Adv. Mater.* **16** 341

[39] Tétreault N, Mihi A, Miguez H, Rodriguez I, Ozin G A, Meseguer F and Kitaev V 2004 *Adv. Mater.* **16** 346

[40] Tétreault N, Arsenault A C, Mihi A, Wong S, Kitaev V, Manners I, Miguez H and Ozin G A 2005 *Adv. Mater.* **17** 1912

[41] Fleischhaker F, Arsenault A C, Kitaev V, Peiris F C, Freymann G von, Manners I, Zentel R and Ozin G A 2005 *J. Am. Chem. Soc.* **127** 9318

[42] Zhao Y, Wostyn K, Schaetzen G de, Clays K, Hellemans L, Persoons A, Szekeres M, Schoonheydt R A 2003, *App. Phys. Lett.* **82** 3764

[43] Wostyn K, Zhao Y, Schaetzen G de, Hellemans L, Matsuda N, Clays K and Persoons A 2003 *Langmuir* **19** 4465

[44] Masse P, Reculusa S, Clays K, Ravaine S 2006 *Chemi. Phys. Lett.* **422** 251

[45] Pradhan R D, Tarhan I Inanç and Watson G H 1996 *Phys. Rev B* **54** 13721

Alicyclic polyimides - a colorless and thermally stable polymer for opto-electronic devices

Toshihiko Matsumoto, Daisuke Mikami, Takahiro Hashimoto, Motoaki Kaise, Risa Takahashi and Shuichi Kawabata

Center for Nano Science and Technology, Tokyo Polytechnic University, 1583 Iiyama, Atsugi, Kanagawa 243-0297, Japan

E-mail: matumoto@nano.t-kougei.ac.jp

Abstract. Alicyclic polyimides were prepared from polyalicyclic monomers using a two-step method. Light absorption of polyimides was elucidated from quantum chemistry. The alicyclic polyimide films were colorless, and the transparencies in the visible region were over 85%. The semi-aromatic polyimide films had an average refractive index range of 1.599 to 1.617, and that of a full-alicyclic polyimide was 1.522. The birefringencies were nearly zero. Semi-aromatic polyimides were soluble in aprotic polar solvents like DMAc. However, the non-aromatic (full-alicyclic) polyimides were insoluble except sulfuric acid. All the alicyclic polyimides have the 5 % weight loss temperatures in nitrogen over 450 °C, and 400 °C in air. They possess glass transition temperatures over 280 °C. The alicyclic polyimide films have a tensile modulus range of 1.3-3.0 GPa, and a tensile strength range of 38-145 MPa. Especially semi-aromatic polyimide films display high strength. They have an elongation at break range 5-30 %. All the polyimides exhibited amorphous patterns and broad dispersive peaks of diffraction around at 16° (2θ).

Keywords: Alicyclic polyimides, colorless, thermally stable, opto-electronic devices, flexible organic substrates.

1. Introduction

Aromatic polyimides are often insoluble and intractable in their fully imidized form, thus presenting serious processing difficulties. Much endeavor has been focused on synthesizing tractable and processable polyimides with no reasonable sacrifice of their own excellent properties [1]. One of the success strategies is to incorporate pendant alkyl and phenyl groups [2] or fluorine [3], bromine [4], unsymmetrical [5] and heterocyclic units [6] in the polymer backbone. Most of aromatic polyimides intensely absorb visible light and are colored reddish yellow or pale yellow, so they can be hardly used in areas where absence of coloration is an important requirement. By now the origin of the coloration is known to be caused by their highly conjugated structures and charge-transfer (CT) between the diamine (donor) and the dianhydride (acceptor) moieties. The first mention of CT in aromatic polyimides was made by Dine-Hart et al in 1971 [7]. Kotov et al. confirmed the occurrence of CT in aromatic polyimides in a study of the UV-VIS spectra of several poly(pyromellitimide)s and their model compounds [8]. The first example of colorless polyimides prepared from hexafluoroisopropylidene-bridged diamine and dianhydride was reported as a patent by Rogers about 45 years ago [9]. St. Clair et al. focused much attention on colorless polyimides from the viewpoint of

© 2009 IOP Publishing Ltd

the potential applications on space components such as multilayer insulation blankets, solar cell, and thermal control coating systems [10]. They demonstrated that polyimides containing the hexafluoroisopropylidene group exhibited high transparency in the visible region. Many other examples have been reported where dianhydrides with a biphenyl structure, sulfone linkaged diamines and fluorinated monomers were employed [11].

Recently, soluble and colorless polyimides using monomers with alicyclic structures, so-called "alicyclic polyimides" have been reported. The introduction of an alicyclic unit into the polyimide backbone would facilitate less polymer-polymer interaction and enhance the solubility in organic solvents. The lack of color is generally associated with the absence or inhibition of intra- and/or intermolecular CT interactions. These polyimides show excellent thermal stability and the high-temperature stability can also be explained by the introduction of an alicyclic structure, which would foster less probability of main chain scission because of the presence of multibonds and would increase main chain rigidity. Alicyclic polyimides have attracted much attention in recent years because of their potential applications including use as liquid crystal orientation layers, flexible solar cell substrates, organic EL substrates, light guide and low dielectric materials [12].

This article deals with the alicyclic polyimides consisting of each alicyclic dianhydride or alicyclic diamines and those of the both.

2. Experimental

2.1. Reagents and monomer synthesis

Bicyclo[2.2.1]heptane-2,3,5,6-tetracarboxylic dianhydrides was prepared as follows: bismethoxycarbonylation of commercial *endo*-type nadic anhydride with carbon monoxide in methanol in the presence of catalytic amount of 5 % Pd-C and stoichiometric amounts of $CuCl_2$ afforded the tetramethyl ester. The dianhydrides having different configurations (BHDAdx and BHDAxx) were obtained from the ester by an acid- and a base-catalyzed hydrolyses followed by dehydration ring-closure [13]. Cyclopentane-1,2,3,4-tetracarboxylic 1,2;3,4-dianhydride (CpDA) was prepared from *cis,cis,cis,cis*- 1,2,3,4-cyclopentanetetracarboxylic acid by thermal dehydration reaction then sublimation at 170-190 ^{o}C. Bicyclo[2.2.2]octane-2,3,5,6-tetracarboxylic dianhydrides with 2-*endo*, 3-*endo*, 5-*exo*, 6-*exo*-configuration (BODAdx) and its all-*exo* isomer (BODAxx) were synthesized according to our previous paper [14]. Two kinds of tetracyclic teracarboxylic dianhydrides, (4a*rH*,8a*cH*)-decahydro- 1*t*,4*t*:5*c*,8*c*-dimethanonaphthalene-2*t*,3*t*,6*c*,7*c*-tetracarboxylic 2,3:6,7-dianhydride (DNDAxx) [15] and the 2*c*,3*c*,6*c*,7*c*-tetracarboxylic 2,3:6,7-dianhydride (DNDAdx) [16], were prepared according to the previous report. An alicyclic dianhydride having a spiro carbon atom, 5-*endo*-carboxymethylbicyclo[2.2.1]-heptane-2-*exo*,3-*exo*,5-*exo*-tricarboxylic acid 2,3:5,5-dianhydride (BSDA) was prepared according to the literature [17]. 5-(2,5-Dioxotetrahydro-3-furanyl)-3-methyl-3-cyclohexene-1,2-dicarboxylic anhydride (MCTC, Epiclon® B-4400) was supplied from Dainippon Ink & Chemicals Incorporation, Japan.

Bis(aminomethyl)bicyclo[2.2.1]heptane (BBH) was donated by Mitsui Chemical Co. Ltd, and used without further purification. BBH consists of four isomers; 2-*exo*,5-*exo*- (30 wt%), 2-*endo*,5-*exo*- (35 wt%), 2-*exo*,6-*exo*- (20 wt%), 2-*endo*,6-*exo*- (15 wt%). 4,4'-Methylenebis(2-methylcyclohexyl-amine) (MCHM) were purchased from Aldrich Chemical Co. Inc. and Tokyo Kasei Kogyo Co. Ltd., respectively, and used as received. *N,N*-dimethylacetamide (DMAc), *N*-methyl-2-pyrrolidone (NMP), 1,3-dimethyl-2-imidazolinone (DMI), and hexamethylphosphoramide (HMPA) were fractionally distilled over CaH_2 under reduced pressure and stored over molecular sieve 4A.

2.2. General polymerization and film preparation

In a 30 mL three necked flask containing DMAc was placed the diamine (purified by sublimation) while a slow stream of nitrogen was maintained. The tetracarboxylic dianhydride was added into the solution and the mixture was stirred mechanically for 2 days in a nitrogen stream. The resulting poly(amic acid) solutions were cast onto glass plates. The polyimide films were obtained by heating the plates in vacuum.

2.3. Measurements

The ^1H and ^{13}C NMR spectra were obtained using a JEOL JNM-LA500 spectrometer. The proton signal in the ^1H NMR spectrum were assigned in the H,H, C,H-COSY, and NOESY (elimination of J correlation) spectra. Infrared spectra were recorded using a JASCO VALOR III and a JASCO FT/IR 460 PLUS Fourier transform spectrometers. UV-VIS spectra of the polyimide films were recorded on a JASCO V-570 UV/VIS/NIR spectrophotometer. Film thickness and refractive were measured by a Metricon model PC-2010 prism coupler at room temperature in open air. The birefringence of sample films was evaluated by means of ellipsometry. Thermal analyses were carried out using Seiko Instruments SSC 5200-TG/DTA 220 at a heating rate of 10 K/min in a nitrogen atmosphere or in open air. Thermomechanical analyses (TMA) were done using Seiko Instruments TMA/SS 100 equipped with a penetration probe of 1.0 mm diameter and using an applied constant load of 10 g (stress, 0.125 MPa) at a heating rate 10 K/min air. The mechanical properties were examined at room temperature in open air using Seiko Instruments TMA/SS 100 at a drawing rate of 1 mm/min; the sample size (film) was 3 mm long, 2 mm wide, and about 20 μm thick. The wide-angle X-ray diffraction measurements (WAXD) were performed on a Rigaku Rint 2500 X-ray diffractometer with graphite monochromated Cu Kα radiation and a 12 kW (40 kV, 300 mA) rotating anode generator. Inherent viscosities (ηinh) were measured with an Ostwald viscometer in a 0.5 g/dL solution of poly(amic acid) at 30 °C.

2.4. Molecular orbital calculations

MO calculations were done with an IBM ThinkPad X31 personal computer (CPU:Pentium M processor 1400MHz). The structures of model compounds for the polyimides were geometrically optimized using AM1 in a software package of WinMOPAC Ver. 3 (Fujitsu Ltd.) which was transplated from MOPAC2000. The electron absorption wavelengths were calculated based on the semi-empirical INSO/S Hamiltonian with configuration interaction (CI) wavefunctions in a software package of MOS-F Ver. 4.2 (Fujitsu Ltd.). In the excited state expressions of all electronic states (20 x 20 = 400 states) are included; they are generated by singly exciting all electrons in the 20 highest occupied MOs to the 20 lowest unoccupied MOs with respect to the ground state [18].

3. Results and discussion

3.1. Polymer synthesis

The structures of monomers used in the present study are illustrated in figure 1 along with their abbreviations. Alicyclic polyimides were prepared via two steps: the first step includes the poly(amic acid) formation at room temperature, and as the second step, the viscous polymer solutions are deposited onto the glass substrates and thermally cyclodehydrated at elevated temperature to yield the corresponding polyimide. The dianhydrides with a bicyclooctane structure, BODAdx and BODAxx, were less reactive than those with the bicycloheptane structure, BHDAdx and BHDAxx, due to the smaller strain of the ring structure. Therefore, even during the first polymerization step, a higher temperature was needed in order to obtain high molecular weight poly(amic acid)s which could be converted to the polyimides with film formability after the thermal cyclohydration processing.

For polycondensations of an aromatic diamine with aromatic or aliphatic dianhydrides, the addition order of the monomers does not play a significant role, although higher molecular weights are obtained for many poly(amic acid)s when the anhydride is added to the diamine rather than the reverse addition. However, in the reaction using an aliphatic diamine with a polyalicyclic structure such as BBH, the order of addition becomes significant. A viscous poly(amic acid) solution was obtained only when BBH was slowly added to the the dianhydride solution. A precipitate developed when adding the dianhydride to the BBH solution. The result can be explained by salt formation [19].That is, in the latter case, the carbonyl group of poly(amic acid) that was produced by the reaction between the diamine and the dianhydride is easily reacted with an excess of the free diamine to form the salt. Semi-aromatic poly(amic acid)s have an inherent viscosity (ηinh) range of 1.00-0.53 dL/g (0.5 g/dL, 30 °C), and the flexible and free-standing polyimide films were obtained after thermal curing. The ηinh's of the non-aromatic poly(amic acid)s were as low as 0.08-0.23 dL/g, however, they formed flexible films.

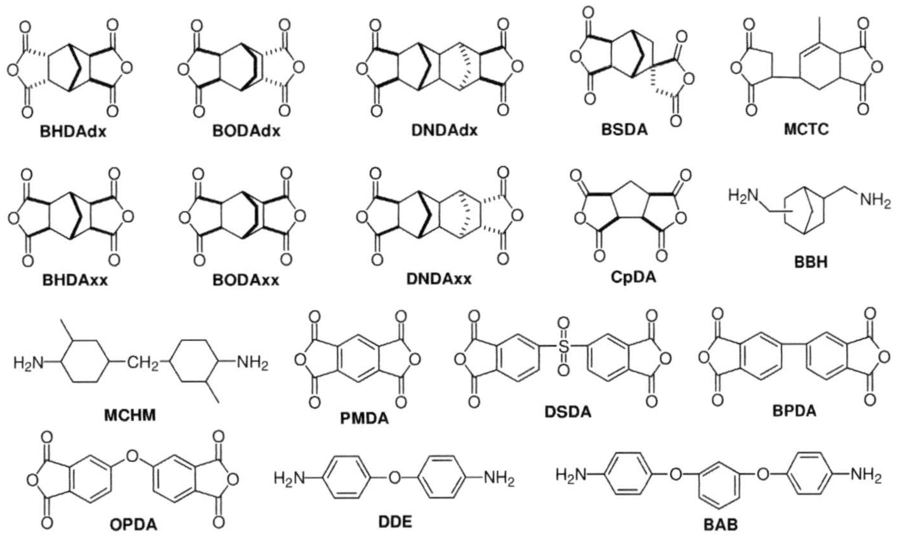

Figure 1. Structures and abbreviations of monomers for polyimide synthesis.

3.2. UV-VIS spectra of polyimide films and quantum chemical consideration

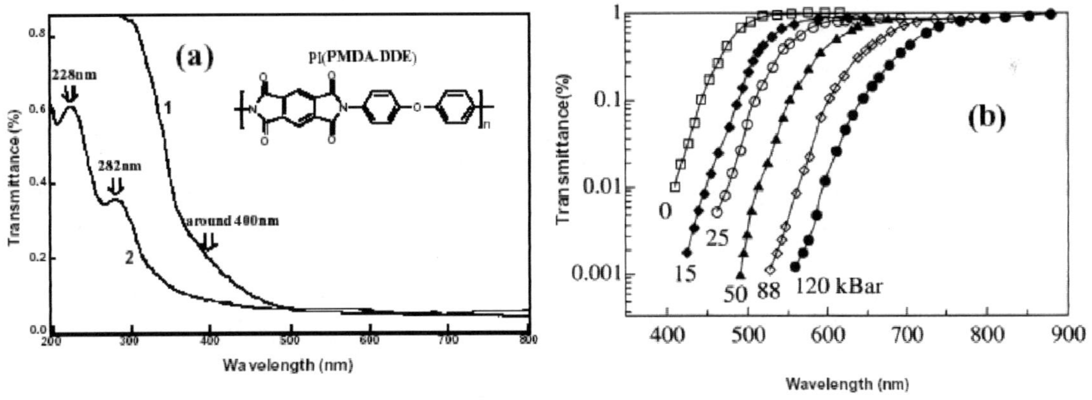

Figure 2. a) Absorption UV-VIS spectra of thin Kapton®-type PI(PMDA-DDE) films (1: ca. 1μm, 2: ca. 0.1 μm) formed on quartz plates [19]. b) The effect of pressure on the light transparency of 7.5 μm thick Kapton® film [21].

The absorption UV-VIS spectra of the Kapton®-type polyimide PI(PMDA-DDE) films formed on quartz plates using a spin-coating technique were determined and are shown in figure 2a. PI(PMDA-DDE) denotes the polyimide obtained from dianhydride PMDA and diamine DDE. The weak absorption is observed around 400 nm as a shoulder, and the peak is considered to the observed λmax of the polyimide. In 1970, at first Bikson directed scientific eye to discoloration aromatic polyimides [20]. Dine-Hart and Kotov reported that the coloration is caused by charge transfer complex [7, 8].

However, it has still not been confirmed whether the CT occurs intermolecularly or intramolecularly. Figure 2b shows the effect of pressure on the light transparency of 7.5 μm thick Kapton® film [21]. Reversible spectral change within 120 kBar is considered to be caused by an intermolecular CT.

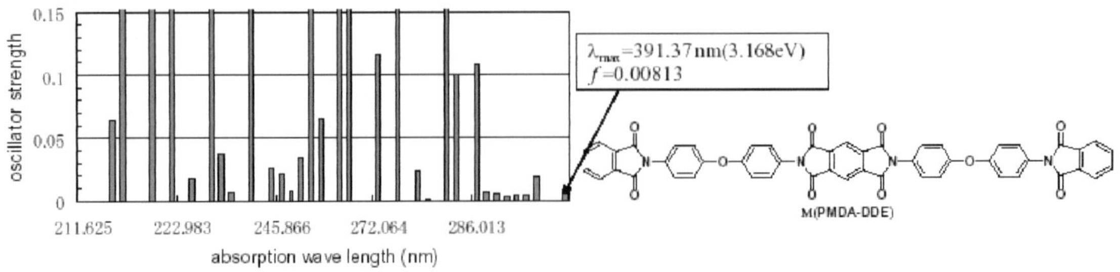

Figure 3. The structure of a model compound M(PMDE-DDE) for Kapton® polyimide and the calculated light absorption spectrum of the model.

Light absorption of aromatic polyimides was elucidated from quantum chemistry that is using the MO calculations for the model compound. The structures were geometrically optimized by means of an AM1 method in MOPAC2000 and the MOs in the ground and excited states for the optimized structures were calculated using INDO/S method in MOS-F software. Figure 3 shows the structure of a model compound M(PMDE-DDE) for Kapton® polyimide and the calculated light absorption spectrum of the model. The calculated λmax is 392 nm, this value is in good agreement with the observed one around 400 nm. The wave function of the first excited state corresponding to 392 nm light absorption for the model, $|\Phi_e\rangle$ is given in equation (1), where $|\Psi_{HOMO\rightarrow LUMO}\rangle$ is monoexcited configuration state function. In this state an electron is excited from the HOMO to the LUMO. The CI coefficients of the first excited state indicate that HOMO-LUMO transition is the major electronic transition, and the calculated contribution of the first excited state is 48 % for the M(PMDA-DDE).

$$|\Phi_{e,PMDA-DDE}\rangle = 0.691\,|\Psi_{HOMO\rightarrow LUMO}\rangle - 0.567\,|\Psi_{HOMO-2\rightarrow LUMO}\rangle - 0.238\,|\Psi_{HOMO-1\rightarrow LUMO+4}\rangle \quad (1)$$

The difference in electron density between the ground and the first excited states was calculated and is illustrated in figure 4. The dark-gray area corresponds to regions of increased electron density in the excited state, while the light gray one corresponds to depleted regions from the ground state. From both the MOs and the charge difference maps, it is found that the benzene ring derived from the diamine DDE acts as an electron donor, and that the five-membered imide and benzene rings derived from the dianhydride PMDA act as electron-withdrawing groups. Therefore, the first excited state is characterized by a strong intramolecular CT.

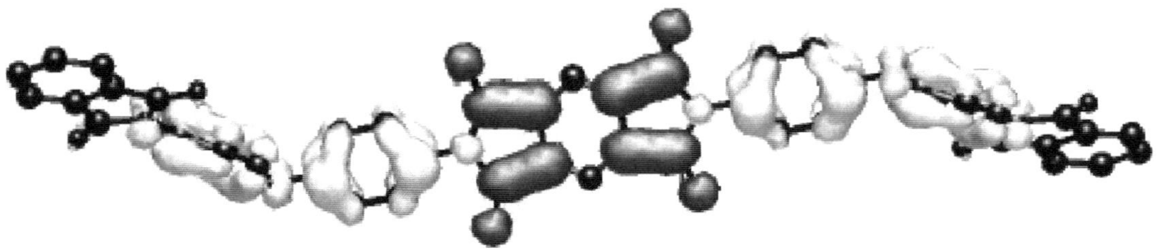

Figure 4. Difference in electron density between the ground state and the first excited states.

Table 1. INDO/S calculated electronic properties of PI model compounds.

model compd.	ω_{ge} [a]	λ_{max} [b]	f [c]	Main SCF [d]		
				MOs	CI coef	Contrib [e]
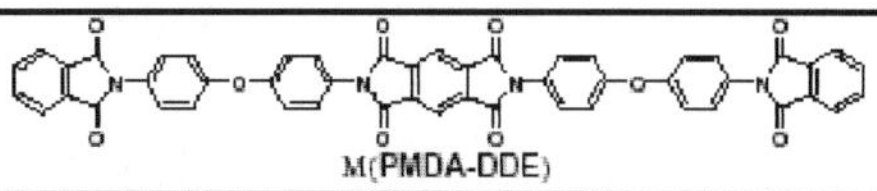 M(PMDA-DDE)	3.17	391	0.008	HOMO → LUMO	0.691	48
				HOMO-2 → LUMO	-0.567	32
				HOMO-1 → LUMO+4	-0.238	6
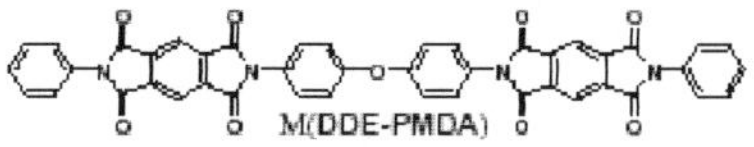 M(DDE-PMDA)	3.20	387	0.010	HOMO → LUMO	0.698	49
				HOMO-1 → LUMO	-0.436	19
				HOMO → LUMO+4	0.258	7
				HOMO-3 → LUMO	-0.246	6
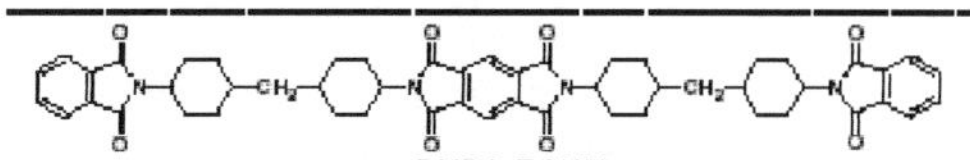 M(PMDA-DCHM)	3.38	366	0.002	HOMO-12 → LUMO	-0.700	49
				HOMO-16 → LUMO+4	-0.502	25
				HOMO-15 → LUMO+7	0.279	8
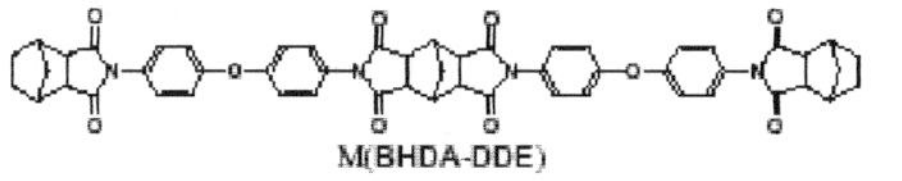 M(BHDA-DDE)	3.76	330	0.006	HOMO-10 → LUMO	-0.600	36
				HOMO-11 → LUMO+1	0.461	21
				HOMO-16 → LUMO+4	0.319	10
				HOMO-19 → LUMO+5	0.222	5
				HOMO-19 → LUMO+13	-0.221	5
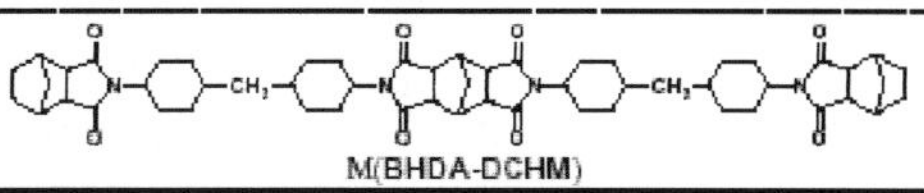 M(BHDA-DCHM)	3.79	327	0.003	HOMO-6 → LUMO	0.621	39
				HOMO-9 → LUMO+1	-0.505	25
				HOMO-10 → LUMO+4	-0.396	16
				HOMO-11 → LUMO+5	-0.363	13

[a]transition energy (eV) [b]transition wavelength (nm) [c]oscillator strength [d]monoexcited configuration state function
[e]contribution of the CSF to the excited state which is given as 100(CI coef.)^2 (%)

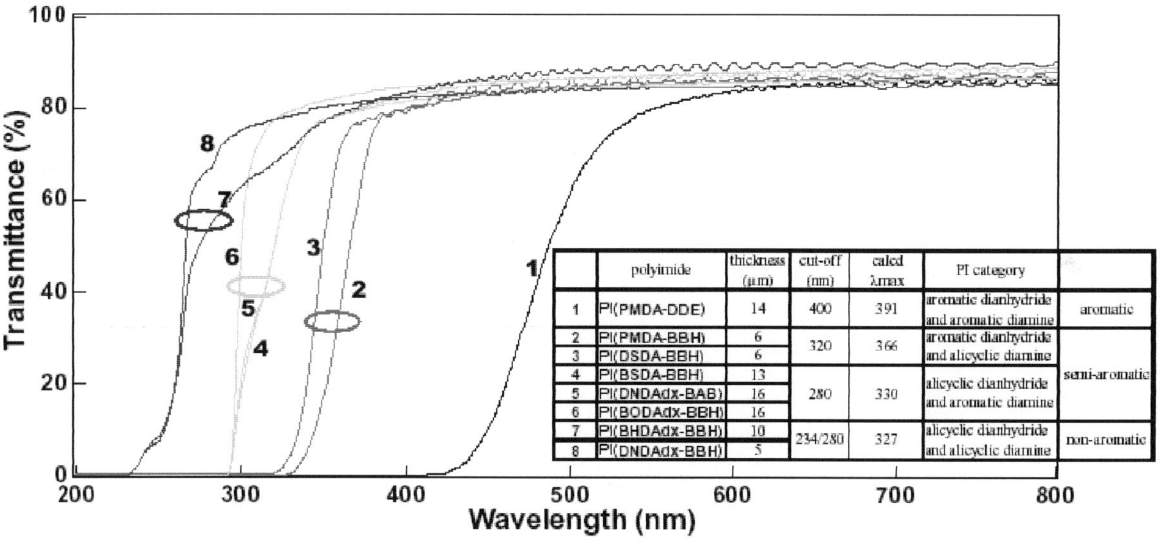

	polyimide	thickness (μm)	cut-off (nm)	calcd λmax	PI category	
1	PI(PMDA-DDE)	14	400	391	aromatic dianhydride and aromatic diamine	aromatic
2	PI(PMDA-BBH)	6	320	366	aromatic dianhydride and alicyclic diamine	semi-aromatic
3	PI(DSDA-BBH)	6				
4	PI(BSDA-BBH)	13	280	330	alicyclic dianhydride and aromatic diamine	
5	PI(DNDAdx-BAB)	16				
6	PI(BODAdx-BBH)	16				
7	PI(BHDAdx-BBH)	10	234/280	327	alicyclic dianhydride and alicyclic diamine	non-aromatic
8	PI(DNDAdx-BBH)	5				

Figure 5. Transmission UV-VIS spectra of free-standing polyimide films.

Similarly, the calculated λmax's and configuration state functions of various kinds of model compounds were estimated, and results are summarized in Table 1. Semi-aromatic polyimide model M(PMDA-DCHM) obtained from an aromatic dianhydride and an alicyclic diamine has a λmax at 366nm in UV area. The HOMO-12 denotes the twelfth orbital below HOMO. The HOMO-12 and the LUMO correspond to a non-bonding and a π* orbital, respectively, therefore the main contributing transition is assigned to the n-π* transition. From both the MOs and the difference electron density between the ground and the first excited states, the first excited state is attributable to light-induced polarization in the five-membered imide ring. The other semi-aromatic polyimide model obtained from an alicyclic dianhydride and an aromatic diamine M(BHDA-DDE) has a λmax at 330nm in UV area. Non-aromatic (full-alicyclic) polyimide shows a λmax at 327 nm.

The transmission UV-VIS spectra of free-standing polyimide films are shown in figure 5 compared with calculated λmax. The semi-aromatic polyimide films exhibit a cut-off wavelength around 300 nm and are colorless, whereas an aromatic Kapton-type polyimide PI(PMDA-DDE) has a λmax at 400 nm. A cut-off wavelength is defined here as the point where the transmittance becomes less than 1% in the spectrum. The cut-off wave lengths of full-alicyclic polyimides were observed at 234 nm and the films were entirely colorless. The order in cut-off wavelength of the polyimide films is essentially the same as obtained for calculated λmax of their model compounds.

3.3. Refractive index, birefringence, and optically estimated dielectric constant

The dielectric constant of the material at optical frequencies can be estimated roughly from the refractive index n according to Maxwell's equation $\varepsilon = n^2$. The ε of polyimide around 10 GHz is evaluated to be $\varepsilon = 1.1n^2$, including an additional contribution of approximately 10 % from the orientation and ionic polarization [22]. Refractive index, birefringence, and optically estimated dielectric constant of free-standing polyimide films are summarized in table 2 together with transparency and glass-transition temperatures. Transparency in the visible region was evaluated by averaging the transmittances in the range from 400 to 780 nm in the UV-VIS spectrum. An average refractive index was calculated following the equation, $n_{av} = (2n_{TE} + n_{TV})/3$, where n_{TE} and n_{TV} indicate the in-plane and out-of-plane refractive indices of the sample polyimide films, respectively. Birefringence Δn, the difference between in-plain and out-of-plain refractive indices gives information on the three dimensional structure of polymer, spatial arrangement of the structure units to contribute to a refractive index, e. g. imide groups and benzene rings. The n_{av} of the full-alicyclic polyimide film PI(BHDA-BBH) is as low as 1.52, which translates into a dielectric constant of 2.55. The decreasing aromatic character of the polyimide diminishes the refractive index as exemplified by the comparison of semi-aromatic and aromatic polyimide in agreement with one generalization that aromatic polymers possess high refractive indices. Aromatic polyimides exhibit a large birefringence which indicated the preference of the polymer chains to orient in the plane of the film. The negligibly small birefringence in semi- and non-aromatic (full-alicyclic) polyimides implies that the polymer chains are randomly oriented in the film due to a weak polymer-polymer interaction.

Table 2. Film thickness, optical properties, and glass transition temperatures of polyimide film.

polyimide	d (μm) [a]	T(%) [b]	n_{TE} [c]	n_{TM} [d]	n_{AV} [e]	Δn [f]	ε [g]	Tg(°C) [h]
PI(**BODAdx-DDE**)	4.6	94	1.605	1.588	1.599	0.017	2.81	383
PI(**BODAdx-BAB**)	14.7	92	1.615	1.615	1.615	0.000	2.87	211
PI(**BODAxx-DDE**)	13.1	86	1.605	1.588	1.600	0.017	2.82	385
PI(**BODAxx-BAB**)	11.3	90	1.617	1.617	1.617	0.000	2.88	225
PI(**DNDAdx-BAB**)	12.3	90	1.607	1.594	1.603	0.013	2.83	257
PI(**BSDA-BAB**)	11.8	86	1.616	1.601	1.611	0.015	2.85	227
PI(**BHDAdx-BBH**)	14.8	ND[i]	1.522	1.522	1.522	0.000	2.55	297
PI(**PMDA-DDE**)[j]	6.2	67	1.715	1.636	1.688	0.079	3.13	ND[i]

[a] Film thickness. [b] Average transmittance in the visible region (400-780nm). [c] The in-plane refractiveindices. [d] The out-of-plane refractive indices. [e] Average refractive index; $n_{AV} = (2n_{TE} + n_{TM})/3$. [f] Optical anisotropy (birefringency); $\Delta n = n_{TE} - n_{TM}$. [g] Optically estimated dielectric constant; $\varepsilon = 1.10 n_{AV}^2$. [h] The glass transition temperature by TMA. [i] Not determined. [j] By Noe et al.

3.4. Qualitative solubility of alicyclic polyimides

The qualitative solubility of the alicyclic polyimide films is shown in table 3. The introduction of polyalicyclic structure as a dianhydride or a diamine unit improved the solubility of the polyimides with retention of excellent thermal stability due to the maltibond and rigidity of the polyalicyclic structure. Semi-aromatic polyimides, I and II series, are soluble in aprotic polar solvents like DMAc, however all the full-alicyclic polyimide films are insoluble except sulfuric acid.

Table 3. Qualitative solubility of alicyclic polyimides.

solvent	Ia	Ib	Ic	IIa	IIIa	IIIb	IIIc
N-methyl-2-pyrrolidone	*vs*	*vs*	*vs*	*s*	*sw*	*i*	*i*
N,N-dimethylacetamide	*vs*	*vs*	*vs*	*s*	*sw*	*i*	*i*
1,3-dimethyl-2-imidazolinone	*s*	*vs*	*s*	*s*	*sw*	*sw*	*i*
dimethyl sulfoxide	*s*	*vs*	*s*	*i*	*sw*	*sw*	*i*
pyridine	*s*	*vs*	*vs*	*vs*	*sw*	*i*	*i*
m-cresol	*s*	ND	*vs*	*vs*	*sw*	*i*	*i*
chloroform	*i*	*i*	*vs*	*vs*	*i*	*i*	*i*
dioxane	*i*	*i*	*s*	*i*	*i*	*i*	*i*
benzene	*i*	*i*	*i*	*i*	*i*	*i*	*i*
hexane	*i*	*i*	*i*	*i*	*i*	*i*	*i*
sulfuric acid	*vs*	*vs*	*vs*	*vs*	*s*	*s*	*s*

vs: soluble at r.t.; *s*: soluble on heating , *sw*: swelling, *i*: insoluble even on heating; ND: not determined.

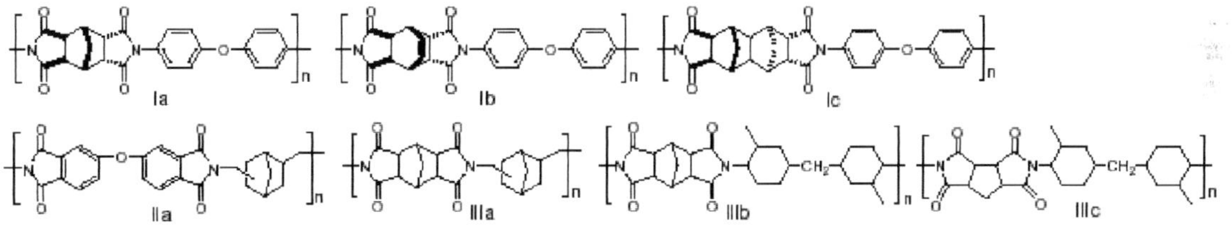

3.5. Thermal properties of alicyclic polyimides
The thermogravimetric analysis (TGA) of polymer films was done at a heating rate of 10 K per minute in nitrogen atmosphere. TGA profiles of the representative alicyclic polyimide films are displayed along with that of polyethylene in figure 6. Full-alicyclic polyimide PI(DNDA-BBH) has good thermal stability with no significant weight-loss up to approximately 350 °C, and the 5 % weight loss temperature, T5, is 459 °C. T5 of semi-aromatic polyimide PI(DNDA-DDE) is as high as 514 °C, which is comparable to that of an aromatic polyimide. The 5 % weight loss temperatures in nitrogen and air and the glass transition temperature of alicyclic polyimides were measured and the results are summarized in table 4 along with Kapton type polyimide PI(PMDA-DDE). All the alicyclic polyimides have the 5 % weight loss temperatures in nitrogen over 450°C, and 400 °C in air. They possess glass transition temperatures over 280 °C. The degradation of polymers in an inert atmosphere is caused by homolytic C-C bond scission. If the polymer backbone consists of a single-chain like that of poly(ethylene), the scission results in direct decreasing of the molecular weight. In most cases, volatile compounds are produced and released as a decomposed gas. On the other hand, in the case of step-ladder type polymers bearing double strands like the alicyclic polyimides, even if the C-C bond scission occurs, it does not lead to a significant decrease in the molecular weight or to the production of volatile compounds. The introduction of a polyalicyclic structure into the polymer backbone would facilitate less polymer-polymer interaction and increased main chain rigidity, and, additionally possibly less probability of main chain scission because of having multibonds.

Journal of Physics: Conference Series **187** (2009) 012005 doi:10.1088/1742-6596/187/1/012005

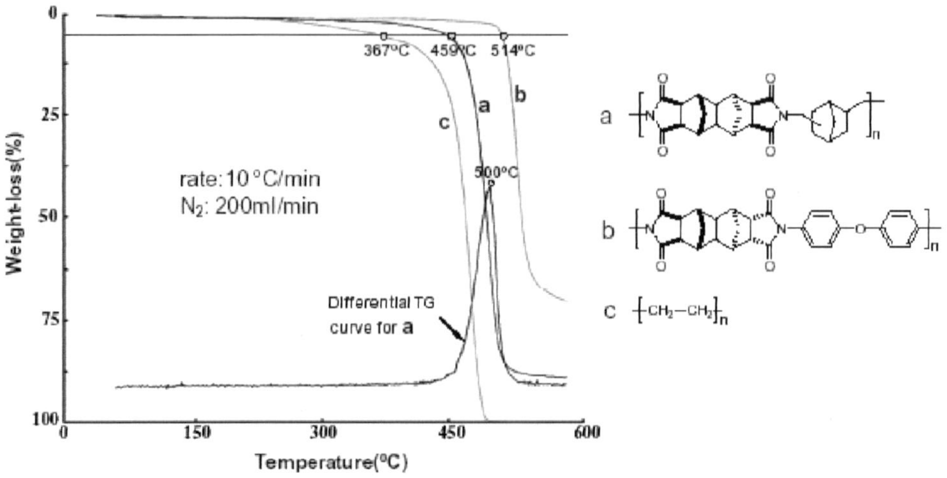

Figure 6. TGA Profiles of alicyclic polyimides [16].

Table 4. Qualitative solubility of alicyclic polyimides.

polymer	T5/N_2(°C) [a]	T5/air(°C) [b]	Tg(°C) [c]
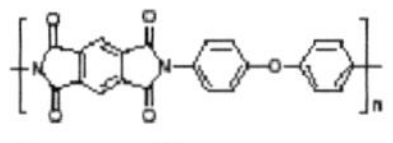	515	480	429
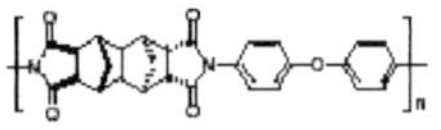	514	470	404
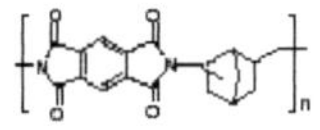	448	406	282
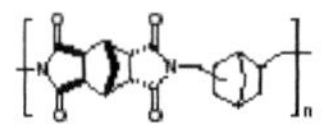	455	411	297
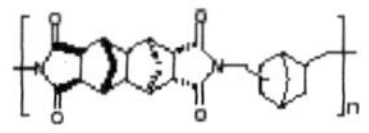	459	416	340
polyethylene	367	352	135

[a] in N_2, [b] in air 5% weight loss temperature [c] by TMA

3.6. Mechanical properties

Mechanical properties of alicyclic polyimide films were examined at room temperature, and the results are listed in table 5. The polyimide films have a tensile modulus range of 1.3-3.0 GPa, and a tensile strength range of 38-145 MPa. They have an elongation at break range 5-30%. The full-alcyclic polyimide film PI(BHDAdx-BBH) possessed a tensile modulus of 2.1 GPa and a tensile strength of 52 MPa, and these values can compete in terms of strength with those of a commercial polycarbonate (PC) [23].

Table 5. Mechanical properties of alicyclic polyimide films.

polymer	TM(GPa) [a]	TS(MPa) [b]	EB(%) [c]
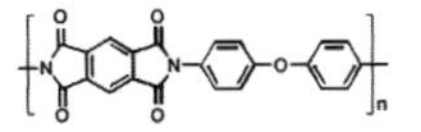	3.5	160	40
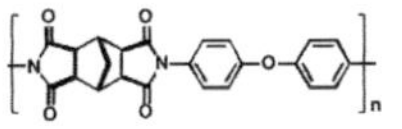	1.3	145	15
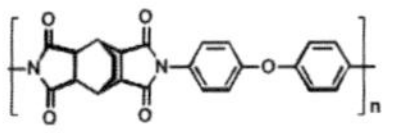	2.0	96	11
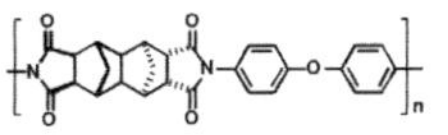	2.3	112	30
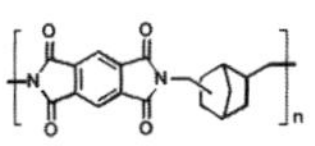	3.0	115	5
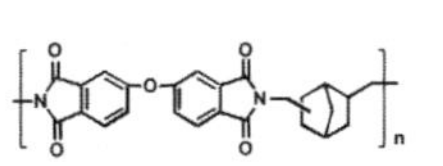	1.6	84	6
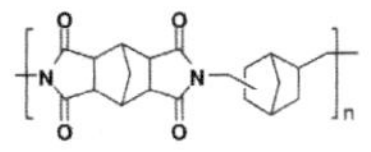	2.1	52	6
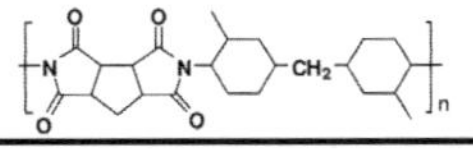	1.3	38	11

[a] Tensile modulus [b] Tensile strength [c] Elongation at break

3.7. Reflection WAXD analysis of alicyclic polyimide films

The crystallinity of the full-alicyclic polyimide films was examined by wide angle X-ray diffractions (WAXD). All polyimides exhibited amorphous patterns and broad dispersive peaks of diffraction around at 16° (2θ). This observation is quite reasonable because the presence of polyalicyclic (bicyclic) structure decrease the intermolecular force between the polymer chains with a consequent decrease in crystallinity.

References

[1] Huang S J, and Hoyt A E 1995 *Trends in Polym. Sci.* **3** 262
[2] a) Tsuda Y, Kojima M, and Oh J M 2006 *Polym. J.* **38** 1043.
b) Huang W, Yan D, Lu Q 2001 *Macromol. Rapid Comm.* **22** 1481
c) Koohmareh G A 2007 *Designed Monomers and Polymers* **10** 517
d) Hsiao S H, Chung C L, and Lee M L 2004 *J. Polym. Sci. Part A: Polym. Chem.* **42** 1008
e) Imai Y, Ishida M, and Kakimoto M 2003 *High. Perform. Polym.* **15** 281
f) Zhou H, Liu J, Qian Z, Zhang S, and Yang S 2001 *J. Polym. Sci. Part A: Polym. Chem.* **39** 2404
g) Hamciuc E, Schulz B, Kopnick T, Kaminorz Y, and Bruma M 2002 *High Perform. Polym.* **14** 63
[3] a) Yang C P, Chen Y P, Woo E M, and Li S H 2006 *Polymer J.* **38** 457
b) Yang C P, Chen R S, and Chen K H 2003 *J. Polym. Sci. Part A: Polym. Chem.* **41** 922
c) Xie K, Zhang S Y, Liu J G, He M H, and S Y Yang 2001 *J. Polym. Sci. Part A: Polym. Chem.* **39** 2581
d) Xie K, Liu J G, Zhou H W, Zhang S Y, and Yang S Y 2001 *Polymer* **42** 7267
e) Chung I S, and Kim S Y 2000 *Macromolecules* **33** 3190
[4] Tsiang R C, and Hung K Y 2003 *Polym. Prepr.* **44** 860
[5] a) Shao Y, Li Y, Zhao X, Ma T, Gong C, and Yang F 2007 *European Polym. J.* **43** 4389

b) Gao C, Wu X, Lu G, Ding M, and Gao L 2004 *Macromolecules* **37** 2754
[6] a) Zhang S J, Li Y F, Wang X L, Zhao X, Shao Y, Ma T 2007 *High. Perform. Polym.* **19** 462
b) Chung I S, Park C E, Ree M, Kim S Y 2001 *Chem. Mater.* **13** 2801
[7] R Dine-Hart A, and Wright W W 1971 *Makromol. Chem.* **143**, 189
[8] a) Gordina T A, Kotov B V, Kolnivov O V, and Pravednikov A N 1973 *Vysokomol. Soyed. Ser. B* **15** 378
b) Kotov B V, Gordina T A, Voishchev V S, Kolninov O V, and Pravednikov A N 1977 *Vysokomol. Soyed. Ser. A* **19** 614
[9] Rogers F E 1964 *US Patent*: 3,356,648
[10] a) Clair A K St, T Clair L St, and Shevket K I 1984 *Polym. Mater. Sci. Eng.* **51** 62
b) Clair A K St, Clair T L St, Slemp W S 1985 *NASA Tech. Memo. NASA-TM-86341, NAS1.*15:86341, 21
c) Clair A K St, Clair T L St, and Slemp W S 1985 *Pro. of the 2nd Inter. Conf. on Polyimide*s (Society of Plastics Engineers Mid Hudson Section, New York) p 333-4
d) Clair A K St, and Slemp W S 1985 *SAMPE Journal* **21** 28
[11] a) Matsuura T, Yamada N, Nishi S, and Hasuda Y 1993 *Macromolecules* **26** 419
b) Mercer F W, McKenzie M T 1993 *High Perform. Polym.***5** 97
c) Matsuura T, Ando S, Sasaki S, and Yamamoto F 1994 Macromolecules **27** 6665
d) Kishi Y, Inoue H, Murata K, Kouzuma S, Morizane M, Shibuya M, Nishiwaki H, and Kuwano Y 1992 *Jpn. J. Appl. Phys.* **31** 12
[12] Hsiao S H, and Chang Y H 2004 *J. Polym. Sci. Part A: Polym. Chem.* **42** 1255
[13] Yamada M, Matsumoto T and Kurosaki T 1992 *J. Org. Chem.* **57** 6075
[14] Matsumoto T, and Kurosaki T 1997 *Macromolecules* **30** 993
[15] Kusama M, Matsumoto T, and Kurosaki T 1994 *Macromolecules* **27** 1117
[16] Matsumoto T 1999 *Macromolecules* **32** 4933
[17] a) Matsumoto T 2001 *J. Photopolym. Sci. Technol.* **14** 695
b) Matsumoto T, and Kurosaki T 1999 *Proc. of 6th Inter. Conf. on Polyimides* (Society of Plastics Engineers Mid Hudson Section, New York) pp 91-115
[18] Abe J, and Shirai Y 1996 *J. Am. Chem. Soc.* **118**, 4705
[19] Matsumoto T 1997 *High Perform. Polym.* **11** 367
[20] Bikson B R, and Freimanis Y F 1970 *Vysokomol. Soed. Ser. A* **12** 69
[21] Erskine D, Yu P Y, and Freilich S C 1988 *J. Polym Sci Part C, Polym. Lett.* **26** 465
[22] Boese D, Lee H, Yoon D Y, Swalen J D, and Rabolt J 1992 *J. Polym. Sci., Part B: Polym. Phys.* **30** 1321
[23] Honma S 2004 *Handbook of plastics and functional polymers* (Tokyo: Sangyo Chosakai)

Acknowledgments

This work was partly supported by a Grant-in-Aid's for Academic Frontier from the Ministry of Education, Culture, Sports, Science and Technology, Japan.

Preparation of nano-thin films from conducting polymer by chemical vapor deposition method and its application for light emitting diodes (LED) and organic solar cells

Ngo Trinh Tung and Nguyen Duc Nghia
Institute of Chemistry, Vietnam Academy of Science and Technology,
18 Hoang Quoc Viet, Cau Giay, Hanoi, Vietnam

E-mail: ducnghia264@fpt.vn

Abstract. It was demonstrated that a series of copolymers consisting of p-phenylenevinylene (PV) and p-phenyleneethylene (PE) units could be prepared from a single monomer, p-(methoxymethyl)benzyl chloride, via chemical vapor deposition polymerization (CVDP) method. The composition of the copolymers can be varied by altering the monomer activation temperature; the higher the temperature, the lower the content of the PV units becomes. The sufficient incorporation of the PE units into the poly(p-phenylenevinylene) backbone caused a significant blue shift in both photoluminescence and electroluminescence. The luminescence properties of the copolymer strongly depend on the amount of the PE units incorporated. The external quantum efficiency of the electroluminescence devices having the configuration of ITO/PEDOT-PSS/copolymer/Al-Li increased as the content of the PV units increased in the copolymer, which can be ascribed to the improved confinement of excitons. For the solar cell based on PPV derivative, the effect of the PCBM concentration on the performance of the solar cell is investigated. The results show that there is no interaction between the MEH-PPV and PCBM phase and the photoluminescence of MEH-PPV is quenched by the incorporation of PCBM. With increasing PCBM concentration, the short circuit current density and the overall energy conversion efficiency of the solar cell have been strongly improved.

Keyword: Copolymers, CVDP, organic solar cell.

1. Introduction

π-Conjugated polymers such as poly(p-phenylenevinylene) (PPV) and its derivatives [1,2] and polyfluorenes and their copolymers [3-5] have attracted much interest over the past decade due to their potential usages in optoelectronic devices such as light-emitting diodes (LED) [6-9] , lasers [10] and solar cell [11-13]. The first green LED of PPV were reported by the Cambridge group [6]. Color tuning of such devices has been realized via many approaches: increasing or decreasing the band gap energy (E_g) of a light-emitting polymer by increasing or decreasing the effective conjugation length of the main chain, which causes red shift or blue shift in the emitted light. Decreasing the E_g value can be achieved by placing electron-donating substituents on polymer backbone whereas the E_g value increases when the polymer backbone [14, 15] is forced to twist out of plane with bulky pendants [16] or by disrupting the conjugated structure of the polymer through copolymerization with a second unconjugated monomer [17]. Although the solution polymerization methods are widely utilized to modify the structure of a polymer, their limitation lies in the problem of side reactions and possible contamination from the solvents used.

The alternative method to the preparation of the insoluble PPV in various shapes is the chemical

© 2009 IOP Publishing Ltd

vapor deposition polymerization method (CVDP). The potential advantages of CVDP compared to the solution processing method include good control of film uniformity, easy elimination of side products, and the potential to control the film architecture and composition on the molecular level. So far, the main starting materials for CVDP preparation of PPV are α, α'-dihalo –p-xylene. But the use of these starting materials suffers the evolution of hydrogen halide gases, which can affect the indium-tin oxide (ITO) electrode and cause worsened device efficiency [18, 19].

In this work, nano-thin film from PPV and its derivative have been prepared by CVDP method for LED application and by spin-coating for solar cell application. The optoelectronic properties of the materials were investigated and discussed.

2. Experimental

The monomer for the CVDP (p-(methoxyethyl)benzyl chloride) and MEH-PPV was synthesized in our Lab. The other chemicals were purschased from Merck Company and were used without purification.

Chemical vapor deposition polymerization and characterization

The experimental setup for the CVDP process is practically the same as described earlier by us [21]. The monomer (0.2 ml) placed on a tungsten boat was vaporized at the temperature of 80-85°C. The CVD-process was performed under a steady stream (8 ml/min) of argon gas at the pressure of 1.0 Torr. The vaporized monomer was allowed to pass through the pyrolysis zone preheated varying from 650°C to 800°C. The substrate temperature was maintained at the temperature of 50°C.

The IR,UV-VIS and photoluminescence (PL) spectra of the materials were recorded by using BOMEM, MB-104 Instrument, HEWLETT PACKARD 8453 Instrument and AMINCO Bowman Series 2 Instrument, respectively. The excited wavelength for PL study was 360 nm. The differential scanning calorimetry (DSC) (Mettler–Toledo DSC 821 Instrument) and the thermogravimetric analysis (TGA) (Mettler-Toledo TC 15 Instrument) were performed under the stream of N_2 at the heating rate of 10°C/min. The device configuration of LED was ITO / PEDOT-PSS (40 nm) / polymer (90 ± 5 nm) / Al-Li (100 nm). The Al-Li alloy contained 0.3 wt % of lithium. The characterization of the devices was performed by assembly instrument consisting of a PC-based DC power supply (HP6623A), digital multimeter (HP 34401) and a light power meter (Newport Instrument, Model 818-UV). The electroluminescence spectra of the devices were measured at room temperature (AMINCO Bowman Series 2 Instrument).

3. Results and Discussion

3.1. Structure analysis

We first examined the IR spectra of the intermediates and the final polymer obtained in the process of CVDP of p-(methoxymethyl)benzyl chloride. Figure 1 shows the changes of IR spectra during the CVDP process after activating the monomer at 650 °C.

It was noted that some side products in the form of liquid were observed in the deposition zone after the CVDP process. The sample obtained by the CVDP process was carefully washed with methylene chloride. By comparing the IR spectrum of the sample obtained right after the CVDP process (figure 1b) with that after washing (figure 1c), we clearly see that two strong peaks at 1701 and 1609 cm^{-1} and many other peaks in the region from 1400 cm^{-1} to 600 cm^{-1} in the spectrum 1b disappeared after washing. The peak at 1701 cm^{-1} is assigned to the carbonyl C=O stretching vibration and the one at 1609 cm^{-1} to an aliphatic C=C stretching vibration mode. It seems that the side products are the same as those formed in the CVDP of α, α'-dialkoxy-p-xylene reported in our previous paper [23]. The IR-spectrum of the washed precursor polymer exhibits absorption peaks characteristic of sp^2 C-H stretching (3017 cm^{-1}), sp^3 C-H stretching (2926 and 2853 cm^{-1}), aromatic C=C stretching (1519, 1452 and 1415 cm^{-1}) and phenylene C-H bending (823 cm^{-1}) modes. The absorption peak of C-Cl stretching mode appears as broad peak at 727 cm^{-1}. And then, the sample was subjected to thermal treatment at 250°C under vacuum for 5 hours converting the precursor polymer to the final copolymer. A new absorption peak at 965 cm^{-1} was observed in the IR-spectrum after thermal treatment, which is from the characteristic out-of-plane C-H bending mode of trans vinylene units newly formed and, thus, the absorption peak of the C-Cl stretching mode simultaneously disappeared (figure 1d). The characteristic feature of the IR spectrum of the final polymer is basically the same as that of the copolymers consisting on PV and PE units as described earlier by Vaeth and Jensen [24] and a superposition of PPX and PPV spectra.

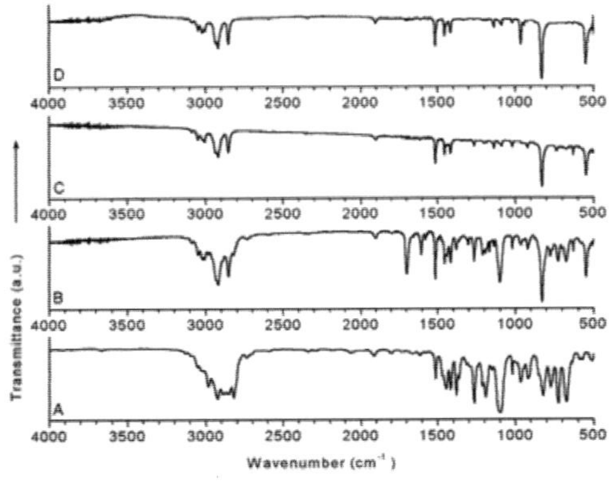

Figure 1. Comparing the IR spectra of monomer, precursor polymer and the final polymer prepared after activating the monomer at 650°C. (a): Monomer; (b): precursor polymer; (c): after washing the precursor polymer with methylene chloride; and (d): after thermal treatment of the washed precursor polymer at 250°C for 5 hours.

We studied the thermal properties of the precursor polymer obtained from the present CVDP followed by washing with CH_2Cl_2. The DSC and TGA thermograms obtained under N_2 atmosphere at the heating rate of 10 °C/min are given in figure 2.

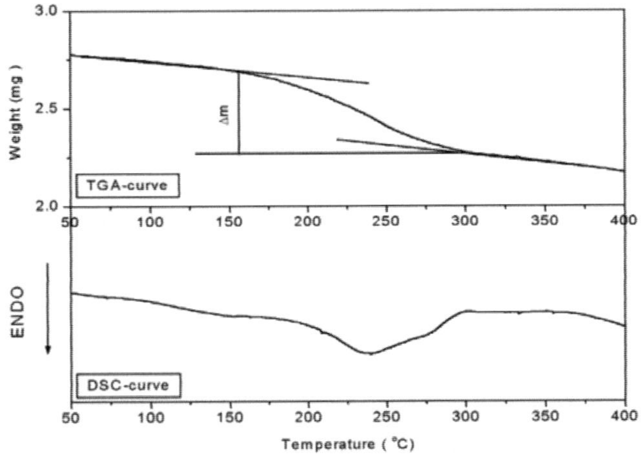

Figure 2. DSC and TGA thermograms of the precursor polymer obtained after the CVDP process followed by washing with CH_2Cl_2.

On the DSC curve, a broad endothermic peak is observed in the region of 160°C to 300°C. A concurrent weight loss is observed over the same temperature range in the TGA thermogram in. The result indicates that an endothermic reaction takes place in this temperature region. This behavior of the precursor polymer is similar to the thermal dehydrochlorination of the PPV precursor as observed in the preparation of PPV from α, α'-dihalo –p-xylene and also as described by Iwatsuki et al. [26]. Assuming all the weight loss mentioned about, is due to elimination of HCl, we estimate that this polymer contains 52 wt % of the PV units. Therefore, the IR spectroscopic and thermal analysis results confirm that the deposited polymer is a copolymer consisting PV and PE units.

The possible mechanism for the formation of the precursor and the final copolymer is given below:

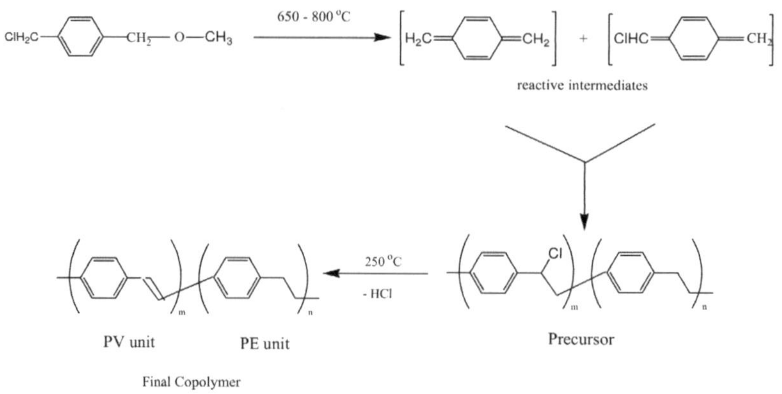

Thermolysis or thermal activation of the monomer at an elevated temperature produces two different quinodimethanes, one with chlorine and the other without chlorine. Condensation of the two actives species on a substrate surface will result in a copolymerization reaction to form a precursor copolymer consisting of two different repeating units, one of which is to be further converted to the final PV units on thermal treatment under vacuum. The side products, although not completely identified, can be removed by washing with methylene chloride.

3.2. Photo-electroluminescent properties

Figure 3 shows the UV-VIS and photoluminescence (PL) spectra of PPV, the precursor obtained after activating the monomer at 650°C and the final copolymer.

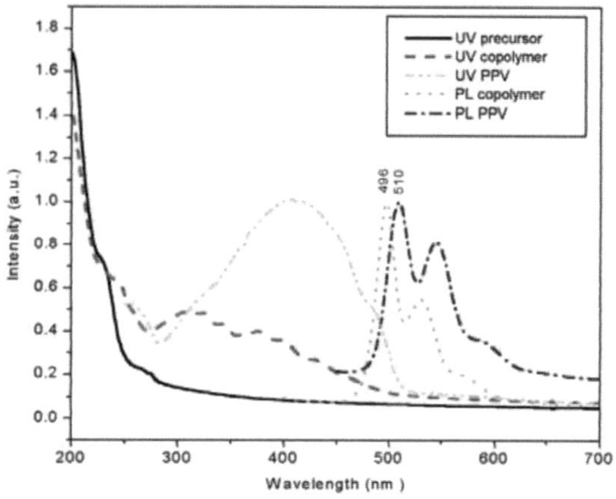

Figure 3. Comparison of UV-VIS absorption and photoluminescence spectra of the precursor, the final copolymer and PPV.

For the sake of comparison, PPV was prepared in the same condition as described earlier by us [21]. Comparing the UV-VIS spectrum of the final polymer to the UV-VIS spectrum of the precursor polymer, it is clearly seen from figure 3 that a new strong broad absorption bands appeared at 324, 400 and 434 nm. The UV-VIS spectrum of the copolymer is basically identical to the UV-VIS spectrum of the partially converted PPV precursor, or not fully conjugated polymer, reported by Gagnon [27]. The fully conjugated PPV has a broad continuous absorption band with the maximum absorption peak at 410 nm. It is interesting to note that the absorption edge of PPV at 512 nm (optical band gap Eg = 2.4 eV) blue-shifted to 497 nm (optical band gap Eg = 2.5 eV) by the copolymerization. The present copolymer exhibits light emission at three positions: 496, 530, and 573 nm at the excitation wavelength of 360 nm. The PL spectrum of the copolymer when compared to the PL spectrum of PPV is significantly blue- shifted. For example, the $S_{0\text{-}0}$ transition peak shifted from 510 to 496 nm. Since poly(p-phenyleneethylene) or poly(p-xylylene) is not a luminescent polymer in the visible region, the incorporation of the unconjugated PE units along the backbone shortens the effective conjugation length of the resulting polymer causing a blue-shift in UV-VIS absorption and also in PL behavior.

Figure 4 compares the PL spectra of the copolymers prepared at different monomer activation temperatures.

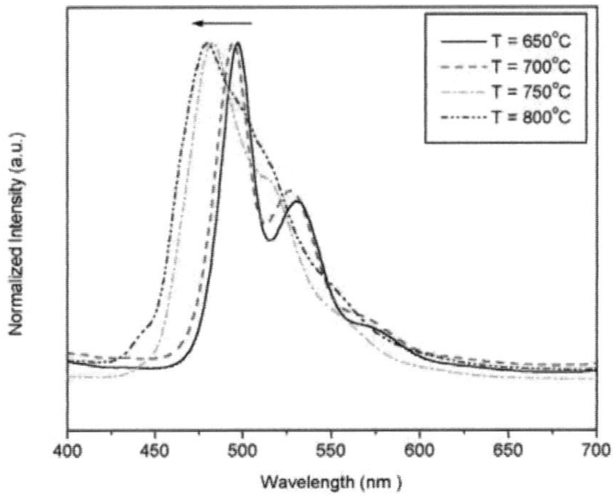

Figure 4. Comparison of PL spectra of the copolymers prepared at different monomer activation temperatures.

It is clearly seen from the figure that with increasing monomer activation temperature from 650 to 800°C, the $S_{0\text{-}0}$ transition peak shifted from 496 nm to 479 nm. We observed a similar trend in their UV-VIS absorption spectra. The PV contents estimated from TGA due to the weight loss are 52, 35, 31 and 26 wt.% respectively for the monomer activation temperature of 650, 700, 750 and 800°C. The estimated PV contents after TGA analysis are lower than the estimated PV contents after analysis of the transmittance intensity by IR spectra proposed by Vaeth and Jensen [20] about 5 wt. % till 20 wt. % but the tendency is the same. This indicates that the formation of the unsubstituted quinodimethane intermediate is favored over the formation of the chlorine-substituted activated species as one increases the monomer activation temperature. Increasing of the PE content in the copolymer would shorten the conjugation length of the polymer backbone resulting in a blue-shift in PL as we have observed in this investigation.

Figure 5 shows the electroluminescence (EL) spectra for the devices of ITO / PEDOT-PSS (40nm) / polymer (90 ± 5 nm) / Al-Li (100nm).

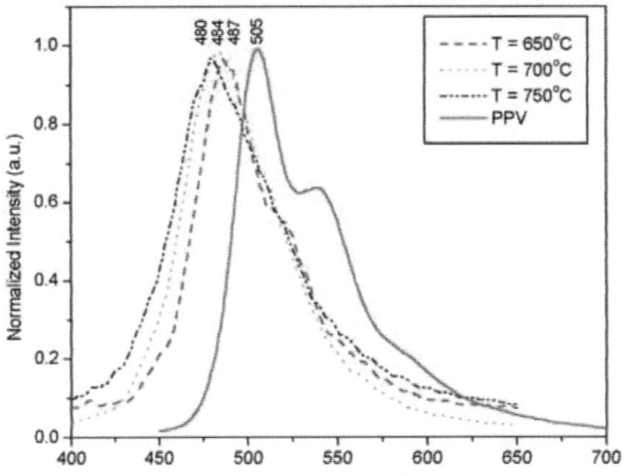

Figure 5. Comparison of the EL spectra of the devices fabricated with PPV and copolymers prepared at different monomer activation temperature.

The general feature of the EL spectra is about the same as the corresponding PL spectra, although less vibronic details are visible in EL spectra. As observed in their PL spectra, the wavelength of the maximum emission intensity decreased from 505 nm for the PPV device to 480 nm for the device of the copolymer obtained after monomer activation at 750°C.

Figure 6 shows the light output vs. electric field and the current density vs. electric field curves for devices.

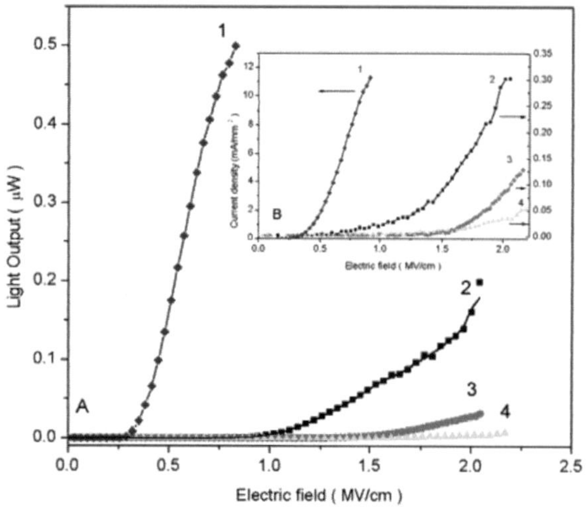

Figure 6. Light output vs. electric field (a) and current density vs. electric field (b) curves for the devices of ITO / PEDOT-PSS (40 nm) / polymer (90 ± 5 nm) / Al-Li (100 nm).

The numbers represent PPV (1), and polymers (2, 3 and 4) prepared after activating the monomer at 650, 700 and 750°C, respectively. The turn-on electric field of the devices fabricated with PPV and copolymers prepared after monomer activation 650, 700 and 750 °C are steadily increased from 0.28 to 0.84, 1.38 and 1.62 MV/cm. The device with polymer 4 showed very little electroluminescent. The same tendency was also observed for the current density of the devices (figure 6b). The enhancement of the turn-on electric field by higher monomer activation temperature is to be expected because of the increased incorporation of the saturated PE units in the backbone leading to diminished electrical conductivity of the resulting copolymers. The external quantum efficiencies of the copolymers devices, however, were higher (6×10^{-3}, 3.4×10^{-3} and 1.1×10^{-3} % respectively for the copolymers prepared at the activation temperature of 650, 700 and 750°C) than that (8×10^{-4} %) of the device fabricated with PPV. The confinement of excitons in the conjugative structure disrupted by the saturated PE units in the polymer backbone appears to be responsible for this observation [19]. It is also noted that the external efficiency drops down in a regular fashion as the content of the PE units in the copolymers increases, which can be ascribed to the decreased concentration of the light emitting PV units in the backbone.

3.3. Solar cell based on PPV derivative

Figure 7 shows the UV-VIS spectra of MEH-PPV, PVBM and their composite in different PCBM concentration.

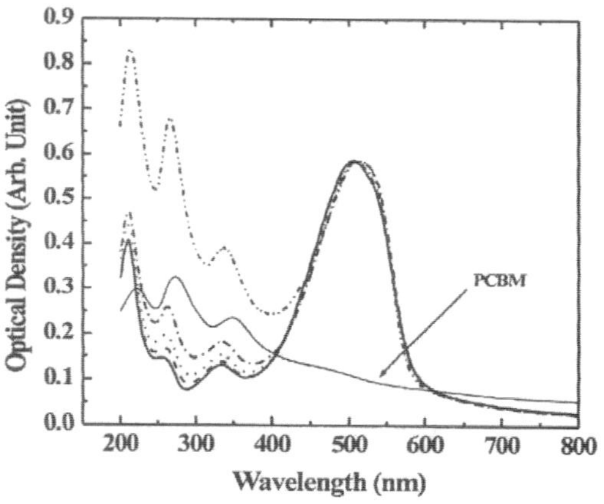

Figure 7. UV-VIS spectra of MEH-PPV, PCBM and their composite.

The characteristic π-π*absorption peak of MEH-PPV is observed around 510 nm in the pure MEH-PPV and its composites. From the UV-VIS spectra, we can see that the three characteristic peaks in UV range (at 339, 260 and 211) of PCBM are overlapped with the absorption peaks of pure MEH-PPV (335, 260 and 211 nm). The absorption peaks of the composite are just superposition of the characteristic peaks of pure MEH-PPV and PCBM. This result confirms that there is no significant interaction between the two materials in the ground state.

Evidence of photoinduced charge transfer in MEH-PPV/PCBM composite can be provided by PL quenching. Figure 8 shows the PL spectra of MEH-PPV and its composite. For the pure MEH-PPV, the characteristic PL peaks are observed around 589 and 625 nm. With increasing PCBM concentration this peaks are significantly quenched.

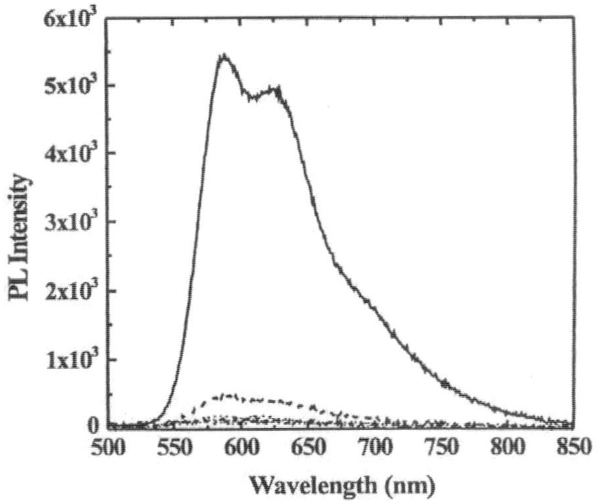

Figure 8. PL spectra of MEH-PPV and its composite (Excitation wavelength: 488 nm).

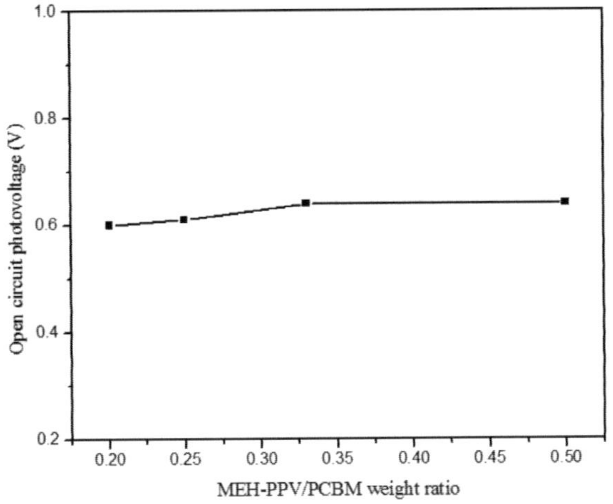

Figure 9. Dependence of the open photovoltage on the MEH-PPV/PCBM ratio.

Figure 9 shows the dependence of the open photovoltage on the MEH-PPV/PCBM ratio.With increasing MEH-PPV/PCBM ratio, the open circuit photovoltage of the solar cell only slightly increases from 0.6 V to 0.64 V. In contrast, with increasing MEH-PPV/PCBM ratio, the short circuit current density strongly decreases from 4.3 mA/cm^2 by the ratio of 0.2 to 1.7 mA/cm^2 by the ratio of 0.5 (figure 10).

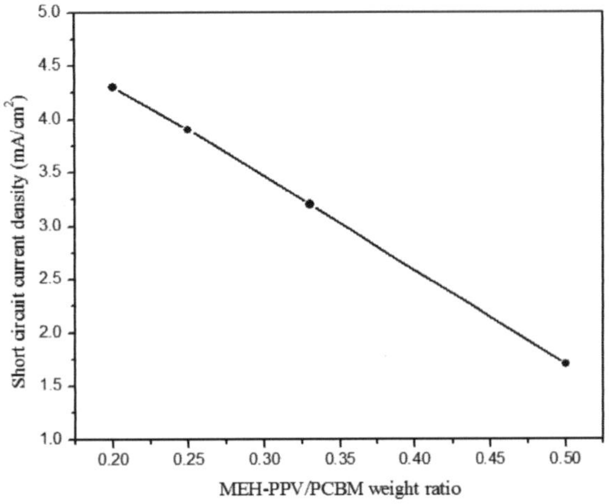

Figure 10. Dependence of the short circuit current density on the MEH-PPV/PCBM ratio.

According to [11], higher concentration of PCBM favors the formation of a phase separated interpenetrated network with sizable domains, which in turn favors effective charge separation. Thus, the short circuit current density is enhanced by higher PCBM concentration or lower MEH-PPV/PCBM ratio. The same tendency comparing to the short circuit current density has been observed by the overall energy conversion efficiency depending on the MEH-PPV/PCBM ratio (figure 11). By the MEH-PPV/PCBM ratio of 0.2, the solar cell reaches the highest overall energy conversion efficiency of 1.1%. This result indicates that the MEH-PPV/PCBM composite has significantly improved the performance of the solar cell comparing to the solar cell using only MEH-PPV as active material.

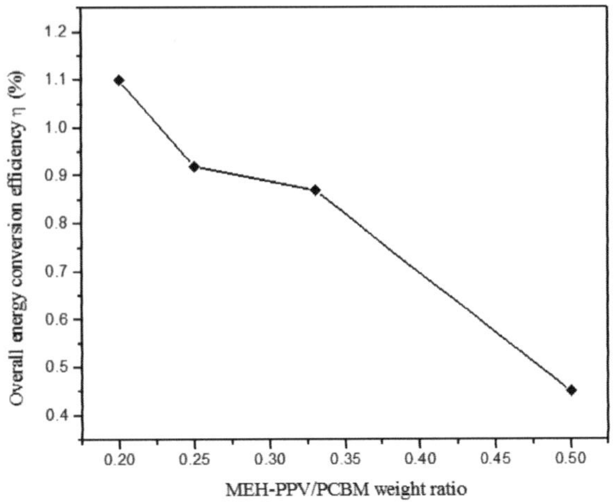

Figure 11. Dependence of the overall energy conversion efficiency on the MEH-PPV/PCBM ratio.

4. Conclusion

It was demonstrated that a series of copolymers, poly(p-phenylenevinylene –co- p-phenyleneethylene), could be prepared from a single monomer, p- (methoxymethyl) benzyl chloride, via the CVDP method after activating the monomer at the different temperatures. Incorporation of p-phenyleneethylene units in the polymer backbone resulted in a blue-shift in both photoluminescence and electroluminescence when compared with those PPV. The luminescence properties of the copolymers strongly depended on the CVDP condition, especially, on the monomer activation temperature, which controlled the amount of the PE units incorporated in the copolymer backbone. A higher monomer activation temperature led to inclusion of a higher level of the PE unit in the deposited copolymer resulting in bigger blue shift in EL and PL spectra.

We have also successfully fabricated the organic solar cell based on the MEH-PPV/PCBM composite. Through the PCBM phase, the photoluminescence of the MEH-PPV was strongly quenched due to the photoinduced charge transfer from MEH-PPV onto fullerene. The incorporation of PCBM phase into MEH-PPV phase had strong effect on both the short circuit current density and the overall energy conversion efficiency of the solar cell. With increasing PCBM concentration, the two parameters above steadily increased. The effect of the PCBM concentration on the open photovoltage was insignificant.

References

[1] Kang H S, Kim K H, Kim M S, Park KT, Kim K M, Lee T H, Joo J, Kim K, Lee DW and Jin J I 2001 *Appl. Phys.* **1** 443
[2] Park S J, Choi E S, Oh E J and Lee, K W 2001 *Synth. Met.* **117** 95
[3] Zheng S J, Goto H, Akagi K 2003 *Synth. Met.* **135** 125
[4] Swensson M, Zhang F, Inganas O and Andersson M R 2003 *Synth. Met.***135** 137
[5] Charas A, Morgado J, Alcacer L, Martinho J M G and Cacialli F 2003 *Synth.Met.* **137** 1039
[6] Burroughes J H, Bradley, D D C, Brown A R, Marks R N, Mackay K, Friend R H, Burn P L, Holmes A B 1990 *Nature* **347** 539
[7] Zhang C, Braun D, Heeger A J 1993 *J. Appl. Phys.* **73** 5177
[8] Shaefer O, Greiner A, Pommerehne J, Guss W, Vestweber H, Tak H Y, Bassler H, Schmidt C, Lussem G, Schartel B, Stumpfen B, Wendorff J H, Spiegel S, Moeller C, Spiess H W 1996 *Synth. Met.* **82** 1
[9] Staring E G J, Braun D, Rikken, Demandt G L J A, Kessener R J C E, Bauwmans Y A R R, Broer M D 1994 *Synth. Met.* **67** 71
[10] Vaeth K M, Jensen K F 1997 *Appl. Phys. Lett.* **71** 2091
[11] Jin H, Hou Y B, Meng X G, Feng F 2007 *Mater. Sci. and Eng. B* **137** 5
[12] Lem S, Bettignies R, Nunzi J.-M, Cariou M, 2004 *Appl. Phys. Let.* **12** 2178
[13] Ltaief A, Davenas J A, Bouazizi R B, Chaabane P, Alcouffe H , Ouada B 2005 *Mater. Sci. and Eng.* **25** 67
[14] Askari S H, Rughooputh S D, Wuld F, Heeger A J 1989 *Polym. Prepr.* **30** 157
[15] Braun, D, Heeger AJ 1991 *J. Appl. Phys. Lett.* **58** 1982
[16] Opfermann J, Scharf U, Raabe D, Nowotny J, Horhold H 1986 *Angw. Makromol. Chem.* **142** 91
[17] Ahn T, Jang M S, Shim H K, Hwang D H, Zyung T 1999 *Macromolecules* **32** 3279
[18] Yang Z, Sokolik I, Karasz F E 1993 *Macromolecules* **26** 1188
[19] Hide F, Diaz-Garcia M A, Schwartz B J, Andersson M R, Pei Q, Heeger A J 1996 *Science* **273** 1833
[20] Vaeth K M, Jensen K F 2000 *Macromolecules* **33** 5336
[21] Kim K, Jin J.-I. 2001 *Nano Letter* **1** 631
[22] Hsieh B R, Antoniadis H, Abkowitz MA, Stolka M 1992 *Polym. Prepr. Am. Chem. Soc. Div. Polym. Chem.* **33** 414
[23] Tung N T, Ju Y J, Kim K, Joo S.-H, Jin J.-I, 2004 *Bulletin of the Korean Chemical Society* **25** 31
[24] Ramakrishna V, Swaroop K, Seshu B D 2001 *Surface Engineering Series:Chemical Vapor Deposition, ASM International* **2** 243
[25] Fou A, Onitsuka O, Ferreira M, Rubner M F, Hsieh B R 1996 *J. Appl. Phys.* **79** 7501
[26] Iwatsuki S, Kubo M, Kumeuchi T 1991 *Chem. Lett.* **6** 1071
[27] Garnon D R 1986 *Dissertation* University of Massachusetts

Journal of Physics: Conference Series **187** (2009) 012007 doi:10.1088/1742-6596/187/1/012007

Tunnelling magnetoresistance in nanometer granular perovskite systems

Bach Thanh Cong, Pham Huong Thao and Nguyen Tien Cuong

Faculty of Physics, Hanoi University of Science, VNUH,
334 NguyenTrai Street, Hanoi, Vietnam

E-mail: congbt@vnu.edu.vn

Abstract. In this contribution the phenomenological theory for the tunnelling magnetoresistance phenomenon observed in granular perovskite manganese systems is developed using Landauer ballistic transport concept. It was shown that the field dependence, magnitude and derivative of magnetoresistance ratio observed experimentally are well reproduced by the presented theory.

Keyword: Tunnelling magnetoresistance, perovskite.

1. Introduction

Tunneling magnetoresistance (TMR) in various spintronic materials including manganese perovskites $Ln_{1-x}A_xMnO_3$ (Ln and A are rare earth and alkaline ions) and structures has involved much attention of researchers during the last decade (see the review [1]). It was shown in Ref. [2] that spin polarized tunneling transport is the dominated mechanism of magnetoresistance effect in nanometer perovskite maganates. In earlier work [3], Hwang et al. suggested that magnetoresistance in polycrystalline perovskite $La_{0.7}Sr_{0.3}MnO_3$ originates from the following two sources: (i) an intrinsic part arising from the double exchange mechanism between two neighboring manganese ions and (ii) the intergranular spin-polarized tunneling at the grain boundaries. Both these factors were considered in the theory developed in Ref. [4] for spin-polarized tunneling in granular perovskites. It was shown in many experiments that spin polarized tunneling gives rise to a sharp drop in resistance at low fields and low temperature. Inoue and Maekawa [5] had applied the Landauer ballistic transport concept [6] for description of tunneling conductance G, where it is given by

$$G = \frac{e^2}{h}|T|^2 \tag{1}$$

In this equation e, h are the electron electric charge and Planck constant, and $|T|^2$ is transmission coefficient of an electron wave crossing intergranular potential barrier (see figure 1). Let χ be the angle between the magnetizations of nearest-neighbor grains (see figure 1), one can write the electron transmission coefficient in the following form (see Ref. [4])

$$|T|^2 \propto \frac{1}{2}\left(n_\uparrow + n_\downarrow\right)^2\left[1 + P^2\cos\chi\right] \tag{2a}$$

$$P = \frac{n_\uparrow - n_\downarrow}{n_\uparrow + n_\downarrow} \tag{2b}$$

© 2009 IOP Publishing Ltd

where P is the electron polarization. Since the electrostatic grain charging energy $E_C = e^2/2C$ (C is the grain capacitance) and the distance s between grains, also influence on the electron tunneling the conductance between grains can be expressed as [5]

$$G_{ij} \propto \left(1+P^2\cos\chi\right)\exp(-\frac{E_c}{k_BT}-2ks) \tag{3a}$$

$$k=\left(2m_e\phi/\hbar^2\right)^{-1/2} \tag{3b}$$

In (3b) m_e, ϕ are the effective mass of electron and the potential barrier height, respectively. Influence of the intergranular distance and charging Coulomb energy on the electron tunneling were discussed also in [7], [8].

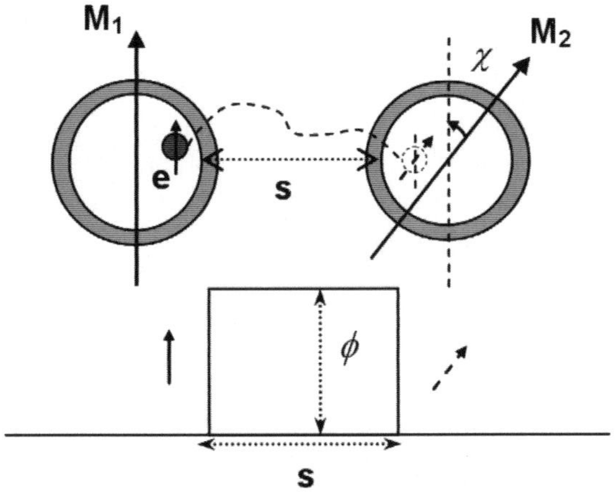

Figure 1. The potential barrier model for the electron tunneling with the spin rotation occurred between two spherical magnetic grains.

Helman and Abeles [8] showed that the product of sE_C is constant for the certain systems

$$sE_c = c \tag{4}$$

Here c is constant. Taking into account the relation (4) one should average the tunneling conductivity (3a) over directions of grain magnetizations and distribution of distance s between grains as follows [5]

$$G_{av}\left(H\right)=\iiint \mathrm{d}s\mathrm{d}\Omega_1\mathrm{d}\Omega_2 f\left(s\right)g\left(\theta,H\right)\left(1+P^2\cos\chi\right)\exp(-\frac{c}{sk_BT}-2ks) \tag{5a}$$

where $\mathrm{d}\Omega=\sin\theta\mathrm{d}\theta\mathrm{d}\varphi$ is the element of solid angle. $f\left(s\right)$, $g\left(\theta,H\right)$ are specific distribution functions for the intergranular distance and grains orientations in the external magnetic field H. Its explicit forms will be given later. The function $\cos\chi$ in (5a) – the cosine of the angle between grain magnetization $\vec{M}_1(M,\varphi_1,\theta_1)$ and $\vec{M}_2(M,\varphi_2,\theta_2)$ is expressed through the spherical coordinate angles as:

$$\cos\chi=\sin\theta_1\sin\theta_2\cos\left(\varphi_1-\varphi_2\right)+\cos\theta_1\cos\theta_2 \tag{5b}$$

In [5] the electron polarization P was considered simply as a parameter of the theory. In more realistic way, the authors [4] calculated P within the molecular field approximation for the localized spins inside the grains and P was equal to the electron relative magnetization m.

The aim of this contribution is to develop further a phenomenological theory for the TMR in granular perovskite systems basing on Inoue- Maekawa theory [5]. We will show that in the case of perovskites, when there is strong local Hund interaction (J_H) between electron and localized spins ($J_H >> t$, t is an electron hopping integral), the electron polarization P can reach almost saturated values. After that, we will calculate conductivity (formula (5a)) with reasonable proposed $f(s)$, $g(\theta,H)$ functions and estimate the TMR effect for nanometric granular perovskite systems.

This theory also can be applied for different perovskite manganese systems when the condition of strong coupling between tunneling electron and localized spin inside nanometer grains is satisfied.

2. Electron polarization

In order to describe the electron polarization in perovskite manganese grains one can use the Kondo lattice model as follows

$$H=\sum_{\langle ij\rangle\lambda} t_{ij}a_{i\lambda}^{+}a_{j\lambda}-J_H\sum_{j,\alpha\beta}\vec{S}_j a_{j\alpha}^{+}(\vec{\sigma})_{\alpha\beta}\,a_{j\beta}-\frac{1}{2}\sum_{\langle i,j\rangle}I_{ij}\vec{S}_i\vec{S}_j \tag{6}$$

where $a_{i\lambda}^{+}(a_{j\lambda})$ is the creation (annihilation) electron operator at site i with spin projection $\lambda=(\uparrow,\downarrow)$ or ± 1. t_{ij} is the nearest-neighbor hopping energy of the e_g electron. J_H is the local ferromagnetic Hund's rule coupling between the e_g electron spin $\vec{\sigma}/2$ and the t_{2g} localized spin $\vec{S}_j$ at i-th site. The last term in (6) corresponds to the antiferromagnetic exchange between nearest neighbor t_{2g} spins with the strength I_{ij}. Because of the strong onsite Hund exchange interaction between the tunneling electron and the localized t_{2g} spin S = 3/2, then the shift of the electron band caused by this interaction is primary important comparing with a grain size effect. For the bulk materials, it was shown by our temperature dependence Green function method calculation [9] that the electron dispersion law in the strong coupling limit ($J_H >> t$) is given approximately by

$$E_{\vec{k}\lambda}\approx-\lambda J_H\left(S+\frac{1}{2}\right)+\left(1+\frac{1}{2S}\right)t_{\vec{k}} \tag{7}$$

It is obviously that when $J_H >> t$, the second term in (7) can be neglected. The electron number $n_\lambda=\frac{1}{N}\sum_{\vec{k}}n_{\vec{k}\lambda}$ figured in the expression for the electron polarization P is derived according to the common formula:

$$n_{\vec{k}\lambda}=\left\langle a_{\vec{k}\lambda}^{+}a_{\vec{k}\lambda}\right\rangle=i\lim_{\gamma\to0}\int_{-\infty}^{\infty}\mathrm{d}E\left\{\left\langle\left\langle a_{\vec{k}\lambda}\middle|a_{\vec{k}\lambda}^{+}\right\rangle\right\rangle_{E+i\gamma}-\left\langle\left\langle a_{\vec{k}\lambda}\middle|a_{\vec{k}\lambda}^{+}\right\rangle\right\rangle_{E-i\gamma}\right\}f(E) \tag{8}$$

where $f(E)=\left\{\exp\left[(E-\mu)/k_BT\right]\right\}^{-1}$ is the Fermi-Dirac distribution function and $\left\langle\left\langle a_{\vec{k}\lambda}\middle|a_{\vec{k}\lambda}^{+}\right\rangle\right\rangle_{E\pm i\gamma}$ is the retarded Green function given in the energy representation. The electron polarization is derived from (7), (8) and (2b):

$$P=\tanh\left[\frac{J_H}{k_BT}\left(S+\frac{1}{2}\right)\right] \tag{9}$$

It follows from (9) that at low temperature, where the electron tunneling phenomenon in perovskites is more interest, the electron polarization P may have values closed to the saturated one (hundred percents).

3. Tunneling magnetoresistance

The tunneling magnetoresistance ratio (MR) according to (5a) is defined by the equation

$$MR=\frac{\rho(H)-\rho(0)}{\rho(0)}=\frac{G_{av}(0)-G_{av}(H)}{G_{av}(H)} \tag{10}$$

where $G_{av}(H)$ ($G_{av}(0)$) is electron tunneling conductivity in the external field H (without the external field H) and $G_{av}(H)=\rho^{-1}(H)(G_{av}(0)=\rho^{-1}(0))$. The distribution function for the intergranular distance is proposed to be:

$$f(s)=\frac{s}{s_0}\exp(-\frac{s}{s_0}) \tag{11}$$

where s_0 is the distance between grains which causes maximum of the electron tunneling conductivity. The grain orientation at finite temperature in the actual field is defined with statistical factor $g(\theta,H)$

$$g(\theta,H)=\frac{1}{A}\exp\left(\frac{M(H-N_dM)(\cos\theta_1+\cos\theta_2)}{k_BT}\right) \tag{12}$$

here N_d means the demagnetization factor for spherical shape grains. N_d equals to $4\pi/3$ (0.33) in CGS (SI) system of units. For small ferromagnetic grains, especially in nanometer range, the demagnetization field effect should be very important and an actual field inside grains acting on electron may be reduced strongly comparing with the applied field H. In the equation (12), the demagnetization field is written as $H_d=-N_dM$ and A stands for a normalized constant

$$G_{av}(H)=\frac{1}{A}\int_0^{\infty}f(s)\mathrm{d}s\int_0^{2\pi}\mathrm{d}\varphi_1\int_0^{2\pi}\mathrm{d}\varphi_1\int_0^{\pi}\mathrm{d}\theta_1\int_0^{\pi}\mathrm{d}\theta_2\sin\theta_1\sin\theta_2\exp\left(\frac{M(H-N_dM)(\cos\theta_1+\cos\theta_2)}{k_BT}\right)$$
$$\left(1+P^2\cos\chi\right)\exp(-\frac{c}{sk_BT}-2ks) \tag{13}$$

The normalized constant A can be calculated exactly and one get

$$A=\int_0^{2\pi}\mathrm{d}\varphi_1\int_0^{2\pi}\mathrm{d}\varphi_1\int_0^{\pi}\mathrm{d}\theta_1\int_0^{\pi}\mathrm{d}\theta_2\sin\theta_1\sin\theta_2\exp\left(\frac{M(H-N_dM)(\cos\theta_1+\cos\theta_2)}{k_BT}\right)$$

$$A=4\pi^2\left[\frac{\sinh\left(\frac{M(H-N_dM)}{k_BT}\right)}{\frac{M(H-N_dM)}{k_BT}}\right]^2 \tag{14}$$

Inoue and Maekawa [5] used the rough approximation when $\overline{\cos\chi}$ is replaced by the square of relative magnetization i.e. $\overline{\cos\chi}=M^2$. In the work [4], $\overline{\cos\chi}$ was considered as an adjust parameter. Here we can calculate directly without these assumption and get the following expression for the average tunneling conductance

$$G_{av}(H)=\frac{2cG_0K_2\left(2\left(\frac{c(2ks_0+1)}{s_0k_BT}\right)^{-1/2}\right)}{s_0k_BT(2ks_0+1)}\left\{1+P^2\left[\coth\left(\frac{M(H-N_dM)}{k_BT}\right)-\frac{k_BT}{M(H-N_dM)}\right]^2\right\} \tag{15}$$

where $K_2(x)$ is McDonal's function.

The magnetoresistance ratio is obtained straightforwardly from (15) and (10):

$$MR=-\frac{P^2\left[\coth\left(\frac{M(H-N_dM)}{k_BT}\right)-\frac{k_BT}{M(H-N_dM)}\right]^2}{1+P^2\left[\coth\left(\frac{M(H-N_dM)}{k_BT}\right)-\frac{k_BT}{M(H-N_dM)}\right]^2} \tag{16}$$

Proposing that at low fields, the magnetization of the nanometric perovskite grains system has the linear dependence on the field intensity, $M=\kappa H$, where κ is the magnetic susceptibility, we get

$$MR = -\frac{\tanh^2\left(\dfrac{\alpha}{H_0^2}\right)\left[\coth\left(\dfrac{H}{H_0}\right)^2 - \left(\dfrac{H_0}{H}\right)^2\right]^2}{1+\tanh^2\left(\dfrac{\alpha}{H_0^2}\right)\left[\coth\left(\dfrac{H}{H_0}\right)^2 - \left(\dfrac{H_0}{H}\right)^2\right]^2} \quad (17)$$

where $\alpha = \dfrac{J_H\left(S+\dfrac{1}{2}\right)}{\kappa(1-N_d\kappa)}$ and $H_0 = \left(\dfrac{k_B T}{\kappa(1-N_d\kappa)}\right)^{1/2}$.

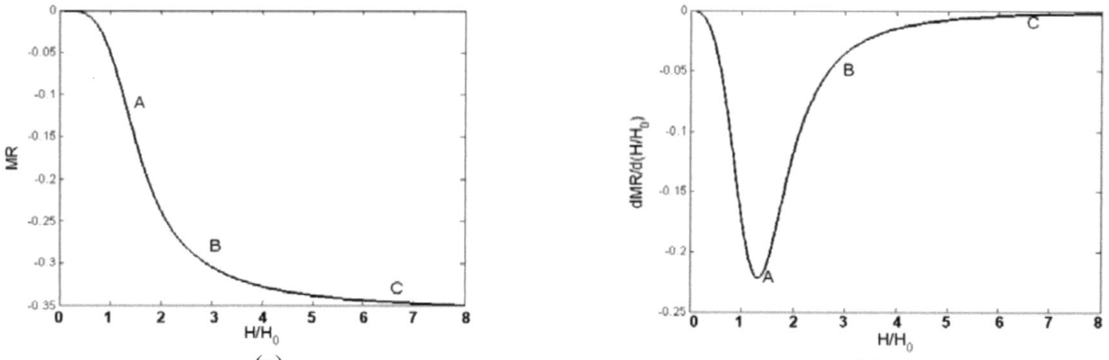

(a) (b)

Figure 2. (a) The magnetoresistance ratio MR as a function of the reduced external magnetic field; (b) The field dependence of the derivative of MR. Here $\alpha / H_0^2 = 0.96$

Figure 2a shows the negative MR given by the formula (17) as a function of inverse relative field strength. We noted that a shape of this curve and a saturated value of MR were the same as observed in experiments [2, 4]. Figure 2b demonstrates the field dependence of the derivative of the MR ratio. The ABC parts of the plots given in figure 2 have the same shape compared with experimental curves obtained in many works, and the 0A parts were not analyzed experimentally. Figure 3 shows the comparison between our theory and experimental results for low field MR (figure 3a), and its derivative (figure 3b) registered in nanometric granular perovskite $La_{0.67}Sr_{0.33}MnO_3$ at temperature 80, 150K (see Dey et al. [10]). One can find here an excellent agreement between the theoretical and experimental results.

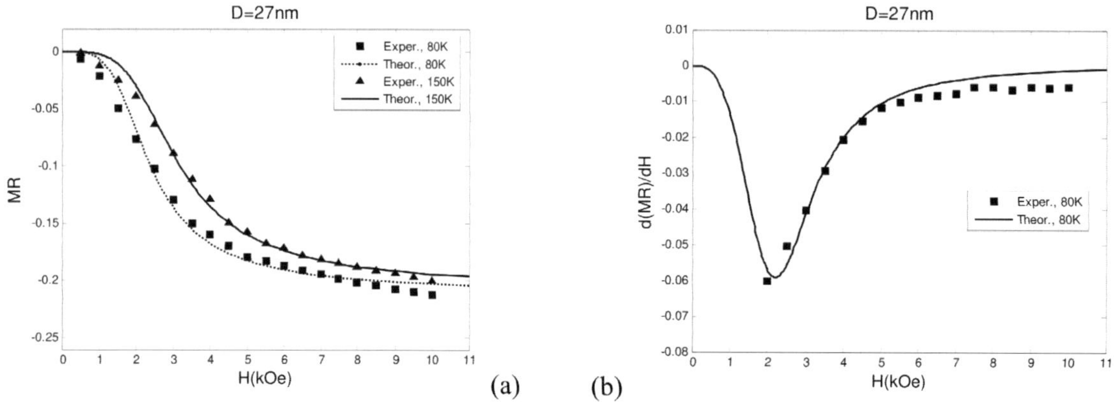

(a) (b)

Figure 3. Comparison between theory and experimental results for low field MR (a) and its derivative (b) given in Ref. [10]. Here the parameter $\alpha = 1.17; H_0 = 1.43$ and $\alpha = 2.00; H_0 = 1.89$ for T = 80 K and T = 150 K, respectively. An average diameter of grains is D = 27 nm.

In conclusion, we have studied the low field tunneling spin transport in magnetic nanometer granular perovskite systems and have shown that the strong interaction between the electron spin and localized atomic spins inside grains leads to the high polarization degree of tunneling electrons. The field dependence of the TMR in nanometer perovskite, its magnitude and derivative can be described properly in the framework of our theory.

References

[1] Zutic, Fabian J and Sarma S Das 2004 *Rev. Mod. Phys.* **76** 323
[2] Dey P and Nath T K 2006 *Appl. Phys. Lett.* **89** 163102
[3] Hwang H Y, Cheong S W, Ong N P and Batlogg B 1996 *Phys. Rev. Lett.* **77** 2041
[4] Raychaudhuri P, Sheshadri K, Taneja P, Bandyopadhyay S, Ayyub P, Nigam A K and Pinto R 1999 *Phys. Rev.* B **59** 13919
[5] Inoue J and Maekawa S 1996 *Phys. Rev.* B **53** R11927
[6] Landauer R 1970 *Philos. Mag.* **21** 863
[7] Sheng P, Abeles B and Arie Y 1973 *Phys. Rev. Lett.* **31** 44
[8] Helman J S and Abeles B 1976 *Phys. Rev. Lett.* **37** 1429
[9] Cong B T, Minh D L, Hieu V T, Vu L V and Chau N 1999 On the reentrant magnetism, spin polaron in colossal magnetoresistance perovskites *Superconductivity, magneto-resistive materials and strongly correlated quantum systems* ed by Nguyen Van Hieu, Tran Thanh Van and Giang Xiao (VNU press) pp 309
[10] Dey P and Nath T K, Kumar U and Mukhopadhyay P K 2005 *J. Appl. Phys.* **98** 014306

Acknowledgments

The authors thank the VNUH research program QGTD 09-05 for support.

Invitro toxicity test and searching the possibility of cancer cell line extermination by magnetic heating with using Fe_3O_4 magnetic fluid

Pham Hoai Linh[1], Nguyen Chi Thuan[1], Nguyen Anh Tuan[1], Pham Van Thach[1], Tran Cong Yen[2], Nguyen Thi Quy[2], Hoang Thi My Nhung[2], Phi Thi Xuyen[2], Nguyen Xuan Phuc[1] and Le Van Hong[1]

[1]Institute of Materials Science, Vietnam Academy of Science and Technology, 18 Hoang Quoc Viet Road, Cau Giay Distr., Hanoi, Vietnam
[2]Biology Faculty, Hanoi University of Science, 334 Nguyen Trai Road, Thanh Xuan, Hanoi, Vietnam

E-mail: honglv@ims.vast.ac.vn

Abstract. A Fe_3O_4 based magnetic fluid with different concentrations ranged between 0.15 ng/cell to 10 ng/cell (nano gram/cell) was used in the *in vitro* toxicity test on several cancer cell lines, Sarcoma 180, HeLa and H358. It shows that the fluid with a concentration of Fe_3O_4 below 1.2 ng/cell is completely non-toxic for these cell lines. Even through in the presence of the highest concentration of 10 ng/cell, the cell viability still reaches more than 60%. The magnetic fluid with Fe_3O_4 concentration of about 0.1 ng/cell was also used to search *ex-vivo* the possibility of Sarcoma 180 extermination by magnetic heating with an AC field of 120Oe and 184 KHz. The result shows that after a heat treatment for 30 min., 40% of Sarcoma 180 cells was killed.

Keywords: Magnetic fluid, hyperthermia, cancer treatment.

1. Introduction

The use of magnetic nanoparticles in cell biology was proposed in the early 1990s in the separation of cells and molecules such as peptides, proteins, and DNA. Magnetic nanoparticles first were used in diagnostics of tumors by using magnetic resonance tomography, in targeting of magnetically–labeled drugs by an external magnetic field [1, 2]. As known, cancer is one of the most dangerous sickness which killed thousands men per year. Humankind has devoted a big endeavor for checking this disaster by using various methods treatment such as ethanol injection [3], radiofrequency ablation [4], laser light and ultrasound waves [5, 6]. In recent years, magnetic nanoparticle hyperthermia has become a potential solution for the treatment of cancers [7]. In this solution tumor cells are selectively killed under heating at the temperature around 45-46^oC by exposing the magnetic nanoparticles in an alternating magnetic field. Basically, when exposing to an alternating magnetic field, the nanoparticles can absorb energy via various physical mechanisms including the magnetic hysteresis loss, Néel relaxation and Brownian loss [8]. This energy is considerably high to make the particles as the local nano-fires for the application of magnetic hyperthermia. Till now the most appropriated material using for magnetic hyperthermia treatment is ferromagnetic nanoparticle, especially Fe_3O_4 magnetite

© 2009 IOP Publishing Ltd

nanoparticle because of its high specific loss power (SLP) and well proven biocompatibility with life cells [9, 10]. The first study on the magnetic heating induced by an alternating magnetic field was performed by Gilchrist *et al.* [11], and its result was later reconfirmed by several groups [12, 13]. The recent investigations showed that the capacity in magnetic hyperthermia of magnetic liquid essentially depends on specific loss power (SLP) of the iron oxide nanoparticles, consequently depends on their structural properties and magnetic behaviors [14]. With the aim of improving SLP of Fe_3O_4 nanoparticle several methods were established and their results showed that the SLP of magnetic Fe_3O_4 nanoparticles has much been improved [15, 16]. The highest value of SLP of 960 W/g at the field of 410 kHz and 10 kA/m was found for magnetosomes produced by magnetotactic bacteria [17]. By now several investigations in magnetic hyperthermia using Fe_3O_4 magnetic nanoparticles were attempted. Hilger et al [18], for example, carried out a hyperthermia treatment for breast cancer using a magnetic fluid. In this paper we report recent new results of our study on the toxicity of Fe_3O_4 magnetic nanoparticles tested on several cancer cell lines and also on an ex-vivo test of cancer cell extermination by using magnetic heating. Three following cell lines were used in our study: (1) Hela cells, an epithelial cell line originally derived from human cervix; (2) a human bronchioalveolar carcinoma H358 cells; (3) mouse ascites sarcoma cells named Sarcoma 180. These cell lines were obtained from American Type Culture Collection (ATCC). Hela and Sarcoma 180 cells were maintained in DMEM (Dulbecco's modified Eagle's) 1g/l of Glucose (Invitrogen). H358 cells were maintained in RPMI (Roswell Park Memorial Institute) (Gibco). These medium were supplemented with 10% foetal bovine serum (Invitrogen), penicillin-streptomycine 100 IU/ml (Invitrogen), L – Glutamine 2mM (Invitrogen). Cells were grown in a humidified chamber in the presence of 5% CO_2, at 37°C.

2. Experimental

2.1. Synthesis of Fe_3O_4 nanoparticles

Fe_3O_4 nanoparticles were prepared by the chemical co-precipitation method. Aqueous solution of $FeCl_3$, $FeCl_2$ salts in their respective stoichiometry of $Fe^{2+}/Fe^{3+} = 1/2$ was mixed and kept at 40°C. NaOH was added to the solution under vigorous mechanical stirring for 30 minutes. The precipitation and formation of Fe_3O_4 nanoparticles were taken by a conversion of the metal salts into hydroxides and, in turn, immediately transformed into ferrites. During the reactions, magnetic-stirring speed was kept at 1000 round/min.

2.2. Coating particles

Starch was used for coating the Fe_3O_4 nanoparticles. The coating process was carried out as follows: Starch was dissolved in water with a concentration of 0.1 g/ml at 80°C. After the good dissolving of starch, the solution was immediately placed in a 60°C water bath and dropped into the obtained aqueous of dispersed Fe_3O_4 particles after a co-precipitation reaction at 80°C in 2 h. Around 50 wt.% of water was evaporated during the boiling. The remaining suspension was then cooled to 0°C within 12 h. The formed gels were washed with deionized water until the pH is reduced to less than 8.5. Excess salts and ions were then removed by washing. For cleaving glycoside bond and shortening the polymer chain, starch-coated iron oxide particles were oxidized by H_2O_2 solution

2.3. Characterization

The structure and the morphological characterization of magnetic nanoparticles were taken by using an X ray diffractometer D5000 of SIEMEN and a field-emission scanning electron microscope (FESEM S-4800 of the Hitachi Company). The magnetic properties were checked by using a home made vibrating sample magnetometer (VSM) and a Quantum Design PPMS 6000.

The magnetic heating experiments were performed under an alternating magnetic field with the frequency changed in a wide range from 50 kHz to 400 kHz and its field strength can increase up to maximal value of 12 kA/m. The magnetic field is generated by a 5.5-loop induction coil with a 4.5 cm diameter and a 6.3 cm length. During the heating experiment, the temperature of the sample was monitored by using a Copper-Constantan thermocouple. The magnetic heating experimental setup is shown in figure 1.

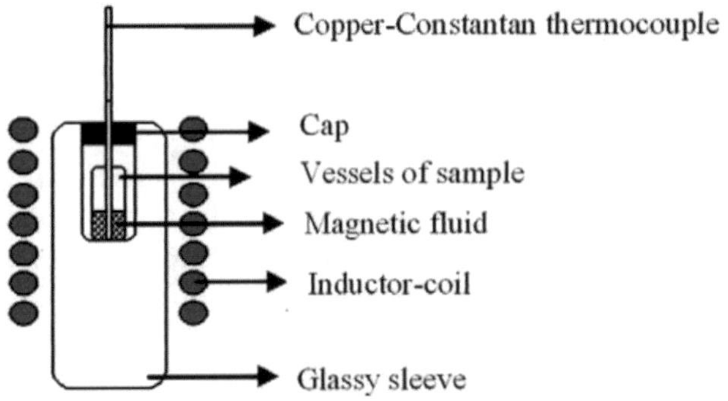

Figure 1. The experimental set up for SLP values measurements.

All the samples used in magnetic heating experiment contain 50 million cells of Sarcoma 180 line and 3 mg/ml of Fe_3O_4 nanoparticles. It means the concentration of Fe_3O_4 nanoparticles in all the samples is about 0.3 ng/cell.

2.4. Cytotoxicity Assay

Sulforhodamine-B (SRB) assay is performed to assess cell survival. SRB is a water-soluble dye that binds to the basic amino acids of the cellular proteins. Thus, colorimetric measurement of the bound dye provides an estimate of the total protein mass that is related to the cell number. Cells were seeded in 96-well plates with the density of $1x10^4$ cells/well. After 24 h, the culture medium was replaced by the new medium plus different concentrations of magnetic particles. 72 h after the incubation, cultures fixed with 10% trichloroacetic acid were stained for 30 minutes at 4°C with 0.4% (wt/vol) sulforhodamine B (SRB) dissolved in 1% acetic acid. Unbound dye was removed by four washes with 1% acetic acid, and protein-bound dye was extracted with 10 mM unbuffered Tris base [tris (hydroxymethyl) aminomethane] for determination of optical density in a computer-interfaced, 96-well microtiter plate reader (SpectraMax 250). The absorbance was measured at 540 nm [19, 20]

2.5. Determination of cells viability by trypan blue

Place a suitable volume of a cell suspension (20-200 μl) in appropriate tube add an equal volume of 0.4% trypan blue and gently mix, let stand for 5 minutes at room temperature. Place 10-20 μl of stained cells in a hemacytometer and count the number of viable (unstained) and dead (stained) cells. The percentage of viable cells is the number of viable cells divided by the number of dead and viable cells [21, 22].

2.6. Determination of cell death rate after the hyperthermia Sarcoma 180 cells ex vivo

Sarcoma 180 cells were collected from Swiss mice bearing 10 days of Sarcoma 180 ascite. Cells then were washed three times with PBS 1X and centrifuged to a pellet. Magnetic nanoparticles were added on the cells. The mixes of cells and nanoparticles were subjected to the hyperthermia treatment either with different times of heating or different concentrations of particles. Cells viability was determined by trypan blue [23, 24].

3. Result and discussion

Our research is an interdisciplinary science mission in physics, biology and medicine, therefore the research was concentrated on both the preparation of magnetic liquid containing Fe_3O_4 nanoparticles as well as the test research on its biocompatibility and the possibility in cancer cell lines treatment by using magnetic heating. Preparation of the magnetic liquid containing Fe_3O_4 nanoparticles has been presented in details elsewhere in other our papers. Here in this paragraph we would like to show its magnetic characterization, especially its response to an alternating magnetic field, possibility of controlling the heating temperature. So we can see in figure 2 which presents the heating curve by using the liquid with various concentrations of Fe_3O_4 nanoparticles exposed in an alternating magnetic field with a frequency of 184 kHz, and an intensity of 12 kA/m. As seen in figure 3 the initial heating rate and the saturation temperature of heating process can be controlled by changing concentration of Fe_3O_4 nanoparticles in the liquid.

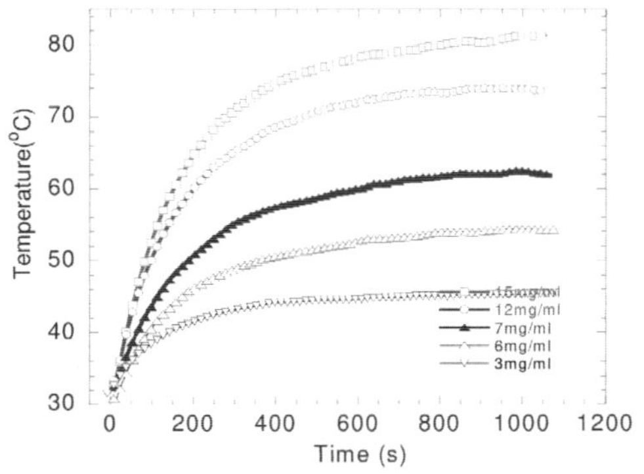

Figure 2. The heating curves of the magnetic fluid with various particles concentration.

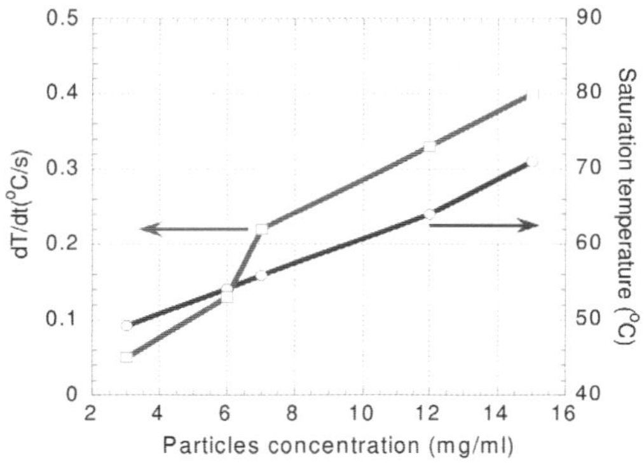

Figure 3. The dependence of the initial heating rate and saturation temperature on particles concentration.

The most decisive parameter in verifying the application possibility of the magnetic liquid in cancer cell treatment by using magnetic heating is a specific loss power (SLP), which was defined as follows [8]:

$$SLP = (C * m_s * \frac{dT}{dt}) / m_i$$

where m_s is the total mass of magnetic liquid, dT/dt is the initial heating rate, which was defined at the moment of about some seconds after starting the heat, C is the specific heat capacity of magnetic liquid, which was accepted equal to that of water, because of the mass of Fe_3O_4 nanoparticles containing in the liquid m_i is much smaller than the total mass m_s. The estimated SLP values are presented in table 1.

Table 1. The initial heating rate, SLP and saturation temperature values of different particles concentration of magnetic fluid.

Concentration of particles (mg/ml)	dT/dt (°C/s)	Satutration temperature (°C)	SLP values (W/g)
3	0.092	45	129
6	0.141	53	97
7	0.159	62	94
12	0.240	73	83
15	0.310	80	86

As mentioned in the Introduction, several research groups devoted their endeavour for looking for the magnetic liquid having high SLP. R. Muller et al [15] using a glass crystallization method received a mix of iron oxides with a SLP of 100 W/g, Q. Zeng *et al.* [16] carried out manufacture the monodispersed Fe/Fe_3O_4 composite nanoparticles with an improved SLP of 345 W/g, and recently R.

Hergt et al received very high SLP of 960 W/g by using bacterial magnetosomes [17]. According to Hergt [25] the ferrofluids with a SLP of 50 W/g are suitable in application for cancer tumors not smaller than about 4 mm diameter. Conclusively the ferrofluid containing Fe_3O_4 nanoparticles manufactured by us could be applicable in hyperthermia applications for cancer cell treatment.

According to the law of health care we have first to take researches on toxicity test and biocompatibility of the ferrofluid containing Fe_3O_4 nanoparticles. Two cell lines, Sarcoma 180 and H358 were taken in the toxicity test with a ferrofluid containing Fe_3O_4 nanoparticles coated with starch. Using the trypan blue method for Sarcoma 180 and SRB method for cell line H358 the proliferation rate of the cell lines was estimated and presented in table 2.

Table 2. Proliferation rate of the cancer cells after 72 hours of growth compared to the control one

Cell lines	Concentration of Fe_3O_4 nanoparticles (ng/cell)				
	0.1	0.5	1.0	5.0	10.0
Sarcoma 180	95.5 ± 2	94.5 ±5	96.3 ± 6	92.0 ± 3	89.5 ± 5
H358	84.5 ± 5	82.5 ± 7	85.0 ± 4	Not determined	78.2 ± 4

As was shown in table 2, the proliferation of two cell lines slightly decreases after 72 h of growth for the test concentration of Fe_3O_4 nanoparticles above 1.0 ng/cell. Fe_3O_4 nanoparticle not equally respects to the cell lines, Sarcoma 180 line has higher proliferation compared with that of the H358 cell line. Probably Fe_3O_4 nanoparticles are more toxic for the adherent cell lines compared to the suspension line. This suggestion has been proven by other experiment which allows us to evaluate the adherent possibility of Fe_3O_4 nanoparticles onto surface of the cancer cells. Observing microscopic images (figure 4) we confirmed that the lower proliferation of the adherent cell line, Hela, is caused because the Fe_3O_4 nanoparticles more strongly adhere to surface of the cells in comparison with the suspension cell line, Sarcoma 180.

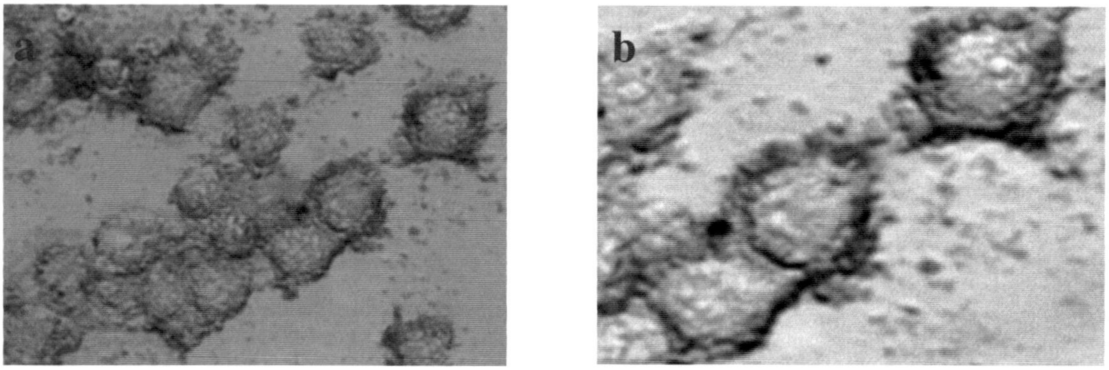

Figure 4. Microscopic images exhibiting the adherence of Fe_3O_4 nanoparticles onto Hela cell membranes (a) and with enlargement (b).

Figure 4 presents microscopic images, exhibiting the adherence of Fe_3O_4 nanoparticles onto the cell membrane of Hela cancer cells. It shows that Fe_3O_4 nanoparticles have tendency to adhere at the surface point where connected to the matrix material, therefore prevents proliferation of the Hela cells.

As shown in table 2, the experimental test with concentration below 1 ng/cell was given higher proliferation for both two cancer cell lines in comparison to other concentrations and actually the proliferation of Sarcoma 180 cell with the Fe_3O_4 nanoparticle concentration below 1 ng/cell was estimated of about 96%.

Next step in searching the application possibility of Fe_3O_4 ferrofluid in biomedicine was an attempt of establishing experiments on cancer treatment by magnetic heating with Fe_3O_4 ferrofluid in an alternating magnetic field. According to the above results we have chosen ferrofluids having Fe_3O_4 concentration below 1 ng/cell for the ex-vivo experiment in hyperthermia. Actually, we set up such an experiment: to observe the effects of the heating time on extermination of Sarcoma 180 cancer cell which was heated with Fe_3O_4 ferrofluid by its exposing to an alternating magnetic field of 12 kA/m in intensity and 184 kHz in frequency. Note here that the intensity of alternating magnetic field, its frequency and Fe_3O_4 concentration in the ferrofluid were chosen how to receive a saturation temperature of heating of about 47°C ± 1°C. As was known at this temperature the cancer cells probably were killed after keeping for a time such that, conversely the life cells were still saving. Figure 5 presents the number of killed Sarcoma 180 cells in dependence on time of heating at 47°C by an alternating magnetic field with an intensity of 12 kA/m and a frequency of 184 kHz. It shows that the Sarcoma 180 cancer cells were successfully killed after being heated for not shorter than100 minutes. Evidently, after stopping of the heating Sarcoma 180 cells were continuously themselves killed. This domino effect in the extermination of cancer cells keeps a time that its length depends on the heating time. For example we present in figure 6 the ratio of killed Sarcoma 180 cells in dependence of time after stopping the heating process.

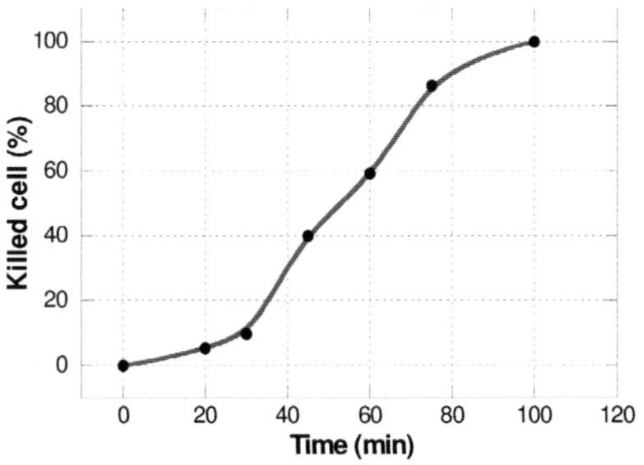

Figure 5. The ratio of the killed cells of Sarcoma 180 line versus time of heating by exposing of Fe_3O_4 magnetic liquid in an alternating magnetic field with an intensity of 12 kA/m and a frequency of 184 kHz.

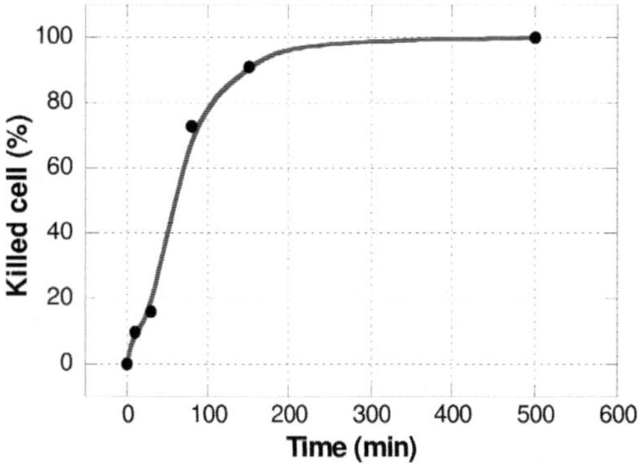

Figure 6. Ratio of the killed cells of Sarcoma 180 line in dependence of time after stopping heating.

For illustrating the progress of the experiment on the extermination of the cancer cells we present microscopic images as shown in figure 7 and 8. Two control samples have almost the same ratio of killed cells of Sarcoma 180 line. It confirmed that the Sarcoma 180 cells survived at least not shorter than 72 hours. In addition it also confirmed that the alternating magnetic field is not harmful for the cancer cell line as well as for the life cells. So it is inferred that the magnetic heating is acceptable method used in vivo cancer treatment for human body. Figure 9 presents two microscopic image of the sample containing 50 millions of Sarcoma 180 cells and Fe_3O_4 nanoparticles after 75 minutes heating by an alternating magnetic field with intensity of 12 kA/m and frequency of 184 kHz. It shows that a large number of cancer cells were killed, shown in the blue colour, resulting from using the trypan blue method in determination of cell viability. The extermination of Sarcoma 180 cells continuously occurred after stopping the heating process. As can be seen in figure 10, number of the killed cells increased very clearly. Among the cells only a few still were alive. This process extended for very long time. As shown in figure 6 the needful time for the completely killed all the Sarcoma 180 cells extended to 500 minutes.

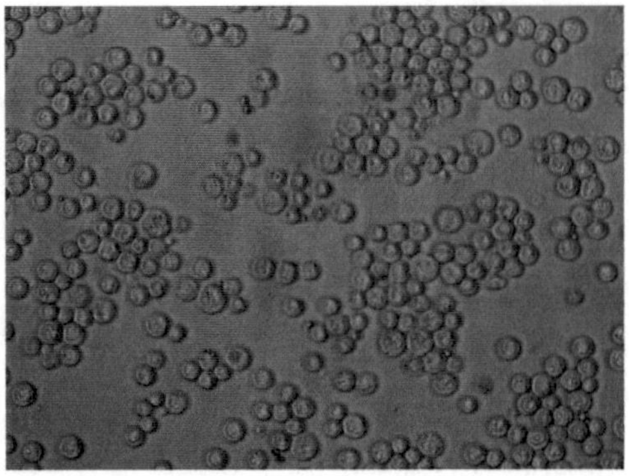

Figure 7. Microscopic images of the sample containing only Sarcoma 180 cancer cells which was kept for time of 72 hours long for using as a control sample.

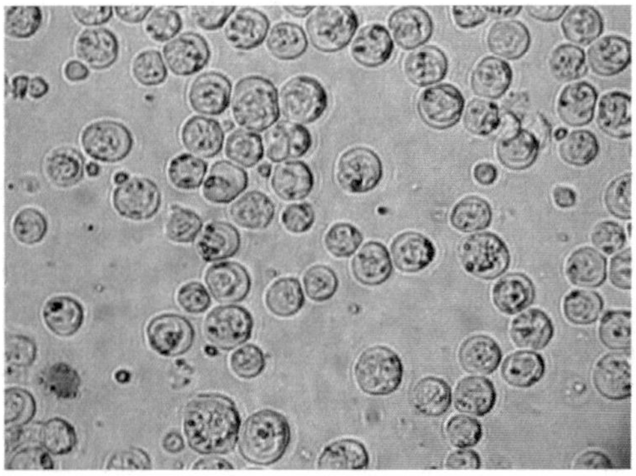

Figure 8. Microscopic images of the sample without Fe_3O_4 nanoparticles after 72 hours exposed in an alternating magnetic field with intensity of 12 kA/m and frequency of 184 kHz.

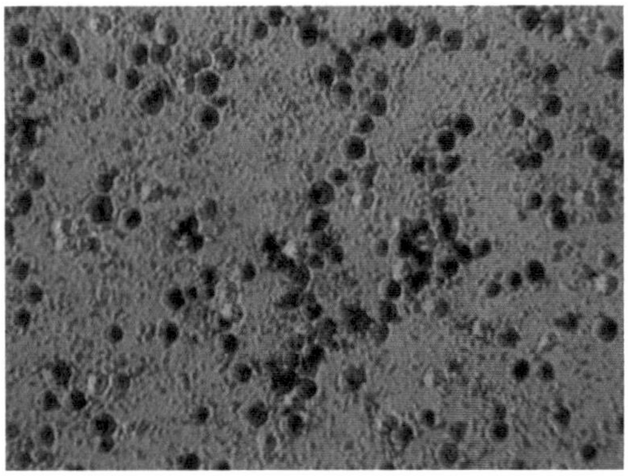

Figure 9. Microscopic image of the sample containing 50 million cells of Sarcoma 180 line and Fe_3O_4 nanoparticles with a concentration of 0.3 ng/cell after heated at 47°C for 75 minutes by an alternating magnetic field with intensity of 12 kA/m and frequency of 184 kHz.

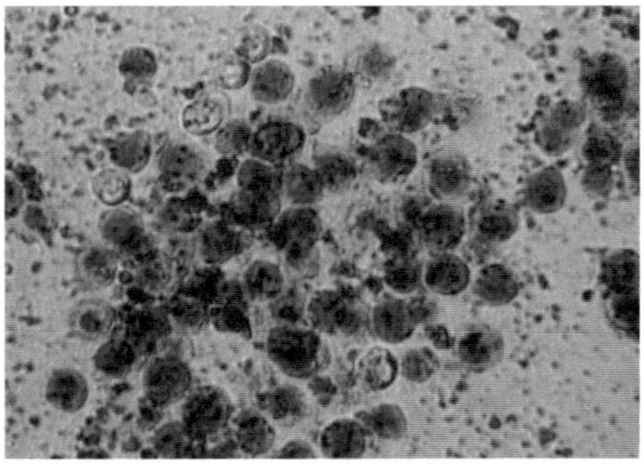

Figure 10. Microscopic images of image of the same sample 45 minutes after stopping the heating process (b).

4. Conclusion

Magnetic fluid containing Fe_3O_4 nanoparticles with different concentrations ranged between 0.15 ng/cell to 10 ng/cell was successfully prepared and used in the in vitro toxicity test on several cancer cell lines, Sarcoma 180, Hela and H358. The fluid with a concentration of Fe_3O_4 below 1.0ng/cell is completely non-toxic for these cell lines. The magnetic fluid with Fe_3O_4 concentration of about 0.3ng/cell was used in ex vivo searching the possibility of Sarcoma 180 extermination by magnetic heating with an AC field of 120 Oe and 219 KHz. The result shows that the heat treatment around 47°C by exposing Fe_3O_4 magnetic liquid in an AC magnetic field is successfully exterminated Sarcoma 180 cells. Probably, after heat treatment in a duration time of 100 minutes almost 100% Sarcoma 180 cells was completely killed. This result opens a large prospect in researching the hyperthermia applications by using magnetic liquid containing Fe_3O_4 nanoparticles.

References

[1] Alexiou C, Arnold W, and Klein R J et al 2000. *Cancer Res.* **60** 6641

[2] Lubbe A S, Bergemann C and Riess H et al 1996 *Cancer Res.* **56** 4686

[3] Yamamoto S, Takatori K, and Oomoto K et al 1991 *Kan Tan Sui.* **22** p 827

[4] Organ L W 1976/1977 *Appl. Neurophysiol.*, **39** 69

[5] Miao Y, Ni Y and Mulier S et al 1997 *J. Surgical Res.* **71** 19

[6] Murakami R, Yoshimatsu S and Yamashita Y et al 1995 *Am. J. Roentgenol* **164** 1159

[7] Mornet S, Vasseur S, Grasset F and Duguet E 2004 *J. Mater. Chem.* **14** 2161

[8] Hergt R, Andra W, d'Ambly C G, Hilger I, Kaiser W A, Richter U, and Schmidt H 1998 *IEEE Trans. Magn.* **34** 3745

[9] Jordan, Scholz R, Wust P, Fahling H and Felix R, Magnetic fluid hyperthermia (MFH) 1999 *J. Mag. Mag. Mater.* **201** 413-419

[10] Hergt R, Hiergeist R, Zeisberger M, Schuler D, Heyen U, Hilger I and Kaiser W A 2005 *J. Mag. Mag. Mater.* **293** 80

[11] Gilchrist R K, Medal R and Shorey W D 1957 *Ann. Surgery* **146** 596

[12] Moroz P, Jones S K and Gray B N 2002 *J. Surgical Oncology.* **80** 149

[13] Moroz P, Jones S K, and Winter J et al. 2001 *A. J. Oncology.* **78** 22

[14] Hergt R and Andra W 2007 *Magnetism in medicine*: a handbook, second edition, Willey, Berlin,

[15] Muller R, Hergt R, Zeisberger M, Gawalek W 2005 *JMMM* **289** 13

[16] Zeng Q, Baker I, Loudis J A, Liao Y, Hoopes P J, and Weaver J B 2007 *Appled Phys. Lett.* **90** 233112

[17] Hergt R, Hiergeist R, Zeisberger M, Schuler D, Heyen U, Hilger I, and Kaiser W A 2005 *JMMM.* **293** 80

[18] Hilger I, Hergt R and Kaiser W A, 2005 *IEE Proc.-Nanobiotechnol.* **152** (1) 33

[19] Skehan P, Storeng R, Scudiero D, Monks A, McMahon J, Vistica D, Warren J T, Bokesch H, Kenney S, Boyd MR 1990 *J Natl Cancer Inst.* **82(13)** 1107

[20] Loberg L I, Engdahl W R, Gauger J R. and McCormick D L 2000 *Radiation Research*, Vol. **153** (5) 725

[21] Choi SH, Jin SE, Lee MK, Lim SJ, Park JS, Kim BG, Ahn WS, Kim CK 2007 *Eur J Pharm Biopharm.* 2008 Mar **68**(3) 545

[22] Sigounas G, Sallah S, Sigounas V Y 2004 *Cancer letters* **214** 171

[23] Mitsugu Y, Masashige S, Hiroyuki H, Toshihiko W, Jun Y and Takeshi K 1998 *Jpn. J. Cancer. Res.* **89** 463

[24] Brusentsov N A, Gogosov V V, Brusentsova T N, Sergeev A V, Jurchenko N Y, Kuznetsov A A, Kuznetsov O A and Shumakov L I 2001 *J. of Magnetism and Magnetic Material* **225** 113

[25] R Hergt, Hiergeist R, Hilger I, Kaiser W A, Lapatnikov Y, Margel S and Richter U 2004 *JMMM.* **270** 345

Acknowledgments

This work was done under the support of Grant 4.107.06 in the Fundamental Research Program of Vietnam (2006-2008) and Project of Vietnam Academy of Science and Technology (VAST). The technique support of The National Key Laboratory for Electronic Materials and Devices is acknowledged.

Preparation of magnetic nanoparticles embedded in polystyrene microspheres

Nguyen Hoang Hai, Nguyen Hoang Luong, Nguyen Chau and Ngo Quy Tai

Center for Materials Science, Faculty of Physics, Hanoi University of Science, Vietnam National University, Hanoi, 334 Nguyen Trai, Hanoi, Vietnam

E-mail: nhhai@vnu.edu.vn

Abstract. Superparamagnetic particles are widely used for biological applications such as cell separation. The size of the particles is normally in the range of 10 – 20 nm which is much smaller than the size of a cell. Therefore small particles create small force which is not strong enough to separate the cells from solution. Superparamagnetic nanoparticles embedded in Polystyrene microspheres (magnetic beads) are very useful for cell separation. Magnetic beads have been prepared by solvent evaporation of an emulsion. The beads with size of 0.2 μm – 1.0 μm have a saturation magnetization of 10 – 25 emu/g. The change of the amount of surfactants, volatile solvent, magnetic particles resulted to the change of size, magnetic properties of the magnetic beads.

Keyword: Fe_3O_4, nanoparticles, superparamagnetism, magnetic beads, bio application.

1. Introduction

Superparamagnetic iron oxide nanoparticles embedded in polymer microspheres with appropriate surface chemistry have been widely used experimentally in numerous biological applications such as magnetic resonance imaging contrast enhancement, tissue repair, immunoassay, detoxification of biological fluids, hyperthermia, drug delivery and cell separation and labelling [1-6]. All these biomedical and bioengineering applications require that these nanoparticles have a superparamagnetic behaviour with high magnetization values, a narrow particle size distribution, a biocompability, and stability under working environments. The typical size of iron oxide nanoparticles is in the range from 10 – 2 nm [7]. Suppose that there is a layer of such nanoparticles coated around a cell, we need a strong external magnetic field gradient to create a strong force to separate the cell from its surrounding medium. One approach to satisfy all requirements in bioapplications is to coat nanoparticles with polymers. Magnetic polymer microspheres should fulfil some criteria to fit further biomedical applications: no sedimentation, uniform size and size distribution, high and uniform magnetic content, superparamagnetic behavior, no toxicity, no iron leaking, high selectivity in case that these particles are used for hyperthermia purposes, and sufficient heat generation at lower frequencies to enhance selective heating. Therefore, magnetite particles homogeneously encapsulated in a hydrophobic polymer which keep away water-soluble components from contacting the magnetite particles are of high interest. Many types of polymers can be used for coating magnetic nanoparticles. Polystyrene is one of the most preferred polymers. There are several reasons to use polystyrene as hydrophobic encapsulation material in biomedical applications, e.g., it is inexpensive and it is a hydrophobic

© 2009 IOP Publishing Ltd

polymer which allows physical adsorption of antibodies or proteins, it can also be functionalized, e.g., by carboxylic groups which enables covalent binding of antibodies, proteins, or cells [8].

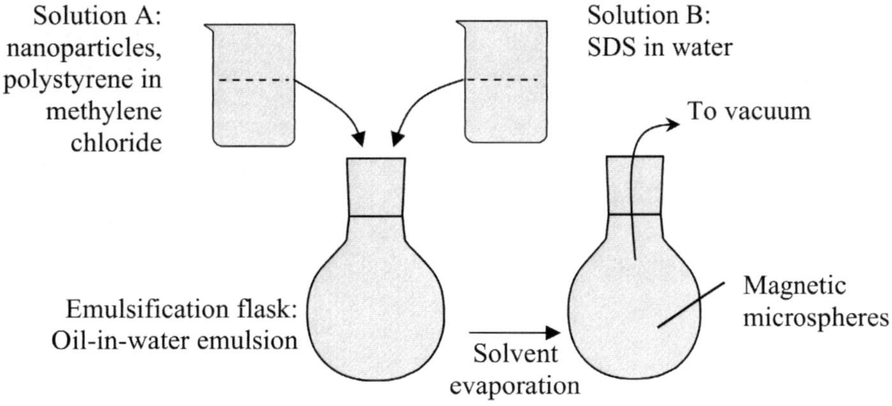

Figure 1. Schematic overview of the four basic steps in microspheres preparation by solvent evaporation.

In Vietnam, nanomaterials, especially nanoparticles, have been paid much attention since 1997. Up to now, dozens of scientific conferences have been organized and hundreds of scientific papers have reported on the preparation, properties and applications of nanomaterials, which have been done by Vietnamese scientists. However, most of the papers focused on the materials with similar properties such as hard/soft nanocomposites [9], CdS/ZnS core/shell structure [10],... This paper will report on another approach to combine nanomaterials with very different properties in one entity – nanoparticles embedded in polymer microspheres. The combination magnetic nanoparticles, quantum dots, metallic nanoparticles, drugs,... in polymer microspheres can create new properties and new applications.

There are various techniques, which feature partly competing, partly complementary characteristics, for obtaining iron oxide nanoparticles embedded in polymer microspheres. Many of them are modifications of the three basic techniques: spray-drying [11], phase separation (coacervation) [12], solvent extraction/evaporation [13]. Spray-drying is simple and of high throughput but must not be used for high temperature-sensitive compounds. Moreover, control of the particle size is difficult, and yields for small batches are moderate. Coacervation is normally impaired by residual solvents and coacervating agents found in the microspheres. In addition, it is not well suited for producing microspheres in the low micrometer size range. Solvent extraction/evaporation neither requires elevated temperatures nor phase separation including agents. Controlled particle sizes in the nano- to micrometer range can be achieved, but careful selection of encapsulation conditions and materials is needed to yield high encapsulation efficiencies and a low residual solvent content. Microsphere preparation by solvent extraction/evaporation consists of four major steps: (i) dissolution or dispersion of the bioactive compound often in an organic solvent containing the matrix forming material; (ii) emulsification of this organic phase in a second continuous (frequently aqueous) phase immiscible with the first phase; (iii) extraction of the solvent from the dispersed phase by the continuous phase, which is optionally accompanied by solvent evaporation; (iv) harvesting and drying of the microspheres.

In this paper, we used the solvent evaporation technique to make iron oxide Fe_3O_4 embedded in polystyrene microspheres. This is one of the first advances in this field in Vietnam.

Journal of Physics: Conference Series **187** (2009) 012009 doi:10.1088/1742-6596/187/1/012009

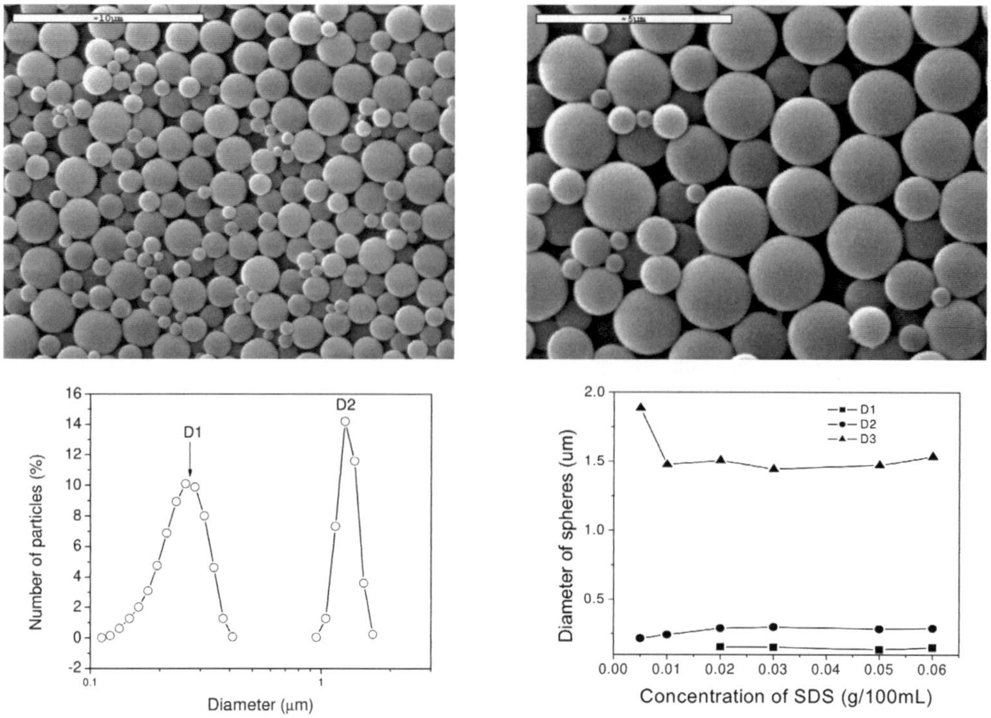

Figure 2. SEM images of the polystyrenes microspheres (top), a typical size distribution of the spheres (bottom left) and the particle diameter of the samples (bottom right).

2. Experimental

Magnetite Fe_3O_4 nanoparticles were synthesized by coprecipitating a Fe^{2+}/Fe^{3+} mixed solution (the molar ratio Fe^{2+}: Fe^{3+} = 1 : 2) with a NH_4OH solution in air [7]. In a typical reaction, 3.98 g of $FeCl_2.4H_2O$ (molar concentration 0.1 M) and 10.82 g of $FeCl_3.6H_2O$ were dissolved in 200 ml of distilled water. An 18 ml solution of NH_4OH 25 % diluted to 100 ml by water, was then added to the Fe^{2+}/Fe^{3+} solution at a rate of one drop per second at room temperature with constant rate. After the reaction, black precipitates of magnetite were washed five times by using distilled water and magnetic decantation to remove all the soluble substances. 0.5 g of magnetic nanoparticles in 20 ml of water were then vigorously mixed with 10 ml of oleic acid (OA, 9-Octadecenoic acid $C_{18}H_{34}O_2$, CAS number: 112-80-1) in order to coat the magnetic nanoparticles with a single surfactant layer of OA (denoted as OA-coated nanoparticles) and to disperse them into a volatile solvent methylene chloride (CH_2Cl_2). The excess OA was washed three times by using magnetic decantation. An amount of 0.1 g of polystyrene was dissolved in 10 ml of methylene chloride by using a magnetic stirrer (IKA RH basic KT/C) at room temperature. The obtained solution was used to disperse 0.01, 0.02, and 0.03 g (correspondingly, three samples with 10, 20, and 30 wt. % of nanoparticles) of OA-coated magnetite nanoparticles to form solution A (figure 1). Solution A was then emulsified with solution B containing water and sodium dodecyl sulfate (SDS) as a stabilizer, which prevents the organic droplets from coalesing. The concentration of SDS was changed from 5 – 60 mg/100 ml. The emulsion process was conducted by mixing solution A and solution B with the help of an overhead stirrer (IKA RW 20) at room temperature. The speed and duration of agitation of stirring affect the size of the droplets, with stronger agitation breaking up the polymer into smaller droplets. Here, we fixed the speed A and the duration time of 2 mins. Methylene chloride was eliminated by using a rotating evaporator (Buchi rotavapor R II) for about 20 mins at 60°C and the microspheres were remained in the flask (figure 1).

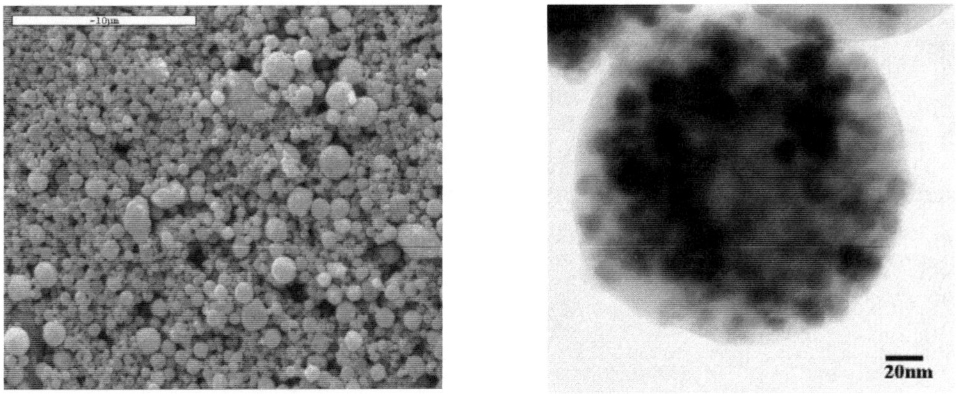

Figure 3. SEM images of the polystyrenes microspheres (top), a typical size distribution of the spheres (bottom left) and the particle diameter of the samples (bottom right).

In the microspheres, magnetite nanoparticles were distributed randomly to form magnetic microspheres (magnetic beads). The microspheres were washed with doubly distilled water and collected by using a centrifuge (Hettich Universal 320, 9000 RPM, 20 mins). Another series of samples in which there was no magnetic nanoparticles have been prepared by a similar method as reference samples. The difference was that, in the solution A, there was only polystyrene dissolved in methylene chloride without magnetic nanoparticles. As a result, the microspheres did not contain magnetic nanoparticles.

The structure of the magnetite nanoparticles was analyzed by using a Bruker D5005 X-ray diffractometer (XRD). Magnetic measurements were conducted by using a DMS-880 sample vibrating magnetometer (VSM) with maximum magnetic field of 13.5 kOe at room temperature. The particle morphology was obtained from a transmission electron microscope (TEM JEM1010-JEOL) and a scanning electron microscope (SEM JEOL 5410 LV).

3. Results and discussion

Figure 2 shows the SEM images of the reference non-magnetic polystyrene microspheres. The shape of the products were almost ideally spherical with very smooth surface. The spherical shape indicated the isotropic property of the oil-in-water emulsion. It is evident that there were two size distributions of the microspheres with different diameters which were denoted by D1 (for small microspheres) and D2 (for big microspheres). In some samples, there were three size distributions with corresponding diameters (denoted by D1, D2, and D3 for the diameter from small to large). The diameters and the size distributions were given in figure 2. Figure 2 (right bottom) presents the diameter of the reference microspheres as a function of SDS concentration. Samples with SDS concentration were smaller than 0.2 had two size distributions and the diameter was too large. Therefor, later on, we chose the concentration of 30 mg of SDS in 100 ml to make magnetic microspheres. When the SDS concentration was equal or larger than 0.2, there were three size distributions and the diameter of the microspheres kept almost constant with SDS concentration. The figure can be explained by the fact that when SDS concentration was high, there was enough SDS to form small oil droplets in the emulsion. After solvent evaporation process, the final microspheres were small.

Figure 3 shows the SEM image of magnetic microspheres. It is not like in the case of the reference microspheres, the shape of microspheres was not ideally spherical with rough surface. This may come from the hydrophobic OA-coated magnetite nanoparticles which were free coated around the microspheres and affected to the surface energy of oil droplets in the emulsion. The TEM image shows the heterogeneous contrast of a microsphere. The dark areas present for the crystalline magnetite nanoparticles and the light areas indicate the amorphous polystyrene.

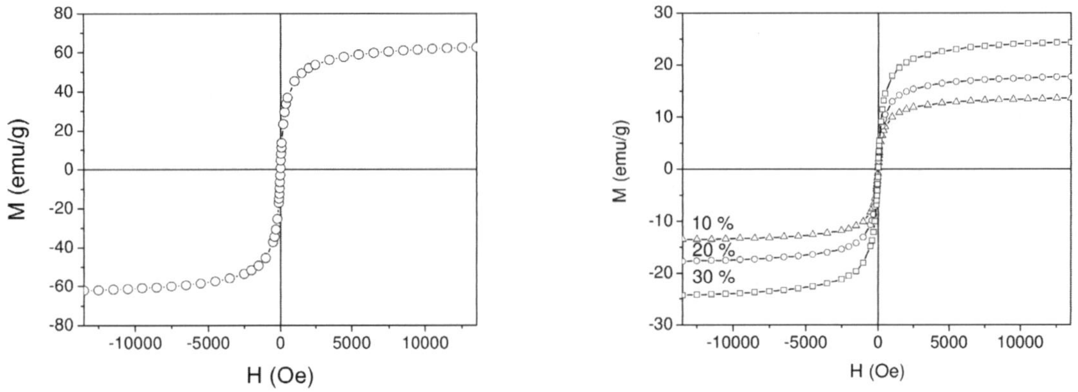

Figure 4. Magnetic curve of magnetite nanoparticles (left) and magnetic microspheres with 10, 20, and 30 wt. % (right) the nanoparticles embedded in polystyrenes microspheres.

Magnetic curves of the magnetite nanoparticles are shown in figure 4. The curve of magnetite nanoparticles shows a superparamagnetic behavior with the saturation magnetization, M_s, of 60 emu/g. The superparamagnetic property did not change when the nanoparticles were embedded in the microspheres. But M_s reduced with the ratio of the nanoparticles in the magnetic microspheres. The values of M_s in magnetic microspheres were higher than expected from the ratio of magnetic nanoparticles. This can be ascribed to the preparation process. During washing by water and magnetic decantation, a commercial magnet was used. The field gradient of the magnet may be not strong to collect microspheres with low magnetization. Therefore, a portion of polystyrene was lost which made the saturation magnetization of the expected values calculated from the ratio smaller than that of magnetic nanoparticles in the microspheres.

4. Conclusion

We have fabricated magnetite nanoparticles embedded in polystyrene microspheres by solvent evaporation techniques. By changing the preparation conditions we could control the diameter, size distribution and saturation magnetization of the final magnetic microspheres. This technique can be applied to make other types of nanoparticles such as metal and semiconducting materials embedded in polymer microspheres.

References

[1] Khuat N T, Nguyen V A T, Phan T N, Thach C V , Hai N H, Chau N 2008 *J. Korean Phys. Soc.* **52** 1323
[2] Gupta A K, Gupta M 2005 *Biomaterials* **26** 3995
[3] Moghimi S M, Hunter A C H 2001 *Pharm. Rev.* **53** 283
[4] Curtis A S G, Wilkinson C 2001 *Trends Biotech.* **19** 97
[5] Wilkinson J M 2003 *Med. Device Technol.* **14** 29
[6] Panyam J, Labhasetwar V 2003 *Adv. Drug Del. Rev.* **55** 329
[7] Thach C V, Hai N H, Chau N 2008 *J. Korean Phys. Soc.* **52** 1332
[8] Ramirez L P, Landfester K 2003 *Macromol. Chem. Phys.* **204** 22
[9] The N D, Hai N H, Anh H D, Hiep V V, Vu L V, Huu C X, Chau N 2008 *J. Korean Phys. Soc.* **52** 1415
[10] Lien V T K, Ha C V, Thuy T T T, Hanh V T H, Cuong P T, Ha L T, Ca V X, Nga P T 2007 *Proc. Natl. Conf. Solid Stat. Phys.* 264
[11] Johansen P, Merkle H P, Gander B 2000 *Eur. J. Pharm. Biopharm.* **50** 413

[12] Thomasin C, Johansen P, Alder R, Bemsel R, Hottinger G, Altorfer H, Wright A D, Wehrli E, Merkle H P, Gander B 1998 *Eur. J. Pharm. Biopharm.* **42** 1233
[13] Freitas S, Merkle H P, Gander B 2005 *J. Control. Release* **102** 313

Acknowledgments
This work is supported by the Vietnamese Fundamental Research Program for Natural Science (Project 406506) and the Key Project of the Vietnam National University, Hanoi (QGTD.08.05). The authors would like to thank Mr. Nguyen Dang Phu and Mr. Nguyen Minh Hieu for the experimental help.

Quantum dynamics of a two-atom-qubit system

Nguyen Van Hieu[1,2], Nguyen Bich Ha[1,2] and Le Thi Ha Linh[2]

[1]Max-Planck Institute for the Physics of Complex Systems,
Noethnitzer Str. 38, D-01187 Dresden, Germany
[2]Institute of Materials Science, Vietnam Academy of Science and Technology,
18 Hoang Quoc Viet, Cau Giay, Hanoi, Vietnam

E-mail: nvhieu@iop.vast.ac.vn

Abstract. A physical model of the quantum information exchange between two qubits is studied theoretically. The qubits are two identical two-level atoms, the physical mechanism of the quantum information exchange is the mutual dependence of the reduced density matrices of two qubits generated by their couplings with a multimode radiation field. The Lehmberg–Agarwal master equation is exactly solved. The explicit form of the mutual dependence of two reduced density matrices is established. The application to study the entanglement of two qubits is discussed.

Keywords: Qubit, quantum information, transmission, density matrix.

1. Introduction

Quantum information processing devices and systems usually contain two-level quantum systems called qubits (quantum bits). The quantum information (QI) encoded into a qubit is its wave function (for a pure state) or its density matrix (for a mixed state), and the QI transfer or transmission between two qubits is the state transfer or transmission between them [1–6]. The physical origin of the state transfer between two qubits may be their direct coupling or their interactions with some intermediate physical fields while the QI transmission between two distant qubits is performed through the intermediary of some physical fields.

The present work is devoted to the theoretical study of the QI exchange in a two-qubit system consisting of two identical two-level atoms coupled to a multimode radiation field whose modes are initially in the vacuum state [7–11]. The exact solution of rate equations is established in section 2. The interpretation of this solution as the explicit form of quantum information transmission between two qubits is presented in section 3. The application to study the entanglement sudden death and revival is discussed in section 4. Section 5 is the conclusion.

2. Rate equations and their solution

Denote ρ the density matrix of this two-qubit system, E – the difference of two energy levels in each atom, $\sigma_+(\sigma_-)$ and $\tau_+(\tau_-)$ – the dipole rising (lowering) operators in two-level atoms in the form of the Pauli matrices. Together with $\sigma_\pm$ and $\tau_\pm$ introduce also corresponding Pauli matrices σ_3 and τ_3 as well as the

© 2009 IOP Publishing Ltd

unit matrices $\sigma_0 = \tau_0 = 1$. As a 4×4 matrix ρ is expressed in terms of 15 generators of the SU(4) group. There are different choices of these SU(4) generators [12,13]. We use following operators

$$\Gamma_{(\alpha\beta)} = \left(\sigma_\alpha \otimes \tau_\beta\right), \qquad \alpha,\beta = 0,+,-,3, \quad (\alpha\beta) \neq (00), \tag{1}$$

and express density matrix ρ in terms of them

$$\rho = \frac{1}{4} + \sum_{(\alpha\beta)\neq(00)} \Gamma_{(\alpha\beta)} \rho_{(\alpha\beta)} \,. \tag{2}$$

The time evolution of density matrix ρ is determined by the Lehmberg–Agarwal [7, 8, 11] master equation

$$\frac{d\rho}{dt} = -i\left[H_S, \rho\right] + L\rho \ , \tag{3}$$

with

$$H_S = \frac{E}{2}\left[(\sigma_3 \otimes 1) + (1 \otimes \tau_3)\right] + J\left[(\sigma_+ \otimes \tau_-) + (\sigma_- \otimes \tau_+)\right], \tag{4}$$

where J is the dipole–dipole coupling constant of two atoms, and

$$\begin{aligned} L\rho = \gamma\Bigg\{&\left[(\sigma_- \otimes 1)\rho(\sigma_+ \otimes 1) - \frac{1}{2}(\sigma_+\sigma_- \otimes 1)\rho - \frac{1}{2}\rho(\sigma_+\sigma_- \otimes 1)\right] + \\ &+\left[(1 \otimes \tau_-)\rho(1 \otimes \tau_+) - \frac{1}{2}(1 \otimes \tau_+\tau_-)\rho - \frac{1}{2}\rho(1 \otimes \tau_+\tau_-)\right]\Bigg\} + \\ +\chi\Bigg\{&\left[(\sigma_- \otimes 1)\rho(1 \otimes \tau_+) - \frac{1}{2}(\sigma_- \otimes \tau_+)\rho - \frac{1}{2}\rho(\sigma_- \otimes \tau_+)\right] + \\ &+\left[(1 \otimes \tau_-)\rho(\sigma_+ \otimes 1) - \frac{1}{2}(\sigma_+ \otimes \tau_-)\rho - \frac{1}{2}\rho(\sigma_+ \otimes \tau_-)\right]\Bigg\}, \end{aligned} \tag{5}$$

where γ is spontaneous decay rate of each atom caused by its coupling with the radiation field, and χ is the collective damping constant of two atoms. The expression of J and χ in terms of γ was given in reference [11].

The master equation (3) has following exact solution:

$$\begin{aligned} \rho_{(+0)}(t) = \frac{1}{2}e^{-iEt}\Bigg\{&\frac{e^{-(\gamma+\chi)t/2}}{2J+i\gamma}\left[\left(J + i\frac{\chi}{2} + i\gamma\right)e^{-iJt} + \left(J - i\frac{\chi}{2}\right)e^{iJt}e^{-\gamma t}\right] + \\ &+\frac{e^{-(\gamma-\chi)t/2}}{2J-i\gamma}\left[\left(J + i\frac{\chi}{2} - i\gamma\right)e^{iJt} + \left(J - i\frac{\chi}{2}\right)e^{-iJt}e^{-\gamma t}\right]\Bigg\}\rho_{(+0)}(0) + \\ +\frac{1}{2}e^{-iEt}\Bigg\{&\frac{e^{-(\gamma+\chi)t/2}}{2J+i\gamma}\left[\left(J + i\frac{\chi}{2} + i\gamma\right)e^{-iJt} + \left(J - i\frac{\chi}{2}\right)e^{iJt}e^{-\gamma t}\right] - \\ &-\frac{e^{-(\gamma-\chi)t/2}}{2J-i\gamma}\left[\left(J + i\frac{\chi}{2} - i\gamma\right)e^{iJt} + \left(J - i\frac{\chi}{2}\right)e^{-iJt}e^{-\gamma t}\right]\Bigg\}\rho_{(0+)}(0) - \\ -\frac{1}{2}e^{-iEt}\left(J - i\frac{\chi}{2}\right)\Bigg\{&\frac{e^{-(\gamma+\chi)t/2}}{2J+i\gamma}\left[e^{-iJt} - e^{iJt}e^{-\gamma t}\right] + \frac{e^{-\frac{1}{2}(\gamma-\chi)t}}{2J-i\gamma}\left[e^{iJt} - e^{-iJt}e^{-\gamma t}\right]\Bigg\}\rho_{(+3)}(0) - \end{aligned} \tag{6}$$

$$-\frac{1}{2}e^{-iEt}\left(J-i\frac{\chi}{2}\right)\left\{\frac{e^{-(\gamma+\chi)t/2}}{2J+i\gamma}\left[e^{-iJt}-e^{iJt}e^{-\gamma t}\right]-\frac{e^{-\frac{1}{2}(\gamma-\chi)t}}{2J-i\gamma}\left[e^{iJt}-e^{-iJt}e^{-\gamma t}\right]\right\}\rho_{(3+)}(0),$$

$$\begin{aligned}\rho_{(+3)}(t)=&\frac{1}{2}e^{-iEt}\left\{\frac{e^{-(\gamma+\chi)t/2}}{2J+i\gamma}\left[\left(J-i\frac{\chi}{2}\right)e^{-iJt}+\left(J+i\frac{\chi}{2}+i\gamma\right)e^{iJt}e^{-\gamma t}\right]+\right.\\&\left.+\frac{e^{-(\gamma-\chi)t/2}}{2J-i\gamma}\left[\left(J-i\frac{\chi}{2}\right)e^{iJt}+\left(J+i\frac{\chi}{2}-i\gamma\right)e^{-iJt}e^{-\gamma t}\right]\right\}\rho_{(+3)}(0)+\\&+\frac{1}{2}e^{-iEt}\left\{\frac{e^{-(\gamma+\chi)t/2}}{2J+i\gamma}\left[\left(J-i\frac{\chi}{2}\right)e^{-iJt}+\left(J+i\frac{\chi}{2}+i\gamma\right)e^{iJt}e^{-\gamma t}\right]-\right.\\&\left.-\frac{e^{-(\gamma-\chi)t/2}}{2J-i\gamma}\left[\left(J-i\frac{\chi}{2}\right)e^{iJt}+\left(J+i\frac{\chi}{2}-i\gamma\right)e^{-iJt}e^{-\gamma t}\right]\right\}\rho_{(3+)}(0)-\\&-\frac{1}{2}e^{-iEt}\left\{\frac{e^{-(\gamma+\chi)t/2}}{2J+i\gamma}\left(J+i\frac{\chi}{2}+i\gamma\right)\left[e^{-iJt}-e^{iJt}e^{-\gamma t}\right]+\right.\\&\left.+\frac{e^{-(\gamma-\chi)t/2}}{2J-i\gamma}\left(J+i\frac{\chi}{2}-i\gamma\right)\left[e^{iJt}-e^{-iJt}e^{-\gamma t}\right]\right\}\rho_{(+0)}(0)-\\&-\frac{1}{2}e^{-iEt}\left\{\frac{e^{-(\gamma+\chi)t/2}}{2J+i\gamma}\left(J+i\frac{\chi}{2}+i\gamma\right)\left[e^{-iJt}-e^{iJt}e^{-\gamma t}\right]-\right.\\&\left.-\frac{e^{-(\gamma-\chi)t/2}}{2J-i\gamma}\left(J+i\frac{\chi}{2}-i\gamma\right)\left[e^{iJt}-e^{-iJt}e^{-\gamma t}\right]\right\}\rho_{(0+)}(0),\end{aligned}\tag{7}$$

$\rho_{(-0)}(t)$ and $\rho_{(-3)}(t)$ have similar expressions with the change of the signs of E and J, $E\to -E$, $J\to -J$, and the substitution $\rho_{(+0)}(0)\to\rho_{(-0)}(0)$, $\rho_{(+3)}(0)\to\rho_{(-3)}(0)$,

$$\begin{aligned}\rho_{(30)}(t)=&\frac{1}{4}\left[\frac{\gamma-\chi}{\gamma+\chi}e^{-(\gamma-\chi)t}+\frac{\gamma+\chi}{\gamma-\chi}e^{-(\gamma+\chi)t}-4\frac{\chi^2}{\gamma^2-\chi^2}e^{-2\gamma t}+2e^{-\gamma t}\cos 2Jt\right]\rho_{(30)}(0)\\&+\frac{1}{4}\left[\frac{\gamma-\chi}{\gamma+\chi}e^{-(\gamma-\chi)t}+\frac{\gamma+\chi}{\gamma-\chi}e^{-(\gamma+\chi)t}-4\frac{\chi^2}{\gamma^2-\chi^2}e^{-2\gamma t}-2e^{-\gamma t}\cos 2Jt\right]\rho_{(03)}(0)\\&-\frac{1}{8}\left[e^{-(\gamma-\chi)t}-e^{-(\gamma+\chi)t}-2ie^{-\gamma t}\sin 2Jt\right]\rho_{(+-)}(0)\\&-\frac{1}{8}\left[e^{-(\gamma-\chi)t}-e^{-(\gamma+\chi)t}+2ie^{-\gamma t}\sin 2Jt\right]\rho_{(-+)}(0)\\&-\frac{1}{2}\left[\frac{\chi}{\gamma+\chi}e^{-(\gamma-\chi)t}-\frac{\chi}{\gamma-\chi}e^{-(\gamma+\chi)t}+2\frac{\chi^2}{\gamma^2-\chi^2}e^{-2\gamma t}\right]\rho_{(33)}(0)\\&+\frac{1}{8}\left[\frac{\gamma}{\gamma+\chi}e^{-(\gamma-\chi)t}+\frac{\gamma}{\gamma-\chi}e^{-(\gamma+\chi)t}-2\frac{\chi^2}{\gamma^2-\chi^2}e^{-2\gamma t}-2\right],\end{aligned}\tag{8}$$

$$
\begin{aligned}
\rho_{(+-)}(t) = & \frac{1}{4}\left[e^{-(\gamma-\chi)t} + e^{-(\gamma+\chi)t} + 2e^{-\gamma t}\cos 2Jt\right]\rho_{(+-)}(0) + \frac{1}{4}\left[e^{-(\gamma-\chi)t} + e^{-(\gamma+\chi)t} - 2e^{-\gamma t}\cos 2Jt\right]\rho_{(-+)}(0) \\
& -\frac{1}{2}\left[\frac{\gamma-\chi}{\gamma+\chi}e^{-(\gamma-\chi)t} - \frac{\gamma+\chi}{\gamma-\chi}e^{-(\gamma+\chi)t} + 4\frac{\chi\gamma}{\gamma^2-\chi^2}e^{-2\gamma t} - 2ie^{-\gamma t}\sin 2Jt\right]\rho_{(30)}(0) \\
& -\frac{1}{2}\left[\frac{\gamma-\chi}{\gamma+\chi}e^{-(\gamma-\chi)t} - \frac{\gamma+\chi}{\gamma-\chi}e^{-(\gamma+\chi)t} + 4\frac{\chi\gamma}{\gamma^2-\chi^2}e^{-2\gamma t} + 2ie^{-\gamma t}\sin 2Jt\right]\rho_{(03)}(0) \\
& +\left[\frac{\chi}{\gamma+\chi}e^{-(\gamma-\chi)t} + \frac{\chi}{\gamma-\chi}e^{-(\gamma+\chi)t} - 2\frac{\chi\gamma}{\gamma^2-\chi^2}e^{-2\gamma t}\right]\rho_{33}(0) - \\
& -\frac{1}{4}\left[\frac{\gamma}{\gamma+\chi}e^{-(\gamma-\chi)t} - \frac{\gamma}{\gamma-\chi}e^{-(\gamma+\chi)t} + 2\frac{\chi\gamma}{\gamma^2-\chi^2}e^{-2\gamma t}\right],
\end{aligned} \tag{9}
$$

$\rho_{(0+)}(t)$, $\rho_{(3+)}(t)$, $\rho_{(0-)}(t)$, $\rho_{(3-)}(t)$, $\rho_{(03)}(t)$ and $\rho_{(-+)}(t)$ are obtained from the expression of $\rho_{(+0)}(t)$, $\rho_{(+3)}(t)$, $\rho_{(-0)}(t)$, $\rho_{(-3)}(t)$, $\rho_{(30)}(t)$ and $\rho_{(+-)}(t)$, respectively, by the interchange $\rho_{(\pm 0)}(0) \leftrightarrow \rho_{(0\pm)}(0)$, $\rho_{(\pm 3)}(0) \leftrightarrow \rho_{(3\pm)}(0)$, $\rho_{(30)}(0) \leftrightarrow \rho_{(03)}(0)$ and $\rho_{(-+)}(0) \leftrightarrow \rho_{(+-)}(0)$,

$$
\rho_{(++)}(t) = e^{-\gamma t}e^{-2iEt}\rho_{(++)}(0), \tag{10}
$$

$\rho_{(--)}(t)$ is obtained from $\rho_{(++)}(t)$ by the change of the sign of E, $E \to -E$, and the substitution $\rho_{(++)}(0) \to \rho_{(--)}(0)$, and

$$
\begin{aligned}
\rho_{(33)}(t) = & \left[-\frac{1}{2}\frac{\gamma-\chi}{\gamma+\chi}e^{-(\gamma-\chi)t} - \frac{1}{2}\frac{\gamma+\chi}{\gamma-\chi}e^{-(\gamma+\chi)t} + \frac{\gamma^2+\chi^2}{\gamma^2-\chi^2}e^{-2\gamma t}\right]\left[\rho_{(03)}(0) + \rho_{(30)}(0)\right] \\
& +\frac{1}{4}\left[e^{-(\gamma-\chi)t} - e^{-(\gamma+\chi)t}\right]\left[\rho_{(+-)}(0) + \rho_{(-+)}(0)\right] + \\
& +\left[\frac{\chi}{\gamma+\chi}e^{-(\gamma-\chi)t} - \frac{\chi}{\gamma-\chi}e^{-(\gamma+\chi)t} + \frac{\gamma^2+\chi^2}{\gamma^2-\chi^2}e^{-2\gamma t}\right]\rho_{(33)}(0) - \\
& -\frac{1}{4}\left[\frac{\gamma}{\gamma+\chi}e^{-(\gamma-\chi)t} + \frac{\gamma}{\gamma-\chi}e^{-(\gamma+\chi)t} - \frac{\gamma^2+\chi^2}{\gamma^2-\chi^2}e^{-2\gamma t}\right] + \frac{1}{4}.
\end{aligned} \tag{11}
$$

3. Quantum Information Exchange

Quantum information encoded into each qubit, called σ-qubit or τ-qubit, is its reduced density matrix $\rho^{\sigma} = Tr_{\tau}\rho$ or $\rho^{\tau} = Tr_{\sigma}\rho$, where Tr_{τ}or Tr_{σ} denotes the trace with respect to the spin indices of spin-qubit τ or σ. We have

$$
\rho^{\sigma} = \frac{1}{2} + \sum_{\alpha}\sigma_{\alpha}\rho_{\alpha}^{\sigma}, \quad \rho_{\alpha}^{\sigma}(t) = 2\rho_{(\alpha 0)}(t), \quad \rho^{\tau} = \frac{1}{2} + \sum_{\alpha}\tau_{\alpha}\rho_{\alpha}^{\tau}, \rho_{\alpha}^{\tau}(t) = 2\rho_{(0\alpha)}(t). \tag{12}
$$

Suppose that at the initial time t = 0 the quantum states of two spin-qubits are independent, and therefore

$$\rho_{(\alpha 0)}(0) = \frac{1}{2}\rho_\alpha^\sigma(0), \quad \rho_{(0\alpha)}(0) = \frac{1}{2}\rho_\alpha^\tau(0),$$
$$\rho_{(\alpha\beta)}(0) = \rho_\alpha^\sigma(0)\rho_\beta^\tau(0), \quad \alpha \neq 0, \beta \neq 0. \tag{13}$$

Then, from the expressions of $\rho_{(\pm 0)}(t)$ and $\rho_{(30)}(t)$ presented in preceding Section we derive following formulae

$$\begin{aligned}
\rho_\pm^\sigma(t) = {} & \frac{1}{2}e^{\mp iEt}\left\{\frac{e^{-(\gamma+\chi)t/2}}{2J \pm i\gamma}\left[\left(J \pm i\frac{\chi}{2} \pm i\gamma\right)e^{\mp iJt} + \left(J \mp i\frac{\chi}{2}\right)e^{\pm iJt}e^{-\gamma t}\right] + \right. \\
& \left. + \frac{e^{-(\gamma-\chi)t/2}}{2J \mp i\gamma}\left[\left(J \pm i\frac{\chi}{2} \mp i\gamma\right)e^{\pm iJt} + \left(J \mp i\frac{\chi}{2}\right)e^{\mp iJt}e^{-\gamma t}\right]\right\}\rho_\pm^\sigma(0) + \\
& + \frac{1}{2}e^{\mp iEt}\left\{\frac{e^{-(\gamma+\chi)t/2}}{2J \pm i\gamma}\left[\left(J \pm i\frac{\chi}{2} \pm i\gamma\right)e^{\mp iJt} + \left(J \mp i\frac{\chi}{2}\right)e^{\pm iJt}e^{-\gamma t}\right] - \right. \\
& \left. - \frac{e^{-(\gamma-\chi)t/2}}{2J \mp i\gamma}\left[\left(J \pm i\frac{\chi}{2} \mp i\gamma\right)e^{\pm iJt} + \left(J \mp i\frac{\chi}{2}\right)e^{\mp iJt}e^{-\gamma t}\right]\right\}\rho_\pm^\tau(0) \\
& - \frac{1}{2}e^{\mp iEt}\left(J \mp i\frac{\chi}{2}\right)\left\{\frac{e^{-(\gamma+\chi)t/2}}{2J \pm i\gamma}\left[e^{\mp iJt} - e^{\pm iJt}e^{-\gamma t}\right] + \frac{e^{-(\gamma-\chi)t/2}}{2J \mp i\gamma}\left[e^{\pm iJt} - e^{\mp iJt}e^{-\gamma t}\right]\right\}\rho_\pm^\sigma(0)\rho_3^\tau(0) - \\
& - \frac{1}{2}e^{\mp iEt}\left(J \mp i\frac{\chi}{2}\right)\left\{\frac{e^{-(\gamma+\chi)t/2}}{2J \pm i\gamma}\left[e^{\mp iJt} - e^{\pm iJt}e^{-\gamma t}\right] - \frac{e^{-(\gamma-\chi)t/2}}{2J \mp i\gamma}\left[e^{\pm iJt} - e^{\mp iJt}e^{-\gamma t}\right]\right\}\rho_3^\sigma(0)\rho_\pm^\tau(0),
\end{aligned} \tag{14}$$

$$\begin{aligned}
\rho_3^\sigma(t) = {} & \frac{1}{4}\left[\frac{\gamma-\chi}{\gamma+\chi}e^{-(\gamma-\chi)t} + \frac{\gamma+\chi}{\gamma-\chi}e^{-(\gamma+\chi)t} - 4\frac{\chi^2}{\gamma^2-\chi^2}e^{-2\gamma t} + 2e^{-\gamma t}\cos 2Jt\right]\rho_3^\sigma(0) + \\
& + \frac{1}{4}\left[\frac{\gamma-\chi}{\gamma+\chi}e^{-(\gamma-\chi)t} + \frac{\gamma+\chi}{\gamma-\chi}e^{-(\gamma+\chi)t} - 4\frac{\chi^2}{\gamma^2-\chi^2}e^{-2\gamma t} - 2e^{-\gamma t}\cos 2Jt\right]\rho_3^\tau(0) - \\
& - \frac{1}{4}\left[e^{-(\gamma-\chi)t} - e^{-(\gamma+\chi)t} - 2ie^{-\gamma t}\sin 2Jt\right]\rho_+^\sigma(0)\rho_-^\tau(0) - \\
& - \frac{1}{4}\left[e^{-(\gamma-\chi)t} - e^{-(\gamma+\chi)t} + 2ie^{-\gamma t}\sin 2Jt\right]\rho_-^\sigma(0)\rho_+^\tau(0) \\
& - \left[\frac{\chi}{\gamma+\chi}e^{-(\gamma-\chi)t} - \frac{\chi}{\gamma-\chi}e^{-(\gamma+\chi)t} + 2\frac{\chi^2}{\gamma^2-\chi^2}e^{-2\gamma t}\right]\rho_3^\sigma(0)\rho_3^\tau(0) \\
& + \frac{1}{4}\left[\frac{\gamma}{\gamma+\chi}e^{-(\gamma-\chi)t} + \frac{\gamma}{\gamma-\chi}e^{-(\gamma+\chi)t} - 2\frac{\chi^2}{\gamma^2-\chi^2}e^{-2\gamma t} - 2\right],
\end{aligned} \tag{15}$$

and similar ones with the interchange $\sigma \leftrightarrow \tau$. They show that reduced density matrix $\rho^\sigma(t)$ of σ–qubit at $t > 0$ depends not only on its initial state $\rho^\sigma(0)$, but also on the initial state $\rho^\tau(0)$ of τ–qubit, and vice-versa. This mutual dependence of reduced density matrices of two qubits exhibits a physical mechanism of the QI exchange between them: read-out QI of σ–qubit at $t > 0$ reflects in-put QI encoded onto τ–qubit at $t = 0$, and vice-versa. Note that even in the special case with $\rho_\alpha^\tau(0) = 0,\ \alpha = \pm 3$, the time evolution of reduced density matrix $\rho^\sigma(t)$,

$$\rho_\pm^\sigma(t) = \frac{1}{2} e^{\mp iEt} \left\{ \frac{e^{-(\gamma+\chi)t/2}}{2J \pm i\gamma} \left[\left(J \pm i\frac{\chi}{2} \pm i\gamma \right) e^{\mp iJt} + \left(J \mp i\frac{\chi}{2} \right) e^{\pm iJt} e^{-\gamma t} \right] + \right.$$
$$\left. + \frac{e^{-(\gamma-\chi)t/2}}{2J \mp i\gamma} \left[\left(J \pm i\frac{\chi}{2} \mp i\gamma \right) e^{\pm iJt} + \left(J \mp i\frac{\chi}{2} \right) e^{\mp iJt} e^{-\gamma t} \right] \right\} \rho_\pm^\sigma(0), \tag{16}$$

$$\rho_3^\sigma(t) = \frac{1}{4} \left[\frac{\gamma - \chi}{\gamma + \chi} e^{-(\gamma-\chi)t} + \frac{\gamma + \chi}{\gamma - \chi} e^{-(\gamma+\chi)t} - 4\frac{\chi^2}{\gamma^2 - \chi^2} e^{-2\gamma t} + 2e^{-\gamma t} \cos 2Jt \right] \rho_3^\sigma(0) +$$
$$+ \frac{1}{4} \left[\frac{\gamma}{\gamma + \chi} e^{-(\gamma-\chi)t} + \frac{\gamma}{\gamma - \chi} e^{-(\gamma+\chi)t} - 2\frac{\chi^2}{\gamma^2 - \chi^2} e^{-2\gamma t} - 2 \right], \tag{17}$$

is different from that of a single qubit

$$\rho_\pm^q(t) = e^{\mp iEt} e^{-\gamma t} \rho_\pm^q(0), \tag{18}$$

$$\rho_3^q(t) = e^{-\gamma t} \rho_3^q(0) + \frac{1}{2}\left(e^{-\gamma t} - 1\right). \tag{19}$$

In general, by comparing the time evolution of reduced density matrix $\rho^\sigma(t)$ with that of a single qubit $\rho^q(t)$, we can discover the presence of τ-qubit and also determine its initial state $\rho^\tau(0)$.

4. Entanglement

For studying the entanglement between two qubits we can apply reasonnings of reference [11] and use the collective Dick states [14]

$$|e\rangle = |11\rangle, \qquad |g\rangle = |22\rangle,$$
$$|s\rangle = \frac{1}{\sqrt{2}}[|21\rangle + |12\rangle], \qquad |a\rangle = \frac{1}{\sqrt{2}}[|21\rangle - |12\rangle] \tag{20}$$

as the basis vectors. In this basis ρ has following matrix elements

$$\rho_{ee}(t) = \rho_{ee}(0) e^{-2\gamma t}, \tag{21}$$

$$\rho_{eg}(t) = \rho_{eg}(0) e^{-\gamma t} e^{-2iEt}, \tag{22}$$

$$\rho_{ge}(t) = \rho_{ge}(0) e^{-\gamma t} e^{2iEt}, \tag{23}$$

$$\rho_{gg}(t) = 1 + \frac{1}{2}\left[e^{-(\gamma-\chi)t} + e^{-(\gamma+\chi)t}\right] \rho_{gg}(0) +$$
$$+ \frac{1}{2}\left[2\frac{\gamma^2 + 3\chi^2}{\gamma^2 - \chi^2} e^{-2\gamma t} + \frac{3\chi - \gamma}{\gamma + \chi} e^{-(\gamma-\chi)t} - \frac{3\chi + \gamma}{\gamma - \chi} e^{-(\gamma+\chi)t} \right] \rho_{ee}(0) + \tag{24}$$
$$+ \frac{1}{2}\left[e^{-(\gamma-\chi)t} - e^{-(\gamma+\chi)t}\right]\left[\rho_{ss}(0) - \rho_{aa}(0)\right] - \frac{1}{2}\left[e^{-(\gamma-\chi)t} + e^{-(\gamma+\chi)t}\right],$$

$$\rho_{ss}(t)=\frac{1}{2}\left[\frac{\gamma-3\chi}{\gamma-\chi}e^{-(\gamma+\chi)t}+\frac{\gamma+\chi}{\gamma-\chi}e^{-2\gamma t}\right]\rho_{ss}(0)+ +\frac{1}{2}\frac{\gamma+\chi}{\gamma-\chi}\left[e^{-2\gamma t}-e^{-(\gamma+\chi)t}\right]\left[\rho_{aa}(0)-\rho_{ee}(0)+\rho_{gg}(0)-1\right], \tag{25}$$

$$\rho_{sa}(t)=\rho_{sa}(0)e^{-\gamma t}e^{-2iJt}, \tag{26}$$

$$\rho_{as}(t)=\rho_{as}(0)e^{-\gamma t}e^{2iJt}, \tag{27}$$

$$\rho_{aa}(t)=\frac{1}{2}\left[\frac{\gamma+3\chi}{\gamma+\chi}e^{-(\gamma-\chi)t}+\frac{\gamma-\chi}{\gamma+\chi}e^{-2\gamma t}\right]\rho_{aa}(0)+ +\frac{1}{2}\frac{\gamma-\chi}{\gamma+\chi}\left[e^{-2\gamma t}-e^{-(\gamma-\chi)t}\right]\left[\rho_{ss}(0)-\rho_{ee}(0)+\rho_{gg}(0)-1\right], \tag{28}$$

$$\rho_{es}(t)=\rho_{es}(0)e^{-(3\gamma+\chi)t/2}e^{-i(E-J)t}, \tag{29}$$

$$\rho_{se}(t)=\rho_{se}(0)e^{-(3\gamma+\chi)t/2}e^{i(E-J)t}, \tag{30}$$

$$\rho_{ea}(t)=\rho_{ea}(0)e^{-(3\gamma-\chi)t/2}e^{-i(E+J)t}, \tag{31}$$

$$\rho_{ae}(t)=\rho_{ae}(0)e^{-(3\gamma-\chi)t/2}e^{i(E+J)t}, \tag{32}$$

$$\rho_{gs}(t)=e^{-(\gamma+\chi)t/2}e^{i(E+J)t}\rho_{gs}(0)-i\frac{\gamma+\chi}{2J-i\gamma}e^{-(\gamma+\chi)t/2}e^{iEt}\left(e^{iJt}-e^{-\gamma t}e^{-iJt}\right)\rho_{se}(0), \tag{33}$$

$$\rho_{sg}(t)=e^{-(\gamma+\chi)t/2}e^{-i(E+J)t}\rho_{sg}(0)+i\frac{\gamma+\chi}{2J+i\gamma}e^{-(\gamma+\chi)t/2}e^{-iEt}\left(e^{-iJt}-e^{-\gamma t}e^{iJt}\right)\rho_{es}(0), \tag{34}$$

$$\rho_{ga}(t)=e^{-(\gamma-\chi)t/2}e^{i(E-J)t}\rho_{ga}(0)-i\frac{\gamma-\chi}{2J+i\gamma}e^{-(\gamma-\chi)t/2}e^{iEt}\left(e^{-iJt}-e^{-\gamma t}e^{iJt}\right)\rho_{ae}(0), \tag{35}$$

$$\rho_{ag}(t)=e^{-(\gamma-\chi)t/2}e^{-i(E-J)t}\rho_{ag}(0)+i\frac{\gamma-\chi}{2J-i\gamma}e^{-(\gamma-\chi)t/2}e^{-iEt}\left(e^{iJt}-e^{-\gamma t}e^{-iJt}\right)\rho_{ea}(0). \tag{36}$$

Consider a special class of quantum states satisfying conditions

$$\rho_{es}=\rho_{se}=\rho_{ea}=\rho_{ae}=\rho_{gs}=\rho_{sg}=\rho_{ga}=\rho_{ag}=\rho_{sa}=\rho_{as}=0 \tag{37}$$

and

$$\rho_{ge}=\rho_{eg}^{*} \tag{38}$$

at $t=0$. From the explicit expressions of the elements of density matrix ρ, equations (21)–(36), it follows that the same conditions still hold at any $t>0$. For this class the formulae of the concurrence [15, 16] were derived by Ficek and Tanas [11]. Similarly, if the elements of ρ satisfy conditions

$$\rho_{eg}=\rho_{ge}=\rho_{es}=\rho_{se}=\rho_{ea}=\rho_{ae}=\rho_{ga}=\rho_{ag}=\rho_{sa}=\rho_{as}=0 \tag{39}$$

and

$$\rho_{sg}=\rho_{gs}^{*} \tag{40}$$

at $t=0$, then these conditions still hold at any $t>0$. For this class, the formulae of the concurrence were derived by Yu and Eberly [17]. Using formulae given in references [11, 17] it can be shown that for suitable initial values of elements of ρ there may take place the entanglement sudden death and revival, as this phenomenon was discovered in references [11, 17] in the cases of some specific states.

5. Conclusion

A model two-atom-qubit system was investigated. It consists of two identical two-level atoms coupled to a multimode radiation field whose modes are initially in the vacuum state. Its quantum dynamics is described by a density matrix satisfying the Lehmberg-Agarwal equation, which was exactly solved. Its solution demonstrates a physical mechanism of the QI exchange between two qubits. It would be also useful for the study of the entanglement sudden death and revival in two wide classes of two-qubit systems satisfying definite initial conditions.

References

[1] Cirac J I, Zoller P P, Kimble H J and Mabuchi H 1997 *Phys. Rev. Lett.* **78** 3221
[2] Duan L-M, Cirac J I, Zoller P and Polzik E S 1999 *Phys. Rev. Lett.* **85** 5158
[3] Browne D E, Plenio M B and Huelga S F 2004 *Phys. Rev. Lett.* **92** 076401
[4] Subrahmanyam V 2004 *Phys. Rev.* A **69** 034304
[5] Li Y, Shi T, Chen B, Song Z and Sun C-P 2005 *Phys. Rev.* A **71** 022301
[6] Zhang J, Rajendran N, Peng X and Suter D 2007 *Phys. Rev.* A **76** 012317
[7] Lehmberg R H 1970 *Phys. Rev.* A **2** 883
[8] Agarwal G S 1974 *in Quantum Statistical Theories of Spontaneous Emission and their Relation other Approaches* ed. Höhler G (Berlin) Springer Tracts in Modern Physics (Springer–Verlag) Vol. 70
[9] Fizek Z and Swain S 2005 *Quantum Interference and Coherence: Theory and Experiments* (New York:Springer)
[10] Fizek Z and Tanas R 2003 *J. Mod. Opt.* **50** 2765
[11] Fizek Z and Tanas R 2006 *Phys. Rev.* A **74** 024304
[12] Rau A R P, Selvaraj G and Uskov D 2005 *Phys. Rev.* A **71** 062316
[13] Zhang J, Vala J, Sastry S and Whaley K B 2003 *Phys. Rev.* A **67** 042313
[14] Dicke R H 1954 *Phys. Rev.* **93** 99
[15] Hill S and Wootters W K 1997 *Phys. Rev. Lett.* **78** 5022
[16] Wootters W K 1998 *Phys. Rev. Lett.* **80** 2245
[17] Yu T and Eberly J H 2004 *Phys. Rev. Lett.* **93** 140404; 2006 *Phys. Rev. Lett.* **97** 140403
[18] Yu T and Eberly J H 2007 *Quantum Inf. Comput.* **7** 459

Acknowledgments

This work was initiated in Max-Planck Institute for the Physics of Complex Systems MPI-PKS, Dresden, Germany. Two of us (N. V. H. and N. B. H.) would like to express their gratitude to MPI-PKS and to Prof. Peter Fulde for the hospitality. The support of Vietnam Academy of Science and Technology is appreciated.

Journal of Physics: Conference Series **187** (2009) 012011 doi:10.1088/1742-6596/187/1/012011

Raman scattering study of perovskite manganites

Nguyen Van Minh

Center for Nano Science and Technology, Hanoi National University of Education, 136 Xuan-Thuy Road, Hanoi, Vietnam

E-mail: minhsp@yahoo.com

Abstract. Raman spectroscopy is shown to be a useful tool for the study of phase transitions, oxygen disorder and local vibrational modes in lanthanum manganese compounds of the perovskite type. In this work, we present results of the Raman scattering and x-ray diffraction studies of $La_{0.7}Sr_{0.3}Mn_{1-x}M_xO_3$ manganites (M=Fe, Co, Ni, Cr). A part of Mn atoms was substituted by M with x in the range of 0.0-0.60. Raman spectra of the $La_{0.7}Sr_{0.3}Mn_{1-x}M_xO_3$ exhibit a significant dependence on the amount and kind of the dopant M, and they are sensitive to the structure and transitions driven by both doping and temperature. X-ray diffraction (XRD) analysis of these samples indicates a change in the cell parameters vs. doping content. Our main focus is the Raman peak at high frequency (>630 cm^{-1}), that is somewhat controversial. The Raman spectra exhibit an additional high frequency mode attributed to the local breathing vibration of oxygen surrounding the substituent ion. The frequency and intensity of Raman spectra in these manganites is very sensitive to the type and content of substituent. From above results, a critical content of substituent at which the structural phase changed was shown.

Keyword: Raman spectroscopy, perovskite, doping.

1. Introduction

The magnetoresistance effect has been known for over 100 years. The discovery of "Giant Magneto Resistance" (GMR) in 1986 [1] started a new area of research in magnetic materials in view of their potential incorporation in efficient technological devices, particularly those used in magnetic recording heads [2]. More recently it was found that another class of materials exhibits much greater magnetoresistance properties, and the effect was accordingly dubbed "Colossal Magneto Resistance" (CMR) [3]. Most representative of the CMR materials are the manganite perovskites $La_{1-x}R_xMnO_3$ where the lanthanide cation is trivalent (La^{3+}) and R is a divalent alkaline earth cation (R = Ca, Sr, Ba). The end members of this series, with x = 0 ($LaMnO_3$) and x = 1 (e.g., $CaMnO_3$) are antiferromagnetic insulators in their ground states. The CMR effect is accompanied by a ferromagnetic to paramagnetic phase transition largest and is near the transition temperature T_c. The various aspects of CMR are described in a variety of publications and reviews [4, 5]. While much of works is driven by potential applications, there is also considerable intrinsic interest in the study of the basic physics and chemistry of the doped perovskite manganites. Many different experimental techniques are used alongside theoretical models in this undertaking. Among these, the technique of Raman spectroscopy has proven itself to be of considerable importance in the study of solid state materials in general. It is the purpose of this presentation to report on a series of Raman spectroscopic experiments on doped in the B site.

© 2009 IOP Publishing Ltd

2. Experiment

The samples of $La_{0.7}Sr_{0.3}Mn_{1-x}M_xO_{3+\delta}$ (x = 0.1, 0.2, 0.3, 0.4, 0.5, 0.6) used in the present study were prepared by solid state reaction method [6]. The room temperature X-ray diffraction (XRD) patterns were collected using a Rigaku X-ray diffractometer. Raman measurements were performed in a back scattering geometry using Jobin Yvon T 64000 triple spectrometer equipped with a cryogenic charge-coupled device (CCD) array detector, using 514.5 nm and 647.1 nm line of Ar and Kr ion lasers, respectively.

3. Result and discussion

3.1. Manganite structure and selection rules for optical vibrational modes

The $La_{1-x}R_xMnO_3$ system may exist in one of the several phases: paramagnetic and ferromagnetic insulators and metals, canted antiferromagnetic insulator and antiferromagnetic insulator phases, depending on the doping value *x*, and the temperature. Moreover, early structural studies of manganite and related systems have shown that this material may be have cubic [7], rhombohedral [8] or orthorhombic [9] symmetries depending on the temperature and value of doping *x*. The activities of vibrational modes in optical spectra are obtained by the methods of group theory. For the three commonly reported symmetries of $La_{1-x}R_xMnO_3$ materials these are summarized below. The perfect perovskite structure belongs to the cubic space group O_h^1 (m3m), and a unit cell contains one formula unit. The optical vibrations are classified as:

$$\Gamma = 3F_{1u}(IR) + F_{2u} ,$$

where the three F_{1u} species are allowed (active) in infrared (IR) absorption while the F_{2u} species are inactive ("forbidden") in both IR and Raman scattering (RS). The IR active modes of $La_{1-x}R_xMnO_3$ materials were analyzed within this assumed symmetry [7]. No Raman modes are allowed in the first order spectra. However, the crystal lattice of most real $La_{1-x}R_xMnO_3$ materials is distorted, for example due to the Jahn-Teller (JT) and other effects. The existing lattice distortions therefore permit the appearance of a first order Raman spectrum. At moderate values of doping, e.g., $x \approx 0.3$, the rhombohedral symmetry $D_{3d}^6(R\overline{3}c)$ is expected. For this structure the unit cell contains two formula units. Therefore the following activity of optical vibrational modes is expected:

$$\Gamma = A_{1g}(RS) + 4E_g(RS) + 3A_{2g} + 2A_{1u} + 3A_{2u}(IR) + 5E_u(IR) .$$

So far Raman spectra for the rhombohedral structure of $La_{1-x}R_xMnO_3$ single crystals were considered. However, no significant difference with the spectra from the orthorhombic phase was detected. The symmetry of the parent $LaMnO_3$ material, as well as those lightly doped (x < 0.125) is usually described by the space group $D_{2h}^{16}(mmm)$. In this case the unit cell has four formula units and the optical vibrations are classified as:

$$\Gamma = 7A_g(RS) + 5B_{1g}(RS) + 7B_{2g}(RS) + 5B_{3g}(RS)$$
$$+ 8A_u + 9B_{1u}(IR) + 7B_{2u}(IR) + 9B_{3u}(IR)$$

Accordingly, the Raman spectrum is expected to be composed of seven A_g, five B_{1g}, seven B_{2g}, and five B_{3g} modes. Since the early study by Geller [10], $LaMnO_3$ is known to possess a distorted perovskite orthorhombic structure with both $\sqrt{2} \times \sqrt{2}$ doubling of the unit cell in the *a-c* plan and additional doubling along the *b* axis. This structure was studied by Raman spectroscopy in [7, 11].

3.2. X-ray data

XRD patterns showed that all of the samples are single phase when x < 0.3, and multiphase in case x > 0.3. All of the reflection lines are successfully indexed with a rhombohedral ABO_3-type perovskite structure. The cell parameters decrease as increasing substituted content in all cases (see figure 1).

However, it monotonously decreases in the range of x < 0.3. At higher content, the tendency seems more complex. The reason might be related to the existence of some secondary phases in the samples. An insight into the differences among the substituents can be obtained by analyzing the X-ray data [4]. A particularly important information is provided by the B–O distance, i.e., the mean distance between the central ion in the perovskite ABO_3 structure and neighbouring oxygens. Since the average valence of the Mn ions is kept constant and their radius is thus fixed, the B–O distance should depend on the radii of the M ions.

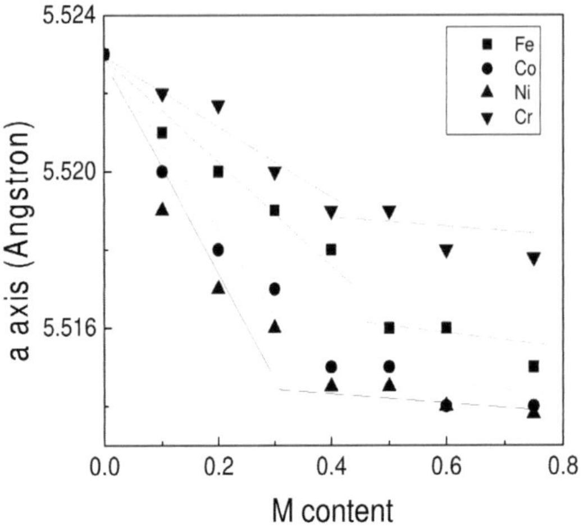

Figure 1. Cell parameters vs. M content.

3.3. Raman scattering

Our samples crystallize in rhombohedral structure, space group D^6_{3d}, Z = 2. This structure can be obtained from the simple cubic perovskite by the rotation of the adjacent MnO_6 octahedra in the opposite directions around the [111]c cubic direction. Of the total of 20 -point modes (A_{1g} + $3A_{2g}$ + $2A_{1u}$ + $4A_{2u}$ + $4E_g$ + $6E_u$), five (A_{1g} + $4E_g$) are Raman active, eight ($3A_{2u}$ + $5E_u$) are infrared active, and five ($2A_{1u}$ + $3A_{2g}$) are silent. We assign the 427 cm^{-1} line to the E_g mode with the predicted frequency of 468 cm^{-1}. The frequencies of the two modes are close to the results obtained on a single crystal (199 and 426 cm^{-1}, respectively). Podobedov *et al.* [11] did not assign the weak band at 663 cm^{-1} to the manganite because its intensity varies substantially from measurement to measurement. They also assumed that it was due to a small proportion of an manganese oxide; e.g. Mn_3O_4 has a strong Raman line at about this frequency.

Figure 2 shows the Raman spectra of the M (M=Fe, Co, Ni, Ti) substituted series. With increasing M concentration, the spectra exhibit the following main features: a fairly strong mode appears at about 670 cm^{-1}, the mode at 427 cm^{-1} gradually disappears while two broad bands appear at around 500 cm^{-1} and 620 cm^{-1}. For Fe and Co doped samples, when increasing doping content, the peak at higher than 600 cm^{-1} up-shifts, whereas the peak downshifts in case of Ti and Ni doped.

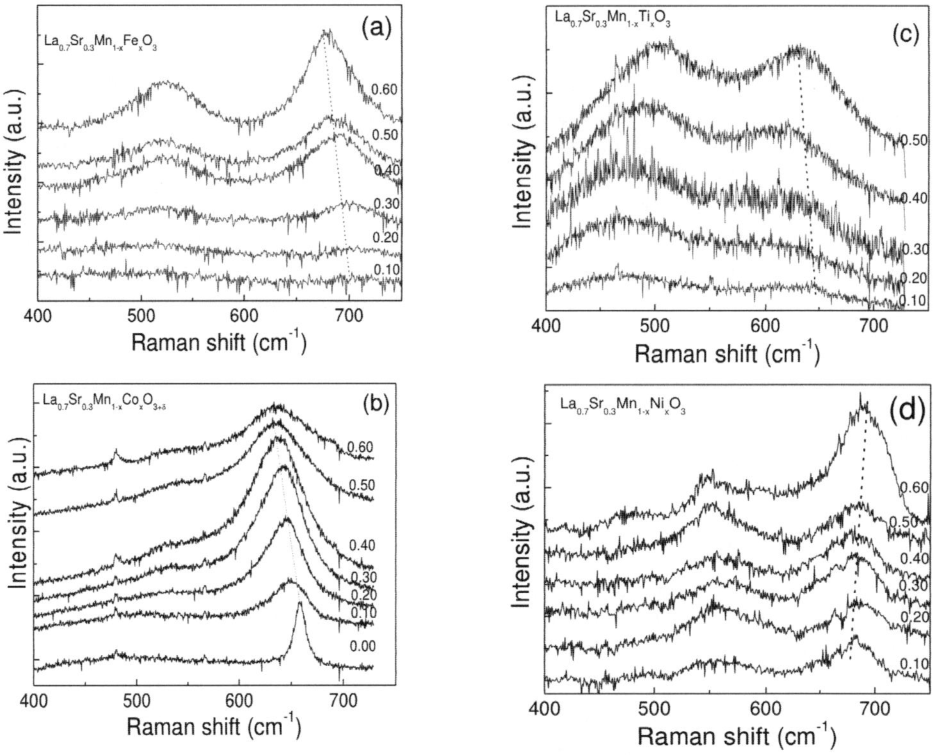

Figure 2. Raman spectra of $La_{0.7}Sr_{0.3}Mn_{1-x}M_xO_3$ (a): Fe, (b): Co; (c): Ti; (d): Ni.

3.4. The additional A_g-like high frequency mode

Let us discuss the peak observed at the highest frequencies of 630-760 cm^{-1}. Its common characteristics are, (a) its intensity increases with the concentration of substituents, (b) its frequency depends rather strongly on the type of the substituent. We assign the peak to a local breathing (in-phase stretching) mode of oxygens in a close vicinity of the substituent ion. This interpretation is further supported by the symmetry of the mode: the suggested eigenvector corresponds to a nondegenerate vibration, i.e., it has the A_g-like symmetry. Moreover, these modes are not infrared active which implies that the eigenvector should have a center of symmetry. Filho et al. [12] observed a similar peak at about 707 cm^{-1} in $La_{0.70}Sr_{0.30}Mn_{1-x}Fe_xO_3$ and interpreted it as the silent A_{2g} mode activated by a symmetry breaking or as the phonon density of state feature.

Evidently, the frequency does not exhibit a simple mass effect. This is not surprising, because the substitutional ions do not take part in the vibration. An important role could be again played by the radius of the substituent ions. The oxygens vibrate along the Mn–O–M bonds. If the ion M is large, it creates an additional pressure on the oxygen, increasing the frequency of the vibration. Some other factor must play an important role. It could be, e.g., the degree of covalency of the M–O bond, which would imply a dependence of the frequency on the position of M atom in the periodic table. The peak of Fe-substituted compound observed in the data of Filho et al. [12] has also a very large intensity as compared with the other phonons. Another factor affecting this frequency can be the valence of the substituent that influence electrostatically the stiffness of the $M–O^{2-}$ bond. This effect would lower the frequencies corresponding to the substituents with valencies 2+ with respect to their hypothetical 3+ state. However, a clear understanding of the frequency and intensity variations remains a challenge for future studies. The properties of the high-frequency mode are even more complex.

4. Summary and conclusion

We have used Raman spectroscopy measurements and X-ray analysis to study the influence of Mn-site substitutions in optimally doped manganites. Our results support the idea that the main factor influencing the oxygen disorder and electric and magnetic properties of these compounds is the radius of the substituents and that their electronic and magnetic structure plays only a minor role. The substitution made local stress and produced by the substitutional ion modifies the Mn–O distances around the substitutional ion. This leads to changes of the energies of the e_g orbitals, and to a localization of the itinerant electrons which competes with the double-exchange interaction. Consequently, the Jahn-Teller disorder is enhanced, d.c. resistivity increases, and Tc decreases. The Raman spectra exhibit an additional A_g-like mode that we attribute to the local breathing mode of oxygen in a close vicinity of the substituent ion. Its frequency and intensity is very sensitive to the type of substituent. They seem to depend on the position of the substituent ion in the periodic table suggesting that the electronic characteristics of the M–O bond are important. However, the origin of these effects remains an open question for future studies.

References

[1] Grunberg P, Schreiber R, Pang Y, Brodsky M B, Sowers H 1986 *Phys. Rev. Lett.* **57** 2442
[2] Parkin S S P 1995 *Annu. Rev. Mater. Sci.*, **25** 357
[3] Jin S, Tiefel T H, McCormack M, Fastnacht R A, Ramesh R, Chen LH 1994 *Science* **264** 413
[4] Salamon M B, Jaime M 2001 *Reviews of Modern Physics*, **73** 583
[5] Haghiri-Gosnet A M, and Renard J P 2003 *Journal of Physics D: Applied Physics* **36** R127
[6] Minh N V, Kim S J and Yang I S 2003 *Physica B* **327** 208
[7] Kim K H, Gu J Y, Choi H S, Park G W, Noh T W 1996 *Phys. Rev. Lett.* **77** 1877
[8] Kawano H, Kajimoto R, Kubota M, Yoshizawa Y, 1996 *Phys. Rev. B* **53** R14709.
[9] Elemans J B A A, van Laar B, van der Veen K R, Loopstra B O 1971 *J. Solid State Chem.* **3** 238
[10] Geller S. 1956 *J. Chem. Phys.* **24** 1236
[11] Podobedov V B, Weber A, Romero D B, Rice J P, Drew HD 1998 *Phys. Rev. B* **58** 43
[12] Souza Filho A G 2003 *Phys. Rev.* B **67** 052405

Acknowledgments

This work is supported in part by the NAFOSTED of Vietnam.

High Q-factor micro-cavity laser: Fabrication and lasing emission properties

Pham Van Hoi [1,3], Bui Van Thien[2], Ha Xuan Vinh[1], Chu Thi Thu Ha[1], Tran Thi Cham[1] and A. N. Gruzintsev[4]

[1] Institute of Materials Science, Vietname Academy of Science and Technology 18 Hoang Quoc Viet Road, Cau Giay District, Hanoi, Vietnam
[2] Faculty of Natural Science, College of Medicine, Thai Nguyen, Vietnam
[3] Faculty of Engineering Physics and Nano Technology, College of Technology, VNUH., Cau Giay District, Hanoi, Vietnam
[4] Institute of Microelectronics Technology and High Purity Materials, Russian Academy of Sciences (IPMT-RAS)

E-mail: hoipv@ims.vast.ac.vn

Abstract. In this article the fabrication method and lasing emission properties of High-Q micro-cavity lasers based on High-concentration Erbium-doped silica-alumina glasses are presented in detail. The configurations of micro-cavities were spherical and/or modified toroidal forms. The lasing threshold of micro-cavity laser pumped by laser diodes was of hundred micro-watts and Q-factor of cavity had been achieved up-to 10^8 in experiment. The emission power of one whispering-gallery-mode (WGM) lasing from micro-cavity laser was of 0.05-0.5 mW that would be enough for applying in the quantum information and optical sensor techniques. The modified toroidal micro-cavity permits to decrease the polar-mode of WGMs, which help to obtain the single-mode emission from micro-cavity lasers.

Keywords: Micro-cavity laser, whispering-gallery-mode emission, high-concentration Er-doped materials.

1. Introduction

Dielectric micro-cavities have attracted much attention due to their ability to entrap photons in the visible and NIR regions of the electromagnetic spectrum via total internal reflection [1]. Such photon confinement can be used for various purposes, ranging from fundamental studies on Quantum Electro-Dynamics (QED) [2] to more applied fields, such as the development of microscopic laser sources, tunable filters, and transducer mechanisms for optical sensing [3]. Due to the total internal reflection condition, photon entrapment works best for those cavity modes that propagate along the circumference of the cavity. Such modes are often referred to as Whispering-Gallery-Modes (WGMs) in analogy to an acoustic phenomenon discovered by Raleigh more than a hundred years ago [4]. As a glass spherical microcavity providing Q-factors in excess of 10^8, silica toroidal microcavity resonators provide both long photon storage times and improved spatial confinement over microsphere resonators of comparable size [5]. Furthermore, with highly ideal, tapered-fiber coupling, efficient pumping and laser emission extraction are possible [6]. Among the solid-state microcavity lasers with ultrahigh-Q performance, Rare-earth doped glass microsphere lasers have been successfully demonstrated in last

© 2009 IOP Publishing Ltd

decade [7-10]. The optical modes of spherical dielectric cavity can be calculated by solving Helmholtz equations in spherical coordinates. A significant simplification occurs if the sphere consists of a homogeneous dielectric, and if the optical modes reflect with grazing incidence upon the dielectric-air boundary, such that the polarization can be assumed to be constant along the optical trajectories. Under this assumption the optical modes can be solved by the scalar wave equation approximation and solutions fall into TE *n, m, l,* and *p* and TM *n, m, l,* and *p,* where *n* is radial mode number, *m* is azimuthally mode number, *l* is polar mode number, and *p* is polarization state of mode [11]. In an ideal sphere cavity the optical modes possess ($2l+1$) degeneracy with respect to the azimuthally mode number *m*. The toroidal microcavity, which creates dense arrays of ultrahigh-Q microcavity structures in equatorial planes, can reduce modal spectra in comparison with microsphere cavity. In particular, single-mode operation was possible in these devices [12]. Recently, a toroidal microcavity has enabled access to Q-factors is excess of 10^8, which was the highest value reported on a planar substrate. These toroidal microcavities were silica based and fabricated on silicon chips combining standard silicon technology with ultrahigh-Q fabrication techniques [13].

In this paper, we used surface-tension-induced structure of High Er-doped silica-alumina glass made by molten method to form spherical and/or modified toroidal microcavity laser and study the lasing characteristics when these devices are coupled to a half-tapered optical fiber. Our experiments are focused on high Q-factor microcavity (with ultra-narrow spectra of laser emission) and reduction of lasing-modes in the Er-doped silica-alumina glass modified toroidal lasers in comparison with microsphere lasers in previous studies [10, 14].

2. Experiments

2.1. Fabrication of Er-doped silica-alumina glass microcavity lasers

The microcavity lasers have been fabricated by multi-component Er-doped silica-alumina glasses ($90SiO_2$ - $6Al_2O_3$: $4Y_2O_3$: xEr_2O_3, where x = 0.065-0.3 is a molar percent of Er_2O_3). This kind of glass material was made by sol-gel method, that we had shown in our previous paper [14]. We formed Er-doped silica-alumina glass microcavity in the form of spheres and/or modified toroid by the following steps: first step is making Er-doped glass microsphere with diameters in the range of 30–120μm by molten method using electrical arc discharge or by CO_2-laser beam and second step is forming modified toroidal cavity by irradiation of CO_2- laser beam with pressure on microspheres via the silicon substrate. From high thermal conductive coefficient of silicon, the area of Si-substrate is an important factor for making glass toroid. In our experiment, the area of Si-substrate was 1-2 mm^2.

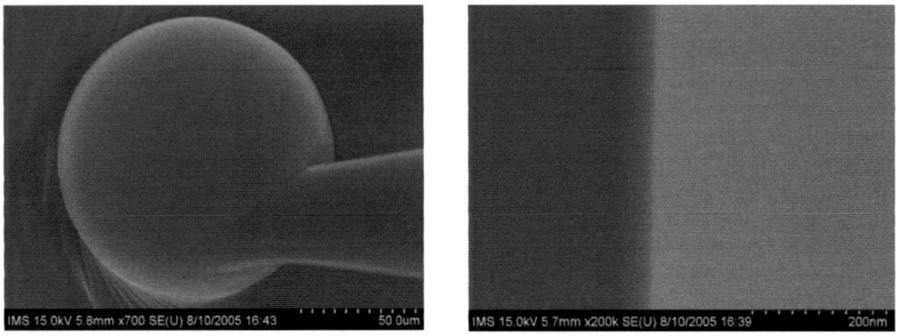

Figure 1. FE-SEM images of glass microsphere (left) and dielectric boundary of microsphere (right) made by thermal molten method.

Using molten method, the solid glass spheres and/or toroids rely upon surface tension to create perfect smooth dielectric boundary, along which the WGMs are confined. Figure 1 shows the dielectric boundary with surface perfection up to nanometer scale. The diameters of spheres were of

30-120μm and the equatorial diameters and thickness of glass modified toroids was of 90-160μm and 20-30μm, respectively.

2.2. Excitation and extraction of WGMs in micro-cavity
The pump laser beam and the collected lasing WGMs are coupled to two different half tapered fibers. Efficient optical coupling to the glass micro-cavity both for pumping and for laser output extraction was performed with optical fiber tapers with waist diameter of 1÷4 μm, which were made from the standard telecommunication single-mode fiber by chemical treatment in 15 mol% -HF solution. The diameters at the waist of fiber tapers were chosen to optimize phase matching and coupling to the fundamental WGM ($n = 1$, $m = l$). We choose 976nm- laser diode adjusted the output power from 0 to 170 mW in single-mode emission (SDLO-2564-170) for excitation of Erbium ions. To couple light efficiently in and out of the micro-cavity laser, the half tapered optical fibers were aligned in the equatorial plane of micro-cavity. The micro-cavity laser was fixed and the half tapered fibers were attached to a micro-positioner with piezoelectric stage that allowed for their precise positioning with respect to the micro-cavity. The collected laser output fiber can go forward or backwards with respect to the pumping direction. The non-absorbed pump beam and the luminescence or laser emission corresponding to the WGMs of the micro-cavity are separated with a de-multiplexing coupler 980nm/ 1550nm. Thus we can analyze simultaneously the pump and the laser signals. The spectral characteristics of the emission around 1550 nm were analyzed with a 0.01 nm resolution optical spectrum analyzer (OSA: Advantest Q8384) directly after the de-multiplexing coupler.

3. Results and discussion
The input threshold pump power can be obtained directly from experiment by setting the pump power from zero. In our Er-doped silica-alumina glass micro-cavity laser, we only obtained supper luminescent emission, when the pump power at wavelength of 976 nm was below 2.0 mW. The laser oscillation modes (WGMs) of the Er-doped glass micro-cavity are in the wavelength range from 1510 nm to 1610 nm (see figure 2). Depending on the gain spectral region of the highly Er-doped silica-alumina glass, the lasing wavelength of the micro-cavity lasers varied within 1540-1610 nm range, which is in the both C- and L-bands. In addition, for any equatorial diameters of the sphere and/or toroid we obtained an enhancement of the laser intensity when increasing the Er-concentration from 1,250 to 3,000 ppm in the silica-alumina glass matrix. The lasing emission could be obtained by forward and backward coupling configuration. Figure 3 shows the spectra of single-mode lasing at laser threshold obtained from 90 μm-diameter and from 110 μm – equatorial diameter of spherical and modified toroidal glass micro-cavity laser doped with 2,500 ppm of Er^{3+} ions, respectively.

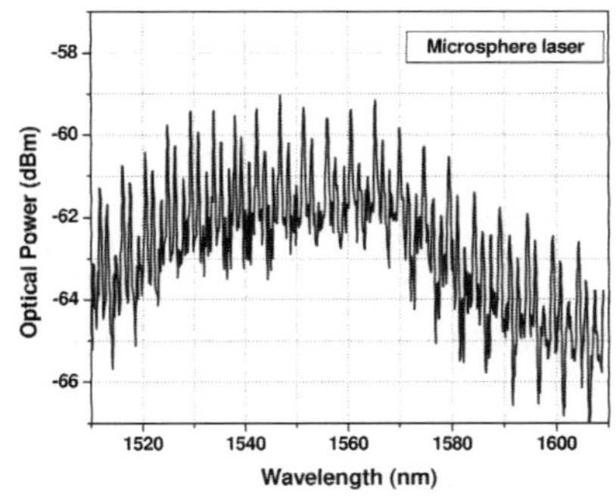

(a)

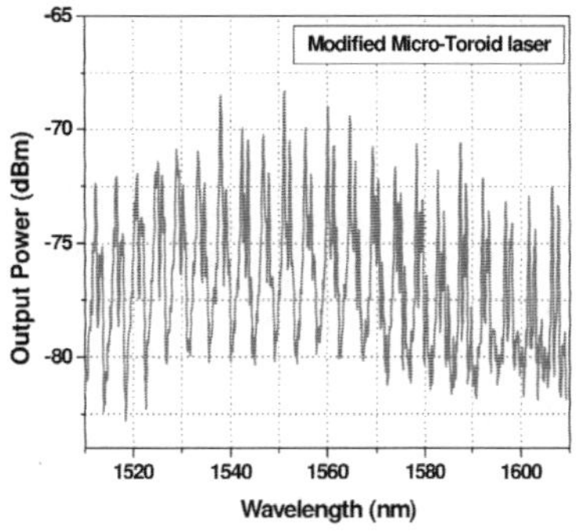

(b)

Figure 2. Superluminescent spectra of micro-sphere (a) and micro-toroidal (b) lasers pumped by 976 nm-laser beam with light power of 2 mW (below threshold)

The WGM lasing spectra rounding 1600 nm-wavelength gives good evidence for larger gain in L-band in highly Er-doped sol-gel silica-alumina glasses. By increasing the pump power the number of lasing modes was increased and the wavelength of lasing peak could be shifted to short range.

Figure 4 presents multi-mode lasing spectra from the same modified toroidal micro-cavity lasers, when pump optical power increased up to 3 mW. We can see, the lasing mode of toroidal micro-cavity laser at wavelength 1599.349 nm was the same at threshold, and most all of the new lasing modes were shifted to the shorter band. The pump transmission was estimated about ~12% in this measurement, so the actual coupled threshold pump power is estimated to be approximately 0.3 mW. As we know, the lasing threshold is a sensitive function of the cavity configuration and coupling gap width between glass surface and tapered fiber. In our experiment, we can remark, that threshold optical pump power in the case of toroidal micro-cavity is the same value in comparison with micro-spherical cavity.

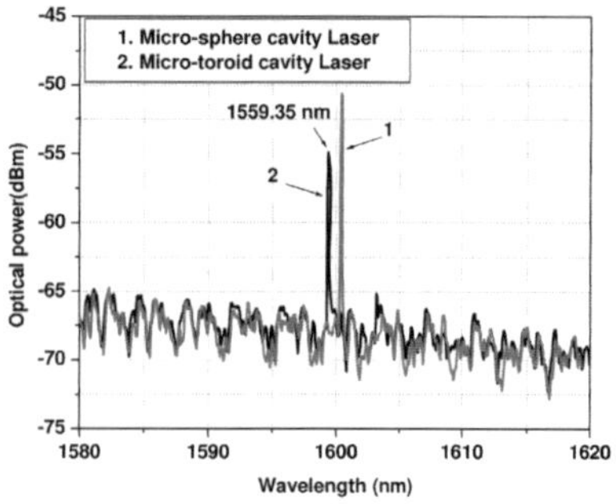

Figure 3. Single-mode WGM lasing from micro-sphere (1) and micro-toroid (2) lasers pumped at threshold by 976 nm-laser beam with 2.2 mW and 2.7 mW, respectively.

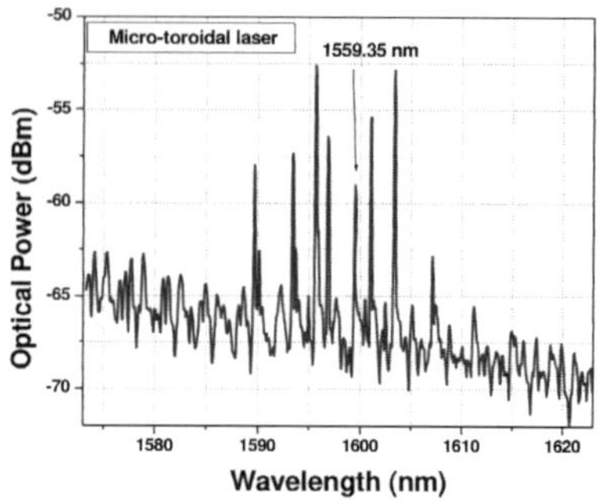

Figure 4. Multi-mode WGM lasing from micro-toroid cavity lasers pumped by 976 nm-laser beam with 3 mW.

Experiments show, that eventually lasing at 1600 nm can no longer be supported with increasing pump power and/or gap width because they cannot be compensated by the relatively small gain at this wavelength range. When pump power was more than 10 mW the lasing spectra of micro-cavity lasers, in general, would be shifted to short-wavelength range. In our case, the WGMs of modified micro-toroidal laser had been distributed in the wavelength range 1550-1580 nm and the number of WGMs of this one was decreased in contradiction with large WGM spectra in range of 1540-1610 nm from micro-sphere laser at the same pumped condition (see figure 5). Figure 6 demonstrates the highest WGM power lasing from micro-sphere laser with diameter of 110 μm and the WGM amplified by EDFA. The maximum lasing output power of one WGM, we achieved, are of -3.2 dBm and of -5.5 dBm from the Er-doped silica-alumina glass spherical and/or toroidal micro-cavity lasers, respectively. The high output power of lasing WGMs in our micro-cavity lasers may be caused by good tapered-micro-cavity coupling at 976 nm – pump and a homogeneous distribution of Er-ions in the silica-alumina glasses.

The number of collected WGMs and the laser output power depend upon many parameters such as loading conditions (i.e. waist diameter of half-taper fiber, gap width between taper fiber and cavity), diameter and eccentricity of sphere, Er-ion concentration. In the case of well satisfying phase condition, single-mode operation can occur. In our previous results [14], the single-mode lasing spectra was obtained when the gap width was changed in the range 0.2-1μm. Figure 7 presents a series of WGM lasing wavelengths under varying gap width between taper fiber and surface of toroidal micro-cavity laser. The multi-mode (figure7a) and single-mode lasing spectra (figure7b) with output power of -19 dBm at the same wavelength of 1561.45 nm were obtained from modified toroidal micro-cavity laser, when the gap widths were changed from 1.5 to 0.5μm, respectively.

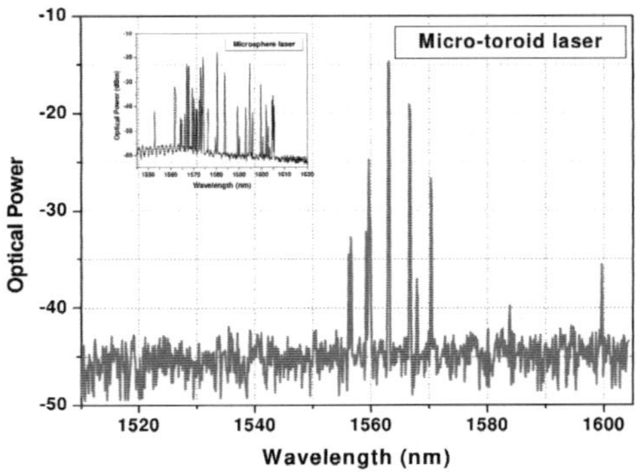

Figure 5. WGMs measured by the fiber taper coupling, when pump power is 90 mW (Lasing peak was at 1563.09 nm). Inset: WGM spectra from microsphere laser at same pump.

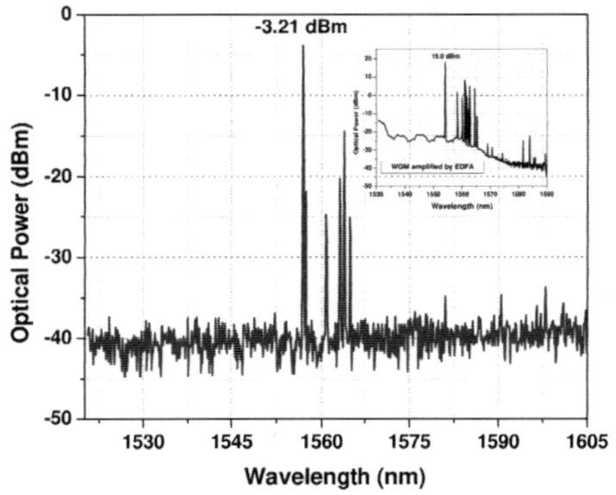

Figure 6. Highest power of -3.2 dBm of WGMs from micro-sphere cavity laser. Inset: WGM signal of 19 dBm amplified by EDFA.

In practice, there are a series of peaks in the single-mode obtained from OSA-Advantest Q8384 with resolution of 0.01 nm and these peaks should be analyzed by Fabry-Perot Etalon (FPE). Using formula $Q= \omega/\Delta\omega$ [13] for calculation of Q-factor, we obtained $Q= 5.10^7$-10^8 for our spherical and/or modified toroidal micro-cavities. This result was in good agreement with the other one in the papers [6-9]. Figure 8 demonstrates the pump-laser emission characteristics of microcavity laser with diameter of 90μm. Threshold of laser was about 2.5 mW of pumping power, that mean the inserted pumping power at threshold was of 0.3 mW. Figure 9 shows the spectra of amplified via EDFA optical signals lasing from DFB-laser with band-width of 0.1 nm and from spherical mcirocavity laser with band-width less than 10^{-5} nm. The saturated output power of EDFA for common DFB-laser emission, which used in the fiberoptic communication network, is of 14 +/- 0.2 dBm. In our experiment, when the WGM's power of about -5 dBm from microcavity laser inserted into EDFA, the amplified output power was achieved up-to 18.5 dBm. This is clear evidence that the band-width of WGM was ultra-narrow in comparison with DFB lasing emission.

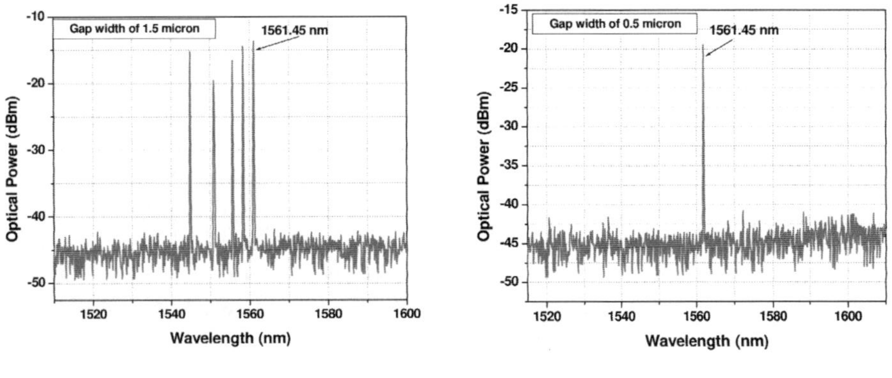

(a) (b)

Figure 7. WGM lasing spectra obtained from toroid micro-cavity laser under varying gap width changed from 1.5 micron (a) to 0.5 micron (b). The WGM single-mode power was achieved up-to -19 dBm.

For DWDM networks, the dispersion phenomenon plays a very important role. The optical signal pulse expansion $\Delta\tau$ by dispersion is given by

$$\Delta\tau = D.L.\Delta\lambda \ , \quad (1)$$

where D is dispersion parameter of fiber at the working wavelength, L is the fiber length and $\Delta\lambda$ is band-width of optical signal. When we use WGM with the band-width $\Delta\lambda < 10^{-5}$ nm, the $\Delta\tau$ would be decreased in four orders in comparison with band-width of 0.1 nm from common DFB lasers. That means we can increase the number of optical channels in one fiber a hundred times. On the other side, the technique of double optical transmission signals with very-closed wavelengths is used in the practical high-security optical network. The marking signal should have ultra-narrow band-width, low optical power and spacing between carrier-signal and marking signal less than 1 GHz. We propose that the WGM lasing from microcavity laser can be used as a marking signal in the high-security optical network. The high-security optical network used WGM lasing from microcavity laser will be studied next time.

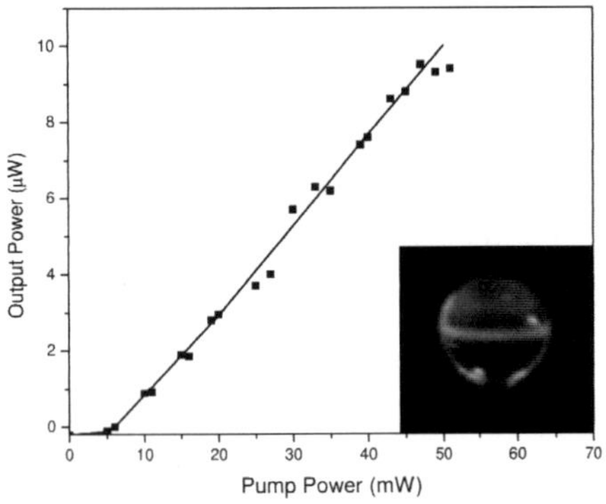

Figure 8. Pump-signal characteristics of micro-sphere cavity laser based on Er-doped silica-alumina glasses. Inset: Up-conversion emission on the equatorial plane of microcavity laser

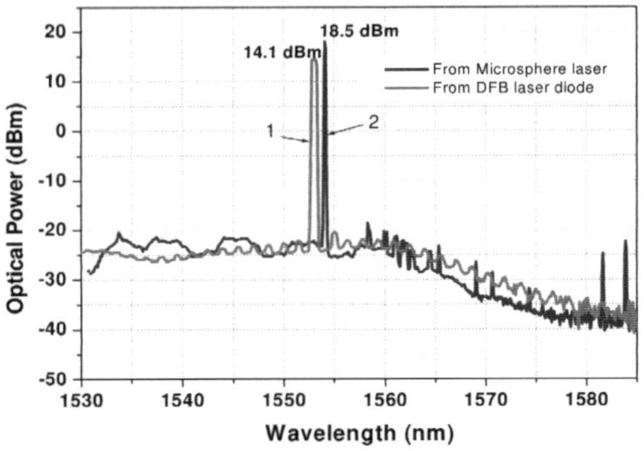

Figure 9. Amplified output power via EDFA from DFB laser with input power of 2.1 dBm (1) and from microcavity laser with input power of -5 dBm (2).

4. Conculusion

We have reviewed a method of fabrication and characterization of High Q-factor micro-cavity lasers based on Er-doped silica-alumina glasses, which has been lasing ultra-narrow spectra at 1550 nm-wavelength range. The lasing output powers of one whispering-gallery-mode were reached up to -3.2 and -5.5 dBm for spherical and modified toroid micro-cavity lasers, respectively. The Q-factor of glass micro-cavities was achieved up to 5.10^7-10^8 and the reduction of lasing WGMs was obtained in the modified toroid micro-cavity laser in comparison with spherical micro-cavity under the same diameter and pumping conditions. We have successfully controlled the number of lasing WGMs by changing the gap width between taper fiber and cavity and by waist diameter of taper. The single-mode spectra of Er-doped glass micro-cavity laser has output power and ultra-narrow line-width, to the best of the authors' knowledge, suitable for applications in the field of DWDM networks, quantum informatics and optical sensor technique.

References

[1] Vahala K J 2003 Optical microcavity *Nature* **424** 839

[2] Gérard J M, Sermage B, Gayral B, Legrand B, Costard E, and Thierry-Mieg V 1998 Enhanced Spontaneous Emission by Quantum Boxes in a monolithic optical microcavity *Phys. Rev. Lett.* **81** 1110-13

[3] Vollmer F, Braun D, Libchaber A, Khoshsima M, Teraoka I, and Arnold S 2002 Protein detection by optical shift of a resonant microcavity *Appl. Phys. Lett.* **80** 4057
[4] Weller A and Himmelhuasa M 2006 Radial cavity mode excitation in fluorescent core-shell microcavities *Appl. Phys. Lett.* **89**, 241105
[5] Armani D K, Kippenberg T K, Spillane S M, and Vahala K J, 2003 Ultra-high-Q toroid microcavity on a chip *Nature* **421** 925
[6] Cai M, Painter O, and Vahala K J 2000 Observation of critical coupling in a fiber taper to a silica-microsphere Whispering-Gallery- Mode system *Phys. Rev. Lett.* **85** 74
[7] Cai M, Painter O and Vahala K J 2000 Fiber-coupler microsphere laser *Opt.Lett.* **25**, 260
[8] Yang L, and Vahala K J, Gain functionalization of silica microresonator 2003 *Opt.Lett.* **28** 592
[9] Peng X, Song F, Jiang S, Peyghambarian N, and Kuwata-Gonokami M 2003 Fiber-taper-coupler L-band Er3+-doped telluride glass microsphere laser *Appl. Phys. Lett.* **82** 1497
[10] Pham V H, Chu T T H, and Hoang Q H 2005 L-band emission of micro-sphere lasers based on Er-doped sol-gel silica-alumina glasses *Appl. Phys. Lett.* **87** 161110
[11] Little B E, Laine J P, and Haus H A 1999 Analytic theory of coupling from tapered fibers and half-blocks into microsphere resonator *J. Lightwave Technol.* **17 (4)** 704
[12] Yang L, Armani D K, and Vahala K J 2003 Fiber-coupler Erbium microlasers on a chip *Appl. Phys. Lett.* **83** 825
[13] Min B, Kippenberg T J, Yang L, and Vahala K J 2004 Erbium-implanted high-Q silica toroidal microcavity laser on a silicon chip *Phys. Rev.A* **70** 033803
[14] Chung L N, Ha C T T, Trang N T, Nga P T, Hoi P V, and Thien B V 2006 High-power micro-cavity lasers based on highly Erbium-doped sol-gel aluminosilicate glasses *J. Materials Sci & Engineer B* **131** 27

Acknowledgments

This work was financially supported by National Natural Science Program in Physics of Vietnam 2006-2008 under Grant No.CB14 and by National Science-Technology Council KC-01 of Ministry of Science and Technology of Vietnam under Contracts KC-01-06/06-10. The authors would like to thank Prof. Acad. Nguyen Van Hieu for his fruitful advice and active support to this work. The authors express our appreciation to the State Key Lab. on Electronic Materials and Devices for testing equipment.

GaN film growth on $LiNbO_3$ surfaces using molecular beam epitaxy

Man Hoai Nam[1, 3], Son Chul Goo[2], Moon Deock Kim[2] and Woochul Yang[1]

[1] Department of Physics, Dongguk University, Seoul 100-715, Korea
[2] Department of Physics, Chungnam National University, Daejeon 305-764, Korea
[3] Institute of Materials Science, Vietnam Academy of Science and Technology, 18 Hoang Quoc Viet, Cau Giay, Hanoi, Vietnam

E-mail: wyang@dongguk.edu

Abstract. GaN has been used for high power, high frequency electronic and optoelectronic devices such as light emitting diodes and laser diodes. Most of the GaN films grow on sapphire (0001) and silicon (111) substrates. In these work, we are trying to grow GaN films on lithium niobate ($LiNbO_3$) substrates using molecular beam epitaxy (MBE). As a ferroelectric materials, $LiNbO_3$ has a spontaneous polarization which may provide excellent control of polarity of GaN. The growth of GaN films on $LiNbO_3$ has been performed after the $LiNbO_3$ substrates was annealed in air at 1000°C for 2 hours. The annealed substrates allowed us to prepare atomically flat surfaces and improve adhesion of GaN on $LiNbO_3$. The AlN buffer layer was deposited to get a smaller lattice mismatch with the GaN films. Compared with GaN films grown without a AlN layer, the crystal qualities of GaN films with AlN buffer layers are extremely improved. The surface morphology of $LiNbO_3$ substrates and the grown GaN films were characterized by atomic force microscopy (AFM), and the crystal structures were studied by X-ray diffraction (XRD).

Keywords: Lithium niobate, gallium nitride, molecular beam epitaxy, polarity.

1. Introduction

Gallium nitride (GaN) and its related compounds have attracted much attention because of their excellent optical and electrical properties for device applications such as light emitting diodes, laser diodes, and high power devices [1-3]. However, it is well known that the epitaxial growth of GaN films suffers from the formation of a high density of crystalline defects. The defects are attributed to the large mismatch of 16.1% between the lattice constants of GaN (a = 0.3189 nm) and sapphire (a = 0.4765 nm), which is widely used for substrates. To solve this problem, the use of an alternative substrate that exhibits a smaller lattice mismatch with GaN is highly desirable [4–10]. Since the (0001) plane of lithium niobate ($LiNbO_3$) has a lattice mismatch of 6.8% with respect to GaN (0001) plane, we might expect to grow higher-quality GaN films on $LiNbO_3$ substrates.

$LiNbO_3$ has a rhombohedral atomic structure, the c-plane surface atomic arrangement of $LiNbO_3$ (LN) has a hexagonal symmetry in an unit cell with a-axis and c-axis dimensions of [12] and 3.8631 Å, respectively [18]. The surface of ($+z$) - LN has hexagonal symmetry in structure similar to sapphire, while LN has different surface oxygen bond lengths around the hexagonal oxygen sub lattice. There

© 2009 IOP Publishing Ltd

are two bonds each of length 2.719, 3.362, and 2.879 Å for LN. Thus, LN has average surface oxygen - oxygen spacing of 2.981 Å, and have a lattice mismatch to GaN $a = 3.185$ Å of 6.8% [12]. A similar analysis of LN mismatch to AlN $a = 3.1114$ Å results in a lattice mismatch of 4.4%.

In addition, the use of $LiNbO_3$ substrates is attractive because large-diameter wafers (> 4 inch) are commercially available at reasonable cost. It should be noted that $LiNbO_3$ has also been used for optoelectronic applications such as optical switching devices and second harmonic generators due to its large nonlinear-optical coefficients [11]. Hence, the successful epitaxial growth of GaN on $LiNbO_3$ could lead to the fabrication of optoelectronic integrated circuits that utilize both GaN lasers and $LiNbO_3$ optical switches [15]. Although Doolittle *et al* [12-16] have succeeded in the epitaxial growth of GaN films on $LiNbO_3$ (0001) substrates, the diffusion of Li atoms into the GaN films and the formation of $LiNb_3O_8$ were observed at the interface between GaN and $LiNbO_3$. The interfacial $LiNb_3O_8$ layers leads to some degradation of the electrical and optical properties of the GaN [15]. To probe the formation of the $LiNb_3O_8$ layer, a reduction in growth temperature and the use of an AlN buffer layer are highly desirable [14]. The other advantage of using an AlN buffer layer lies in its smaller lattice mismatch (4.4%) with $LiNbO_3$ (0001).

In this study, we report on the first successful epitaxial growth of GaN films on ferroelectric materials of lithium niobate ($LiNbO_3$) substrates.

2. Experimental

The GaN film was grown on $LiNbO_3$ substrates with a AlN buffer layer by plasma-assisted molecular beam epitaxy (MBE). The system is equipped with reflection high energy electron diffraction (RHEED) for monitoring the growth. In order to obtain atomically flat substrates, the as-received Lithium niobate ($LiNbO_3$) substrates are cleaned with de-ionized (D.I) water, then the $LiNbO_3$ were put on a quartz board and annealed in the air at 1000°C for two hours in quartz tube of furnace system. On the cleaned substrate backside (-*z* surface), Ti (2000 Å in thickness) was deposited by sputtering to enhance radiation absorption from the heater.

After Ti deposited, LN substrates were cleaned with organic solvents EKC 830 (Posistrip positive photoresist remover filtered to 10 microns) and rinsed with de-ionized D.I water again. Next, the samples were outgassed in vacuum at 200°C for 1 hour in a preparation chamber before being loaded into the growth chamber of the MBE system. Prior to growth of GaN, a AlN buffer layer of 100 nm in thickness [11, 12] was grown on (+*z*)-LN substrates at 600°C with 0.8 SCCM (standard cubic centimeter per minute) nitrogen at nitrogen plasma power of 250 W, that will provides for predominantly Ga-polar GaN films. The GaN bulk film was grown at 650°C with 0.5 sccm nitrogen. The thickness of the grown GaN film was ~ 400 nm. During the growth, a reflection high energy electron diffraction (RHEED) pattern was monitored to examine the film growth mode. Each sample was analyzed *ex situ* by atomic force microscopy (AFM) and x-ray diffraction measurement using a Phillips X'pert Pro MRD to investigate surface morphology and structure.

3. Results and discussions

As-received lithium niobate (LN) substrate has a lot of surface damages such as scratches and corrugations in the nanometer scale range due to mechanical polishing, which decrease the adhesion of GaN to the substrate, as shown in figure 1(a). To produce atomically flat surfaces, furnace annealing at 1000°C in dry air environment [11, 12, 14, 18] have been performed. After annealing, the surface damages of $LiNbO_3$ have been drastically changed.

Journal of Physics: Conference Series **187** (2009) 012013 doi:10.1088/1742-6596/187/1/012013

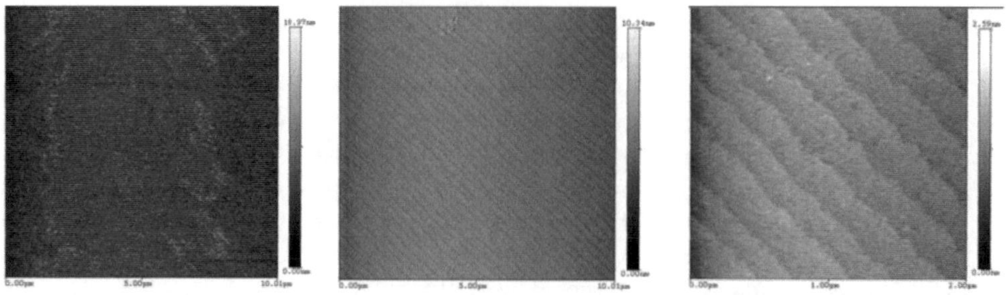

(a) (b) (c)

Figure 1. AFM images of the $LiNbO_3$ substrate (a) as-received substrate (without annealing), (b) after 2 hours of annealing (10 μm x 10 μm, scaled view), and (c) after 2 hours of annealing (2 μm x 2 μm, scaled view).

Figure 1 show AFM images of $LiNbO_3$ substrate surfaces of an as-received and after furnace annealing at 1000°C. Figure 1a shows many scratches on the surface of the as-received $LiNbO_3$ that are not suitable for GaN growth. The surface roughness was 0.32 nm in root mean square (rms), for a 2μm x 2μm area. The AFM image of $LiNbO_3$ substrate after 2 hours of annealing at 1000°C in air was shown in the figure 1b. It showed that, all the surface damage was removed, indicating a unit-cell height step and terrace structures on the surface after annealing at high temperature. The step terrace width is about 380 nm, and the surface roughness is 0.19 nm, for a 2μm x 2μm area, in figures 1b and 1c.

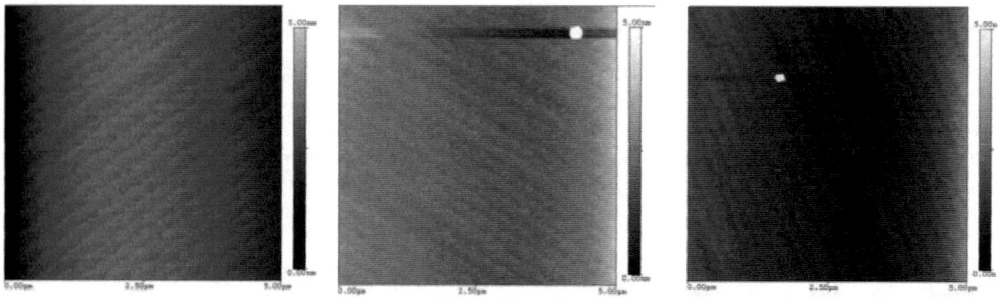

(a) (b) (c)

Figure 2. AFM images of the surface of $LiNbO_3$ substrate (5 μm x 5 μm scaled view): (a) after 3 hours of annealing), (b) after 4 hours of annealing, and (c) after 6 hours of annealing.

However, longer annealing produces a facet coarsening surface which might increase surface roughness. Figure 2 shows the AFM images of the $LiNbO_3$ surfaces after annealing at various times. After 3 hours annealing the surface is very similar to one annealed for 2 hours. The $LiNbO_3$ substrate produces atomically flat terraces of 400 nm width, and the surface roughness of 0.49 nm, as shown in figure 2a. However, after annealing for 4 hours and 6 hours, the surface shows a slightly porous morphology that increases surface roughness, as shown in the figure 2b and figure 2c. The surface roughnesses are 0.65 nm, and 0.51 nm for annealing samples for 4 hours and 6 hours, respectively. From these results, we can conclude that 2–3 hours annealing gives the best surface smoothness.

When the $LiNbO_3$ substrate was annealed at high temperature, the topmost surface is thermodynamically unstable and is transformed to the equilibrium crystal surface by the rearrangement of the surface atoms [14]. The improvement of $LiNbO_3$ surface quality has been attributed to diffusion and re-growth processes, leading to atomic flatness and removal of scratches from the surface.

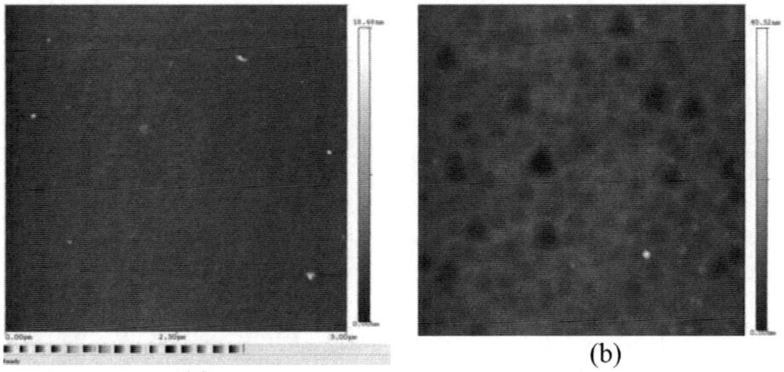

(a) (b)

Figure 3. AFM morphologies of the etched with HF acid for 15 minutes. (a) as-received $LiNbO_3$ substrate without furnace annealing, (b) after furnace annealing at 1000°C for 2 hours

To check the presence of ferroelectric inversion domains on $LiNbO_3$, a chemical etching was performed using HF for 15 minutes. Before high temperature thermal treatment, as-received (+*z*)-cut $LiNbO_3$ was not etched, indicating a single domain. However, after high-temperature thermal treatment at 1000°C in air followed by etching, (+*z*)-cut $LiNbO_3$ shows faceted pits, indicating repolarization on the surface, as shown in figure 3b. Therefore, high-temperature thermal treatment gives a trade-off between adhesion and control of polarity of GaN epitaxial layer on $LiNbO_3$.

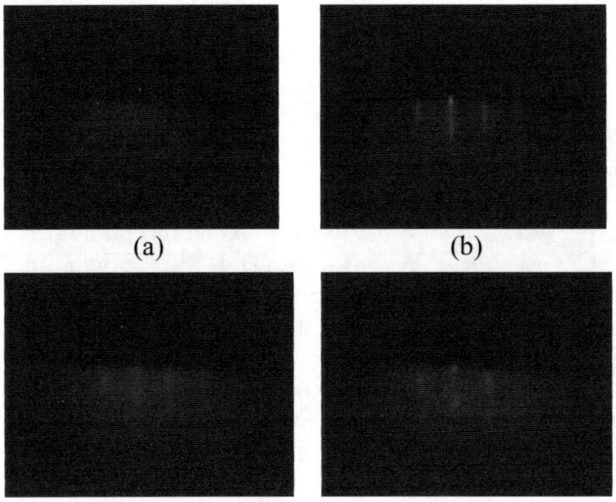

(a) (b) (c) (d)

Figure 4. In situ RHEED patterns of (a) $LiNbO_3$ substrate at 200°C, (b) $LiNbO_3$ substrate at 500°C, (c) AlN on $LiNbO_3$ at 500°C after 2 minutes, and (d) AlN on $LiNbO_3$ at 500°C after 2 minutes

During the GaN growth, the *in situ* RHEED pattern was monitored. Figures 4a and 4b show the RHEED patterns of $LiNbO_3$ substrates at high temperatures. The RHEED patterns of $LiNbO_3$ substrate at 200°C is a dim. However, when the substrate temperature increases to the growth temperature at 500°C, the pattern degrades and exhibits transmission features. The RHEED patterns were streaky and clearly, as shown in figure 4b. This indicates that the corrugation of the surface has increased significantly, suggesting initiation of $LiNbO_3$ is decomposing at high temperature.

The RHEED patterns during the growth of the AlN buffer layer consists of streaks and spots, as shown in figures 4c and 4d, which indicates that single-crystal AlN grows epitaxially on $LiNbO_3$, and that the surface morphology is slightly roughened, probably due to stress in the film. Careful interpretation of the RHEED pattern allows us to conclude that AlN (0001) grows on $LiNbO_3$ (0001) with an in-plane epitaxial alignment.

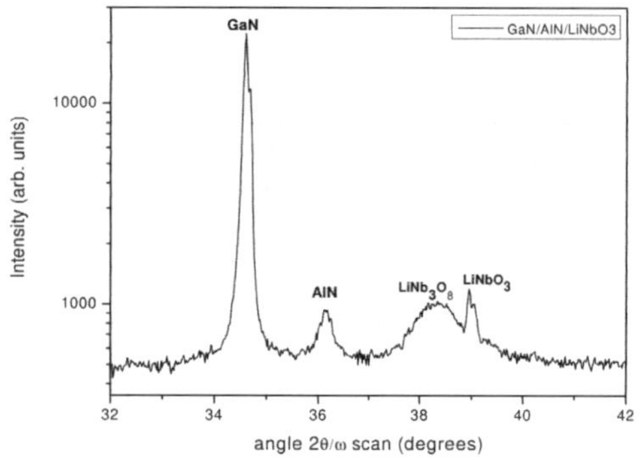

Figure 5. X-ray diffraction of GaN film grown on AlN buffer layer grown on $LiNbO_3$ substrate at 600°C for 50 minutes.

The structural quality of the GaN /AlN/$LiNbO_3$ was assessed by X-ray diffraction study. Figure 5 shows the omega – two theta X-ray diffraction spectrum for a GaN film on lithium niobate. The dominant peak of GaN (0002) is located at 34.56 degrees of a two theta, indicating alignment of the GaN (0001) direction along the (0001) direction of lithium niobate substrate. This result shows the growth of a well-oriented GaN film on the $LiNbO_3$ substrate. The peaks at $2\theta = 36.124$ degrees and $2\theta = 39.1$ degrees are due to the reflections of AlN buffer layer and $LiNbO_3$ substrate, respectively. The additional peak is found to be located at two theta of 38.3 degrees, which peak is related to lithium triniobate, $LiNb_3O_8$, a stable monoclinic phase in the lithium oxide/niobate oxide system resulting from preferential desorption of LiO_2 near the sample surface [15]. This preferential desorption results in an imbalance state in the lithium oxide/niobium oxide ratio which becomes 1:3 LiO_2:Nb_2O_5 in the near surface regions.

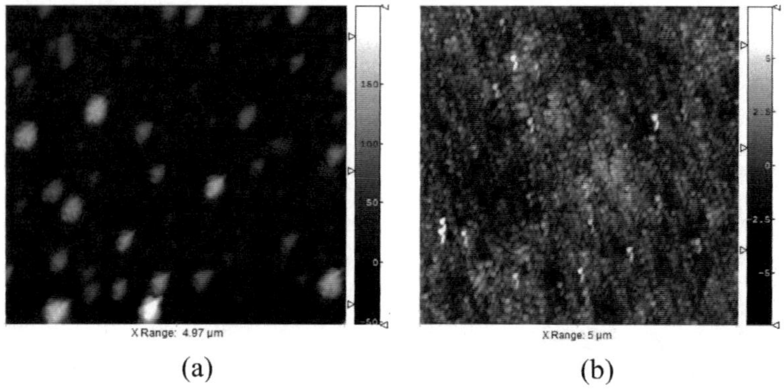

Figure 6. AFM morphologies of, (a) GaN film grown on $LiNbO_3$ substrates (without an AlN buffer layer), at 600°C, and (b) GaN thin film grown on $LiNbO_3$ substrates with an AlN buffer layer. The scan size is 5 µm x 5 µm.

Figures 6a and 6b show the surface morphology of GaN epitaxial layer grown on $LiNbO_3$ without AlN buffer layer and GaN films with a AlN buffer layer of 100 nm thickness, respectively. For a GaN film grown on $LiNbO_3$ without AlN buffer layer, the surface roughness is 28.40 nm, while for the GaN film grown on $LiNbO_3$ with AlN buffer layer, the surface roughness is 1.22 nm. The surface roughness of GaN/AlN/$LiNbO_3$ is much smoother than GaN/$LiNbO_3$. It is inferred that the AlN buffer layer was used to get a smaller lattice mismatch with the GaN films surface. Therefore the surface of GaN films with AlN buffer layers is extremely improved.

4. Conclusions

$LiNbO_3$ can be a promising substrate for the growth of GaN film due to its slightly lattice mismatch and the structural similarity between GaN and $LiNbO_3$. The high temperature thermal treatment has been used to obtain atomically flat surface and to remove the surface damage from mechanical polishing. The different annealing conditions gave rise to different of surface smoothness. The annealing at 1000°C for 2 hours produces the optimal surface smoothness for adhesion of GaN to $LiNbO_3$ substrates. We have first successfully epitaxial growth of decent quality GaN layer on lithium niobate ($LiNbO_3$) substrates by molecular beam epitaxy. The surface morphology was also improved by the insertion of the AlN buffer layer of 100 nm thickness, It give the better surface smoothness compared with GaN film grown on $LiNbO_3$ without AlN buffer layer.

References

[1] Nakamura S, Senoh M, Nagahama S, Iwasa N, Yamada T, Matsushita T, Kiyoku H, Sugimoto Y, Kozaki T, Umemoto H, Sano M and Chocho K 1997 *Jpn. J. Appl. Phys.* **36** L1568
[2] Akaski I and Amano H 1996 *Jpn. J. Appl. Phys.* **36** 5393
[3] Jain S C, Willander M, Narayan J and Overstraeten R V 2000 *J. Appl. Phys.* **87** 956
[4] Ito S, Fujioka H, Ohta J, Takahashi H and Oshima M 2003 *Thin Solid Films* **435** 215
[5] Ohta J, Fujioka H, Takahashi H and Oshima M 2002 *Appl. Surf. Sci.* **483** 197
[6] Takahashi H, Ohta J, Fujioka H, Oshima M and Kimura M 2002 J. *Cryst. Growth* **237**, 1158
[7] Ohta J, Fujioka H, Ito S and Oshima M 2003 *Thin Solid Films* **435** 218
[8] Ito S, Ohta J, Fujioka H and Oshima M 2002 *Appl. Surf. Sci.* **197** 384
[9] Ohta J, Fujioka H, Takahashi H and Oshima M 2001 *Phys. Stat. Sol. (a)* **188** 497
[10] Takahashi H, Ohta J, Fujioka H and Oshima M 2002 *Thin Solid Films* **407** 114
[11] Yousuke T, Atsushi K, Ohta J, Fujioka H and Oshima M 2005 *Phys. Stat. Sol. (a)* **202** R145
[12] Lee K, Namkoong G, Doolittle W A, Losurdo M, Bruno G and Dieter H J 2006 *J. Vac. Sci. Technol. B* **24**, 2093
[13] Namkoong G, Doolittle W A, Brown A S, Losurdo M, Giangregorio M and Bruno G 2003 *Journal of Crystal Growth* **252** 159
[14] Namkoong G, Lee K, Shannon M, Henderson W, Ralph S E and Doolittle W A 2005 *Appl. Phys. Lett.* **87** 171107
[15] Doolittle W A, Namkoong A, Carver A, Henderson W, Dieter J and Brown A S 2003 *Mat. Res. Soc. Symp. Proc.* **743**, L1.4.1
[16] Lee K, Namkoong G, Shannon M, Stephen R, Doolittle W A, Losurdo M, Bruno G and Cho H K 2007 *Materials Science and Engineering B* **140** 203
[17] Nakamura K, Ando H and Shimizu H 1987 *Appl. Phys. Lett.* **50** 1413
[18] Matsubara K, Fons P, Yamada A, Watanabe M and Niki S 1999 *Thin Solid Films* **347** 238
[19] Kang T W, Park S H, Song H, Kim T W, Yoon S G and Kim C O 1998 *J. Appl. Phys.* **84** 2082
[20] Teisseyre H 2000 *Phys. Rev. B* **62** 10151
[21] Kucheyev S O, Toth M, Phillips M R, Williams J S and Jagadish 2001 *Appl. Phys. Lett.* **79** 2154
[22] Yue J S, Oliver B, Jahn U, Tian Y L, Achim T, Sven C, Dhar S and Klauus H P 2002 *J. Appl. Phys.* **92**, 5714
[23] Rieger W, Dimitrov R, Brunner D, Rohrer E, Ambacher O and Stutzmann M 1996, *Phys. Rev. B* **54** 17596
[24] Eckey L 1996 *Appl. Phys. Lett.* **68** 415
[25] Wetzel C, Fischer S, Kruger J, Haller E, Molnar R J, Moustakas T D, Mokhov E N and Baranov P G 1996 *Appl. Phys. Lett.* **68** 2556

Acknowledgments

This work is supported by the Korea Research foundation grant funded by the Korean Government (KRF-2007-331-C00082).

Time-resolved photoluminescence measurements of InP/ZnS quantum dots

Pham Thi Thuy[1,2], Ung Thi Dieu Thuy[1,3], Tran Thi Kim Chi[1], Le Quang Phuong[1], Nguyen Quang Liem[1,*], Liang Li[3] and Peter Reiss[3]

[1] Institute of Materials Science, VAST, 18 Hoang Quoc Viet, Cau Giay, Hanoi, Vietnam
[2] Baria-Vungtau Commune College, National Road 51C, 11 Dist., Vungtau city, Vietnam
[3] CEA Grenoble, DSM/INAC/SPrAM (UMR 5819 CEA-CNRS-Université Joseph Fourier)/ LEMOH, 17 rue des Martyrs, 38054 Grenoble cedex 9, France

E-mail: liemnq@ims.vast.ac.vn

Abstract. This paper reports the results on the time-resolved photoluminescence study of InP/ZnS core/shell quantum dots. The ZnS shell played a decisive role to passivate imperfections on the surface of InP quantum dots, consequently giving rise to a strong enhancement of the photoluminescence from the InP core. Under appropriate excitation conditions, not only the emission from the InP core but also that from the ZnS shell was observed. The emission peak in InP core quantum dots varied as a function of quantum dots size, ranging in the 600 – 700 nm region; while the ZnS shell showed emission in the blue region around 470 nm, which is interpreted as resulting from defects in ZnS.

Keyword : Time-resolved photoluminescence, quantum dots, InP/ZnS.

1. Introduction

Colloidal semiconductor quantum dots (QDs) or nanocrystals (NCs) are a class of materials intermediate between single molecules and bulk solid-state crystals. Due to their unique optical properties, they are of considerable interest for both fundamental research and technological applications. Among them, II-VI semiconductor QDs such as CdSe and CdTe have been mainly studied due to their strong luminescence in the visible spectral range [1-5]. Applications of the mentioned semiconductor QDs is however problematic due to the intrinsic toxicity of Cd. One of the alternative choices for Cd-free semiconductor QDs with similar emission properties is InP. Within the past decade, NCs of this material have been synthesized using various methods [6-9]. InP is a direct band gap semiconductor with a band gap energy of 1.27 eV. Similar as II-VI semiconductors, InP exhibits quantum size effects and size-dependent optical spectra at the nanometer scale. InP QDs and can be used as fluorescent probes in biological labeling, as light emitters or absorbers in opto-electronic devices such as light emitting diodes and solar cells. The as-prepared InP NCs generally show weak luminescence because of the existence of non-radiative channels originating from surface states. After appropriate surface treatments, e.g., with HF etching [10] or by overgrowth with a ZnS shell [9-12], InP NCs can become highly luminescence materials. Using the steady-state photoluminescence (PL) technique, the overall PL spectra generally show various spectral components

© 2009 IOP Publishing Ltd

originating from different recombinations/transitions. The PL intensity of each spectral component depends on the corresponding recombination/transition rate and its density of states. When looking at the PL spectra of InP/ZnS NCs in the literature, only the emission component of InP has been reported. It clearly showed the quantum size effect with the PL wavelength varying from green to near infrared as the core InP NCs size increased from around 2 to 6 nm [11-13]. The observed high fluorescence quantum yield (Q.Y.) also showed that an efficient passivation of trap states was achieved by the ZnS shell.

The lack of PL from the ZnS shell is questionable, particularly when exciting the InP/ZnS NCs with light of short wavelength, say below 400 nm. In such a case the corresponding excitation photon energy is well above the transition energy of the InP core NCs and also high enough to excite ZnS trap sttates generated from lattice imperfections like Zinc-vacancy or certain donors. The self-activated (SA) PL band peaking around 470 nm is well-known in ZnS [14]. It originates from the recombination of donor-acceptor pairs, in which the donor is usually generated from the replacement of S by a Group-VII atom (e.g. Cl) or the replacement of Zn by a Group-III atom (e.g. Al), while the acceptor is preferably the Zn-vacancy. The SA band in ZnS bulk crystals is efficiently excited by 365 nm photons [14].

In time-resolved photoluminescence (TRPL) measurements, the PL intensity corresponding to each spectral component that is detected at a certain moment delayed from the start of the excitation laser pulse provides a real intensity in the dynamic regime. Therefore, TRPL is a good tool to investigate the PL from the ZnS shell beside that from the InP core, even if it is not revealed in the overall PL spectra. Furthermore, the recombination/transition dynamics after excitation of the InP/ZnS QDs can be followed by measuring their TRPL spectra.

In the present work, the optical characteristics of InP/ZnS core/shell QDs were investigated by means of steady-state and time-resolved PL spectroscopies. The experimental results showed that the ZnS shell has efficiently passivated the surface states of the InP QDs core, consequently strongly increasing their PL Q.Y. While InP QDs emitted luminescence at various spectral ranges depending on their size, the ZnS shells of different InP QDs emitted at the same wavelength characteristic to defects in ZnS.

2. Experimental

Two different InP/ZnS core/shell QDs samples were used in the study. The detailed syntheses of these QDs are described elsewhere [8]. Briefly, InP QDs were synthesized from the high temperature (250°C) reaction of the In and the P precursors in 1-octadecene (ODE). The In precursor was indium myristate, formed *in situ* from indium acetate and myristic acid, while the P precursor was provided in the form of PH_3 gas, generated by the reaction of finely ground Ca_3P_2 or Zn_3P_2 powder with H_2SO_4 in a separate flask, connected to the synthesis flask. The PH_3 gas flow was controlled by the flux of Ar carrier gas. The size of the InP NCs was controlled by the concentration of the precursors and by the growth time. As-synthesized InP NCs could easily be dispersed in a number of solvents, including hexane, toluene, or chloroform. The ZnS shell was made by the subsequent slow addition of the air-stable ZnS precursors (zinc stearate and zinc ethylxanthate) to the crude InP core NCs dispersion. The high quality of the synthesized InP core and InP/ZnS core/shell QDs was confirmed by TEM images (not shown in this paper) and by their UV-vis and PL characteristics.

In the steady-state PL measurements, several excitation sources were used, such as 370 nm or 460 nm LEDs. The excitation power density was weak, about 100 $\mu W/cm^2$ in all cases. For the TRPL measurements, the 355 nm pulsed light from a frequency-tripled Nd:YAG laser was used as the excitation source (with the excitation power density ~1 kW/cm^2, pulse duration of 5 ns). The PL signals were dispersed by using a 0.6-m grating monochromator (Jobin-Yvon HRD1 or Horiba iHR550) and were then detected using a thermoelectrically cooled Si-CCD camera (Synapse) for the steady-state PL measurement and a fast photomultiplier (Hamamatsu model H733, with the rise time of 700 ps) for the TRPL measurements. Averaging the multi-pulses at each spectral point using a 1.5 GHz digital oscilloscope (LeCroy 9362) strongly improved the signal-to-noise ratio.

3. Results and discussion

As-synthesized InP NCs without a ZnS shell did not exhibit detectable photoluminescence. In this case surface states act as non-radiative channels for the relaxation of the excited charge carriers. The ZnS shell on the InP core resulted in a strong increase of the PL intensity of the InP NCs, which is attributed to the passivation of their surface states and confinement of the photogenerated carriers inside the InP core.

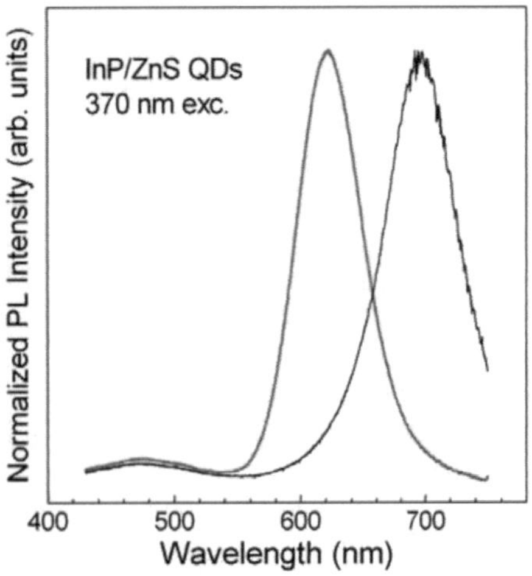

Figure 1. Steady-state PL spectra from two different-sized InP/ZnS core/shell NC samples under the 370 nm excitation.

Figure 1 shows the normalized PL spectra of the two different-sized InP/ZnS core/shell samples investigated under 370 nm excitation. For the InP core sizes of 2.7 and 3.6 nm, strong emission bands at 620 and 690 nm were observed, respectively. This size-dependence is a direct consequence of the quantum size effect active on the InP QDs, namely on their exciton transitions. For easier identification of each sample hereafter, we have named the InP/ZnS core/shell QDs emitting at 620 nm as S-QDs and the other as L-QDs. Along with the PL band originating from the exciton transition in the InP QDs (core), one can see another PL band peaking at shorter wavelength around 470 nm. The latter signal is very similar in difference InP/ZnS core/shell QDs systems, regardless of the size of core InP QDs. We attribute the 470 nm band to lattice defect related emissions from ZnS. In fact, ZnS single crystals emit intensive blue luminescence under excitation with the 365 nm line from a high-pressure mercury lamp [14]. The 365 nm light is below the band gap of ZnS, however it can excite defect-related centers like SA to emit luminescence.

If we suppose that the PL spectrum taken from InP/ZnS QDs contains two spectral components, in the steady-state PL measurement the PL intensity of each component appears to be proportional to the product of the recombination/transition rate and its density of states (with the same absorbance). This product is much larger for the InP QDs than for the ZnS shell, giving rise to the relative intensities of the PL signal observed in figure 1.

The TRPL technique, by tracing the PL intensity at certain moments delayed from the excitation, enables us to determine exactly the instant PL intensity of each spectral component at a certain time interval after the pulsed excitation. It therefore allows us to determine the role of each relevant transition resulting in the overall emission. Figure 2 presents the TRPL spectra of the two InP/ZnS core/shell QDs, in which the 470 nm band was clearly observed, very prominent as compared to the proper emission from the InP QDs core. The intensity evolution of both bands as a function of the delay time (figure 2a) is possibly related to the different transfer rate of carriers from the ZnS shell to the InP core.

The dynamics of photo-generated carriers can be studied by analyzing the decay curves corresponding to each emission band. Figure 3 presents the decay characteristics of the two InP/ZnS core/shell QD systems. Decay curves at various PL wavelengths were measured showing similar decay behavior of the 470 nm PL band for both sizes of InP/ZnS QDs, S- and L-QDs. This is not the case for the evolution of the carriers in the InP cores: The S-QDs exhibit the same decay behavior at different emission wavelengths belonging to the same excitonic transition (580, 630, 660 nm).

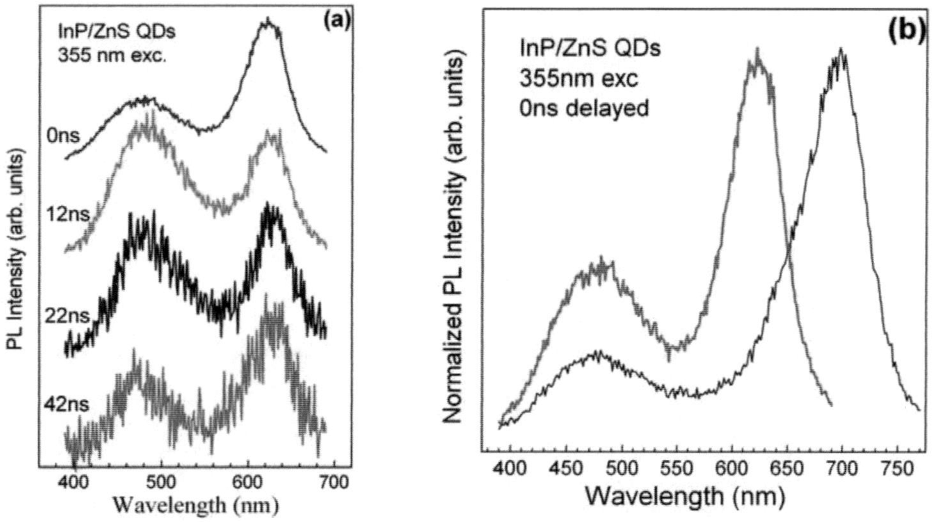

Figure 2. TRPL spectra from InP/ZnS core/shell QDs (S-QDs) under 355 nm excitation at different delay times (a), and those from S-QDs (red) and L-QDs (black), at the 0 ns-delayed time (b) showing the two emission bands originating from the InP cores (at 620 and 690 nm) and the ZnS shell (at 470 nm).

For the L-QDs, the decay behavior is more complicated: there are at least two decay times for each transition, one is shorter comparable to the instrumental response and the other is in the tens of nanoseconds range. In addition, the 630 nm spectral component, which appears as a small shoulder in the PL spectra, shows a much longer decay time than the longer wavelength components (690, 750 nm). A detailed study of the carrier dynamics in InP/ZnS core/shell QDs is in progress with the goal to achieve a better understanding of the optical processes in these systems.

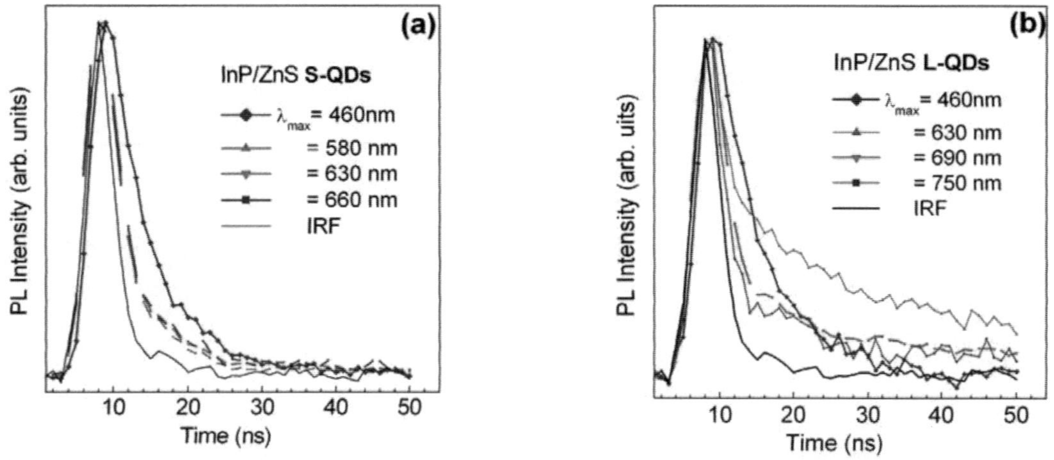

Figure 3. Decay times at various PL wavelengths (indicated on the figures) for the two different InP/ZnS core/shell QDs, (a) S-QDs and (b) L-QDs, showing a similar decay behavior of the 470 nm PL band. Concerning the other spectral components, the L-QDs exhibit a much longer decay time than the S-QDs for the 630 nm band, which appears as a small shoulder in the PL spectra of the former.

4. Conclusion

In conclusion, InP/ZnS core/shell QDs were studied using PL and TRPL techniques. The emission peak related to the InP core varied as a function of the QD size in the 600–700 nm range, while the ZnS shell showed emission in the blue region around 470 nm, arising from defect states. The results from TRPL measurements demonstrated that the ZnS shell emitted on a comparable timescale to that from the InP core, in the nanosecond range. On the one hand, the ZnS overcoating played a decisive role in the passivation of imperfections at the surface of InP QDs, consequently increasing the photoluminescence intensity related to the excitonic transition in the InP core. On the other hand, our study reveals that, beside the optical transitions in the InP core, the optical processes within the ZnS shell strongly influence the dynamics of carriers' population and evolution after photo-generation.

References

[1] Murray C B, Norris D J, Bawendi M G 1993 *J. Am. Chem. Soc.* **115** 8706
[2] Dabbousi B O, Rodriguez-Viejo J, Mikulec F V, Heine J R, Mattoussi H, Ober R, Jensen K F and Bawendi M G 1997 *J. Phys. Chem. B* **101** 9463
[3] Qu L, Peng Z A, Peng X 2001 *Nano Lett.* **1** 333
[4] Reiss P, Bleuse J, Pron A 2002 *Nano Lett.* **2** 781
[5] Gu Z, Zou L, Fang Z, Zhu W, Zhong X 2008 *Nanotechnology* **19** 135604
[6] Micic O I, Curtis C J, Jones K M, Sprague J R, Nozik A J 1994 *J. Phys. Chem.* **98** (19) 4966
[7] Guzelian A A, Katari J E B, Kadavanich A V, Banin U, Hamad K, Juban E, Alivisatos A P, Wolters R H, Arnold C C, Heath J R 1996 *J. Phys Chem.* **100** 7212
[8] Liu Z, Kumbhar A, Xu D, Zhang J, Sun Z, Fang J 2008 *Angew. Chem. Int. Ed.* **47** 1
[9] Li L, Protière M, Reiss P 2008 *Chem. Mater.* **20** 2621
[10] Talapin D V, Gaponik N, Borchert H, Rogach A L, Haase M, Weller H 2002 *J. Phys. Chem. B* **106** (49) 12659
[11] Xie R, Battaglia D, and Peng X 2007 *J. Am. Chem. Soc.* **129** (50) 15432
[12] Xu S, Ziegler J, and Nann T 2008 *J. Mater. Chem.* **18** 2653
[13] Adam S, Talapin D V, Borchert H, Lobo A, McGinley C, de Castro A R B, Haase M, Weller H, and Möller T 2005 *J. Chem. Phys.* **123** 084706
[14] Liem N Q, Quang V X, Thanh D X 1996 *Acta Physica Polonica A* **89** 717

Acknowledgements

This work was partially supported by the Basic Research Programme in Natural Science (MOST Vietnam). U.T.D.T. benefits from the Evariste Galois fellowship of the French Ministry of Foreign Affairs.

Metamaterials: Detailed parametric studies on cut-wire pair and combined structure

Vu Dinh Lam[1,2], Nguyen Thanh Tung[1], Le Van Hong[2] and Young Pak Lee[1]

[1] Quantum Photonic Science Research Center and Department of Physics, Hanyang University, Seoul 133-791, Korea

[2] Institute of Materials Science, Vietnam Academy of Science and Technology, Vietnam

E-mail: lamvd@hanyang.ac.kr

Abstract. We studied the electromagnetic behavior of cut-wire pair structure, which is considered as a magnetic metamaterial for designing the left-handed material. It was found that the lattice constant of cut-wire pair structure provides subtle effects not only on the electric resonance but also on the magnetic one. In addition, we also investigated on how this affects the left-handed behavior of combined structures of cut-wire pair and continuous wire. These structures were designed, fabricated, and measured in the microwave frequency regime.

Keywords: Metallic nanoparticles, gold nanoparticles, chemical reduction method, photochemical method.

1. Introduction

Materials with simultaneously negative permittivity and negative permeability have been first theoretically studied in 1968 by Veselago [1]. More than 30 years have elapsed until the first left-handed material (LHM) was conceived and demonstrated experimentally by Smith et al., [2] which was a combination of split-ring resonators (SRRs) and continuous wires. Recently, such materials were investigated both experimentally and theoretically more and more because of the potential roles in physical phenomena, novel applications such as superlens [3], filters [4], resonator [5], invisibility cloak [6, 7], and biosensor [8]. In general, the LHM is an artificially composite structure of electric and magnetic components that have effective properties. To apply the uniform effective medium theory in determining the effective permittivity ε and permeability μ for these structures, the size of the unit cell should be much smaller than the wavelength. Hence, nano-fabrication technologies are required to obtain the LHM working at optical frequencies. Therefore, the development of geometry and fabrication techniques is still an area of significant effort. The design and the construction of magnetic and electric components play a central role for the electromagnetic response of LHMs.

It is well-known that, to obtain the left-handed (LH) behavior, these materials should have $\mu< 0$ and $\varepsilon< 0$simultaneously over a certain frequency range. A negative permittivity medium can be easily obtained by a periodic array of wires, however, obtaining $\mu< 0$is still a major challenge for researchers, since it occurs only in a narrow frequency band. Interestingly, Koschny *et al.* [9] and Zhou *et al.* [10] showed that the magnetic components such as SRR and cut-wire pair provide not only a magnetic resonance but also an electric resonance. Thus, the electric resonance of the magnetic component is added to the electric response of the continuous wire, resulting in a new effective plasma frequency of combined structure, which is much lower than that of continuous wire alone [11, 12]. To obtain $\mu< 0$

© 2009 IOP Publishing Ltd

and ε< 0simultaneously over a certain frequency range, one of the most important thing is that the effective plasma frequency of combined structure should be higher than the magnetic resonance frequency of the magnetic component [12, 13]. Therefore, when the magnetic component is combined with the continuous wire, the electric resonance of the magnetic component plays an important role in determining the effective plasma frequency of the combined structure.

For the potential applications of LHMs, it would be desirable to design the LHMs that exhibit a controlled active, dynamical [14, 15], and/or tunable response [16-19]. One of the main characteristics of LHMs is the existence of both LH and right-handed (RH) transmission bands [13, 20, 21]. Two pass bands are normally separated by a certain stop band, which strongly depends on the electromagnetic behavior of the combined structure. Thus, to exploit the potential applications of LHMs, the most important tasks is to separate the LH and the RH transmission bands. Therefore, one has to avoid the crossing region where the LH behavior is destroyed or overlaps with the RH one.

In this research, we studied experimentally the electromagnetic response of cut-wire pair structure, which could be utilized as the magnetic metamaterial for constructing the LHMs. The results were compared with the previous theoretical studies. In addition, we also investigated how this affects the LH behavior of combined structures of cut-wire pair and continuous wire.

2. Experiment

The cut-wire pair and the combined structures were fabricated using the conventional printed circuit board process with copper patterns (0.036 mm thick) on both sides of a dielectric printed circuit board (0.4 mm thick) with a dielectric constant of 4.8. Further detail on the fabrication of the cut-wire pair and the combined structures are given in reference [21]. The geometrical parameters of cut-wire pair are defined in figure 1a. The length and the width of cut-wire pair are 5.5 and 1.0 mm, respectively. To obtain the LH behavior, the cut-wire pairs were combined with the continuous wires. The geometrical parameters of combined structure are depicted in figure 1b. We performed the transmission measurements in free space by using a Hewlett-Packard E8362B network analyzer connected to the microwave standard-gain horn antennas.

3. Results and discussion

It is known that the cut-wire pair exhibits both electric and magnetic resonances as the SRR and can be explained by LC model [10]. If the cut-wire pair is electrically shorted at the end of the cut-wire pair through the dielectric spacer, it eliminates the capacitor of the cut-wire pair, and therefore the magnetic resonance will be destroyed. By this way we can easily identify the magnetic and electric resonance frequencies of the cut-wire pair.

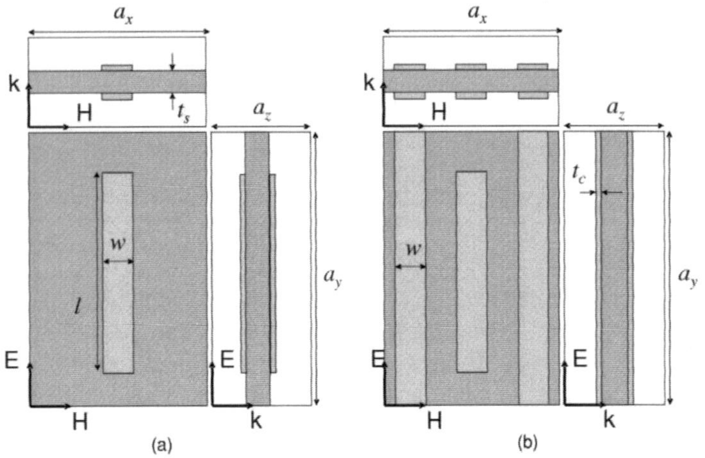

Figure 1. a) Geometry of cut-wire pair with a length l of 5.5 mm and a width w of 1.0 mm;
b) Unit-cell of combined structure.

Figure 2 presents the measured transmission spectra of lattices of shorted cut-wire pair and cut-wire pair. As can be seen, there are two band gaps in the transmission spectra of the cut-wire pair medium (dotted line): the first band gap between 13.8 and 14.3 GHz and the second band gap starting at 17.5 GHz. Whereas for the case of shorted cut-wire pair medium there exists only one band gap starting at 17.5 GHz and the first band gap is quite destroyed (solid line). This result confirms that the first band gap between 13.8 and 14.3 GHz in the transmission spectrum of the cut-wire pair medium is due to the magnetic resonance, providing a negative magnetic permeability, and the second band gap beginning at 17.5 GHz is due to the electric resonance, providing a negative electric permittivity.

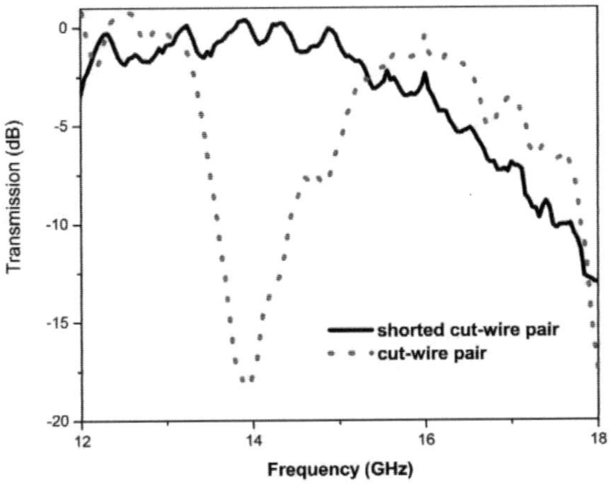

Figure 2. Measured transmission spectra of lattices of cut-wire pairs and shorted cut-wire pairs.

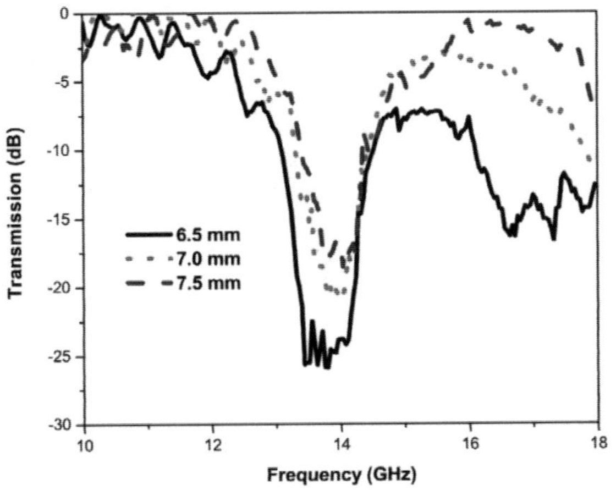

Figure 3. Transmission spectra of several cut-wire pair structures, where the lattice constants a_x and a_z are kept constant to be 3.5 and 1.4 mm, respectively, while in the E direction the lattice constant is varied from a_y = 6.5 to 7.5 mm.

Figure 3 presents the transmission spectra of several cut-wire pair media, where the periodicities of the cut-wire pair in the H-k plane is kept constant to be a_x = 3.5 mm and a_z = 1.4 mm, while a_y along the E direction is varied from 6.5 mm to 7.5 mm. As can be seen, the magnetic resonance band comes to be narrower when a_y is increased. Interestingly, the low-frequency edge of electric resonance band is strongly shifted to a lower frequency as a_y is decreased, but the high-frequency edge of magnetic resonance band is nearly unchanged. Therefore, the separation between two resonances is reduced as at a shorter a_y. In the case of a_y = 6.5 mm, this separation is narrowed (solid line). If a_y keeps being decreased until a certain one, two resonances might be overlapped, and then the cut-wire pair structures exhibits the LH behavior. However, it is not too easy to overlap them, since the distance between cut-wire pairs along E direction is required to be much smaller than the length of cut-wire. This observation is in good agreement with the previous theoretical study and is explained by the simple LC equivalent-circuit proposed by Zhou *et al.* [10]. This model reports that, besides the capacitances formed by the upper and the lower halves of the cut-wire pair, there exists another capacitance established by neighboring pairs along the E direction, which strongly depends on their distance, in other word, lattice constant a_y. This capacitance is added to that of cut-wire pair, hence, affecting the electric-resonance frequency. This influences the LH behavior of combined structure when the cut-wire pairs are combined with the continuous wires, as indicated in figure 4.

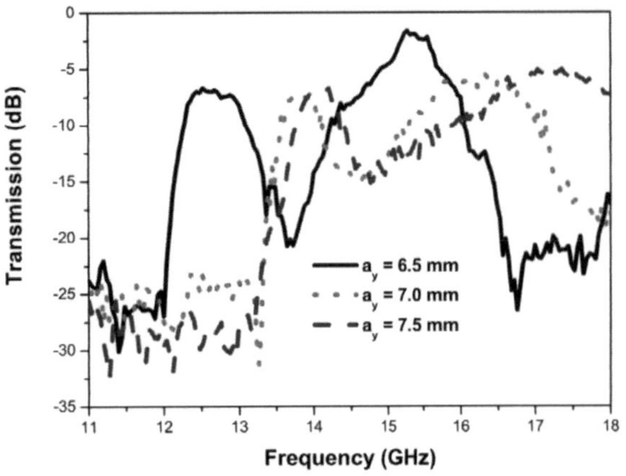

Figure 4. Transmission spectra of several combined structures, where a_x and a_z are 6.5 and 1.4 mm, respectively, while a_y is varied from 6.5 to 7.5 mm.

Figure 4 shows the measured transmission spectra of various combined structures, in which a_x and a_z are kept constant to be 6.5 and 1.4 mm, respectively, while a_y varies from 6.5 to 7.5 mm. As seen in figure 4, two pass bands exist in the transmission spectra of the combined structure with a_y = 7.0 mm, which are separated by a shallow dip (dotted line). The first pass band between 13.4 and 14.5 GHz exhibits the LH character, while the second pass band starting from 17 GHz is for the RH behavior [21]. When a_y is increased to be 7.5 mm, two pass bands are shifted to higher frequencies, but the first pass band is less moved. This is quite reasonable because both the magnetic and the electric resonance frequency bands of cut-wire pair are moved to higher frequencies according to a_y as shown in figure 3.

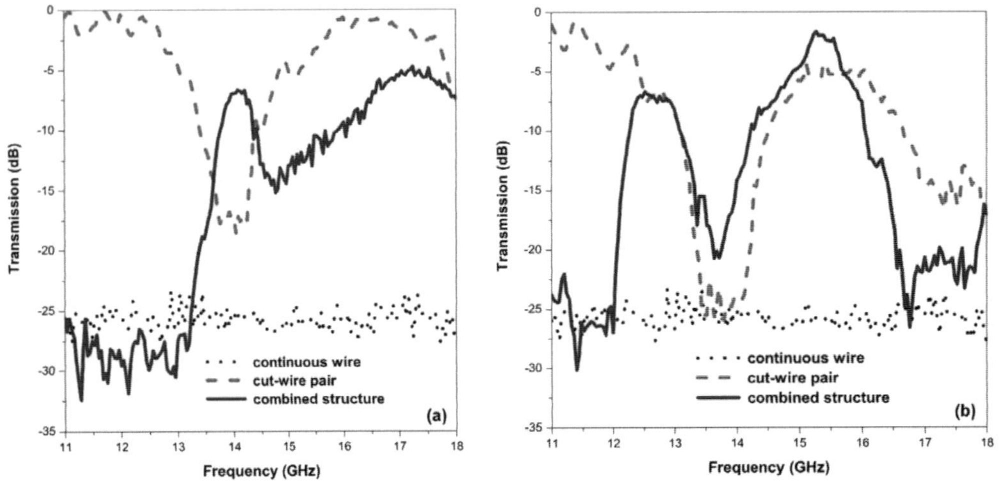

Figure 5. Measured transmission spectra of the continuous wire, the cut-wire pair and the combined structures, where a_x and a_z are 6.5 and 1.4 mm, respectively, while a_y is 6.5 mm for (a) and 7.5 mm for (b).

Interestingly, the first pass band still coincides with the stop band of the cut-wire pair structure as shown in figure 5a, indicating the LH behavior. In the case of a_y = 6.5 mm, two pass bands are strongly shifted to lower frequencies (solid line in figure 4). However, the first pass band of combined structure does not coincide with the stop band of the cut-wire structure as shown in figure 5b, hence the first pass band is not the LH transmission band but the RH one.

The observed result can be explained by reducing the electric-resonance frequency of cut-wire pair, resulting in a decrease in the effective plasma frequency of combined structure as a_y is decreased.

Journal of Physics: Conference Series **187** (2009) 012015 doi:10.1088/1742-6596/187/1/012015

Therefore, the plasma frequency is lower than the magnetic-resonance frequency of the cut-wire pair structure, and then the LH transmission is completely destroyed.

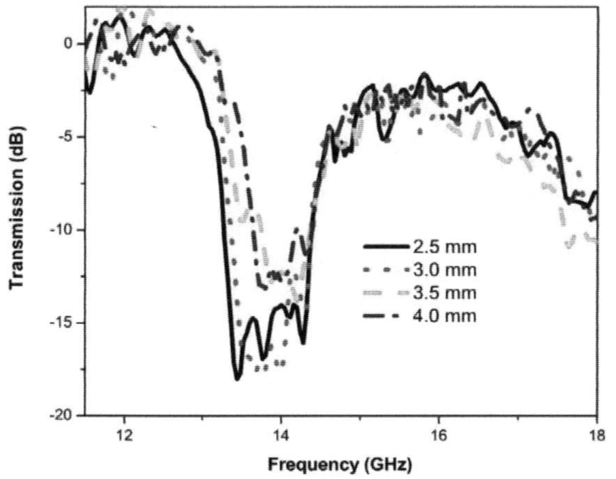

Figure 6. Transmission spectra of various cut-wire pair structures where a_y and a_z are kept constant to be 7.0 and 1.4 mm, respectively, while the periodicity of cut-wire pairs along the H direction is varied from a_x 2.5 to 4.0 mm.

To study the influence of the distance between cut-wire pair along the H direction, in other word, the lattice constant a_x on the resonance frequencies, the periodicity a_x is varied from 2.5 to 4.0 mm while, in the (E, k) plane, the periodicities of cut-wire pair is kept constant to be $a_y = 7.0$ mm and $a_z = 1.4$ mm, respectively. The transmission spectra of these media were measured as in figure 6. Clearly, the high edge frequency of magnetic resonance band is nearly unchanged as increasing the distance between cut-wire pairs, while the low-frequency edge is slightly increased. This means that the magnetic-resonance frequency band comes to be narrower at the bigger distance between cut-wire pairs. Another considerable change here is that the magnetic-resonance frequency band becomes stronger with decreasing the distance between cut-wire pairs. These effects can be explained by the reduced coupling between cut-wire pairs as increasing the distance between cut-wire pairs [22]. Interestingly, the electric-resonance frequency is nearly unchanged with changing the distance between cut-wire pairs, hence this does not influence on the plasma frequency of combined structure by changing a_x.

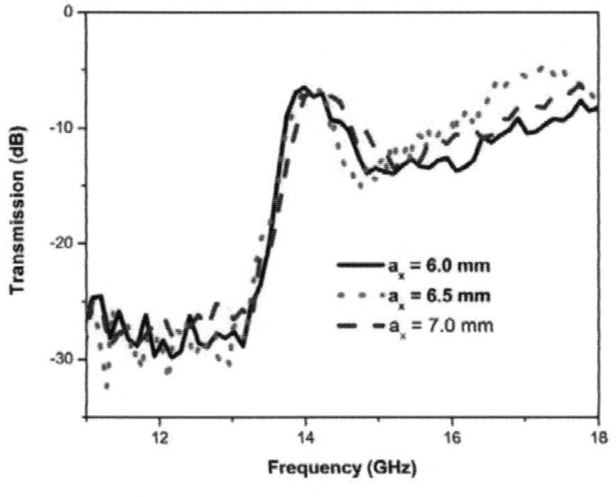

Figure 7. Transmission spectra of several combined structure media. The periodic along the H direction is varied from $a_x = 6.0$ to 7.0 mm, but those in the (k, E) plane is kept constant to be $a_z = 1.4$ mm and $a_y = 7.0$ mm.

Figure 7 gives the measured transmission spectra of several combined structures, where a_y and a_z are 7.0 and 1.4 mm, respectively, while a_x varies from 6.0 to 7.0 mm. Clearly, the first pass band between 13.8 and 14.5 GHz, which exhibits the LH behavior remains unchanged according to a_x. In addition, the separation between the LH and RH pass band is also maintained. This means that the effective plasma frequency of combined structure is not affected by the a_x. It is consistent with the phenomenon observed in figure 6.

The influence of distance between cut-wire pair layers, in other words, the a_z lattice constant along the k direction on the resonance frequencies was studied both theoretically and experimentally detail in reference [23]. It was found that a_z provides a subtle effects not only on the electric resonance but also on the magnetic one. Therefore, this also affects the LH behavior of combined structure.

4. Conclusions

We studied the electromagnetic behavior of cut-wire pair structure in the microwave frequency regime. It was found that the lattice constant of cut-wire pair structure provides subtle effects not only on the electric resonance but also on the magnetic one. In addition, we also investigated on how this affects the LH behavior of combined structures of cut-wire pair and continuous wire. This is of fundamental importance in understanding the electromagnetic response of combined structure of cut-wire pair and continuous wire in the low frequency region, and in designing the LHMs working at the high frequencies.

References

[1] Veselago V G 1968 *Sov. Phys. Usp.* **10** 509
[2] Smith D, Padilla W, Vier D, Nemat-Nesser S and Chultz S 2000 *Phys. Rev. Lett.* **84** 4184
[3] Pendry J B 2000 *Phys. Rev. Lett.* **85** 3966
[4] Lu J and He S 2003 *Microwave Opt. Technol. Lett.* **37** 292
[5] Engheta 2002 *N IEEE Antennas Wireless Propag. Lett.* **1** 10
[6] Pendry J B, Schurig D and Smith D R 2006 *Science* **312** 1780
[7] Schurig D, Mock J J, Justice B J, Cummer S A, Pendry J B, Starr A F and Smith D R 2006 *Science* **314** 977
[8] Lee H J and Yook J G 2008 *Appl. Phys. Lett.* **92** 254103
[9] Koschny T, Kafesaki M, Economou E N and Soukoulis C M 2004 *Phys. Rev. Lett.* **93** 107402
[10] Zhou J, Economon E N, Koschny T and Soukoulis C M 2006 *Opt. Lett.* **31** 3620
[11] Aydin K, Guven K, Kafesaki M, Zhang L, Soukoulis C M and Ozbay E 2004 *Opt. Lett.* **26** 2623
[12] Guven K, Caliskan M D and Ozbay E 2006 *Opt. Express* **14** 8685
[13] Katsarakis N, Koschny T, Kafesaki M, Economou N E, Ozbay E and Soukoulis C M 2004 *Phys. Rev. B* **70** 201101
[14] Lim S, Caloz C and Itoh T 2004 *IEEE Trans. Microw. Theory Tech.* **52** 2678
[15] Padilla W J, Taylor A J, Highstrete C, Lee M and Averitt R D 2005 *Phys. Rev. Lett.* **96** 107401
[16] Shadrivov I V, Morrison S K and Kivshar Y S 2006 *Opt. Express* **14** 9344
[17] Sheng Z and Varadan V V 2007 *J. Appl. Phys.* **101** 014909
[18] Driscoll T, Andreev G O, Basov D N, Palit S, Cho S Y, Jokerst N M and Smith D R 2007 *Appl. Phys. Lett.* **91** 062511
[19] Aydin K and Ozbay E 2007 *J. Appl. Phys.* **101** 024911
[20] Zhou J, Zhang L, Tuttle G, Koschny T and Soukoulis C M 2006 *Phys. Rev. B* 73 041101
[21] Lam V D, Kim J B, Lee S J and Lee Y P 2008 *J. Appl. Phys.* **103** 033107
[22] Gay-Balamz P and Martin O J F 2002 *J. Appl. Phys.* **92** 2929
[23] Lam V D, Kim J B, Tung N T, Lee S J, Lee Y P and Rhee J Y 2008 *Opt. Express* **16** 5934

Acknowledgements

This work was supported by the Ministry of Education, Science and Technology (MEST) through the Quantum Photonic Science Research Center, Korea.

Preparation, photoluminescence and time-resolved luminescence of $ZnS:Mn^{2+}$ nanophosphors

Nguyen Vu[1], Pham Thi Minh Chau[2], Dinh Xuan Loc[1], and Tran Kim Anh[1]

[1] Institute of Materials Science, Vietnam Academy of Science and Technology, 18 Hoang Quoc Viet Road, Cau Giay Distr., Hanoi, Vietnam
[2] Institute of Geography, Vietnam Academy of Science and Technology, 18 Hoang Quoc Viet Road, Cau Giay Distr., Hanoi, Vietnam

E-mail: nguyenvu@ims.vast.ac.vn

Abstract. Nanocrystals of $ZnS:Mn^{2+}$ were prepared by chemical synthesis. The samples were characterized by scanning electron microscopy, photoluminescence, and time-resolved luminescence measurements. The emission spectra of $ZnS:Mn^{2+}$ nanoparticles under 325-nm excitation consist of an intense broad band at 585 nm, ascribed to the $^4T_1$-$^6A_1$ transition of the Mn^{2+} ions. No emission in the blue spectral region (about 425 nm) from the ZnS host is observed. The dependence of PL intensity on Mn^{2+} concentration has been investigated.

Keywords: Photoluminescence, time-resolved, nanophosphor, ZnS:Mn.

1. Introduction

In the past 20 years, nanostructure materials have been widely studied [1-5]. Application of nanometer phosphors for display has gathered great interest, since Bhargava *et al.* reported the optical properties of $ZnS:Mn^{2+}$ nanocrystals with high quantum efficiency in 1994 [2]. ZnS is a typical II-VI semiconductor. It has been commercially used as a phosphor as well as in thin film electroluminescent devices, especially if doped with divalent manganese ions [6]. Recent years, Mn-doped ZnS nanocrystals have attracted much attention in many groups for preparation, physical properties and potential applications [7-11]. $ZnS:Mn^{2+}$ nanocrystals have possibility to enhance reliability of organic electroluminescent devices due to the stability of $ZnS:Mn^{2+}$ which has been established already in commercial use as an inorganic electroluminescent [7, 8]. Mohagheghpour [9] reported on synthesis and characterization of Mn-doped ZnS nanoparticles which could be used with avidin bioconjugated antibodies. Luminescence kinetics and time-resolved luminescence of $ZnS:Mn^{2+}$ nanocrystals were also discussed [8, 10]. In time-resolved luminescence measurements, the photo luminescence intensity corresponding to each spectral component which is detected at a certain moment delayed from the start of the excitation laser pulse provides a real intensity in the dynamic regime.

Red and green nanophosphors were successfully synthesized in our group [12, 13]. Many different techniques such as, co-precipitation, sol–gel, hydrothermal, sonochemical… have been used in preparation of $ZnS:Mn^{2+}$ nanoparticles. In the present work, we describe our procedure concerning the synthesis of $ZnS:Mn^{2+}$ nanoparticles (orange nanophosphors) using the wet chemical method. Structural phase and micro image of nanoparticles are presented. Photoluminescence (PL) and time-resolved luminescence properties of $ZnS:Mn^{2+}$ nanoparticles are reported.

© 2009 IOP Publishing Ltd

Journal of Physics: Conference Series **187** (2009) 012016 doi:10.1088/1742-6596/187/1/012016

2. Experimental

2.1. *Synthesis of ZnS:Mn^{2+} nanocrystals*

The ZnS:Mn^{2+} nanocrystals were synthesized using the wet chemical method, involving a propagation step of ZnS which was terminated by the adsorption of thiolate on nanoparticle surface. The crystal size is controlled by the relative rates of the propagation and termination reaction of initiating (sulphide) and terminating (thiolate) species.

Five stock solutions were prepared: (I) 20 vol.% 0.1 M succinic acid ($C_4H_60_4$) in 80 vol.% methanol (solvent 1); (II) 0.1 M S^{2-} from Na_2S in methanol and de-ionized water; (III) 0.1 M Zn^{2+}, and 0.1 M Mn^{2+} were dissolved by Zinc acetate with stoichiometric calculated concentration and Manganese (II) chloride with concentrations 0.5, 1.5, 2 mol.% in solvent 1 (solution A); (IV) 0.1 M thiosemicarbasite (CH_5N_3S) in methanol; (V) Mixture of the 50 ml (II) with 50 ml (IV) (solution B). To form nanoparticles, the solution A was added drop-wise during continuous stirring. A white cloudy solution was stirred for one hour then washed by methanol then centrifuged (5000 rpm, 10 min). The sedimentation was dried in vacuum oven at 105°C for one hour, and then calcimined at 500°C in vacuum for 30 min.

2.2. *Characterizations*

The ZnS:Mn^{2+} powder was analyzed by the X-ray diffractometer (XRD, Siemens, D-5000) with Cu Kα radiation. The morphology of ZnS:Mn^{2+} was observed by using a transmission electron microscopy (TEM, JEM-1010) and field emission scanning electron microscopy (FESEM, Hitachi, S-4800). The photoluminescence measurements were performed at room temperature by using a cw He-Cd laser (325 nm) as the excitation source, a Spectrapro 2300i monochromator (Acton) as the dispersive unit, and a Pixis 256 CCD (Acton) as the detector. Time-resolved emission spectra were recorded by using a HRD1 monochromator (Jobin-Yvon) as the dispersive unit, a pulsed nitrogen laser (337 nm) and a fast photomultiplier (Hamamatsu model H733) as the excitation source and the detector, respectively. Averaging the multi-pulses at each spectral point using a 1.5 GHz digital oscilloscope (LeCroy 9362) strongly improved the signal-to-noise ratio.

3. Results and discussion

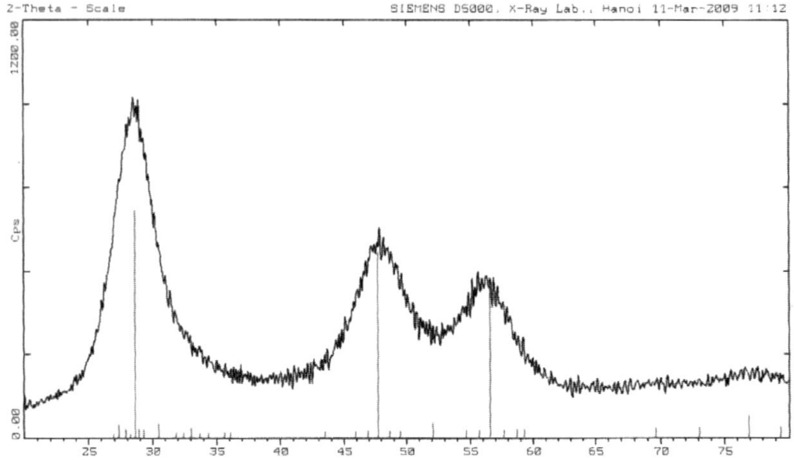

Figure 1. XRD pattern of ZnS:Mn^{2+} nanoparticles (main lines) and ZnS from JCPDS card (vertical lines).

The powder XRD pattern of typical ZnS:Mn^{2+} nanoparticles is given in figure 1. It is noted that all of the diffraction peaks could be well indexed to the zinc blende crystal structure, having prominent (111), (220), and (311) *d*-spacings. In this Mn-doped sample no additional phase was observed, indicating that manganese has been successfully doped in the ZnS crystalline lattice. However, the width of the diffraction lines is broadened because of the small size of the crystallites. The full width at half maximum (FWHM) of the diffraction peak at $2\theta = 8.487^{o}$ is 3.6218^{o}. The XRD patterns also agree with that of the ZnS:Mn^{2+} nanoparticles reported by Mohagheghpour [9].

Figure 2 shows typical TEM and FE-SEM images of the as prepared ZnS:Mn^{2+} nanoparticles. From the image, one can see that the mean size of the Mn-doped ZnS is about 20 - 30 nm. These images also reveal the possibility of powders in somewhat agglomerated forms.

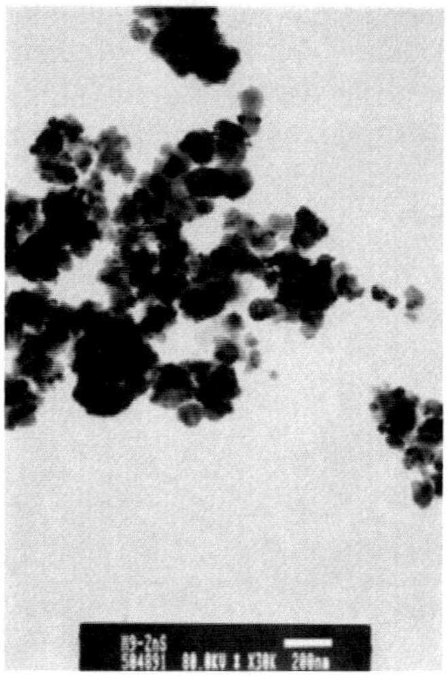

Figure 2. TEM (left) and FESEM (right) images of the as prepared ZnS:Mn^{2+} nanoparticles.

Figure 3 shows the PL spectra of ZnS:Mn^{2+} (0.5, 1, and 2 mol.%) nanoparticles upon 325-nm excitation. The spectra consist of a very intense emission Gaussian broad band with a maximum at 585 nm, ascribed to the $^4T_1$ - $^6A_1$ transition of Mn^{2+} [2]. This result is consistent with the emission of Mn^{2+} in ZnS nanoparticles [2-4]. But no emission in the blue spectral region (~ 425 nm), which ascribed to radiative recombination at defects in the ZnS nanoparticles [10], is observed. It was reported that in Mn^{2+}-activated ZnS nanocrystals in which the Mn^{2+} ions are distributed outside the ZnS nanocrystals, the PL is totally different from that of ZnS:Mn^{2+} nanocrystals in which the Mn^{2+} is incorporated within the nanocrystals. When the Mn^{2+} is incorporated within the nanocrystals, both the 435-nm blue emission of ZnS and the orange Mn^{2+} emission at 590 nm are observed [3]. However, in the Mn^{2+}-activated ZnS nanocrystals in which the Mn^{2+} ions are distributed outside the ZnS nanocrystals, no orange emission at 590 nm is observed; a new peak at ~350 nm appears and the blue 435-nm emission of ZnS is quenched considerably and shifted to 390 nm. Comparison of these observations with our own suggests that in our material the Mn^{2+} ions are incorporated within the ZnS nanoparticles.

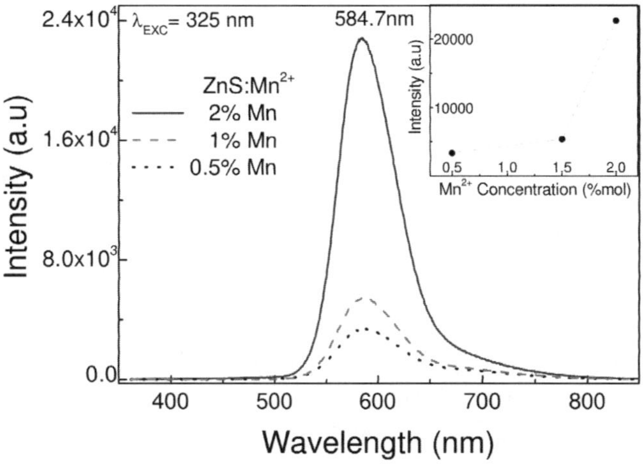

Figure 3. PL spectrum of ZnS:Mn^{2+} (2 mol.%) nanoparticles upon 325-nm excitation.
Inset: PL intensity (at 585 nm) of ZnS:Mn^{2+} nanoparticles with different Mn^{2+} concentrations.

In order to study the Mn-concentration dependence of the luminescence, we prepared a series samples with different concentration ratios of Mn^{2+} (0.5, 1, and 2 mol.%) to Zn^{2+} in the precursor solution under the same conditions. Figure 3 (inset) presents the relative PL intensity (at 585 nm) of $ZnS:Mn^{2+}$ nanoparticles as a function of the manganese concentration. This shows the maximum intensity at a 2 mol.% Mn concentration. Some researchers [3, 13] also observed the highest intensity peak for 2 mol.% doping.

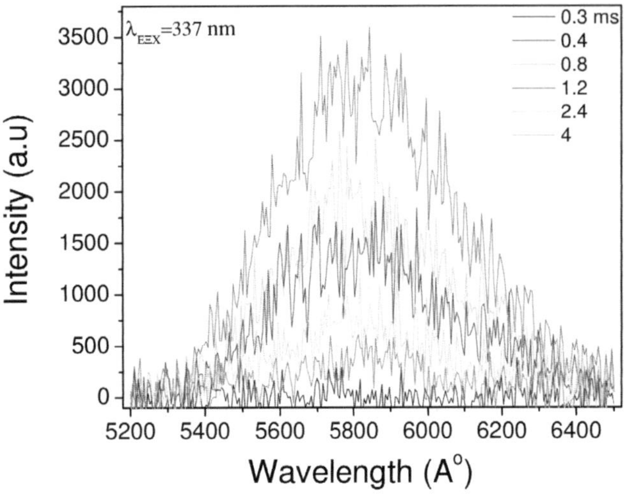

Figure 4. Time-resolved emission spectra of the $ZnS:Mn^{2+}$ (2 mol.%) nanoparticles, upon 337-nm excitation, with variable time delays.

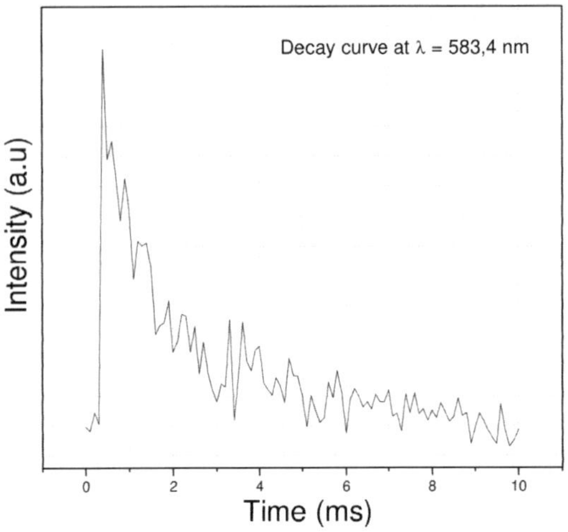

Figure 5. PL decay curve of $ZnS:Mn^{2+}$ (2 mol.%) nanoparticles at 583.4 nm, upon 337-nm excitation. This curve was recorded at room temperature.

Figure 4 presents the room temperature time-resolved emission spectra of the $ZnS:Mn^{2+}$ (2%) nanoparticles upon 337-nm excitation and with variable time delays. The maximum intensity correspond to the $^4T_1$ - $^6A_1$ transition of Mn^{2+} is observed in a time delays of 0.4 ms. As the time delays increases from 0.4 ms the emission intensity correspond to the $^4T_1$ - $^6A_1$ transition of Mn^{2+} decreases. This is due to the fact that the decay time of the $^4T_1$ - $^6A_1$ transition of Mn^{2+} ion is in the ms range. The decay curve at 583.4 nm of the $ZnS:Mn^{2+}$ (2 mol.%) nanoparticles is shown in figure 5. The obtained decay curve can be fitted by a exponential function as $I = I_0 exp(-t/\tau)$ (where I_0 is the initial intensity at $t = 0$, τ is the 1/e lifetime). For the 2 mol.% Mn^{2+}-doped sample, the 1/e decay time is about 1.2 ms which is attributed to $^4T_1$ - $^6A_1$ transition of Mn^{2+} in $ZnS:Mn^{2+}$.

Journal of Physics: Conference Series **187** (2009) 012016 doi:10.1088/1742-6596/187/1/012016

4. Conclusions

The Mn^{2+} -doped ZnS nanoparticles were prepared by chemical synthesis. The mean size of the Mn^{2+}-doped ZnS is about 20 - 30 nm. The emission spectra of $ZnS:Mn^{2+}$ nanoparticles upon 325-nm excitation consists of a very intense emission broad band at 585 nm, ascribed to the $^4T_1$ - $^6A_1$ transition of Mn^{2+} ions. The maximum PL intensity of $ZnS:Mn^{2+}$ nanoparticles was observed in the case of 2 mol.% Mn-doped sample. The room temperature time-resolved emission spectra of the $ZnS:Mn^{2+}$ nanoparticles upon 337-nm excitation and with variable time delays indicates that the 1/e decay time is about 1.2 ms for the 2 mol.% Mn^{2+}-doped sample.

References

[1] Nalwa H S 1999 *Handbook of Nanostructured Materials and Nanotechnology* (New York: Academic Press)

[2] Bhargava R N, Gallagher D, Hong X, and Nurmikko A 1994 *Phys. Rev. Lett.* **72** 416

[3] Sooklal K, Cullum B S, Angel S M, Murphy C J 1996 *J. Phys. Chem.* **100** 4551

[4] Chen W, Li G, Malm J O, Huang Y, Wallenberg R, Hanb H, Wang Z, Bovin J O 2000 *J. Lumin.* **91** 139

[5] Anh T K, Benalloul P, Barthou C, Giang L T K, Vu N, and Minh L Q 2007 *J. of Nanomaterials* **Vol. 2007** Article ID 48247

[6] Shionoya S, Yen W M 1999 *Phosphor Handbook* (CRC Boston-London-New York-Wasington, D.C, CRC. Press LLC)

[7] Hwang J M, Oh M O, Kim I, Lee J K, Ha C S 2005 *Curr. Appl. Phys.* **5** 31

[8] Mahlik S, Zalewska M, Grinberg M, Klonkowski A M, Godlewski M 2008 *J. Lumin.* **128** 921

[9] Mohagheghpour E, Rabiee M, Moztarzadeh F, Tahriri M, Jafarbeglou M, Bizari D, Eslami H 2009 *Mater. Sci. and Eng.: C* In Press

[10] Donega C de Mello, Bol A A, Meijerink A 2002 *J. Lumin.* **96** 87

[11] Yang P and Bredol M 2008 *Res. Lett. in Mater. Sci.* **Vol. 2008** Article ID 506065

[12] Vu N, Anh T K, Yi G C, and Strek W 2007 *J. Lumin.* **122-123** 776

[13] Vu N, Quan N H, Anh T K, Khoi N T and Liem N Q 2008 *J. Korean Phys. Soc.* **52** 1514

[14] Sohling U, Jung G, Saenger D U, Lu S, Kutsch B and Menning M 1998 *J. Sol-Gel Sci. Technol.* **13** 685

Acknowledgments

This work is supported in part by Vietnam Academy of Science and Technology (VAST). A part of the work was done in the National Key Laboratory for Electronic Materials and Devices, Institute of Materials Science, VAST. We would like to thank MSc. Tran Quang Huy, MSc. Do Hung Manh, and MSc. Tran Dang Thanh for their helps on TEM, SEM, and XRD measurements, respectively.

Synthesis of $SrAl_2O_4$: Eu^{2+}, Dy^{3+} phosphorescence nanosized powder by combustion method and its optical properties

Nguyen Manh Son[1], **Le Thi Thao Vien**[2], **Le Van Khoa Bao**[3] **and Nguyen Ngoc Trac**[1]

[1] Department of Physics, College of Sciences, Hue University, Hue, Vietnam
[2] Department of Physics, Qui Nhon University, Qui Nhon, Vietnam
[3] Department of Natural Sciences, Duy Tan University, Da Nang, Vietnam

E-mail: manhson03@yahoo.com

Abstract. Eu^{2+}, Dy^{3+} codoped strontium aluminate ($SrAl_2O_4$) nanosized phosphorescent powder with high brightness and long afterglow were prepared by urea-nitrate solution combustion method at 540^{0}C for 5 minutes. The average particle size of the powders was about 80 nm. The photoluminescent and thermoluminescent properties have been studied. The broad band photoluminescence of $SrAl_2O_4$: Eu^{2+}, Dy^{3+} were observed with maximun wavelength λ_{max} = 516 nm due to transitions from the $4f^65d^1$ to $4f^7$ configuration of the Eu^{2+} ions. The main peak of the emission spectra shifted to the short wavelength compared with phosphorescence obtained by the solid state reaction method. The decay time of the afterglow for nanosized phosphorescence was shorter than that obtained by the solid state reaction method.

Keyword: phosphorescent, photoluminescent, thermoluminescent, nanosized.

1. Introduction

Green emiting $SrAl_2O_4$ co-doped with Eu^{2+}, Dy^{3+} ions has been considered one of the best and long lasting phosphorscent materials. Until recent decade, strontium aluminate phosphors doped with europium and dysprosium ions have attracted much attention since they show excellent properties [2]. Compared with classical sulfide phosphorescent phosphors, aluminates have several valuable properties [3-6]: high radiation intensity, color purity, longer afterglow, chemically stabilization, safe and no radioactivity, etc. Together with the development of scientific technologies on materials, several chemical synthesis techniques, such as: co-precipitation, sol-gel and combustion synthesis methods have been applied to prepare $SrAl_2O_4$: Eu^{2+}, Dy^{3+}. In this paper, the phosphor synthesized by combustion method using urea as reducer and fuel, nitrate precursors as oxidizer at 540°C for 5 minutes. The combustion process to prepare this powder, however, is very facile and only takes few minutes which have been extensively applied to the preparation of various nanosized oxide materials. The $SrAl_2O_4$: Eu^{2+}, Dy^{3+} phosphor resulted from combustion method has optical properties like this phosphor resulted from the other method but the sintering temperature of the sample is much lower than that prepared by solid state reaction or sol-gel method.

2. Experimental

The starting materials included: $Al(NO_3)_3.9H_2O$, $Sr(NO_3)_2$, Eu_2O_3, Dy_2O_3, B_2O_3, $CO(NH_2)_2$ and HNO_3 (1.4 g/ml). Small quantities of B_2O_3 were used as a flux; amounts of urea were added as reducer and

© 2009 IOP Publishing Ltd

fuel, metal nitrates were used as oxidizer. The choice of oxidizer and fuel in the stoichiometric ratios are dictated by the propellant chemistry principles.

The procedure used to prepare $Sr_{0.97}Al_2O_4$: $Eu^{2+}_{0.01}$, $Dy^{2+}_{0.02}$ had the following stages. Firstly, $Al(NO_3)_3.9H_2O$, $Sr(NO_3)_2$, B_2O_3 and $CO(NH_2)_2$ were dissolved enough into deionised water to obtain transparent solution. Eu_2O_3 and Dy_2O_3 were dissolved into concentrated nitric acid to form nitrate solution, then mixed the two solutions together and stirred at 70°C for 4 h. The process of heat treatment was studied by DTA which was presented in figure 1. After that, the precursor solution was introduced into a Lenton furnace maintained at 540°C. Initially, the solution boiled and underwent dehydration, followed by decomposition with escaping large amounts of gases (oxides of carbon, nitrogen and ammonia) and then spontaneous ignition occurred and underwent smouldering combustion with enormous swelling, producing white foamy and voluminous ash, the whole process was over within less than 5 minutes, after the product was cooled to room temperature, milled the ash slightly. The $SrAl_2O_4$: Eu^{2+}, Dy^{3+} was obtained.

Crystalline structure and phase composition of the sample were checked by X-ray diffractometer (Bruker D8, Germany) using Cu Kα radiation, $\lambda = 1.5406$ Å. The morphology and particle size of the prepared powders were observed by scanning electron microscopy (SEM). The excitation and emission spectra were measured at room temperature on the luminescent spectrophotometer. The decay time and the glow-curve of the sample were recorded to determine afterglow characteristics.

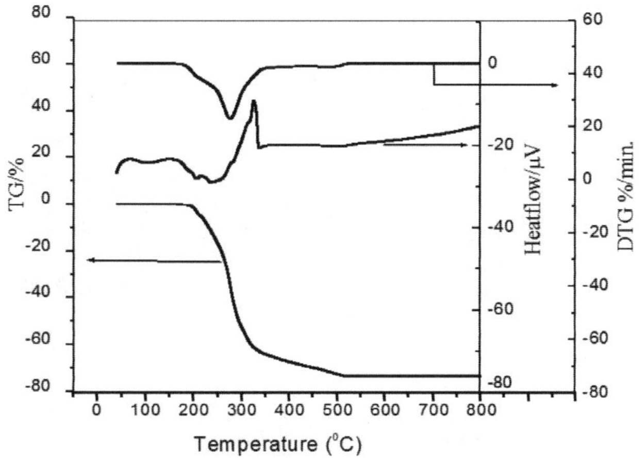

Figure 1. Diagram DTA of $SrAl_2O_4$: Eu^{2+}, Dy^{3+}.

3. Results and disscusions

3.1. Structure

A typical X-ray diffraction pattern of the resultant $SrAl_2O_4$: Eu^{2+}, Dy^{3+} powders are shown in figure 2. The pure monoclinic phase diffraction peaks of $SrAl_2O_4$ are predominant in the XRD pattern, which are similar with the sample resulted from the other method. $SrAl_2O_4$ powders are all low-temperature monoclinic phase (α-phase), lattice constants are $a = 8.442$ Å, $b = 8.822$ Å, $c = 5.160$ Å, $\beta = 93.415°$. It was found that the single phase $SrAl_2O_4$ was formed at 540^0C, which is lower than that required for the sol-gel method [8] and the solid-state reaction method [1]. In fact, the volatile combustible gases ignite and burn with a flame and thus provide conditions for formation of phosphor lattice with dopant. No other products or starting materials were observed, implying that the little amount of doped rare earth ions have almost no effect on the $SrAl_2O_4$ phase composition [10]. In fact, α-$SrAl_2O_4$ can melt with metal ions in a wide range; this gives enough space for doped rare earth ions to enter.

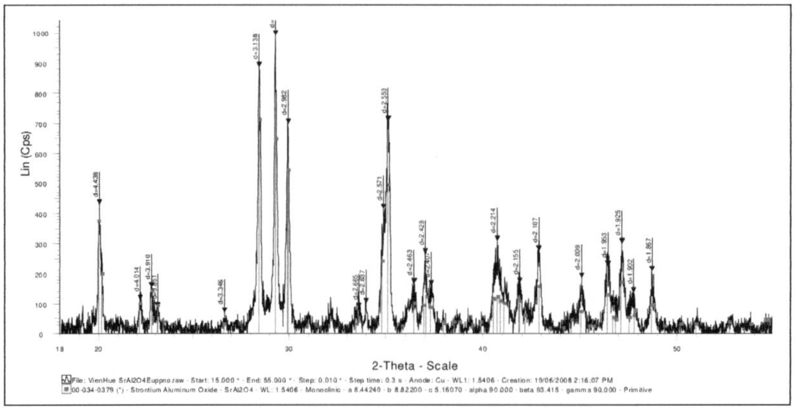

Figure 2. XRD diagram of $SrAl_2O_4$: Eu^{2+}, Dy^{3+} annealing at 540^{0}C.

3.2. Phosphor morphology

In order to determine the morphology and size of the sample, typical scanning electron microscopy (SEM) measurements were carried out. Figure 3 displays a typical SEM image of the prepared strontium aluminate crystal under combustion at 540°C for 5 minutes.

Figure 3. SEM image of $SrAl_2O_4$: Eu^{2+}, Dy^{3+}.

The phosphor powders are irregular spherical particles, the surfaces of the foams show a lot of cracks, voids and pores formed by the escaping gases during combustion reaction. It reveals that the mean size is about 80 nm. Compared with the phosphor prepared via the traditional ceramic method, the grain size is greatly smaller. In fact, the large amount of escaping gases dissipates heat and thereby prevents the material from sintering and thus provides conditions for formation of nanocrystalline phase. It is ascribed to that the solid state reaction requires a high calcinating temperature, which induces sintering and aggregation of particles, and it is an advantage for perfect crystal formation.

3.3. Excitation and emission spectra

The emission spectra at room temperature of $SrAl_2O_4$: Eu^{2+}, Dy^{3+} was shown in the figure 4 at excitation wavelength of 365 nm, it consists of a broad band and the emission peak was 516 nm which is attributed to the typical $4f^65d^1 \rightarrow 4f^7$ transition of Eu^{2+} ion [1]. The excitation spectra at the emission wavelength 516 nm are shown in the figure 5. It is observed that the excitation spectrum has a broad band with the main peak at 410 nm.

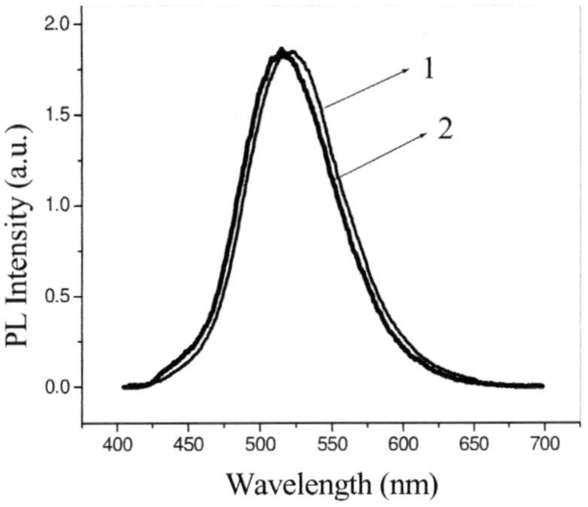

Figure 4. PL spectra of $SrAl_2O_4$: Eu^{2+}, Dy^{3+} prepared by the solid state method (1), the combustion method (2).

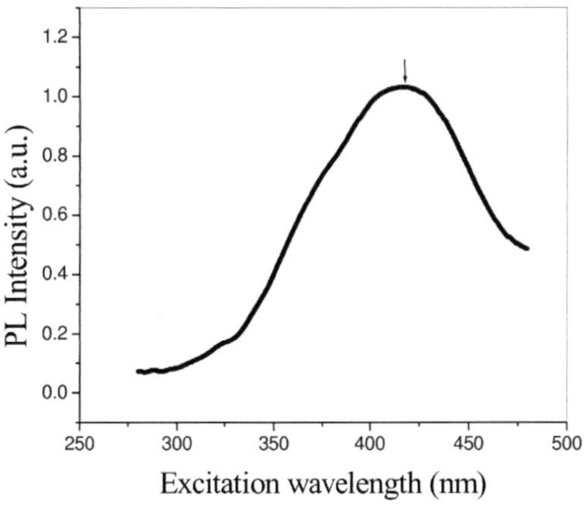

Figure 5. Excitation spectra of $SrAl_2O_4$: Eu^{2+}, Dy^{3+} at λ_{em} = 516 nm prepared by combustion method.

Compared with the phosphor resulted from solid state reaction method, the emission maximum of the phosphor that was prepared by the combustion method shifts to shorter wavelength (from 520 nm to 516 nm), the slight blue shift in the emission band may be attributed to the changes of the crystal field around Eu^{2+}. Since the excited $4f^65d^1$ configurations of Eu^{2+} ion is extremely sensitive to the change in the lattice environment, the 5d electron may couple strongly to the lattice [7]. Hence, the mixed states of 4f and 5d configuration will be split by the crystal field, which may lead to the blue shift of its emission peak.

3.4. Decay time and afterglow characteristics

Figure 6 shows the decay times of $SrAl_2O_4$: Eu^{2+}, Dy^{3+} phosphors that prepared by two techniques using the excitation wavelength 365 nm for 1 minute. After the irradiation light source was cut off, both of them showed a rapid decay at first and subsequently long lasting phosphorescence. The rate of decay is comparatively faster in case of solid-state reaction samples to those observed in the sample made by combustion method. The observed enhancement of afterglow intensity and lengthening of decay time of the sample by combustion method can be attributed to increase the number of defects due to rapid reaction combustion process and a higher concentration of Eu^{2+} generated in the process.

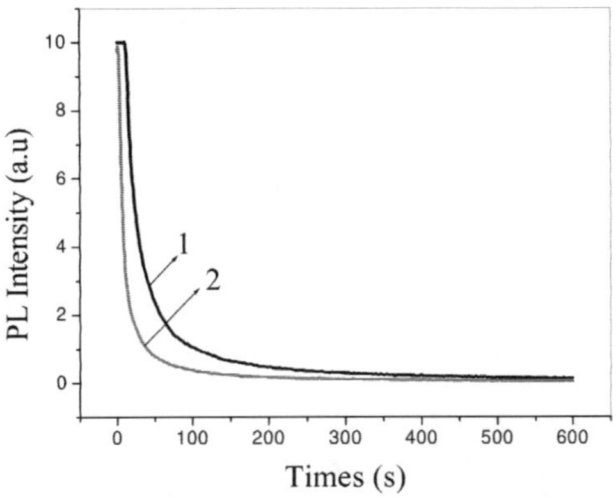

Figure 6. The decay time of $SrAl_2O_4$: Eu^{2+}, Dy^{3+} prepared by the solid-state method (1), the combustion method (2).

Using exponential fit, the decay curve of both phosphors were simulated by following exponential equation: $I = I_0+A_1\exp(-t/\tau_1)+A_2\exp(-t/\tau_2)$, where I presents the phosphorescence intensity, I_0, A_1 and

A_2 are the constants, t is the time, and τ_1, τ_2 are the decay constants of the phosphor. Significant and varied values of τ_1 and τ_2 clearly indicate the concentration of shallow and deeper traps, respectively. The decay (1) has $\tau_1 = 26$ s, $\tau_2 = 247$ s, and the decay (2) has $\tau_1 = 7$ s, $\tau_2 = 97$ s. It shows that the sample by combustion method has shorter decay times and so deeper trap density than those by solid state reaction method.

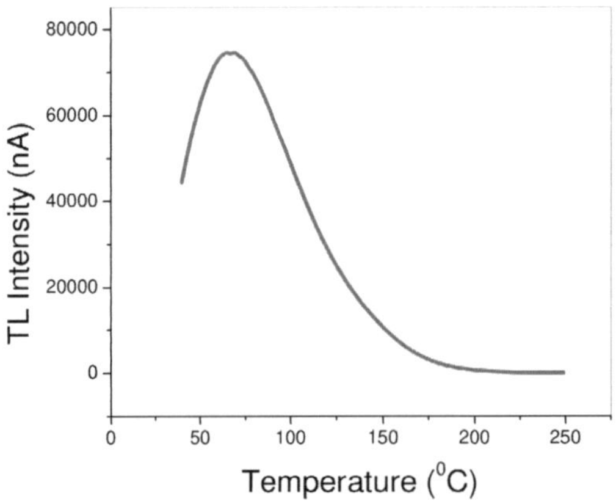

Figure 7. Glow curve of $SrAl_2O_4$: Eu^{2+}, Dy^{3+}.

The glow-curve of the phosphor was measured by the HARSHAW TLD-3500 reader. Figure 6 shows the glow-curve of $SrAl_2O_4$: Eu^{2+}, Dy^{3+} phosphor which was prepared by combustion method using D_2 lamp as ionization radiation, heating rate $\beta = 1^{o}C/s$.

As can be seen from the figure, the glow-curve of the sample shows a single peak with the maximum temperature at 86°C, and it is the second order kinetics thermoluminescent peak. Activation energy was analyzed by R. Chen mothod with E = 0.68 eV.

It is well known that in $SrAl_2O_4$: Eu^{2+}, Dy^{3+} phosphor, Eu^{2+} ions are the luminescent centers, the luminescence is considered to be due to the transition from 5d level to 4f level of Eu^{2+}, and holes were presented by Dy^{3+} ions. In the present system, it is the hole trapped-transported-detrapped process that results in the properties of long afterglow of $SrAl_2O_4$: Eu^{2+}, Dy^{3+} phosphor. Therefore, Dy^{3+} ions play an important role and work as hole traps, the trap depth was about 0.68 eV from the valence band. Upon the exposure to the ultraviolet lights, electrons of Eu^{2+} ion in the 4f level transfer to 5d level, and holes are produced. Some of free holes released thermally to the valence band and are captured last by the Dy^{3+} hole traps. When the excitation source was removed, the excitated Eu^{+} is relaxed to the stable state $(Eu^{+})^*$. The trapped holes are released thermally to the valence band and migrate to recombine with the excess electron in the metastable state $(Eu^{+})^*$ site and lead to the long afterglow. The long phosphorescence depends on the depth and the density of the hole trap.

4. Conclusion

The long afterglow phosphor $SrAl_2O_4$: Eu^{2+}, Dy^{3+} was successfully synthesized by combustion method. Analytical results implied that the phosphor powders have monoclinic phase with the average particles size of about 80 nm. The emission color is green and the blue shift of main emission peak of sample prepared by combustion method, compared with the solid state reaction method, occurred. The synthesized phosphor has bright initial phosphorescence intensity and the decay time is long.

References

[1] Matsuzawa T, Aoki Y and Takeuchi N *et al.* 1996 *J. Electrochem. Soc.* **8** 2670
[2] Jia W, Yuan H and Lu L *et al.* 1999 *J. Cryst. Growth* **200** 179
[3] Blasse G, Wanmaker W L and Bril A 1968 *Philips. Res. Rep.* **23** 201
[4] Groppi G, Cristiani C and Forzatti P 1994. *Mater. Sci.* **29** 3441

[5] Abanti N and Kutty T R N 2003 *J. Alloys Comp.* **354** 221
[6] Peng M, Pei Z, Hong G and Su Q 2003 *Chem. Phys. Lett.* **371** 1
[7] Hölsä J, Högne J, Lastusaari M and Niittykoski J 2001 *J. Alloys Comp.* **323** 326
[8] Yu X, Zhou C, He X, Peng Z and Yang S 2004 *Mater. Lett.* **58** 1087
[9] Kurihara L K and Suib S L 1993 *J. Chem. Mater.* **5** 609

Acknowledgments
This work is supported by basic research program of the Vietnam Ministry of Science and Technology, No 40.91.06.

Luminescence properties of II/VI semiconductor colloidal nanocrystals at collective and single scales

Céline Vion[1], Carlos Barthou[1], Laurent Coolen[1], Paul Bennaloul[1], Vu Duc Chinh[2], Pham Thuy Linh[2], Vu Thi Bich[2], Pham Thu Nga[2] and Agnès Maître[1]

[1]Institut des NanoSciences de Paris, Unité Mixte de Recherche-CNRS 7588, Université Pierre et Marie Curie, Paris, France
[2]Institute of Materials Science, Vietnam Academy of Science and Technology, Hanoi, Vietnam

E-mail: celine.vion@insp.jussieu.fr

Abstract. Colloidal nanocrystals are crystalline spheres of semiconductors of a few nanometers, obtained by chemical synthesis. At this size scale, lower than Bohr radius of the exciton, emission properties are dominated by quantum confinement effects and depend crucially on the nanocrystal radius, which can be controlled by adjusting the synthesis parameters. Nanocrystals present high photostability and good quantum efficiency, even at room temperature. Their emission wavelength can be tuned over the whole visible range, making them very attractive solid state light sources which are already used in optoelectronic devices or for biological labeling. The luminescence properties of CdSe colloidal nanocrystals synthesized at the Institute of Materials Science in Hanoi are presented. At collective scale, the emission properties reveal the synthesis quality. Temperature effects from ambient to 4 K on spectra and decay rates will be presented and analyzed in terms of emitting level fine structure. The study of CdSe colloidal quantum dots at the single emitter scale is of great interest as it reveals properties which are hidden by collective studies, such as luminescence "blinking", a random switching from a fluorescent to a non fluorescent state, which is closely related to the crystalline defects of a nanocrystal and its interaction with its environment. We will present the blinking properties of the prepared nanocrystals, and relate them to the nanocrystals synthesis quality and shell quality.

Keywords: Nanocrystals, CdSe, quantum dot, single scale emitter, luminescence.

1. Introduction

Colloidal CdSe nanocrystals (NCs) have attracted growing attention due to their size-tunable optical properties and ease of chemical manipulation into structures of varied complexity using a bottom-up approach. They can be synthesized with a narrow size distribution (<5% rms) and the diameters are tunable during the synthesis (between 1.5 and 10 m). In this size range, where the nanocrystal is smaller than the exciton Bohr diameter (11.2 nm), the confinement energy of the electron and hole is stronger than their Coulomb interaction, making them true zero-dimensional structures. The spectral properties and the photostability at room temperature make them promising light sources for a large range of applications, including quantum cryptography [1], optoelectronic devices [2, 3], or biological labeling [4]. The study of their optical properties has greatly benefited from the advent of single

© 2009 IOP Publishing Ltd

molecule techniques. When observed individually, NCs have, indeed, proved to be more complex than ensemble-averaged studies could have inferred. Studies at single scale have revealed striking phenomena such as fluorescence “blinking” [5], spontaneous spectral shifts [6], and lifetime fluctuations [7]. Most of these phenomena are due to the crystalline defects of a nanocrystal and strong interactions with their surrounding environment. In the last few years the synthesis quality has largely improved to limit crystalline defects and partially removes the interactions between the core and the environment by adding a passivating shell of ZnS or CdS.

In the first part, we present study of CdSe colloidal nanocrystals, synthesized at the Institute of Materials Science in Hanoi, at collective scale. Temperature effects from ambient to 4 K on spectra and decay rates will be presented and analyzed in terms of emitting level fine structure. The effect of a ZnS shell will be shown. In the second part, the properties of CdSe colloidal quantum dots at the single emitter scale will be presented and discussed. Single scale studies such as luminescence “blinking”, a random switching from a fluorescent to a non fluorescent state will be present and discussed. The blinking properties of the prepared nanocrystals will be related to the synthesis quality.

2. Ensemble measurements

Two different types of nanocrystals are compared in this article: CdSe nanocrystal in toluene covered by trioctylphosphine oxide (TOPO), to prevent from oxidation and solubilized them in organic solvents (574 nm peak at 300 K), and ZnS-coated CdSe NCs in toluene (585 nm peak emission at 280 K).

We first study the temperature dependence of nanocrystals luminescent spectra and lifetimes in the range 4 - 300 K.

Samples of nanocrystals were prepared by depositing a solution of NCs and PMMA (2% in mass) in toluene on a clean glass substrate. The low temperature experiments were carried out in a Janis Supertran-VP cryostat with a variable temperature controlled between 4 and 300 K.

The photonic excitation at 400 nm was obtained with a dye laser pumped by a pulsed nitrogen laser (0.14 mJ energy per pulse, pulse width 0.6 ns). The visible emitted light from the sample, collected by an optical fiber, was analyzed with a Spectrometer and a multichannel CCD detector (2000 pixels). The decays were analyzed with a Photo Multiplier Hamamatsu R5600U and a scope Tektronix TDS 784A with a time resolution of the order of 1 ns.

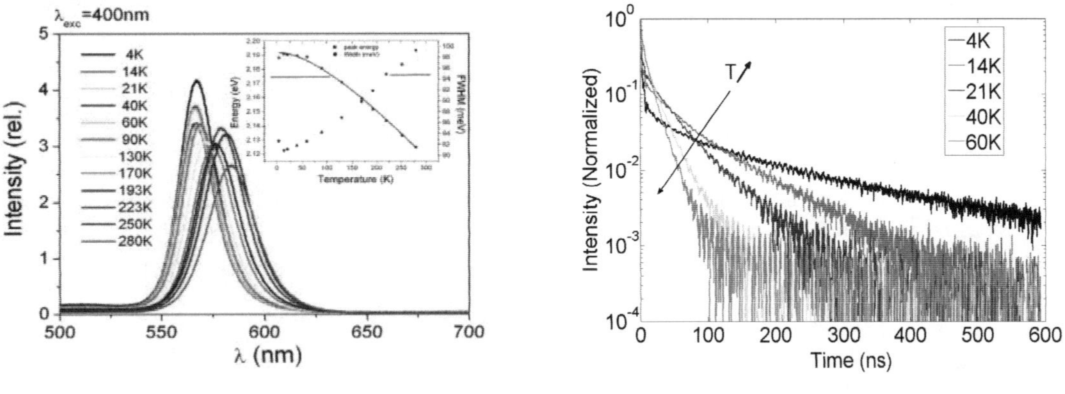

(a) (b)

Figure 1. a) Luminescence spectra of CdSe/ZnS nanocrystals at different temperature and variation of the position of the maximum of luminescence with the temperature. b) Luminescent decays curves of CdSe/ZnS nanocrystals.

The figure 1 presents the luminescent spectra and decay variation with temperature of the CdSe/ZnS nanocrystals. The maximum of the luminescent spectra stays stable until 60 K and is then linearly red-shifted as the temperature increases. The FWMH decreases when the temperature decreases as phonons broadening are limited at low temperature. These results are in agreement with previous studies on temperature dependence luminescence of CdSe/ZnS emitters [8]. The NCs emission wavelength presents a redshift of about 17 nm. This temperature dependence is directly related to the temperature dependence of the energy gap of bulk CdSe semiconductors as given by Varshni [9]. The temperature coefficient is found to be (4±0.3) 10^{-4} eV/K as given in literature for bulk CdSe.

The excited-state lifetime T_1 increases as the temperature decreases, from 17.5 ns at 300K to more than 200 ns at 4 K (see figure 1b). This variation is well known, and understood considering the fine structure of the emitting level. The emitting level ($1S_{3/2}1S_e$), having electron spin 1/2 and hole momentum 3/2, is eightfold degenerate. This degeneracy is known to be due to three perturbative effects: the hexagonal crystalline structure (Wurtzite), the slighty prolate shape of the nanocrystal, and the exchange term in the electron-hole coulombian interaction, which is enhanced by confinement. A full mathematical treatment of the band edge exciton degeneracy can be found in ref [10]. The figure 2a presents the results for a CdSe/ZnS nanocrystal of diameter 3.5 nm. The exciton lowest sublevel has a total spin projection on the crystal hexagonal axis J = ±2 (twofold degenerate), cannot emit light through a linear process and is located a few meV below the lowest-energy optically active (J = ±1L) exciton states.

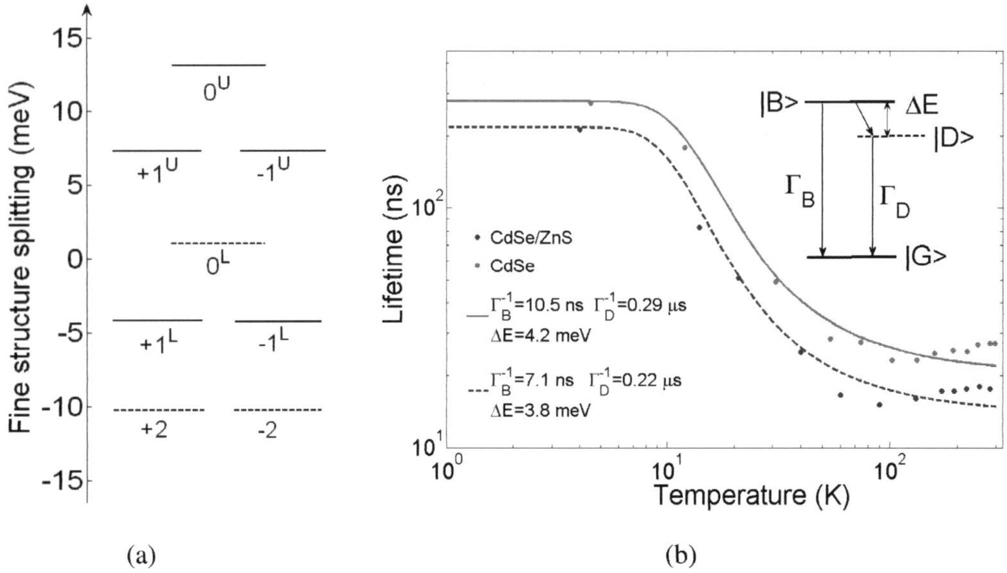

Figure 2. a) Fine structure of the emitting ($1S_{3/2}1S_e$) level of a 1.8 nm radius CdSe nanocrystal calculated from [8]; b) Variation of lifetime with temperature (log-log scale) for NCs of CdSe/TOPO: red line and CdSe/ZnS: dotted blue line. Inset: The three states model as proposed in [9].

The scheme is simplified on the basis of a three level system (see inset in figure 2b): a zero exciton ground states |G>, the |B> state representing the bright (J = ±1L) exciton, and the |D> state representing the "dark" (J = ±2) exciton. The energy splitting ΔE between the dark and bright states is in a few meV ranges, depending on the size and the shape of the QDs. After excitation, the system relaxes rapidly to the dark exciton state; it is later excited to the bright state through the phonon bath, and emits a photon [11]. This process is faster at high temperature, which explains the observed dependence of the excited states lifetime T_1 as a function of the temperature.

Figure 2b presents the variation of decays with temperature for the CdSe/ZnS and CdSe nanocrystals. The temperature dependence is modeled by the three level system as done in reference

[11]. The experimental results are well fitted leading to the parameters ΔE, Γ_B, and Γ_D respectively the decay rates of the dark and the bright sates in good agreement with previous studies [11, 12]. From the model, two characteristic times are revealed, a short one, present only at low temperature, and a long one. This model reproduces then satisfactorily the decay curves presented on figure 1b. At low temperature (<30 K) two characteristic decay times are found, one lower than 2 ns and one larger than 100 ns. The short decay disappears with growing temperature.

Figure 1b presents the luminescence decay of an ensemble of nanocrystals recorded at 4 K and 280 K. It presents a multi-exponential behavior distributed over a large time scale. This can be due either to an inhomogeneous distribution of lifetimes or emitters containing a large amount of defects. To discriminate between these two causes, experiments at single emitter scale are essential as they can reveal fundamental features otherwise masked in ensemble experiments.

3. Single scale characterisations

CdSe/ZnS core-shell nanocrystals are spin coated on silica layer with a very low density (0.1/µm^2), adjusted by different dilutions. Indeed, the distance between each nanocrystal should be higher than 1 µm to be excited and detected at single scale. The nanocrystals are then covered of a polymethyl (methacrylate) PMMA layer (50 nm thick). The optical emissions at single scale were characterized by micro-photoluminescence resolved in time at room temperature [13]. The excitation was made using a 400 nm pulsed laser diode focused onto a single NC via an oil objective lens (1.4 NA, x100). The fluorescence photons were collected by the same objective lens and sent to an avalanche photodiode (PerkinElmer SPCM-AQR-14) followed by a time-resolved photon counting card (Picoquant, Picoharp 300) (see the scheme in figure 3a). The photoluminescence intensity and decay were recorded in time-resolved acquisition mode.

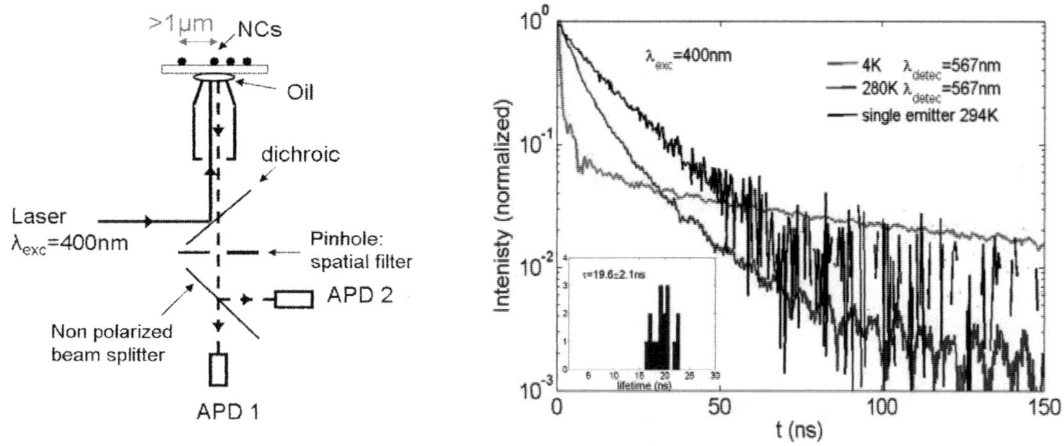

(a) (b)

Figure 3. a) Schematic overview of the optical setup for collecting single NC fluorescence. (b) An ensemble measurement at 280 K gives a multi-exponential decay (blue line). The decay is bi-exponential at 4 K (red line): the short time is within our time resolution (<2 ns) and the long time decay is around 240 ns. Single NC decay (black line). The mean lifetime is found to be 19.5±2 ns (mean from 10 measurements).
Inset: single NCs lifetimes distribution.

On figure 3b, the decay of single NC (black line) can be compared to the ensemble decay at 280 K. The decay is monoexponential at single scale while it is multi-exponential at collective scale. The monoexponential decay suggests that there are few defects and the NC is of good quality. The multi-exponential decay at collective scale is then due to an inhomogeneous distribution of lifetimes.

Actually, the distribution of lifetimes measured on 16 single NCs gives a mean lifetime of 19.5 ns with a standard deviation of ±2 ns (see inset figure 3b).

It was seen for the first time in 1996 [5] that the light emission from single nanocrystals under continuous excitation turns on and off intermittently.

Indeed, QDs significantly interact with their local environment due to their high surface to volume ratio. The interaction of charges with their surroundings directly translates to fluctuations of their properties, particularly in the form of fluorescence intermittency. Most blinking-statistic models associate dark states with the presence of a charge in the dot's core, which quenches the fluorescence emission by Auger processes [14]. On figure 4a, the fluorescent blinking of single CdSe/TOPO and CdSe/ZnS emitters are compared. Statistics made over more than 20 NCs shows that the CdSe NCs are in the "on" state 5 to 30% of the total time, whereas 70% of "on" states are recorded on CdSe NCs with ZnS shell. ZnS shell acts as an effective barrier to ionization and passivate potential surface trap sites [14]. This blinking behaviour is a good indication of the core/shell NCs synthesis quality.

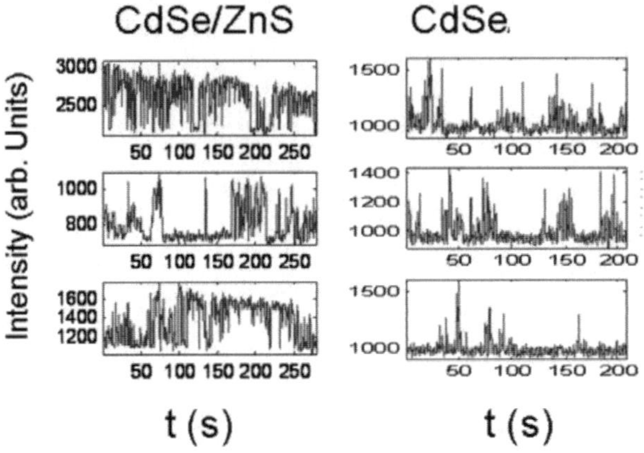

(a)

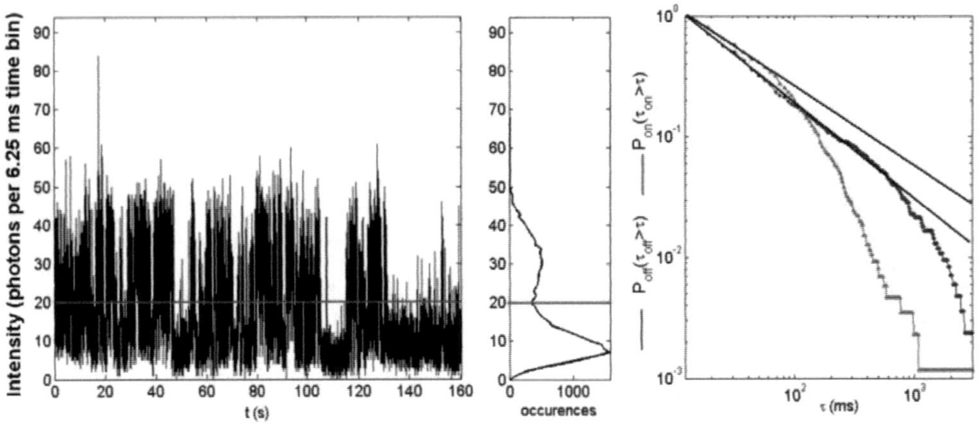

(b)

Figure 4. a) Fluorescence intensity versus time trace of a CdSe nanocrystal compared to ZnS covered nanocrystal: three typical trace for each type of nanocrystal recorded with CCD camera; b) From left to right: Intensity fluctuations of single CdSe/ZnS NC measured with avalanche photodiodes. Intensity distribution for the NC trace: the two peaks correspond to off and on intensities; the threshold used is 20 photons per 6.25 ms time bin. Cumulative distributions of on (red line) and off times (blue line) in a logarithmic scale.

Journal of Physics: Conference Series **187** (2009) 012018 doi:10.1088/1742-6596/187/1/012018

Due to the binary nature of the switching process, we can consider the time trace of each NC as a sequence of on and off times from which the statistical distributions P_{on}(□) and P_{off}(□) can be determined. The figure 4b presents the cumulative distributions of the on and off times obtained for one of our CdSe/ZnS emitters. The on and off durations do not follow an exponential statistics, Kuno and al. confirmed this observation in 2001 showing that off states are power law distributed [15]. The cumulative probability density P(□$_{on,off}$> □) decreases in time according to a power law decay p(□) □□-$^{□}$, with a power exponent □$_{on,off}$= □ around 1/2. For our CdSe/ZnS NCs, the values of □$_{off}$ = 0.76 and □$_{on}$ = 0.67 are found in good agreement with previous studies. This kind of distribution is called "broad distribution" (or Lévy distribution) because it yields formally infinite averages for □$_{on}$ and □$_{off}$; in other words, the signal is dominated by a few rare long events [16]. Accounting for this surprising behavior in nanocrystals remains challenging [17], although some satisfying models were recently proposed [18, 19]. The cutoff (i.e: the truncation of the power-law statistics for long □) for "off" states, as it can be seen on figure 4b, is an artifact due to the finite duration of the acquisition, whereas the cutoff for on states at shorter times is a "true" cutoff, as already seen in references [20] and [21]. This implies that NC will be off most of the time although the exponents of on and off states are the same. The exponent □$_{off}$ and □$_{on}$ are independent of the power of the continuous optical excitation, of temperature, or of eventual NCs capping layer, whereas the cutoffs strongly depends on the experimental conditions [20, 21].

Recently, CdSe-CdS core-shell NCs with thick crystalline shells have shown a complete absence of blinking when observed at acquisition rates of 33 Hz, suggesting that well designed shells are a key parameter for obtaining non-blinking NCs [22].

It was shown in 2000 [23] that single nanocrystals, like single molecules and other nano-emitters, exhibit emission antibunching : their emission intensity correlations vanish at zero delays, meaning that photons are emitted one by one, and never two photons at the same time. Under pulsed excitation, a single nanocrystal constitutes a highly efficient triggered source of these so-called "single photons", one laser pulse leading to one single-photon emission [1], which could be used in quantum cryptography protocols and other quantum optics experiments.

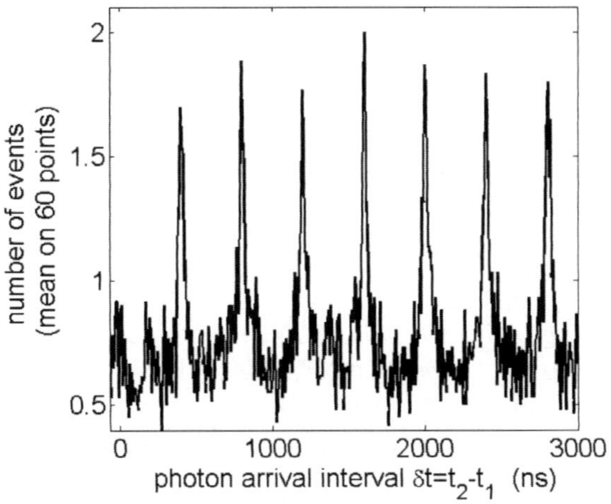

Figure 5. Histogram of the time intervals between detection events on start and stop detectors. No peak can be seen at time delay □t = 0.

Photon antibunching is evidenced using a Hanbury-Brown and Twiss-type photon correlation setup [24]. We realized such an experiment with our CdSe/ZnS nanocrystals. The luminescence collected by the objective is separated by a beamsplitter into two beams, each one detected by a single-photon detector, connected respectively to the start (t_1) and stop (t_2) inputs of the time-interval counter (see scheme figure 3a. This counter builds a histogram of the time intervals □t = t_2 - t_1 between start and stop detection events. Figure 5 presents a typical histogram; it consists of a series of peaks separated by the period of pump cycle (400 ns). The absence of the correlation peak at □t = 0 is the signature of single-photon emission.

Journal of Physics: Conference Series **187** (2009) 012018 doi:10.1088/1742-6596/187/1/012018

4. Conclusions

To conclude, we underline the importance of characterizing the properties of emission at single scale, such as blinking statistics or decay profiles, as they are strongly related to synthesis quality. The CdSe/ZnS nanocrystals synthesized at the Institute of Materials Science in Hanoi are found to be of good quality at ensemble and single scales. Indeed, the same behaviors as in previous studies were found for experiments at collective scale: variation of lifetimes with temperature from 4 to 300 K, and single scale: single scale decays profiles, statistics of on and off states, photon antibunching. The synthesis of good quality nanocrystals, especially the suppression of blinking, is an important goal for bio-optics use as fluorescent labels, as well as for single-photon quantum optics experiments and applications.

References

[1] Brokmann X, Giacobino E, Dahan M, and Hermier J P 2004 *Appl. Phys. Lett.* **85** 5

[2] Schmidt R, Scholz U, Metzner C, Wieck A, Zrenner A. et al. 2006 *Appl. Phys. Lett.* **88** 121115

[3] Hikmet R A M, Chin P T K, Talapin D V and Weller H 2005 *Adv. Mater.* **17** 1436

[4] Dahan M, Lévi S, Luccardini C, Rostaing P, Riveau B, Triller A 2003 *Science* **302** 442

[5] Nirmal M, Dabbousi B O, Bawendi M G, Macklin J J, Trautman J K, Harris T D and Brus L E 1996 *Nature* **383** 802

[6] Empedocles S A, Norris D J, and Bawendi M G 1996 *PRL* **77** 3873

[7] Schelgel G, Bohnonberger J, Potapova I, and Mews A 2002 *PRL* **88** 137401

[8] Valerini D, Creti A, Lomascolo M, Manna L, Cingolani R and Anni M 2005 *Physical Review B* **71** 235409

[9] Varshni Y P 1967 *Physica (Amsterdam)* **34** 149

[10] Efros A L, Rosen M, Kuno M, Nirmal M, D Norris J and Bawendi M 1996 *PRB* **54** 4843

[11] Labeau O, Tamarat P and Lounis B. 2003 *Phys. Rev. Lett* **90** 2575404

[12] Crooker S A, Barrick T, Hollingsworth J A and Klimov V I 2003 *Appl. Phys. Lett.* **82** 2793

[13] Tamarat Ph, Maali A, Lounis B, and Orrit M 2000 *The Journal of Physical Chemistry A* **104** 1

[14] Cichos F, Borczyskowski C von, Orrit M 2007 *Current Opinion in Colloid and Interface Science* **12** 272

[15] Kuno M, Fromm D P, Hamann H F, Gallagher A and Nesbitt D J 2000 *J. Chem. Physics* **112** 3117

[16] Brokmann X, Hermier J P, Messin G, Desbiolles P, J Bouchaud P and Dahan M 2003 *PRL* **90** 120601

[17] Kuno M, Fromm D P, Johnson S T, Gallagher A and Nesbitt D J 2003 *PRB* **67** 125304

[18] Tang J and Marcus R A 2005 *The J. Chem. Phys.* **123** 054704

[19] Frantsuzov P A and Marcus R A 2005 *PRB* **72** 155321

[20] Chung and Bawendi M 2004 *Phys. Rev. B* **70** 165304

[21] Chung, Witkoskie J B, J Cao and Bawendi M 2006 *Phys. Rev. E* **73** 011106

[22] Malher B, Spinicelli P, Buil S, Quelin X, Hermier J P, and Dubertret B 2008 *Nature Materials* **7** 659

[23] Michler P, Imamoglu A, Mason M.D, Carson P.J, Strouse G F, and Buratto K 2000 *Nature (London)* **406** 968

[24] Hanbury-Brown R and Twiss R Q 1956 *Nature (London)* **178** 1447

Preparation and optical properties of ZnO, ZnO: Al nanomaterials

Tran Kim Anh[1], Man Hoai Nam [1], Dinh Xuan Loc[1], Nguyen Vu[1], Wieslaw Strek[2] and Le Quoc Minh[1,3]

[1] Institute of Materials Science, Vietnam Academy of Science and Technology, 18 Hoang Quoc Viet Road, Cau Giay Distr., Hanoi, Vietnam
[2] Institute of Low Temperature and Structure Research, 2 Okolna, Wroclaw, Poland
[3] College of Technology, National University Hanoi, Hanoi, Vietnam

E-mail: kimanh1949@gmail.com

Abstract. ZnO nanorods were prepared by different methods of MOCVD, CVD and ZnO: Al nanomaterials were synthesized by the hydrothermal method. Their structure was investigated using X-ray diffraction XRD and field emission scanning electron microscopy FESEM. The diameter of the nanorods is from 40 to 90 nm and the length is from 500 nm to 1000 nm dependent on the growth conditions. Optical properties were studied by photoluminescent with different excited of 325 and 266 nm. A dominant photoluminescent peak 380 nm was observed at room temperature. At low temperature, five peaks were observed at 3.347, 3.321, 3.301, 3.208, and 3.111 eV. The nature of luminescent peaks will be investigated. ZnO:Al nanomaterials have strong luminescent intensity in visible range dependent on Al concentration, optimal molar concentration is 7%.

Keywords: ZnO nanomaterials, CVD, MOCVD, hydrothermal, photoluminescence.

1. Introduction

ZnO nanomaterials have attractive properties such as wide-band gap (3.37eV), large exciton binding energy (60 meV versus 28 meV for GaN), direct and wide band-gap energy, high optical gain, thermal stabilities and luminescent properties. ZnO nanorods and nanowires have been used in nanodevices such as light-emitting device arrays, logic gate, transistors for short-wavelength optoelectronic-device applications like solar cell, piezoelectric devices, electro acoustic transducers, laser diodes and semiconductor laser [1-3]. One dimensional ZnO such as nanotubes, nanowires, nanorods, nanobelts have been synthesized by various methods, including arc discharge, metal organic chemical vapor deposition, laser vaporization, pyrolysis, electrodeposition, sol-gel [4-10]. Recently ZnO nanowires were site - selectively deposited on the catalytic gold - patterned substrates using a vapor transport process at 600°C [7]. A facile route to ZnO nanorod arrays using wet chemical method was reported in [8]. The fine structures electron-phonon coupling and temperature effect on green luminescence band of ZnO was presented in [9]. The synthesis, characterization and applications ZnO nanorods were also studied in [10]. ZnO nanorods have attracted attention as potential optical components integrated in nanometer sized electronic devices, applications and study nature of luminescent peaks are important [11, 12]. In this paper, we report results of the preparation of ZnO nanorods by MOCVD and CVD

© 2009 IOP Publishing Ltd

method as well as ZnO, ZnO:Al prepared by the hydrothermal method and the study of luminescent spectra.

2. Experimental

2.1. *Synthesis of ZnO nanorod*

ZnO nanorod arrays were grown on Al_2O_3(0001) and Si (100) substrates by metal-organic chemical vapor deposition (MOCVD) system. Before ZnO nanorod growth, 20 nm thick ZnO buffer layers were deposited at room temperature, which significantly improved the vertical alignment of ZnO nanorod. During buffer layer growth, oxygen and diethyl zinc (DEZn) flow rates were in the ranges of 10 to 100 sccm at DEZn bubbler temperature of –15°C. The reactor pressure was kept at 5.0 torr. After ZnO buffer layer growth, the flow of DEZn vented while the oxygen was kept flowing into the reactor to prevent decomposition of the buffer layer. Then, the reactor pressure was decreased to 0.5 torr while the temperature was increased to a growth temperature of 500 - 800°C with a ramp rate of temperature 50°C/min. The typical growth time was 1-2 hours.

ZnO nanorods have been grown by CVD method from Zn powders on the Si wafer (111) covered by Au or Pt. The Carbollite furnace was used for ZnO nanostructure growth systems. The schematic illustration of furnace system used in this experiment is shown in figure 1. The evaporation process was performed in the inner quartz tube. First, the alumina boat and Au-or-Pt-coated silicon wafer were washed by alcohol and acetone. Pure Zn nano powders and silicon were put on an alumina boat and the boat into the quartz tube heated under the argon flow of 0.5 to 1 litre/min. Heating temperature increased from room temp to 800°C, it was kept constant for 10- 60 minutes. This heating processing was controlled by a computer.

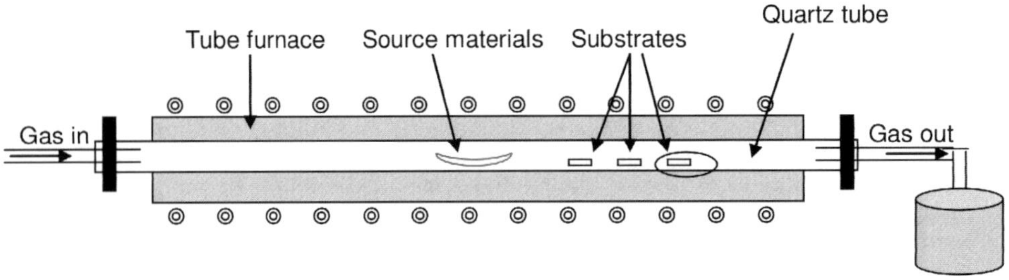

Figure 1. Schematic illustration of furnace system for ZnO nanostructure growth.

2.2. *ZnO:Al nanomaterials*

ZnO: Al nanorods were prepared by hydrothermal method: $ZnSO_4.7H_2O$ and $CO(NH_2)_2$ are used as precursor's materials. The amount of $Al_2(SO_4)_3.18H_2O$ was calculated for doping with different concentration of Al such as 1, 3, 5, 7 and 10%. The solution was stirred for 30 minutes at room temperature, NaOH was slowly added to the solution of this salt, and the solution was poured into a Teflon lined stainless steel autoclave. The micro autoclave instrument (MMM Medcenter Einrichtungen GmbH) allows all of heating process can be controlled fully by programmer in time and temperature; the minimum increment/decrement is 1°C. Temperature increased to 180°C for 4h. The clear transparence solution was received. Adding the alcohol, the solution were separed by the centrifugation followed by washing with distilled water and alcohol. The fine powder of ZnO: Al was received.

2.3. *Characterizations*

The crystal orientation of samples was characterized by X-ray diffraction (XRD) (Philips, X'pert PRO MRD or Siemens D5000). General morphology was analyzed using a Philips XL30SFEG field emission scanning electron microscope (FESEM) equipped with 4 an axis motorized stage. The morphology of the obtained product has also been investigated by FESEM (Hitachi).

Photoluminescence (PL) spectra were measured by Spectrapro 2300i, CCD. The PL spectra were measured at room temperature in the wavelength range of 350–600 nm by using excited He–Cd laser of 325 nm as the excitation source and by 266 nm of Ti sapphire laser.

3. Results and discussion

The crystal structure and orientation of ZnO nanorods were investigated measuring XRD. Figure 2 shows the results of XRD measurement of the specimen prepared at 500°C for 1 hour. From the XRD, θ-2θ scan data of the ZnO nanorods, three peaks at ZnO (0002), ZnO (0004) and ZnO (0006) were observed at 34.39°, 72.53°, and 125.15°, respectively.

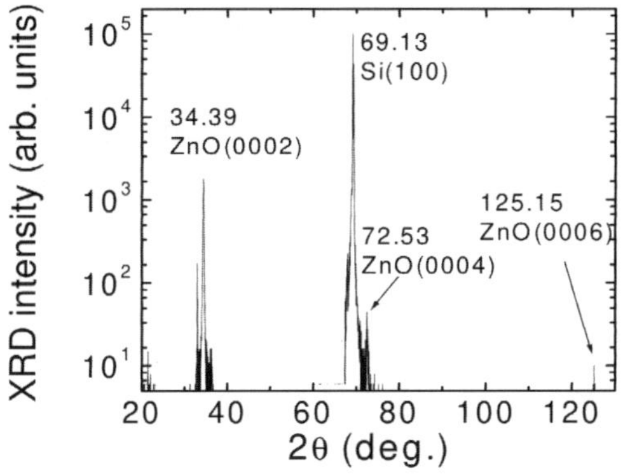

Figure 2. XRD pattern of ZnO nanorods grown at 500°C, 1 hour.

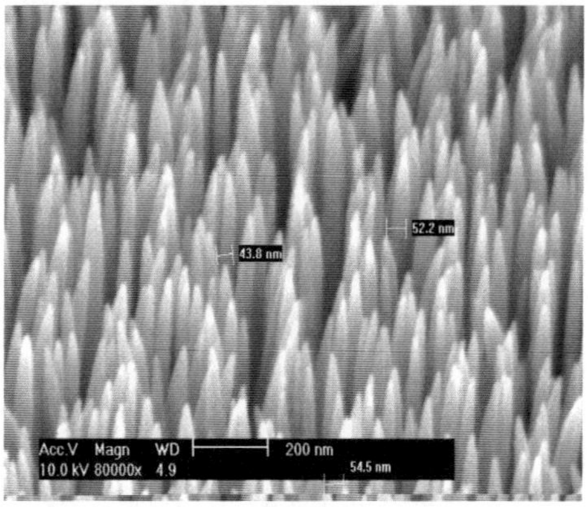

Figure 3. FESEM image of ZnO nanorods grown at 500°C, 2 hours.

Influence of technology conditions such as the method of MOCVD, CVD or hydrothermal, the temperature, the time, the distance from source and selective excited were studied. Figure 3 presents FESEM image of ZnO nanorods grown at 500°C, 2 hours by MOCVD method, diameter is about 43.8-52.2 nm. One can notice that the MOCVD method is excellent one to prepare ZnO nanorods. Figure 4 shows FESEM images of ZnO nanorods grown at 500°C, 1 hour by MOCVD method. It shows that the rods have diameters of 40 nm and lengths of 900 nm.

Figure 4. FE SEM images of ZnO nanorods: plan-view (left) and cross-sectional (right). Sample was grown at 500°C, 1 hour by MOCVD method.

Figure 5 presents FESEM images of ZnO nanorods grown at 450°C for 2 hours by MOCVD method. It shows that the rods have lengths of 500 nm and diameters of about 20 nm.

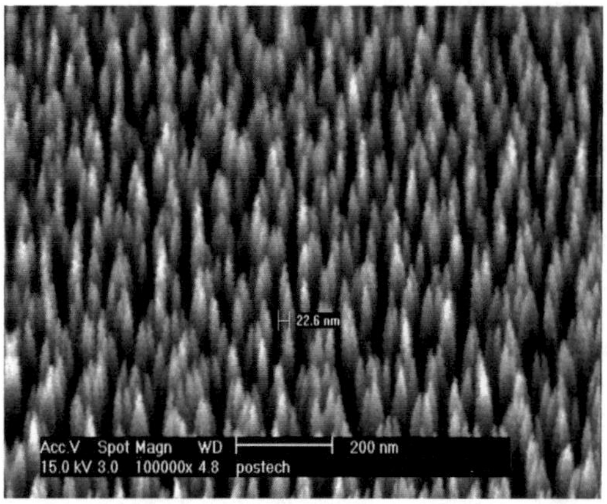

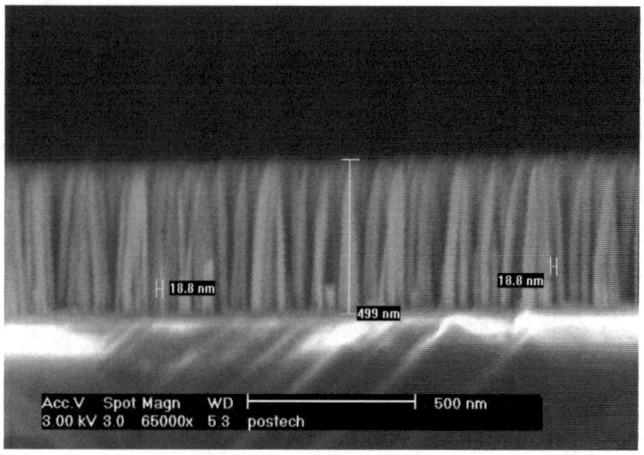

Figure 5. FESEM images of ZnO nanorods: plan-view (left) and cross-sectional (right). Sample was grown at 450°C, 2 hours by MOCVD method.

Figure 6 shows XRD pattern of ZnO nanorods prepared by CVD method at 800°C after 30 minutes annealing, argon flow was 1litre/min. X-ray diffraction pattern (XRD) shows that ZnO is monophase with one diffraction peak at $2\theta = 34.39^0$ corresponding to (002) plane of ZnO. Verticals lines in figure 6 are corresponding to the JCPDS-ICDD 36-1451 card of ZnO Zincite (Crystal system: Hexagonal, space group P63mc, a = 3.24982 and c = 5.20661). FESEM image of ZnO nanorods is presented in figure 7. It can be seen that the product is primarily composed of nanorods with mean diameters of 50-100 nm and lengths of 800 nm.

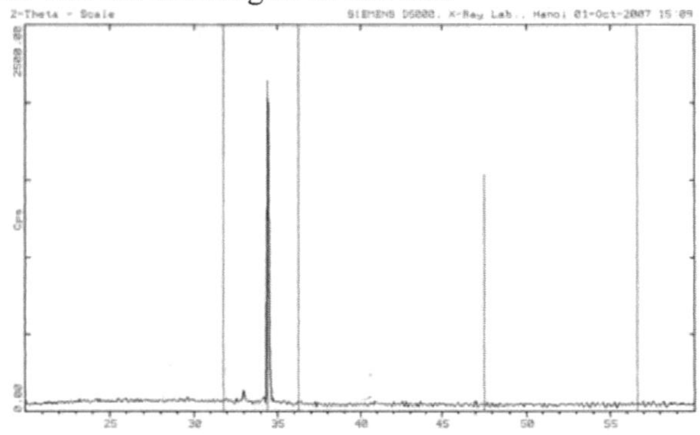

Figure 6. X-ray diffraction pattern of the ZnO prepared by CVD method (800°C, 30 minutes, argon 1 l/min).

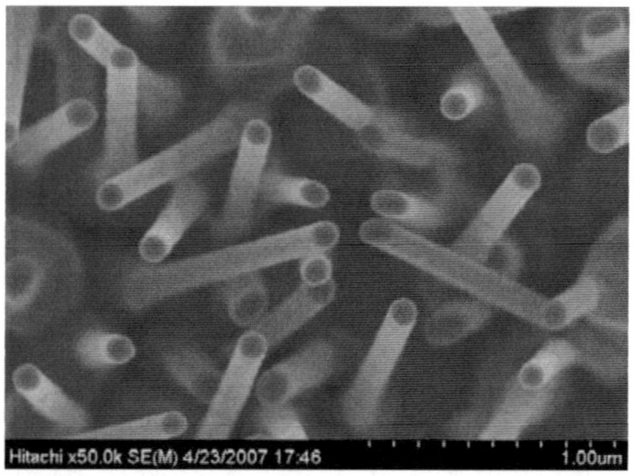

Figure 7. FESEM image of ZnO nanorods prepared by CVD method (800°C, 30 minutes, argon 0.8l/ min).

Figure 8 presents a typical XRD pattern of the ZnO sample prepared by hydrothermal method. It shows that ZnO samples have some diffraction peaks of ZnO with 2θ at 31.77° (100), 34.39° (002), 36.25° (101), 47.54° (102), 56.6° (110), 62.86° (103), 66.38° (200), 67.96° (112), and 69.1 (201). Diffraction peaks in figure 8 could be readily index as hexagonal phase of ZnO, the lattice parameters of which were all consistent with report values (corresponding to the JCPDS-ICDD 36-1451 card of ZnO Zincite). FESEM image of ZnO nanorods prepared by hydrothermal method is shown in the figure 9. By the hydrothermal method which shows that they were random flower like ZnO structure. Figures 10 and 11 present a typical XRD pattern and FESEM image of the ZnO:Al (7%) prepared by the hydrothermal method, respectively. They have also a random flower like structure. Properties of transparent conductive ZnO:Al films prepared by co-sputtering was studied in [13].

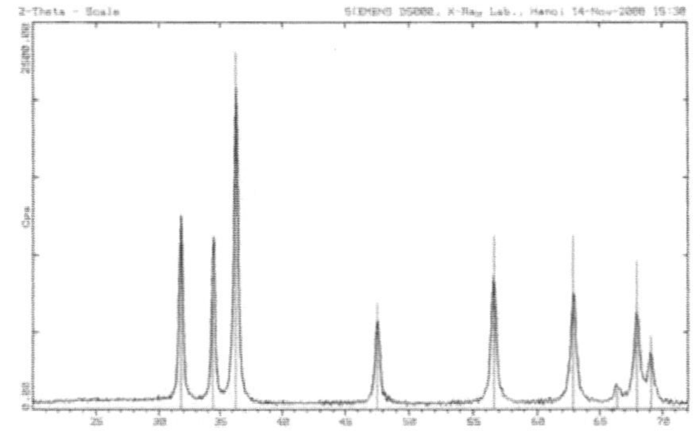

Figure 8. X-ray diffraction pattern of the ZnO sample prepared by hydrothermal method.

Figure 9. FESEM image of ZnO sample prepared by hydrothermal method.

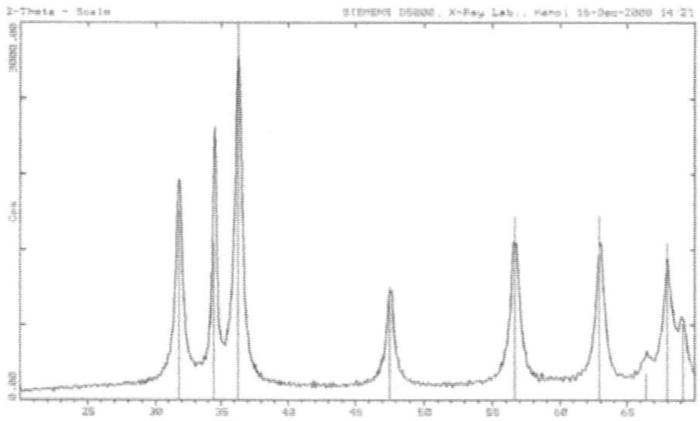

Figure 10. X-ray diffraction pattern of the ZnO:Al (7%) sample prepared by hydrothermal method.

Figure 11. FESEM image of ZnO: Al (7%) sample prepared by hydrothermal method.

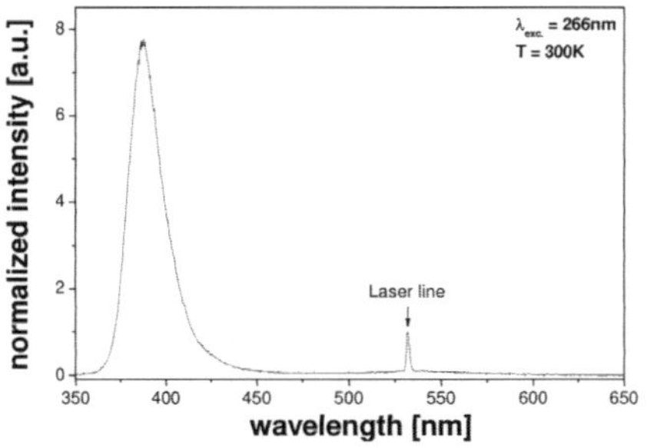

Figure 12. Luminescence spectra of ZnO nanorods (CVD, for distance from source of 7 cm) under 266 nm excited by Ti-sapphire laser.

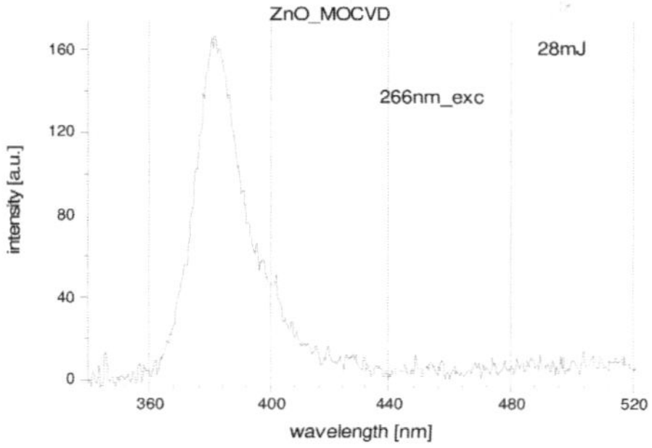

Figure 13. Luminescence spectra of ZnO nanorods (MOCVD) under 266 nm excited by Ti-sapphire laser excited.

The CVD method is a good one to prepare ZnO nanorods and the hydrothermal method is a simple and cheap method to prepare ZnO nanomaterials. The luminescence spectra of ZnO were measured at room and low temperature. Figure 12 shows PL spectra of ZnO nanorods prepared by CVD method under 266-nm excitation, the optimal distance from Zn nano source to the substrate of is 7 cm. Figure 13 presents PL spectra of ZnO nanorods prepared by MOCVD method under 266 nm excited by Ti-sapphire laser (266 nm, 28 mJ). Influence of excited power was studied: from 26 to 32 mJ, PL intensity at 380 nm increase linearly with excited power. It clearly reveals that under 266 nm laser

Journal of Physics: Conference Series **187** (2009) 012019 doi:10.1088/1742-6596/187/1/012019

excitation for the ZnO samples prepared by CVD method only one peak is observed at 380 nm like the sample prepared by MOCVD.

PL spectra of ZnO:Al nanorods prepared by hydrothermal method observed at 685-688 nm (orange region) were presented in the figure 14. It shows that the influence of the Al concentration doped on the PL intensity was observed. PL intensity is strongest at 7 mol % concentration. Full width at haft maximum (FWHM) are 203, 197 and 206 nm, respectively for the case of 5%, 7% and 10 % Al-doped ZnO. Nanosized flower-like ZnO synthesized by a simple hydrothermal method and applied as matrix for horseradish peroxidase immobilization for electro-biosensing presented in [11] open the interesting application. PL spectra of ZnO were measured at room and low temperatures.

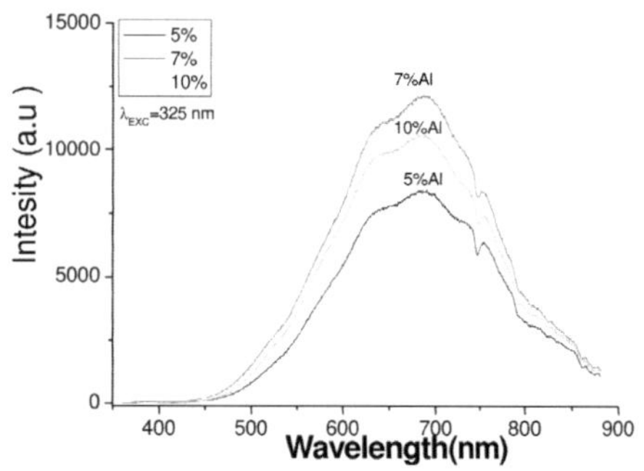

Figure 14. PL spectra of ZnO:Al (5, 7, 10%) prepared by hydrothermal method, 325 nm excited.

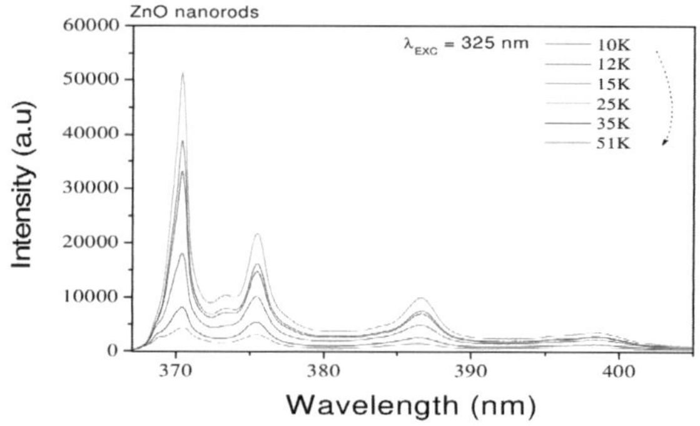

Figure 15. PL spectra of nanorod ZnO in low temperature of 10-51K, from top to bottom, 325 nm excited.

PL spectra of ZnO were studied in detail in the temperature dependent and presented in figure 15. They are quite different at low temperatures. From 10-51K the structure of spectra separate between 5 peaks 370.4, 373.3, 375.5, 386.5, and 398.5 nm (3.347, 3.321, 3.301, 3.208, and 3.111 eV). Sharp peak of 3.347 eV is assigned to an exciton bound to a neutral donor (D^oX). The broad peak of donor-acceptor pair is around 3.208 eV (DAP). The nature of peaks was detailed discussed by G C Yi: main peak at UV emission is attributed to the exciton transition bound to neutral acceptors or deep donor. For example, nanorod vertically oriented and nanowires random orientations are obtained respectively at 500 to 800°C according to Mezy *et al.* prepared by MOCVD method [12]. The nature of luminescent peaks were investigated by time resolved photoluminescence spectra at low temperature photoluminescence has been performed using 325 nm line He-Cd laser as the excitation source. The time resolved PL (TRPL) set up uses the frequency-tripled radiation (266 nm) from a pulsed Ti-sapphire mode-locked laser cavity, with a typical pulse width of 2 ps, to excite the PL of the sample. 3.361 eV (D_oX_1) 3.354 eV (D_0X_2), 3.208 eV line is attributed to donor-acceptor pair (DAP) recombination [12]. In order to apply ZnO nanomaterials, new subjects such as synthesis and field emission of novel ZnO nanorod chains [14], upright-standing ZnO nano-sheets growth using wet chemistry [15] were studied, PPV/PVA/ZnO nanocomposite prepared by complex precursor method and its photovoltaic application [16], high PL-efficiency ZnO nanocrystallites/PPV composite nanofibers [17], optical and structural properties of the sol–gel prepared ZnO thin films and their effect on the photocatalytic activity [18] high-performance ethanol sensing based on an aligned assembly of ZnO nanorods [19] and synthesis and red-shifted photoluminescence of single-crystalline ZnO nanowires [20] were presented. The high quality of nanorods can be prepared in the National CRI Center for Semiconductor Nanorods POSTECH, Pohang Korea open many interesting applications.

4. Conclusions

ZnO nanomaterials, specials nanorods were successfully prepared by MOCVD, CVD and ZnO:Al by hydrothermal methods. The X-ray diffraction results show that our samples are good quality. Luminescence spectra are detail studied, they have strong intensity. ZnO nano rods have diameter of 40 - 90 nanometers, some hundred nanometers of length. The promising materials for imaging technology, excited by He-Cd laser, influence of Al^{3+} concentrations were investigated, optimal concentration of Al is 7 mol %. The influence of temperature and concentration as well as the technology conditions were detail investigated and compare in order to find the optimal condition to prepare nanorods. Luminescent intensity is strong in the UV 380 nm for ZnO and visible at 687 nm for ZnO:Al. Selective excited of 266 nm by laser indicated that only one peak at 380 nm was observed in ZnO prepared by CVD or MOCVD. By the hydrothermal method, ZnO and ZnO:Al were prepared with strong luminescence. They are promising materials for application in optoelectronic and lighting technique.

References

[1] Lee W, Jeong M C and Myoung J M 2004 *Acta Materialia* **52** 3949
[2] Xu C L, Qin D H, Li H, Xu T, Li H L 2004 *Materails letters* **58** 3976
[3] Park W I, Jun Y H, Jung S W and Yi G C 2003 *Applied physics letter* **82** 964
[4] Li C, Fang G, Fuhaisu, Li G, Wu X G and Zhao X Z 2006 *Nanotech.* **16** 3740
[5] Kim H J, Lee C H, Kim D W and Yi G C 2006 *Nanotech.* **17** S327
[6] Shvalagin V V, Stroyuk A L and Kuchmii S Y 2007 *J. Nanoparticle Research* **9** 427
[7] Shin J H, Song J Y, Park H M 2009 *Materials Letters* **63** 145
[8] Chen Z, Gao L 2006 *J.Cryst. Growth* **293** 522
[9] Shi S L, Li G Q, Xu S J, Zhao Y and Chen G H 2006 *J. Phys. Chem B* **110** 10575
[10] Yi G C, Wang C and Park W I 2005 *Semicond. Sci. Technol* **20** S22
[11] Liu Y L, Yang Y H, Yang H F, Liu Z M, Shen G L, Yu R Q 2005 *J. Inorganic Biochemistry* **99** 2046
[12] Mezy A, Anceau S, Bretagnon T, Lefebvre P, Taliercio T, Yi G C, Yoo J 2006 *Supertattices and Microstructures* **39** 358
[13] Oh BY, Jeong MC, Lee W, Myoung J M 2005 *J. Crystal Growth* **274** 453
[14] Cui Q, Huang Y, Zhu Z 2009 *Current Applied Physics* **9** 426
[15] Min S K, Mane R S, Joo O S, Ganesh T, Cho B W, Han S H 2009 *Current Applied Physics* **9** 492
[16] Wang M, Lian Y, Wang X 2009 *Current Applied Physics* **9** 189
[17] Jiang Z , Huang Z, Yang P, Chen J, Xin Y, Xu J 2008 *Composites Science and Technology* **68** 3240
[18] Delgado G T, Romero C I Z, Hernández S A M, Pérez R C, Angel O Z 2009 *Solar Energy Materials and Solar Cells* **93** 55
[19] Yang Z, Li L M, Wan Q, Liu Q H, Wang TH 2008 *Sensors and Actuators B: Chemical* **13** 57
[20] Deng Y, Wang G, Li N, Guo L 2009 *J. Luminescence* **129** 55

Acknowledgments

We would like to thank Prof. Acad. Nguyen Van Hieu, Prof. Gyu Chul Yi (POSTECH, Pohang, Korea), Prof. Tamio Endo (Mie University, Japan) for their helps and financial support from the Basic research programmer, Laboratory of Quantum Optoelectronic (PAS, Wroclaw, Poland), Optoelectronic Materials Lab. and National Key Laboratory of Electronic Materials and Devices, Institute of Materials Science, Vietnam Academy of Science and Technology.

Synthesis of undoped and M-doped ZnO (M = Co, Mn) nanopowder in water using microwave irradiation

Nguyen Viet Tuyen[1], Ta Dinh Canh[1], Nguyen Ngoc Long[1], Nguyen Xuan Nghia[2], Bui Nguyen Quoc Trinh[3] and Zhongrong Shen[3]

[1] Faculty of Physics, College of Science, Vietnam National University, Hanoi
[2] Institute of Materials Science – VAST
[3]Shimoda Nano-Liquid Process Project, ERATO, Japan Science and Technology Agency (JST) 2-5-3 Asahidai, Nomi, Ishikawa 923-1211

E-mail: canhtd@vnu.edu.vn

Abstract. Undoped and M-doped ZnO (M = Co, Mn) nanopowders have been successfully synthesized by microwave-assisted approach using $Zn(NO_3)_2.2H_2O$ and NaOH. The results obtained from X-ray diffraction (XRD) analysis and the transmission electron microscopy (TEM) showed that the mean particle size was about 10-15 nm. The XRD analysis of the synthesized samples also showed a single phase ZnO structure without any additional phase. Magnetic property studies of the as-prepared M-doped ZnO (M = Co, Mn) samples showed them to be paramagnetic material. However, thermal annealing in air of particular samples at 800°C for 3 h resulted in transforming the samples into a room temperature ferromagnetic material.

Keyword: Nanopowder, undoped and M-doped ZnO (M = Co, Mn), microwave irradiation.

1. Introduction

Zinc oxide (ZnO) nanoparticles are used in a variety of applications such as ultraviolet (UV) light emitting device [1], catalyst [2], photo-catalyst [3], electro-acoustic wave device, and UV detector [4].

ZnO technologically is an important material due to its wide range of optical and electrical properties as well as semiconductor crystal with a large exciton binding energy (60 meV) and wide band gap (3.37 eV). The stable structure of ZnO is wurtzite, in which four atoms of oxygen locate in tetrahedral coordination surround each atom of zinc. Recently, Co-, Mn-doped ZnO have been investigated for possible application as spintronic material [6]. Synthesis of these materials is often accomplished by sputtering [7], chemical vapor deposition [8] and sol-gel technique [9, 10].

In this paper, we report a simple method for preparation of hexagonal undoped and M-doped (M = Co, Mn) by using microwave irradiation. Microwave-solvothermal synthesis is generally faster, more simply and energy efficient.

2. Experimental

The ZnO nanopowders were synthesized under the microwave irradiation in a tube water cooling systems (figure 1). The ZnO nanopowders were prepared according to the following equations:

$$Zn(NO_3)_2 + NaOH \rightarrow Zn(NO_3)(OH) + NaNO_3 \ , \quad (1)$$

$$Zn(NO_3)OH + NaOH \rightarrow ZnO + NaNO_3 + H_2O \ , \quad (2)$$

© 2009 IOP Publishing Ltd

and the $NaNO_3$ was removed by washing the sample with water.

Figure 1. Image of microwave irradiation system.

Synthesis of ZnO nanopowders was achieved by dissolving 1.89 g zinc nitrate (Merk, 99% purity) in ethanol to make a 50 ml solution of 0.2 M $Zn(NO_3)_2$ (Merk, 99% purity). While stirring, 50 ml of 0.4 M NaOH (Merk, 99% purity) was added dropwise. The resulting solution (the final pH of the solution was 12) was then placed in a conventional microwave. The microwave power was set to 150 W and the reaction time was 5 minutes. During the microwave irradiation the temperature of the solution reached up 80°C. After 5 minutes, a white product precipitated, the precipitator was washed several times with absolute ethanol and distilled water. This process was repeated until the pH of the solution was 7. Finally, the product was dried at 70°C for 4 hours. For doping, appropriate amounts of metal nitrates such as $Co(NO_3)_2$, $Mn(NO_3)_2$ (Merk, 99% purity) were added to zinc nitrate solution until the concentration of the dopant was 5%.

The particle morphologies and structures of the products were investigated by a transition electron microscope (TEM JEM 1010-JEOL), high-resolution transmission electron microscope (HRTEM JEM 2010-JEOL) and X-ray diffractometer (Bruker-AXS D5005). Raman scattering spectra at room temperature in the energy region between 100 and 1000 cm^{-1} were recorded by a micro-Raman spectrograph LABRAM-1B equipped with a He-Ne laser (λ = 632,8 nm) with a power of 11 mW. Photoluminescence (PL) measurement at room temperature was carried out on a 325 nm He-Cd laser. A UV-vis spectrophotometer PU 8700 was used to record the UV-visible absorption spectra. Magnetic measurement was conducted by using a DMS-880 sample vibrating magnetometer (VSM). Magnetization was measured as a function of the static externally applied field at room temperature.

3. Results and discussion

X-ray diffraction is used to investigate the changes of phase structure and crystallite size of the undoped ZnO powders and the 5% transition metal-doped ZnO samples. Figure 2 shows the XRD patterns of the undoped ZnO samples and Co- and Mn-doped ZnO samples, which demonstrates that all the ZnO samples have wurtzite hexagonal structure [4]. As indexed in the figure, all the diffraction peaks match those of wurtzite ZnO with lattice constants of a = 3.250 Å and c = 5.207 Å. The strong diffraction peaks appear at 31.8, 34.3 and 36.5°, which correspond to (100), (002) and (101) planes of wurtzite ZnO, respectively. The average crystallite size of the samples, evaluated by the Scherrer formula [5], is about 13 nm. This value agreed well with TEM observations of the ZnO nanopowders. It should be pointed that in the XRD patterns except the diffraction peaks of ZnO, no peak of additional phase was observed.

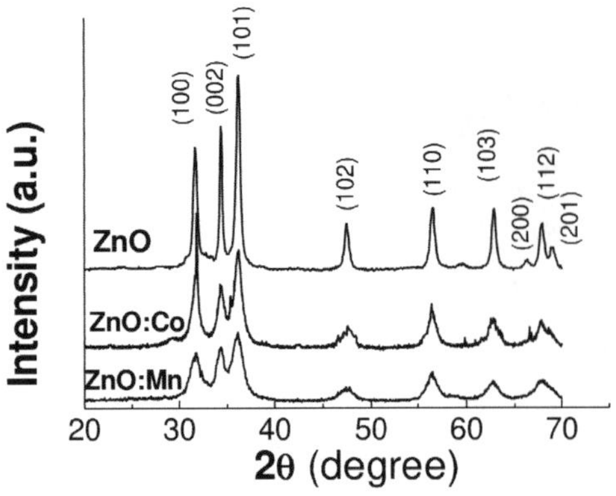

Figure 2. XRD patterns of the undoped and Co, Mn-doped ZnO nanoparticles.

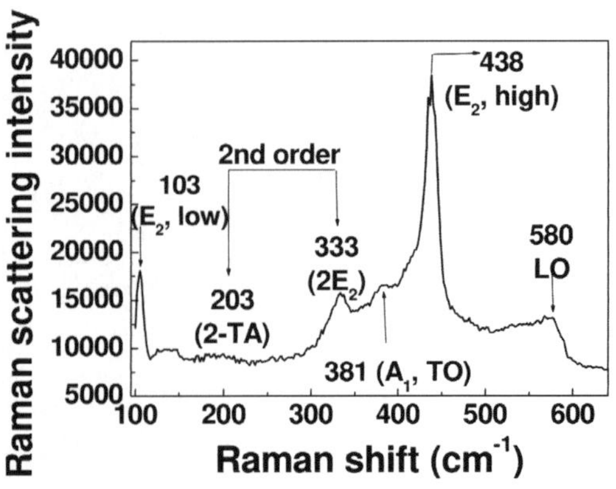

Figure 3. Raman scattering spectrum of the undoped ZnO nanoparticles.

Further, Raman scattering, due to its sensitivity to the crystallization, structural disorders and defects in nanostructures, was measured for the ZnO nanoparticles. Figure 3 shows the Raman spectrum of ZnO nanoparticles. ZnO has a wurtzite crystal structure and belongs to C_{6v} group. According to group theory analysis, the $A_1+E_1+2E_2$ modes are Raman active. The two higher peaks at 103 and 438 cm^{-1} can be assigned to E_2 modes, characteristic of the wurtzite lattice. The much weaker peak at 379 cm^{-1} is attributed to the transverse optical modes of A_1. The other two weaker and broader peaks at 203 and 333 cm^{-1} can be assigned to the secondary Raman scattering arising from zero-boundary phonons 2-TA (M) and 2-E_2 (M), respectively [8]. The presence of the E_1 (LO, 580 cm^{-1}) mode of oxygen deficiency indicates that there are oxygen vacancies in our ZnO nanoparticles.

The EDS elemental analysis is shown in figure 4. This result indicates that ZnO nanoparticles were only composed of Zn and O. This suggested a high purity of the ZnO nanopowder.

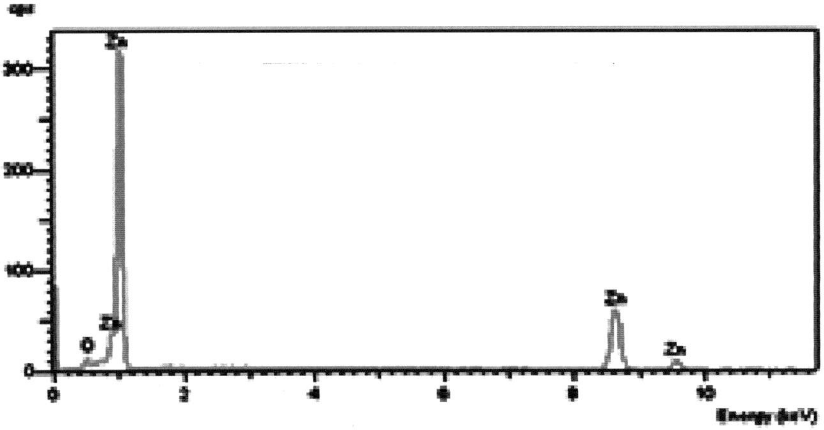

Figure 4. Energy dispersion spectrum of ZnO nanopowder.

TEM is used to observe the particle size and morphology of the samples. Figure 5 shows the TEM photographs of the undoped and the 5% transition metal-doped ZnO samples.

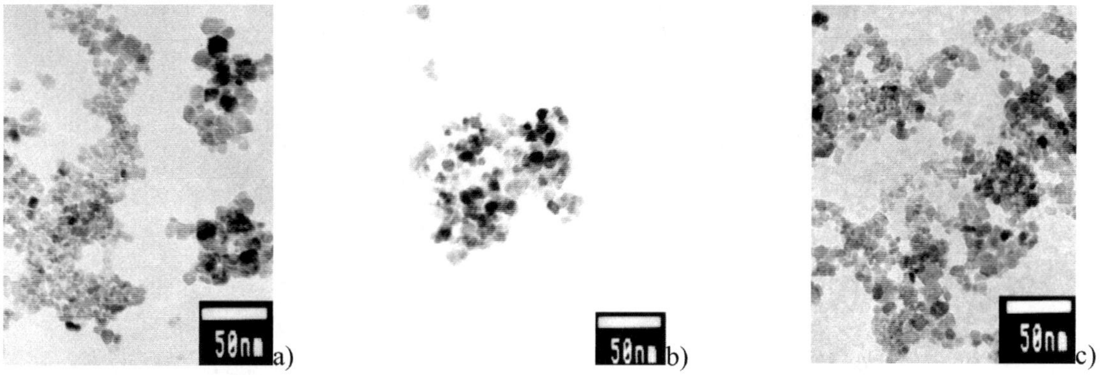

Figure 5. TEM images of the undoped and doped ZnO nanoparticles: a) ZnO; b) Co-doped ZnO; c) Mn-doped ZnO, prepared with the microwave power of 150 W.

In figure 6 the morphology of the ZnO nanoparticles prepared with different powers of the microwave is presented. It can be seen that decreasing power of the microwave results in an aggregation of the nanoparticles.

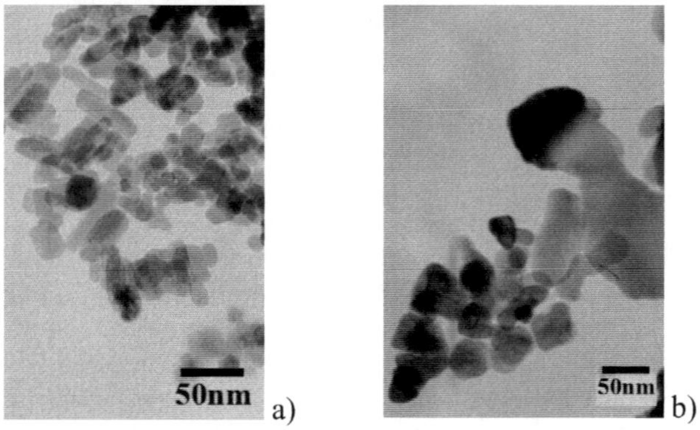

Figure 6. TEM images of the undoped ZnO nanoparticles, prepared with different powers of the microwave a) 120 W and b) 80 W.

The morphology and structure of powders were further investigated by TEM. TEM gives us more details about microstructure of the ZnO nanopowders, as shown in figure 7. The nanopowders are of good transparency to the electron beam. The particles appeared to be well separated from each other. Figure 7a shows the morphologies of ZnO nanoparticles containing mainly spherical particles typically with diameters ranging from 10-15 nm. The micrographs revealed that the particles were nearly spherical in shape (figure 7b). Figure 7c shows an electron diffraction (ED) pattern of the produced nanopowder. The ED pattern shows ring patterns without any additional spots and rings of secondary phases revealing their highly crystalline ZnO wurtzite structure. Three fringe patterns were observed with plane distances of 2.79, 2.58, and 2.44 Å in the electron diffraction pattern which corresponds to (100), (002), and (101) planes of pure wurtzite hexagonal structure of ZnO. The fringe spacing is about 0.28 nm, corresponding to the (100) crystal planes of ZnO (figure 7d). The ED patterns illustrate crystalline, which is in good agreement with the XRD data. The ZnO nanoparticles can be observed as particles having the hexagonal and spherical shapes.

Photoluminescence and absorption spectra of the undoped and Co- and Mn- doped ZnO nanoparticles are shown in figure 8 and figure 9, respectively.

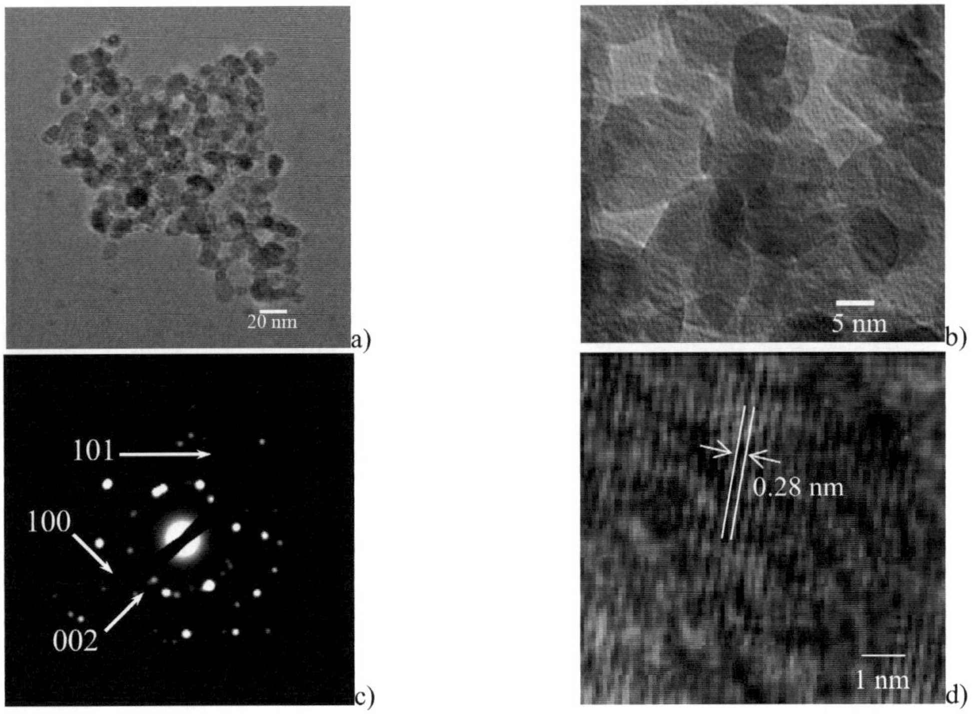

Figure 7. (a) TEM image of ZnO nanopowder; (b) magnified TEM image of ZnO nanopowder; (c) corresponding electron diffraction pattern, and (d) HRTEM image of single ZnO nanoparticle showed (100) crystalline planes.

The absorption spectrum of ZnO nanoparticles dispersed in ethanol solution is shown in figure 8. Compared with bulk ZnO, the blue shift observed in the ZnO nanostructures is due to the quantum size effect.

The green emission band is attributed to the radiative recombination of photogenerated holes with electrons belonging to singly ionized oxygen vacancies on the surface and subsurface. The observation of strong green band emission relative to bulk ZnO indicates the existence of oxygen vacancies concentrated on nanoparticles surface.

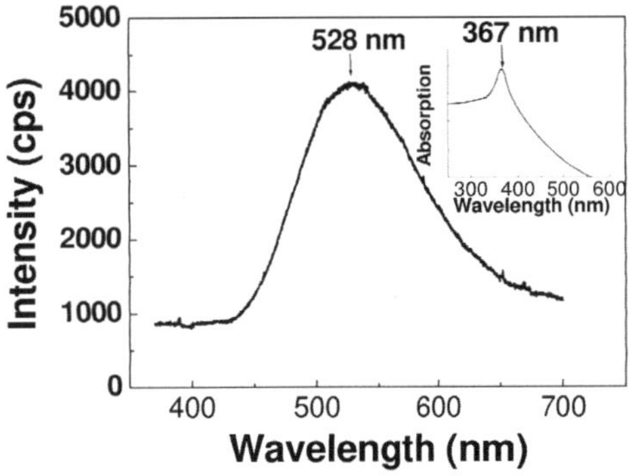

Figure 8. The PL spectrum of undoped ZnO nanopowder. The inset shows UV-vis spectrum of the ZnO nanopowder.

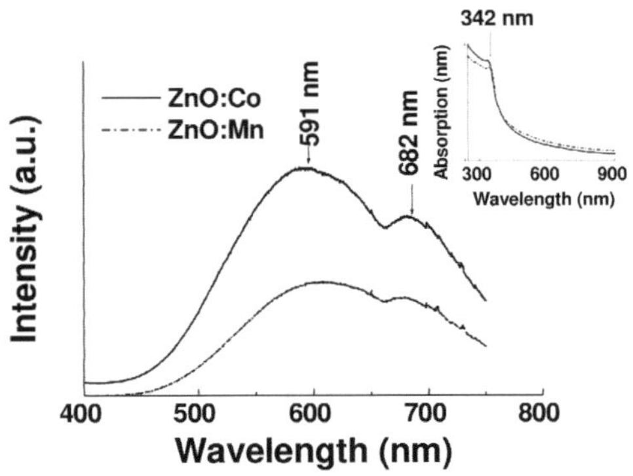

Figure 9. The PL spectrum of the 5% Co- and 5% Mn-doped ZnO samples annealed at 800°C in air for 3 hours. The inset shows UV-vis spectra of Co- and Mn-doped ZnO nanopowder.

The PL spectra of the 5% Co- and 5% Mn-doped ZnO samples annealed at 800°C in air for 3 h (figure 9) are similar to that of undoped ZnO nanopowder. However, besides the broad band at 591 nm, it is noted that there is another peak at 682 nm with a full width at half maximum of 110 meV, which is attributed to the 3d-3d electron transitions. The authors [11, 12] assert that this radiation band corresponding to the $^4T_1 \rightarrow {}^4A_2$ transition of Co^{2+} or Mn^{2+} in the crystalline network ZnO.

The as-prepared samples of 5% Co-doped ZnO and 5% Mn-doped ZnO display a paramagnetic behavior, but after annealing at 800°C in air for 3 h, these samples display room temperature ferromagnetism as can be seen from the hysteresis loop shown in figure 10.

The remanence (M_r) and coercive field (H_c) data for the annealling in air samples displaying ferromagnetism are $M_r = 0.35\times10^{-4}$ emu/g, $H_c = 88$ Oe for Co-doped ZnO samples and $M_r = 0.29\times10^{-4}$ emu/g, $H_c = 29$ Oe for Mn-doped ZnO samples, respectively. So we strongly believe that when ZnO crystallites were doped with cobalt or manganese, Co^{2+} ions or Mn^{2+} ions substituted for Zn^{2+} ions.

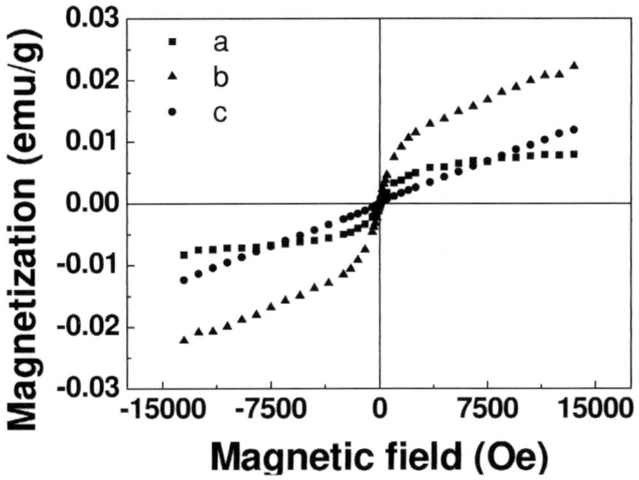

Figure 10. Magnetization as a function of magnetic field for a) 5% Mn-ZnO samples annealed at 800°C in air for 3 hours; b) 5% Co-ZnO samples annealed at 800°C in air for 3 hours; and c) as-prepared samples.

4. Conclusion

The easy and very fast microwave-assisted approach was used for preparation of the undoped and M-doped ZnO (M = Co, Mn) nanoparticles. The XRD measurements showed the formation of single phase compound without any additional phase and the magnetic measurements showed that the as-prepared cobalt- or manganese-doped zinc oxide is paramagnetic. However, thermal annealing of these samples at 800°C in air for 3 hours transforms the magnetic nature from paramagnetic to ferromagnetic. The particle sizes of 10-15 nm determined by using Scherer formula were in good agreement with TEM observations. This microwave-assisted technique may possibly be extended for synthesis of other transition metal-doped oxides.

References

[1] Glaspell G, Dutta P and Manirannan A. 2005 *J. Cluster Sci.* **16** 523
[2] Khrenow V, Schwager F, Klapper M, Koch M and Mulen K. 2007 *Polymer Bull.* **58** 799
[3] Moghaddam F M, Saeidim H. 2007 *Mater. Sci. Eng. B* **139** 265
[4] Hamedani N F and Farzaneh F. 2006 *Journal of Science Islamic Republic of Iran* **17** 231
[5] Cullity B D. 1978 *Elements of X-ray diffraction* (M A: Addison-Wesley) 102
[6] Liqiong J, Dejun W, Baiqui W, Shudan L, Baifu X, Honggang F and Jiazhong S. 2006 *J. Mol. Cat. A, Chemical* **244** 193
[7] Zhu B L, Zeng D W, Wu J, Song W L and Xie C S 2003 *J. Mater. Sci., Material in electronics* **14** 521
[8] Shao M, Li Q, Xie B, Wu J and Qian Y. 2002 *Mater. Chem. Phys.* **78** 288
[9] Cao B, Cai W, Li Y, Sun F and Zhang L 2005 *Nanotechnology* **16** 1734

[10] Singh P, Kumar A,Kaushal A, Kaur D, Pandey A and Goyal R N 2008 *Bull.Mater.Sci.* **31** 573
[11] Denisov A, Volk Y V, Malyarevich A M, Yumashev K V, Dymshits O S, Zhilin A A, Kang U and Lee K 2003 *J.Appl.Phys.* **93** 3827
[12] Yang S G, Pakhomov A B, Hung S T and Wong C Y 2002 *IEEE Transactions on Magnetics* **38** 2877

Acknowledgments
This work is supported by the Fundamental Research Program, Ministry of Science and Technology of Vietnam (Project No 405506) and by Vietnam National University, Hanoi (Project QG 09 05).

Photoluminescence and Raman studies of GaN films grown by MOCVD

Luong Tien Tung[1,3], K L Lin[1], E Y Chang[1], W C Huang[1], Y L Hsiao[1] and C H Chiang[2]

[1] CSD Lab, Department of Materials Science and Engineering, National Chiao Tung University, Taiwan R. O. C.
[2] Department of Electrophysics, National Chiao Tung University, Taiwan R.O.C.
[3] Faculty of Physics, Hanoi National University of Education, Vietnam

E-mail: tungluongtien@gmail.com

Abstract. The photoluminescence (PL) of high quality GaN epitaxial layer grown by MOCVD was investigated for various excitation power and temperatures from 8.3 to 300K. The PL at 8.3 K and with relatively low excitation power of GaN film grown on c-plane sapphire by using MOCVD shows clearly free-exciton A, B and exciton A bound to neutral donors (D^oX) at 3.502 eV, 3.509 eV, and 3.496 eV, respectively. The full width at half maximum (FWHM) and binding energy of exciton A of the high quality GaN film were evaluated as small as 3.7 meV and 27.9 ± 0.5 meV, respectively. In addition, PL and Raman scattering of GaN films grown on r-plane sapphire and (111) Si substrates by using MOCVD were examined. The residual stress effect was detected in all films.

Keyword: Metalorganic chemical vapour deposition, GaN, photoluminescence, binding energy, Raman.

1. Introduction

Gallium nitride (GaN) and its alloys with InN and AIN have emerged as important semiconductor materials with applications to green, blue, and ultraviolet portions of the spectrum as emitters and detectors and to high-power/temperature radio frequency electronic devices. GaN is probably the most important semiconductor since silicon due to their direct and wide band gap, high saturation velocity and high breakdown field. Metal-organic chemical vapour deposition (MOCVD) is currently the most widely used technology. Usually, all optoelectronic commercial device structures are fabricated using MOCVD. Recently, the high quality GaN grown on sapphire or on low cost Silicon substrate by MOCVD have been investigated with low density of threading dislocations (TDs) suitable for Blue LEDs and UV LEDs, which require dislocation densities lower than 10^8 cm^{-2} [1]. Photoluminescence (PL) and Raman are the most powerful methods to study material. The quality of GaN material relates to pure free exciton transition. However, some experimental data appear conflicting, an example of very low value for binding energy of free exciton A (FX_A) in GaN layer grown on sapphire substrate [2 - 4], which is expected to be larger than 27 meV in unstrained GaN. It could be due to the incorrect labelling of free exciton peaks or a mistake of temperature range of pure exciton transition.

In this paper, photoluminescence (PL) and Raman spectra of GaN grown on sapphire and silicon substrate by MOCVD have been examined. The re-combination mechanism in different ranges of

© 2009 IOP Publishing Ltd

temperature was clearly made and the residual strain in GaN layers was deduced from detailed analysis of the energy shift observed on the Raman spectra.

2. Experimental

The 1-μm thick GaN layers were grown at 1080°C by MOCVD on c–plane, r–plane sapphire and (111) oriented Silicon substrates with different buffer layers, low temperature GaN buffer layers, high-low-high temperature (H-L-H Temp.) GaN or AlN layers for instance. The PL measurements were carried out using a Spectra Pro PL system with a He–Cd laser operating at 325 nm as an excitation source. Temperature was controlled between ~ 8 K and 325 K using a closed-cycle cryostat. All the Raman spectra have been recorded with a double grating spectrometer in backscattering geometry, at room temperature; the He–Ne laser (632.8 nm) was used for the excitation. The diameter of the spot in the sample was typically 1 μm.

3. Results and discussion

Figure 1 presents PL spectra of GaN on c-plane sapphire measured at 14 K. The PL spectra exhibits the dominant transition of near band-edge (NBE) band at 3.5013 eV, together with shoulder at higher energy of 3.509 eV and another at lower energy of 3.496 eV. The inset exhibits three emission lines identifying as the shallow neutral donor-bound (D^oX) and free exciton (FX_A and FX_B), the bound-exciton (D^oX) level exist 5.7 meV below the free exciton (FX_A) energy. The splitting of FX_A and FX_B levels is 6.6 meV. Our results agree with [5, 8], and the linewidths of D^oX, FX_A and FX_B as small as 4.0, 3.9 and 6.9 meV, respectively, show the high quality crystal of GaN layer [5, 6, 7]

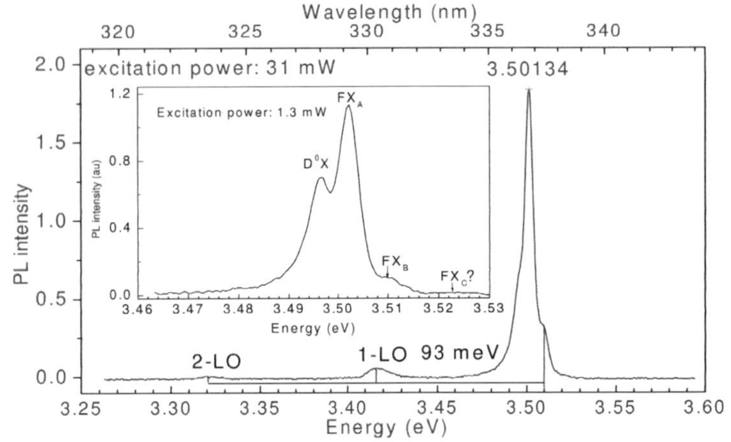

Figure 1. (a) PL spectra of GaN on sapphire at 14 K and 31 mW of excitation power. Inset: near band-edge band of GaN at 14 K and 1.3 mW excitation power.

The temperature dependence of GaN film was investigated. The redshift and broadening of all emission lines with increasing temperature were observed. The temperature dependence of the linewidth is attributed to phonon-induced band broadening [9]. The characteristic redshift with increasing temperature is attributed to band gap reduction from lattice expansion and to electron-phonon interactions [10]. The first contribution has major play at low temperature. In figure 2a, the temperature dependence of the energy position of lines (D^oX, FX_A, and FX_B), therefore, is similar and the energy level difference between at 14 K and at 85 K is 4.9 meV. As the temperature increases, the quick decrease in PL intensity was observed and the shallow neutral donor-bound (D^oX) transition is extinguished above 100 K. In figure 3b, we show PL intensity variation as a function of temperature and the fit curves were described by formula:

$$I(T) = \frac{I(0)}{1 + C_1 \exp\left(-\frac{\Delta E_1}{kT}\right) + C_2 \exp\left(-\frac{\Delta E_2}{kT}\right)} , \quad (3)$$

where, ΔE is activation energy. The thermal quenching of the FX_A exciton is satisfactorily fitted with two nonradiative processes with activation energies of 5.7 ± 0.5 meV and 27.9 ± 0.5 meV, respectively. The first process is attributed to the change of the exciton from the lower branch of the *A* exciton to the branch of the *B* exciton [11]. The increase of the population in the *B* branch is also evidenced by the enhancement of FX_B intensity in the temperature range lower 25 K (figure 3b). At higher temperature (> 40 K) the PL intensity of FX_A and FX_B quench with the same activation energy (27.9 meV), it has been reported that the binding energy of the A and B exciton are almost the same and the excitons binding energy of GaN/c-plane sapphire are slightly larger than in unstrained GaN (~ 26 – 27 meV) [11-15].

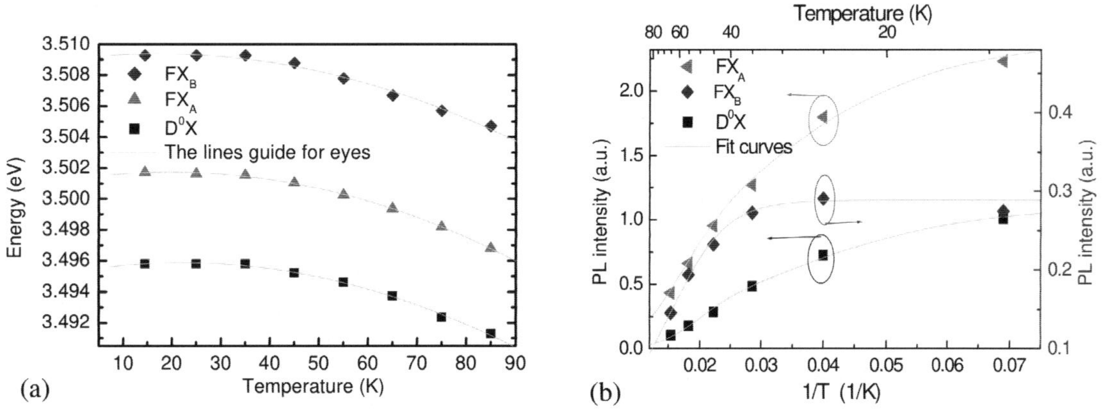

Figure 3. The temperature dependence of energy position of GaN/sapphire (a) and PL intensity various 1/T; (b) of FX_B (♦-diamond), FX_A (▲-triangle), and D°X (■-square). The fit curves were described by formula (1).

The quenching of the donor-bound exciton D°X emission is also described by two nonradiative processes. The first process is assigned for thermal destruction of shallow donor–exciton bound with activation energy of 5.3 ± 0.5 meV. This value is in a good agreement with the difference of energy level between D^oX and FX_A. The second process with activation energy of 32.3 ± 0.5 meV can be attributed to the ionization of the donors or to a simultaneous exciton delocalization and exciton dissociation [11, 16]. This value is in excellent agreement with the binding energies of Si_{Ga} and O_N donors, 30.18 ± 0.1 and 33.2 ± 0.1 meV, respectively [17].

At low temperatures and low excitation densities, free carriers can form free exciton states or become bound to impurities so that the luminescence spectra are dominated by these states rather than by band-to-band recombination. However, the band–to–band transition cannot be ignored at higher temperature. The excitation power-dependent photoluminescence has been widely used for determining the origin of light emission in semiconductors, $I_{PL} = \mu I_{laser}^{\alpha}$(2) [18, 19]. In this relation, it is predicted that for free-exciton recombination $\alpha \approx 1$; for band gap recombination, i.e., free-carrier electron-hole bimolecular recombination, $\alpha \approx 2$; and for the intermediate case involving free exciton and free carriers $1 < \alpha < 2$. Table 1 shows the α values and recombination mechanism of transitions in different ranges of temperature.

Table 1. The α values in different ranges of temperature. The coefficient α is obtained by fitting the measured power dependence of the luminescence lines by $I_{PL} = \mu I_{laser}^{\alpha}$ formula.

	D^oX	FX_A	NBE	NBE
Temp. range	8.3 K	8.3 – 75 K	100 – 300 K	375 – 450 K
α	~ 1[a]	~ 1	1.5 – 1.73	~ 2
Recombination mechanism	Donor bond exciton	Pure free exciton	Both free exciton and free carriers band-to-band	Free carriers band-to-band

[a] This value is obtained at low range of excitation power (< 5 mW).

In addition, the dependence of PL intensity of D^oX on excitation power at 8.3 K is shown in figure 4a. It saturates at about 12 mW. The decrease of PL intensity and the redshift of energy of FX_A as the excitation power increases higher than 15 mW because of the inelastic exciton-exciton interaction are presented in figure 4a and 4b. It implies that the accuracy of calculating values must be based on the measurements at sufficiently low excitation power and low range temperature (3.6 mW and lower 85 K in our experiments).

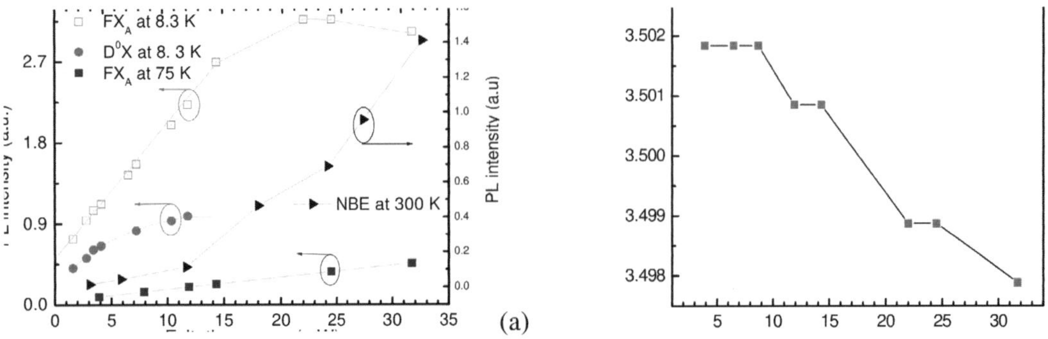

Figure 4. (a) The PL relative intensity of GaN/sapphire: FX_A at 8.3 K (□ open square), D^oX at 8.3 K (● circle), FX_A at 75 K (■ solid square), and near band-edge emission (NBE) (▶ triangle) and (b) energy of FX_A at 8.3 K at various excitation power.

Growth of GaN on r-plane sapphire substrate is expected to improve the efficiency of light emitting diodes (LEDs) by removing the spontaneous polarization and the piezoelectric fields to the quantum wells active layer based on heterostructure [21-24]. In contrast with GaN/c-plane sapphire substrate, the GaN/r-plane sapphire has significantly higher density of the basal plane (0001) stacking faults (BSFs) because of the planar anisotropic nature of the growing surface [25]. The BSFs play role as luminescence active centers in GaN layer.

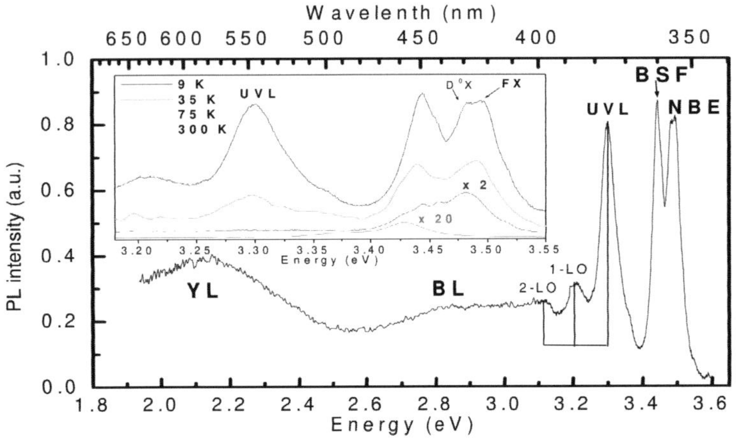

Figure 5. Typical PL spectra of GaN/r-plane sapphire substrate grown by MOCVD measured at 9K. Inset: PL spectra at different temperatures. Note PL intensity is magnified by 2 at 75 K and by 20 at 300 K.

Figure 5 shows the typically PL spectra of GaN grown on r-plane sapphire by MOCVD measured at 9 K. NBE emission band involving free exciton FX transition energy of 3.494 eV and D^0X transition energy of 3.484 eV was observed. The band labelled BSF at 3.443 eV belongs to basal plane stacking faults [26, 27]. It is a specific band in GaN on non-c-plane sapphire substrate and dominant at low temperature. In examining GaN/c-plane sapphire, the broad band at 3.300 eV which followed by LO replica sidebands separated by 91 meV contributes to both donor-acceptor-pair (DAP) and free-electron-acceptor (e-A) emission, called ultra-violet luminescence band (UVL) [14]. In particular, in GaN/r-plane sapphire, UVL also is contributed by the transitions related to prismatic stacking faults

formed on {11$\bar{2}$0} planes [26, 28, 29]. In inset of figure 5, PL spectra at different temperature reveal very fast thermal quenching of UVL band due to primarily related ionization of acceptors and donors by heating. While BSFs act as the quantum-well-like region confine electrons. Thus, BSFs emission can be observed at elevated temperatures [26]. Two very broad bands, a blue band (BL) at about 2.85 eV and a yellow band (YL) at about 2.1 meV, have been detected. In undoped GaN, the BL band is caused by transition from shallow donors (at low temperature) or conduction band to relatively deep acceptor (Zn_{Ga}), from Zn contamination by impure gases used in MOCVD, while the YL band relates to V_{Ga} or V_{Ga}-complex, such as $V_{Ga}O_N$ or $V_{Ga}Si_N$ [14].
Although the spontaneous polarization and the piezoelectric fields can be eliminated from AlGaN/GaN heterostructure, the high density of basal plane stacking faults (BSFs) is a big challenge in growth GaN on r-plane sapphire substrate by MOCVD, so far.

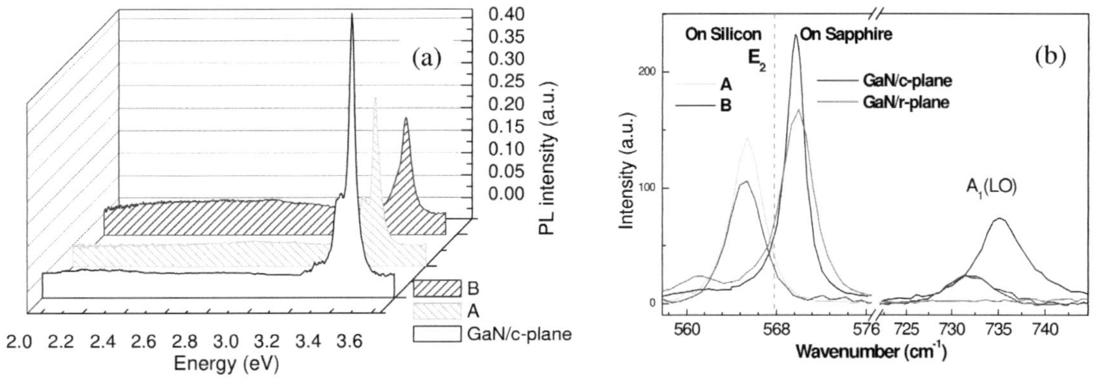

Figure 6. (a) Room temperature PL spectra of GaN grown on c-plane sapphire and on silicon substrates; (b) Raman spectra of GaN grown on silicon substrates (A, B); on c-plane, c-plane sapphire substrate at room temperature. The vertical dotted line denotes E_2 mode frequency position of strain-free GaN of 567.6 cm^{-1} at room temperature.

GaN grown on silicon substrate lowers cost and uses better heat dissipation in comparison with sapphire substrates but it is not as good as GaN grown on sapphire substrate. In figure 6a, we show room temperature PL of two samples, called A and B grown on silicon substrates using the same buffer layer structure H-L-H temperature AlN; and GaN grown on c-plane sapphire substrate. The PL spectra reveal that both samples A and B have the NBE band at the same energy position of 3.40 eV dominantly attributed to band-to-band transition. The NBE band is more broaden, however, and the blue band (BL) and yellow band (YL) relate to the defects are clearly observed in the B sample. It is indicated that there is higher defects dislocation in the B sample than in the A sample. In the GaN grown on c-plane sapphire substrate, the NBE band at 3.43 eV is very strong and sharp, while YL band cannot hardly be nearly observed. The NBE band shift can be caused not only by strain but also strongly by the Burstein-Moss effect. Therefore, we move to study Raman measurement of the E_2 Raman mode that is known to be shifted by stress only [30]

Raman is a powerful method to examine quality and to monitor residual stress of the GaN film. The width of the strongest mode E_2 (high) can be used to analyze lattice defects [5, 31], whereas its frequency shift can be well used to monitor residual stress [5, 32, 33]. The residual stress due to mismatch of the lattice constants and the thermal expansion coefficients of the substrate and epitaxial film significantly effects the optical and electrical properties of epitaxial film. The differences in thermal expansion coefficients between layers and substrates dominantly contribute to the residual strain [31]. Usually, growth GaN films on sapphire substrates induce compressive strains and on silicon substrates induce tensile strains. The E_2 mode of strain-free GaN is known to be 567.6 cm^{-1} at room temperature [31, 34].

In investigating, Raman spectra of all samples were measured at room temperature. The measured Raman spectra of all samples are presented in figure 6b. The upshift of the E_2 peak frequencies is corresponding to compressive strain, GaN grown on sapphire substrates, whereas the tensile strain in GaN grown on silicon substrates is observed by the decrease in E_2 mode frequency in comparison to strain-free GaN. While E_2 mode is only affected by residual stress and crystalline quality. The A_1(LO) mode depends not only on the residual stress in GaN film but also on the coupling to free carriers (plasmon) of LO mode [35, 36]. It is beyond the scope of study in this work.
The frequencies position and linewidth of E_2 peaks at room temperature of all samples given in table 1. The Raman shift and linewidth (Γ, full width at half maximum) are determined by fitting Lorentzian functions to the experimental data.

Table 2. Energy position, linewidth of NBE bands, frequencies position and linewidth of E_2 peaks at room temperature of all samples.

Sample	E_2 (cm^{-1})	Γ of E_2 (cm^{-1})
GaN/c-plane sapphire	569.68	1.98
GaN/r-plane sapphire	569.86	3.25
A (GaN/silicon)	565.40	3.40
B (GaN/silicon)	565.22	3.70
References	567.6 [31,34]	1.75 [5]; 2.0 [31]

Linewidth of B sample is larger than that of A sample to indicate that B sample has higher defects density than A sample, this result is good in agreement with above PL examination. The linewidth of E_2 peak is as narrow as 1.98 cm^{-1} compares well with previously published reports [5, 7, 31] indicating very low defects density in GaN grown on c-plane sapphire sample. In comparison to A and B samples, the influence of defects on the shift of E_2 peak is small. Thus, we can neglect the hydrostatic stress component, which can arise from point defects in low quality GaN or even from dopants in doped GaN layers. For pure biaxial stress, the frequency shift of the E_2 phonon with respect to unstrained GaN is related to stress-induced changes the E_2 phonon of 4.2 cm^{-1} GPa^{-1} [37]. The upshifted E_2 frequency in GaN film grown on c-plane sapphire corresponds to a compressive stress of 0.5 GPa and higher values in the other samples.

4. Conclusion

GaN films grown by MOCVD on different substrates were examined by Photoluminescence (PL) and Raman measurement. We show that in the high quality crystal GaN grown on c-plane sapphire, the PL linewidth of free exciton A transition at 8.3 K and Raman E_2 mode line are as small as 3.7 meV and 1.98 cm^{-1}, respectively. Thermal quenching of the different transitions was observed by study dependent temperature PL. It reveals very fast thermal quenching of the transitions relating to exciton bond to acceptors, donors, or dislocations. The binding energy of free exciton A and B in compressive strain GaN/c-plane sapphire are the same of 27.9 meV and slightly higher than that in strain-free GaN. The pure free exciton recombination is just at a very low temperature. We also point out that for accuracy of above calculations the measurements necessarily have to carry out at sufficiently low range of temperatures and excitation power. In comparison with GaN/c-plane, the GaN films grown on r-plane and silicon substrates are higher residual stress and lower quality crystal.

References

[1] Gibart P 2004 *Pep. Prog. Phys.* **67** 667-715
[2] Smith M, Chen G D, Lin J Y, Jiang H X, Asif-Khan M, Sun C J, Chen Q, Yang J W 1996 *J. Appl. Phys.* **79**, 7001-7004
[3] Smith M, Chen G D, Li J Z, Lin J Y, Jiang H X, Salvador A, Kim W K, Aktas O, Morkoc A 1995 *Appl. Phys. Lett.* **67**, 3387-3389

[4] Reynolds D C, Look D C, Kim W, Aktas O, Botchkarev A, Salvador A, Morkoc H 1996 *Appl. Phys.* **80**, 594-596
[5] Song D Y, Basavaraj M, Nikishin S A, Holtz M, Soukhoveev V, Usikov A and Dmitriev V 2006 *J. Appl. Phys.* **100** 113504
[6] Keller S, Keller B P, Wu Y-F, Heying B, Kapolnek D, Speck J S, Mishra U K and DenBaars S P 1996 *Appl. Phys. Lett.* **68** (11) 1525
[7] Martínez O, Avella M, Jimenez J, Gerard B, Cusco R and Artus L 2004 *J. Appl. Phys.* **96** 3639
[8] Shan W, Schmidt T J, Yang X H, Hwang S J, Song J J and Goldenberg B 1995 *Appl. Phys. Lett.* **66** 985
[9] Rudin S, Reinecke T L and Segall B 1990 *Phys. Rev.* B **42** 11218
[10] Capaz R B, Spataru C D, Tangney P, Cohen M L and Louie S G 2005 *Phys. Rev. Lett.* **94** 036801
[11] Paskov P P, Paskova T, Holtz P O and Monemar B 2004 *Phys. Rev.* B **70** 035210
[12] Monemar B, Bergman J P, Buyanova I A, Li W, Amano H and Akasaki I 1996 *Material Research Soc. Internet Journal of Nitride Semiconductor Research* **1** 2
[13] Eckey L, Holst J.-Chr, Hoffmann A, Broser I, Detchprohm T and Hiramatsu K 1997 *Material Research Soc. Internet Journal of Nitride Semiconductor Research* **2** 1
[14] Reshchikov M A and Morkoç H 2005 *J. Appl. Phys.* **97** 061301
[15] Rodina A V, Dietrich M, Göldner A, Eckey L, Hoffmann A, Efros Al L, Rosen M and Meyer B K 2001 *Phys. Rev.* B **64** 115204
[16] Leroux M, Grandjean N, Beaumont B, Nataf G, J Massies and Gibart P 1999 *J. Appl. Phys.* **86** 3721
[17] Moore W J, Freitas J A, Lee S K, Park S S and Han J Y 2002 *Phys. Rev.* B **65** 081201
[18] Fouquet J E and Siegman A E 1985 *Appl. Phys. Lett.* **46** 280
[19] Bergman L, Cheng X B, Morrison J L and Huso J 2004 *J. Appl. Phys.* **96** 675
[20] Chen X B, Huso J, Morrison J L and Bergman L 2006 *J. Appl. Phys.* **99** 046105
[21] Bernardini F, Fiorentini V and Vanderbilt D 1997 *Phys. Rev.* B **56** R10024
[22] Lefebvre P *et al.* 2001 *Appl. Phys. Lett.* **78** 1252
[23] Nishida T and Kobayashi N 1999 *Phys. Stat. Sol.* (a) **176** 45
[24] Fiorentini V, Bernardini F, Sala F.D, Carlo A D and Lugli P 1999 *Phys. Rev.* B **60** 8849
[25] Paskov P P, Paskova T, Holtz P O and Monemar B 2001 *Phys. Rev.* B **64** 115201
[26] Paskov P P, Schifano R, Monemar B, Paskova T, Figge S and Hommel D 2005 *J. Appl. Phys.* **98** 093519
[27] Wu Z H, Fischer A M, Ponce F A, Bastek B, Christen J, Wernicke T, Weyers M and Kneissl M 2008 *Appl. Phys. Lett.* **92** 171904
[28] Liu R, Bell A, Ponce F A, Chen C Q, Yang J W and M Khan A 2005 *Appl. Phys. Lett.* **86**, 021908
[29] J Mei, S Srinivasan, Liu R, Ponce F A, Narukawa Y and Mukai T 2006 *Appl. Phys. Lett.* **88** 141912
[30] Perlin P, Jauberthie-Carillon C, Itie J P, Miguel A S, Grzegory I and Polian A 1992 *Phys. Rev.*B **45** 83
[31] Kitamura T, Nakashima S, Nakamura N, Furuta K and Okumura H 2008 *Phys. Stat. Sol.* (c) **5** No. 6 1789
[32] Inoue T, Toda Y, K Hoshino, Someya T and Arakawa Y 2003 *Phys. Stat. Sol.* (c) **0** No. 7 2428
[33] Park H J, Park C, Yeo S, Kang S.W, Mastro M, Kryliouk O, and Anderson T J *Phys. Stat. Sol.* (c) **2**, No. 7, 2446 (2005).
[34] Davydov V Yu, Kitaev Yu E, Goncharuk I N, Smirnov A N, Graul J, Semchinova O, Uffmann D, Smirnov M. B, Mirgorodsky A. P. and Evarestov R. A. 1998 *Phys. Rev.* B **58** 12899
[35] Kozawa T, Kachi T, Kano H, Taga Y, and Hashimoto M, Koide N and Manabe K 1994 *J. Appl. Phys.* **75** (2) 1098
[36] Kuball M 2001 *Surf. Interface Anal.* **31** 987–999

[37] Kisielowski C, Kruger J, Ruvimov S, Suski T, Ager J W III, Jones E, Liliental-Weber Z, Rubin M, Weber E R, Bremser M D and Davis R F 1996 *Phys. Rev.* B **54** 17745

Acknowledgments

This work was supported by the Ministry of Economic Affairs and the National Science Council of the Taiwan under the contracts: NSC 96-2752-E-009-001-PAE and 96-EC-17-A-05-S1-020.

Controlled synthesis and luminescence of Eu doped ZnO nanowires and nanorods via hydrothermal method

Trinh Thi Hang[1], Trinh Xuan Anh[2] and Pham Thanh Huy[1]

[1] Hanoi Advanced School of Science and Technology (HAST)
[2] Faculty of Chemistry and Technology (FCT),
Hanoi University of Technology, 1 Dai Co Viet road, Hanoi, Vietnam

E-mail: trinhxuananh-fct@mail.hut.edu.vn

Abstract. Undoped and Eu^{3+} doped ZnO nanostructures were prepared by single-step mild hydrothermal conditions. The nanostructures were deposited on a Si or glass substrate without template. The reaction was carried out for 10–20h which resulted in morphology and aspect ration of the final products depend on the concentration of the precursors and reaction time. The concentration of dopant Eu^{3+} in the synthesized nanostructures was below 4 at.%. The crystal structure and the microstructure of the doped nanostructures were studied and compared with undoped ZnO. Photoluminescence measurements of the nanowires exhibit a strong band gap emission at room temperature. In partcular, relative intensity of band gap and defect emission changes drastically in ZnO:Eu and ZnO nanostructures with different morphology. The highly crystalline ZnO nanorods, nanowires can be a good one-dimensional candidate to study optical and electronic properties

Keyword: luminescence, nanowires, nanorods, nanostructures.

1. Introduction

One-dimensional nanoscale materials have stimulated considerable interest recently because of their unique electronic, optical and mechanical properties as well as their potential application in nanodevices. Many attempts have been made to synthesize one-dimensional nanostructured materials. One of the widely synthesized materials is ZnO, which has important electronic and photonic properties. The ZnO is of interest in many applications including electrodes for solar cells [1], transparent conductive coatings [2], gas sensors [3] and electro- and photoluminescent materials [4]. Many synthetic methodologies for one-dimensional ZnO nanostructures have been developed, including vapour phase methods such as thermal evaporation [5], chemical vapor deposition [6] and metal–organic chemical vapor deposition [7]... However, controlled synthesis of large scale ZnO nanostructures still needs further investigation.

In recent years, rare-earth (RE) doped ZnO nanostructured materials have attracted great attention in both fundamental studies and applications [8]. ZnO is a wide band gap (3.37 eV) semiconductor material at room temperature [9]. This property makes it a good candidate as host materials for the visible and infrared emission of various rare-earth ions. On the other hand, rare-earth ions are good luminescence centers due to their narrow and intense emission lines originating from 4f intrashell transitions.

In this work, we present the effects of precursor concentration, reaction time on morphology of undoped and Eu^{3+} doped ZnO nanostructures prepared by mild hydrothermal conditions. The

© 2009 IOP Publishing Ltd

photoluminescence (PL) of undoped and Eu^{3+} doped ZnO nanorods, nanowires are studied and discussed.

2. Experimental

The samples of undoped and Eu^{3+} doped ZnO nanostructures were prepared on substrates (glass slides or silicon wafers) by hydrothermal methods. The substrates were placed in chemical batches containing equimolar ratio of zinc nitrate and hexamine. The growth process was carried out in a laboratory oven at 95°C for 10-24h. For the experiments studying the effect of precursor concentration (0.0001–0.1M), the ratio of Zn–amine remained constant at 1 : 1. Subsequently, the final products were washed with distilled water and then dried at 95°C for 60 minutes to remove any residual salts and organic materials. The crystal phase of the as-synthesized nanostructures were analyzed using an X-ray diffractometer (XRD). The morphology of the studied samples were characterized using a scanning electron microscope (FESEM 4800 Hitachi, secondary electron imaging) and PL properties were measured using photoluminescence system (Aston, Winspec) mapping system with a He–Cd laser (325 nm).

3. Results and discussion

The typical XRD pattern of the as-prepared ZnO nanotubes and nanowires is shown in figure 2. All the diffraction peaks of both nanotubes and nanowires can be indexed to the hexagonal ZnO. No characteristic peaks were observed for the other impurities.

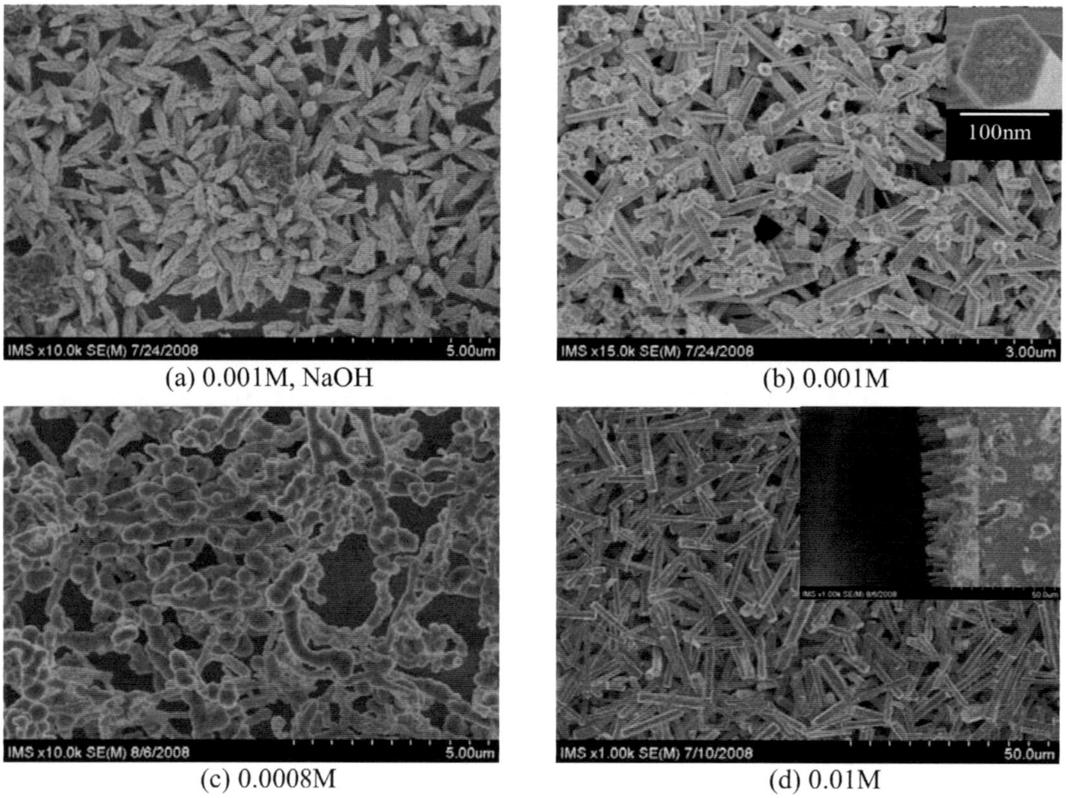

Figure 1. FESEM images of ZnO nanostructures growth at different precursor concentration and reaction conditions. (a) 0.001M with presence NaOH for 10h, (b) 0.001M for 10h, (c) 0.0008M for 20h, (d) 0.01M for 12h (with presence of Mn^{2+} as in inset figure).

Figure 1a shows a FE-SEM micrograph of ZnO nanostructures grown on the substrate at a precursor concentration of 0.001M with presence of NaOH after a 10h hydrothermal process. The morphology shows bead shapes of nano-particles aligned along the substrate. The rough surface of particles is due to speed up of hydrothermal process when NaOH was added. Figure 1b shows the morphology after undergoing a hydrothermal process for 10h; the ZnO nanorods have hexagonal structures with narrow distribution of diameter of ~100 nm and a length of ~2 μm. ZnO nanowires (figure 1c), it is observed that the grown nanowires are well aligned along the substrate the nanowires is a result of intergrowth and branching of nanospheres. The microrods were formed after hydrothermal process of 0.01M precursors concentration for 12h, these microrods also aligned along the substrate. However, the microrods vertically aligned on the substrate when Mn^{2+} was added into initial solution as in inset figure of figure 1d. The morphology of ZnO nanostructures strongly depended on the precursor concentration and reaction conditions.

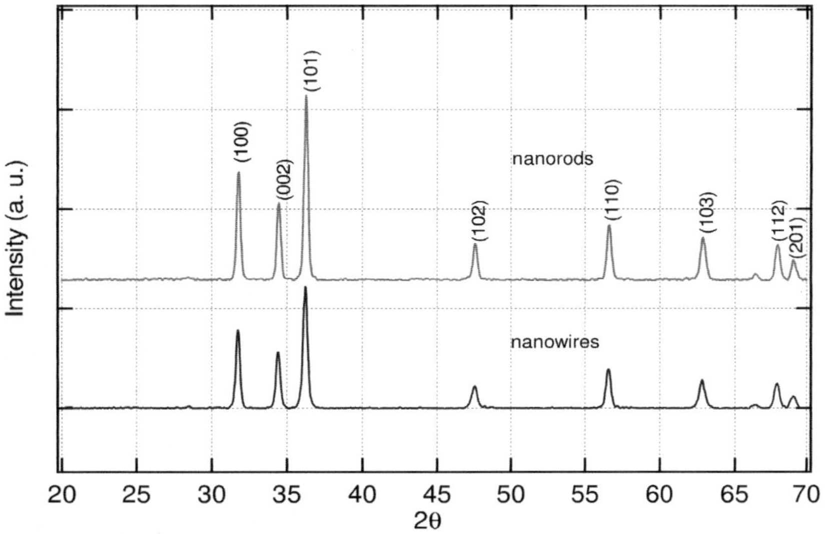

Figure 2. X-ray diffraction patterns of ZnO nanorods and nanowires.

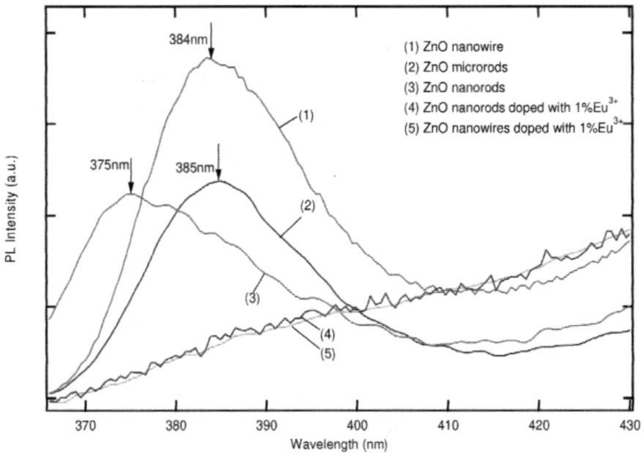

Figure 3. The UV emission of undoped and Eu^{3+} doped ZnO nanostructures.

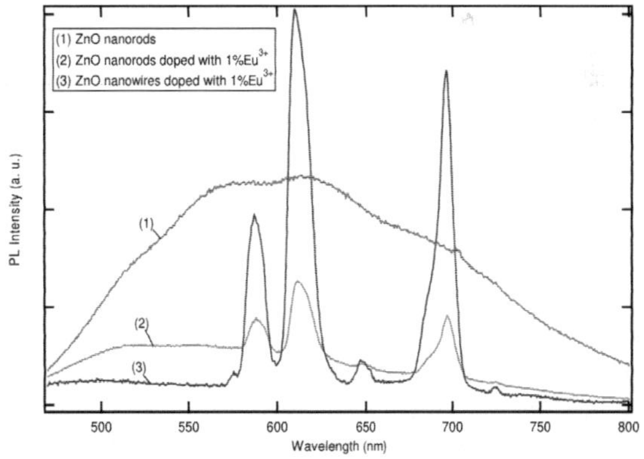

Figure 4. The visible emission of undoped and Eu^{3+} doped ZnO nanostructures.

The photoluminescence of ZnO nanostructures were measured using excitation 325 nm He-Cd laser at room temperature. The undoped ZnO samples show the band-edge emission around 380 nm and a broad peak located between 450 and 800 nm which can be attributed to free-exciton annihilation and deep level emission. The UV and visible peaks are familiar in the photoluminescence spectrum of ZnO nanostructures. As the concentration of the precursor or rather the diameters of the ZnO nanorods increased, it is observed that peak red-shifting of band-edge emission occur. The nanorods exhibit a band-edge emission at 375 nm as compared to the microrods, which has a peak at 385 nm as shown in the figure 3. In view of this, the spectral shift of the nanorods with increasing diameter is attributed to a decrease in crystal defects [10] and not the confinement effect, which is likely to be observed at low temperature [11].

The UV and visible peaks in the photoluminescence spectrum of ZnO are drastically decreased in Eu^{3+} doped ZnO samples as show in figure 3 and figure 4. The visible broad peaks of ZnO are decreased when ZnO doped with Eu^{3+}. This result implies that the defects are decreased by the doping of Eu^{3+} ions into ZnO nanorods and nanowires. The UV peaks of ZnO are disappeared in samples ZnO nanorods and nanowire doped with Eu^{3+}. This result indicates an energy transfer from the host ZnO nanostructures to Eu^{3+} ions. Several sharp-peaks in the photoluminescence spectrum are observed in ZnO nanorods and nanowire doped with Eu^{3+}, and the energies of all these peaks agree well with the energies of intra-4f transitions of Eu^{3+} ions. However, the intensity of these peaks in ZnO nanowires is much stronger than in ZnO nanorods due to effect of energy transfer from the host ZnO nanowires to Eu^{3+} ions is better than from the host ZnO nanorods. The ratio between PL peaks of Eu^{3+} in ZnO one-dimension nanostructures (nanorods and nanowires) are different compared to the nanospheres as show in figure 5. The transitions of $5D_o \rightarrow 7F_1$ and $5D_o \rightarrow 7F_4$ are enhanced in 1D-nanostructures. This results implies that the structures of host ZnO have effects on transition random from $5D_o$ to $7F_J$ ($J = 1 \rightarrow 4$) in Eu^{3+} ions.

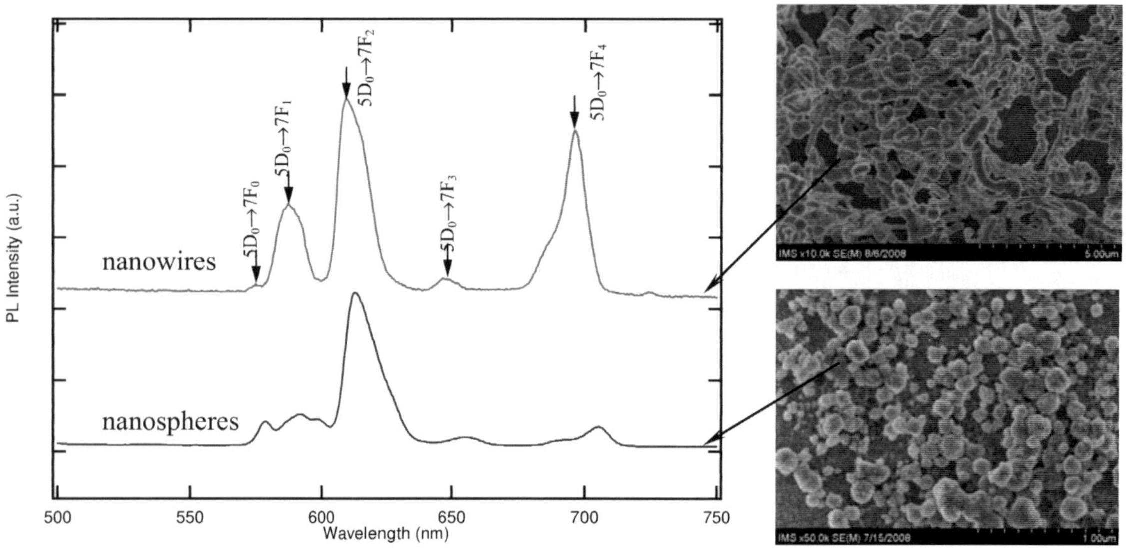

Figure 5. PL spectra of Eu^{3+} doped ZnO nanowires and nanospheres.

4. Conclusions

In summary, we have realized a high-yield synthesis of ZnO nanostructures via mild hydrothermal reaction at low temperature. We can obtain various morphology of ZnO nanostructure through the control of nucleation events by an organic additive and modification of the amount of Zn concentration. The synthesized ZnO nanostructures exhibit hexagonal wurtzite structure, high crystallinity, and narrow distribution. We have studied PL properties of undoped and Eu-doped ZnO nanorods and nanowires. The ZnO:Eu 1D-nanostructures exhibit the strong PL peaks of Eu^{3+} due to energy transfer from the host to Eu^{3+} ions.

References

[1] Law M, Greene L, Johnson J.C, Saykally R, Yang P, 2005 *Nat. Mater.* **4** 455
[2] Minami T, Vac J 1999 *Sci. Technol. A* **17** 1765
[3] Wan Q, Li Q H, Chen Y J, Wang T H, He X L, Li J P, Lin C L, 2004 *Appl. Phys. Lett.* **84** 3654
[4] Mo C M, Li Y H, Lin Y S, Zhang Y, Zhang L P 1998 *J. Appl. Phys.* **83** 4389
[5] Huy P T, An T T, Chien N D, Kim D 2008 *Adv. Mate. Res.* **31** 68

[6] Wu J J, Liu S C 2002 *Adv. Mater.* **14** 215
[7] Zhang B P, Binh N T, Segawa Y, Wakatsuki K, Usami N 2003 *Appl. Phys. Lett.* **83** 1635
[8] Kossanyi J, Kouuyate D, Pouliquen J, Ronfard-Haret J C, Valat P 1990 *J. Lumin.* **46** 17
[9] Zamfirescu M, Kavokin A, Gil B, Malpuech G, Kaliteevski M 2002 *Phys. Rev. B* **65** 161
[10] Wang X, Ding Y, Summers C J, Wang Z L 2004 *J. Phys. Chem. B* **108** 8773
[11] Koyano M, Bao P Q, Binh L T T, Ha L H, Long N N, Katayama S 2002 *Phys. Stat. Solid A* **193** 125

Acknowledgments
This work is supported in part by the National Basic Research Program Grant.

Effect of electrochemical etching solution composition on properties of porous SiC film

Dao Tran Cao[1], Cao Tuan Anh[2], Nguyen Thi Thu Ha[1], Huynh Thi Ha[1], Bui Huy[1], Pham Thi Mai Hoa[1,3], Pham Hong Duong[1], Nguyen Thi Thanh Ngan[1] and Ngo Xuan Dai[4]

[1]Institute of Materials Science, Vietnam Academy of Science and Technology, 18 Hoang Quoc Viet Road, Cau Giay Distr., Hanoi, Vietnam
[2] Institute of Physics, Vietnam Academy of Science and Technology, 10 Dao Tan Street, Ba Dinh Distr., Hanoi, Vietnam
[3] Delft University of Technology, Mekelweg 4, 2628 CD Delft, the Netherlands
[4] College of Science, Vietnam National University – Hanoi, 334 Nguyen Trai Road, Thanh Xuan Distr., Hanoi, Vietnam

E-mail: dtcao@ims.vast.ac.vn

Abstract. Porous amorphous SiC (a-SiC) layer with pore size in the nanometer region was fabricated on the a-SiC/Si substrates by the electrochemical etching method using HF/H_2O/surfactant solution. Systematic study showed that the HF concentration in the etching solution (in the 1-73% region) strongly affects the structure (both the pore size and the pore density) of the porous a-SiC layer. It was also observed the changing of the photoluminescence properties of the porous a-SiC layer when its structure has been changed.

Keywords: Porous SiC, thin film, anodization, photoluminescence.

1. Introduction

Silicon carbide (SiC) has many excellent properties, such as wide band gap, high breakdown voltage, high-saturated electron drift velocity, and high thermal conductivity. Moreover, it is stable at high temperature and in hostile environments. Its wide band gap also renders it applicable for blue light emitting devices. Unfortunately, its indirect band gap makes SiC hardly suited for optical device application although it has been used for blue light emitting diodes. The discovery of porous silicon (PSi) emitting strong light have inspired research on the blue light emission of porous SiC (PSiC). Furthemore, the interest to PSiC has grown up, because unlike the red luminescence in PSi, which is reduced very quickly in air due to oxidation, many authors have found that the blue luminescence in PSiC degrades very little in air at room temperature [1, 2].

In a previous work [3], we have concentrated our efforts to find out a way for making a porous SiC layer on the nondoped high-thickness amorphous SiC (a-SiC) films fabricated in the DIMES, T.U. Delft (The Netherlands). This was a difficult task due to the high resistivity character of the nondoped high-thickness a-SiC films used. Nevertheless, we have solved this task by using high anodizing voltages (up to 500 V for the 6 μm thick a-SiC films) and aqueous 73% HF electrolyte containing 1% Triton X-100 surfactant. Continuing the research in this direction, in this work we report our study

© 2009 IOP Publishing Ltd

results concerning the role of the Triton X-100 surfactant in the anodic etching process and the effect of the HF concentration in the electrolyte on the structural and photoluminescence properties of the porous a-SiC layer. There are two reasons why we have chosen porous a-SiC as the object of study. Firstly, porous a-SiC is not sufficiently studied, in literature we could find only a few works concerning porous a-SiC [4-6], but with the thickness of the a-SiC films much lower and in addition, in general, the a-SiC films were p-doped. Secondly, a-SiC films are rather easy-fabricated and low-cost films, but at the same time, in our opinion, they are perspective for many applications.

2. Experimental, results and discussion

2.1. a-SiC anodic etching

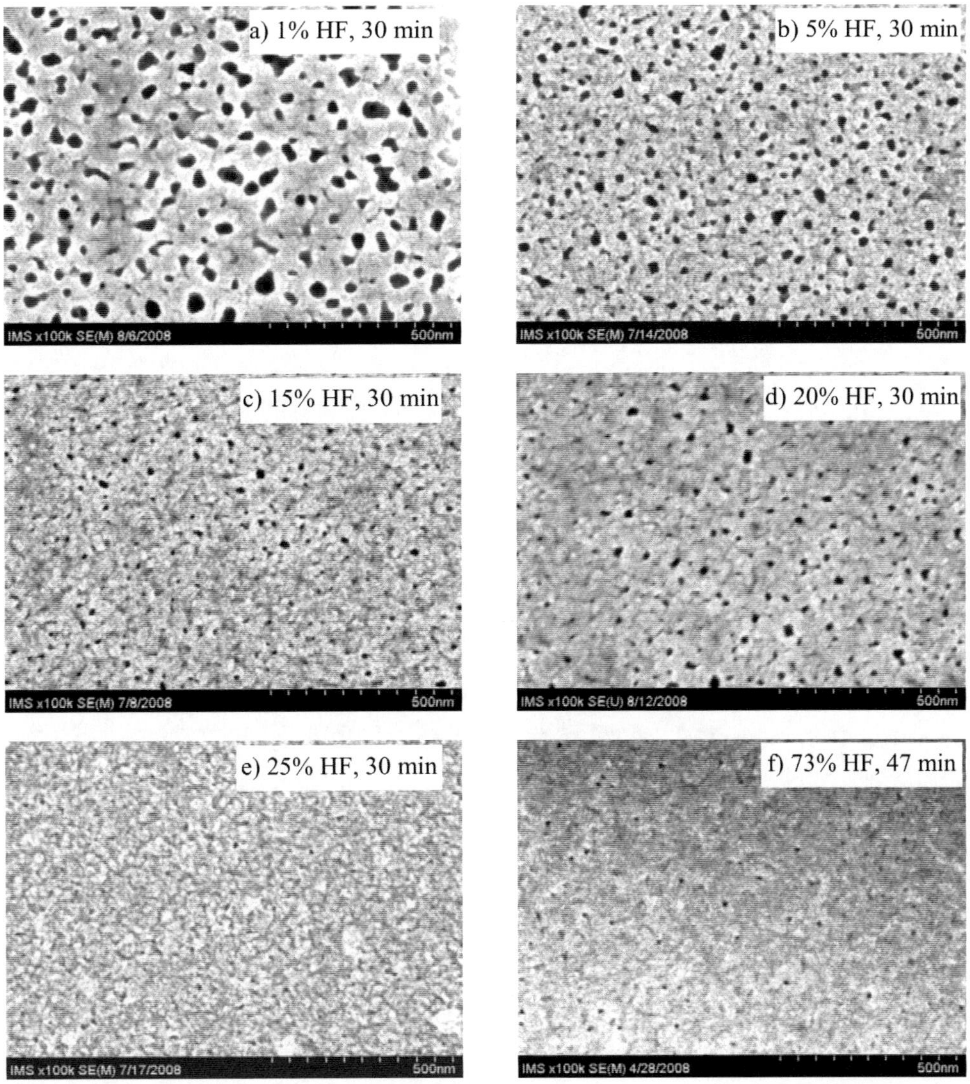

Figure 1. SEM images of a-SiC samples after anodization (the darker points present the etched pores). The anodization conditions are the same for all samples except the HF concentration of the electrolytes (and the anodization time for the case f).

SiC original samples are the undoped amorphous SiC (a-SiC) thin films with the thickness of 3 μm fabricated by plasma-enhanced CVD deposition at 400°C without doping on the standard Si substrate. We made a-SiC film porous by electrochemical etching/anodization. For anodization, first of all, Al was evaporated on the backside of the samples to turn the SiC sample to one electrode, whereas another electrode is made from Pt. The second step is protection of the backside and the edges of the samples by a special protection layer to ensure that only the a-SiC surface is exposed to the electrolyte during anodization. As mentioned in our previous work [3], a novel aspect of our anodizing method was to use the protection layer made from polystyrene instead of black wax normally used in such processes. The polystyrene protection layer has the advantage that it can be mechanically removed from the samples. It means that there is no need of using solvents to dissolute the protection layer after etching and as a consequence, there is not any risk of chemical contamination of the sample.

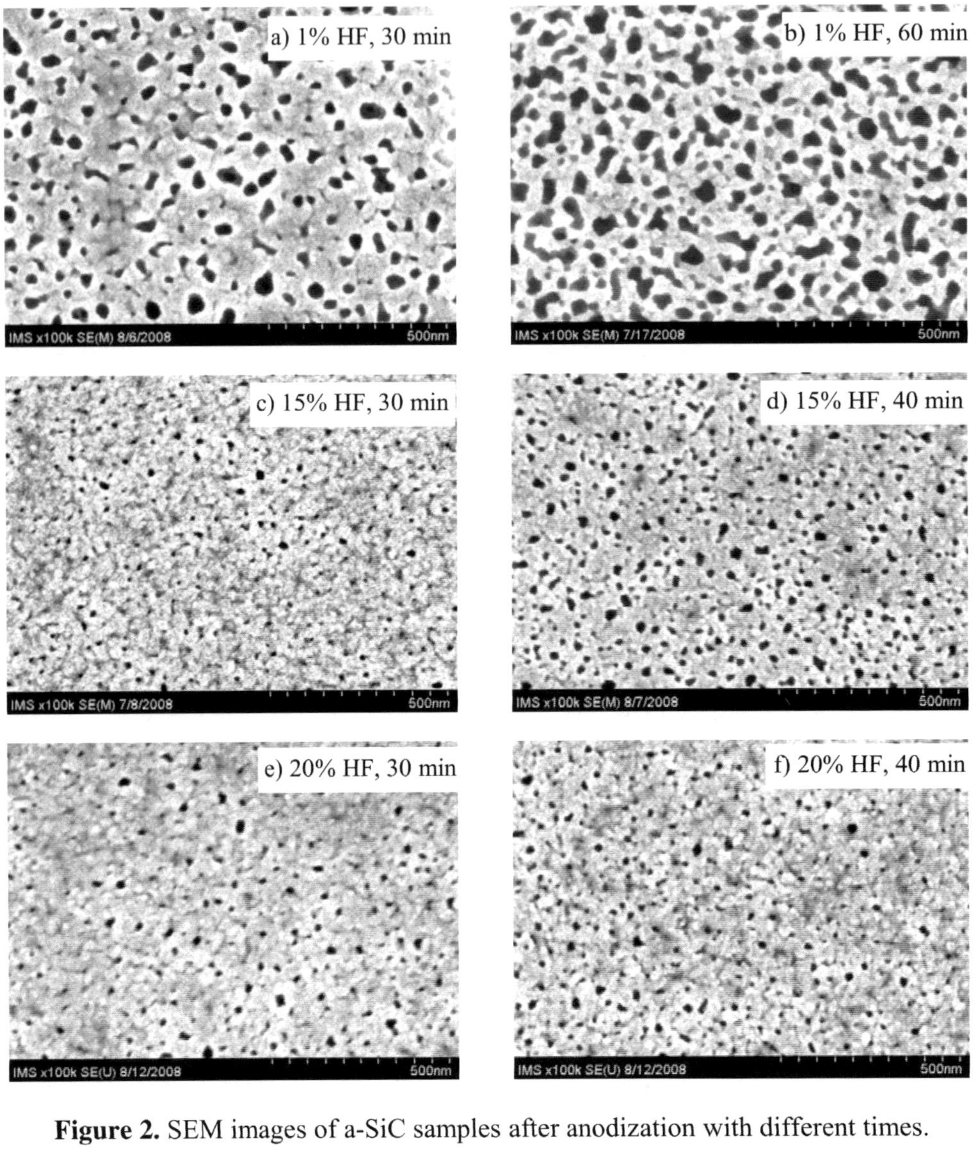

Figure 2. SEM images of a-SiC samples after anodization with different times.

In our previous work [3], following the experiences of the group DIMES [5, 6], we have used the electrolyte with very high concentration of HF ($HF:H_2O$ = 73:27). But now, when we are attempting to carry out a systematical study of the effect of the HF concentration in the electrolyte on making the nondoped a-SiC film porous, to our surprise, we have found that the dilute HF etchant turns the nondoped a-SiC film to porous (in nanometer scale) much more strongly than the concentrated HF etchant. The results of our study are shown in figure 1 for the samples, which were electrochemically etched in the same regime except the HF concentration in the electrolyte. The SEM images of figure 1 were obtained on the S-4800 Field Emission Scanning Electron Microscope (Hitachi). It is worth pointing out that all electrolytes contain 1% Triton X-100 serving as surfactant. This is because our previous experiments with 73% HF etchant showed that without Triton X-100 activation, the anodizing process could not occur even at very high anodizing voltages [3]. All samples have been anodized using a voltage constant source, which output voltage (250 V for all cases of the figure 1) is applied to the load consisting of a sample connected in series with a 14 kΩ resistor. We would like to emphasize that in our present work, due to the high resistivity character of the nondoped high-thickness (3 μm) a-SiC film used, we have to use much higher values of anodizing voltage than the values reported in literature [4]. As etching proceeds, normally we have observed the increasing of the anodizing current. In general, at the end of the etching process, the anodizing current density was in the range of 3.8-5.0 mA/cm^2. The case (a) (1% HF electrolyte) is an exceptional one, because there the anodizing current changed very little in the whole etching process and at the end the current density has attained only 0.1 mA/cm^2.

It can be seen clearly from figure 1 that the porosity (including both the pore size and the pore density) of the a-SiC layer formed on the original a-SiC films after etching depends strongly on the HF concentration. In particular, the more dilute HF etchant will give the larger pore size and also the denser pore distribution. It seems that this is not true for the case (a), because there although the pores are very large in size, the pore density is not very high. But this can be explained by the fact that the anodizing current density in this case is much lower than in the remaining cases. Indeed, our experiments of etching with the same etchant, under the same conditions, but varying the time of etching (for giving to the samples the opportunity to be etched either at higher current density or longer at high current density) brought us to the results shown in figure 2. From this figure we can see that increasing of the etching time only affects the pore density, making it higher, while the pore size remains almost unchanged. Nevertheless, it is worth emphasizing that with 1% HF electrolyte, the anodic etching already took place although there the anodizing current density was only 0.1 mA/cm^2.

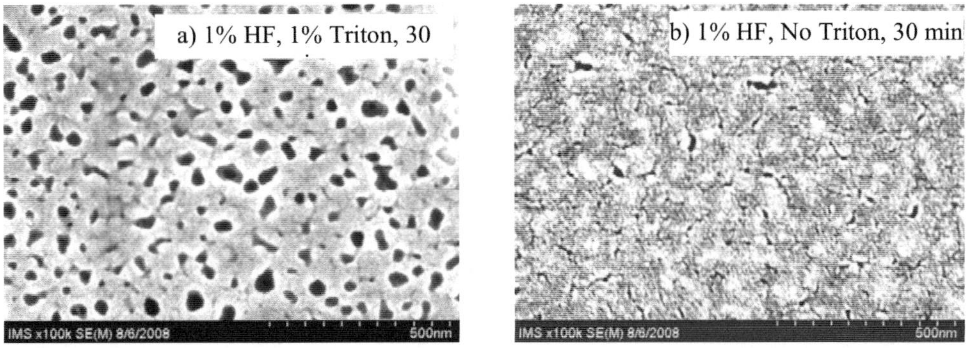

Figure 3. SEM images of a-SiC samples after anodization in the 1% HF electrolyte with and without Triton X-100.

Our further investigations showed that without Triton X-100 surfactant in the electrolyte, the anodic etching process could also occur, provided that the etchant is HF-dilute enough. Additionally, for the case of the electrolyte without surfactant, the anodizing voltage applied to the sample must be higher than for the case of the electrolyte with surfactant. For example, for the 1% HF etchant without Triton X-100, if the anodizing voltage was 250 V sharing between the sample and a 14 kΩ resistor, there was no etching, but if we used 300 V sharing between the sample and a 3.4 kΩ resistor, we could observe a good etching process. The results of this experiment are shown in figure 3. It is worth noting that although the anodic etching process could take place in the electrolyte without surfactant, the morphology of the obtained porous a-SiC layer is totally different in comparison with the case of the electrolyte with 1% Triton X-100. In particular, the pore shape is not roundish anymore and becomes prolonged, and in addition, the pore density is much lower.

2.2. Photoluminescence from porous a-SiC

The photoluminescence (PL) measurements of the obtained porous a-SiC layers have been carried out using either a He-Cd laser (325 nm excitation) or a commercial spectrofluorometer FL3-22 (Jobin Yvon-Spex) equipped with a xenon lamp. First of all, we observed that all samples etched in the electrolyte containing 1% Triton X-100 surfactant exhibit broad photoluminescence (PL) spectra in the 400-900 nm spectral regions with the PL intensity increasing with the porosity of sample (figure 4).

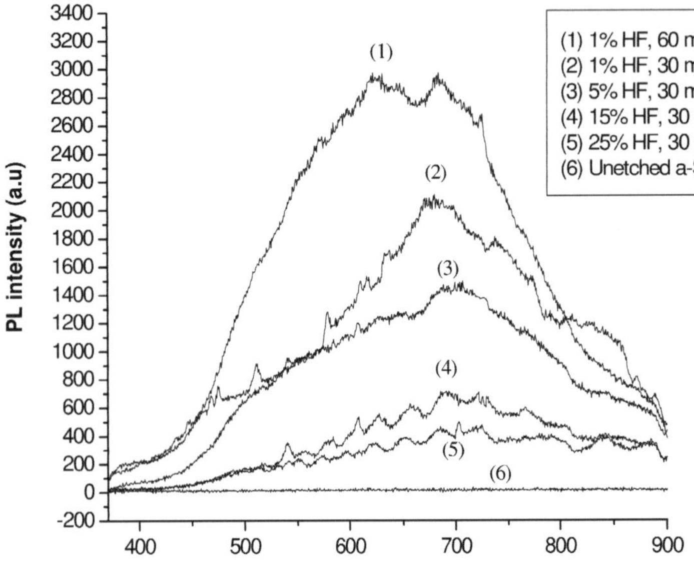

Figure 4. PL spectra of a-SiC samples after anodization in different HF-concentration electrolytes containing 1% Triton X-100 excited at 325 nm.

The tendency of PL intensity increasing with porosity is an important result because it shows the role of the porosity in the PL emission of a-SiC. Concerning the PL spectral region of porous SiC, the results of different research groups are diverging. Our results with broad PL in the 400-900 nm spectral region peaking in the 600-700 nm region are similar with what reported in [7, 8]. Some other groups of authors have received the broad PL spectra in the 350-600 nm regions peaking in the 400-500 nm regions [1, 9]. The third kind of PL spectrum is reported in [10] for porous p-type 6H-SiC, where the spectral range spans from 350 nm to 800 nm with a nearly flat band in the 470-650 nm region. In summary, PL properties of porous SiC depend very much on the kind and the properties of the SiC material to be etched and also on the etching conditions. Relating to the origin of luminescence from porous SiC, nowadays it is generally believed to stem from surface or defect states [11]. No explicit evidence of quantum confinement has been observed. The main reasons are that there are many surface or defect states which dominate the luminescence and that the SiC nanocrystals fabricated in the etching processes are too large. Our further investigations showed that Triton X-100 surfactant weakens the photoluminescence from porous a-SiC, especially the photoluminescence in the blue region. This result is illustrated by figure 5 for two samples, which have been etched anodically in similar conditions except in one electrolyte there was 1% Triton X-100, whereas in another there was no Triton X-100 at all. It means that while improving the porosity of the a-SiC, Triton X-100 surfactant badly affects its PL properties.

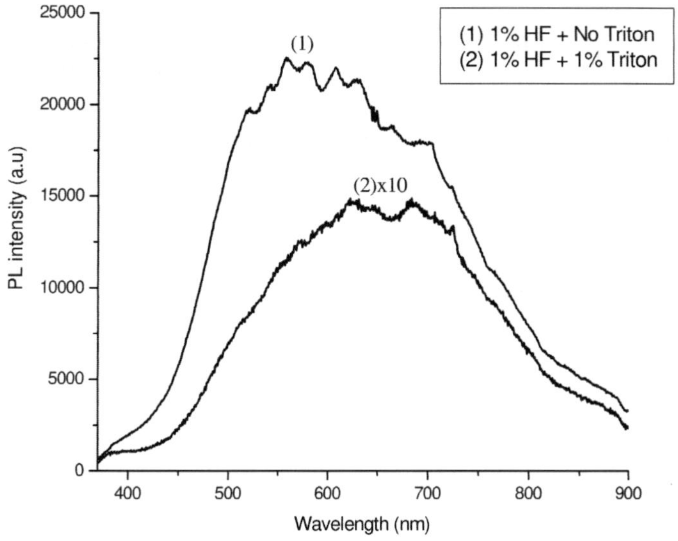

Figure 5. PL spectra of a-SiC samples after anodization in 1% HF electrolytes with and without Triton X-100 excited at 325 nm.

To verify why our samples do not strongly emit the light in the blue region we have paid our attention to the intense blue light (visible to the naked eye) emitting from some macropores which have been etched through the a-SiC to the Si-substrate when samples were excited by the 325 nm light from the He-Cd laser. Sometimes such reach-through macropores appeared on the surface of the etched a-SiC samples due to the poor quality of the original PE-CVD fabricated a-SiC film. The PL spectra of some porous a-SiC samples with 325 nm He-Cd laser beam concentrated either to said macropores or to their edges are shown in figure 6. From this figure we can see that the macropores emit really intense blue light peaking approximately at 500 nm. Note that the intensity of this peak is not the same for all macropores but varies from one to another. Furthermore, we have observed that not every macropore will emit the blue light, i.e. the blue light will be emitted from the macropores only in the cases where special requirements have been satisfied. From figure 6 we can see also that when the laser beam is concentrated not directly to the macropore, but to its edge, the PL in the 600-900 nm will reappear. The origin of the blue luminescence from said macropores is unclear. In our opinion, it is probably that this blue luminescence has the same origin as the intense blue luminescence from carbon-doped silicon-rich silicon oxide reported in reference [12].

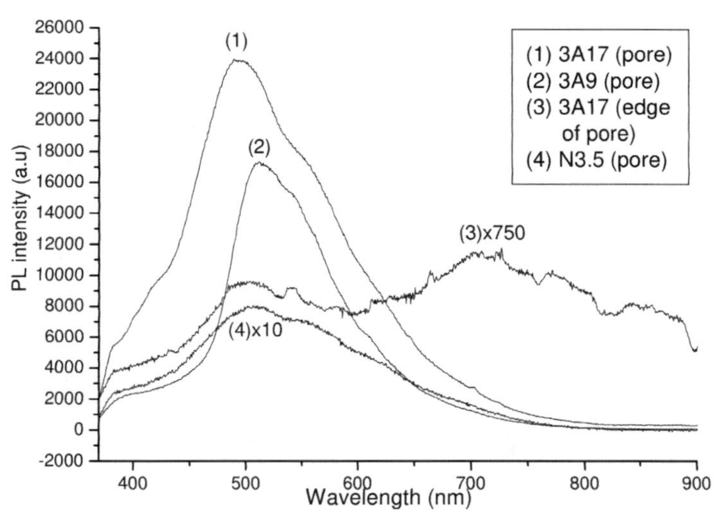

Figure 6. PL spectra of different porous a-SiC samples excited at 325 nm from the He-Cd laser with the laser beam concentrated either to the reach-through macropores or to their edges.

3. Conclusion

i. The Triton X-100 surfactant has very pronounced effect on the anodic etching of the a-SiC material. With assistance of the Triton X-100, the a-SiC material can be easier etched anodically, i.e. the required anodizing voltage is lower. Furthermore, with assistance of the Triton X-100 in the etching process, the pore shape becomes more roundish and the pore density becomes higher.
ii. When a-SiC anodic etching is carried out using the aqueous HF-electrolyte containing 1% Triton X-100, the more HF-dilute etchant will result in the larger pore sizes and denser pore distribution, i.e. the a-SiC porosity depends strongly on the HF concentration.
iii. Porous a-SiC samples exhibit broad photoluminescence (PL) spectra in the 400-900 nm spectral region with the PL intensity increasing with the porosity of the sample.
iv. Although the porosity of the a-SiC layer is greatly improved by adding the Triton X-100 surfactant to the electrolyte, Triton X-100 weakens its PL, especially the PL in the blue region.

References

[1] Matsumoto T, Takahashi J, Tamaki T, Futagi T, Mimura H, and Kanemitsu Y 1994 *Appl. Phys. Lett.* **64** 226
[2] Rittenhouse T L, Bohn P W, and Adesida H 2003 *Sol. St. Comm.* **126** 245
[3] Cao D T, Ha N T T, Thanh V T, Ha H T, Huy B, Hoa P T M, and Sarro P M 2007 *Proc. 1th Int. Workshop on Nanotechnology and Application (Vungtau, Vietnam, 15-17 November 2007)* p 176
[4] Parkhutik V P, Namavar F, and Andrade E 1997 *Thin Solid Films* **297** 229
[5] Connolly E J, Pham H T M, Groeneweg J, Sarro P M and French P J 2004 *Sensors and Actuators B* **100** 216
[6] Connolly E J, Timmer B, Pham H T M, Groeneweg J, Sarro P M, Olthuis W, and French P J 2005 *Sensors and Actuators B* **109** 44
[7] Stecki A J, Su J N, Xu J, Li J P, Yuan C, Yih P H, and Mogul H C 1994 *Appl. Phys. Lett.* **64** 1419
[8] Hassen F, M'Ghaieth R, Maaref H, and Madar R 2001 *Mat. Sci. and Eng. C* **15** 113
[9] Rittenhouse T L, Bohn P W, Hossain T K, Adesida H, Lindsay J and Marcus A 2004 *J. Appl. Phys.* **95** 490
[10] Jessensky O, Muller F and Gosele U 1997 *Thin Solid Films* **297** 224
[11] Fan J Y, Wu X L and Chu P K 2006 *Progr. Mat. Sci.* **51** 983
[12] Seo S-Y, Cho K-S and Shin J H 2004 *Appl. Phys. Lett.* **84** 717

Acknowledgements

The authors would like to thank MSc. Do Hung Manh for the SEM measurements. A part of the work was done with the help of the National Key Laboratory for Electronic Materials & Devices, Institute of Materials Science.

Iron(III) and manganese(II) substituted hydroxyapatite nanoparticles: characterization and cytotoxicity analysis

Yan Li, Chai Teck Nam and Chui Ping Ooi

Division of Bioengineering, School of Chemical and Biomedical Engineering, Nanyang Technological University, Singapore 639798

E-mail: ascpooi@ntu.edu.sg

Abstract. Calcium hydroxyapatite (HA) is the main inorganic component of natural bones and can bond to bone directly *in vivo*. Thus HA is widely used as coating material on bone implants due to its good osteoconductivity and osteoinductivity. Metal ions doped HA have been used as catalyst or absorbents since the ion exchange method has introduced new properties in HA which are inherent to the metal ions. For example, Mn^{2+} ions have the potential to increase cell adhesion while Fe^{3+} ions have magnetic properties. Here, Fe(III) substituted hydroxyapatite (Fe-HA) and Mn(II) substituted hydroxyapatite (Mn-HA) were produced by wet chemical method coupled with ion exchange mechanism. Compared with pure HA, the colour of both Fe-HA and Mn-HA nanoparticles changed from white to brown and pink respectively. The intensity of the colours increased with increasing substitution concentrations. XRD patterns showed that all samples were single phased HA while the FTIR spectra revealed all samples possessed the characteristic phosphate and hydroxyl adsorption bands of HA. However, undesired adsorption bands of carbonate substitution (B-type carbonated HA) and H_2O were also detected, which was reasonable since the wet chemical method was used in the synthesis of these nanoparticles. FESEM images showed all samples were elongated spheroids with small size distribution and of around 70 nm, regardless of metal ion substitution concentrations. EDX spectra showed the presence of Fe and Mn and ICP-AES results revealed all metal ion substituted HA were non-stoichiometric (Ca/P atomic ratio deviates from 1.67). Fe-HA nanoparticles were paramagnetic and the magnetic susceptibility increased with the increase of Fe content. Based on the extraction assay for cytotoxicity test, both Fe-HA and Mn-HA displayed non-cytotoxicity to osteoblast.

Keywords: Hydroxyapatite, nanoparticles, ion-exchange, paramagnetic, cytotoxicity.

1. Introduction

Population ageing is a global phenomenon. It is expected that by the year 2050, the number of people aged 60 years and above will be the first time in history to exceed the number of young people (under 15 years of age). Osteoporosis is common among older people because of the reduction in bone density, leading to increased risk of bone fracture, which requires prosthesis implantations (usually metallic implants) for total recovery. However, due to the bioinert property of metallic implants, the slow osseointegration process tends to cause failure of implantations. This problem is more serious for older persons who have lower proliferation and differentiation rates of bone cells [1].

Hydroxyapatite (HA, $Ca_{10}(PO_4)_6(OH)_2$) attracts great interest due to its good osteoconductivity and

© 2009 IOP Publishing Ltd

Journal of Physics: Conference Series **187** (2009) 012024 doi:10.1088/1742-6596/187/1/012024

osteoinductivity. However, pure HA implants feature inferior mechanical characteristics, which limit its use to low load bearing applications [2]. Therefore, a better alternative will be to use composite structures of thin HA layer coated metallic implants, since these implants possesses both good mechanical strengths of metal and bioactivity of HA. HA coated metallic implants promote osseointegration that enhances tight-fit fixation and reduces motion damage to surrounding tissues [3-5]. Hence, the adhesion of osteoblast to HA coating is a crucial step for subsequent osteoblast functions. Mn^{2+} ions increase ligand binding affinity of integrate and activate cell adhesion [6]. His-Chin Wu *et al.* [7] reported that iron (Fe^{2+}) substituted HA nanoparticles were superparamagnetic and showed good biocompatibility. This work aims to synthesize Mn^{2+} and Fe^{3+} doped HA and investigates their bioactivity compared to pure HA.

Wet chemical method coupled with ion exchange mechanism was used to produce pure HA, Mn^{2+} substituted HA (Mn-HA) and Fe^{3+} substituted HA (Fe-HA). All samples were characterized by X-ray Diffraction (XRD), Fourier Transform Infrared Spectroscopy (FTIR), Field Emission Scanning Electron Microscopy (FESEM) coupled with Energy Dispersive X-ray Spectroscopy (EDX) and Inductively Coupled Plasma Atomic Emission Spectrometry (ICP-AES). The magnetic property of Fe-HA was tested using a Vibrating Sample Magnetometer (VSM). Cytotoxicity tests for all samples were performed on osteoblast cells by extraction method according to ISO 10993 part 5 and 12.

2. Experiment

2.1. Synthesis of Pure Hydroxyapatite and Metal Ion Substituted Hydroxyapatite

$Ca(OH)_2$ (Greenrich Chemical Enterprise, 96.0 wt.% pure with less than 4% $CaCO_3$), H_3PO_4 (Merck), $MnCl_2.4H_2O$ (Riedel-de Haën) and $FeCl_3.6H_2O$ (Sigma-Aldrich) were used as sources of Ca, P, Mn and Fe. Pure HA was produced using the following chemical reaction:

$$10Ca(OH)_2 + 6H_3PO_4 \rightarrow Ca_{10}(PO_4)_6(OH)_2\downarrow + 18H_2O\ .$$

3.71g $Ca(OH)_2$ was suspended in 250 ml DI water (Mili-Q unit from Milipore). The suspension temperature was kept at 98.5°C and stirred at 500 rpm for 30 minutes. 2 ml 85 wt.% H_3PO_4 was diluted into 250 ml DI water and added into the $Ca(OH)_2$ suspension at a rate of 4 ml/min until the pH of mixture decreased to the desired values. Ion exchange was used for Mn-HA and Fe-HA. After the addition of H_3PO_4 was stopped, metal salt solution (different molar quantity according to the different exchange ratio X_M/X_{Ca}) was added to the mixture. The pH of the mixture was kept at desired values with NH_4OH. The mixture was left stirring at 500 rpm and aged at 98.5°C for 2 hours. After 24 hours at room temperature, the supernatant was replaced by DI water three times to wash away non-reacted substances. The samples were finally dried at 100°C for 12 h and grounded to powder for further characterization

2.2. XRD Characterization

Bruker D8 Advance X-ray diffractometer with Cu-Kα radiation was used for crystal phase analysis. The diffractometer was operated at 40 kV and 40 mA. Data was collected at step size of 0.04 degree and counting time of 15 seconds per step in a 2θ range of 20-80°.

2.3. FTIR Characterization

FTIR (PerkinElmer, Spectrum GX) characterization was carried out under classic KBr pellet technique. The samples were diluted by KBr at ratio 1/300 (by weight). The FTIR spectrum was the average of five scans with resolution of 1.0 cm^{-1} in the range of 400-4000 cm^{-1}.

2.4. FESEM and EDX Characterization

FESEM (JEOL JSM-6700F) was used to study the morphology of samples. To get the best view under FESEM, samples were slightly pressed into pellets at 0.5 ton-load. The elements present in each sample were also characterized by EDX, which is coupled with FESEM.

2.5. ICP-AES Analysis
The element concentrations in each sample were quantified by ICP-AES (Prodigy-Teledyne Instruments Leeman Lab). ICP-AES was calibrated by standard solutions for elements interested and samples were prepared by dissolving 40 mg powder in 250 ml 1 wt.% HNO_3.

2.6. Magnetic property test
The magnetic property of Fe^{3+} doped HA was studied at room temperature in the range of magnetic field 0-10 kOe using VSM (Lake Shore 7404).

2.7. Cytotoxicity Test
Cytotoxicity test was based on ISO 10993-5 and ISO 10993-12. Human osteoblast cell line hFOB 1.19 ATCC was used. Specimens were prepared by pressing 120 mg powder in a 13 mm die (Specac®, PT. No. 3006 BDH) at 1 ton-load. The pellets were autoclaved and placed into 35 mm petri dishes. 2.12 ml DMEM/F12 medium for each pellet was added and incubated at 37°C for 24 hours in an incubator. The extraction medium may contain leachable substances from samples. Before the extraction, cells were seeded into 24-well plates at a density of 2×10^4 cell/ml and incubated for one day. The medium in each well was then replaced by the extraction medium (4 replicates for each composition) and incubated for another 24 hours. The cells cultured with DMEM/F12 medium acted as the control. MTT method was used to measure the metabolic activity of living cells. After the cells were incubated with extraction medium for 24 hours, the medium was replaced with 0.5 ml MTT solution (1 mg/ml) and incubated for another 2 hours. 0.5 ml Dimethyl Sulfoxide (DMSO) was then added and incubated for another 2 hours to dissolve the precipitate. The absorbance was measured by the Multiwell Scanning Spectrophotometer (ELISA reader) at 490 nm.

3. Results and discussion
Pure HA, Mn-HA and Fe-HA nanoparticles were successfully synthesized by wet chemical method. Pure HA was white, while the Mn-HA nanoparticles were pink and the Fe-HA nanoparticles were brown. As substitution concentration increased, the colour intensity increased accordingly.

3.1. XRD Pattern

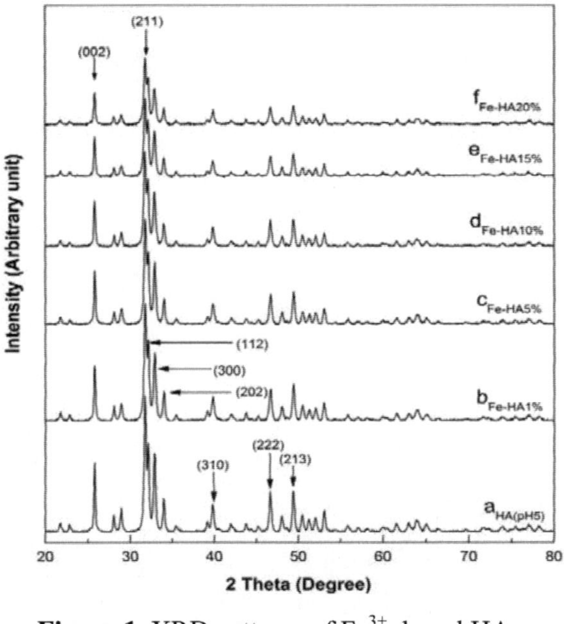

Figure 1. XRD patterns of Fe^{3+} doped HA.

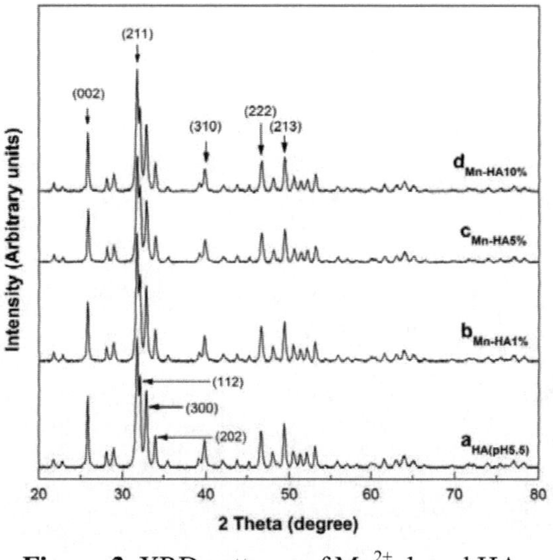

Figure 2. XRD patterns of Mn^{2+} doped HA.

Journal of Physics: Conference Series **187** (2009) 012024 doi:10.1088/1742-6596/187/1/012024

Figures 1 and 2 show the XRD patterns of all samples. The peaks present agree well with ICSD file no. 26204 which is hexagonal hydroxyapatite and none of the patterns displayed extra peaks indicating that all samples were single phased HA. Compared with pure HA, all metal ion substituted HA showed comparable peaks without significant shifting of peak positions, regardless of substitution concentrations. This suggested that the ion exchange process did not greatly modify the structure of HA. For Fe-HA nanoparticles, the peak intensities decreased as substitution concentration increased (figure 1), which may be due to the high substituted concentration of Fe (20 wt.%) decreasing the crystallinity. For Mn-HA nanoparticles, the peak intensities did not decrease significantly as substitution concentration increased (figure 2).

3.2. FTIR Spectra

Figures 3 and 4 show the FTIR spectra of Mn-HA and Fe-HA nanoparticles. The detection of PO_4^{3-} (υ_1 to υ_4) and OH^- adsorption bands indicated that all samples possessed fundamental apatite structures [8]. $PO_4^{3-}(\upsilon_2)$ peak at around 462 – 472 cm^{-1} was relatively harder to observe compared to the other PO_4^{3-} bands in all spectra. In addition, carbonate CO_3^{2-}, and water, H_2O, adsorption bands were noticeable within all spectra. This was reasonable in wet chemical synthesis which required long maturing time for precipitation of hydroxyapatite. The carbonate was from atmospheric carbon dioxide and/or $CaCO_3$ impurities contained in the reactants. The carbonate groups partially replacing the PO_4^{3-} groups in HA lattice, which was B-type carbonated HA. The presence of hydrogen phosphate, such as HPO_4^{2-} and $H_2PO_4^-$ adsorption bands were not visible in all spectra.

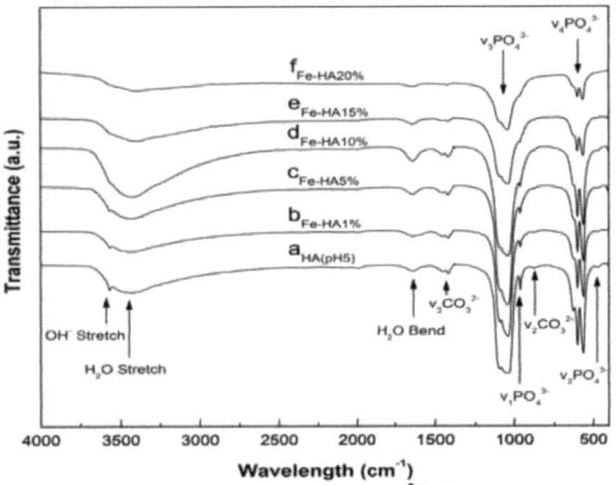

Figure 3. FTIR spectra of Fe^{3+} doped HA.

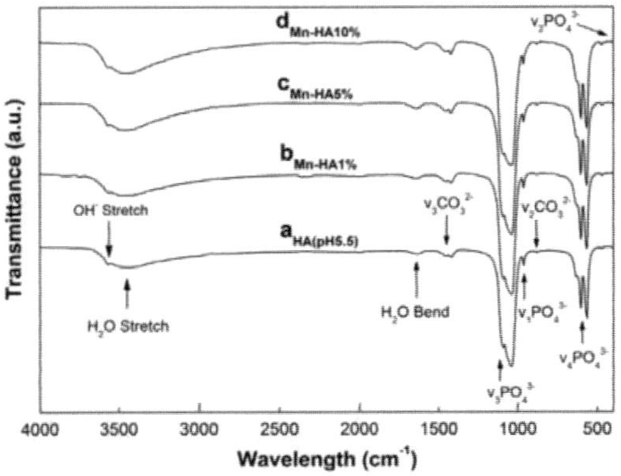

Figure 4. FTIR spectra of Mn^{2+} doped HA.

3.3. FESEM and EDX Characterization

Figure 5. The FESEM image of Fe-HA 20 wt.%.

The FESEM image of Fe-HA 20 wt.% is shown in figure 5. This image was representative of pure HA, Mn^{2+} and Fe^{3+} doped HA, where all samples had similar crystal morphology. All samples were of elongated spheroids shape of around 70 nm and with a small size distribution. The elements present in HA phase, such as C, O, P, Ca and Mn or Fe, were detected by EDX for each sample. The representative EDX spectra are shown in figure 6. For both Mn-HA and Fe-HA nanoparticles, no Cl^- ions were detected although $MnCl_2$ and $FeCl_3$ were used for synthesis.

3.4. Chemical Analysis

From ICP-AES data, Ca/P, Metal/Ca and (Ca+Metal)/P atomic ratios were tabulated in table 1. For HA (pH = 5) and HA (pH = 5.5), Ca/P atomic ratio was higher than the stoichiometric value of 1.67, which indicated that the synthetic pure HA was non-stoichiometric. This may be due to the carbonate groups partially replacing the phosphate groups in the crystal lattice or the unreacted amorphous $Ca(OH)_2$. For Mn-HA and Fe-HA nanoparticles, Ca/P ratio decreased while M/Ca ratio increased with the increase of initial metal ion concentration, indicating that more Ca ions in the lattice were replaced by the metal ions. Compared with the theoretical value of Ca/P*, ICP-AES result of Ca/P was higher. This was because not all metal ions added in the suspension were involved in the ion exchange process. The result indicated that Fe^{3+} was more active than Mn^{2+} in replacing Ca^{2+} in HA lattice, which was consistent with the ICP-AES result of M/Ca ratio and theoretical value of M/Ca*, as shown in Table 1. For Mn-HA nanoparticles, ICP-AES result of Mn/Ca ratio was lower than the theoretical value, especially for Mn-HA 10 wt.%; while for Fe-HA nanoparticles, ICP-AES result of Fe/Ca ratio was comparable with the theoretical value. This confirmed that the exchange ratio of Mn/Ca was 1/1 while the exchange ratio of Fe/Ca was 2/3 to maintain the charge balance [9]. Thus the (Ca+Fe)/P ratio decreased with the increase of Fe content while for Mn-HA nanoparticles, the (Ca+Mn)/P ratio was comparable.

Table 1. The atomic ratios for metal ion doped HA from ICP-AES result.

Samples	Ca/P	Ca/P *	(Ca+M)/P	M/Ca	M/Ca *
HA (pH = 5.5)	1.681	-	-	-	-
Mn-HA 1%	1.660	1.650	1.675	0.010	0.010
Mn-HA 5%	1.620	1.584	1.681	0.041	0.050
Mn-HA 10%	1.599	1.500	1.683	0.053	0.100
HA (pH = 5)	1.713	-	-	-	-
Fe-HA 1%	1.661	1.642	1.678	0.010	0.010
Fe-HA 5%	1.589	1.542	1.677	0.055	0.054
Fe-HA 10%	1.471	1.417	1.643	0.117	0.118
Fe-HA 15%	1.334	1.292	1.591	0.193	0.194
Fe-HA 20%	1.217	1.167	1.566	0.287	0.286

Ca/P*: theoretical atomic ratio of Ca to P after Ca replaced by all the M ions added at the corresponding ratio. For Mn^{2+} doped HA, the exchange ratio of Mn/Ca was 1:1 and for Fe^{3+} doped HA, the exchange ratio of Fe/Ca was 2/3 to maintain the charge balance.

M/Ca*: theoretical atomic ratio of M to Ca after Ca replaced by all the M ions added at the corresponding ratio.

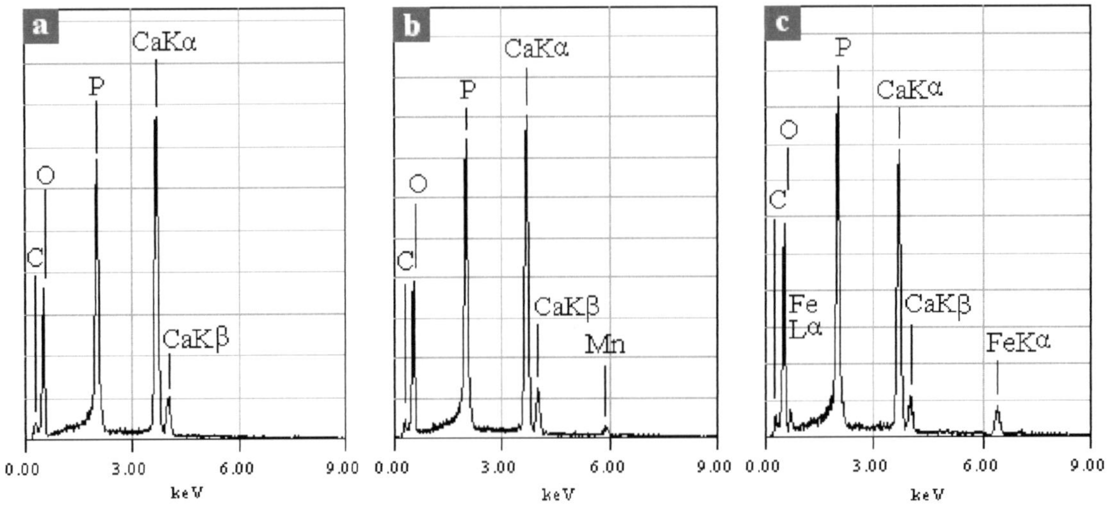

Figure 6. The EDX spectra of (a) pure HA, (b) Mn-HA 10% and (c) Fe-HA 20%.

3.5. Magnetic property of Fe^{3+} substituted HA

The magnetization of the powders versus external magnetic field is shown in figure 7. The magnetization increased linearly with external field, indicating that Fe^{3+} doped HA was paramagnetic. The slope, which represents the magnetic susceptibility (χ_g), increased with Fe concentration. Here, for Fe-HA 10 wt.%, Fe-HA 15 wt.% and Fe-HA 20 wt.%, the χ_g was 4.4×10^{-6}, 7.5×10^{-6}, 1.0×10^{-5} emu.g^{-1}.Oe^{-1}, respectively. However, His-Chin Wu *et al.* [7] reported that Fe^{2+} substituted HA nanoparticles were superparamagnetic and the highest saturation magnetization (M_s) was 20.92 emu/g. This may be due to the different valance of Fe. Here, Fe^{3+} was used instead of Fe^{2+}, as reported by His-Chin Wu *et al.* [7].

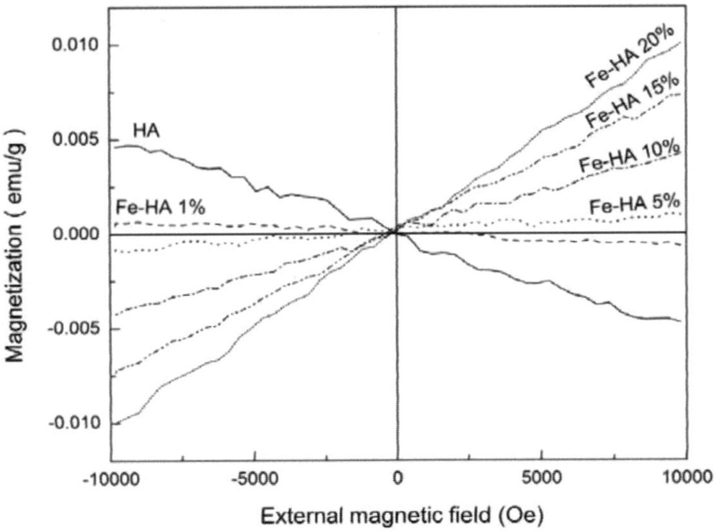

Figure 7. The magnetization curves of Fe^{3+} doped HA measured at room temperature.

3.6. Cytotoxicity analysis

The relative absorbance value for Mn-HA and Fe-HA nanoparticles after MTT test are shown in figures 8 and 9. The colorimetric value presented the quantity of vital osteoblasts. Compared with pure HA, all Mn-HA and Fe-HA nanoparticles had comparable absorbance values, which suggested that all samples were non-cytotoxic. As Mn-HA nanoparticles had slightly higher absorbance values compared with pure HA, further cell adhesion experiment will be carried out to confirm its increase in bioactivity compared to pure HA.

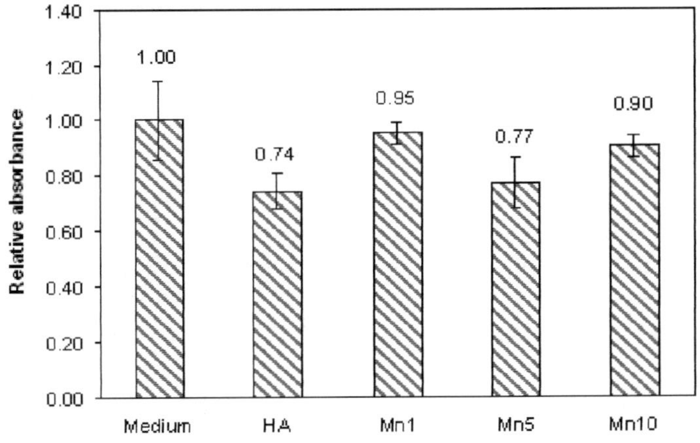

Figure 8. The relative absorbance value for Mn^{2+} doped HA after MTT test.

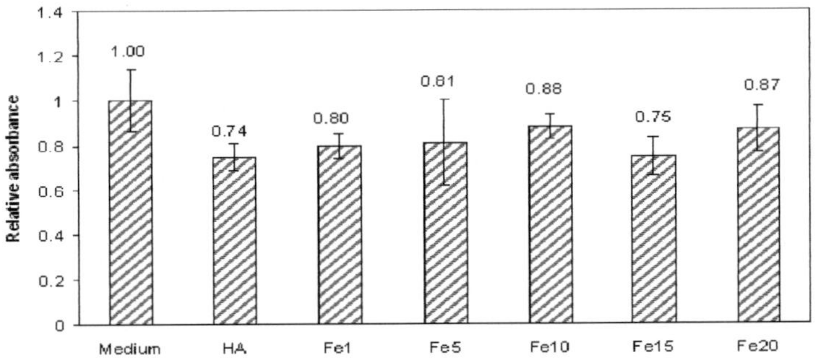

Figure 9. The relative absorbance value for Fe^{3+} doped HA after MTT test.

4. Conclusion

HA substituted with manganese (II) and iron (III) were successfully synthesized by wet chemical method coupled ion exchange mechanism. All products were of single-phase and the lattice structure was not affected by the ion exchange process. For Mn-HA nanoparticles, the crystallinity did not change significantly with the increase of Mn content while for Fe-HA nanoparticles, the crytallinity decreased with increase in Fe quantity. All samples were characterized as carbonated HA by FTIR, which was B-type. The metal ion substituted HA nanoparticles were elongated spheroids and uniform in size (70 nm). From the ICP-AES results, it can be concluded that the synthetic pure HA was non-stoichiometric and Fe^{3+} was more active than Mn^{2+} in the exchange of Ca^{2+} ions in HA lattice. To maintain the charge balance for Fe-HA nanoparticles, the vacancies of Ca positions in the lattice increased with the increase of Fe concentration. All powders were tested as non-cytotoxic to osteoblast based on extraction assay. According to the magnetization curves, Fe^{3+} doped HA was paramagnetic and the magnetic susceptibility increased with the increase of Fe content.

References

[1] Ratner B D, Hoffman A S, Schoen F J, Lemons J E 2004 *Biomaterials Sciences-an introduction to materials in medicine* (London: Elsevier academic press)
[2] Oktar F N 2007 *Cera. Int.* **33** 1309
[3] Mihailescu I N, Torricelli P, Bigi A, Mayer I, Iliescu M, Werckmann J, Socol G, Miroiu F, Cuisinier F, Elkaim R, G Hildebrand 2005 *Appl. Surf. Sci.* **248** 344
[4] Weiser E, Tsyganov I, Matz W, Reuther H, Oswald S, Pham T, Richter E. 1999 *Surf. Coat. Technol.* **111** 103
[5] Feng Q L, Kim T N, Wu J, Park E S, Kim J O, Lim D Y, Cui F Z 1998 *Thin Solid Films* **335**

214
[6] Mayer I, Jacobsohn O, Niazov T, Werckmann J, Iliescu M, Richard-Plouet M, Burghaus O and Reinen D 2003 *Eur. J. Inorg. Chem.* **2003** 1445
[7] Wu H C, Wang T W, Sun J S, Wang W H and Lin F H 2007 *Nanotechnology* **18** 165601
[8] Kumar R, Prakash K H, Cheang P and Khor K A 2004 *Langmuir* **20** 5196
[9] Wakamura M, Kandori K and Ishikawa T 2000 *Colloid Surf. A-Physicochem. Eng. Asp.* **164** 297

Acknowledgements
We thank Mr. Patrick Lai from the School of Materials Science and Engineering, Nanyang Technological University (NTU), Singapore for helping us to finish the magnetic property test using VSM and Ms. Tang Meng Lu, Ms. Saraswati and Assoc. Professor Philip Cheang from School of Chemical and Biomedical Engineering, NTU, Singapore for their help in cytotoxicity analysis.

Studies on gold atom chains and lead nanowires on silicon vicinal surfaces

Hoang Vu Chung[1], Markus Klevenz[1], Robert Lovrincic[1], Frank Neubrech[1], Olaf Skibbe[1], Annemarie Pucci[1], Pawel Nita[2], Mieczyslaw Jalochowski[2] and Tadaaki Nagao[3]

[1] Kirchhoff Institute of Physics, Heidelberg University,
Im Neuenheimer Feld 227, D-69120 Heidelberg, Germany
[2] Institute of Physics, Maria Curie-Sklodowska University,
Pl. M. Curie-Sklodowskiej 1, PL-20031 Lublin, Poland
[3] International Center for Materials Nanoarchitectonics (MANA),
National Institute for Materials Science, 1-1 Namiki, Tsukuba 305-0044, Japan

E-mail: pucci@kip.uni-heidelberg.de

Abstract. The Si(557) vicinal surface is produced from Si(111) by cutting the crystal at a small angle of 9.45° along the $[\bar{1}\bar{1}2]$ direction. By a special cleaning process performed under ultra-high vacuum conditions, Si(557) displays a perfect grating structure consisting of parallel narrow terraces, each of them composes of a silicon plane with half unit cell of the Si(111)-7x7 reconstruction and a silicon row. Such an atomic grating makes it possible to get self-assembled atomic wires by decorating the surface with an optimal amount of 0.2 monolayer (ML) of Au. In the first step of the study, Si(557) is used as a template for fabricating of Pb nanowires by self-assembling in which the steps on the Si(557) surface provide the high-oriented diffusion channels for Pb wires growth. By means of polarized infrared spectroscopy the optical response of the Pb nanowires is investigated. As a major consequence of the interaction between infrared light and nanostructures, an antenna-like resonance appears. We found the shift of the antenna resonance frequency to lower wavenumbers due to the development of the nanowires in length under parallel excitation, whereas there is no significant change due to perpendicular excitation observed.

Keywords: Infrared spectroscopy, Pb nanowires, Au chains, vicinal surface.

1. Introduction

Si(557) and Si(557)-Au are the most popular vicinal surfaces derived from Si(111) and has attracted so much attention in the last few years since these structures are ideally suited for basic research on atomic wires with respect to Luttinger-Tomonaga-liquid behaviour and Peierls distortion occurring at low temperature due to electron-phonon coupling, e.g. [1, 2]. In general, the clean Si(557) surface consists of the parallel and well-ordered steps but they are only stable at the temperature above 600°C [3], hence, it is necessary to stabilize the surface. As it known from previous work [4], to totally stabilize the surface with gold, one has to obey the condition that every unit cell of the Si vicinal surface must be occupied by one Au atom. In case of Si(557), the optimal amount of Au is 0.2

© 2009 IOP Publishing Ltd

monolayers (ML) and the reconstructed surfaces possess the co-appearance of atomic chains made of Au and Si parallel to each other on one terrace.

As a fully decorated surface, the Si(557)-Au composes of parallel diffusion channels that is an ideal substrate to grow parallel nanowires by self-assembling. The metallic nanowires made by this technique are single crystals with defect-free, highly oriented structures [4]. Under the appropriate excitation conditions by the external infrared (IR) electromagnetic field, these nanowire arrays show antenna-like response which can be assigned to the surface plasmon excitation [5-7]. The enhancement of the electrical field at the ends of the nanowires can be exploited for molecular sensing by tailoring the quantum interference between the resonance range of the antenna and the vibrational features of the molecules [6].

Within this short contribution, we would like to explain the experimental process to fabricate the clean Si(557) surface as well as Si(557)-Au where self-assembled Au chains display perfect one-dimensional atomic wires. Based on such careful surface preparation studies, the Si(557) surface is used to prepare Pb nanowires by taking advantage of the natural diffusion properties of Pb atoms, *i.e.* of the high mobility along the parallel channels. The shift of the antenna resonance frequencies under IR light makes it possible to control the development of Pb nanowires in length *in situ* [5].

2. Experimental detail

The experiments were performed under the ultra-high vacuum (UHV) conditions with a base pressure better than $1x10^{-10}$ mbar. The set up consists of a chamber connected with a nitrogen purged Fourier-transform infrared spectrometer (FTIR) (Bruker Tensor 27 with Mercury Cadmium Telluride - MCT detector) and a reflection high energy electron diffraction (RHEED) gun. Scanning tunnelling microscopy measurements (STM) were performed in a separated UHV chamber with a similar set up. Atomic force microscopy measurement (AFM) is done immediately after sample transfer to air.

Si(557) wafers with dimensions of 18x5x0.38 mm^3 and a specific resistivity about 0.5 Ωcm were used as substrates. Prior to transferring into vacuum, the sample was prepared by *ex situ* cleaning with ethanol, acetone, and deionised water in the ultrasonic bath for 15 minutes for each solvent. This process removed dust from the wafer, but the natural silicon dioxide was still remaining. The cleaning process in UHV was performed by applying a direct heating current passing through the sample with the heating current parallel to the steps to avoid the electron migration effects (see [8] and references therein).

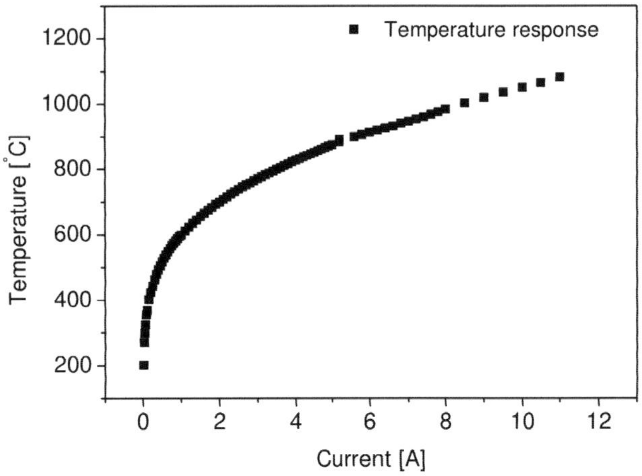

Figure 1. The temperature response of the silicon wafer (18x5x0.38 mm^3, about 0.5 Ωcm) to a flashing current.

In order to determine the temperature response of the silicon sample to the direct heating current, a thermocouple (type K) was mounted on the Si sample with two spacers made of sapphire plates, mainly to insulate the thermocouple from the heating current. As can be seen in figure 1, at temperatures higher than 600°C, where the phonon-induced resistivity is dominant, an almost linear response is measured.

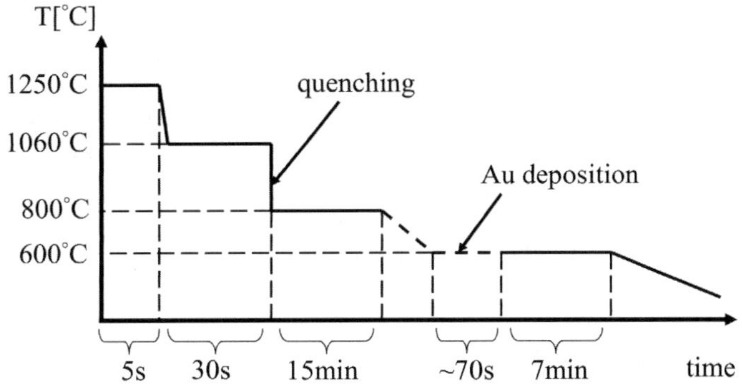

Figure 2. Heat treatment process of the Si(557)-Au fabrication.

Figure 2 shows the heating diagram that had been used to treat the silicon wafer in UHV and to decorate that sample with 0.2 monolayer of Au. The sample was first heated to 600°C and kept at that temperature for 20 hours. Afterwards, the wafer was heated up several times to 1200°C by heating pulses (so called flashing) with 5s for each, to remove the native silicon dioxide and the residual silicon carbide from the surface. The proper flashing current can be extrapolated from the response behaviour seen in figure 1. Within 30 s the sample was then cooled to 1060°C at which the step structures are most stable [3].

The most critical part in the sample preparation is to quench the surface from 1060°C to 830°C to avoid the formation of silicon carbide which can be produced on the fresh Si surface at temperatures higher than 900°C. The sample subsequently was kept at 830°C for 15 minutes to undergo the full transition from 1x1 phase to 7x7 reconstruction and then slowly cooled down to room temperature for the measurements or kept at 600°C for Au decoration, respectively. To capture the STM images, a special care had been taken to avoid the thermal drift by waiting at least one hour. Au and Pb evaporation were carried out with ceramic crucibles and heating filaments. Within the present study, one ML of Au corresponds to one ML of Si(111) in the topmost layer, consists of 7.84×10^{14} atoms/cm^2. One ML of Pb is estimated for Pb(111) layers the thickness which is 0.286 nm. The evaporated average thickness was estimated via the deposition rate measured with a water-cooled quartz crystal microbalance. Surface preparation steps were controlled by RHEED observations.

3. Result and discussion

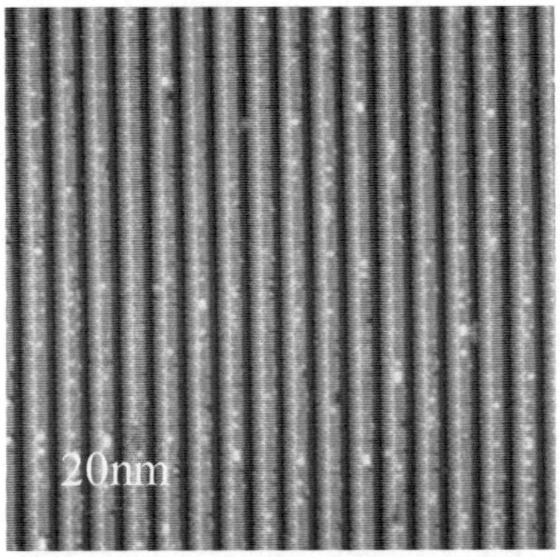

Figure 3. A 100x100 nm^2 STM image of the Si(557) surface after the flashing and annealing procedure which was explained in figure 2. The achieved structure reveals parallel terraces with perfect periodicity of 5.9 nm within the accuracy of our measurement. The image was recorded at room temperature with a tunnelling current I = 0.2 nA and a sample bias U = -2.193 V.

The result of this publication came from the initial step of our co-operation project and it can be divided into two main parts: the first part dealing with the STM observation of the Si(557) surface and

Si(557) decorated by 0.2 ML of Au; the second part mentions the IR studies of self-assembled Pb nanowires on the bare Si(557) surface.

Figure 3 shows the STM image of a clean Si(557) surface, captured at room temperature with a tunnelling current of 0.2 nA and a bias voltage of -2.193 V. The surface reveals well-oriented steps with perfect grating structure at atomic scale. Every terrace consists of one silicon plane made of the halves of the Si(111)-7x7 unit cell. We found that as longer keeping the sample at 830°C as more complete is the transition from the 1x1 to the 7x7 reconstruction. This is the main factor to form a vast and stable area of steps. Accordingly, the RHEED observation displayed very clear diffraction pattern (not shown here) of the 7x7 character.

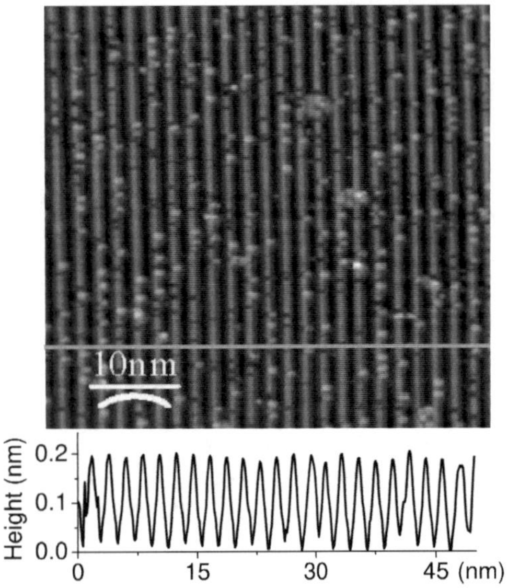

Figure 4. STM image of Si(557)-Au with height profile (along the blue line) of the surface decorated by 0.2 ML Au. The new Au-induced surface structure with the periodicity of 2.1 nm was formed. The image was recorded at room temperature with the tunnelling current I = 0.2 nA and the bias voltage U = -2.154V.

In an earlier work, Krawiec & Jalochowski claimed the existence of the double non-equivalent chain structure on one terrace forming the Si(557)-Au [9], as predicted before by a theoretical calculations by J. Crain *et al.* [10]. It is known that the electron wave function of the chains localizes at the end-atoms, so-called end states that influence the obvious chain length under STM current [11], depending on whether the scanning regime is for filled states or for emptied ones. Figure 4 shows the Au chains on the Si(557)-Au, recorded at room temperature with a tunnelling current of 0.2 nA and a voltage of -2.514 V. The height profile reveals that the periodicity of the structure was 2.1 nm, smaller than in the case of bare surface with 5.9 nm, which means that the surface had been reconstructed by the presence of Au chains. The bright spots that were distributed randomly on the surface originate from silicon adatoms. The co-appearance of the Au and Si chains can be identified via changing the tunnelling bias to see the emptied states.

As known, IR spectroscopy enables us to measure the optical conductivity of the metallic nanostructures by looking at the response of the free electron gas [7, 12, 13]. In the case of the nanowires, the appropriate polarized IR radiation with the optical frequencies could excite collective oscillation of the free electron gas in the range of the skin depth. The nanowires act as optical nanoantennae. As a consequence, it is possible to approximately describe such a scenario by the antenna relation at the fundamental resonance, i.e.

$$< L >= \frac{\lambda(\mu\mathrm{m})}{2n_{\mathrm{eff}}} \tag{1}$$

This rough estimation does not take into account the influence of the detailed shape, and of the material of the nanowires. In equation (1), $<L>$ is the average length of the nanowire array and λ the excitation wavelength, and n_{eff} the effective refractive index of the surrounding environment (here 2.52 [5]), respectively. For more precise relations see [5] and references therein.

The IR measurements in figure 5 show the development of the relative transmittance (reference is the substrate before Pb deposition) coming along with the shift of the antenna resonance during Pb evaporation. Already at 2 ML Pb, the resonance appears at 2500 cm^{-1}. The shift of the antenna frequency to lower wavenumbers (longer wavelength) reveals the corresponding growth of nanowires in length.

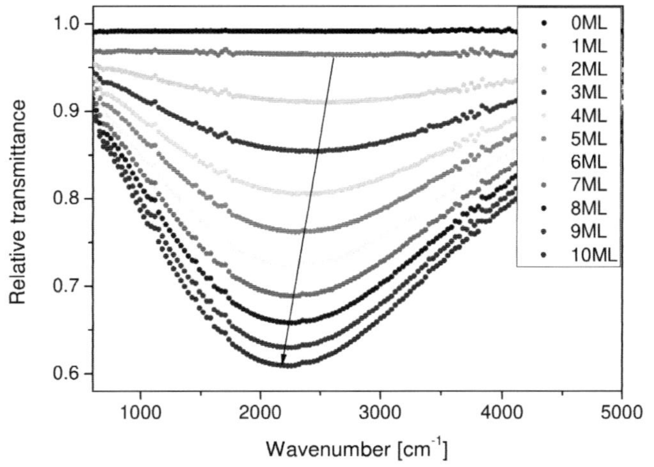

Figure 5. IR spectra recorded in situ during Pb evaporation from 0 to 10 ML. The shift to smaller wavenumbers is indicated. The antenna-like resonance was measured with parallel polarized electric field (parallel to the steps). 0 ML is the start of evaporation when the shutter of the Pb crucible was opened and 15 ML is the end of the deposition process when the shutter was closed.

The nucleation process at room temperature of the Pb nanowires was started via forming a wetting layer with the thickness of 1.3 ML (not shown here). Beyond this first initial stage the nucleation and aggregation of the Pb wires were developing. Hence, it can be inferred from figure 5 where at 2 ML the nanowires start to be excited by the external field that only 0.7 ML of Pb contributed to the formation of nanowires. A rough estimate gives an the average length reaching of about 800 nm which means that there existed very long nanowires with very high aspect ratio. In a former work on Au-Si(335), a resonance length of only 500 nm at 2 ML is found under similar experimental conditions [5]. Compared to that situation, Si(557) seems to provide better ordered terraces with larger areas. If not taking into account the effects by different thermal treatment procedures in sample preparation, one can notice that the different intrinsic structure gives the differences in Pb diffusion and thus they influence the final length of the array via corner rounding diffusion and step separation.

It is also possible to estimate wire lengths from figure 5, like done in [5]. From figure 5 it turned out, that only length of nanowires from 800 nm (2500 cm^{-1} at 2 ML) to 900 nm (2200 cm^{-1} at 10 ML) occurs. This means also that after nucleation the growth rate in the width of the wires was faster than that in length. Since the final structure is defined from the very beginning stage of the nucleation process, it is reasonable to notice that both the surface energy and the sample preparation (including evaporation rate, substrate temperature) defined the nucleation density, and therefore influenced also the final average length in the nanowire arrays.

Figure 6 displays the different responses of the lead nanowires to the incoming electrical field for two different orientations: parallel and perpendicular to the long axis of the nanowires. The transmittance spectrum in case of perpendicular polarization looks similar to spectra of metal island films regarding the lower transmittance (*i.e.* higher extinction) at the higher wavenumbers. The antenna resonance peak clearly appears for parallel excitation. The inset shows the AFM morphology of the lead wires (taken right after the sample had been transferred to the air).

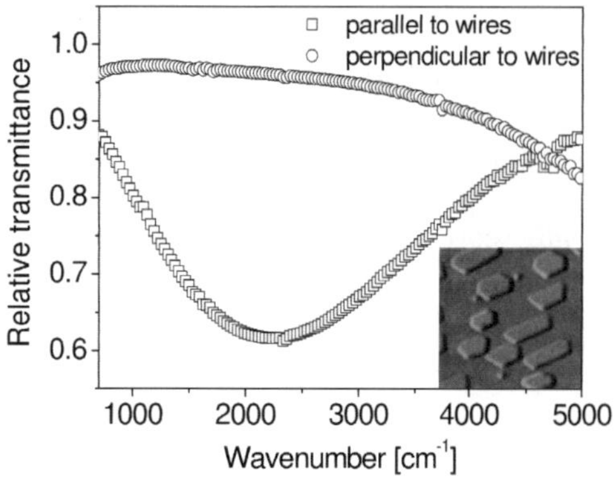

Figure 6. Comparison of IR transmittance of lead wires on Si(557) for the two different polarization directions. Inset: AFM image of such lead rods grown at room temperature, scan size 1.25 μm x 1.25 μm, tapping mode.

4. Conclusion

In summary, we demonstrated the first stage of our studies on Au atom-chains and Pb nanowires on the Si(557) surface by STM and IR measurements. Si(557) and Si(557)-Au revealed nearly ideal atomic chain structures which are well-suited for fundamental research on low dimensional system such as atom chains and nanowires. The Pb nanowires grown on such surfaces behave like IR optical antennae under the appropriate excitation regime.

References

[1] T Nagao, Yaginuma S, Inaoka T, Sakurai T 2006 *Phys. Rev. Lett.* **97** 116802
[2] Yeom H W, Takeda S, Rotenberg E, Matsuda I, Horikoshi K, Schaefer J, Lee C M, Kevan S D, Ohta T, Nagao T and Hasegawa S 1999 *Phys. Rev. Lett.* **82** 4898
[3] Phaneuf R J and Williams E D 1990 *Phys. Rev. B* **41** 5
[4] Segovia P, Purdie D, Hengsberger M and Baer Y 1999 *Nature (London)* **402** 504
[5] Klevenz M, Neubrech F, Lovrincic R, Jalochowski M and Pucci A 2008 *Appl. Phys. Lett.* **92** 133116
[6] Neubrech F, Pucci A, Cornelius T. W, Karim S, Garcia-Etxarri A and Aizpurua J 2008 *Phys. Rev. Lett.* **101** 157403
[7] Neubrech F, Weber D, Lovrincic R, Pucci A, Lopes M, Toury T and Lamy M de L C 2008 *Appl. Phys. Lett.* **93** 163105
[8] Viernow J, Lin J L, Petrovykh D Y, Leibsle F M, Men F K and Himpsel F J 1998 *Appl. Phys. Lett.* **72** 8
[9] Krawiec M and Jalochowski M 2006 *Phys. Rev. B* **73** 075415
[10] Crain J N, McChesney J L, Zheng F, Gallagher M C, Snijders P C, Bissen M, Gundelach C, Erwin S C 2004 *Phys. Rev. B* **69** 125401
[11] Crain J N and Pierce D T 2005 *Science* **307** 703
[12] Pucci A, Kost F, Fahsold G and Jalochowski M 2006 *Phys. Rev. B* **74** 125428
[13] Fedorov D V, Fahsold G, Pucci A, Zahn P and Mertig I 2007 *Phys. Rev. B* **75** 245427

Acknowledgments

CVH would like to acknowledge the financial support by the Daimler Benz Foundation and the Heidelberg Graduate School of Fundamental Physics (HGSFP), Heidelberg University. The project is supported by the Deutsche Forschungsgemeinschaft (PU193/9-1) and by the Japan Science and Technology Agency.

Synthesis and optical properties of colloidal gold nanoparticles

Nguyen Ngoc Long, Le Van Vu, Chu Dinh Kiem, Sai Cong Doanh, Cao Thi Nguyet, Pham Thi Hang, Nguyen Duy Thien, Luu Manh Quynh

Faculty of Physics, Hanoi University of Science, 334 Nguyen Trai, Thanh Xuan, Hanoi, Vietnam

E-mail: longnn@vnu.edu.vn

Abstract. Colloidal gold nanoparticles (spheres) have been prepared from $HAuCl_4$ containing aqueous solution by using X-ray irradiation and by chemical reduction method. Gold nanorods were synthesized according to the seed-mediated growth method. The colloidal gold nanoparticles were characterized by using transmission electron microscopy, X-ray diffraction, and UV-VIS absorption spectroscopy. It was found that the concentration of the precursors affects the size of the nanoparticles. In the chemical reduction approach the size of nanoparticles can be controlled by varying amount of trisodium citrate, but in the photochemical method the size of nanoparticles can been controlled by varying the ratio of $HAuCl_4$ to TX-100 and X-ray irradiation duration. Gold nanorods have been synthesized according to the seed-mediated growth method with two steps. The effect of silver acetate and CTAB on formation of gold nanorods has been studied.

Keywords: Metallic nanoparticles, gold nanoparticles, chemical reduction method, photochemical method.

1. Introduction

Metal nanoparticles, in general, and gold nanoparticles, in particular, are very attractive because of their size- and shape-dependent properties [1]. For example, gold nanospheres have a characteristic red colour, but anisotropic gold nanorods have a dramatically changed colour. More recent treatments have shown that the colour is due to the collective oscillation of the electrons in the conduction band, known as the surface plasmon oscillation [2]. The oscillation frequency is usually in the visible region for gold giving rise to the strong surface plasmon resonance absorption. The plasmon resonance absorption of gold nanospheres has an absorption coefficient orders of magnitude larger than strongly absorbing dyes. Anisotropic shapes like nanorods have plasmon resonance absorptions that are even stronger, leading to increased detection sensitivity. Gold nanoparticles generate enhanced electromagnetic fields that affect the local environment. The field is determined by the geometry of the nanoparticle and can enhance fluorescence of the gold itself, the Raman signal of a molecule on the surface, and the scattering of light. Because of the mentioned above factors, many applications in biomedicine of gold nanoparticles became possible [3, 4]. In the existing literature Au nanoparticles have been prepared using various methods such as chemical reduction [5, 6], photochemical using UV irradiation [7, 8], sonochemical [9, 10], sonoelectrochemical, etc [11, 12]. Gold nanorods have been synthesized by the seed-mediated growth method [13-15].

© 2009 IOP Publishing Ltd

In this work we present the size- and shape-controlled synthesis of gold nanoparticles by chemical reduction method and in particular, we first time report on synthesis of gold nanoparticles by photochemical approach using X-ray irradiation. Structural property, morphology and absorption spectra of the gold nanoparticles are discussed.

2. Experimental

Using chemical reduction method, synthesis of gold nanospheres was carried out as follows: 50 ml of 0.01 wt.% chlorauric acid ($HAuCl_4.3H_2O$) solution was heated to boiling while stirring in a 100 ml beaker. Then a few hundred microlitres (μl) of 1 wt.% trisodium citrate dihydrate ($Na_3C_6H_5O_7.2H_2O$) solution was quickly added to the auric solution. The solution changed color within several minutes from yellow to black and then to red or purple color depending on the size of the nanoparticles.

Gold nanoparticles were also synthesized via a photochemical approach by using X-ray irradiation. In this approach, a few μl of 1 mM triton X-100 (TX-100) solution was mixed with 3 ml of 1 mM $HAuCl_4.3H_2O$ solution. Pure $HAuCl_4$ solution without TX-100 addition was also prepared for comparison. The X-ray radiation was Cu-Ka radiation of an X-ray diffractometer Siemens D5005 with the parameters of 40 kV and 40 mA. The X-ray irradiation time was set up to 40 or 60 minutes to guarantee the complete reduction of gold precursor solution.

Gold nanorods were synthesized according to the seed-mediated growth method [13, 14]. The synthesis of gold nanorods was divided into two steps.

Step 1: Preparation of seed solution: 2.5 ml of 1 mM chlorauric acid solution was mixed with 7.5 ml of 0.2 M cetyltrimettyammonium bromide (CTAB: $C_{19}H_{42}NBr$) solution and stirred. To the stirred solution, 0.6 ml of ice-cold 0.01 M sodium borohydride ($NaBH_4$) solution was added. The final solution had a brownish yellow color. Stirring the seed solution was continued for 2 minutes. After that, the seed solution was kept at room temperature.

Step 2: Growth of nanorods: 5 ml of 0.2 M CTAB solution was added to 5 ml of 1 mM chlorauric acid solution. Then a few hundred μl of 4 mM silver acetate ($AgC_2H_3O_2$) solution was added and gently mixed. After that, to this solution, 70 μl of 0.08 M ascorbic acid ($C_6H_8O_6$) solution and 12 μl of the seed solution were added (see table 1). By changing the amount of silver acetate, the aspect ratio defined as the length of the major axis divided by the width of the minor axis of the nanorods can be varied. To separate nanorods from nanospheres, the solution was centrifuged at 3000 rpm for 10 min.

Table 1. Preparation of gold nanorods

Chemicals	Sample 1	Sample 2	Sample 3	Sample 4	Sample 5
CTAB	5 ml	5 ml	5 ml	5 ml	5 ml
$AgC_2H_3O_2$	50 μl	100 μl	150 μl	200 μl	250 μl
$HAuCl_4$	5 ml	5 ml	5 ml	5 ml	5 ml
Ascorbic acid	70 μl	70 μl	70 μl	70 μl	70 μl
Seed solution	12 μl	12 μl	12 μl	12 μl	12 μl

The obtained products were characterized by X-ray diffraction (XRD) by using a Brucker D5005 diffractometer and transmission electron microscopy (TEM) by using a JEOL JEM1010 electron microscope. UV-VIS absorption spectra of the nanoparticle solutions were collected with a Shimadzu UV 2450 PC spectrometer.

3. Results and discussion

We have prepared the gold nanoparticles by adding different amounts (200, 240, 280, 320, 360, and 400 μl) of 1 wt.% trisodium citrate solution to 50 ml of 0.01 wt.% chlorauric acid solution while heating to boiling.

Figure 1 illustrates a typical XRD pattern of the gold nanoparticles obtained from the solution with 400 μl of trisodium citrate. It can be seen from the figure that all the diffraction peaks corresponding to the diffraction planes (111), (200), and (220) were indexed to the gold metal with face centered cubic structure. Lattice constant determined from XRD pattern is a = 4.078 Å in good agreement with the standard diffraction pattern of cubic gold metal (Pattern 4-784).

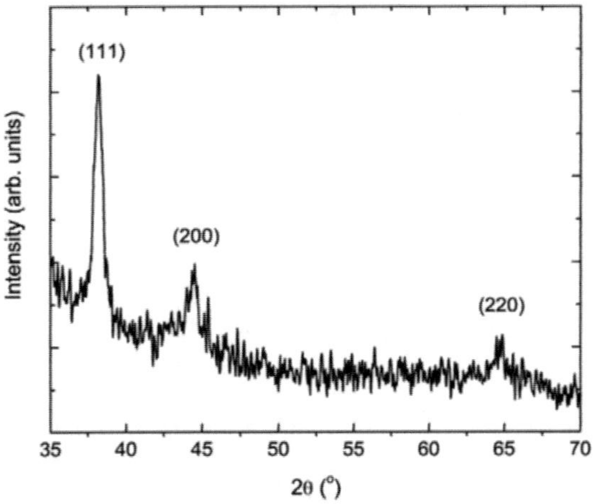

Figure 1. XRD pattern of the gold nanoparticles prepared by chemical reduction technique.

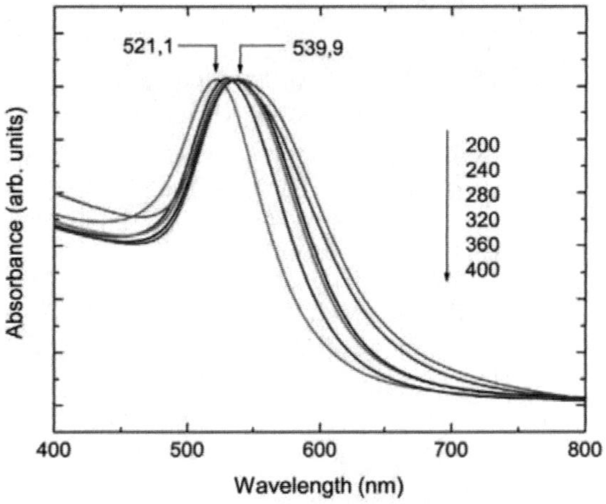

Figure 2. Surface plasmon absorption spectra of spherical gold nanoparticles prepared with different amounts of 1 wt.% trisodium citrate solution: 200, 240, 280, 320, 360, and 400 μl.

Surface plasmon absorption spectra of spherical gold nanoparticles prepared with different amounts of 1 wt.% trisodium citrate solution are shown in figure 2. As seen from the figure, the maximum of the absorption spectrum is shifted to the short wavelength side (from 539.9 to 521.1 nm) and the full width at half maximum (FWHM) of the spectrum is decreased with increasing amount of trisodium citrate from 200 to 400 μl. Thus, the amount of citrate solution determines the size of the nanoparticles. The faster the capping of the nanoparticles by the citrate, the smaller the resulting nanoparticles.

Typical TEM images of the gold nanoparticles prepared with different amounts of 1 wt.% trisodium citrate solution are shown in figure 3. From the figure one can see that when the amount of trisodium citrate increases from 200 to 400 μl, the size of the synthesized gold nanoparticles decreases from 60 to 30 nm. That means the more amount of trisodium citrate, the smaller obtained nanoparticles are.

Gold nanoparticles were also synthesized via a photochemical approach by using X-ray irradiation as mentioned above. Figure 4*a* shows the TEM image of the gold nanoparticles for the sample prepared from $HAuCl_4$ solution without TX-100 with the corresponding selected area electron diffraction (SAED) pattern in the inset. It can be seen that the gold nanoparticles with the size about 7-10 nm are aggregated. From the SAED pattern one can see that the diffraction rings were indexed to (111), (220), and (311) diffraction planes of the cubic structure of gold metal in agreement with the XRD pattern (not shown here). The TEM image of the gold nanoparticles for the sample prepared from the mixed solution of $HAuCl_4$ and TX-100 is illustrated in figure 4*b*. As seen from this figure, the gold nanoparticles are spherical and well dispersed.

It is found that the addition of TX-100 behaving as a particle stabilizer and reducing agent into $HAuCl_4$ solution tremendously affected the formation and growth of the gold nanoparticles. Figure 5 shows the surface plasmon absorption spectra of gold nanoparticles as a function of amount of $HAuCl_4$ solution. When the amount of $HAuCl_4$ is increased from 3.6 to 10.8 μl, while the amount of TX-100 is kept constant, the intensity of absorption band is much stronger.

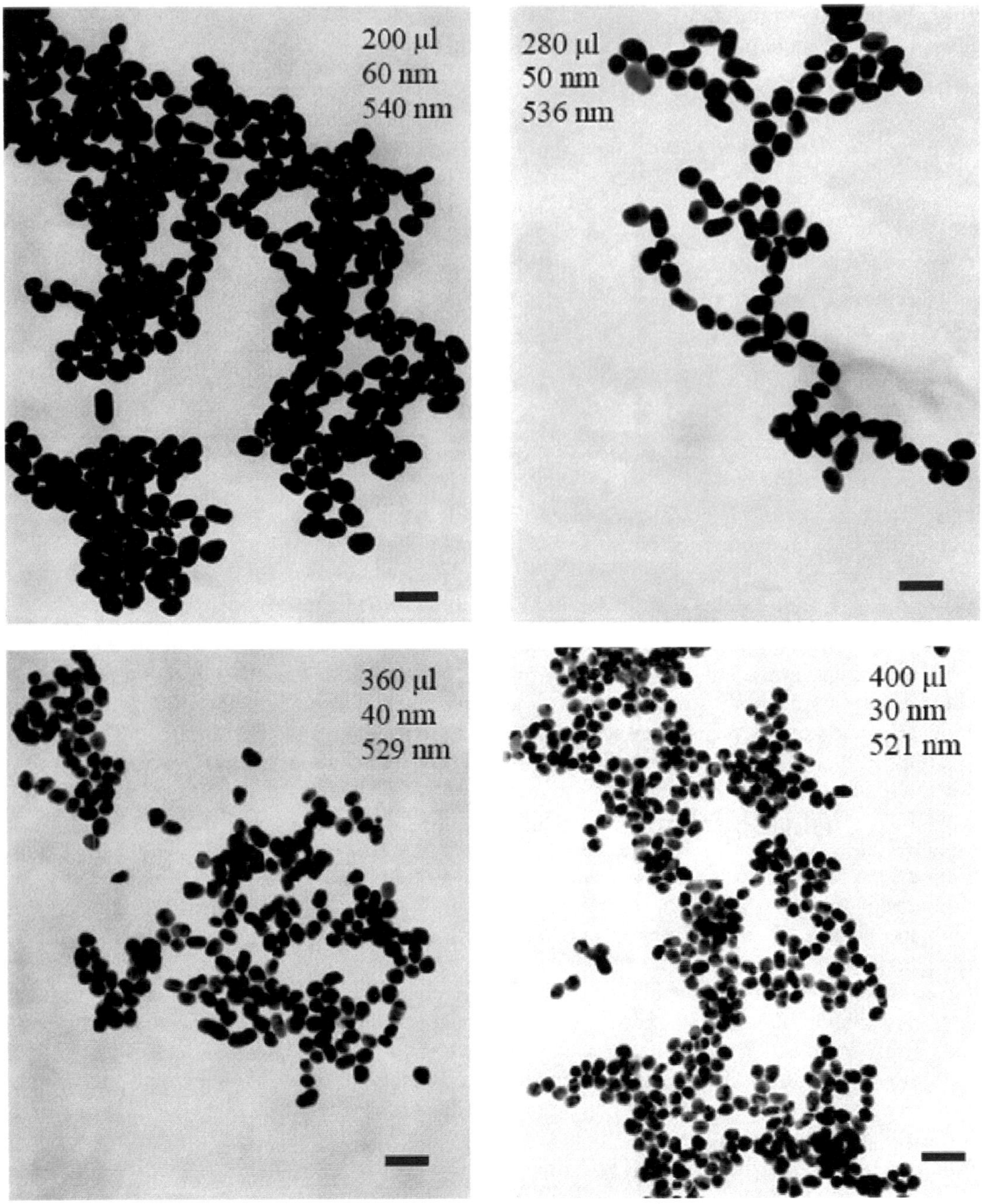

Figure 3. TEM images of the gold nanoparticles prepared with different amounts (200, 280, 360, and 400 µl) of 1 wt.% trisodium citrate solution. All the scale bars are 100 nm. On the pictures are shown the size of the nanoparticles and the wavelength of the maximum of surface plasmon absorption spectra.

When trivalent gold ions $Au^{III}Cl_4^-$ are irradiated by an intensive X-ray beam, the reduction occurs as follows: Under the effect of the X-ray radiation, hydrated electrons produced from water molecules contributes to the reduction of $AuCl_4^-$ ions [16]. The free radical ($H^\bullet$) and solvated electrons (e_{aq}) are powerful reducing agents and easily reduce metallic ions to zero valence gold through the below reactions:

$$H_2O \xrightarrow{h\nu} e_{aq}^- + H^{\bullet} + OH^{\bullet} + H_2O_2 + H_2 \quad (1)$$

$$Au^{3+} + 3e_{aq}^- \rightarrow Au \quad (2)$$

$$Au^{3+} + 3H^{\bullet} \rightarrow Au \quad (3)$$

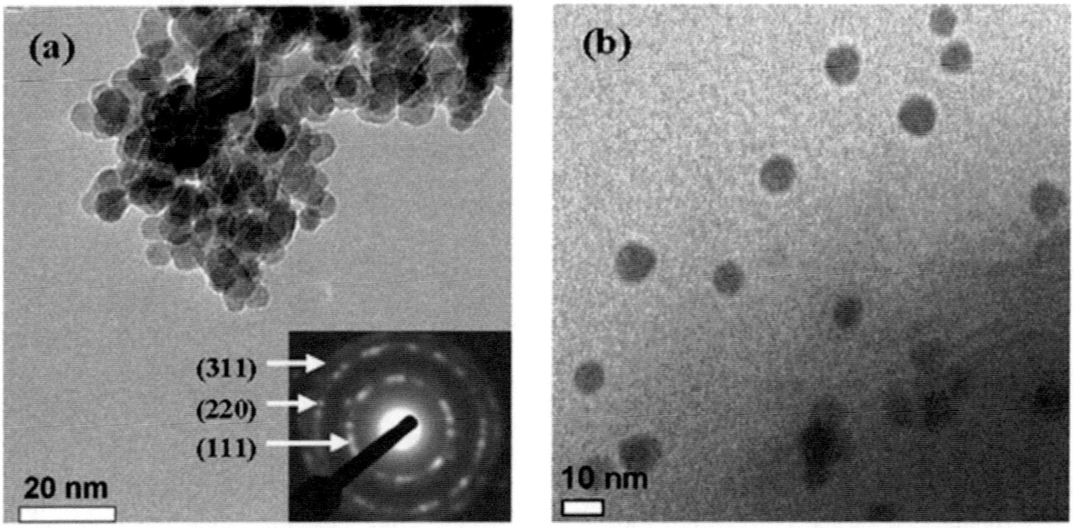

Figure 4. TEM images of the gold nanoparticles synthesized by photochemical method using X-ray irradiation: *a*) sample prepared from $HAuCl_4$ solution without TX-100, inset is electron diffraction pattern revealing the cubic structure of gold metal; *b*) sample prepared from the mixed solution of $HAuCl_4$ and TX-100.

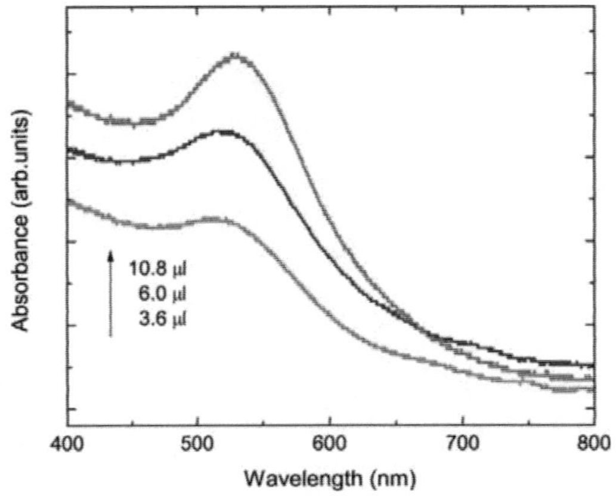

Figure 5. Surface plasmon absorption spectra of gold nanoparticles as a function of the amount of $HAuCl_4$, (the amount of TX-100 is constant).

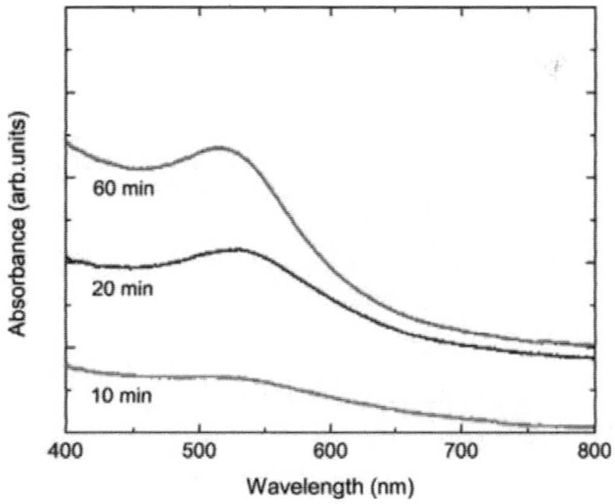

Figure 6. Surface plasmon absorption spectra of gold nanoparticles as a function of different X-ray irradiation time.

According to these equations, the amount of formed reducing agents strongly depends on the incident X-ray dose, which was dependent on both the flux of incident beam and the amount of precursor. The relationship between X-ray dose and formation of gold nanoparticles was investigated. Figure 6 shows the the surface plasmon absorption spectra of gold nanoparticles as a function of X-ray irradiation time. The increase of the absorption peak intensity of gold nanoparticles with increasing X-ray irradiation duration from 10 to 60 min is an evident proof about an enhancement of nanoparticle concentration with increasing X-ray dose.

Gold nanorods were synthesized according to the seed-mediated growth method with two steps. By changing silver acetate content (from 50 to 250 μl of 4 mM solution), the aspect ratio of the nanorods can be varied. Table 1 shows the content of the solutions in different samples. Typical XRD pattern of the gold nanorods obtained from the solution with 200 μl of silver acetate solution is shown in figure 7. It can be seen from the figure that all the diffraction peaks were indexed to the gold metal with face centered cubic structure. Lattice constant computed from the diffraction peak (111) is $a = 4.083$ Å.

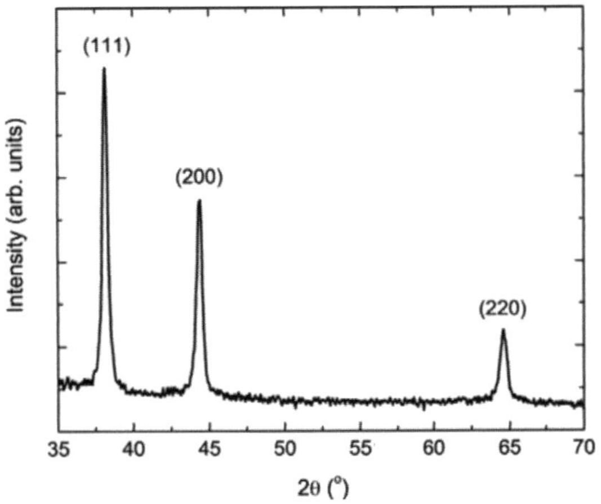

Figure 7. XRD pattern of the gold nanorods prepared by the seed-mediated growth method.

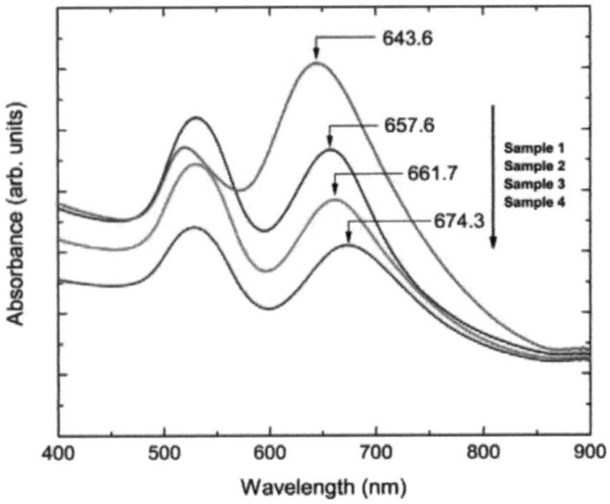

Figure 8. Surface plasmon absorption spectra of the gold nanorods prepared with different amounts (50, 100, 150, and 200 μl) of 4 mM silver acetate solution.

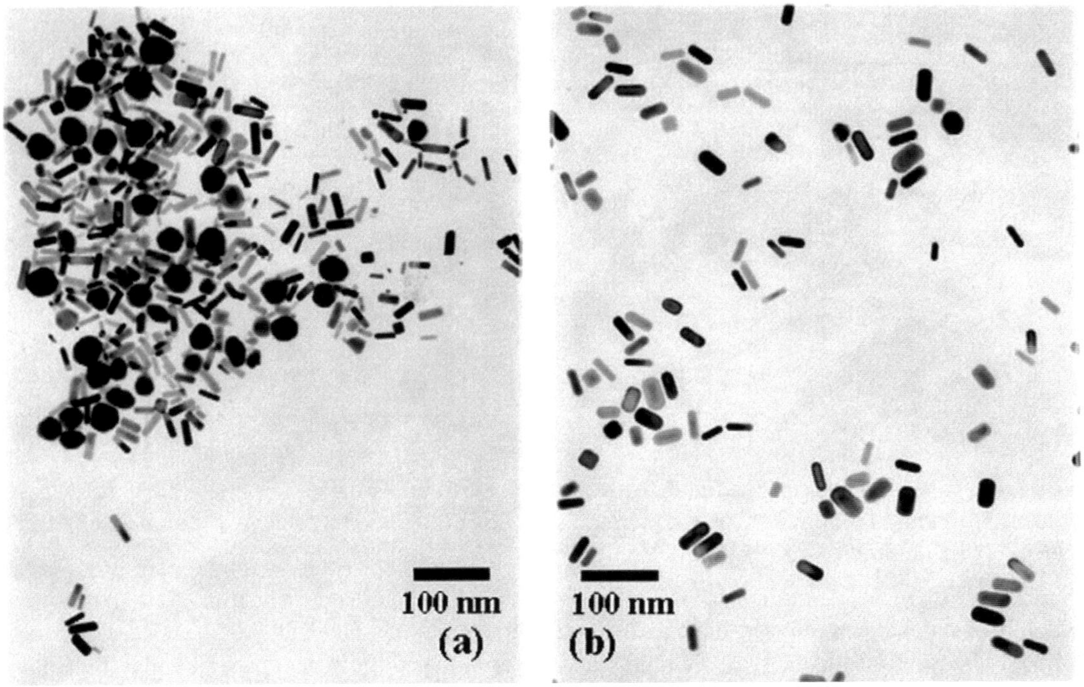

Figure 9. TEM images of gold nanostructures for sample 1:
a) as-synthesized produce; b) produce after centrifugation at 3000 rpm for 10 min.

Journal of Physics: Conference Series **187** (2009) 012026 doi:10.1088/1742-6596/187/1/012026

Surface plasmon absorption spectra of anisotropic gold nanorods prepared with different amounts of 4 mM silver acetate solution are shown in figure 8. As seen from the figure, when the shape of the nanoparticles changes from nanospheres to nanorods, the surface plasmon absorption spectra also change. The surface plasmon absorption spectra of gold nanorods have two bands: one band in the blue-green region around 525 nm due to the transverse oscillation of electrons and another band in the red-near infrared region due to the longitudinal electronic oscillation. In addition, the position of the band due to the transverse electronic oscillation is almost unvaried; meanwhile the position of the band due to the longitudinal oscillation is increased from 643.6 to 674.3 nm with increasing the silver acetate content from 50 to 250 μl.

Typical TEM image of gold nanostructures for the as-prepared sample 1 (table 1) with 50 μl of 4 mM silver acetate solution is shown in figure 9a. Two major components of the sample are spherical nanoparticles and nanorods. After centrifugation at 3000 rpm for 10 min, the nanorods were separated from spherical nanoparticles (figure 9b).

Our experimental results indicated that if no amount of silver acetate was added to the growth solution, no rods were formatted (inset in figure 10). In this case, the increase of CTAB amount made the particle size slightly decrease. It can be seen from figure 10 that for the samples prepared without silver acetate, there are no absorption bands due to the longitudinal oscillation of electrons in the visible spectra.

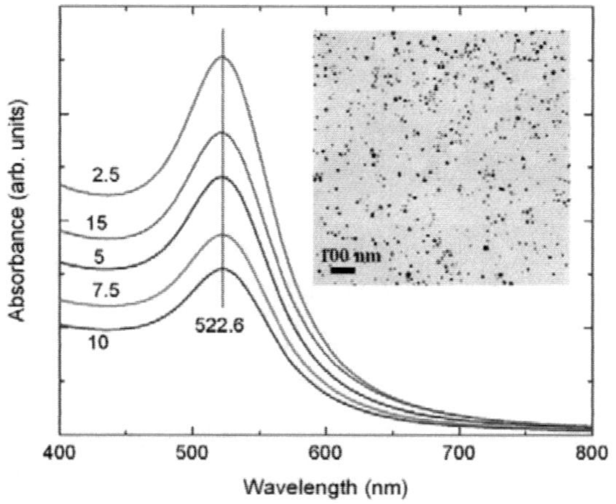

Figure 10. Visible absorption spectra of gold nanoparticles prepared with 2.5, 5, 7.5, 10, 15 μl of 0.05 M CTAB solution in the absence of silver ions. Inset shows a TEM image of the sample containing 15μl CTAB.

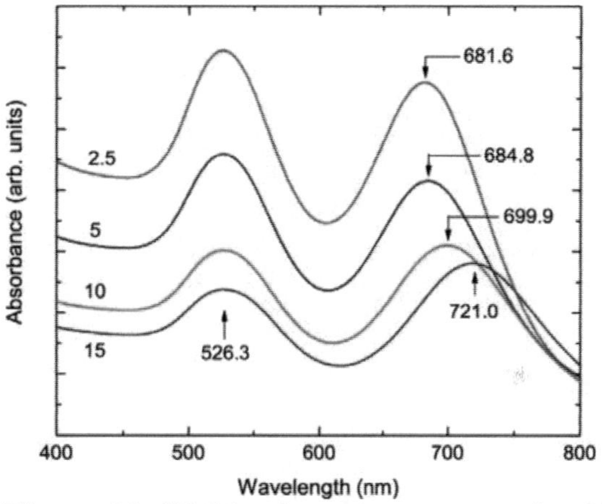

Figure 11. Visible absorption spectra of gold nanoparticles prepared with 2.5, 5, 7.5, 10, 15 μl of 0.05 M CTAB solution and 300 μl of 4 mM silver acetate solution.

In addition, the position of the absorption bands due to the transverse oscillation is almost kept constant. When a certain amount of silver acetate was added to the growth solution, gold nanorods were easily formatted. In this case, the increase of CTAB amount resulted in increasing aspect ratio of nanorods. This fact is clearly illustrated in figure 11. When 300 μl of 4 mM silver acetate solution are added to the growth solution, the absorption bands due to the longitudinal oscillation of electrons appear in the visible spectra, that is obvious evidence on the formation of gold nanorods. In addition, with increasing the content of CTAB from 2.5 to 15 μl of 0.05 M CTAB solution, the position of the longitudinal oscillation absorption bands is shifted from 681.6 to 721.0 nm, corresponding to the enhance of the aspect ratio of gold nanorods.

4. Conclusion

The colloidal gold nanoparticles have been prepared by chemical reduction method and by photochemical method using X-ray irradiation. In the chemical reduction approach the size of nanoparticles can been controlled by varying amount of trisodium citrate, but in the photochemical method the size of nanoparticles can been controlled by varying the ratio between the amount of $HAuCl_4$ and TX-100, and X-ray irradiation duration. The gold nanorods have been prepared by the seed-mediated growth method. The aspect ratio of nanorods can be controlled by varying amount of silver acetate and CTAB.

References

[1] Eustis S and El-Sayed M A 2006 *Chemical Society Review* **35** 209
[2] Liz-Marzan L M 2004 *Materialstoday* **7** 26
[3] Jain P K, El-Sayed I H and El-Sayed M A 2007 *Nanotoday* **2** 18
[4] Rayavarapu R G, Petersen W, Ungureanu C, Post J N, van Leeuwen T G and Manohar S 2007 International Journal of Biomedical Imaging Article ID 29817
[5] Frens G 1973 *Nature* **241** 20
[6] Leff D V, Ohara P C, Heath J R and Gelbart W M 1995 *J. Phys. Chem.* **99** 7036
[7] Kaushik Mallick, Wang Z L and Tarasankar Pal 2001 *Journal of Photochemistry and Photobiology A: Chemistry* **140** 75
[8] Sau T K, Anjali P, Jana N R, Wang Z L and Tarasankar P 2001 *Journal of Nanoparticle Research* **3** 257
[9] Kenji O, Yoshiteru M, Yamamoto T A, Yasuaki M and Yoshio Na 2007 *Materials Letters* **61** 3429
[10] Cuncheng L, Weiping C, Caixia K, Ganhua F and Lide Z 2003 *Materials Letters* **58** 196
[11] Liping W, Wei M, Dandan N, Junwei D, Ying W and Yifeng T 2008 *Electrochemistry Communications* **10** 673
[12] Min Z, Shenhao C, Shiyong Z and Houyi M 2006 *Physica E* **33** 28
[13] Nikoobakht B and El-Sayed M A 2001 *Langmuir* **17** 6368
[14] Sau T K and Murphy C J 2004 *Langmuir* **20** 6414
[15] Jorge P-J, Isabel P-S, Luis M L-M and Paul M 2005 *Coordination Chemistry Reviews* **249** 1870
[16] Gachard E, Remita H, Khatouri J, Keita B, Nadjo L, Belloni J 1998 *New J. Chem.* **22** 1257

Acknowledgements

This work is financially supported by Ministry of Science and Technology of Vietnam (Project 405506 and Task of Protocol with Israel) and Vietnam National University, Hanoi (Key Project QGTD 0805 and QG 07-03).

Influence of Cd:Se precursor ratio on optical properties of colloidal CdSe tetrapods prepared in octadecene

Le Ba Hai, Nguyen Xuan Nghia, Pham Thu Nga, Do Hung Manh, Vu Thi Hong Hanh and Nguyen Thi Thu Trang

Institute of Materials Science, Vietnam Academy of Science and Technology, 18 Hoang Quoc Viet Rd., Hanoi, Vietnam

E-mail: nghianx@ims.vast.ac.vn

Abstract. Colloidal CdSe tetrapods were synthesized by chemical method in octadecene with the initial Cd:Se precursor ratios from 6:1 to 1:10 and the different monomer concentrations. Their stoichiometry and optical properties were investigated using the energy dispersion X-ray, absorption and photoluminescence spectroscopy. It was found that the CdSe tetrapods are Cd rich for all of the obtained sizes. Their stoichiometry was almost not changed during the growth and depends on the initial precursor ratio. The double-peak structure corresponding to the optical transitions in the core and arms of CdSe tetrapods was observed clearly in both the optical absorption and photoluminescence spectra for the Cd:Se precursor ratios from 2:1 to 1:2. The influence of the initial Cd:Se precursor ratio and monomer concentration on the growth of CdSe tetrapods and their optical properties has been analysed and discussed.

Keyword: CdSe tetrapods, Cd:Se precursor ratio, optical properties, monomer concentration.

1. Introduction

Semiconductor nanocrystals (NCs) have attracted great interest over the past years because their properties can be tailored by a judicious control of particle composition, size, and surface [1-2]. These characteristics arise from several phenomena (quantum confinement of charge carriers, surface effects, and geometrical confinement of phonons) and have turned semiconductor NCs into promising materials for many applications, such as light emitting diodes [3], lasers [4], and biomedical tags for fluoroimmunoassays, nanosensors, and biological imaging [5].

The high surface-to-volume ratio of small NCs suggests that the surface properties should have significant effects on their structural and optical properties. It is well-known that the emission efficiency, spectrum, and time evolution are very strongly affected by the surface. This is generally understood as being due to the presence of gap surface states arising from surface nonstoichiometry, unsaturated bond, etc. Control of the surface is in particular the key to highly luminescent NCs.

To realize this wide range of potential applications the luminescent properties (emission colour, photoluminescence quantum yield (PL QY), luminescence lifetime, and stability) of semiconductor NCs must be strictly controlled. At present, the control and reproducibility of the PL QY has remained an elusive issue, only recently some of the factors determining the PL efficiency of NCs have been properly addressed [6-8]. The influence of the initial ratio between the Cd and the Se precursor on the temporal evolution of the ensemble PL QY of CdSe NCs during their growth at 290°C in a coordinating solvent has been investigated by Qu and Peng [6]. The position and temporal width of the

© 2009 IOP Publishing Ltd

"bright point", the highest QY, the growth kinetics, and the sharpness of the PL peak were all reported to be strongly dependent on the initial Cd:Se ratio. The existence of the "bright point" was interpreted as a signature of an optimal surface structure/reconstruction of the NCs grown under a given set of initial conditions.

Although the crucial role of the surface is clearly demonstrated, but the mechanisms underlying the correlation between the surface quality and the growth conditions have yet to be determined. Most importantly, a comprehensive investigation of several parameters (for example, the initial Cd:Se precursor ratio) in a single synthetic system is still lacking. In this work, we report on the influence of initial Cd:Se precursor ratio on the optical properties as well as the growth of colloidal CdSe tetrapods (TPs) prepared in octadecene – a noncoordinating solvent.

2. Experimental

Chemicals: Cadmium oxide (CdO), oleic acid (OA), selenium (Se), trioctylphosphine (TOP), octadecene (ODE).

Preparation: To prepare the stock Se injection solution, Se powder was reacted with TOP, the resulting colorless solution was dispersed in ODE to give a final stock solution and was loaded into a syringe. Meanwhile, a mixture of CdO, OA and ODE was loaded into a three-neck flask, stirred and heated to 200°C under nitrogen gas flow to form an optical clear solution. At this temperature, the TOPSe solution was rapidly injected into the vigorously stirred reaction flask to grow CdSe TPs. The influence of the initial Cd:Se precursor molar ratio was investigated for the different values (6:1, 4:1, 2:1, 1:1, 1:2, 1:4, 1:6, 1:8 and 1:10) chosen to cover the limits from a large excess of Cd precursor to a large excess of Se precursor. The OA:Cd and TOP:Se molar ratio of 6:1 and 4.5:1, respectively, were kept for all of our experiments. At various time intervals, aliquots of the reaction mixture were removed for the further optical measurements. A centrifugation procedure was used to purify the CdSe TPs from unreacted precursors and side products, and then the CdSe TPs were dispersed in toluene.

Measurements: The optical absorption spectra of CdSe TPs were recorded by Jasco V670 UV-Vis-NIR spectrometer. The PL spectra were collected on LABRAM-1B spectrometer, fitted with the Argon ion laser of wavelength 488 nm. To determine the stoichiometry of CdSe TPs the energy dispersion X-ray spectroscopy (EDXS) was performed using Hitachi FE-SEM S4800 at the beam voltage of 20 kV. The elemental analysis was carried out at different places of area 2 mm×2 mm on the surface of solid samples. For EDXS measurements, the purification procedure was repeated 3 times. Transmission electron microscopy (TEM) was carried out using a JEM 1010 microscope fitted with an acceleration voltage of 80 kV.

3. Results and discussion

It is known that the injection of the TOPSe at high temperatures leads to the creation of CdSe TPs. Figure 1 is a TEM image of the CdSe TPs prepared with the Cd:Se precursor ratio of 2:1 and the reaction time of 180 min.

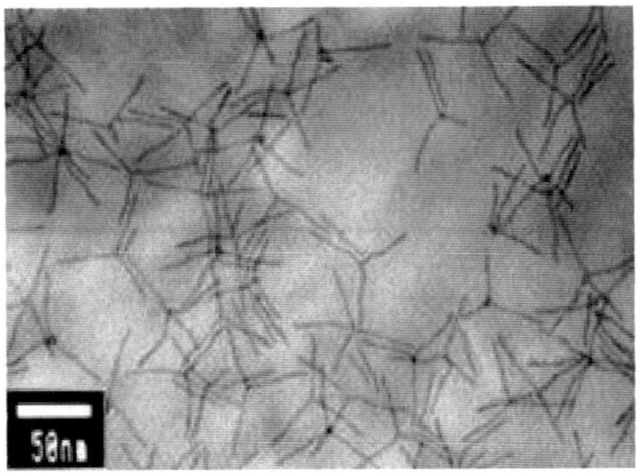

Figure 1. TEM image of CdSe TPs prepared with the initial Cd:Se precursor ratio of 2:1 and the reaction time of 180 min.

The Cd:Se stoichiometry of CdSe TPs with the arm diameters ranging from 4.6 to 5.7 nm is presented in figure 2. Surprising, all of the CdSe TPs are Cd rich for the initial Cd:Se precursor ratios from 6:1 to 1:10. Figure 3 displays the stoichiometry of CdSe TPs prepared with the initial Cd:Se precursor ratios ranging from 6:1 to 1:10 and the reaction time of 20 min. It can be seen that the Cd:Se ratio of CdSe TPs increases with increasing the initial Cd:Se precursor ratio. It is possible that the metallic cadmium was formatted in our preparation conditions. Donega et al. noted that the injection of $Cd(CH_3)_2$ in hot TOPO – HAD (hexadecylamine) under strictly water and oxygen – free conditions leads to the formation of metallic cadmium [9].

Recent experiments have shown that colloidal NCs can be nonstoichiometric. In this case of CdSe NCs, the Cd:Se stoichiometry of the resulting particles are always found to be between 1 and 1.2 [10-11]. Jasieniak et al. have manipulated the CdSe NCs surface, yielding nonstoichiometric quantum dots with a surface enriched in Cd or Se [11]. Guzelian et al. have determined an In:P ratio of 0.86:1 for InP NCs [12]. Recently, Moreels et al. also have shown that PbSe quantum dots prepared in OA-TOP-DPE (diphenil ether) reaction system are nonstoichiometric, consisting of a quasi-stoichiometric core terminated by a Pb surface shell [13].

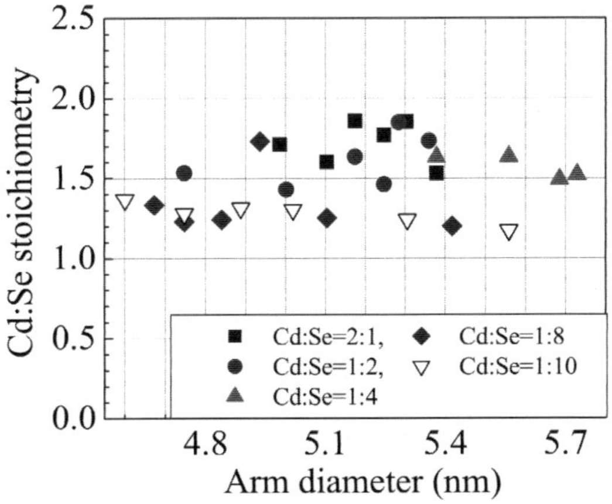

Figure 2. Cd:Se stoichiometry of CdSe TPs prepared with the different initial Cd:Se precursor ratios vs. the arm diameters.

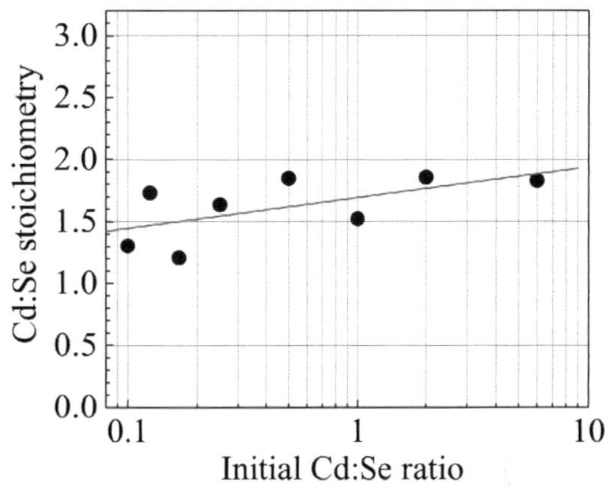

Figure 3. Cd:Se stoichiometry of CdSe TPs drawn at the reaction time of 20 min vs. the initial Cd:Se precursor ratios.

Figure 4 displays the temporal evolution of the optical absorption and PL spectra of CdSe TPs prepared with the initial Cd:Se precursor ratios of 4:1, 1:1 and 1:4. The first absorption peaks and the emission peaks were normalized in the intensity. The growth of the TPs is clearly evidenced in Figure 4 by the shift of both the absorption and PL spectra to long wavelength. It can be seen that the TP size quickly increases at the initial stages of the synthesis. Especially, the appearance of a new emission peak during the first reaction minutes is observed clearly for the initial precursor ratio of 1:1. The long-wavelength peak is attributed to the emission of TP core, and the short-wavelength peak to the emission from TP arms. These two emission peaks are separated gradually and the emission intensity of the arms increases in comparison with that of the core as the TP size increases.

It seems that the PL spectra exhibit a one-peak structure for the initial precursor ratios of 4:1 and 1:4. However, the second derivative spectra of absorbance confirm the formation of CdSe TPs in these cases. Although the PL spectra of CdSe TPs prepared with the Cd:Se precursor ratios of 4:1 and 1:4 exhibit one-peak structure, but we believe that the broad of PL spectra to long wavelength is due to the contribution of the emission from TP core. It should be noted that the previous reports on their optical properties of CdSe TPs are very different. The double-peak structure was observed clearly only in the absorption spectra [14], or in the PL spectra [15], even this double-peak structure was not observed in both the absorption spectra and the PL spectra [16]. Our experimental results show that the emission peak of TP core is separated gradually from that of the arms and quenched rapidly with increasing the TP size. It is possible that the appearance of the double-peak structure relates to the surface chemistry of colloidal CdSe TPs, namely, to the initial Cd:Se precursor ratio.

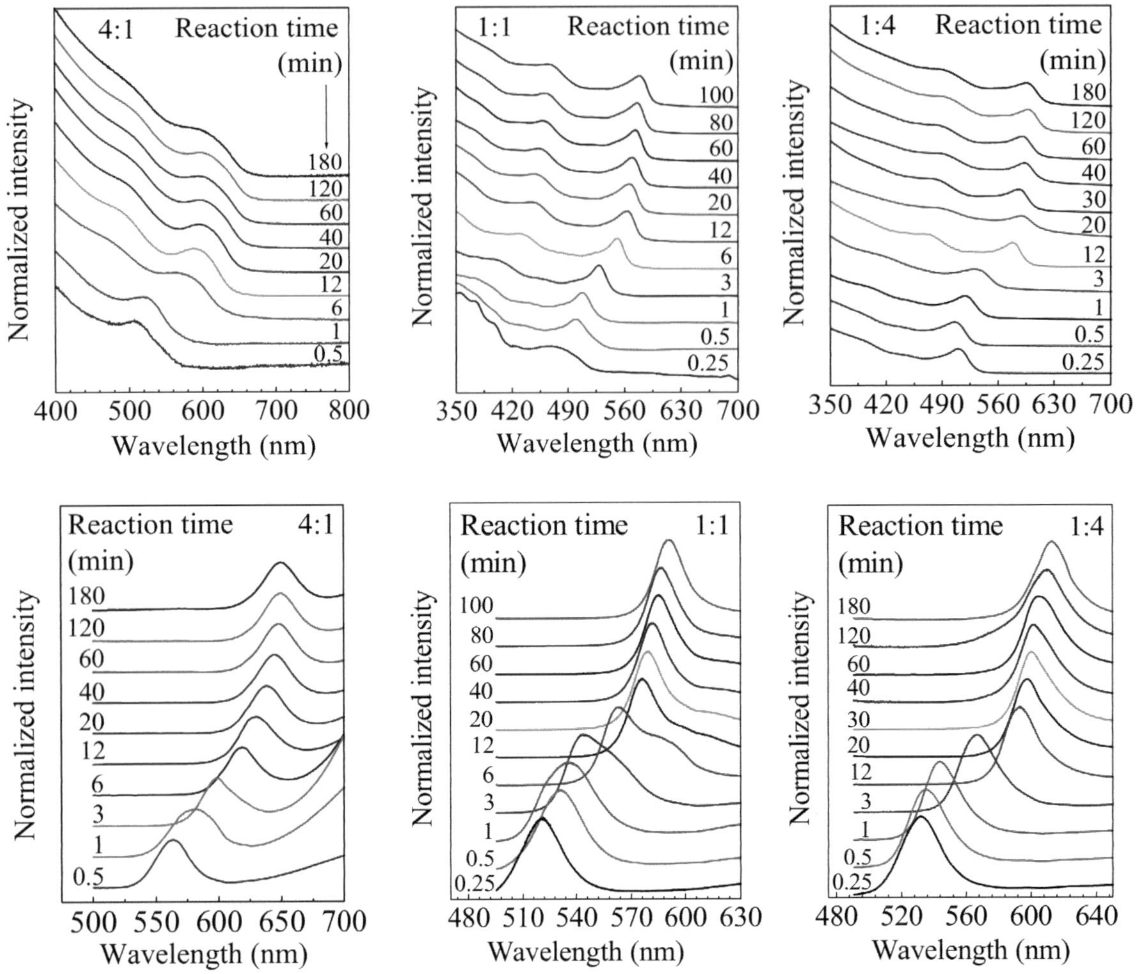

Figure 4. Temporal evolution of the absorption and PL spectra of CdSe TPs prepared with the initial Cd:Se precursor ratios of 4:1, 1:1, and 1:4.

Remarkably, the temporal evolution of the PL spectra of CdSe TPs prepared with the initial precursor ratio of 1:4 is relatively regular. Meanwhile, the temporal evolution of their absorption spectra is unusual. The absorption peak shifts to short wavelength at the reaction time of 0.5, 30, 60, and 180 min. After 6 min from the moment of the TOPSe-solution injection into the reaction flask the temperature in the reaction mixture is stable at 200 ± 3°C. Therefore, it is possible that the unusual temporal evolution of the absorption spectra is due to the unstable growth of CdSe TPs. The unstable growth of CdSe TPs also is observed for the Cd:Se precursor ratios of 6:1, 1:6, 1:8, and 1:10. It is well-known that only a limited amount of Cd monomer is stable under the conditions prevalent during the synthesis. A large excess of Cd precursor can lead to uncontrolled growth of large CdSe NCs [9]. The preparation of CdSe NCs in trioctylphosphine oxide (TOPO), in which the particles were grown stably with the Cd:Se precursor ratios from 2:1 to 1:10, was reported in [6]. However, our experiments show that the growth of CdSe TPs in ODE is more stable only for the initial precursor ratios ranging from 2:1 to 1:2, indicating the important role of the reaction media.

In order to determine the influence of monomer concentration on the optical properties of CdSe TPs, we prepared the CdSe TPs with the initial Cd:Se ratio of 2:1and Cd monomer concentrations in ODE of 0.033, 0.017, and 0.011 M. The temporal evolution of the absorption and PL spectra of CdSe TPs is presented in figure 5. The double-peak structure was observed clearly in both the absorption and PL spectra of the sample series. The temporal evolution of the absorption and PL spectra is similar to that in the case of the Cd:Se ratio of 1:1. It can be seen that the monomer concentrations almost do not change the spectroscopic characteristics of the CdSe TPs.

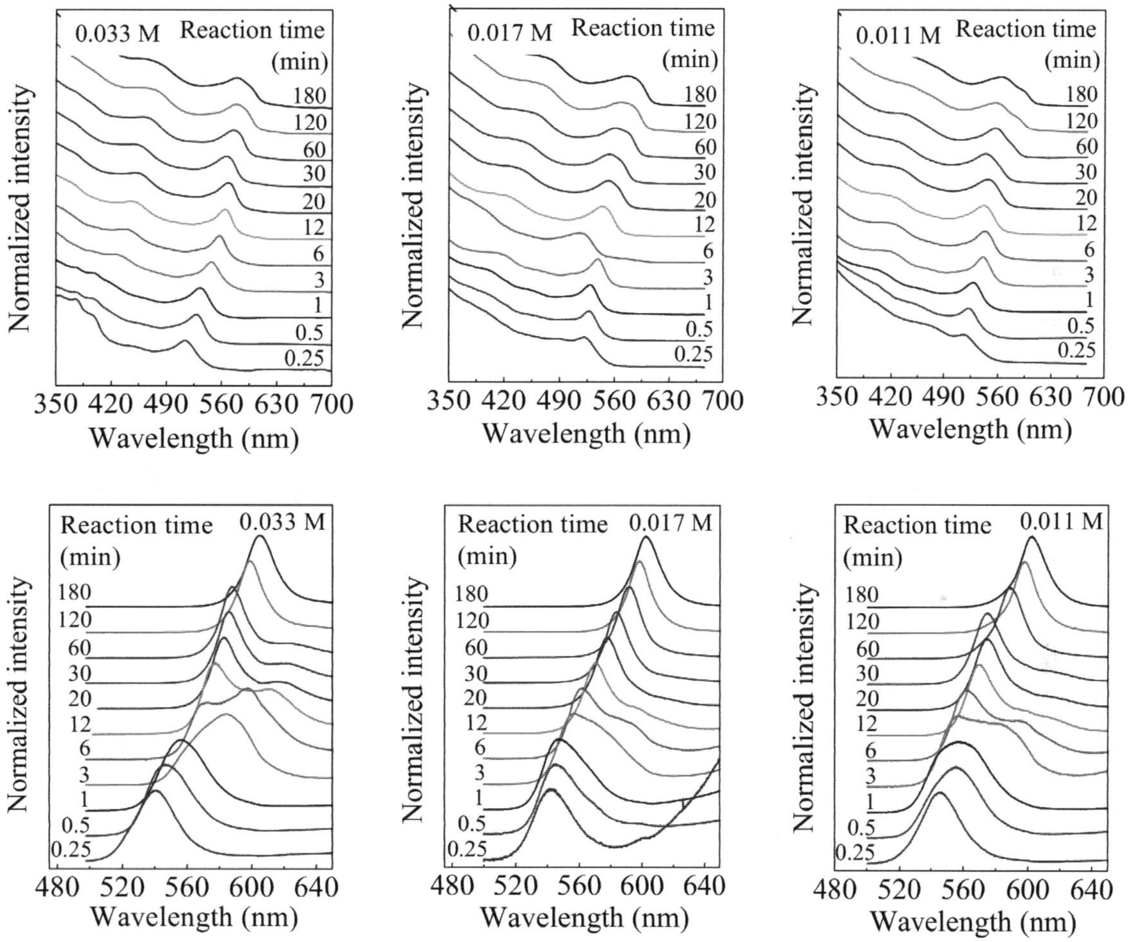

Figure 5. Temporal evolution of the absorption and PL spectra of CdSe TPs prepared with the initial Cd:Se ratio of 2:1 and Cd monomer concentrations in ODE of 0.033, 0.017, and 0.011 M.

Figure 6 presents the temporal evolution of the TP size during the growth. The diameters of the spherical core and columnar arms are estimated according to Brus expression [17] and the method of Gu *et al.* [18], respectively. The positions of the first two exciton peaks are determined from the second derivative spectra of absorbance with respect to the energy. It should be noted that: (i) the diameter of TP core is larger than that of the arm; (ii) the core size increases more quickly in comparison with the arm diameter; and (iii) the low monomer concentration leads to the decrease in the diameter of core and arms.

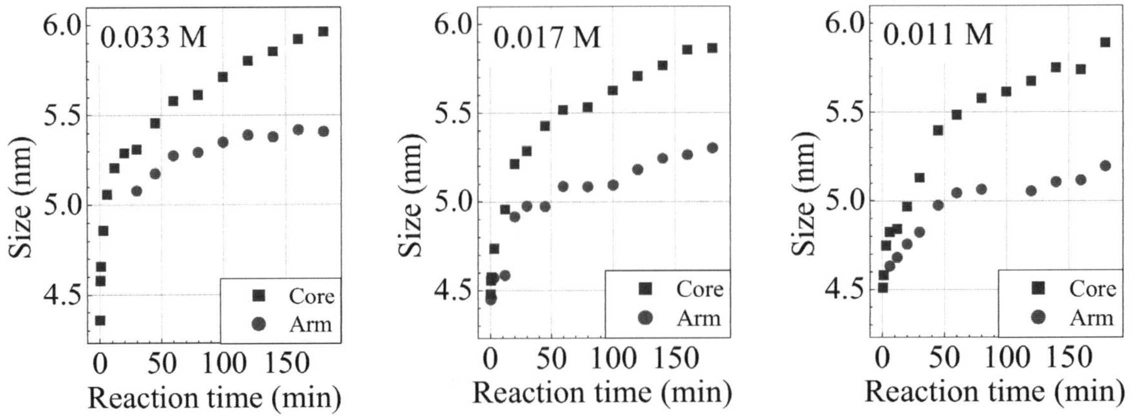

Figure 6. Temporal evolution of the size of CdSe TPs prepared with the initial Cd:Se ratio of 2:1and Cd monomer concentration of 0.033, 0.017 and 0.011 M.

The PL spectra of CdSe TPs prepared with three different monomer concentrations and drawn at the same reaction time are compared in figure 7. It can be seen that the low monomer concentrations lead to the rapid quenching of the luminescence from the TP core, and therefore the spectral shape becomes more symmetric.

Figure 8 shows the temporal evolution of PL spectra of CdSe TPs prepared with the Cd monomer concentration of 0.017 M in the wavelength region from 500 to 830 nm. The growth of CdSe TPs during the first reaction minutes leads to not only the appearance of a second peak, but also increase in the intensity of the emission band at ~ 750 nm. The intensities of both the emission from TP core and long wavelength band decrease in the next reaction time, and then they disappear.

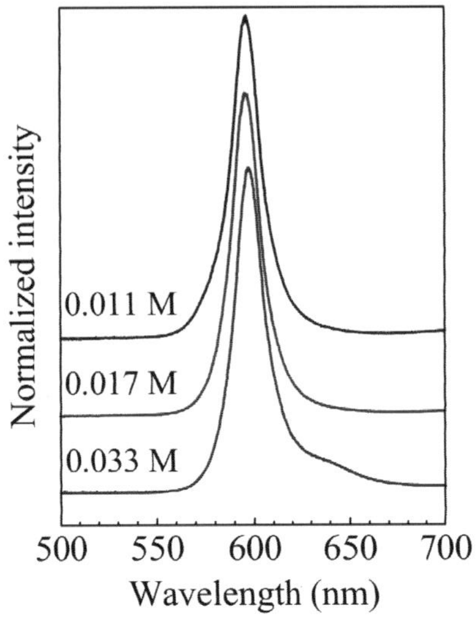

Figure 7. PL spectra of CdSe TPs prepared with the Cd monomer concentrations of 0.033, 0.017, and 0.011 M. The samples were drawn at the reaction time of 100 min.

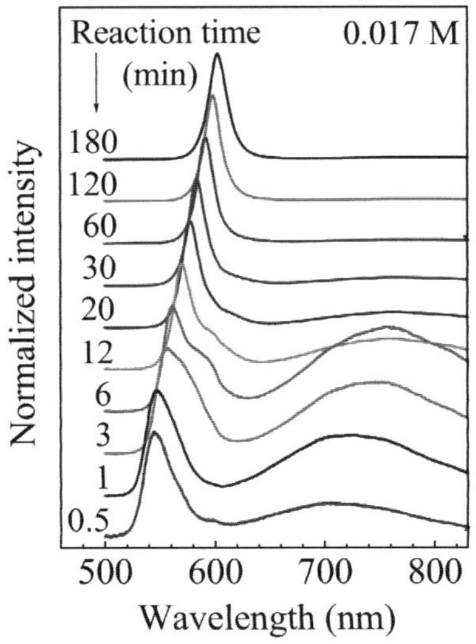

Figure 8. Temporal evolution of PL spectra of CdSe TPs prepared with the Cd monomer concentration of 0.017 M.

4. Conclusions

Colloidal CdSe TPs were synthesized by chemical method in ODE with the initial Cd:Se precursor ratios from 6:1 to 1:10 and different monomer concentrations. It was found that the CdSe TPs are Cd rich for all of the obtained sizes. Their stoichiometry depends on the initial precursor ratio.

The double-peak structure corresponding to the optical transitions in the core and arms of CdSe TPs was observed clearly in both the optical absorption and PL spectra for the Cd:Se precursor ratios from 2:1 to 1:2. These two emission peaks are separated gradually and the emission intensity of the arms increases in comparison with that of the core as the TP size increases. The rapid decrease in the emission intensity of the TP core with increasing its size is unclear. The temporal evolution of the optical absorption spectra shows that the growth of CdSe TPs in ODE is stable only for the initial precursor ratios ranging from 2:1 to 1:2. Moreover, the Cd monomer concentration in the range of 0.011 – 0.033 M almost does not change the spectroscopic characteristics of the CdSe TPs.

References

[1] Wang Y and Herron N 1991 *J. Phys. Chem.* **95** 525
[2] Alivisatos A P 1996 *J. Phys. Chem.* **100** 13226
[3] Gao M, Lesser C, Kristein S, Mohwald H, Rogach A L and Weller H 2000 *J. Appl. Phys.* **87** 2297
[4] Klimov V I, Mikhailovsky A A, Xu S, Hollingsworth J A, Leatherdale C A, Eisler H J and Bawendi M G 2000 *Science* **290** 314
[5] Gerion D, Pinaud F, Williams S C, Warak W J, Zanchet D, Weiss S and Alivisatos A P 2001 *J. Phys. Chem.* B **105** 8861
[6] Qu L and Peng X 2002 *J. Am. Chem. Soc.* **124** 2049
[7] Talapin D V, Rogach A L, Shevchenko E V, Kornowski A, Haase M and Weller H 2002 *J. Am. Chem. Soc.* **124** 5782
[8] Ebenstein Y, Mokari T and Banin U 2002 *Appl. Phys. Lett.* **80** 4033
[9] Donega C M, Hickey S G, Wuister S F, Vanmaekelbergh D and Meijerink A 2003 *J. Phys. Chem* B **107** 489
[10] Taylor J, Kippeny T and Rosenthai S J 2002 *J. Cluster Science* **12** 571
[11] Jasieniak J and Mulvaney P 2007 *J. Am. Chem. Soc.* **129** 2841
[12] Guzelian A A, Katari J E B, Kadavanich A V, Banin U, Hamad K, Juban E, Alivisatos A P, Wolters R H, Arnold C C and Heath J R 1996 *j. Phys. Chem.* **100** 7212
[13] Moreels I, Lambert K, Muynck D D, Vanhaecke F, Poelman D, Martins J C, Allan G and Hens Zeger 2007 *Chem. Mater.* **19** 6101
[14] Mohamed M B, Tonti D, Salman A A and Chergui M 2005 *Chem. Phys. Chem.* **6** 2505
[15] Tari D, Giorgi M D, Sala F D, Carbone L, Krahne R, Manna L and Cingolani R 2005 *Appl. Phys. Lett.* **87** 224101
[16] Pang Q, Zhao L, Cai Y, Nguyen D P, Regnault N, Wang N, Yang S, Ge W, Ferreira R, Bastard G and Wang J 2005 *Chem. Mater.* **17** 5263
[17] BrusL E 1986 *J. Phys.Chem.* **90** 2555
[18] Gu Y, Kuskovsky L, Yin M, O'Brien S and Neumark G F 2004 *Appl. Phys. Lett.* **85** 3833

Acknowledgments

This work was supported by the National Fundamental Research Foundation of Vietnam.

Optical properties of CdS and CdS/ZnS quantum dots synthesized by reverse micelle method

Vu Thi Kim Lien[1], Chu Viet Ha[1], Le Tien Ha[2] and Nguyen Nhu Dat[3]

[1] Thai Nguyen University of Education (TUE), Thai Nguyen, Vietnam
[2] Hanoi University of Mining and Geology, Hanoi, Vietnam
[3] Institute of Physics, Vietnam Academy of Science and Technology, Hanoi, Vietnam

E-mail: vkliensptn@yahoo.com

Abstract. The CdS and CdS/ZnS semiconductor quantum dots have been synthesized by reverse micelle method using sodium bis (2-ethylhexyl) sulfosuccinate (AOT) surfactant agent. The quantum dot diameter is about 2.5 to 4 nm depending on the concentration of the surfactant agent. It is interesting that, in contrast to other colloidal methods, the size of quantum dots does not depend on the growth time. The absorption spectra of CdS quantum dots show the narrow size distribution. The photoluminescence (PL) spectra include two bands, the intrinsic emission of CdS nanocrystals and the emission of surface states. There is noticeable increase of the PL intensity and subsequent photostability of CdS/ZnS core-shell quantum dots in comparison with CdS quantum dots without the ZnS shell.

Keyword: CdS, CdS/ZnS, quantum dots, semiconductor, reverse micelle method.

1. Introduction

Nowadays semiconductor nanocrystal quantum dots have been investigated in different aspects from physical properties of low-dimensional systems to their potential applications [1]. Efficient luminescent quantum dots form an important and interesting class of luminescent materials. They have demonstrated excellent optical properties and higher photochemical stability than most organic emitters [1, 2]. Their broad absorption spectrum and narrow emission band would be tunable by changing their size. One of the most interesting applications is the use of quantum dots as luminescent labels for optical sensors or fluorescent label sources for biological imaging applications [3-5].

Among approaches to prepare nanocrystals, the direct solution chemical routes have widely been used for synthesis of semiconductor quantum dots because these methods allow us to create the quantum dots with a narrow size distribution [1]. In the reverse micelle method, a solution is created by three components, surfactant molecules, water and a non-polar organic solvent. The polar heads of the surfactant molecules are directed toward the interior of a water-containing sphere, whereas the aliphatic tails are oriented toward the non-polar organic phase. The size of the quantum dots obtained by the method can be controlled by changing the ratio of water to the surfactant agent. In this work, we have studied the optical properties of CdS and CdS/ZnS core/shell semiconductor nanocrystal quantum dots synthesized by the reverse micelle method. The core/shell nanocrystal structures show the enhanced luminescence and photostability.

© 2009 IOP Publishing Ltd

2. Experimental

The CdS quantum dots were synthesized by the reverse micelle method with sodium bis (2-ethylhexyl) sulfosuccinate (AOT) used as the surfactant agent and heptane used as the solvent [1, 2, 6]. To obtain two micellar solutions with the aqueous solution encapsulated within the reverse micelles formed by the surfactant in oil, we prepared two identical mixtures of heptane and AOT and then added an aqueous solution of $Cd(CH_3COO)_2{\cdot}2H_2O$ to one of the mixtures and an aqueous solution of $Na_2S{\cdot}9H_2O$ to the other. After addition of one micellar solution containing Cd^{2+} ions into the other micellar solution containing S^{2-} ions in a vigorously stirring condition, CdS nanocrystal quantum dots were yielded in the nanometer-sized water pool within the reverse micelles. The size of the nanocrystals was changed by the molar ratio $W = [H_2O]/ [AOT]$. The value of W was chosen of 2.5, 5, 7.5 and 10.

To synthesize CdS/ZnS core/shell quantum dots, a ZnS shell layer was grown on CdS core nanocrystals by using two aqueous solutions of Zn^{2+} and S^{2-} ions which were obtained by dissolving the compounds $Zn(CH_3COOH)_2.2H_2O$ and $Na_2S{\cdot}9H_2O$ into water. The aqueous solutions of Zn^{2+} and S^{2-} were mixed separately with two identical AOT/heptane stock solutions. Then two micellar solutions containing Zn^{2+} and S^{2-} ions were injected at a very slow rate into the CdS nanocrystal micellar solution which had been prepared before. The obtained solution was then stirred vigorously, so that ions Zn^{2+} and S^{2-} could hang on the drops containing CdS quantum dots, forming the ZnS shell layer. The CdS/ZnS core/shell quantum dots were dispersed in organic solvents such as methanol and heptane.

Optical (UV-VIS) absorption measurements for nanocrystal samples were performed in the wavelength range of 300 - 750 nm using a Jasco V-600 (Japan). Photoluminescence (PL) and photoluminescence excitation (PLE) measurements were carried out at room temperature with a monochromatized light source (200 nm $< \lambda_{exc} <$ 900 nm) from a 450 W Xe lamp of a FS 920 Edinburgh (England).

3. Results and discussion

The prepared CdS and CdS/ZnS core/shell quantum dots were clusters which were protected by the surfactant agent (AOT) layer and dispersed in the heptane solvent. Figure 1 presents the absorption spectra of CdS quantum dots prepared with different values of the molar ratio of water to surfactant agent (W). There is an evident shift of the absorption peak towards lower wavelengths (or higher energy) compared to that of bulk CdS crystal ($\lambda_{bulk} \sim$ 500 nm corresponding to $E_g \approx 2.482$ eV at room temperature [7]). This blue shift is in connection with a decrease of the particle size.

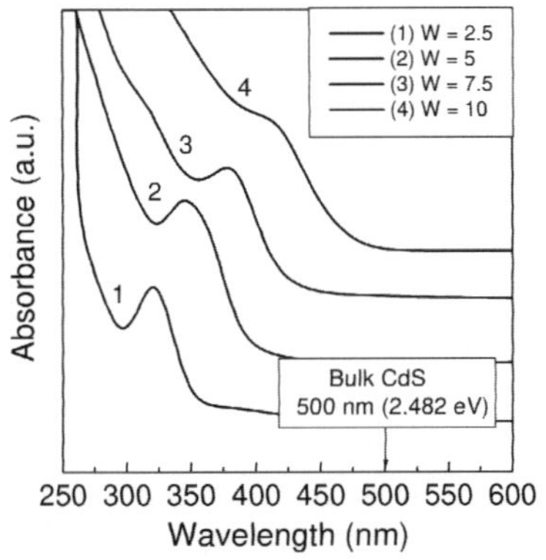

Figure 1. Absorption spectra of CdS quantum dots prepared with different values of the molar ratio of water to surfactant agent.

Various theoretical approaches have been employed to account for the variation in the electronic structure of nanocrystal quantum dots as a function of their size [3]. In the strong confinement regime, when the quantum dots radius R is much smaller than the Bohr exciton radius a_B, the electron-hole individual confinement model is applied. We used the Brus' formula [8] to estimate the mean radius of the CdS nanocrystals from the absorption peak energy. The observed peak values are 3.875 eV (320 nm), 3.594 eV (345 nm), 3.38 eV (378 nm) and 3.024 eV (410 nm) for the samples with the value W of 2.5, 5, 7.5 and 10, respectively. The mean radius of CdS quantum dots are estimated to be 1.2, 1.4, 1.6 and 2.0 nm, respectively. It is evident that increasing the molar ratio W of water to the surfactant agent leads to the increase of the mean quantum dots radius.

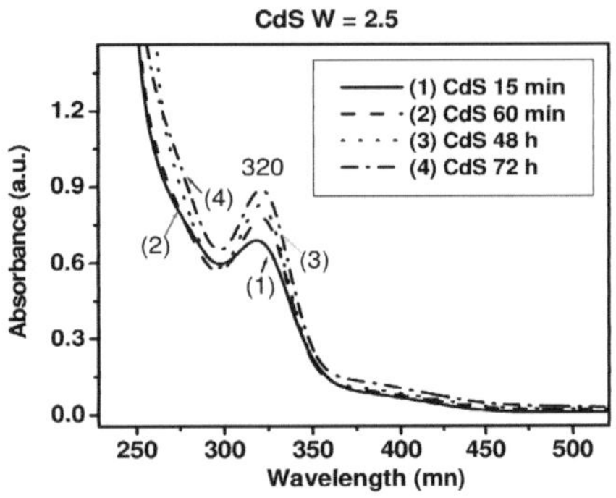

Figure 2. Absorption spectra of CdS quantum dots in the condition of W = 2.5 with different growth times.

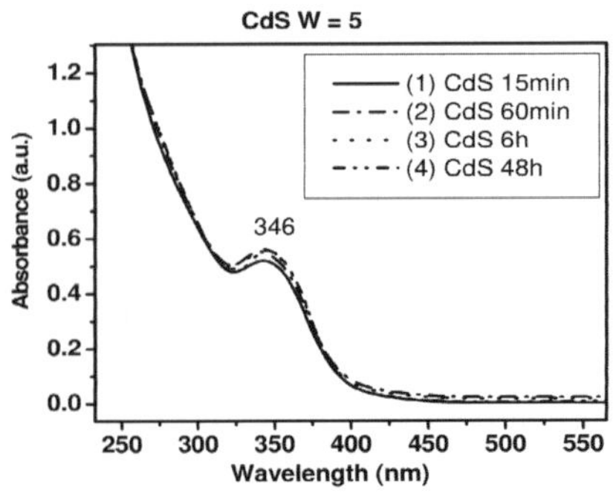

Figure 3. Absorption spectra of CdS quantum dots in the condition of W = 5 with different growth times.

Figure 2 shows the absorption spectra of the CdS quantum dots samples with different growth times in the condition of W = 2.5. Curves 1, 2, 3 and 4 correspond to growth times of 50 min, 60 min, 48 hours and 72 hours, respectively. The position of the absorption peak is the same (320 nm) for the samples of different growth times. This indicates the role of surfactant agent AOT in preventing further aggregation of nanocrystals. This is different from other colloidal methods, for example, the synthesis of nanocrystals by using organometallic precursors in a standard airless condition at high temperature [3], in which the size of nanoparticles increases with increasing growth time. Figures 3, 4 and 5 present the absorption spectra of the CdS quantum dots grown in different periods of time with the ratio W = 5, 7.5 and 10, respectively. These figures exhibit the same feature as figure 2. Hence, we believe that the method applied in our experiment can be used to control the size of semiconductor quantum dots by choosing a relevant value of W.

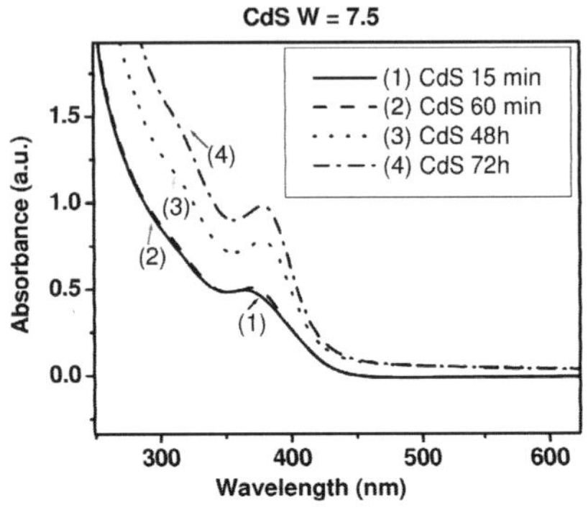

Figure 4. Absorption spectra of CdS quantum dots in the condition of W = 7.5 with different growth times.

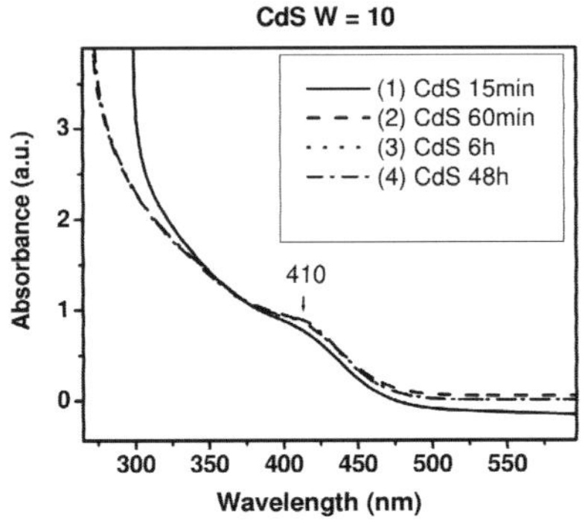

Figure 5. Absorption spectra of CdS quantum dots in the condition of W = 10 with different growth times.

The observation of the clear absorption peaks indicates the narrow size distribution of the CdS quantum dots. The size distribution width σ can be estimated from the line shape analysis. Following Kim *et al.* [1], the total absorption spectrum $F_{abs}(E)$ is given by the contributions from all quantum dots, *i.e.*, $F_{abs}(E) = \int A(R, E) N(R) dR$, where A(R, E) is the absorption of an individual quantum dot with a given radius R, N(R) represents a quantum dot size distribution assumed to be Gaussian. The shape of A(R, E) as a function of R is also assumed to be Gaussian to take into account the inhomogeneous broadening due to differences in the shape and/or surface morphology of the quantum dots.

To calculate the width σ, the spectral width Γ of A(R, E) has to be referred to the experimental results of single-dot spectroscopy, but there have been no reported data at room temperature. Here, we assume the peak area to be Gaussian and determine the full width at half maximum (FWHM).

Figure 6 presents the absorption spectra of CdS quantum dots with Gaussian fitting of the peak area (circle-symbol curves). The FWHM of the curves is 40, 55, 45 and 52 nm for W = 2.5, 5, 7.5 and 10, respectively. The size distribution width σ is estimated to be about 10-15%, showing the narrow quantum dot size distribution.

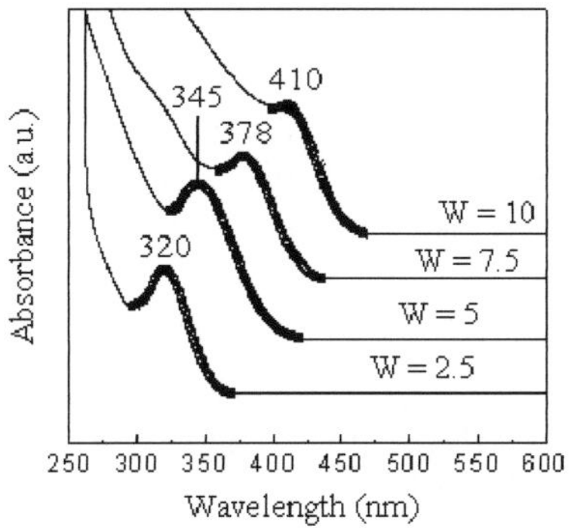

Figure 6. Absorption spectra of CdS quantum dots with Gaussian fitting of the peak area.

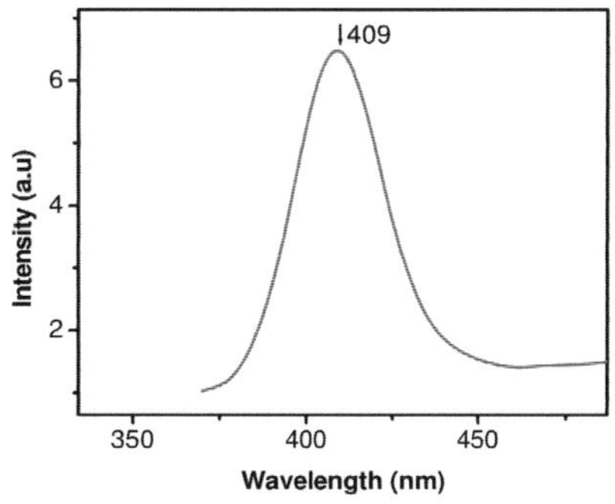

Figure 7. PL spectrum of CdS nanocrystals under excitation of 350 nm at room temperature.

The narrow size distribution of the CdS quantum dots is also shown in photoluminescence (PL) spectra. Figure 7 presents the room temperature PL spectrum of CdS quantum dots of W = 5 dispersed in a heptane solution. The PL emission is a narrow band peaked at 409 nm and has the FWHM of 26 nm.

Figure 8 exhibits the absorption spectra of the CdS and core/shell nanocrystal dispersed in a solution with the molar ratio of water to surfactant agent W = 5. In the case of CdS/ZnS quantum dots charge carriers can penetrate into the ZnS layer, leading to lowering the confinement energy of the charge carriers in the CdS core. This results in a red shift of the absorption peak in the absorption spectrum of CdS/ZnS quantum dots in comparison with bare CdS quantum dots. This red shift is more pronounced in smaller dots where the leakage of the charge carriers into the ZnS shell has a more dramatic effect on their confinement energies [3]. In our experiment the shift is of about 294 meV (31 nm) for the core size of 2.8 nm (corresponding to W = 5).

The photoluminescence (PL) measurements of the CdS nanocrystal quantum dots were carried out under continuous excitation with 330, 350, and 370 nm wavelengths. Figure 9 shows the PL spectra of CdS and CdS/ZnS quantum dots with *W* = 5. The emission of CdS quantum dots without the ZnS shell (curve 1) includes two bands, a narrow band peaked at 410 nm and a broad band peaked at 620 nm. The 410 nm emission band is attributed to the intrinsic emission of CdS quantum dots (or the band-to-band transition in CdS nanocrystals), while the other band centered at 620 nm is attributed to the emission of surface states due to the small size of nanocrystals. When the CdS quantum dots are overcoated with a ZnS shell, the intensity of intrinsic emission of CdS quantum dots increases significantly and is stronger than the emission of surface states (curve 2). This fact shows the passivation of the ZnS shell. It makes the emission of the surface states decrease, so that the intrinsic emission is enhanced. However, the emission of surface states is quite strong for CdS/ZnS quantum dot samples. This may result from a surplus of some compounds that did not take part in reaction and some CdS quantum dots were not passivated, so that the density of surface states is high.

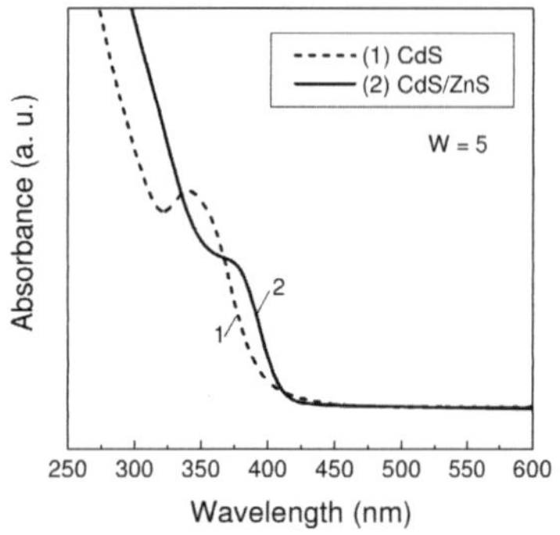

Figure 8. Absorption spectra of CdS and CdS/ZnS quantum dots samples with W=5.

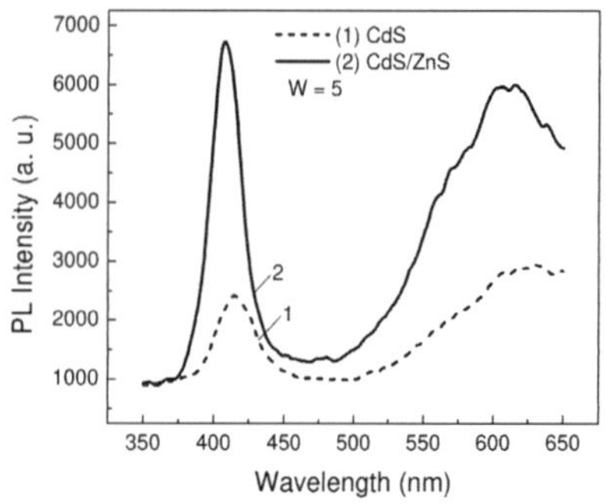

Figure 9. PL spectra of CdS and CdS/ZnS core/shell quantum dots with excitation wavelength of 330 nm at room temperature.

In order to obtain only the intrinsic emission of CdS quantum dots in PL spectra, besides passivating the quantum dots by ZnS shells, the prepared quantum dots were precipitated and then washed and centrifuged by methanol. Figure 10 presents the PL spectrum of CdS/ZnS core/shell quantum dots after washing and centrifugation. It can be seen that the emission of surface states is diminished significantly.

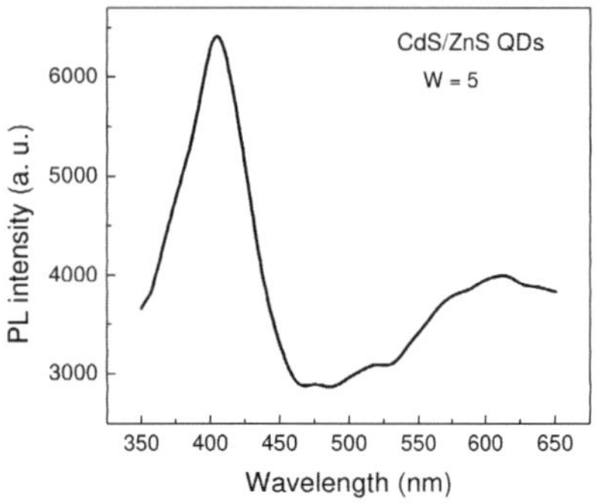

Figure 10. PL spectra of CdS/ZnS core/shell nanocrystals quantum dots after washing and centrifugation with excitation wavelength of 330 nm at room temperature.

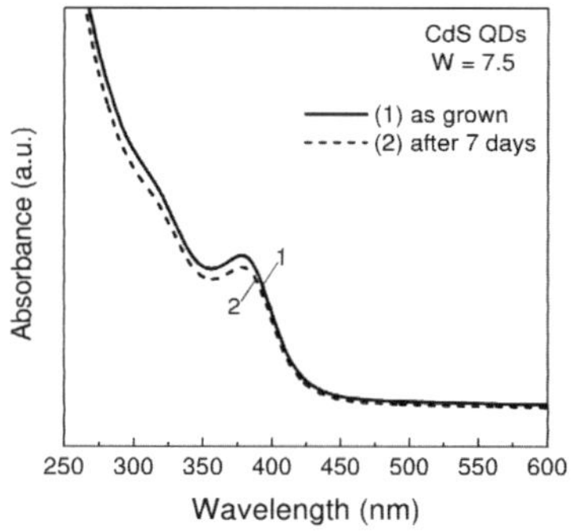

Figure 11. The absorption spectra of CdS quantum dots with W = 7.5 as grown and after the preparation 7 days.

To estimate the stability of CdS quantum dots, the prepared samples were protected by surfactant agent (AOT) layer and dispersed in heptane or methanol solvent. The samples were kept in condition of dark at room temperature. Figure 11 shows the absorption spectra of CdS quantum dots with W = 7.5 measured just after the preparation and after 7 days. The curves seem to be nearly the same: no significant change in the absorption peak energy and the spectrum width is observed. These features are necessary for the practical applications of quantum dots.

4. Conclusion

The CdS and CdS/ZnS nanocrystals quantum dots were synthesized by the reverse micelle method. The diameters of the CdS nanocrystals have a narrow distribution and their mean value is about from 2.5 to 4 nm depending on the molar ratio W of water to surfactant agent. The absorption spectra show that the size of prepared nanocrystals does not depend on their growth time and can be controlled by changing the ratio W. This shows the role of the surfactant agent AOT in preventing the aggregation of nanocrystals.

The photoluminescence spectra include two bands, the intrinsic emission of CdS nanocrystals and the emission of surface states. There is a noticeable increase of the photoluminescence intensity and the photostability of CdS/ZnS quantum dots in comparison with CdS quantum dots without the ZnS shell.

References

[1] Kim D G, Teratani N and Nakayama M 2002 *Jpn. J. Appl. Phys.* **44** 5064
[2] Ishizumi A and Kanemitsu Y 2006 *Adv. Mater.* **18** 1083
[3] Nga P T, Chuc N V, Chinh V D, Nghia N X, Dung P T, Cuong P T, Ha C V, Hanh V T H, Lien V T K, Thuan D N, Barthou C, and Benalloul P 2006 *Proc. IEEE SENSORS (Daegu, Korea, 22-25 October 2006)* pp 22-25
[4] Bailey R E, Smith A M and Nie S 2004 *Physica* E **25** 1
[5] Klimov V I 2000 *J. Phys. Chem.* B **104** 6112
[6] Yang H, Holloway P, Cunningham H G and Schanze K S 2004 *J. Chem. Phys.* **121** 10233
[7] Woggon U 1996 *Optical Properties of Semiconductor Quantum Dots (Springer Tract in Modern Physics* vol 136*)* (Berlin: Springer)
[8] Brus L E 1984 *J. Chem. Phys.* **80** 4403

Acknowledgments

This work is supported in part by the Natural Science Research Project (2006-2008) and the Asia Research Center, Vietnam National University, Hanoi. We thank Prof. Nguyen Van Hieu, Prof. Phan Hong Khoi, and Prof. Vu Ngoc Tu for their encouragement.

Journal of Physics: Conference Series **187** (2009) 012029 doi:10.1088/1742-6596/187/1/012029

Preparation and characterization of nanostructured composite films for organic light emitting diodes

Nguyen Nang Dinh[1], Le Ha Chi[1], Nguyen Thang Long[1], Tran Thi Chung Thuy[2], Tran Quang Trung[3] and Hyung-Kook Kim[4]

[1] College of Technology, Vietnam National University,
144 Xuan Thuy Road, Cau Giay District, Hanoi, Vietnam
[2] Thai Nguyen University of Education, Thai Nguyen City, Vietnam
[3] College of Natural Science, Vietnam National University,
227 Nguyen Van Cu Road, 5 District, Ho Chi Minh, Vietnam
[4] College of Nanoscience and Nanotechnology, Pusan National University,
San 30 Jangjeon-dong Geumjeong-gu Busan 609-735, Korea

E-mail: dinhnn@vnu.edu.vn

Abstract. With the aim to improve the photonic efficiency of an organic light emitting diode (OLED) and its display duration, both the hole transport layer (HTL) and the emitting layer (EML) were prepared as nanostructured thin films. For the HTL nanocomposite films were prepared by spin coating solutions of poly(vinylcarbazole) (PVK) and PEDOT-PSS containing TiO_2 nanoparticles onto low resistivity ITO substrates; for the EML, TiO_2-embedded MEH-PPV (MEH-PPV+nc-TiO_2) conjugate polymers were spin-coated onto the HTL. OLEDs prepared from above mentioned films have the structure of multilayers such as Al/MEH-PPV+nc-TiO_2/PVK+nc-TiO_2/ITO and Al/ MEH-PPV+nc-TiO_2/PEDOT-PSS+nc-TiO_2/ITO. Characterization of the nanocomposite films showed that both the I-V characteristics and the photoluminescence spectra of the nanocomposite materials were significantly enhanced in comparison to the standard polymers.

Keyword: Nanostructure, thin film, nanocomposite films, OLED.

1. Introduction

Organic light emitting diodes (OLED) have been intensively investigating for the last decade, because they have many potential applications such as optoelectronics, urban lighting, screen for TV and cellular phones, large-arc displays, solar flexible cells, etc [1-4]. However to replace the LED based on inorganic semiconducting materials it is necessary to improve both the time of service and efficiency of the OLEDs. It is known that the efficient device operation requires optimization of three factors: (i) equalization of injection rates of positive (hole) and negative (electron) charge carriers (ii) recombination of the charge carriers to form singlet exciton in emitting layer (EL) and (iii) radiative decay of the excitons. Recently, there have been reported different ways to deal with these problems, for instance, to add a hole transport layer (HTL) between the transparent anode and the emitting layer (EML) [5] and/or an electron transport layer (ETL) sandwiched between the EML and cathode [6]. With these solutions one can enhance electroluminescent efficiency of the OLED, but the long-lasting service sometimes is limited. There is another way to enhance both the efficiency and the service

© 2009 IOP Publishing Ltd

duration of the devices. It is the use of nanocomposite films instead of pure-polymers served as HTL and EML. The embedded oxides can substantially influence both the electrical, and optical properties of the polymer, for instance MEH-PPV + nc-TiO_2 composite thin films were studied as photoactive material [7]. By adding a hole transport layer (HTL) and an electron transport layer (ETL) to the three-layer device the equalization of injection rates of hole and electron was improved and the higher electroluminescent efficiency of the OLED was obtained. However, a large difference in the structures of an inorganic material (ITO) and an organic poly(vinylcarbazole) (PVK) or polyethylene (3,4-dioxythiophene) (PEDOT) usually causes a poor interface contact between them. By embedding TiO_2 nanoparticles in PEDOT one can expect to enhance both the contact of HTL with ITO and the working function of both the PVK and PEDOT films.

In this work we present the results of the research on the componential layers in OLEDs. I-V characteristics of the devices with the use of nanocomposite films are also presented.

2. Experimental

The ITO-coated glass substrates used for spincoating nanocomposite films were ultrasonically cleaned in distilled water, followed by ethanol and acetone. To deposit the EML composites films onto ITO, MEH-PPV solution was prepared by dissolving MEH-PPV powder in xylene at a ratio of 10 mg of MEH-PPV in 1 ml of xylene. Then, TiO_2 nanoparticles were embedded in these solutions according to the weight ratio of TiO_2/MEH-PPV ranging from 0.05 to 0.20, further referred to as MEHPPV+nc-TiO_2. To obtain homogenously dispersion of TiO_2 in polymer, the solutions were mixed for 8 hours by using ultrasonic stirring. The conditions for spincoating are following: delay time is 120s, rest spin time is 30s, spin speed is 1500 rpm, accelerate is 500 rpm and finally drying time is 2 min. Then the samples were put in a flow of dried gaseous nitrogen for 12 hours. All the samples were put in a fore vacuum box to await the measurements. For preparing an initial solution of PVK and PEDOT, the corresponding pure polymer powders were solved in cloruaform and PSS, respectively. To get both the PVK+ nc-TiO_2 and the PEDOT+nc-TiO_2 films that served as the HTL, the above mentioned procedure for MEHPPV+TiO_2 was used, with however a different content of nc-TiO_2 in PVK and PEDOT (i.e. 35 wt.% and 20 wt.%, respectively). Aluminum coating was used as the cathode was made by vacuum metallization. Thus, the OLEDs with multilayer layers had the structure of ITO/HTL/EML/Al.

The surface morphology of samples was characterized by using a "Hitachi" Field Emission Scanning Electron Microscope (FESEM). Atomic force microscope (AFM) images on a NT-MDT Atomic Force Microscope employed by a tunnel current mode. Nanocrystalline structures were checked by a Philips D-5005 X-ray diffractometer using filtered Cu Kα radiation (λ = 0.15405 nm). Photoluminescence spectra (PL) were carried-out by using a FL3-2 spectrophotometer and current-voltage (I-V) characteristics were measured on an Auto-Lab Potentiostat PGS-30. The UV-VIS absorption spectra were carried out on a Jasco UV-VIS-NIR V570.

3. Results and discussion

Poly(vinylcarbazole) (PVK) blended is a conducting polymer and possesses a bandgap, i.e. the separated energy gap between the lowest unoccupied molecular orbital (LUMO) and the highest occupied molecular orbital (HOMO) of about 3.6 eV [8]. It is transparent and possesses a workfunction higher than that of ITO, so it is often used for the HTL in OLEDs. In this work PVK composite films were prepared also for HTL. Firstly, the PVK+nc-TiO_2 films (the average size of TiO_2 particles is about 35 nm) were characterized by comparing the surface morphology and PL spectra with that of pure PVK. The surface of PVK+nc-TiO_2 nanocomposite films was examined by the SEM technique. Figure 1 shows SEM images of a pure PVK and a composite sample with embedding of 35 wt.% nanocrystalline titanium oxide particles (35 nm in size). The surface of the pure PVK exhibited much smoother than the composite films with large percentages and large TiO_2 particles.

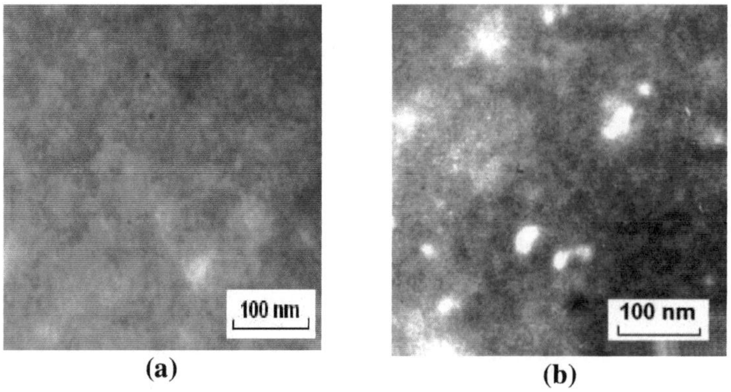

Figure 1. SEM of PVK (a) and PVK+nc-TiO_2 (30 wt.% of TiO_2) annealed in vacuum at 150°C (b).

The influence of the heat treatment on the morphology of the films is weak, i.e. no noticeable differences of the surface were observed in samples annealed at 120°C, 150°C and 180°C in the same vacuum. But the most suitable heating temperature for other operations like I-V characteristics or PL spectra was of 150°C. In the observed sample, the distribution of TiO_2 nanoparticles is not uniform, there are big bright points indicating clusters of the nanoparticles with larger size than that of initial nc-TiO_2. This proved that by using the ultrasonic stir the nanoparticles were not completely dispersed. However, the photoluminescence (PL) spectra of the composite exhibited quite different from the one obtained for the pure PVK (figure 2). Under an excitation of a large-energy photonic beam (nitrogen laser of 337.1 nm) the PL intensity of the PVK+nc-TiO_2 increased about 3.5 times in comparison with the PL intensity of the pure PVK under the same excitation source. From Raman spectra, it is known that nc-TiO_2 particles embedded into PVK did not make the molecular structure of PVK change. Thus the PVK+nc-TiO_2 can be seen as a composite material, where nc-TiO_2 particles were surrounded by PVK creating numerous TiO_2/PVK interfaces inside the composite.

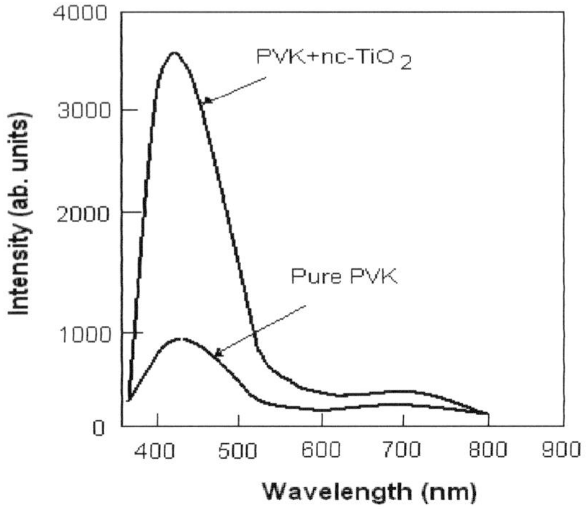

Figure 2. PL spectra of a pure PVK (bottom) and PVK+nc-TiO_2 (top) under excitation of laser beam with $\lambda = 337.1$ nm.

The mechanism of the increase of PL intensity for the composite is complicated. It is strongly dependent on the size and concentration of nc-TiO_2 embedded in. In this case the TiO_2 particles had the size of 35 nm and some large clusters were formed, so the effect in enhancement of PL intensity of the composite may be attributed to the presence of interfaces between large oxide particles and polymer. Under the excitation electron from the valence band jumped to the conduction band of TiO_2, then diffused to the PVK at the interface, because the Fermi level of TiO_2 is higher than the LUMO level of PVK. From the LUMO this electron transferred to the lower energy level in the HOMO of PVK emitting a photon. This resulted in the improvement of the photoemission.

Let us see the PEDOT case. It is known that PEDOT is also used for the HTL in OLED, because it has a high transmission in the visible region, good thermal stability and high conductivity [9-10]. To enhance the interface contact between ITO and PEDOT, TiO_2 nanoparticles were embedded into PEDOT.

Figure 3 shows the AFM of the pure PEDOT and the composite with the percentage of 20 wt.% TiO_2 nanoparticles (5 nm in size). With such a high resolution of the AFM one can see a clear round distribution of nanoparticles in the polymer due to the spin-coating process. For the pure PEDOT, the surface is smooth. The TiO_2 nanoparticles made the composite surface rougher in nanoscale and created numerous TiO_2/PEDOT boundaries in the composite film.

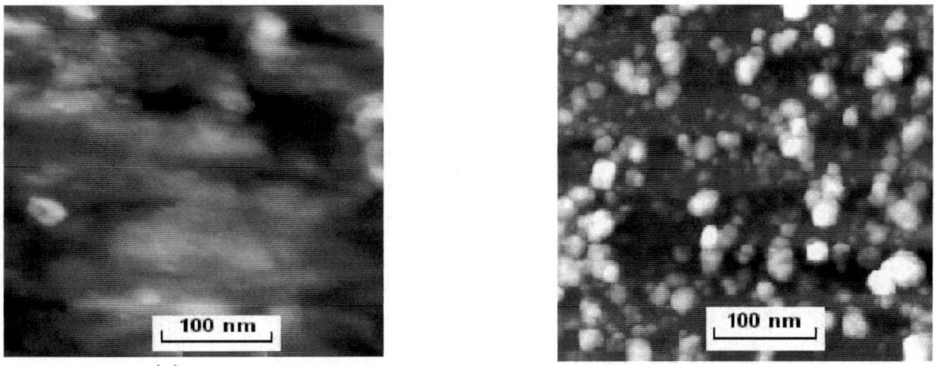

(a) (b)

Figure 3. AFM of a pure PEDOT (a) and a PEDOT composite film with 50 wt. % of nc-TiO_2.

Transmittance spectra of a pure PEDOT and a nanocomposite film plotted in figure 3 shows the AFM of the pure PEDOT and the composite with the percentage of 20 wt.% TiO_2 nanoparticles (5 nm in size). With such a high resolution of the AFM one can see a clear round distribution of nanoparticles in the polymer due to the spin-coating process. For the pure PEDOT, the surface is smooth. The TiO_2 nanoparticles made the composite surface rougher in nanoscale and created numerous TiO_2/PEDOT boundaries in the composite film (figure 3 shows the AFM of the pure PEDOT and the composite with the percentage of 20 wt.% TiO_2 nanoparticles (5 nm in size). With such a high resolution of the AFM one can see a clear round distribution of nanoparticles in the polymer due to the spin-coating process. For the pure PEDOT, the surface is smooth. The TiO_2 nanoparticles made the composite surface rougher in nanoscale and created numerous TiO_2/PEDOT boundaries in the composite film (figure 4). From this figure one can see that the PEDOT+nc-TiO_2 was strongly absorbed in the near UV range and weaker in the near IR range. The transmittance decreased with the increase of the embedded TiO_2 concentration, above 20 wt.% the transmittance lowered to a value smaller than 60%. The HTL film with such a low transmittance can not be utilized in OLEDs. At the range of the emission light of the MEH-PPV, namely from 540 to 600 nm, the transmittance of the 20 wt.% TiO_2 embedded PEPOT reached a value as high as 75%. This transmittance is a bit lower, but still comparable to the transmittance of the ITO anode. So we used this film for the HTL in the multilayer OLEDs. The electrical conductivity of the PEDOT blend is high; it reaches up to 80 S/cm [11]. But the presence of TiO_2 nanoparticles in PEDOT may result in the cleavage of the polymer conjugation pathway, consequently the decrease in conductivity of the film. However, a negligible decrease in conductivity does not much affect the performance of a OLED when using the composite as a hole transport layer.

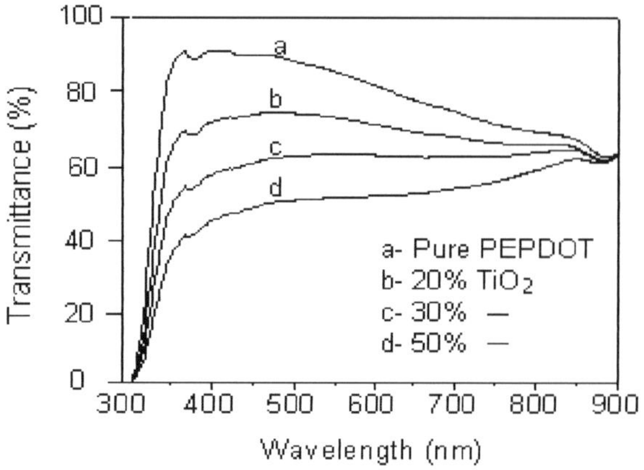

Figure 4. Transmittance spectra of the PEDOT composite vs. concentration of TiO_2.

Similar results of PL measurements for the PVK+nc-TiO_2, when the MEHPPV+TiO_2 nanocomposite excited at a short wavelength, the PL intensity of the MEH-PPV composite increased about 4 times larger than the one of pure MEH-PPV (figure 5). The photoluminescence spectra were measured on a FL3-2 spectrophotometer with a He-Ne laser as an excitation source (λ = 325 nm). Notice that, with such a short wavelength excitation both the polymer and composite emitted only one broad peak of wavelengths. From this figure, it is seen that photoemission of the composite film exhibits much higher luminescence intensity than that of the pure MEH-PPV. A blue shift of a peak from 580.5 to 550.3 nm was observed similarly to the result in [12] reporting on PL spectra of a hybrid MEH-PPV/nc-MoO_3 film. This result is consistent with currently obtained results on the polymeric nanocomposite [13], where the blue shift was explained by the reduction of the chain length of polymer, with embedded TiO_2 nanoparticles. The fact that the blue shift was not observed for the PVK composite proved the size of the TiO_2 particles embedded in PVK was larger than that in MEH-PPV. Thus, for such a small size of 5 nm, nc-TiO_2 particles can be seen as the nanocrystalline grains located in the polymer, and then the increase of PL intensity for this MEH-PPV nanocomposite film can be explained by the large absorption coefficient for TiO_2 grains. This similar phenomenon was explained due to the non-radiative Förster resonant energy transfer (FRET) [14] from TiO_2 nanocrystals to polymer with excitation of wavelength less than 350 nm.

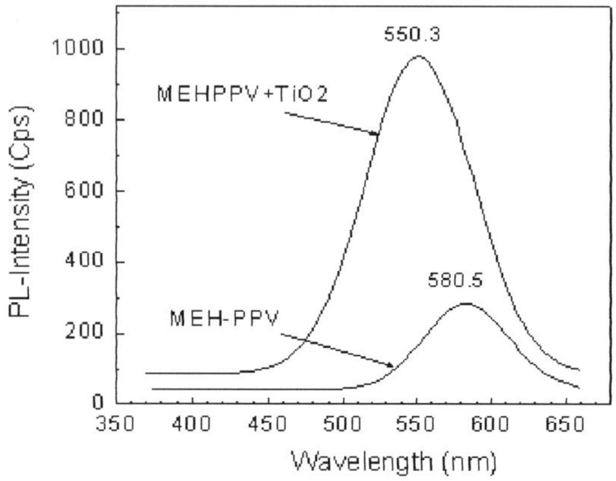

Figure 5. PL spectra of MEH-PPV+nc-TiO_2 with 15 wt. % of TiO_2. Excitation beam with λ = 325 nm.

To have clear understanding of the effect of nanoparticles in composites films used for both the EML and HTL, the OLEDs were made from different layers, such as a single pure EML diode (ITO/MEH-PPV/Al, adopted as SPD0), two diodes with double polymers (ITO/PVK/MEH-PPV/Al and ITO/PEDOT/MEH-PPV/Al, adopted as DPD1 and DPD2, respectively), a double polymeric composite layers, where the MEH-PPV+TiO_2 composite was used as EML and the PVK- or PEDOT-composite film was used as HTL (ITO/PVK+TiO_2/MEH-PPV+TiO_2/Al and ITO/PEDOT+TiO_2/MEH-PPV+TiO_2/Al or PKCD and PECD, respectively). The I-V characteristics of these devices are shown in figure 6. From this figure one can notice the following:

(i) The turn-on voltage for the diodes lowered from SPD0 to DPD1, DPD2 and PKCD. The smallest value of the turn-on voltage was obtained for the PECD device. This indicates that the equalization of injection rates of holes and electrons was improved due to both the HTL and the ETL added to the OLED.

(ii) The PEDOT film served as the HTL is better than the PVK composite. It favours the hole injection from ITO into the organic composite EML deposited on the HTL, resulting in an enhancement of the I-V characteristic. Thus the turn-on voltage decreased from 4.2 to 2.6 V (for PVK) and to 2.4 V (for PEDOT) (see curves "a, b and c" in figure 6). Nanoparticles in both the EML and HTL films have contributed to significantly lowering the turn-on voltage of the device (see the curve "d and e" of PKCD and PECD diodes).

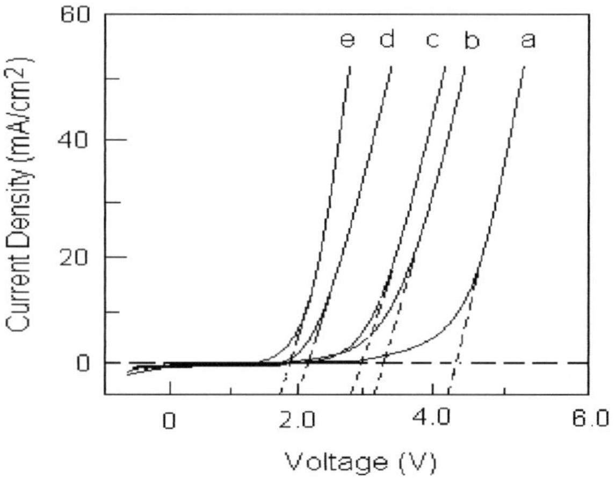

Figure 6. I-V characteristics of OLED with different laminated structure:
(a) – Single MEH-PPV, SPD0
(b) – with PVK layer, DPD1
(c) – with PEDOT layer, DPD2
(d) – with both EML and PVK composites
(e) – with both EML and PEDOT composites.

It is known that the effect of both the HTL and ETL on the enhancement of the I-V characteristic was explained, without argumentation due to the equalization of injection rates of holes and electrons. But the question why the nanoparticles can improve the device performance is still open. For instance, in [15] the authors attributed this enhancement to stimulated emission of optically-pumped MEH–PPV films when TiO_2 particles were embedded in. Whereas, in [16] by their obtained results the author indicates that no evidence of line narrowing or changes in the line shape are observed at different voltages implying that the mechanism for improved performance is distinctly different from that found in optically-pumped TiO2/MEH–PPV films. The authors concluded that optical scattering phenomenon is not causing the increase in performance. Another possible explanation is that the nanoparticle surfaces increase the probability of electron-hole recombination; however, again this would result in a change in the external quantum efficiency, rather than the current density as is observed [16].

From the data of PL spectra for the PVK and MEH-PPV nanocomposites, as well as the transmittance for the PEDOT composite, we have observed the improvement in PL intensity for the first two composites and the stronger absorption for the third one (see figures 2, 3 and 5). Similar phenomena obtained for nanohybrid layers were explained due to the TiO_2/polymer boundaries causing differences in band gap between oxide nanoparticles and the conjugate polymer [17]. Based on these results we take an explanation for the improved performance which supports the suggestion by Carter *et al.* [16]. It is suggested by the change in device morphology caused by the incorporation of nanoparticles into the solution. During the spinning process in the spin-coating technique, the nanoparticles can stick by strong electrostatic forces to the HTL and to themselves, and capillary forces can then draw the MEH–PPV solution around the nanoparticles into cavities without opening up pinholes through the device. This will result in a rough surface over which the Al-cathode is evaporated and subsequently a large interface between the cathode and the electroluminescent composite material. Thus at a low voltage, charge-injection into MEH–PPV is expected to be cathode limited; the very steep rise in the I–V curves for the composite diodes suggests that more efficient injection at the cathode is occurring which could be caused by the rougher interface of the nanocomposites. At the higher-voltage, transport in MEH–PPV appears to be space-charge limited. This possibly is also the reason why using the PVK composite as the HTL was less efficient in comparison with using PEDOT composite, where the nanoparticles (35 nm in size) embedded in the first are much larger than the one (5 nm in size) embedded in the second.

The electroluminescence quantum efficiency can be calculated by using a simple expression obtained by Tsutsui and Saito in [18], as follows:

$$\eta_{\phi} = \gamma \cdot \eta_r \cdot \phi_r \qquad (1)$$

where γ is a double charge injection factor which is dependent on the processes of carrier injection and is maximal ($\gamma = 1$) if a balanced charge injection into the emission layer of the device is achieved, i. e. the number of injected negative charge (electron) equals the number of injected positive charge (hole); η_r quantifies the efficiency of the formation of a singlet exciton from a positive and a negative polaron, and ϕ_f is the photoluminescence quantum efficiency. From the PL spectra and I-V characteristics obtained one can see that γ for the PECD is the largest due to the addition of the best HTL like PEDOT+TiO_2 film. The η_r and ϕ_f of the PECD may be improved due to the embedding of TiO_2 nanoparticles into both the emission and hole transport layers. Thus the electroluminescence quantum efficiency of the multilayer polymeric composite diodes can be qualitatively evaluated from equation (1) and suggested much larger than for the single polymeric layer device. For the accurate estimation of the electroluminescence quantum efficiency of a multilayer OLED we shall focus our forthcoming research work on the electroluminescence properties of such laminated structures.

4. Conclusion

Both the hole transport layer and the emitting layer were prepared as nanostructured thin films by embedding nc-TiO_2 particles into PVK, PEDOT and MEH-PPV. The HTL nanocomposite films were spin-coated onto low resistivity ITO substrates. TiO_2-embedded MEH-PPV conjugate polymers were spin-coated onto the HTL. Two types of the multilayer OLEDs of the laminated structure of Al/MEH-PPV+nc-TiO_2/PVK+nc-TiO_2/ITO and Al/MEH-PPV+nc-TiO_2/PEDOT+nc-TiO_2/ITO were prepared for I-V measurements. Characterization of the nanocomposite films showed that both the photoluminescence and I-V characteristics of the nanocomposite materials were significantly enhanced in comparison with the standard polymers. From these results one can expect a large electroluminescent efficiency of the OLEDs made from these layers.

References

[1] Salafsky J S 1999 *Phys. Rev.* B **59** 10885
[2] Burlakov V M, Kawata K, Assender H E, and Briggs G A D, Ruseckas A and Samuel I D W 2005 *Phys. Rev. B* **72** 075206
[3] Petrella A, Tamborra M, Cozzoli P D, Curri M L, Striccoli M, Cosma P, Farinola G M, Babudri F, Naso F and Agostiano A 2004 *Thin Solid Films* **451- 452** 64
[4] Spanggaard H and Kerbs F C 2004 *Sol. Energy. Mat. Sol. Cells* **83** 125
[5] Choulis S A, ChoongV E, Mathai M K and So F 2005 *Appl. Phys. Lett.* **87** 113508
[6] Markov D E and Blom P W M 2005 *Phys. Rev.* B **72** 161401(R)
[7] Kawata K, Burlakov V M, Carey M J, Assender H E, Briggs G A D, Ruseckas A and Samuel I D W 2005 *Sol. Energy Mat. Sol Cells* **87** 715

[8] Kolosov D, Adamovich V, Djurovich P, Thompson M E and Adachi C 2002 *J. Am. Chem. Soc.* **124** 9945
[9] Quyang J, Xu Q, Chu C-W, Yang Y, Li G and Shinar 2004 *J Polymer* **45** 8443
[10] Tehrani P, Kanciurzewska A, Crispin X, Robinson N D, Fahlman M and Berggren M 2007 *Solid State Ionics* **177** 3521
[11] Quyang J, Chu C-W, Chen F-C, Xu Q and Yang Y 2005 *Advanced Functional Mat.* **15** 203
[12] Dinh N N, Chi L H, Thuy T T C, Thanh D V and Nguyen T P 2008 *JKPS* **53** (2) 802
[13] Yang S H, Nguyen T P, Le Rendu P and Hsu C S 2005 *Composites Part A: Appl. Sci. Manufact.* **36** 509
[14] Heliotis G, Itskos G, Murray R, Dawson M D, Watson I M and Bradley D D C 2006 *Adv. Mater.* **18** 334
[15] Scott J C, Kaufman J, Brock P J, DiPietro R, Salem J and Goitia J A 1996 *J. Appl. Phys.* **79** 2745
[16] Carter S A, Scott J C and Brock P J 1997 *Appl. Phys. Lett.* **71** 1
[17] Thuy T T C, Chi L H and Dinh N N 2009 *JKPS* **54** (1) 291
[18] Tsutsui T and Saito S 1993 *NATO ASI Ser. E: Appl. Sci.* **246** 123

Acknowledgments

This work is supported in part by the Protocol in Scientific Cooperation between Vietnam and South Korea during period of 2009 - 2010.

Analytical determination of Kondo and Fano resonances of electron Green's function in a single-level quantum dot

Nguyen Bich Ha[1] and Nguyen Van Hop[2]

[1]Institute of Materials Science, Vietnam Academy of Science and Technology, 18 Hoang Quoc Viet Road, Cau Giay Dist., Hanoi, Vietnam
[2]Hanoi National University of Education, Hanoi, Vietnam

E-mail: bichha@iop.vast.ac.vn

Abstract. The Kondo and Fano resonances in the two-point Green's function of the single-level quantum dot were found and investigated in many previous works by means of different numerical calculation methods. In this work we present the derivation of the analytical expressions of resonance terms in the expression of the two-point Green's function. For that purpose the system of Dyson equations for the two-point nonequilibrium Green's functions in the complex-time Keldysh formalism was established in the second order with respect to the tunneling coupling constants and the mean field approximation. This system of Dyson equations was solved exactly and the analytical expressions of the resonance terms are derived. The conditions for the existence of Kondo or Fano resonances are found.

Keywords: quantum dot, Green's function, nonequilibrium, resonance.

1. Introduction

The electrons transport through a single-level quantum dot (QD) connected with two conducting leads has been the subject for theoretical and experimental studies in many works since the early days of nanophysics [1–22]. Two observable physical quantities, which can be measured in experiments on electron transport, are the electron current through the QD and the time-averaged value of the electron number in the QD. Both can be expressed in terms of the single-electron Green's functions. In the pioneering theoretical works [1, 3] on the electron transport through a single-level QD, the differential equations for the real-time Green's functions were derived with the use of the Heisenberg equations of motion for the electron destruction and creation operators. Due to the presence of the strong Coulomb interaction between electrons in the QD the differential equations for the single-electron Green's functions contain multi-electron Green's functions, and all the coupled equations for these Green's functions form an infinite system of differential equations. In order to have a finite closed system of equations, one can assume some approximation to decouple the infinite system of equations. Moreover, since the electron transport is a nonequilibrium process, one should work with the Keldysh formalism of nonequilibrium complex-time Green's functions [23, 24].

In the study of the nonequibrium complex-time Green's functions by means of the perturbation theory with respect to the Coulomb interaction, one usually retains some chain of ladder diagrams and assumes noncrossing approximation (NCA). The systems of equations for the Green's functions were solved by means of different numerical methods, for example the Quantum Monte Carlo technique [13] and the numerical renormalization group method [10, 12, 14, 17]. Electron two-point Green's functions were shown to have the resonance related to the Kondo effect. Beside of this Kondo

© 2009 IOP Publishing Ltd

resonance, the Fano quasi-bound state in the energy spectrum of the electron system of the QD and the leads might also give some resonant contribution.

In this work, the exact analytical expressions of the Kondo and the Fano resonance terms are derived by explicitly solving the equations. From these analytical expressions we obtain the whole set of resonances and the conditions for their existence. In particular, we shall demonstrate the distinction between the Kondo and the Fano resonances, if they do exist.

2. System of Dyson Equations

Consider the nanosystem consisting of a single level QD connected to two leads through the potential barriers with the total Hamiltonian

$$\begin{aligned} H = E\sum_{\sigma} c_\sigma^+ c_\sigma + U N_\uparrow N_\downarrow + \sum_{\mathbf{k}}\sum_{\sigma}\left\{E_a(\mathbf{k}) a_\sigma^+(\mathbf{k}) a_\sigma(\mathbf{k}) + E_b(\mathbf{k}) b_\sigma^+(\mathbf{k}) b_\sigma(\mathbf{k})\right\} \\ + \sum_{\mathbf{k}}\sum_{\sigma}\left\{V_a(\mathbf{k}) a_\sigma^+(\mathbf{k}) c_\sigma + V_a(\mathbf{k})^* c_\sigma^+ a_\sigma(\mathbf{k}) + V_b(\mathbf{k}) b_\sigma^+(\mathbf{k}) c_\sigma + V_b(\mathbf{k})^* c_\sigma^+ b_\sigma(\mathbf{k})\right\} \end{aligned} \tag{1}$$

where $a_\sigma(\mathrm{k}), b_\sigma(\mathrm{k})$ and $a_\sigma^+(\mathrm{k}), b_\sigma^+(\mathrm{k})$ are the annihilation and creation operators of the electrons with momenta **k** in the two leads,

$$N_\sigma = c_\sigma^+ c_\sigma,$$

and denote $G_{\sigma\sigma'}(t-t')_C$ the complex time single-electron Green's function

$$G_{\sigma\sigma'}(t-t')_C = \delta_{\sigma\sigma'} G(t-t')_C = -i\left\langle T_C[c_\sigma(t) c_{\sigma'}^+(t')]\right\rangle \tag{2}$$

which is a set of four functions

$$G_{\sigma\sigma'}(t-t')_{\alpha\beta} = \delta_{\sigma\sigma'} G(t-t')_{\alpha\beta}\,, \qquad \alpha, \beta = 1, 2 \tag{3}$$

with the Fourier transforms

$$\widetilde{G}_{\sigma\sigma'}(\omega)_{\alpha\beta} = \delta_{\sigma\sigma'} \widetilde{G}(\omega)_{\alpha\beta}\,. \tag{4}$$

Due to the presence of the Coulomb interaction term in the Hamiltonian (1) the differential equation for the Green's function (2) contains a new Green's function $H(t-t')_C$ defined as

$$H_{\sigma\sigma'}(t-t')_C = \delta_{\sigma\sigma'} H(t-t')_C = -i\left\langle T_C[N_{-\sigma} c_\sigma(t) c_{\sigma'}^+(t')]\right\rangle.$$

From the Heisenberg equations of motion for the operators $c_\sigma(t)$, $a_\sigma(\mathrm{k},t)$ and $b_\sigma(\mathrm{k},t)$

$$i\frac{dc_\sigma(t)}{dt} = Ec_\sigma(t) + UN_{-\sigma}(t) c_\sigma(t) + \sum_{\mathbf{k}}\left\{V_a(\mathbf{k})^* a_\sigma(\mathbf{k},t) + V_b(\mathbf{k})^* b_\sigma(\mathbf{k},t)\right\}, \tag{5}$$

$$i\frac{da_\sigma(\mathbf{k},t)}{dt} = E_a(\mathbf{k}) a_\sigma(\mathbf{k},t) + V_a(\mathbf{k}) c_\sigma(t), \tag{6}$$

$$i\frac{db_\sigma(\mathbf{k},t)}{dt} = E_b(\mathbf{k}) b_\sigma(\mathbf{k},t) + V_b(\mathbf{k}) c_\sigma(t), \tag{7}$$

we can derive the differential equations for the Green's functions. Applying the mean field approximation to the products of four operators and neglecting the terms of the fourth and higher orders with respect to the tunnelling coupling constants $V_a(\mathbf{k})$ and $V_b(\mathbf{k})$, we obtain the system of Dyson equations:

$$\left(i\frac{d}{dz} - E\right) G(t-t')_C = \delta(t-t')_C + UH(t-t')_C + \int_C dt'' \Sigma^{(1)}(t-t'')_C G(t''-t')_C, \tag{8}$$

$$\begin{aligned} \left(i\frac{d}{dt} - E - U\right) H(t-t')_C = n\delta(t-t')_C + \rho(t-t')_C + \\ + \int_C dt'' \Sigma^{(2)}(t-t'')_C H(t''-t')_C - \int_C dt'' \Sigma^{(3)}(t-t'')_C G(t''-t')_C, \end{aligned} \tag{9}$$

where

$$\Sigma^{(1)}(t-t')_C = \sum_{\mathbf{k}} \left\{ |V_a(\mathbf{k})|^2 S^{E_a(\mathbf{k})}(t-t')_C + (a \to b) \right\}, \tag{10}$$

$$\Sigma^{(2)}(t-t')_C = \sum_{\mathbf{k}} \left\{ |V_a(\mathbf{k})|^2 \left[2S^{E_a(\mathbf{k})}(t-t')_C + S^{2E+U-E_a(\mathbf{k})}(t-t')_C \right] + (a \to b) \right\}, \tag{11}$$

$$\Sigma^{(3)}(t-t')_C = \sum_{\mathbf{k}} \left\{ n_a(\mathbf{k}) |V_a(\mathbf{k})|^2 \left[S^{E_a(\mathbf{k})}(t-t')_C + S^{2E+U-E_a(\mathbf{k})}(t-t')_C \right] + (a \to b) \right\}, \tag{12}$$

$$\rho(t-t')_C = \sum_{\mathbf{k}} \left\{ l_a(\mathbf{k}) |V_a(\mathbf{k})|^2 [S^{E_a(\mathbf{k})}(t-t')_C - S^{2E+U-E_a(\mathbf{k})}(t-t')_C] + (a \to b) \right\}, \tag{13}$$

$$n_a(\mathbf{k}) = \left\langle a_\sigma^+(\mathbf{k}) a_\sigma(\mathbf{k}) \right\rangle = \frac{e^{-\beta E_a(\mathbf{k})}}{1+e^{-\beta E_a(\mathbf{k})}}, \tag{14}$$

$$n = \left\langle c_\sigma^+ c_\sigma \right\rangle = -iG(-0)_{11}, \tag{15}$$

$$l_a(\mathbf{k}) = \frac{1}{Z} \left\{ \frac{e^{-\beta E} - [1+e^{-\beta E}] n_a(\mathbf{k})}{E - E_a(\mathbf{k})} + e^{-\beta E} \frac{e^{-\beta(E+U)} - [1+e^{-\beta(E+U)}] n_a(\mathbf{k})}{E+U-E_a(\mathbf{k})} \right\}, \tag{16}$$

$$Z = 1 + 2e^{-\beta E} + e^{-(2E+U)}, \tag{17}$$

and $S^E(t-t')_C$ is the complex time Green's function of the free electrons at the single energy level E. It satisfies the differential equation

$$\left[i\frac{d}{dt} - E \right] S^E(t-t')_C = \delta(t-t')_C. \tag{18}$$

The terms containing $S^{2E+U-E_a(\mathbf{k})}(t-t')_C$ in the r.h.s. of equations (11)–(13) are the crossing terms. In the NCA they are omitted.

The system of Dyson equations (8) and (9) is more rigorous than those derived previously by many other authors. In particular, in the derivation of equations (8) and (9), the contributions of all crossing terms have been taken into account: there was no necessity to use NCA. Moreover, in all previous works on the Green functions of quantum dots in the Keldysh formalism, the solution of the system of equations was solved by numerical methods. In the present work we derive the analytical expressions for the exact solution of the system of Dyson equations.

In terms of the 2×2 matrices $\hat{G}(\omega)$, $\hat{H}(\omega)$, $\hat{\Sigma}^{(i)}(\omega)$ and $\hat{\rho}(\omega)$ with the matrix elements being the Fourier transforms of the functions $G(t-t')_{\alpha\beta}$, $H(t-t')_{\alpha\beta}$, $\Sigma^{(i)}(\omega)_{\alpha\beta}$ and $\rho(\omega)_{\alpha\beta}$, the system of equations (8) and (9) becomes that of the matrix equations

$$\hat{G}(\omega) = \hat{S}^E(\omega) + U\hat{S}^E(\omega)\hat{\eta}\hat{H}(\omega) + \hat{S}^E(\omega)\hat{\eta}\hat{\Sigma}^{(1)}(\omega)\hat{\eta}\hat{G}(\omega), \tag{19}$$

$$\hat{H}(\omega) = n\hat{S}^{E+U}(\omega) + \hat{S}^{E+U}(\omega)\hat{\eta}\hat{\rho}(\omega) + \hat{S}^{E+U}(\omega)\hat{\eta}\hat{\Sigma}^{(2)}(\omega)\hat{\eta}\hat{H}(\omega) - \hat{S}^{E+U}(\omega)\hat{\eta}\hat{\Sigma}^{(3)}(\omega)\hat{\eta}\hat{G}(\omega), \tag{20}$$

where

$$\hat{\eta} = \begin{pmatrix} 1 & 0 \\ 0 & -1 \end{pmatrix}. \tag{21}$$

From these two matrix equations we derive two similar systems of algebraic ones, each of which consists of four equations of $G(\omega)_{\alpha 1}$ and $H(\omega)_{\alpha 1}$ or $G(\omega)_{\alpha 2}$ and $H(\omega)_{\alpha 2}$, α = 1, 2. For investigating the electron transport through a QD it is necessary to study the function $G(\omega)_{11}$. By solving the system of equations for the functions $G(\omega)_{\alpha 1}$ and $H(\omega)_{\alpha 1}$, we have derived the expression of this function in the form

$$G(\omega)_{11} = \frac{Z(\omega)}{Y(\omega)}, \tag{22}$$

where $Z(\omega)$ and $Y(\omega)$ are the polynomials of $S^{E}(\omega)_{\alpha\beta}$, $S^{E+U}(\omega)_{\alpha\beta}$, $\Sigma^{(i)}(\omega)_{\alpha\beta}$ and $\rho(\omega)_{\alpha\beta}$. The explicit forms of the rather cumbersome numerator $Z(\omega)$ and denominator $Y(\omega)$ in equation (22) were found by exactly solving equations (19) and (20) and presented in the Appendix.

3. Kondo and Fano Resonances

Now we consider the resonances of the Green's function (22). The expression of the denominator $Y(\omega)$ consists of the terms of two types: finite terms which do not contain self-energies $\Sigma^{(i)}(\omega)_{\alpha\beta}$ and those proportional to $\Sigma^{(i)}(\omega)_{\alpha\beta}$ or their products. Because $\Sigma^{(i)}(\omega)_{\alpha\beta}$ are the linear combinations of the squared moduli of small tunnelling coupling constants $V_a(\mathbf{k})$ and $V_b(\mathbf{k})$ they give small contributions to $Y(\omega)$ at their regular points. However some of them may be divergent at some points, and near these divergence points $Y(\omega)$ may vanish: there appear the resonances. The study of the resonances of Green's functions requires the determination of the divergence points of functions $\Sigma^{(i)}(\omega)_{\alpha\beta}$. From equations (10)–(12) it follows that $\Sigma^{(i)}(\omega)_{12}$, $\Sigma^{(i)}(\omega)_{21}$ and the imaginary parts of $\Sigma^{(i)}(\omega)_{11}$ and $\Sigma^{(i)}(\omega)_{22}$, i = 1, 2, 3, are always finite and only the real parts of $\Sigma^{(i)}(\omega)_{11}$ and $\Sigma^{(i)}(\omega)_{22}$ may be divergent at some points. Introducing the spectral functions

$$\Gamma^{(p)}_{a,b}(\omega) = \pi \sum_{\mathbf{k}} \left[\frac{e^{-\beta E_{a,b}(\mathbf{k})}}{1+e^{-\beta E_{a,b}(\mathbf{k})}} \right]^{p} \left| V_{a,b}(\mathbf{k}) \right|^{2} \delta[\omega - E_{a,b}(\mathbf{k})], \tag{23}$$

$p = 0, 1, 2,$

we can express the functions

$$\Sigma^{(i)}(\omega) = \mathrm{Re}\,\Sigma^{(i)}(\omega)_{11} = -\mathrm{Re}\,\Sigma^{(i)}(\omega)_{22} \tag{24}$$

in the form convenient for the investigation of their divergence

$$\Sigma^{(1)}(\omega) = \frac{1}{\pi} P \int d\omega' \frac{1}{\omega - \omega'} \left[\Gamma^{(0)}_{a}(\omega') + \Gamma^{(0)}_{b}(\omega') \right], \tag{25}$$

$$\Sigma^{(2)}(\omega) = \frac{1}{\pi} P \int d\omega' \left[\frac{2}{\omega - \omega'} + \frac{1}{\omega - 2E - U + \omega'} \right] \left[\Gamma^{(0)}_{a}(\omega') + \Gamma^{(0)}_{b}(\omega') \right], \tag{26}$$

$$\Sigma^{(3)}(\omega) = \frac{1}{\pi} P \int d\omega' \left[\frac{1}{\omega - \omega'} + \frac{1}{\omega - 2E - U + \omega'} \right] \left[\Gamma^{(1)}_{a}(\omega') + \Gamma^{(1)}_{b}(\omega') \right]. \tag{27}$$

Near the divergence points of the functions $\Sigma^{(i)}(\omega)$ we can neglect $\Sigma^{(i)}(\omega)_{12}$ and $\Sigma^{(i)}(\omega)_{21}$ as well as the imaginary parts of $\Sigma^{(i)}(\omega)_{11}$ and $\Sigma^{(i)}(\omega)_{22}$, i = 1, 2, 3, and have following asymptotic expression of function (22):

$$G_{11}(\omega) = \frac{A(\omega)}{B(\omega)} \tag{28}$$

with

$$A(\omega) = \omega - E - U[1 - n - \rho(\omega)] - \Sigma^{(2)}(\omega), \tag{29}$$

$$B(\omega) = (\omega - E - U)(\omega - E) - (\omega - E - U)\Sigma^{(1)}(\omega) - (\omega - E)\Sigma^{(2)}(\omega) + U\Sigma^{(3)}(\omega) + \Sigma^{(1)}(\omega)\Sigma^{(2)}(\omega), \tag{30}$$

where

$$\rho(\omega) = \frac{1}{\pi} P \int d\omega' \left[\frac{1}{\omega - \omega'} - \frac{1}{\omega - 2E - U + \omega'} \right] \left[\lambda_a(\omega) + \lambda_b(\omega) \right], \tag{31}$$

$$\lambda_{a,b}(\omega) = \pi \sum_{\mathbf{k}} l_a(\mathbf{k}) \left| V_{a,b}(\mathbf{k}) \right|^2 \delta[\omega - E_{a,b}(\mathbf{k})]. \tag{32}$$

The functions $\Sigma^{(i)}(\omega)$ and $\rho(\omega)$ contain the dispersion integrals with the spectral functions $\Gamma^{(p)}_{a,b}(\omega)$ and $\lambda_{a,b}(\omega)$. Denote μ_a, μ_b and Ω_a, Ω_b the chemical potentials and the tops of the energy bands of the systems of conducting electrons in the leads "a" and "b", respectively. Because

$$E_{a,b}(\mathbf{k}) = E^{(0)}_{a,b}(\mathbf{k}) - \mu_{a,b} ,$$

where $E^{(0)}_{a,b}(\mathbf{k})$ are the kinetic energies of the conducting electrons in the leads, $0 \le E^{(0)}_{a,b}(\mathbf{k}) \le \Omega_{a,b}$, the spectral functions $\Gamma^{(p)}_{a,b}(\omega)$ and $\lambda_{a,b}(\omega)$ vanishes for $\omega < -\mu_{a,b}$ and $\omega > \Omega_{a,b} - \mu_{a,b}$. For the study of the divergence of $\Sigma^{(i)}(\omega)$ and $\rho(\omega)$ we set $\mu_a = \mu_b$, $\Omega_a = \Omega_b$, $\Gamma^{(0)}_a(\omega) = \Gamma^{(0)}_b(\omega)$, $\lambda_a(\omega) = \lambda_b(\omega)$ and approximately replace the values of these functions in the interval $-\mu < \omega < \Omega - \mu$ by some constants Γ and λ, respectively. Then we have following divergent asymptotic expressions of functions $\Sigma^{(i)}(\omega)$ and $\rho(\omega)$:

$$\Sigma^{(1)}(\omega) \approx \frac{1}{2}\Sigma^{(2)}(\omega) \approx \Sigma^{(3)}(\omega) \approx \frac{\Gamma}{\lambda}\rho(\omega) \approx \frac{2\Gamma}{\pi} \ln\left|\frac{\omega + \mu}{\Omega}\right| \text{ at } T \approx 0 \text{ and } \omega \approx -\mu , \tag{33}$$

$$\Sigma^{(3)}(\omega) \approx -\frac{2\Gamma}{\pi} \ln\left|\frac{\omega}{\mu}\right| \text{ at } T = 0 \text{ and } \omega \approx 0 , \tag{34}$$

$$\Sigma^{(2)}(\omega) \approx \Sigma^{(3)}(\omega) \approx -\frac{\Gamma}{\lambda}\rho(\omega) \approx -\frac{2\Gamma}{\pi} \ln\left|\frac{\omega - 2E - U - \mu}{\Omega}\right| \text{ at } T \approx 0 \text{ and } \omega \approx 2E + U + \mu , \tag{35}$$

$$\Sigma^{(3)}(\omega) \approx \frac{2\Gamma}{\pi} \ln\left|\frac{\omega - 2E - U}{\mu}\right| \text{ at } T = 0 \text{ and } \omega \approx 2E + U . \tag{36}$$

The formulae (33) and (34) contain some unknown finite constant Ω of the order of the width of the energy band of the conducting electrons in the leads. However we shall formulate the conclusion which does not depend on the concrete value of this constant.

Consider now the behaviour of function (28) in the neighbourhoods of the divergence points of functions $\Sigma^{(i)}(\omega)$, neglecting the finite very small terms of the second order with respect to the tunnelling coupling constants.

a) At $\omega \to -\mu$ and low temperature $T \approx 0$, all functions $\Sigma^{(i)}(\omega)$ and also $\rho(\omega)$ tend to $-\infty$. Function (28) has following asymptotic expression in the neighbourhoods of this point:

$$G(\omega)_{11} \approx -\left[1 - n + \frac{1}{2}(E + \mu)\frac{\lambda}{\Gamma}\right] U\left(E + 2U + \mu\right)^{-1} \left[\frac{1}{2}(E + \mu) + \frac{2\Gamma}{\pi}\ln\left|\frac{\omega + \mu}{\Omega}\right| + 2i\Gamma\right]^{-1} - \\ - E + \mu + \left[1 + n - (E + U + \mu)\frac{\lambda}{\Gamma}\right] U\left(E + 2U + \mu\right)^{-1} \left[(E + U + \mu) + \frac{2\Gamma}{\pi}\ln\left|\frac{\omega + \mu}{\Omega}\right| + 2i\Gamma\right]^{-1} . \tag{37}$$

If $E + \mu > 0$, then $G(\omega)_{11}$ has two resonances at two points

$$\omega_1^{(\pm)} = -\mu \pm \Omega e^{-\frac{\pi}{4\Gamma}(E+\mu)} \tag{38}$$

and two resonances at two points

$$\omega_2^{(\pm)} = -\mu \pm \Omega e^{-\frac{\pi}{2\Gamma}(E+U+\mu)}. \tag{39}$$

Between these four resonances there are the dips. If $E+\mu<0$ but $E+U+\mu>0$, then $G(\omega)_{11}$ has only two resonances at the points $\omega_2^{(\pm)}$. If $E+U+\mu<0$, then function $G(\omega)_{11}$ has no resonance in the neighbourhood of the point $\omega=-\mu$. All four points $\omega_1^{(\pm)}$ and $\omega_2^{(\pm)}$ are very close to the point $\omega=-\mu$ and the resonances at $\omega_1^{(\pm)}$ and $\omega_2^{(\pm)}$ look like a resonance at $\omega=-\mu$. The origin of these resonances is the presence of the Fano quasi-bound state at the lower edge of the energy band of the conducting electrons. If they exist, they would be called the Fano resonances.

b) At $\omega\to 0$ and $T=0$, functions $\Sigma^{(1)}(\omega)$, $\Sigma^{(2)}(\omega)$ and $\rho(\omega)$ are bounded but the functions $\Sigma^{(3)}(\omega)$ tends to $+\infty$. Function $G(\omega)_{11}$ has following asymptotic expression:

$$G(\omega)_{11} \approx -[E+(1-n)U]\left[E(E+U)+\frac{2\Gamma U}{\pi}\ln\left|\frac{\mu}{\omega}\right|+2i(3E+2U)\Gamma\right]^{-1}. \tag{40}$$

If $E(E+U)<0$, then $G(\omega)_{11}$ has two resonances at the points

$$\omega_3^{(\pm)} = \pm\mu e^{-\frac{\pi}{2\Gamma}\frac{|E(E+U)|}{U}}, \tag{41}$$

which are very close to the point $\omega=0$. At $\omega=0$ and $0<T<T_K$,

$$T_K = \frac{1}{k}\mu\exp\left\{-\frac{\pi}{2}\cdot\frac{|E(E+U)|}{\Gamma U}\right\}, \tag{42}$$

where k is the Boltzmann constant. Instead of formula (40) we have

$$G(0)_{11} \approx \frac{\pi}{2}\cdot\frac{E+(1-n)U}{\Gamma U}\cdot\frac{1}{\ln|T/T_K| - i\pi(3E+2U)\Gamma}. \tag{43}$$

The resonances in the neighbourhood of the point $\omega=0$ have the same physical origin as the Kondo effect in the scattering of electrons by a magnetic impurity. They are the Kondo resonances.

c) *At $\omega\to 2E+U$ and $T=0$* the functions $\Sigma^{(1)}(\omega)$, $\Sigma^{(2)}(\omega)$ and $\rho(\omega)$ are bounded, but the function $\Sigma^{(3)}(\omega)$ tends to $+\infty$. Function $G(\omega)_{11}$ has following asymptotic expression:

$$G(\omega)_{11} \approx (E+nU)\left[E(E+U)-\frac{2\Gamma U}{\pi}\ln\left|\frac{\mu}{\omega-2E-U}\right|-2iE\Gamma\right]^{-1}. \tag{44}$$

Therefore if $E(E+U)>0$, then $G(\omega)_{11}$ has also two resonances at the points

$$\omega_4^{(\pm)} = 2E+U\pm\mu e^{-\frac{\pi}{2\Gamma}\frac{E(E+U)}{U}}, \tag{45}$$

which are very close to the point $\omega=2E+U$. At $\omega=2E+U$ and $0<T<T_K$,

$$T_K = \frac{1}{k}\mu\exp\left\{-\frac{\pi}{2}\cdot\frac{E(E+U)}{\Gamma U}\right\}. \tag{46}$$

instead of equations (42) we have

$$G(\omega)_{11} \approx \frac{\pi}{2}\cdot\frac{E+nU}{\Gamma U}\cdot\frac{1}{\ln|T/T_K| - i\pi E\Gamma}. \tag{47}$$

The resonances in the neighbourhood of the point $\omega=2E+U$ are the Kondo resonances of the crossing terms.

d) At $\omega \to 2E+U+\mu$ and low temperature $T \approx 0$, the functions $\Sigma^{(1)}(\omega)$, $\Sigma^{(2)}(\omega)$ tend to $-\infty$ and $\rho(\omega)$ tends to $+\infty$. Function $G(\omega)_{11}$ has following asymptotic expression:

$$G(\omega)_{11} \approx -U(E+\mu)^{-1}\left[1-n+(E+U+\mu)\frac{\lambda}{\Gamma}\right]\left[E+U+\mu+\frac{2\Gamma}{\pi}\ln\left|\frac{\omega-2E-U-\mu}{\Omega}\right|-2i\Gamma\right]^{-1}. \quad (48)$$

If $E+U+\mu>0$, then $G(\omega)_{11}$ has two resonances at the points

$$\omega_5^{(\pm)} = 2E+U+\mu \pm \Omega e^{-\frac{\pi}{2}\cdot\frac{E+U+\mu}{\Gamma}}, \quad (49)$$

which are very close to the point $\omega = 2E+U+\mu$. They are the Fano resonances of the crossing terms.

4. Conclusion

By means of the equation of motion method the system of Dyson equations for the complex-time non-equilibrium electron Green's functions of the system consisting of a single-level QD connected with two conducting leads was derived in the mean field approximation with respect to the products of four creation and destruction operators of the electron and in the second order with respect to the effective tunnelling coupling constants. All the crossing terms are included into the equations. The exact solution of the system of Dyson equations may have the resonances of four types in the dependence on the physical parameters of the system: the Kondo resonances at the Fermi surface, whose origin is similar to that of the Kondo effect in the scattering of electrons on magnetic impurities, the Fano resonances due to the presence of the electron quasi-bound state at the lower edge of the energy band of the conducting electrons, the Kondo resonances in the crossing terms and the Fano resonances in the crossing terms. The analytical asymptotic expressions of the single-electron Green function at these resonances were derived.

The results of the present study are the complement of and agree with the numerical calculations in previous works [6–19] on the electron Green's functions in QD.

5. Appendix

$$\begin{aligned} Z(\omega) = &\left\{B(\omega)[1+\Omega^{(1)}(\omega)_{22}]-UD(\omega)_{22}\right\}\left\{B(\omega)S^E(\omega)_{11}+UC_1(\omega)\right\}- \\ &-\left\{B(\omega)\Omega^{(1)}(\omega)_{12}-UD(\omega)_{12}\right\}\left\{B(\omega)S^E(\omega)_{21}+UC_2(\omega)\right\}, \end{aligned} \quad (A.1)$$

$$\begin{aligned} Y(\omega) = &\left\{B(\omega)[1-\Omega^{(1)}(\omega)_{11}]+UD(\omega)_{11}\right\}\left\{B(\omega)[1+\Omega^{(1)}(\omega)_{22}]-UD(\omega)_{22}\right\}+ \\ &+\left\{B(\omega)\Omega^{(1)}(\omega)_{12}-UD(\omega)_{12}\right\}\left\{B(\omega)\Omega^{(1)}(\omega)_{21}-UD(\omega)_{21}\right\}, \end{aligned} \quad (A.2)$$

$$B(\omega) = [1+\Omega^{(2)}(\omega)_{22}][1-\Omega^{(2)}(\omega)_{11}]+\Omega^{(2)}(\omega)_{12}\Omega^{(2)}(\omega)_{21}, \quad (A.3)$$

$$\begin{aligned} C_i(\omega) = &\left\{S^E(\omega)_{i1}[1+\Omega^{(2)}(\omega)_{22}]-S^E(\omega)_{i2}\Omega^{(2)}(\omega)_{21}\right\}\lambda(\omega)_{11}- \\ &-\left\{S^E(\omega)_{i1}\Omega^{(2)}(\omega)_{12}+S^E(\omega)_{i2}[1-\Omega^{(2)}(\omega)_{11}]\right\}\lambda(\omega)_{21}, \\ &i = 1, 2. \end{aligned} \quad (A.4)$$

$$\begin{aligned} D(\omega)_{ij} = &\left\{S^E(\omega)_{i1}[1+\Omega^{(2)}(\omega)_{22}]-S^E(\omega)_{i2}\Omega^{(2)}(\omega)_{21}\right\}\Omega^{(3)}(\omega)_{1j}- \\ &-\left\{S^E(\omega)_{i1}\Omega^{(2)}(\omega)_{12}+S^E(\omega)_{i2}[1-\Omega^{(2)}(\omega)_{11}]\right\}\Omega^{(3)}(\omega)_{2j}, \end{aligned} \quad (A.5)$$

$$\Omega^{(1)}(\omega)_{ij} = S^E(\omega)_{i1}\Sigma^{(1)}(\omega)_{1j}-S^E(\omega)_{i2}\Sigma^{(1)}(\omega)_{2j},$$

$$\Omega^{(2)}(\omega)_{ij} = S^{E+U}(\omega)_{i1}\Sigma^{(2)}(\omega)_{1j} - S^{E+U}(\omega)_{i2}\Sigma^{(2)}(\omega)_{2j}, \tag{A.6}$$
$$\Omega^{(3)}(\omega)_{ij} = S^{E+U}(\omega)_{i1}\Sigma^{(3)}(\omega)_{1j} - S^{E+U}(\omega)_{i2}\Sigma^{(3)}(\omega)_{2j},$$

$$\lambda(\omega)_{ij} = nS^{E+U}(\omega)_{ij} + S^{E+U}(\omega)_{i1}\rho(\omega)_{1j} - S^{E+U}(\omega)_{i2}\rho(\omega)_{2j}, \tag{A.7}$$

$$i = 1,2, \quad j = 1,2.$$

References

[1] Meir Y, Wingreen N S and Lee P A 1991 *Phys. Rev. Lett.* **66** 3048
[2] Hershfield S, Davies J H and Wilkins J W 1991 *Phys. Rev. Lett.* **67** 3720
[3] Meir Y, Wingreen N S and Lee P A 1993 *Phys. Rev. Lett.* **70** 2061
[4] Ng T K 1993 *Phys. Rev. Lett.* **70** 3635
[5] Yeyati A L, Martin-Rodero A and Flores F 1993 *Phys. Rev. Lett.* **71** 2991
[6] Inoshita T, Shimizu A, Kuramoto Y and Skaki H 1993 *Phys. Rev.* B **48** 14725
[7] Wingreen N S and Meir Y 1994 *Phys. Rev.* B **49** 11040
[8] Craco L and Kang K 1999 *Phys. Rev.* B **59** 12244
[9] Fujii T and Ueda K 2003 *Phys. Rev.* B **68** 155310
[10] Choi M.-S, Sánchez D and López R 2004 *Phys. Rev. Lett.* **92** 056601
[11] Takagi O and Saso T 1999 *J. Phys. Soc. Jpn.* **68** 1997, **68** 2894
[12] Izumida W, Sakai O and Shimizu Y 1997 *J. Phys. Soc. Jpn.* **66** 717; 1998 *J. Phys. Soc. Jpn.* **67** 2444
[13] Sakai O, Suzuki S, Izumida W and Oguri A 1999 *J. Phys. Soc. Jpn.* **68** 1640
[14] Izumida W, Sakai O and Suzuki S 2001 *J. Phys. Soc. Jpn.* **70** 1045
[15] Torio M E, Hallberg K, Ceccatto A H and Proetto C R 2002 *Phys.Rev.* B **65** 085302
[16] Swirkowicz R, Barnas J and Wilczyuski M 2003 *Phys.Rev.* B **68** 195318
[17] Costi T A, Hewson A C and Zlatic V 1994 *J. Phys.: Condens. Matter* **6** 2519
[18] Pustilnik M and Glasman L 2004 *J. Phys.: Condens. Matter* **16** R513
[19] Swirkowicz R, Wilczynski M and Barnas J 2006 *J. Phys.: Condens. Matter* **18** 2297
[20] Hieu N V and Ha N B 2005 *Adv. Nat. Sci.* **6** 1
[21] Konig J and Gefen Y 2005 *Phys.Rev.* B **71** 201308 (R)
[22] Suidel M, Silva A, Oreg Y and von Delft J 2005 *Phys. Rev.* B **72** 125318
[23] Keldysh L V 1965 *Jh. Eksp. Teor. Fiz.* **20** 1018
[24] Chou K -C, Su Z -B, Hao B -L and Yu L 1985 *Physics Reports* **118** 1

Acknowledgments

This work is a part of a research project in the Vietnam National Research Programme in Natural Sciences. We thank Ministry for Science and Technology of Vietnam for the support.

Exact solution of rate equations for a two-spin-qubit system

Le Thi Ha Linh[1,2] **and Nguyen Bich Ha**[1]

[1] Institute of Materials Science, Vietnam Academy of Science and Technology, 18 Hoang Quoc Viet Road, Cau Giay District, Hanoi, Vietnam
[2] Thai Nguyen University, Thai Nguyen, Vietnam

E-mail: linhlth@ims.vast.ac.vn

Abstract. The quantum dynamics of a system of two interacting spin-qubits is studied for elaborating the physical mechanism of the quantum information transfer between them. A simple model with their Heisenberg XY exchange interaction is investigated. The rate equations are established. The analytical expressions of their solution are exactly derived. They explicitly demonstrate, how the quantum information encoded into a spin-qubit at the initial time $t = 0$ is transfered to other one at any time $t > 0$.

Keywords: qubit, spin-qubit, quantum information transfer.

1. Introduction

The basic element of any quantum information processing system is the quantum bit or qubit – a two-state quantum system. The physical processes related to the interaction of a qubit with a dissipative environment were widely investigated [1–10]. Any device or system to be used for the quantum information processing always consists of two or many qubits interacting one with another. Therefore, for elaborating the physical mechanisms of the quantum information processing operations, it is essential to study the quantum dynamics of two-qubit or multi-qubit systems. At present, the related problems are widely investigated [11–20].

An electron at two Zeeman states, splitted from a discrete energy level of an atom or a quantum dot due to the interaction of its magnetic moment with a constant magnetic field, is a two-state system, which can be used in quantum information technology as a qubit, called spin-qubit. The purpose of this work is to study the quantum dynamics of a two-spin-qubit system. The interaction between two spin-qubits may have different physical origins and, therefore, the interaction Hamiltonian may have different expressions. One of them is the spin-spin Heisenberg XY exchange Hamiltonian. Following references [11–14] we shall use the interaction Hamiltonian of this kind. A quantum statistical state of the two-qubit system is a 4×4 density matrix ρ. Due to the dissipative interaction of the environment, the quantum Liouville equation for the density matrix ρ contains a linear operator called Liouvillian superoperator. In the Born-Markov approximation, it is determined by the Lindblad formula [21–23].

In section 2, after the statement of problem, the system of rate equations is established. The exact solution of rate equations is derived in section 3. As the result, we obtain explicit time-dependent expressions of the elements of the density matrix. The discussion on the physical meaning of these expressions is presented in section 4. Section 5 is the conclusion.

© 2009 IOP Publishing Ltd

2. Rate equations

Consider a system of two interacting identical spin-qubits localized one near another in a constant magnetic field, and denote $\sigma_\pm, \sigma_3$ and $\tau_\pm, \tau_3$ the components of the spin operators representing these spin-qubits. Following [11–14] we assume that Hamiltonian of this system without the dissipative interaction of the environment has the form

$$H_S = \frac{E}{2}[(\sigma_3 \otimes 1) + (1 \otimes \tau_3)] + J[(\sigma_+ \otimes \tau_-) + (\sigma_- \otimes \tau_+)]. \tag{1}$$

Its 4×4 density matrix ρ is a linear combination of 15 generators of the SU(4) group. There are different choices of these generators [24]. For our purpose it is convenient to use following representation

$$\Gamma_{(\alpha\beta)} = (\sigma_\alpha \otimes \tau_\beta), \quad (\alpha\beta) \neq (00), \quad \alpha, \beta = 0, +, -, 3, \tag{2}$$

with $\tau_0 = \sigma_0 = 1$. The density matrix ρ is expressed in terms of 15 components $\rho_{(\alpha\beta)}$:

$$\rho = \frac{1}{4} + \sum_{(\alpha\beta)\neq(00)} \Gamma_{(\alpha\beta)} \rho_{(\alpha\beta)} \, . \tag{3}$$

The decoherence of the system due to the dissipative interaction of the environment is represented by some Liouvillian superoperator L acting on the density matrix ρ. The quantum dynamics of the system is determined by the quantum Liouville equation

$$\frac{d\rho}{dt} = -i[H_S, \rho] + L\rho \, . \tag{4}$$

In the Born-Markov approximation, Liouvillian superoperator L is expressed by the Lindblad formula [21–23]. Suppose that the interaction of the environment on a spin-qubit does not depend on the presence of the other. Then, with the choice of $\sigma_+, \sigma_-, \sigma_3$ and τ_+, τ_-, τ_3 to be generators of the corresponding SU(2) groups in the Lindblad formula, Liouvillian superoperator has the form

$$\begin{aligned} L\rho = \sum_{i=\pm,3} \alpha_i \Big\{ & (\sigma_i \otimes 1)\rho(\sigma_i^+ \otimes 1) - \frac{1}{2}(\sigma_i^+\sigma_i \otimes 1)\rho - \frac{1}{2}\rho(\sigma_i^+\sigma_i \otimes 1) \\ & + (1 \otimes \tau_i)\rho(1 \otimes \tau_i^+) - \frac{1}{2}(1 \otimes \tau_i^+\tau_i)\rho - \frac{1}{2}\rho(1 \otimes \tau_i^+\tau_i) \Big\}. \end{aligned} \tag{5}$$

Substituting the expressions (1), (3) and (5) of H_S, ρ and L into equation (4), we derive following system of rate equations

$$\frac{d\rho_{(+0)}}{dt} = -i\left[E\rho_{(+0)} - J\rho_{(3+)}\right] - \gamma_2 \rho_{(+0)} \, , \tag{6}$$

$$\frac{d\rho_{(0+)}}{dt} = -i\left[E\rho_{(0+)} - J\rho_{(+3)}\right] - \gamma_2 \rho_{(0+)} \, , \tag{7}$$

$$\frac{d\rho_{(-0)}}{dt} = i\left[E\rho_{(-0)} - J\rho_{(3-)}\right] - \gamma_2 \rho_{(-0)} \, , \tag{8}$$

$$\frac{d\rho_{(0-)}}{dt} = i\left[E\rho_{(0-)} - J\rho_{(-3)}\right] - \gamma_2 \rho_{(0-)} \, , \tag{9}$$

$$\frac{d\rho_{(30)}}{dt} = -i\frac{J}{2}\left[\rho_{(-+)} - \rho_{(+-)}\right] - \gamma_1 \rho_{(30)} - \frac{\lambda}{4} \, , \tag{10}$$

$$\frac{d\rho_{(03)}}{dt} = i\frac{J}{2}\left[\rho_{(-+)} - \rho_{(+-)}\right] - \gamma_1 \rho_{(03)} - \frac{\lambda}{4} \, , \tag{11}$$

$$\frac{d\rho_{(+3)}}{dt} = -i\left[E\rho_{(+3)} - J\rho_{(0+)}\right] - \gamma_3 \rho_{(+3)} - \lambda \rho_{(+0)} \, , \tag{12}$$

$$\frac{d\rho_{(3+)}}{dt} = -i\left[E\rho_{(3+)} - J\rho_{(+0)}\right] - \gamma_3\rho_{(3+)} - \lambda\rho_{(0+)}\,, \tag{13}$$

$$\frac{d\rho_{(-3)}}{dt} = i\left[E\rho_{(-3)} - J\rho_{(0-)}\right] - \gamma_3\rho_{(-3)} - \lambda\rho_{(-0)}\,, \tag{14}$$

$$\frac{d\rho_{(3-)}}{dt} = i\left[E\rho_{(3-)} - J\rho_{(-0)}\right] - \gamma_3\rho_{(3-)} - \lambda\rho_{(0-)}\,, \tag{15}$$

$$\frac{d\rho_{(++)}}{dt} = -2iE\rho_{(++)} - 2\gamma_2\rho_{(++)}\,, \tag{16}$$

$$\frac{d\rho_{(--)}}{dt} = 2iE\rho_{(--)} - 2\gamma_2\rho_{(--)}\,, \tag{17}$$

$$\frac{d\rho_{(+-)}}{dt} = 2iJ\left[\rho_{(30)} - \rho_{(03)}\right] - 2\gamma_2\rho_{(+-)}\,, \tag{18}$$

$$\frac{d\rho_{(-+)}}{dt} = -2iJ\left[\rho_{(30)} - \rho_{(03)}\right] - 2\gamma_2\rho_{(-+)}\,, \tag{19}$$

$$\frac{d\rho_{(33)}}{dt} = -2\gamma_1\rho_{(33)} - \lambda\left[\rho_{(30)} + \rho_{(03)}\right], \tag{20}$$

where

$$\lambda = \alpha_- - \alpha_+\,,$$
$$\gamma_1 = \alpha_- + \alpha_+\,,\quad \gamma_2 = \frac{1}{2}(\alpha_- + \alpha_+) + 2\alpha_3\,,\quad \gamma_3 = \frac{3}{2}(\alpha_- + \alpha_+) + 2\alpha_3\,. \tag{21}$$

3. Solution of rate equations

The system of linear differential equations (6)–(15) was solved exactly. As the result we obtain the explicit expressions of the components $\rho_{(\alpha\beta)}(t), (\alpha\beta) \neq (00)$. They contain arbitrary initial values $\rho_{(\alpha\beta)}(0)$. For example, two components $\rho_{(+0)}(t)$ and $\rho_{(+3)}(t)$ have following expressions:

$$\begin{aligned}
\rho_{(+0)}(t) = {} & \frac{1}{2}e^{-iEt}\left\{e^{-\left(\frac{\gamma_2+\gamma_3}{2}+\lambda'\right)t}\left[\left(1 - \frac{\gamma_3-\gamma_2}{2}\frac{\lambda'}{J'^2+\lambda'^2}\right)\cos J't + \frac{\gamma_3-\gamma_2}{2}\frac{J'}{J'^2+\lambda'^2}\sin J't\right]\right. \\
& \left. + e^{-\left(\frac{\gamma_2+\gamma_3}{2}-\lambda'\right)t}\left[\left(1 + \frac{\gamma_3-\gamma_2}{2}\frac{\lambda'}{J'^2+\lambda'^2}\right)\cos J't + \frac{\gamma_3-\gamma_2}{2}\frac{J'}{J'^2+\lambda'^2}\sin J't\right]\right\}\rho_{(+0)}(0) \\
& + \frac{i}{2}e^{-iEt}\left\{e^{-\left(\frac{\gamma_2+\gamma_3}{2}+\lambda'\right)t}\left[\left(1 - \frac{\gamma_3-\gamma_2}{2}\frac{\lambda'}{J'^2+\lambda'^2}\right)\sin J't - \frac{\gamma_3-\gamma_2}{2}\frac{J'}{J'^2+\lambda'^2}\cos J't\right]\right. \\
& \left. - e^{-\left(\frac{\gamma_2+\gamma_3}{2}-\lambda'\right)t}\left[\left(1 + \frac{\gamma_3-\gamma_2}{2}\frac{\lambda'}{J'^2+\lambda'^2}\right)\sin J't - \frac{\gamma_3-\gamma_2}{2}\frac{J'}{J'^2+\lambda'^2}\cos J't\right]\right\}\rho_{(0+)}(0) \\
& + \frac{1}{2}e^{-iEt}\frac{J}{J'^2+\lambda'^2}\left\{e^{-\left(\frac{\gamma_2+\gamma_3}{2}+\lambda'\right)t}(J'\cos J't + \lambda'\sin J't) - e^{-\left(\frac{\gamma_2+\gamma_3}{2}-\lambda'\right)t}(J'\cos J't - \lambda'\sin J't)\right\}\rho_{(+3)}(0)
\end{aligned} \tag{22}$$

$$+\frac{i}{2}e^{-iEt}\frac{J}{J'^2+\lambda'^2}\left\{e^{-\left(\frac{\gamma_2+\gamma_3}{2}+\lambda'\right)t}(J'\sin J't-\lambda'\cos J't)+e^{-\left(\frac{\gamma_2+\gamma_3}{2}-\lambda'\right)t}(J'\sin J't+\lambda'\cos J't)\right\}\rho_{(3+)}(0),$$

$$\begin{aligned}\rho_{(+3)}(t)=\frac{1}{2}e^{-iEt}&\left\{e^{-\left(\frac{\gamma_2+\gamma_3}{2}+\lambda'\right)t}\left[\left(1+\frac{\gamma_3-\gamma_2}{2}\frac{\lambda'}{J'^2+\lambda'^2}\right)\cos J't-\frac{\gamma_3-\gamma_2}{2}\frac{J'}{J'^2+\lambda'^2}\sin J't\right]\right.\\
&\left.+e^{-\left(\frac{\gamma_2+\gamma_3}{2}-\lambda'\right)t}\left[\left(1-\frac{\gamma_3-\gamma_2}{2}\frac{\lambda'}{J'^2+\lambda'^2}\right)\cos J't-\frac{\gamma_3-\gamma_2}{2}\frac{J'}{J'^2+\lambda'^2}\sin J't\right]\right\}\rho_{(+3)}(0)\\
+\frac{i}{2}e^{-iEt}&\left\{e^{-\left(\frac{\gamma_2+\gamma_3}{2}+\lambda'\right)t}\left[\left(1+\frac{\gamma_3-\gamma_2}{2}\frac{\lambda'}{J'^2+\lambda'^2}\right)\sin J't+\frac{\gamma_3-\gamma_2}{2}\frac{J'}{J'^2+\lambda'^2}\cos J't\right]\right.\\
&\left.-e^{-\left(\frac{\gamma_2+\gamma_3}{2}-\lambda'\right)t}\left[\left(1-\frac{\gamma_3-\gamma_2}{2}\frac{\lambda'}{J'^2+\lambda'^2}\right)\sin J't+\frac{\gamma_3-\gamma_2}{2}\frac{J'}{J'^2+\lambda'^2}\cos J't\right]\right\}\rho_{(3+)}(0)\\
+\frac{1}{2}\frac{e^{-iEt}}{J'^2+\lambda'^2}&\left\{\lambda\left[e^{-\left(\frac{\gamma_2+\gamma_3}{2}+\lambda'\right)t}(\lambda'\cos J't-J'\sin J't)-e^{-\left(\frac{\gamma_2+\gamma_3}{2}-\lambda'\right)t}(\lambda'\cos J't+J'\sin J't)\right]\right.\\
&\left.+J\left[e^{-\left(\frac{\gamma_2+\gamma_3}{2}+\lambda'\right)t}(J'\cos J't+\lambda'\sin J't)-e^{-\left(\frac{\gamma_2+\gamma_3}{2}-\lambda'\right)t}(J'\cos J't-\lambda'\sin J't)\right]\right\}\rho_{(+0)}(0)\\
+\frac{i}{2}\frac{e^{-iEt}}{J'^2+\lambda'^2}&\left\{\lambda\left[e^{-\left(\frac{\gamma_2+\gamma_3}{2}+\lambda'\right)t}(\lambda'\sin J't+J'\cos J't)+e^{-\left(\frac{\gamma_2+\gamma_3}{2}-\lambda'\right)t}(\lambda'\sin J't-J'\cos J't)\right]\right.\\
&\left.+J\left[e^{-\left(\frac{\gamma_2+\gamma_3}{2}+\lambda'\right)t}(J'\sin J't-\lambda'\cos J't)+e^{-\left(\frac{\gamma_2+\gamma_3}{2}-\lambda'\right)t}(J'\sin J't+\lambda'\cos J't)\right]\right\}\rho_{(0+)}(0),\end{aligned}\tag{23}$$

where

$$J'=\frac{1}{2}\sqrt{\frac{4J^2-(\gamma_3-\gamma_2)^2}{2}+\frac{1}{2}\sqrt{\left[4J^2-(\gamma_3-\gamma_2)^2\right]^2+16\lambda^2J^2}}\,,\tag{24}$$

$$\lambda'=\frac{1}{2}\sqrt{\frac{(\gamma_3-\gamma_2)^2-4J^2}{2}+\frac{1}{2}\sqrt{\left[(\gamma_3-\gamma_2)^2-4J^2\right]^2+16\lambda^2J^2}}\,.\tag{25}$$

The components $\rho_{(-0)}(t)$ and $\rho_{(-3)}(t)$ have similar expressions with the change of the sign of E and J: $E\to -E,\ J\to -J$. The explicit expressions of two components $\rho_{(30)}(t)$ and $\rho_{(+-)}(t)$ are

$$\begin{aligned}\rho_{(30)}(t)=&\frac{1}{2}\left[e^{-\gamma_1 t}+e^{-\frac{\gamma_1+2\gamma_2}{2}t}\left(\cos 2J''t+\frac{2\gamma_2-\gamma_1}{4J''}\sin 2J''t\right)\right]\rho_{(30)}(0)+\\&+\frac{1}{2}\left[e^{-\gamma_1 t}-e^{-\frac{\gamma_1+2\gamma_2}{2}t}\left(\cos 2J''t+\frac{2\gamma_2-\gamma_1}{4J''}\sin 2J''t\right)\right]\rho_{(03)}(0)+\\&+\frac{i}{4}\frac{J}{J''}e^{-\frac{\gamma_1+2\gamma_2}{2}t}\sin 2J''t\left[\rho_{(+-)}(0)-\rho_{(-+)}(0)\right]-\frac{\lambda}{4\gamma_1}\left(1-e^{-\gamma_1 t}\right),\end{aligned}\tag{26}$$

$$\begin{aligned}\rho_{(+-)}(t)=&\frac{1}{2}\left[e^{-2\gamma_2 t}+e^{-\frac{\gamma_1+2\gamma_2}{2}t}\left(\cos 2J''t-\frac{2\gamma_2-\gamma_1}{4J''}\sin 2J''t\right)\right]\rho_{(+-)}(0)+\\&+\frac{1}{2}\left[e^{-2\gamma_2 t}-e^{-\frac{\gamma_1+2\gamma_2}{2}t}\left(\cos 2J''t-\frac{2\gamma_2-\gamma_1}{4J''}\sin 2J''t\right)\right]\rho_{(-+)}(0)+\\&+i\frac{J}{J''}e^{-\frac{\gamma_1+2\gamma_2}{2}t}\sin 2J''t\left[\rho_{(30)}(0)-\rho_{(03)}(0)\right],\end{aligned}\tag{27}$$

where

$$J''=\frac{1}{4}\sqrt{16J^2-(2\gamma_2-\gamma_1)^2}.\tag{28}$$

The formulae of $\rho_{(0+)}(t)$, $\rho_{(3+)}(t)$, $\rho_{(0-)}(t)$, $\rho_{(3-)}(t)$, $\rho_{(03)}(t)$ and $\rho_{(-+)}(t)$ are obtained from the expressions of $\rho_{(+0)}(t)$, $\rho_{(+3)}(t)$, $\rho_{(-0)}(t)$, $\rho_{(-3)}(t)$, $\rho_{(30)}(t)$ and $\rho_{(+-)}(t)$ by the interchanges $\rho_{(\pm 0)}(0)\leftrightarrow\rho_{(0\pm)}(0)$, $\rho_{(\pm 3)}(0)\leftrightarrow\rho_{(3\pm)}(0)$, $\rho_{(30)}(0)\leftrightarrow\rho_{(03)}(0)$ and $\rho_{(+-)}(0)\leftrightarrow\rho_{(-+)}(0)$. For $\rho_{(++)}(t)$, $\rho_{(--)}(t)$ and $\rho_{(33)}(t)$ we have following expressions:

$$\rho_{(++)}(t)=e^{-2iEt}e^{-2\gamma_2 t}\rho_{(++)}(0),\tag{29}$$

$$\rho_{(--)}(t)=e^{2iEt}e^{-2\gamma_2 t}\rho_{(--)}(0),\tag{30}$$

$$\begin{aligned}\rho_{(33)}(t)=&e^{-2\gamma_1 t}\rho_{(33)}(0)-e^{-\gamma_1 t}\left(1-e^{-\gamma_1 t}\right)\frac{\lambda}{\gamma_1}\left[\rho_{(30)}(0)+\rho_{(03)}(0)\right]\\&-e^{-\gamma_1 t}\left(1-e^{-\gamma_1 t}\right)\frac{\lambda^2}{2\gamma_1^2}+\left(1-e^{-2\gamma_1 t}\right)\frac{\lambda^2}{4\gamma_1^2}.\end{aligned}\tag{31}$$

4. Discussion

The expressions of $\rho_{(\alpha 0)}$ or $\rho_{(0\alpha)}$, $\alpha=\pm,3$, determine the reduced density matrix of each spin-qubit $\rho^\sigma=Tr_\tau\rho$ or $\rho^\tau=Tr_\sigma\rho$, where Tr_τ or Tr_σ denotes the trace with respect to the spin indices of the spin-qubit τ or σ, respectively,

$$\begin{aligned}\rho^\sigma&=\frac{1}{2}+\sum_\alpha\sigma_\alpha\rho_\alpha^\sigma, & \rho_\alpha^\sigma(t)&=2\rho_{(\alpha 0)}(t),\\ \rho^\tau&=\frac{1}{2}+\sum_\alpha\tau_\alpha\rho_\alpha^\tau, & \rho_\alpha^\tau(t)&=2\rho_{(0\alpha)}(t).\end{aligned}\tag{32}$$

Suppose that at the initial time $t=0$ the quantum states of two spin-qubits are independent and therefore

$$\rho_{(\alpha 0)}(0)=\frac{1}{2}\rho^{\sigma}_{\alpha}(0),\quad \rho_{(0\alpha)}(0)=\frac{1}{2}\rho^{\tau}_{\alpha}(0),$$
$$\rho_{(\alpha\beta)}(0)=\rho^{\sigma}_{\alpha}(0)\rho^{\tau}_{\beta}(0),\ \alpha\neq 0,\beta\neq 0. \tag{33}$$

Then, from the expressions of $\rho_{(\pm 0)}(t)$ and $\rho_{(30)}(t)$ presented in section 3, we obtain following formulae

$$\begin{aligned}
\rho^{\sigma}_{\pm}(t)=&\frac{1}{2}e^{\mp iEt}\left\{e^{-\left(\frac{\gamma_2+\gamma_3}{2}+\lambda'\right)t}\left[\left(1-\frac{\gamma_3-\gamma_2}{2}\frac{\lambda'}{J'^2+\lambda'^2}\right)\cos J't+\frac{\gamma_3-\gamma_2}{2}\frac{J'}{J'^2+\lambda'^2}\sin J't\right]\right.\\
&\left.+e^{-\left(\frac{\gamma_2+\gamma_3}{2}-\lambda'\right)t}\left[\left(1+\frac{\gamma_3-\gamma_2}{2}\frac{\lambda'}{J'^2+\lambda'^2}\right)\cos J't+\frac{\gamma_3-\gamma_2}{2}\frac{J'}{J'^2+\lambda'^2}\sin J't\right]\right\}\rho^{\sigma}_{\pm}(0)\\
&\pm\frac{i}{2}e^{\mp iEt}\left\{e^{-\left(\frac{\gamma_2+\gamma_3}{2}+\lambda'\right)t}\left[\left(1-\frac{\gamma_3-\gamma_2}{2}\frac{\lambda'}{J'^2+\lambda'^2}\right)\sin J't-\frac{\gamma_3-\gamma_2}{2}\frac{J'}{J'^2+\lambda'^2}\cos J't\right]\right.\\
&\left.-e^{-\left(\frac{\gamma_2+\gamma_3}{2}-\lambda'\right)t}\left[\left(1+\frac{\gamma_3-\gamma_2}{2}\frac{\lambda'}{J'^2+\lambda'^2}\right)\sin J't-\frac{\gamma_3-\gamma_2}{2}\frac{J'}{J'^2+\lambda'^2}\cos J't\right]\right\}\rho^{\tau}_{\pm}(0)\\
&+e^{\mp iEt}\frac{J}{J'^2+\lambda'^2}\left\{e^{-\left(\frac{\gamma_2+\gamma_3}{2}+\lambda'\right)t}(J'\cos J't+\lambda'\sin J't)-e^{-\left(\frac{\gamma_2+\gamma_3}{2}-\lambda'\right)t}(J'\cos J't-\lambda'\sin J't)\right\}\rho^{\sigma}_{\pm}(0)\rho^{\tau}_{3}(0)\\
&\pm ie^{\mp iEt}\frac{J}{J'^2+\lambda'^2}\left\{e^{-\left(\frac{\gamma_2+\gamma_3}{2}+\lambda'\right)t}(J'\sin J't-\lambda'\cos J't)+e^{-\left(\frac{\gamma_2+\gamma_3}{2}-\lambda'\right)t}(J'\sin J't+\lambda'\cos J't)\right\}\rho^{\sigma}_{3}(0)\rho^{\tau}_{\pm}(0),
\end{aligned} \tag{34}$$

$$\begin{aligned}
\rho^{\sigma}_{3}(t)=&\frac{1}{2}\left[e^{-\gamma_1 t}+e^{-\frac{\gamma_1+2\gamma_2}{2}t}\left(\cos 2J''t+\frac{2\gamma_2-\gamma_1}{4J''}\sin 2J''t\right)\right]\rho^{\sigma}_{3}(0)+\\
&+\frac{1}{2}\left[e^{-\gamma_1 t}-e^{-\frac{\gamma_1+2\gamma_2}{2}t}\left(\cos 2J''t+\frac{2\gamma_2-\gamma_1}{4J''}\sin 2J''t\right)\right]\rho^{\tau}_{3}(0)+\\
&+\frac{i}{2}\frac{J}{J''}e^{-\frac{\gamma_1+2\gamma_2}{2}t}\sin 2J''t\left[\rho^{\sigma}_{+}(0)\rho^{\tau}_{-}(0)-\rho^{\sigma}_{-}(0)\rho^{\tau}_{+}(0)\right]-\frac{\lambda}{2\gamma_1}\left(1-e^{-\gamma_1 t}\right),
\end{aligned} \tag{35}$$

and similar ones with the interchange $\sigma\leftrightarrow\tau$. These formulae determine the mutual dependence of the reduced density matrices of two spin-qubits: $\rho^{\sigma}_{\pm}(t)$ depends not only on $\rho^{\sigma}_{\pm}(0)$, but also on $\rho^{\sigma}_{3}(0)$ and $\rho^{\tau}_{\pm}(0)$; $\rho^{\sigma}_{3}(t)$ depends not only on $\rho^{\sigma}_{3}(0)$, but also on $\rho^{\tau}_{3}(0)$, $\rho^{\sigma}_{\pm}(0)$ and $\rho^{\tau}_{\pm}(0)$, and vice-versa. These formulae exhibit the quantum information transfer from a spin-qubit to the other.

Moreover, the expressions of $\rho_{(\alpha\beta)}(t)$, $\alpha\neq 0$, $\beta\neq 0$, presented in section 3, explicitly demonstrate also the entanglement in the two-qubit system. This dynamical entanglement is generated not only by the direct interaction of qubits, but also by their interaction with the environment.

5. Conclusion

The physical mechanism of the quantum information transfer between two spin-qubits due to their

Heisenberg XY exchange interaction was investigated. The rate equations for the components of the density matrix of the two-spin-qubit system interacting with a dissipative environment were exactly solved, and its solution was presented.

The expressions of this solution explicitly demonstrate the physical mechanism of the entanglement of quantum states of two spin-qubits and, in particular, that of the quantum information transfer from a spin-qubit to the other.

References

[1] Witzel W M, de Sousa R and Das Sarma S 2005 *Phys. Rev.* B **72** 161306
[2] Morello A, Stamp P C E and Tupitsyn I S 2006 *Phys. Rev. Lett.* **97** 207206
[3] Nghiem D and Joynt R 2006 *Phys. Rev.* A **73** 032333
[4] Yao W, Liu R-B and Sham L J 2006 *Phys. Rev.* B **74** 195301
[5] Cai J-M, Zhou Z-W and Guo G-C 2006 *Phys. Rev.* A **74** 022328
[6] Deng C and Hu X 2006 *Phys. Rev.* B **73** 241303
[7] Jirari H and Pötz W 2006 *Phys. Rev.* A **74** 022306
[8] Saikin S K, Yao W and Sham L J 2007 *Phys. Rev.* B **75** 125314
[9] Rossini D, Calarco T, Giovannetti V, Montangero S and R. Fazio R 2007 *Phys. Rev.* A **75** 032333
[10] Nguyen Bich Ha 2008 *Adv. Nat. Sci.* **9** 23
[11] Ficek Z and Tanas R 2006 *Phys. Rev.* A **74** 024304
[12] Trif M, Golovach V N and Loss D 2007 *Phys. Rev.* B **75** 085307
[13] Hamdouni Y 2007 *J. Phys. A: Math. Theor.* **40** 11569
[14] Yuan X-Z, Goan H-S and Zhu K-D 2007 *Phys. Rev.* B **75** 045331
Yuan X-Z, Goan H-S and Zhu K-D 2007 *Phys. Rev.* B **75** 085307
[15] Venuti L, Boschi C D E and Roncaglia M 2006 *Phys. Rev. Lett.* **96** 247206
Venuti L, Boschi C D E and Roncaglia M 2007 *Phys. Rev. Lett.* **99** 060401
[16] Storez M J, Hellmann F, Hrelescu C and Wilhelm F K 2005 *Phys. Rev.* A **72** 052314
[17] Rao D D, Ravishankar V and Subramanyam V 2006 *Phys. Rev.* A **74** 022301
[18] Spört A, Schulte-Herbrüggen T, Glaser S J, Bergholm V, Storcz M J, Ferber J and Wilhelm F K 2007 *Phys. Rev.* A **75** 012302
[19] Morello A, Stamp P C E and Tupitsyn I S 2006 *Phys. Rev. Lett.* **97** 207206
[20] Cao X and Zheng H 2008 *Phys. Rev.* A **77** 022320
[21] Lindblad G 1976 *Commun. Math. Phys.* **48** 119
[22] Gorini V, Kossakowski A and Sudarshan E C G 1976 *J. Math. Phys.* **17** 821
[23] Havel T F 2003 *J. Math. Phys.* **44** 534
[24] Rau A R P, Selvaraj G and Uskov D 2005 *Phys. Rev.* A **71** 062316

Acknowledgments

We would like to thank Prof. Nguyen Van Hieu and Vietnam Academy of Science and Technology for the support.

Dynamics in a system of four qubits in two cavities

Duong Hai Trieu

Institute of Materials Science, Vietnam Academy of Science and Technology, 18 Hoang Quoc Viet Road, Cau Giay Dist., Hanoi, Vietnam

E-mail: trieudh@ims.vast.ac.vn

Abstract. We study the dynamics of a closed system of four qubits in two coupled cavities. Each cavity contains two qubits which interact with it's quantum field and photons which may hop between the cavities. The state of the system is first expressed in delocalized modes and then is expressed in the system of local bases. Exact general solutions are obtained for one-excitation state with arbitrary initial condition. We consider in detail two cases of large and small hopping at resonance. In the case of resonance and large hopping the state of the first qubit may also be transferred to the second qubit in the same cavity without excitation of the field. Such transfer of states minimizes decoherence due to photon loss.

Keywords: Cavity QED, qubit, coupled cavities, transfer of states.

1. Introduction

Cavity quantum electrodynamics (QED) is one of the promising approaches for quantum computation. It can be used for quantum information processing [1] or for quantum gate operations [2, 3]. The study of cavity QED may have also applications in circuit QED [4]. For applications of cavity QED we must first understand the physics of different systems of coupled qubits and cavities. In this paper we study the dynamics of a closed system of four qubits in two coupled cavities. Each cavity contains two qubits and two cavities are coupled by an optical fibre. We shall find the general solution for one-excitation state with arbitrary initial conditions and then investigate the important cases of resonance with large and small hopping of photons between cavities. As in [5] we shall find the state of the system in terms of the delocalized field and qubit modes and then express the state in the system of local bases.

The paper is organized as follows. In section 2 we find the one-excitation state for a closed system of two qubits in a cavity. In section 3 we find the one-excitation state of the system of four qubits in two coupled cavities for arbitrary initial condition. In section 4 we consider the cases of resonance with small and large hopping. Conclusions are presented in section 5.

2. System of two qubits in a cavity

To understand the physics of a complicated system we first study a closed system of two qubits in a cavity. The Hamiltonian of this system is ($\hbar = 1$)

$$H = E_e(|e_1\rangle\langle e_1| + |e_2\rangle\langle e_2|) + \omega\gamma^+\gamma + g_1(\gamma|e_1\rangle\langle g_1| + \gamma^+|g_1\rangle\langle e_1|) + g_2(\gamma|e_2\rangle\langle g_2| + \gamma^+|g_2\rangle\langle e_2|), \quad (1)$$

where $|e_i\rangle$, $|g_i\rangle$ represent the excited and ground states of qubit i in the cavity respectively, E_e is the energy of the excited state of qubit, γ is the annihilation operator for photon in the cavity, ω is

© 2009 IOP Publishing Ltd

the angular frequency of the photon, g_i is coupling strength of qubit i with the cavity. The energy of a qubit in the ground state is taken to be zero. Without losing the generality we can take $g_i \geq 0$. The one-excitation eigenstate of the system may be written as

$$|\phi_i\rangle = \sum_{j=1}^{3} \alpha_{ij} |\psi_j\rangle, \tag{2}$$

where $|\psi_1\rangle = |e_1\rangle$ is the state in which qubit 1 is excited, the field and qubit 2 is not excited, $|\psi_2\rangle = |e_2\rangle$ is the state in which qubit 2 is excited, the field and qubit 1 is not excited, $|\psi_3\rangle = |\gamma\rangle$ is the state with one photon and qubits are not excited. The energy of the eigenstates are

$$\varepsilon_1 = \omega + \frac{1}{2}\left[\Delta - \sqrt{\Delta^2 + 4(g_1^2 + g_2^2)}\right],\ \varepsilon_2 = E_e,\ \varepsilon_3 = \omega + \frac{1}{2}\left[\Delta + \sqrt{\Delta^2 + 4(g_1^2 + g_2^2)}\right], \tag{3}$$

where $\Delta = E_e - \omega$ is detuning. The matrix α equals

$$\alpha = \begin{bmatrix} \dfrac{\sqrt{2}g_1}{\sqrt{D(D+\Delta)}} & \dfrac{\sqrt{2}g_2}{\sqrt{D(D+\Delta)}} & -\dfrac{\sqrt{D+\Delta}}{\sqrt{2D}} \\ \dfrac{g_2}{\sqrt{g_1^2+g_2^2}} & -\dfrac{g_1}{\sqrt{g_1^2+g_2^2}} & 0 \\ \dfrac{\sqrt{2}g_1}{\sqrt{D(D-\Delta)}} & \dfrac{\sqrt{2}g_2}{\sqrt{D(D-\Delta)}} & \dfrac{\sqrt{D-\Delta}}{\sqrt{2D}} \end{bmatrix}, \tag{4}$$

where $D = \sqrt{\Delta^2 + 4(g_1^2 + g_2^2)}$.

We note that with adiabatic passage we can transfer the excited state from qubit 1 to qubit 2. Suppose that initially we have $g_1 = 0$, $g_2 \neq 0$ and the system is in the state $|\phi_2\rangle = \alpha_{21}|\psi_1\rangle = |e_1\rangle$. By slowly increasing g_1 and slowly decreasing g_2 (so that $\sqrt{g_1^2 + g_2^2} \neq 0$) the system remains in the state $|\phi_2\rangle$ and finally $|\phi_2\rangle = \alpha_{22}|\psi_2\rangle = -|e_2\rangle$.

Now we use the above eigenstates to find the evolution of the state of the system $|\Psi(t)\rangle$ from an arbitrary one-excitation initial state $|\Psi(0)\rangle = \sum_j A_j |\phi_j\rangle$. We have

$$|\Psi(t)\rangle = \exp(-\mathrm{i}Ht)|\Psi(0)\rangle = \sum_j A_j \exp(-\mathrm{i}\varepsilon_j t)|\phi_j\rangle = \sum_{jk} A_j \exp(-\mathrm{i}\varepsilon_j t)\alpha_{jk}|\psi_k\rangle = \sum_k B_k(t)|\psi_k\rangle, \tag{5}$$

where $B_k(t) = \sum_j A_j \exp(-\mathrm{i}\varepsilon_j t)\alpha_{jk}$. Putting $t = 0$ we have $B_k(0) = \sum_j A_j \alpha_{jk}$, $A_j = \sum_l B_l(0)\alpha_{jl}$, then

$$B_k(t) = \sum_{jl} \alpha_{jk}\alpha_{jl} \exp(-\mathrm{i}\varepsilon_j t) B_l(0). \tag{6}$$

3. System of four qubits in two coupled cavities

3.1. Transformation of Hamiltonian

The Hamiltonian of a closed system of four qubits in two coupled cavities is

$$H = \sum_{i=1}^{2}\sum_{j=1}^{2}[E_e|e_{ij}\rangle\langle e_{ij}| + g_i(\gamma_j|e_{ij}\rangle\langle g_{ij}| + \gamma_j^+|g_{ij}\rangle\langle e_{ij}|)] + \sum_{j=1}^{2}\omega\gamma_j^+\gamma_j + A(\gamma_1^+\gamma_2 + \gamma_2^+\gamma_1), \tag{7}$$

where $|e_{ij}\rangle$, $|g_{ij}\rangle$ represent the excited and ground states of the qubit of type i in cavity j (or qubit ij) respectively, E_e is the energy of the excited state of a qubit, γ_j is the annihilation operator for photon in the cavity j, ω is the angular frequency of the photon, g_i is the coupling strength of the qubit of type i in a cavity and A is cavity-cavity hopping strength. The energy of a qubit in the ground state is taken to be zero. Without losing the generality we can take $g_i \geq 0$ and $A \geq 0$. In the following the state of the system with one excited qubit ij and without photon we also denote as $|e_{ij}\rangle$.

Using the transformation [5]

$$|f_j\rangle = \sum_{k=1}^{2} F_{jk}|\gamma_k\rangle, \; f_j^+ = \sum_{k=1}^{2} F_{jk}\gamma_k^+, \; f_j = \sum_{k=1}^{2} F_{jk}\gamma_k, \tag{8}$$

$$|E_{ij}\rangle = \sum_{k=1}^{2} F_{jk}|e_{ik}\rangle, \tag{9}$$

where the matrix F equals

$$F = \frac{1}{\sqrt{2}}\begin{bmatrix} 1 & -1 \\ 1 & 1 \end{bmatrix}, \tag{10}$$

we can rewrite the Hamiltonian in the form

$$H = \sum_{k=1}^{2}\left\{\sum_{i=1}^{2}[E_e|E_{ik}\rangle\langle E_{ik}| + g_i(f_k|E_{ik}\rangle\langle 0| + f_k^+|0\rangle\langle E_{ik}|) + \omega_k f_k^+ f_k]\right\}. \tag{11}$$

Here $|\gamma_j\rangle$ is the state of the system with one photon in the cavity j, $|f_j\rangle$ is the state of the system with one photon in the delocalized mode j and f_j is the annihilation operator for this photon, $|E_{ij}\rangle$ is the excitation state of the qubit of type i in the delocalized mode j, $|0\rangle$ is the state of the system without photon and without excitation of qubits, $\omega_1 = \omega - A$ and $\omega_2 = \omega + A$. By this transformation we obtain two effectively independent systems ($k = 1, 2$), in each of which we have two effective qubits ($i = 1, 2$) interacting with a delocalized field mode.

3. 2. Eigenstates of the system

The eigenstates of the system $|\phi_{ij}\rangle$ may be expressed in the bases of delocalized modes $|\psi_{1j}\rangle = |E_{1j}\rangle$, $|\psi_{2j}\rangle = |E_{2j}\rangle$, $|\psi_{3j}\rangle = |f_j\rangle$ as

$$|\phi_{ij}\rangle = \sum_k \alpha_{ik,j}|\psi_{kj}\rangle, \tag{12}$$

where $\alpha_{ik,j}$ is the elements of matrix α_j which has the form

$$\alpha_j = \begin{bmatrix} \frac{\sqrt{2}g_1}{\sqrt{D_j(D_j+\Delta_j)}} & \frac{\sqrt{2}g_2}{\sqrt{D_j(D_j+\Delta_j)}} & -\frac{\sqrt{D_j+\Delta_j}}{\sqrt{2D_j}} \\ \frac{g_2}{\sqrt{g_1^2+g_2^2}} & -\frac{g_1}{\sqrt{g_1^2+g_2^2}} & 0 \\ \frac{\sqrt{2}g_1}{\sqrt{D_j(D_j-\Delta_j)}} & \frac{\sqrt{2}g_2}{\sqrt{D_j(D_j-\Delta_j)}} & \frac{\sqrt{D_j-\Delta_j}}{\sqrt{2D_j}} \end{bmatrix}, \tag{13}$$

where $\Delta_j = E_e - \omega_j$, $D_j = \sqrt{\Delta_j^2 + 4(g_1^2 + g_2^2)}$.

The eigenvalues of Hamiltonian are

$$\varepsilon_{1j} = \omega_j + \frac{1}{2}(\Delta_j - D_j),\ \varepsilon_{2j} = E_e,\ \varepsilon_{3j} = \omega_j + \frac{1}{2}(\Delta_j + D_j). \tag{14}$$

3. 3. Evolution of the state of the system

In practice we often work with low excited states. We shall study the one-excitation or zero-excitation states. The initial state of the system may be written as

$$|\Psi(0)\rangle = \sum_{ij} A_{ij} |\phi_{ij}\rangle + A_0 |0\rangle, \tag{15}$$

where A_{ij}, A_0 are constants. The state of the system at t is

$$|\Psi(t)\rangle = \exp(-\mathrm{i}Ht)|\Psi(0)\rangle = \sum_{ij} A_{ij}\exp(-\mathrm{i}\varepsilon_{ij}t)|\phi_{ij}\rangle + A_0|0\rangle = \sum_{ijk} A_{ij}\exp(-\mathrm{i}\varepsilon_{ij}t)\alpha_{ik,j}|\psi_{kj}\rangle + A_0|0\rangle. \tag{16}$$

The term $A_0|0\rangle$ does not change and we shall not regard it. We have

$$|\Psi(t)\rangle = \sum_{kj} B_{kj}(t)|\psi_{kj}\rangle, \tag{17}$$

where $B_{kj}(t) = \sum_i A_{ij}\exp(-\mathrm{i}\varepsilon_{ij}t)\alpha_{ik,j}$. At $t = 0$ we have $B_{kj}(0) = \sum_i A_{ij}\alpha_{ik,j}$, $A_{ij} = \sum_l B_{lj}(0)\alpha_{il,j}$, then

$$B_{kj}(t) = \sum_{li} \alpha_{ik,j}\alpha_{il,j}\exp(-\mathrm{i}\varepsilon_{ij}t)B_{lj}(0). \tag{18}$$

The form of states in equations (17, 18) is not convenient because in practice the initial state is given in local bases. Now we express $|\Psi(t)\rangle$ in local bases. With the notation

$$|\varphi_{1j}\rangle = |e_{1j}\rangle,\ |\varphi_{2j}\rangle = |e_{2j}\rangle,\ |\varphi_{3j}\rangle = |\gamma_j\rangle, \tag{19}$$

we have

$$|\psi_{ij}\rangle = \sum_k F_{jk}|\varphi_{ik}\rangle, \tag{20}$$

$$|\Psi(t)\rangle = \sum_{kl} C_{kl}(t)|\varphi_{kl}\rangle, \tag{21}$$

where $C_{kl}(t) = \sum_j B_{kj}(t)F_{jl} = \sum_{mij} \alpha_{ik,j}\alpha_{im,j}\exp(-\mathrm{i}\varepsilon_{ij}t)B_{mj}(0)F_{jl}$. At $t = 0$ we have $C_{kl}(0) = \sum_j B_{kj}(0)F_{jl}$, $B_{mj}(0) = \sum_n C_{mn}(0)F_{jn}$, then

$$C_{kl}(t) = \sum_{mnij} \alpha_{ik,j}\alpha_{im,j}\exp(-\mathrm{i}\varepsilon_{ij}t)F_{jl}F_{jn}C_{mn}(0). \tag{22}$$

In equations (21, 22) we have obtained the evolution of the one-excitation state of the system from arbitrary initial condition, expressed in local bases.

4. The case of resonance and small or large detuning

4.1. The case of resonance and small detuning

The case of resonance is important and will be considered in this section. First we consider the case of resonance $\Delta = 0$ and small hopping $A << g_1, g_2$. With condition $\Delta = 0$, $A << g_1, g_2$ we have

$$\alpha_1 = \alpha_2 = \begin{bmatrix} \frac{\sqrt{2}g_1}{D} & \frac{\sqrt{2}g_2}{D} & -\frac{1}{\sqrt{2}} \\ \frac{g_2}{\sqrt{g_1^2+g_2^2}} & -\frac{g_1}{\sqrt{g_1^2+g_2^2}} & 0 \\ \frac{\sqrt{2}g_1}{D} & \frac{\sqrt{2}g_2}{D} & \frac{1}{\sqrt{2}} \end{bmatrix}, \tag{22}$$

where notation $D = \sqrt{A^2 + 4(g_1^2 + g_2^2)}$ is used in all section 4. We shall regard two cases of initial condition.

4.1.1. Only qubit of type 1 in cavity 1 is initially excited

Suppose that only qubit of type 1 in cavity 1 is initially excited, $C_{11}(0) = 1$, then in approximation we obtain

$$C_{11}(t) = \frac{g_1^2}{g_1^2+g_2^2}\exp(\text{-i}E_e t)\left(\cos\frac{Dt}{2}\cos\frac{At}{2} + \frac{g_2^2}{g_1^2}\right), \tag{23}$$

$$C_{21}(t) = \frac{g_1 g_2}{g_1^2+g_2^2}\exp(\text{-i}E_e t)\left(\cos\frac{Dt}{2}\cos\frac{At}{2} - 1\right), \tag{24}$$

$$C_{31}(t) = -\frac{\text{i}g_1}{\sqrt{g_1^2+g_2^2}}\exp(\text{-i}E_e t)\sin\frac{Dt}{2}\cos\frac{At}{2}, \tag{25}$$

$$C_{12}(t) = -\frac{\text{i}g_1^2}{g_1^2+g_2^2}\exp(\text{-i}E_e t)\cos\frac{Dt}{2}\sin\frac{At}{2}, \tag{26}$$

$$C_{22}(t) = -\frac{\text{i}g_1 g_2}{g_1^2+g_2^2}\exp(\text{-i}E_e t)\cos\frac{Dt}{2}\sin\frac{At}{2}, \tag{27}$$

$$C_{32}(t) = -\frac{g_1}{\sqrt{g_1^2+g_2^2}}\exp(\text{-i}E_e t)\sin\frac{Dt}{2}\sin\frac{At}{2}. \tag{28}$$

We have oscillation first in cavity 1. In this oscillation the excitation transfers from qubit of type 1 to qubit of type 2 and field (in cavity 1) and then back. The excitation is slowly transferred from cavity 1 to cavity 2 (but not entirely) and then back again.

If we find minimal integers m, n satisfying $D/A = (2m+1)/(2n+1)$, then we can find the period of the evolution of the system $T = 2\pi(2n+1)/A = 2\pi(2m+1)/D$. In this case at the moments $t = kT$, where k is an integer, the excitation returns entirely to qubit of type 1 in cavity 1 and at the moments $(k+(1/2))T$ the probability of excitation of the field in the cavity 2 is maximal and equals $g_1^2/(g_1^2+g_2^2)$.

If we find minimal integers m, n satisfying $D/A = 2m/(2n+1)$, then at the moment $\tau = 2\pi(2n+1)/A = 4\pi m/D$ the state of qubit 11 is entirely transferred to qubit 21 and at the moments $t = \tau + k2\tau$, where k is integer, this excited state returns entirely to qubit 21.

If we find minimal integers m, n satisfying $D/A=(2m+1)/(2n)$, then at the moment $\tau=4\pi n/A=2\pi(2m+1)/D$ the state of qubit 11 is entirely transferred to qubit 21 and at the moments $t=\tau+k2\tau$, where k is integer, this excited state returns entirely to qubit 21.

4.1.2. Only field in cavity 1 is initially excited

Suppose that initially cavity 1 has one photon and all qubits are on the ground state. In this case $C_{31}(0)=1$. In approximation we obtain

$$C_{11}(t)=-\frac{\mathrm{i}g_1}{\sqrt{g_1^2+g_2^2}}\exp(\text{-i}E_e t)\sin\frac{Dt}{2}\cos\frac{At}{2}, \tag{29}$$

$$C_{21}(t)=-\frac{\mathrm{i}g_2}{\sqrt{g_1^2+g_2^2}}\exp(\text{-i}E_e t)\sin\frac{Dt}{2}\cos\frac{At}{2}, \tag{30}$$

$$C_{31}(t)=\exp(\text{-i}E_e t)\cos\frac{Dt}{2}\cos\frac{At}{2}, \tag{31}$$

$$C_{12}(t)=-\frac{g_1}{\sqrt{g_1^2+g_2^2}}\exp(\text{-i}E_e t)\sin\frac{Dt}{2}\sin\frac{At}{2}, \tag{32}$$

$$C_{22}(t)=-\frac{g_2}{\sqrt{g_1^2+g_2^2}}\exp(\text{-i}E_e t)\sin\frac{Dt}{2}\sin\frac{At}{2}, \tag{33}$$

$$C_{32}(t)=-\mathrm{i}\exp(\text{-i}E_e t)\cos\frac{Dt}{2}\sin\frac{At}{2}. \tag{34}$$

If we find minimal integers m, n satisfying $D/A=(2m+1)/(2n+1)$, then we can find the period of the evolution of the system $T=2\pi(2n+1)/A=2\pi(2m+1)/D$. In this case at the moments $t=kT$, where k is an integer, the excitation returns entirely to the field in cavity 1 and at the moments $(k+(1/2))T$ the state of the field in cavity 1 is entirely encoded in qubits in cavity 2.

If we find minimal integers m, n satisfying $D/A=2m/(2n+1)$, then at the moment $\tau=\pi(2n+1)/A=2\pi m/D$ the state of field in cavity 1 is entirely transferred to the field in cavity 2 and at the moments $t=\tau+k2\tau$, where k is integer, this excited state returns entirely to the field in cavity 2.

4.2. The case of resonance and large detuning

Now we study the case of resonance and large detuning $\Delta=0$, $A>>g_1, g_2$. Suppose that initially only qubit of type 1 in cavity 1 is excited, $C_{11}(0)=1$. In approximation we obtain

$$C_{11}(t)=\frac{g_1^2}{g_1^2+g_2^2}\exp(\text{-i}E_e t)\left(\cos\frac{(D-A)t}{2}+\frac{g_2^2}{g_1^2}\right), \tag{35}$$

$$C_{21}(t)=\frac{g_1 g_2}{g_1^2+g_2^2}\exp(\text{-i}E_e t)\left(\cos\frac{(D-A)t}{2}-1\right), \tag{36}$$

$$C_{31}(t)=0, \tag{37}$$

$$C_{12}(t)=\frac{\mathrm{i}g_1^2}{g_1^2+g_2^2}\exp(\text{-i}E_e t)\sin\frac{(D-A)t}{2}, \tag{38}$$

$$C_{22}(t)=\frac{\mathrm{i}g_1 g_2}{g_1^2+g_2^2}\exp(\text{-i}E_e t)\sin\frac{(D-A)t}{2}, \tag{39}$$

$$C_{32}(t)=0. \tag{40}$$

We see that in this case the field is not excited. The evolution of the state of the system has period $T = 4\pi/(D - A)$.

We can find the probability that qubit ij is in excited state $P_{ij} = |C_{ij}|^2$. For simplicity we take $g_1 = g_2$. The functions $P_{ij}(t)$ are presented in the figure below. From the figure we see that with time, excitation is transferred from qubit 11 to other qubits and back without the population of the field. At $t = T/4$, $3T/4$,... the probability of excitation of each qubit equals $1/4$. At $t = T/2$, $3T/2$,... the state of qubit 11 is entirely transferred to qubit 21.

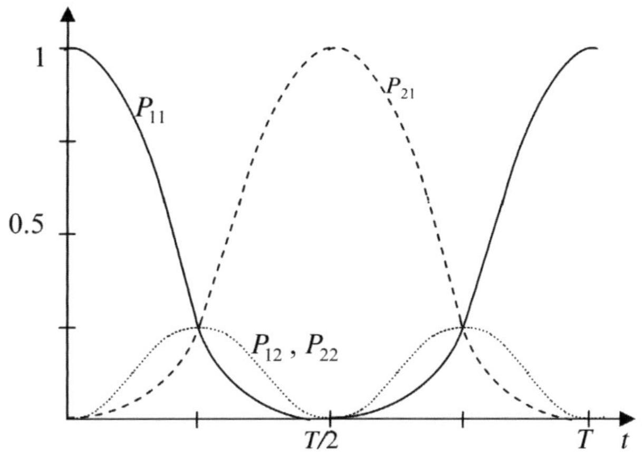

We note that although the transfer of states in this case is slow, but this transfer has the advantage that the absence of field population minimize the decoherence due to photon loss.

5. Conclusions

We have studied the dynamics of a closed system of four qubits in two coupled cavities. The general expression for one-excitation state of the system in arbitrary initial condition is derived. We have investigated the case of resonance and small or large hopping. In the case of resonance and small hopping the transfer of state is accompanied by the population of the field. In the case of resonance and large hopping the qubit state is transferred without the population of the field and such a state transfer minimize the decoherence due to photon loss. If we may turn off the parameters g_1, g_2 and A in an appropriate moments we can perform a desired quantum gate. In application to circuit QED such operations may easily be performed [6].

References

[1] Cirac J I, Ekert A K, Huelga S F and Macchiavello C 1999 *Phys. Rev.* A **59** 4249
[2] Zubairy M S, Kim M and Scully M O 2003 *Phys. Rev.* A **68** 033820
[3] Wallraff A, Schuster D I, Blais A, Frunzio L, Huang R S, Majer J, Kumar S, Girvin S M and Schoelkopf R J 2004 *Nature* (London) **431** 162
[4] Schuster D I, Houck A A, Schreier J A, Wallraff A, Gambetta J M, Blais A, Frunzio L, Majer J, Johnson B, Devoret M H *et al.* 2007 *Nature* (London) **445** 515
[5] Ogden C D, Irish E K and Kim M S 2008 *arXiv*:0804.2882v1 [quant-ph]
[6] Hu Y, Xiao Y F, Zhou Z W and Guo G C 2007 *Phys. Rev.* A **75** 012314

Acknowledgement

This work is supported by the Vietnam National Basic Research Program. The author would like to express his sincere thanks to Prof. Nguyen Van Hieu for the support.

Porous silicon as a low dimensional and optical material

Bui Huy, Pham Van Hoi, Phan Hong Khoi, Do Khanh Van, Pham Thanh Binh and Tran Thi Cham

Institute of Materials Science, Vietnam Academy of Science and Technology, 18 Hoang Quoc Viet Road, Cau Giay District, Hanoi, Vietnam

E-mail: buihuy@ims.vast.ac.vn

Abstract. The considerable and controllable changes in electronic structure and refractive index of porous silicon fabricated by electrochemical anodization make it become a promising material for photonics in comparison with bulk silicon. As the study of PS in terms of a low-dimensional material, we reviewed the effect of the surface passivation of silicon nanocrystals on photoluminescence spectra of such zero-dimensional crystals. In terms of an optical material, we show the fabrication method and optical properties of planar waveguide as well as active waveguide operated in the range of 1.54 μm wavelengths. We have also investigated the effect of energy transfer from silicon nanocrystals to erbium ions based on erbium-doped porous silicon waveguide.

Keyword: Porous silicon, electrochemical, nanocrystals.

1. Introduction

Since 1990 when Canham reported on the efficient visible photoluminescence (PL) from porous silicon (PS) [1] this material has attracted an increasing interest for research as well as applications [2]. The considerable and controllable changes in electronic structure and refractive index of PS fabricated by electrochemical anodization make it become a promising material for photonics in comparison with bulk silicon. In the structure, PS consists of many micropores and silicon residuals with the size and width of the order of a few nm. The exciton Bohr radius in Si is around 4.3 nm, so that quantum confinement can occur and change in electronic structure of those silicon nanocrystals [3]. On the other hand, because the value of porosity (the fraction of Si removed from the substrate) is directly linked to the effective index of refraction of the PS layer, this layer appears as an effective medium, whose refractive index has a tunable value between the index of refraction of bulk Si and that of the air (pores) [2]. The considerable and controllable changes in electronic structure and refractive index of PS as compared to bulk Si make it fascinating in terms of both a low-dimensional material and an optical one.

In the initial work [1], Canham proposed that the upship of the luminescence spectrum into the visible was due to quantum confinement in the silicon crystalline wire structure and hydride passivation of the Si wire was the reason for the high efficiency of the observed PL. For a no-long time after that, spectroscopic studies, particularly those of the polarization of the PL [4, 5] and of features observed under conditions of resonant excitation [6], have provided strong positive proof of the quantum confinement model. However, there were a lot of spectroscopic phenomena that can not be explained in the frame of the simple quantum confinement model. In this embarrassing situation,

© 2009 IOP Publishing Ltd

numerous models have been proposed as alternative explanations of the PL from PS such as hydrogenated amorphous silicon, surface hydrides, defects, molecules, surface states [3]. It is well known that in PS the surface to volume ratio is very large, so the surface effects are expected to play an important role in the material properties, especially optical ones [7]. Because the Si atoms in Si nanocrystals are either at the surface or a few lattice sites away, the arrangement of interfacial atomic bonds, i. e. the passivation with Si-H or Si-O bonds, strongly affects the energy distribution of electronic states [8]. In this paper, we present effects of ageing on the spectral, intensity and lifetime of PL from PS. Our results show that thc effect of the surface passivation on the re-distribution of electronic states and the luminescence efficiency is dependent on the size of nanocrystals in PS.

As study of PS in terms of a low-dimensional material, we focus attentions on strong emission properties and employ PS as materials for light sources, i. e. light emitting diodes and lasers operated in visible region. Concentration on controllable changes in refractive index of PS, we would like to use PS as materials for filters, optical planar waveguides and amplifier waveguides. In the paper, we also present the fabrication method and properties of optical planar and amplifier waveguides operated in the range of 1.54 μm wavelength. An advance of amplifier waveguides based on Er-doped PS is effect of efficient energy transfer from electron-hole pairs generated in the Si nanocrystals to their near erbium ions, which radiatively decay by emitting photon at 1.54 μm. The effective Er excitation cross-section in Er-doped PS is more than two orders of magnitude higher with respect to the Er resonant absorption of a photon [9], so that the pump efficiency in amplifier Er-doped PS waveguides can be very high. The effect of energy transfer has also been explored in elaborated Er-doped PS waveguides.

With the before-mentioned aim this paper consists of sections as following: 2- PS as a low-dimension material, 3- PS as an optical material, and 4- conclusions.

2. Porous silicon as a low-dimension material

In the first of this section we present the affect of surface states on the PL properties of PS based on the ageing processs in the air. In the last, we would like to show the reason for the intense and stable luminescence in the blue region which has been of interest in recent studies [10].

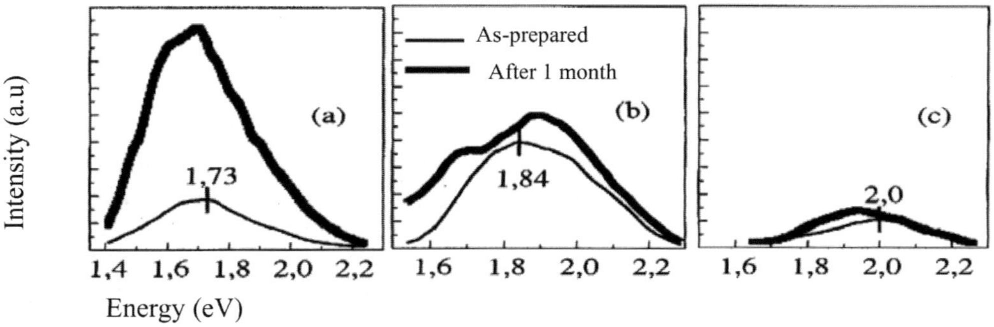

Figure 1. PL spectra of the as-prepared samples and after exposure to air for 1-month; samples, denoted as 1,2 and 3, were prepared by the anodical etching in 20%, 13%, and 10% HF solution, respectively. (a) sample 1, (b) sample 2 and (c) sample 3 [10].

Previous studies on the interaction of oxygen in the air on the as-prepared PS [8] show that i) the samples as-prepared were well passivated by hydrogen and free of oxygen, ii) after exposure to air the samples were gradually pasivated by oxygen and the red-shift of PL spectral occurred as exposure samples to air and was nearly completed after ageing of 24 h. They suggest that the ageing process can be divided into periods: the first one in which the transition of the luminescence mechanism occurs after exposing samples to air for a short time and the second one in which non-radiative centre concentration is changed by oxygen passivation [11].

In order to reveal the dependent of surface passivation effect on the size of Si nanocrystals a series of PS samples, denoted as 1, 2 and 3, were prepared by the anodical etching in 20%, 13% and 10% HF solution, respectively. As seen in figure 1, the PL peaks of the as-prepared samples 1, 2 and 3 have energy of 1.73, 1.84 and 2.00 eV, respectively. It is related with a decrease of particle size in the considered samples. The figure also reveals that the ageing produces a pronounced increase in PL intensity from sample 1 and only a small increase for samples 2 and 3. As seen in figure 2, the decay rate of the as-prepared samples (the curves 1a, 2a and 3a) shows that concentration of non-radiative centers in the sample 1 is higher than those in samples 2 and 3. The pronounced increase in intensity (in figure 1) as well as the pronounced decrease in decay rate (in figure 2) of sample 1 could be caused by the oxygen passivation of non-radiative defects. In the samples 2 and 3 containing smaller particles, the initial passivation degree is higher, therefore the ageing is expected to induce a small change both in intensity and decay rate. The comparison of data from curves 2a and 2c in figure 2 reveals that the modification of emission mechanism has no effect on the decay rate as well as its energy dependence $\tau^{-1}(E)$. This result seems to indicate that the replacement of Si-bond by Si-O one acting as a radiative center has no effect on the lifetime.

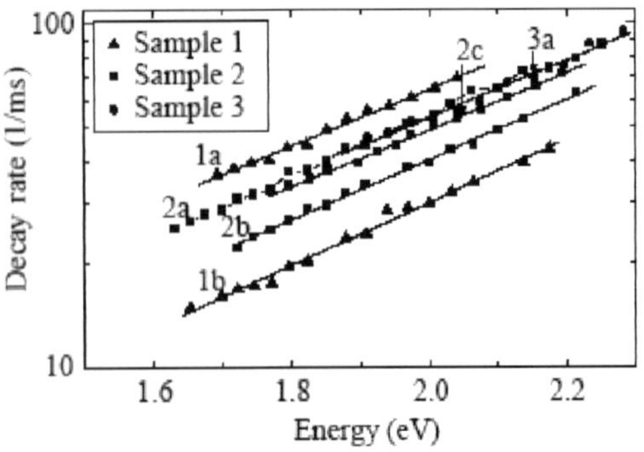

Figure 2. Evolution of decay rate as a funtion of emission energy from sampes after preparation, curves 1a, 2a, 3a and after exposure to mair for 1-month, curves 1b, 2b. Curve 2c coresponds to sample 2 for 24 h [10].

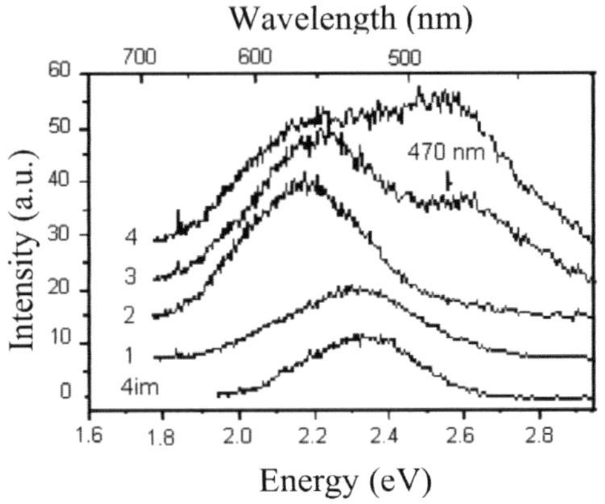

Figure 3. Evolution of PL spectral measured at the end of excitative pulse from a PS sample after different exposure time (1): as-prepared, (2): after 26, 72 and 94 h of exposure to air, respectively, (4im): corresponding to sample exposed to air for 94 h. and then immersed in 5% HF: ethanol solution for 10 sec [12].

From figures 1 and 2 the relation between the size of particles, intensity and decay rate during ageing was received. In the sample containing larger nanocrystals, the change of intensity and decay rate, i.e. the luminescence lifetime, are much larger during ageing.

Figure 3 shows the evolution of PL spectra, measured at the end of an excitative pulse after different exposure time. The figure reveals that the blue zone with the PL emission peaked at 470 nm is only observed after 72 hours of exposure to air. Furthermore, the figure also reveals that the PL intensity increases by increasing the air exposure time. These observations are different as compared to those reported in reference [8] in which the intensity of blue emission from the as-prepared sample containing the small Si particles was shown to decrease as the exposure time increased. This result indicates that blue-light emission observed in the present work does not arise from very small nanocrystals. Curve 4im shows the PL spectrum of a sample, which has exposed to air for 94 hours and then immersed in HF: ethanol solution. Comparing curves 4 and 4im, one can state that the blue zone in the PL spectrum, observed for the sample after 94 hours of exposure to air, is completely quenched. This quenching clearly relates to the fact that the silicon oxide layers in the exposed sample

Journal of Physics: Conference Series **187** (2009) 012033 doi:10.1088/1742-6596/187/1/012033

have been removed. The above results indicate that the intense and stable emission in the blue zone of the PL spectra, which observed in considered samples, relates to defects in silicon oxide layers.

3. Porous silicon as an optical material

In this section, before presentation of the fabrication method and properties of optical planar and amplifier waveguides we would like to show the ways of producing PS multilayer which is the base of elaboration of those waveguides. In the last, effect of energy transfer in prepared samples is shown.

Two main ways of producing PS multilayer have been proposed up to now [2]: by periodically varying the etching parameters, such as for the current density or the light power on the surface of Si under etch, or by using periodically doped substrates and maintaining constant the various etching parameters. In the following we will discuss the first approach, because it is more easily accomplished and it has been the one employed by us. This approach is based onto the following statements:

- the etching process is self-limited (once a porous layer is formed, the electro-chemical etching of this layer stops);
- the etching occurs mainly in correspondence to the pore tips;
- the porosity depends only on the current density once the other etching parameters are kept fixed;
- the refractive index of PS, n, depends on its porosity;

Hence, by varying the current density during the etch process, it is possible to vary the porosity in the etching direction only at the etch front. In this way, the current versus time profile is transferred in the porosity versus depth profile, i.e. n versus depth profile.

Our preparation process for an optical planar waveguide consists of 2 steps: making a PS film containing a core layer and a cladded one and stabilizing the structure of the PS film by thermal annealing at high temperature. In the fabrication of an optical amplifier waveguide, a step of deposition of Er ion into the PS film is added before thermal annealing. The treatment in high temperatures can cause an optical activation of Er ions in PS.

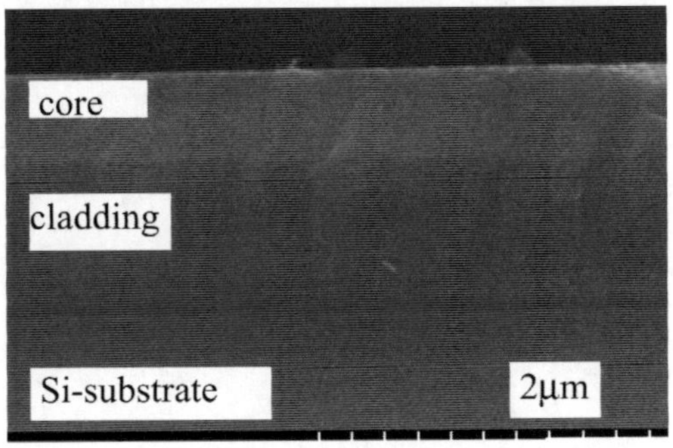

Figure 4. Core-cladding structure of PS film on Si-substrate of a planar waveguide layer.

The PS films were formed by electrochemical etching of 1Ωcm p-type Si wafers in 30%HF: ethanol solution. The top core layer has been fabricated by applying a 15 mA/cm^2 for 7 min. The cladding has been formed in the same way, with a 65 mA/cm^2 current density for 3 min. These conditions allow receiving a porosity of about 60% and 65%, corresponding, to the core and cladding. To deposit Er ions into the PS layers, the film was immersed in a 0.2 M $ErCl_3$ solution and the negative bias, relative to a platinum electrode, was applied to the sample at a constant current density in the range from 175 $\mu A/cm^2$ to 450 $\mu A/cm^2$ for 5 min. For optical activation of Er ions distributed in pores of PS layer, the sample was annealed at 400°C for 2 h, and then at 800°C for 15 min. in dry oxygen.

A cross section of the PS film was characterized by using a Scanning Electron Microscopy (SEM) measurement as presented in figure 4. Based on the different contrast between the core and the cladding due to the difference in porosities, it is observed that the film consisted of two layers in which the core layer thickness is about 4,5 μm, and the cladding about 7 μm. These data consisted with the layer thickness received from the prism-coupling method.

Figure 5 shows the SEM image of surface of the core and cladding before and after the period of thermal annealing. As seen from figures 5a and 5c, the difference in density of the black area presented the pores in the PS layer shows that the porosity in the core layer is lower than that in the cladding. From this figure we also observed the differences in density of the black area and in the contrast between the black area and the while one from the PS layers before (figures 5a and 5c) and after (figures 5b and 5d) the thermal treatment. Those differences appear that the treatment can cause decrease in the size of pores and the porosity of PS layers. The prepared PS layers have been densed and therefore optical properties of the waveguides would be stabilized.

Figure 6 shows the m-line received from the prism-coupling measurement in the prepared samples. As seen in figure 6a, the optical waveguide effect did not occur in the sample containing only a core layer. However, the m-line in figure 6b from the sample having the core-cladding structure showed an existence of the coupling angles at which the waveguide effect appeared. Base on the measured coupling angles, the effective refractive indices of 1.45 and 1.42, respectively, from the core and the cladding layer have been determined. The prepared optical waveguide with a refractive index contract up to 0.03 is suitable in optical integrated devices.

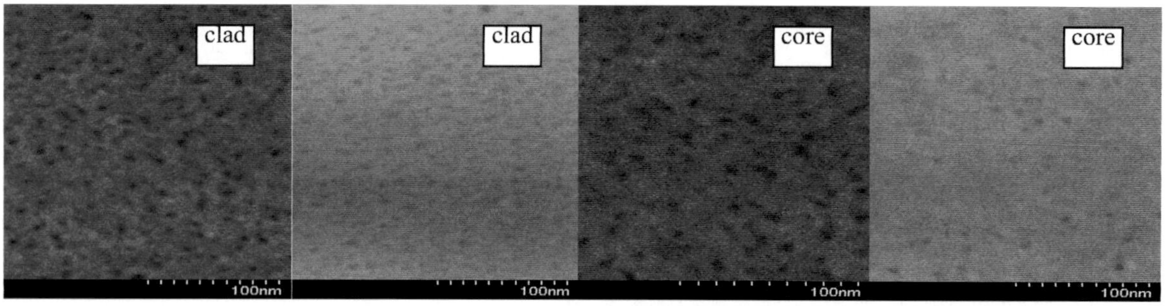

(a) (b) (c) (d)

Figure 5. SEM image of surface of the core (c, d) and a PS layer, which was prepared in the same etching parameters for the cladding (a, b) before (a, c) and after (b, d) thermal annealing in order to estimate their porosities.

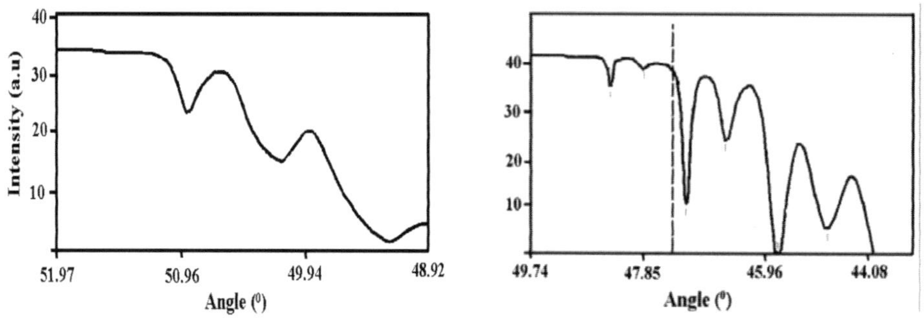

(a) (b)

Figure 6. The m-line from the sample having the single layer structure (a) and the core-cladding structure (b) in order to determine the waveguide modes and effective refractive index of each layer.

As above mentioned, Er ions need to incorporate into the core layer during the fabrication process of the amplifier waveguides. In our method, the Er-ion concentration, which doped into PS, can be controlled by Er concentration in $ErCl_3$ solution or drift current density in electrochemical cell. For the purpose of obtaining high-concentration Er-doped SRO materials (more than 0.1 at.% of Er) without Er-clusters, that will be good candidate for planar-waveguide amplifier, we have carried out very careful study of the distribution of Er-ions along the depth of the prepared PS layers. Figure 7 shows the Er concentration profile as a function of the waveguide depth that has been characterized by EDX method with the SEM technique. The Er-ion concentration was increased from 0.11 at.% from top-surface to 0.2 at.% at the depth of 3.5 μm from the top-surface. But the Er-ion concentration became decreased with the depth inside the samples.

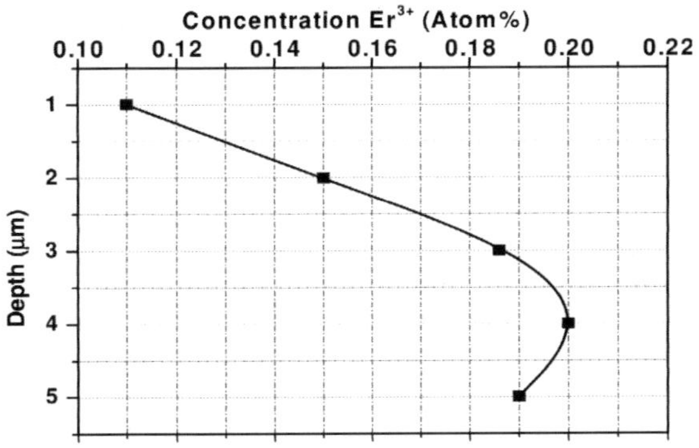

Figure 7. Er concentration profile as a function of the waveguide depth.

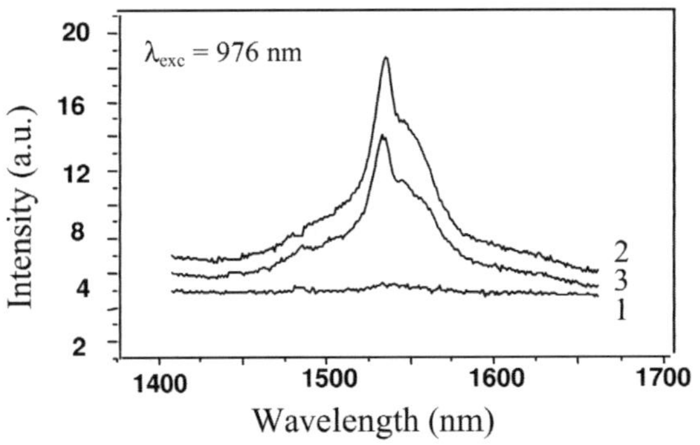

Figure 8. Luminescence spectra from samples 1, 2 and 3 under drift current density of 0.17, 0.25 and 0.45 mA, respectively.

Figure 8 presents the luminescence spectra at 1550 nm-region of the samples with different drift currents from 0.17 to 0.45 $mA.cm^{-2}$ under excitation of 976 nm-laser beam. The 1550 nm-luminescence intensity of all kind of samples has been increased with increasing drift current density from 0.17 to 0.25 $mA.cm^{-2}$, but when the drift current density is more than 0.25 $mA.cm^{-2}$ the luminescence would be slightly decreased with increasing the drift current.

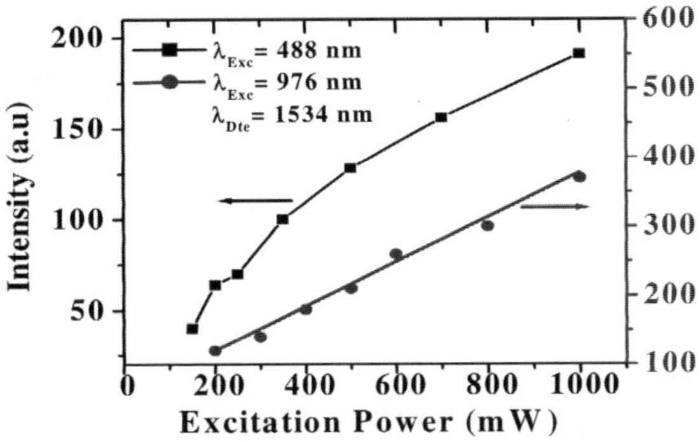

Figure 9. Dependence of luminescence intensity at 1534 nm on the power of the excitation laser at wavelengths of 488 nm and 976 nm.

In general, the intensity of luminescence emission at 1550 nm-region would be increased with increasing the concentration of Er-ion in PS layer. However, as the Er-ions concentration has reached the certain value, the luminescence intensity at 1550 nm-region would be decreased by the quenching effect from Er-ion clusters [13].

Figure 9 demonstrates the photoluminescence intensity at 1550 nm-region as a function of the light power from pumped lasers radiating at wavelengths of 488 and 976 nm. A difference in the intensity dependence on excited power from lasers radiating at 488 and 976 nm was observed in the figure. The pump at 976 nm only causes a direct excitement of Er ions (from $^4I_{15/2}$ to $^4I_{13/2}$ level), whereas the pump at 488 nm causes both a direct excitement of Er ions (from $^4I_{15/2}$ to $^4F_{7/2}$ level) and an indirect one related to the energy transfers from Si-nc to Er ions. Because the effective Er excitation cross-section in Er-doped PS is more than two orders of magnitude higher with respect to the Er resonant absorption of a photon, so that the pump at 976 nm causes a linear dependence of intensity on excited power [14] and the pump at 488 nm causes the non-linear one as seen in the figure.

References

[1] Canham L T 1990 *Appl. Phys. Lett.* **57** 1046
[2] Bettotti P, Cazzanelli M, Negro L Dal, Danese B, Gaburro Z, Oton C J, Prakash G Vijaya and Pavesi L 2002 *J. Phys. Condens. Matter* **14** 8253
[3] Amato G, Delerue C, Bardeleben V 1997 *Structure and Optical Properties of Porous Silicon Nanostructures* (Gorden and Breach Science Publishers)
[4] Starukhin A N, Lebedev A A, Razbirin B S and Kapitonova I M 1992 *Sov. Tech. Phys. Lett.* **18** 535
[5] Koch F, Kovalev D, Aberboukh B, Polisski G, Ben-Chorin M, Efros A L and Rosen M 1996 *J. Lumin.* **70** 320
[6] Calcott P D J, Nash K J, Canham L T, Kane M J, and Brumhead D 1993 *J. Lumin.* **57** 257
[7] Kanemitsu Y, Uto H, Masumoto Y, Futagi T, and Mimura H 1993 *Phys. Rev. B* **48** 2827
[8] Wolkin M V, Jorne J and Fauchet P M 1999 *Phys. Rev. Lett.* **82** 197
[9] Friolo F, Franzo G, Pacifici D, Vinciguerra V, Iacona F and Irrena A 2001 *J. Appl. Phys.* **89** 264
[10] Gorelkinskii Y V, Abdullin K A, Kalykova G K, Kikarin S M and Mukashev B N 2008 *Materials science & engineering B* **147** 258
[11] Huy B, Binh P H, Diep B Q and Luong P V 2003 *Physica E* **17** 134
[12] Huy B, Hoi P V, Binh P H, Chi T T K, Huy L Q and Liem N Q 2006 *J.Phys. IV France* **132** 321
[13] Kit P G and Polman A 2000 *J. Appl. Phys.* **88** 1992

Acknowledgments

This work is supported in part by the Vietnam National Basic Research Program on Physic under Project 408306.

Membrane-based microcalorimetry for thin films and sub-milligram single-crystal

Nguyen Duc Tuyen[1,2], Marco Evangelisti[1], Alessandro Di Bona[1] and Marco Affronte[1,2]

[1] CNR-INFM National Research Center on Nanostructures and Biosystems at Surfaces (S3), 41100 Modena, Italy
[2] Dipartimento di Fisica, Università di Modena e Reggio Emilia, via Campi 213/a, 41100 Modena, Italy

E-mail: tuyen.nguyenduc@unimore.it

Abstract. We report on the design, fabrication and use of micro-machined, membrane-based calorimeters for heat capacity measurements in the 20 K < T < 300 K temperature range. The membrane is made of Si_3N_4 and its thickness is as small as 250 nm. Thermometer and heater are fabricated in Pt by sputtering and optical lithography. Remarkably, the addenda of our devices are three to four orders of magnitude smaller than the most commercially performing available calorimeters. This leads to a high sensitivity, making them suitable for addressing samples in the form of thin films and sub-milligram single-crystals. Automated measurements are performed using a custom-made programme based on the quasi-adiabatic thermal relaxation method. A sample measurement is taken for a Na-Mn-O sub-milligram, and the results are reported so as to compare with the expected signal.

Keyword: Calorimetry, microcalorimeter, low temperature.

1. Introduction

Thermodynamic measurements give a great deal of information on fundamental properties, since they provide direct and quantifiable insight into, e.g., densities of state and phase transitions [1,2]. However, many interesting materials can only be obtained in the form of thin films or sub-milligram single crystals. The signal out of these samples is reduced accordingly to their size, being hence not sufficient for conventional calorimetry. Progress is growing at a pace determined by the development of the necessary experimental tools. In this respect, a primary advancement can be obtained by fabricating and using highly-sensitive calorimeters of reduced size.

The typical calorimeter has a sample holder with attached a thermometer and a heater. This as a whole is weakly linked to an external thermal bath, which provides the cooling power. The measurement technique is "simple": a controlled heat quantity Q is provided by the heater to the sample and the holder block so the response of the system is given in terms of temperature changes ΔT. The heat capacity is thus measured as $C = Q/\Delta T$. In practice, its realization within reduced sizes is very challenging and relatively few works have been carried out so far [3-8].

One simple technical issue that must be taken into consideration when developing any calorimeter, is that its components and surrounding elements are contributing to the measured signal. This unwanted but unavoidable heat capacity is called the "addenda". If a high resolution is the aim, the

© 2009 IOP Publishing Ltd

contribution of the addenda to the total heat capacity must be of the same order of magnitude as the sample or preferably less. Therefore the materials out of which the calorimeter is made, have to be chosen very carefully.

We show herein that micro-electro-mechanical-system (MEMS) techniques, like lithography and dry/wet etching, can be successfully employed for fabricating highly-sensitive micron-sized calorimeters. Specifically, we report on the design, fabrication and use of a membrane-based micro-calorimeters for heat capacity experiments in the 20 K < T < 300 K temperature range. The membrane is made of Si_3N_4 and its thickness is as small as 250 nm, whereas the thermometer and heater are both made of Pt. The addenda is very small (6 μJ/K at room temperature), resulting in sufficient sensitivity for detecting and measuring sub-milligram samples.

2. Principle of measurement and experimental set-up

Although the AC steady state method [9] could be used, we use the relaxation technique described by Bachmann et al. [10]. In this method, the power (P) to heat the sample is applied in the form of rectangular pulses with typical time periods of the order of seconds. After the heat is switched on (or off), the calorimeter relaxes to a new thermal equilibrium situation with a fixed temperature step, which typically amounts to $\Delta T = 3\%$. The heat capacity is then obtained by fitting the time evolution of the temperature of the calorimeter to an exponential. Our experimental set-up is schematized in figure 1. The heating is provided by a DC current source (Keithley 220), whereas the temperature in the calorimeter is monitored by a lock-in amplifier (Stanford Research 830) working in combination with an AC current source (Keithley 6221). Both heater and thermometer are measured using a four-probe configuration. The cooling is provided by inserting our devices into a Quantum Design PPMS cryostat, although they can easily implemented in any commercially available cryostat in view of the reduced size of our calorimeters (see below). All instruments are remotely controlled by a computer through GPIB connections. Automatized measurements are carried out by means of a home-made software developed using National Instruments Lab-View suite (see Section on Measurement procedure).

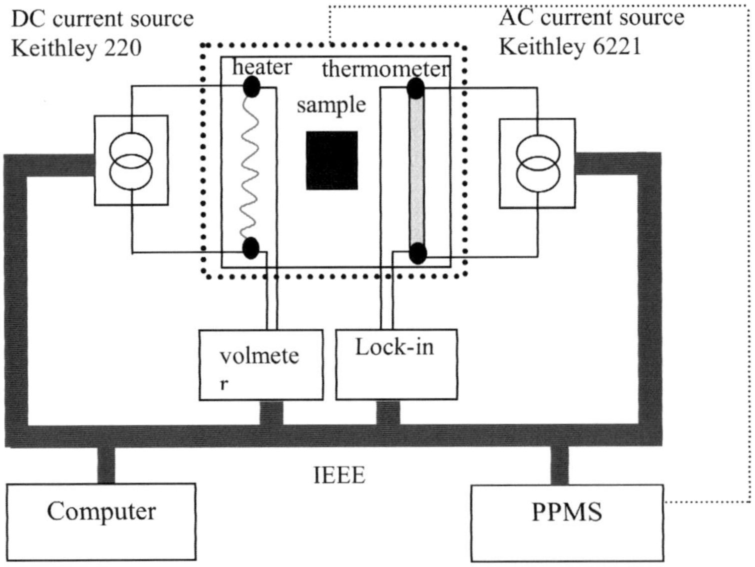

Figure 1. Scheme of the experimental set-up.

For each temperature we determine the characteristic relaxation time of the measurement (τ_e) as well as the thermal conductivity between calorimeter and its environment according to

$$K = \frac{P}{\Delta T}. \tag{1}$$

The damping of the temperature during the heating/cooling cycle is related to the heat capacity by

$$C = K\tau_e. \tag{2}$$

It should be noted that, in order to obtain the proper value of the heat capacity, it is crucial that τ_e is much longer than the internal relaxation time, i.e. the time it takes for the thermometer to be in thermal equilibrium with the sample and addenda. This condition implies that particular care should be taken in designing the calorimeter such that the sensing part should be nearly thermally-isolated from its environment.

3. Design, fabrication and testing

3.1. Design of the calorimeter

A special calorimeter was designed, allowing highly-sensitive specific heat measurements with the relaxation method. As detailed in the Fabrication Section, we start from a Si wafer (double-side polished with Si_3N_4) that offers a relatively high thermal conductivity, and forms the calorimeter frame that will be fixed to the cryostat sample-holder, e.g. a commercial sample puck of the Quantum Design PPMS. The micromachined calorimeter consists of a substrate, a heater and a thermometer. While measuring, these components contribute with their own heat capacity (i.e. addenda) which must be subtracted from the total heat capacity. In order to achieve high sensitivity, the addenda therefore has to be small, hence implying the size-shrinking of all components. The substrate, i.e. the sensing platform, is a thin membrane of Si_3N_4 whose size is 5x5 mm^2 with a thickness as small as 250 nm. Heater and thermometer are lithographically fabricated on top of the membrane. Since the Si_3N_4 thermal conductivity is appreciably smaller, we design the transducers to be parallel and close to each other (figure 2).

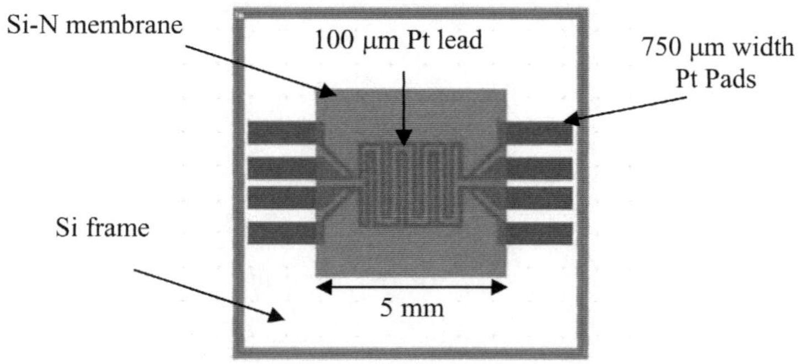

Figure 2. Design of the calorimeter.

We use Pt for fabricating the transducers as well as the leads and pads. The choice of one material only for all electrical components facilitates the fabrication process. The Pt patterning that connects

Journal of Physics: Conference Series **187** (2009) 012034 doi:10.1088/1742-6596/187/1/012034

the pads with the transducers, represents the "heat link" between calorimeter and thermal bath. Its design, therefore, directly influences the characteristic time τ_e of the measurement.

3.2. Microcalorimeter construction

The processing steps to fabricate our calorimeters are detailed in figure 3. We start from a 300 μm-thick Si wafer double-coated with Si_3N_4. In order to selectively remove the Si_3N_4, we pattern the top side with a square photoresist mask of 5x5 mm^2 (figure 3a). The Si_3N_4 is completely removed by Reactive Ion Etching (RIE) in a composition of $9CF_4$:$1O_2$ for four minutes (figure 3b). The region covered by Si_3N_4 is then used as protection for Si etching in a hot KOH solution (30 wt.% of water). The temperature of the solution is kept stable at 80°C, providing an etching rate of ~1.2 μm/min. The complete removal of the Si leads to the suspended Si_3N_4 membrane inside a Si frame (figure 3c). Because the Si_3N_4 membrane is very thin (~ 250 nm), care must be taken in order not to break it when the Si on the borders is etched completely.

Upon completing the fabrication of the membrane, we pattern on the back side a photoresist mask for all electrical components, i.e. heater, thermometer, leads and pads (figure 3d). Hence we sputter a Ti / Pt film (3 nm / 50 nm), whose thickness is calibrated by profilometry measurements (figure 3e). The heater and thermometers are finally obtained after a lift-off process (figure 3f).

A problem we are faced with refers to the low thermal conductivity of the Si_3N_4. In order to guarantee a good thermalization between heater and thermometer, we deposit a thin copper film by shadow mask on the top side. Alternatively, we add a small amount of thermal grease, e.g. Apiezon N.

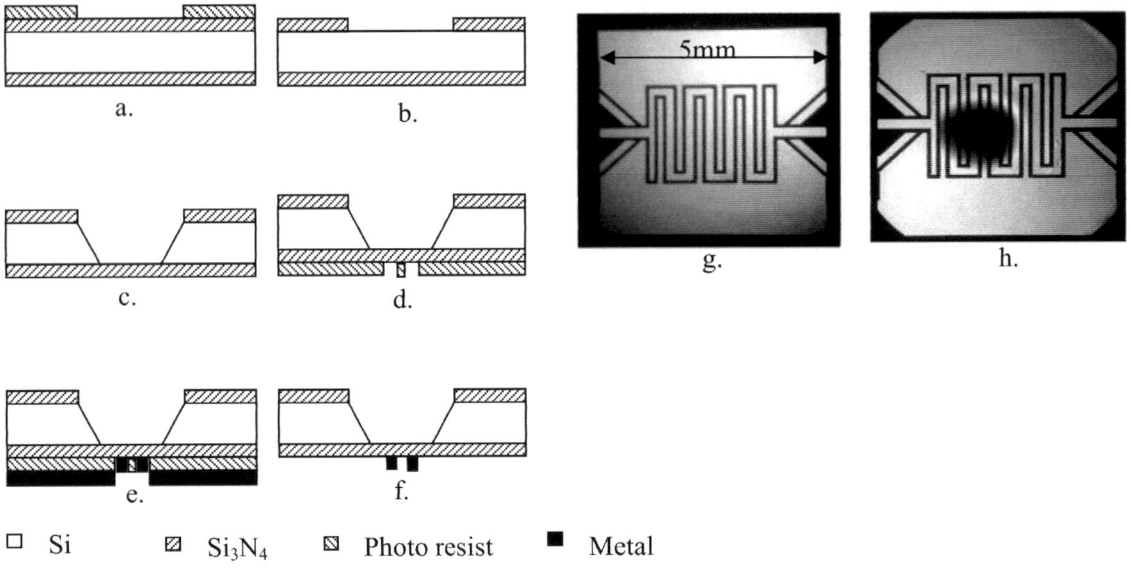

Figure 3. a) Si wafer covered by photoresist mask; b) reactive ion etching of Si_3N_4; c) Si_3N_4 membrane after wet etching of Si; d) photo resist mask for heater and thermometer; e) Pt deposition; f) lift-off to remove photoresist; g) completed device; h) device with sample.

3.3. Measurement procedure

In order to measure the heat capacity as a function of temperature, we install our calorimeter in a sample puck of the Quantum Design PPMS system. The electronics is then set-up as depicted in figure 1. Fully automated measurements are performed using a custom-made programme based on the quasi-adiabatic thermal relaxation method. Figure 4 shows a representative screen capture taken during the running of a measurement.

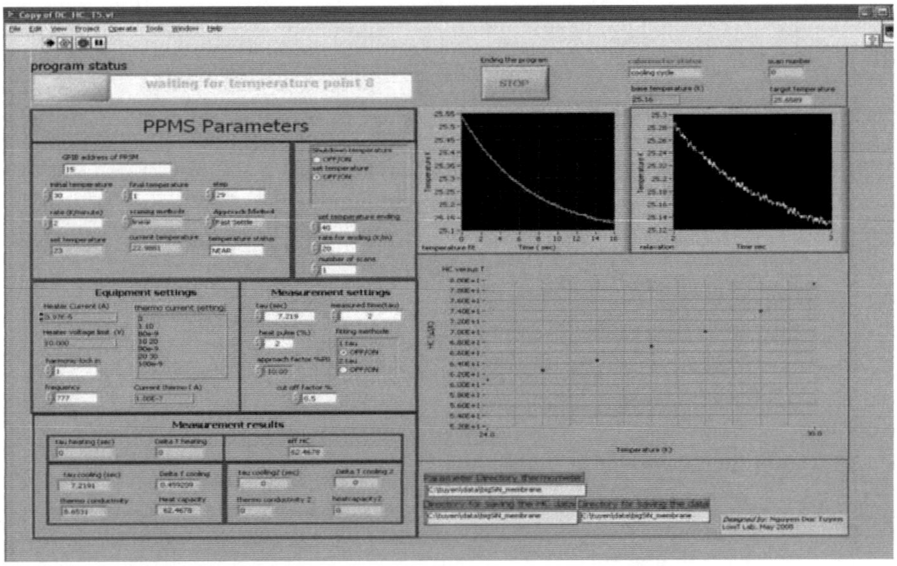

Figure 4. Main control panel of the home-made software.

In a typical measurement, we start by setting the system temperature (corresponding to the initial temperature of the measurement) in the “PPMS parameters” panel (figure 4). Following the stabilization of the system temperature, we check for the temperature in the membrane and wait till we obtain a stable reading. A heat pulse is then provided to the calorimeter according to the parameters that we set in the “Measurement settings” panel (figure 4). The programme automatically looks for that pulse which properly provides the desired ΔT (typically we work with 3% ΔT steps). When this step is accomplished, we start collecting experimental data. In figure 4 top-right panel, one can notice an example temperature relaxation, which follows the characteristic exponential decay. Finally the software performs the fits, and calculates, saves and plots the resulting heat capacity. Before switching to the next temperature, measurements may be repeated to minimize the noise.

3.4. Characterization and example measurement

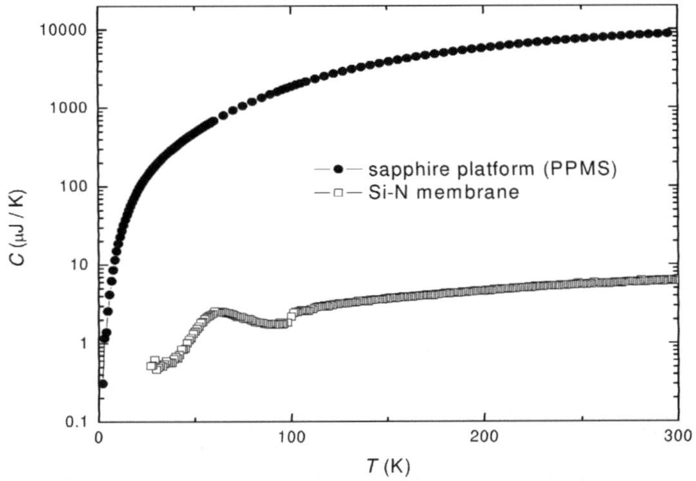

Figure 5. Heat capacity of our empty membrane-based calorimeter together with that of an empty commercial PPMS calorimeter based on a sapphire platform.

The resistance of Pt is known to linearly decrease with temperature down to 20 K, below which it levels off. This value limits the low-temperature applicability of our Pt thermometers. The heat capacity of the present design empty calorimeter is depicted in figure 5, together with that of an empty

PPMS calorimeter for comparison. To measure the addenda depicted in figure 5, we have thermalized heater and thermometer together by adding a 200 nm-thick copper film on the back side of the membrane. Remarkably, one can notice that the addenda of our device are up to three-four orders of magnitude smaller than that of the commercial apparatus. The addenda of this work amounts to 6 μJ/K at room temperature and decreases to 0.5 μJ/K at the lowest temperature. The reason for the broad bump observed around 60 K (figure 5) is unknown yet.

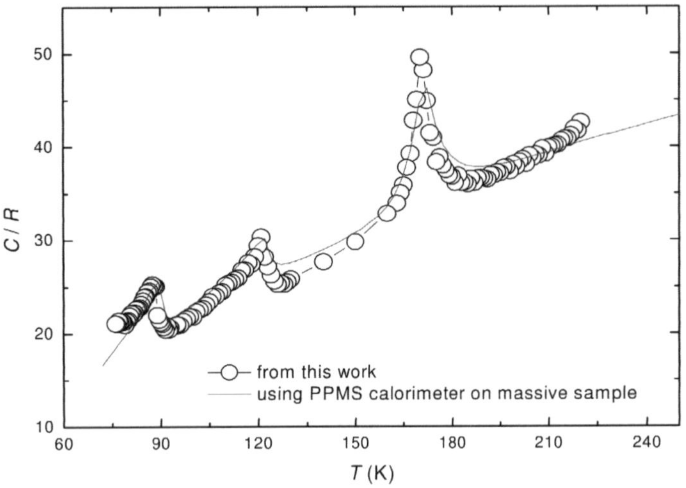

Figure 6. Experimental heat capacity of Na-Mn-O, as collected with a membrane-based calorimeter and a standard PPMS calorimeter.

As a test measurement, we have collected the heat capacity of a m ≈ 200 μg poly-crystal of the double perovskite compound Na-Mn-O [11,12], see figure 3h for a photograph of the mounted sample on the Si_3N_4 membrane. Figure 6 shows the excellent agreement of our data with that measured with a standard (PPMS) calorimeter on a massive (m = 50 mg) equivalent sample. Concerning the three observed features, the peak at 176 K is associated with a structural phase transition, whereas the ones at 125 K and 90 K are due to antiferromagnetic ordering of two different magnetic sublattices, respectively [11, 12].

4. Conclusions

Employing MEMS techniques, we successfully fabricated calorimeters based on a Si_3N_4 membrane having thickness as small as 250 nm. Thin-film heater and thermometer, both made of Pt, are lithographically implemented on the membrane. We experimentally demonstrate that the reduced size of all components leads to a heat capacity of the empty calorimeter that is three-four orders of magnitude smaller than conventional calorimetry. This results in an increased sensitivity making our devices particularly valuable for addressing studies in microgram samples, like thin films and single crystals.

References

[1] Evangelisti M, Luis F, de Jongh L J and Affronte M 2006 *J. Mater. Chem.* **16** 2534
[2] Stewart G R 1983. *Rev. Sci. Instrum.* **54** 1
[3] Denlinger D W, Abarra E N, Allen K, Rooney P W, Messer M T, Watson S K and Hellman F 1994 *Rev. Sci. Instrum.* **65** 946
[4] Lai S L, Ramanath G, Allen L H, Infante P and Ma Z 1995 *Appl. Phys. Lett.* **67** 1229
[5] Fominaya F, Fournier T, Gandit P and Chaussy J 1997 *Rev. Sci. Instrum.* **68** 4191
[6] Doettinger-Zech S G, Uhl M, Sisson D L, and Kspitulnik A 2001 *Rev. Sci. Instrum.* **72** 2398
[7] Zhang Y Y and Tadigadapa S 2005 *Appl. Phys. Lett.* **86** 034101
[8] Suh K S, Kim H J, Park Y D, Kim K H and Cheong S W, Kor J 2006 *Phys. Soc* **49** 1370
[9] P.F. Sullivan and G. Seidel 1968 *Phys. Rev.* **173** 679

[10] Bachmann R, DiSalvo F J, Geballe T H, Greene R L, Howard R E, King C N, Kirsch H C, Lee K N, Schwall R E, Thomas H U and Zubeck R B 1972 *Rev. Sci. Instrum.* **43** 205
[11] Gauzzi A, Gilioli E, Prodi A, Bolzoni F, Licci F, Marezio M, Calestani G L, Affronte M, Huang Q, Santoro A and Lynn J 2005 *J. Supercond.: Incorporating Novel Magnetism* **18** 75
[12] Prodi A, Gilioli E, Gauzzi A, Licci F, Marezio M, Bolzoni F, Huang Q, Santoro A and Lynn J W 2004 *Nature Mater.* **3** 48

Acknowledgments
This work was carried out within the framework of the Network of Excellence "MAGMANet" supported by the EC under contract No. 515767, and the CNR-INFM S3 PFD61101-SEED activity.

The effects of hydrogen dilution on structure of Si:H thin films deposited by PECVD

Tran Quang Trung[1], Stuchlik Jiri[2], Ha Stuchlikova[2], Le Khac Binh[1], Nguyen Nang Dinh[3], Huynh Kim Khuong[1], Phan Thi Nhu Quynh[1] and Nguyen Thi Huynh Nga[1]

[1] University of Natural Science HCM, 227 Nguyen Van Cu Str., 5 Distr., Ho Chi Minh City, Vietnam
[2] Institute of Physics, Na Slovance 2 CZ-182 21 Prague Praha, Czech Republic
[3] College of Technology, VNU, 144 Xuan Thuy, Cau Giay, Hanoi, Vietnam

E-mail: trungvlcr@yahoo.com.sg

Abstract. Si:H thin films were deposited using PECVD method at substrate temperature of 250°C with different hydrogen dilution. Structure investigations and hydrogen concentration estimation were performed using FTIR, Raman and UV-VIS spectroscopy. The IR and Raman spectra show various peak shifting when H dilution is changed. Band gaps of films were evaluated from the strong absorption regions in UV-VIS transmission spectra. The results demonstrate a transition from amorphous to nc/μc structure when hydrogen dilution increased.

Keywords: Plasma-enhanced chemical vapor deposition, hydrogen dilution, a-Si:H, nc/μc-Si:H, para-crystallines.

1. Introduction
Amorphous hydrogenated silicon (a-Si:H) is one of the famous materials for photovoltaics due to its availability in thin film, the high absorption coefficient, and the low cost of manufacturing. However, a-Si:H has some disadvantages such as low conductivity and the instability on light exposure, leading to the degradation of devices in the working environment. In recent years, microcrystalline Si:H (μc-Si:H) is considered as the ideal Si-based semiconductors which can overcome the drawbacks of a-Si:H. Both a-Si:H and μc-Si:H thin films can be deposited by PECVD method in which structures of products are depended on the experimental parameters. It is generally accepted that the crystallinity of thin films is increased with the hydrogen dilution in deposition processes. In this paper, we report our quantitative investigations of this effect. The results in this report are an advantage for preparing thin film PV devices in structures nip or pin with higher efficiency.

2. Experimental
Si:H thin films were prepared using PECVD method with the conventional 13.56 MHz RF plasma. The precursor gases silane (SiH_4) and hydrogen were used as dilution media. Two kinds of substrates (glass and two-side polished silicon) were employed in depositing: samples deposited on glass substrates were used for Raman and UV-VIS measurements whereas those on silicon substrates were for FTIR characterizations. Before depositing, the deposition chamber was desorption at 10^{-7} Torr and 250° C for 2 hours in order to reduce the contamination from oxygen and moisture.

© 2009 IOP Publishing Ltd

To investigate the effects of hydrogen dilution on the films structure, all experimental parameters for various samples were kept constant except for the dilution ratio. The values of RF power, substrate temperature, deposition pressure, and deposition time were 10 W/cm^2, 250°C, 1 Torr, and 180 minutes, respectively. On the other hand, the dilution ratio R (R = flowrate of H_2 / flowrate of SiH_4) was varied to from 1 to 6 in order to investigate the effects of hydrogen on the films structure. In each experiment, the films were deposited both on glass as well as silicon substrates.

Films thickness was measured using Dektak 6M Stylus Profiler. FTIR transmission spectra of films were taken by a Bruker-Equinox spectroscopy in the range 500–6000 cm^{-1}. Raman spectra were investigated using Horiba-Yvon Raman spectrophotometer. UV-VIS transmissions are characterized by a Jasco 530 UV-VIS spectroscopy. The deposition conditions and films thickness are shown in table 1 below.

Table 1. Experimental parameters of samples.

Sample	Dilution ratio **R**	Film thickness (μm)	Average growth rate (A/s)
No. 1	1	3.7	3.42
No. 2	2	3.3	3.05
No. 3	3	1.8	1.67
No. 4	4	0.98	0.91
No. 5	5	0.8	0.7

3. Results and discussions

3.1. Infrared spectra and hydrogen content

The raw IR transmission data of the film is an accumulation between sinusoidal interference fringes and the strong absorption bands resulted from the vibration of Si-H or Si-O bonds. The interference fringes were eliminated from the spectra by normalizing the transmittances to the absorption-free region (6000-2500 cm^{-1}) and absorption spectra was then calculated from the normalized transmittances using the procedure described in Ref. [1]. Figure 1 represents the calculated absorption spectra of films. In all samples, there are absorption bands at 600-660 cm^{-1}, 800-890 cm^{-1} and 2000-2200 cm^{-1} which resulted from dipole vibrations of various SiH_x groups as well as different vibrational modes for every group [1-4]. The strong bands around 1000 cm^{-1} were assigned to Si-O vibrations from oxygen contaminations in our films.

The main absorption bands of Si-H bonds show evident alternations with increasing H dilution. The 640 and 880 cm^{-1} bands tend to low energy side whereas the 2000 cm^{-1} one shows peak shifting and raising of new peaks in the high energy side when the ratio R is increased. These features of IR spectra are the signs of amorphous – crystalline transition in high dilution depositing conditions [2, 3].

Figure 1b clarifies the alternations of the absorption bands around 2000 cm^{-1} with increasing H dilution. At R = 1, the main absorption peak appears around 2000 cm^{-1}, which is often assigned to the vibration of Si-H bond in the amorphous matrix. The shoulder which occurred around 2100 cm^{-1} (when R = 2 and at high H dilution) becomes a peak and a new peak arises at about 2250 cm^{-1}. For higher dilution the peak at 2100 cm^{-1} is lower while the peak at 2250 cm^{-1} becomes more pronounced. The peak at 2100 cm^{-1} is explained as vibration mode of hydrogen bonded to the surface of crystalline phase, i.e. the hydrogen in the grain boundary voids in μc-Si:H [3].

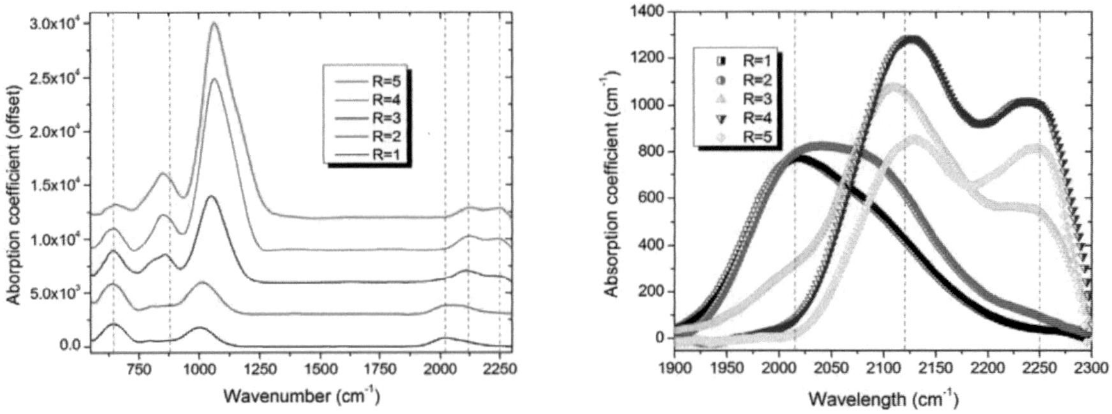

(a) (b)

Figure 1. a) The absorption spectra of the samples on Si substrates. **R** is the dilution ratio. The peak shifts are emphasized by vertical dashed lines. b) The alternations of absorption bands around 2100 cm^{-1}.

In Ref. [11] the 2125 and 2245 cm^{-1} peaks are assigned to the stretching mode of SiH_2 group with one, two or three backbone oxygen atoms. Therefore, at high H dilution, the effects of oxygen contamination are larger. According to our opinion, the peak shifting and raising at the region of 2100 cm^{-1} represents the sum of both oxygen impurity and amorphous – microcrystalline transition.

The hydrogen content of the samples is estimated by using the following formula [2, 3]

$$C_{H,i}(at)=\frac{A}{N_{Si}}\int_{\nu_i}\frac{\alpha(\nu_i)}{\nu}d\nu \ , \qquad (1)$$

where $\alpha(\nu)$ is the film absorption coefficient at the wave number ν and ν_i stands for the wave number region of vibration, N_{Si} is the atomic density of silicon. For different absorption bands, the proportionality constant A_i takes different values. In many references, the rocking-wagging vibrations of all kinds SiH_x group result in the absorption band around 660 cm^{-1} [1, 3, 5]. In addition, the proportionality constant for 660 cm^{-1} band is independent of the preparation conditions and H content [3]. Therefore, this band was employed to estimate hydrogen content evaluation in our samples. The constant was chosen at 1.6×10^{-19} cm^{-2} and the atomic density of Si was taken at the value of the crystalline Si: 5×10^{22} cm^{-3} [3]. The evaluated values of H content are shown in figure 2. Due to various reasons, the H content estimated from this method may have an error larger than 15% [3, 4].

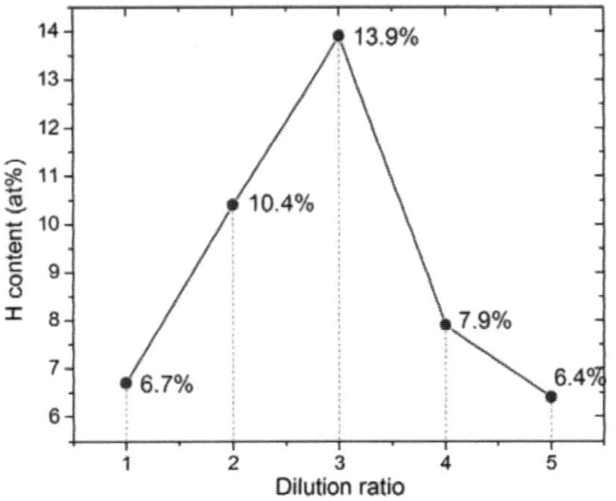

Figure 1. Hydrogen content in the samples as a function of H dilution.

The hydrogen content at first increases with H dilution in the low and medium dilution (**R**=1-3) and has its maximum at **R** = 3. From figure 1b it can be observed that when **R** = 1 or 2, there are only weak shoulders around 2250 cm^{-1} together with the increasing of the H content . As this shoulder becomes peaks and these peaks get stronger (**R** > 3), the H dilution also decreases. The maximum at **R** = 3 can be explained as the "transition region" between μc-Si:H regime and a-Si:H regime of deposition [3]. When the dilution is higher the H content decreases together with the growth of crystalline phase in the films.

3.2. Raman spectra

In figure 3a, Raman spectra of the samples are shown in comparison with the amorphous lines taken from our other work [9]. The features of Raman scattering lines show that the films are mixtures between amorphous, paracrystalline and crystalline phase of various volume fractions. With increasing H dilution, the main peak shifts gradually to 520 cm^{-1}, demonstrating the transition from amorphous regime to microcrystalline regime.

One can observe in the spectra a broad shoulder at 480 cm^{-1}, which is typically represented for amorphous phase, and the peaks around 505 to 518 cm^{-1}, which can be interpreted as TO scattering form of different kinds of silicon grains. In crystalline silicon, the TO phonon scattering results a sharp peak at 520 cm^{-1}. However, due to the finite grain size and internal stress in thin films, this peak has a slight shift to 518 cm^{-1} with a wider peak width [7, 10]. Besides, the peak whose position varies from 500 to 510 cm^{-1} is often assigned to different structures: paracrystalline grains (about 1 nm) [8], wurtzite phase or stacking faults on grain boundaries [7-9]. We therefore decouple the Raman spectra into two Gaussian peaks at 480 and 518 cm^{-1} and one intermediate peak at 500 - 510 cm^{-1}, in order to evaluate the crystallinity volume fraction in each sample (figure 3b). The position of the intermediate peak I_{int} is calculated to minimize the error between the experimental and the fitted data.

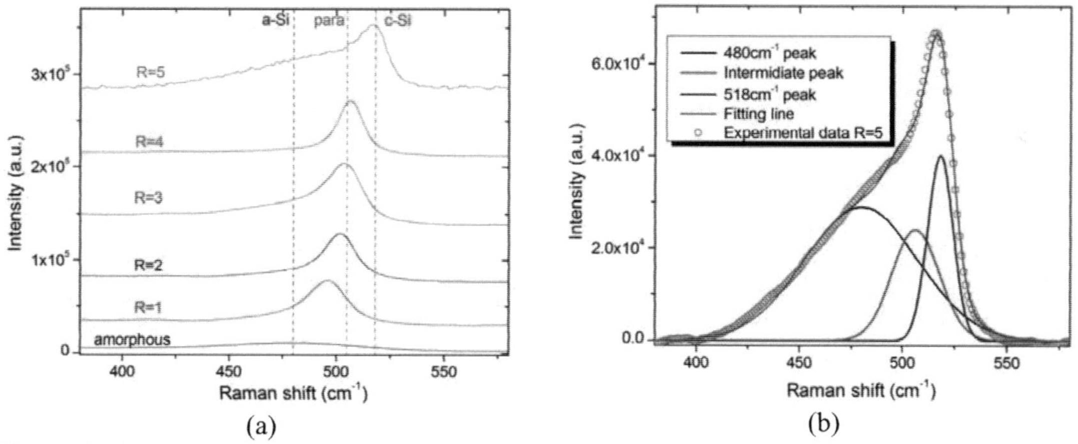

Figure 2. a) Raman spectra of Si:H thin films (on glass) of H dilution from 1 to 5. The amorphous line is added for the ease of comparison. b) The deconvolution of Raman spectra of sample No.5 and fitted lines from three Gaussian peaks at 480, 508, and 518 cm^{-1}.

The total crystallinity volume fraction is then estimated using the following formula:

$$X_C = \frac{I_{518} + I_{Int.}}{I_{518} + I_{508} + I_{480}} \times 100 \ , \qquad (2)$$

where I_{480}, $I_{int.}$ and I_{518} are the integrated intensities of the corresponding peaks. The reduced integrated intensity of the peaks, the position of the intermediate peak, and the calculated crystallinity volume fractions are shown in figure 4, the volume fractions of the intermediate phase and the crystalline phase are shown together with the calculated position of intermediate peak. The volume fraction of crystalline phase and intermediate phase are estimated by comparing the corresponding peak intensity to the sum intensity of three peaks.

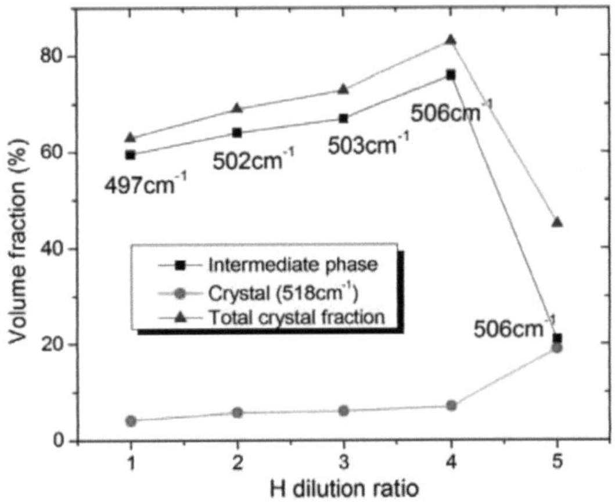

Figure 3. Crytallinity volume fraction of samples. The value noted at each point of the black line is the position of the corresponding intermediate peak.

The results demonstrate a transition from amorphous phase to nanocrystalline phase when H dilution is increased through the shifting of position of the intermediate peak and the intensity of crystalline peaks. When **R** $\leq$ 4, the amorphous and the paracrystalline phase dominate film structures but the position of the intermediate peak is shifted gradually to the high energy side. When **R** > 4, the phase transition take places and the volume fraction of the crystalline phase grows strongly, leading to the decrease of the intermediate phase volume fraction. The ambiguous thing here is that in this high dilution condition (**R** = 5), the total crystallinity volume fraction also decreases from 83 to 45%.

3.3. UV-VIS transmittances

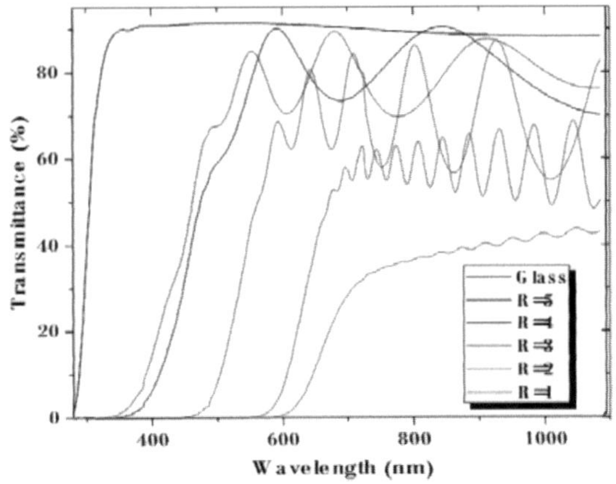

Figure 4. The UV-VIS transmittances of the samples with different H dilution.

The UV-VIS transmissions of the samples on glass are shown in figure 5. Similar as the case of IR transmission, the UV-VIS spectra are a superposition of the inference fringes and the true transmittances of the material. It is therefore difficult to interpret the absorption coefficients from these spectra without a good optical model which accounts for multiple interferences inside the films. In figure 5, however, the effects if H dilution on the band gap of the material is clear: a film of higher H dilution has absorption edges shifted to the high energy side, which indicates larger band gaps. We therefore perform a band gap estimation using the data in strong absorption region of the spectra (table 2). The band gap increase can be explained by two reasons: *(i)* when the H dilution is increased, the film structure transforms gradually to the crystalline phase, which has a larger band gap than the amorphous one, *(ii)* in high H dilution, the oxygen contamination is important, resulting in a large amount of SiO_x phase. Therefore the band gap increased.

Table 2. Estimated band gap values of Si:H samples.

Samples	Band gap (eV)
1	1.8
2	2.1
3	2.5
4	2.8
5	3.0

4. Conclusion

In this report, we have investigated the effects of hydrogen dilution on the structures of Si:H thin films. The results show an increasing in crystallinity and a transition region of H content in samples when H dilution rise, which are agreed with other references. However, the larger oxygen contamination together with higher dilution demonstrates that oxygen can attack the deposition process easier in high dilution regimes, leading to a mixture structure between Si:H and SiO_x in the studied films.

References

[1] Brodsky M H, Cardona M, Cuomo J J 1977 *Phys. Rev. B* **16** (8) 3556
[2] Mahan A H, Yang J, Cuba S, Wiliamson D L 2000 *Phys. Rev. B* **61** (3) 1667
[3] Kroll U, Meier J, Shah A, Mikhailov S, Weber J 1996 *J. Appl. Phys.* **80** (9) 4971
[4] Tolstoy V P, Chenyshova I V, Skryshevsky V A 2003 *Handbook of infrared spectroscopy of ultra thin films* section 5.5
[5] Lucovsky G, Pollard W B 1984 *The Physics of Hydrogenated Silicon II* (Springer-Velag P)
[6] Debajyoti Das 1997 *Bull. Mater. Sci.* **20** (1) 9
[7] Thorsten Dylla 2004 *Electron Spin Resonance and Transient Photocurrent Measurements on Microcrystalline Silicon* Doctoral thesis (Free University of Berlin)
[8] Voyles P M, Zotov N, Nakhmansonc S M, Drabold D A, Gibson J M, Treacy M M J, Keblinski P 2001 *J. Appl. Phys.* **90** (9) 4437
[9] Trung T Q, Jiri S et al. 2008 *Proc. 1st IWNA 2008 (Vietnam)*
[10] Mishra P, Jain K P 2002 *Mater. Sci. Eng. B* **95** 202
[11] Gupta N D, Chaudhuri P 2001 *J. Non. Cryst. Sol.* **289** 168

Acknowledgments

This work is supported in part by VNU-HCM City, the Protocol in Scientific cooperation between Vietnam and South Korea (2009 – 2010).

Preparation and characterization of magnetic nanoparticles with chitosan coating

Doan Thi Kim Dung[1], Tran Hoang Hai[1], Le Hong Phuc[1], Bui Duc Long[1], Le Khanh Vinh[1] and Phan Nha Truc[2]

[1] Ho Chi Minh City Institute of Physics, Vietnam Academy of Science and Technology, Vietnam
[2] Camau Community College, Vietnam

E-mail: dtkdung@vast-hcm.ac.vn

Abstract. Magnetic chitosan nanoparticles were prepared by the suspension cross-linking technique for use in the application of magnetic carrier technology. The Fe_3O_4 magnetic nanoparticles were synthesized by co-precipitation of $FeCl_2$ and $FeCl_3$ solution in base medium for using in the preparation of the magnetic chitosan. The morphological and magnetic properties of the magnetic nanoparticles were characterized by different techniques (TEM, XRD, VSM, FTIR, etc.). The magnetic properties of chitosan – magnetic nanoparticles were analyzed by VSM, and M_S around 15 emu/g.

Keyword: Magnetic nanoparticles, chitosan.

1. Introduction

The recent developments of nanotechnology in synthesizing biocompatible and functionalized magnetic nanoparticles have numerous novel applications in biomedicine as well as diagnosis [1-7]. Especially, Fe_3O_4 ferrites magnetic nanoparticles have been rising as a significant useful material due to their specific properties such as supperparamagnetic, non toxic and small size, etc. Magnetic nanoparticles are used to coat several surfactants to anti – aggregation which was caused by magnetic dipole – dipole attractions between particles [1, 2, 7]. The studies of utilizing various biocompatible and biodegradable polymers which can functionalize and modify the surface of the Fe_3O_4 nanoparticles have been broadly developed [2, 6, 7]. A magnetic nanoparticle is usually composed of the magnetic cores and a polymeric shell having favorable functional groups and features for various applications [8, 9].

Magnetic nanoparticles technology is a significant method for solving decomposition, inactivation or deformation in biochemical processes such as separation, purification, etc. In this technology, magnetic particles are used as the support materials and they can be easily separated from the reaction medium and stabilized in a fluidized reactor by applying a magnetic field [1, 2, 7, 9]. The most popular applications of magnetic carrier technology are bioaffinity chromatography, wastewater treatment, immobilization of enzymes or other biomoleculars and the preparation of immunological assays [9].

Chitosan, a partially acetylated glucosamine, poly(1 $\rightarrow$ 4)-2-amino-2-deoxy-d-glucan, biopolymer which exist in cell walls of some fungi such as the Mucorales, is a polyaminosaccharide with many

© 2009 IOP Publishing Ltd

significant biological (biodegradable, biocompatible and bioactive) and chemical properties (polycationic, hydrogel, reactive groups such as OH and NH_2) [1, 2, 7, 9]. Therefore, it is used in many different types of applications (i.e. wastewater treatment, chromatographic support, enzyme immobilization, drug-delivery systems). Chitosan can be produced by alkaline deacetylation of chitin, and chitin is the second most abundant natural polymer after cellulose, and therefore is a very cheap biopolymer [9].

In this study, we prepared magnetic Fe_3O_4-chitosan nanoparticles by coprecipitation Fe(II) and Fe(III) in the presence of ammonium. And then, the suspension cross-linking technique was used for the preparation of magnetic chitosan nanoparticles and glutaraldehyde was used as a cross-linker. The size and the size distribution, magnetic properties and the functional groups of the magnetic chitosan microspheres were evaluated using selected parameters (such as stirring rate of the suspension medium, molecular weight of the chitosan and the Fe_3O_4/chitosan ratio).

The structure, size and the size distribution, magnetic properties and functional groups of magnetic nanoparticles were evaluated using of a Vibrating Sample Magnetometer (VSM), Transmission Electron Microscopy (TEM), X-ray Diffraction (XRD), Fourier Transforms Infrared (FTIR) spectroscopy.

The magnetic- chitosan nanoparticles in this study were applied in the immobilizing enzyme in biotechnology or removal of heavy metal ions in environment at technology.

2. Experimental

2.1. Materials

Chitosan polymer 150 kDa was obtained from Fluka (Switzerland). Aqueous acetic acid (Aldrich, USA) solution was used as a solvent for the chitosan polymers and glutaraldehyde (Merck, Germany) was used as the cross-linker. The suspension medium was composed of a mixture of paraffin (Merck, Germany), and Span-80 (Merck, Germany) was added as an emulsifier. All chemicals were of analytical grade and no further purification was required.

2.2. Preparation of mangnetic nanoparticles

Fe_3O_4 nanoparticles were prepared by coprecipitaion method with a ferrous complex in presence of NH_4OH. Firstly, $FeCl_2.4H_2O$ and $FeCl_3.6H_2O$ [Fe^{2+}: Fe^{3+}= 1:2] were dissolved in about 50 ml re-dionized water, and stirring this solution under strong ultrasonic agitation while heating solution to 70°C. Next, this iron solution source was added drop-wise into NH_4OH under strong ultrasonic agitation for 30 minutes, and bubbling N_2 gas.

The chemical reaction of Fe_3O_4 precipitation is expected as follows:

$$FeCl_2.4H_2O + 2\ FeCl_3.6H_2O + 8NaOH \rightarrow Fe_3O_4 + 8NaCl + 20H_2O \qquad (1)$$

Black Fe_3O_4 particles were decanted by permanent magnet and cleaned by re-dionized water several times.

2.3. Preparation of magnetic chitosan nanoparticles

The suspension cross-linking technique was used for the preparation of magnetic chitosan nanoparticles. In this specific procedure, a 5% chitosan solution was prepared using a 2% aqueous acetic acid solution containing 0.2 g Fe_3O_4 dry magnetic nanoparticles. And then, this solution was poured, drop-wise, into the dispersion medium, which was composed of 30 ml paraffin and 0.5 ml span-80. During this process, the dispersion medium was stirred with a strong ultrasonic agitation at room temperature. Next, an additional 3 ml 25% glutaraldehyde solution was added to the dispersion medium and then solution was stirred for further 5 h. At the end of this period, the chitosan-magnetite nanocomposite particles were recovered from the reaction mixture by using a permanent magnet; the products were washed with ethanol and dried in a vacuum oven at 500°C for 2 days.

3. Results and discussions

3.1. 3.1. Synthesis of Fe_3O_4 nanoparticles with chitosan coating
The preparation of magnetic chitosan nanoparticles proceeds in two steps by cross-linking methods as follows:
- The synthesis of Fe_3O_4 particles by coprecipitation of equimolar Fe(III) and Fe(II) in the presence of NH_4OH. Schematic representation of the preparation of Fe_3O_4 nanoparticles was shown in equation (1).
- Chitosan and Fe_3O_4 were mixed in appropriate proportion to form the chitosan-magnetite nanocomposite particles with amine group by reverse-phase suspension cross-linking method.

3.2. Particle size and structure of Fe_3O_4- chitosan nanoparticles
The morphology of naked and Fe_3O_4 with chitosan coating were shown in figure 1. The magnetic particles showed a small size in nanoscale before and after being coated by chitosan. The image of Fe_3O_4 nanoparticles, was shown in figure 1, was proved that Fe_3O_4 particles were monodisperse and had a mean diameter of 15-17 nm. The TEM image of Fe_3O_4–chitosan nanoparticles in figure 2b was showed that the structure of CTS-coated magnetite nanoparticles was looser, causing to the increasing in size, the average diameter of such a structure was 23-25 nm.

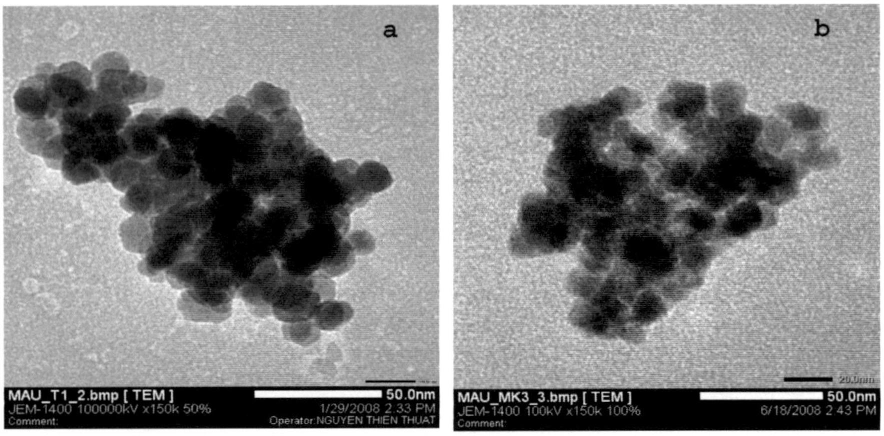

Figure 1: TEM imaged of pure Fe_3O_4 nanoparticles (a) and Fe_3O_4- chitosan nanoparticles (b).

This also represented that the coating process did not significantly effect the agglomeration and the change in size of the particles. However, the aggregative phenomenon still present a little in the Fe_3O_4-chitosan nanoparticles.

3.3. Phase structure of Fe_3O_4 magnetic nanoparticles
The XRD patterns for the naked and Fe_3O_4 nanoparticles coated with chitosan were showed in figure 3. Six characteristic peaks for Fe_3O_4 marked by their indices (2 2 0), (3 1 1), (4 0 0), (4 2 2), (5 1 1) and (4 4 0) were observed for both samples reveal that the resultant nanoparticles were pure Fe_3O_4 with inverse-spinel structure.

That is the reason for the unchangeable pace of Fe_3O_4 during coating process. In the pattern of XRD diffraction, the broad nature of the diffraction bands indicated that Fe_3O_4 have small particle sizes. The particle sizes can also be quantitatively evaluated from the XRD data using the Debye–Scherrer equation which gives a relationship between peak broadening in XRD and particle size:

$$d = (k\lambda/\beta\cos\theta) \quad , \tag{2}$$

where d is the particles size, k is the Debye–Scherrer constant (0.89), λ is the X-ray wavelength (0.15406 nm) and β is the full width at half maximum, θ is the Bragg angle. According to the Debye-Scherrer equation (2), the particles size of the naked and the chitosan coated Fe_3O_4 nanoparticles is 15 and 23 nm, respectively.

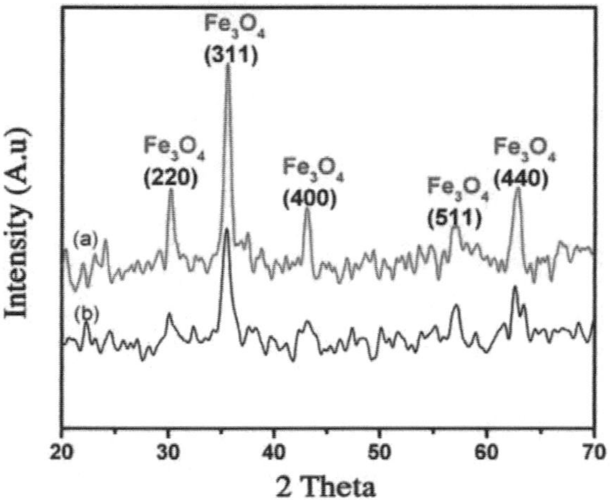

Figure 2: X-ray diffraction of pure Fe_3O_4 nanoparticles (a) and Fe_3O_4- chitosan nanoparticles (b).

3.4. FTIR spectra analysis

Figure 3 shows FTIR spectra of naked Fe_3O_4, chitosan (CTS), and Fe_3O_4- chitosan nanoparticles (CF1) and we can affirm the binding between the other groups. The peak around 3440 cm^{-1} observed in curve CTS and magnetic CTF1 indicates to the OH group. The peak at 543 cm^{-1} relates to Fe–O bond. In the IR spectrum of CTS, the characteristic absorption bands appeared at 1570 cm^{-1} which can be assigned to N–H bending vibration, peaks 1309 cm^{-1} appeared to C–O stretching of primary alcoholic group in chitosan.

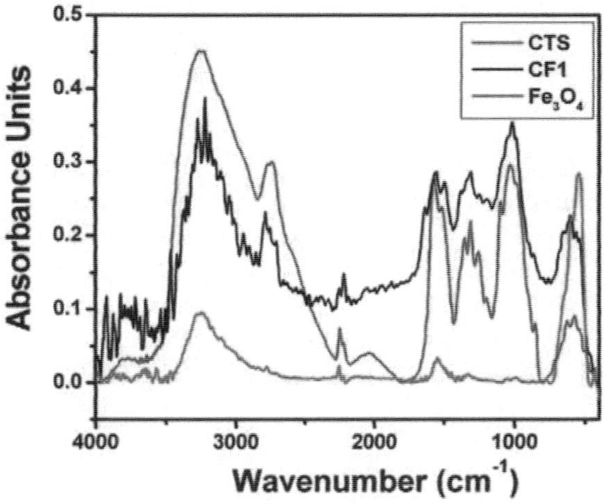

Figure 3: FTIR spectra of pure Fe_3O_4, chitosan (CTS), and Fe_3O_4-chitosan nanoparticles (CF1).

In the spectrum of Fe_3O_4-chitosan nanoparticles (figure 3), compared with the spectrum of CTS, the 1570 cm^{-1} peak of N–H bending vibration shifted to 1558 cm^{-1}, and a new sharp peak 1627 cm^{-1} appears, it indicated that chitosan react with glutaraldehyde to form the Schiff base, and a newsharp peak 577 cm^{-1} relates to Fe–O bond appears. Results indicated that CTS was coated to the magnetic Fe_3O_4 nanoparticles successfully. Because the surface of iron oxide with negative charges has an affinity toward CTS, protonated CTS could coat the magnetite nanoparticles by the electrostatic interaction and chemical reaction through glutaraldehyde cross-linking.

3.5. Magnetic properties of magnetic nanoparticles

Magnetic materials having around a dozen nanometers in size are used to show a superparamagnetic behavior in which coercivity and remanence has zero value. The hysteresis loop of magnetic Fe_3O_4 nanoparticles naked and coated with chitosan at room temperature is showed in figure 4. As could be seen no remainance and coercivity is observed in the magnetic loop. This phenomenon proved that magnetic nanoparticles made from this study are superparamagnetic. Also in figure 4, saturated magnetization M_S of pure magnetic nanoparticles is 70 emu/g while of magnetic chitosan coating is 15 emu/g. This represented that magnetic properties of particles after coating have a content of 25% by comparing the value of naked particles.

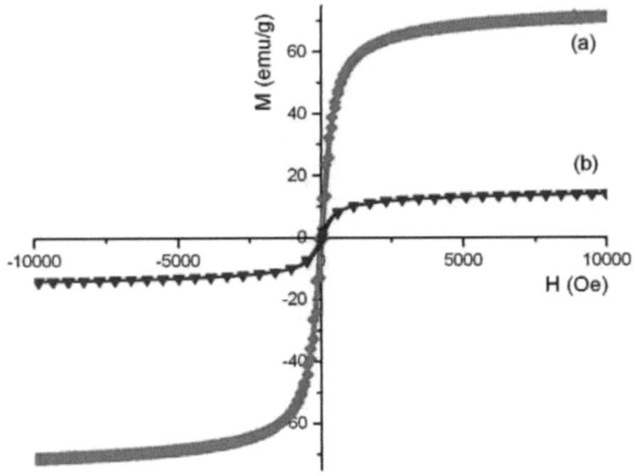

Figure 4: VSM of pure Fe_3O_4, chitosan (CTS), and Fe_3O_4-chitosan nanoparticles (CF1).

The remanence of span-80 on the surface of magnetic nanoparticle could be due to the loss of magnetic properties of magnetic chitosan nanoparticles because it caused a magnetically dead layer. Because any crystalline disorder existed in the surface layer could also cause a significant decrease of saturated magnetization. Ferrite magnetic nanoparticle with special properties- superparamagnetic is an especially important material for applying to magnetic targeting carriers.

4. Conclusions

The magnetic Fe_3O_4-chitosan nanoparticles were synthesized by binding chitosan on the surface of Fe_3O_4 nanoparticles. The results from TEM and XRD proved that the Fe_3O_4-chitosan nanoparticles were inverse- spinel structure, monodisperse and quite sphere in shape with a mean diameter of 23 nm. The coating process had no changing effect in the spinel structure of Fe_3O_4. The saturated magnetization of magnetic chitosan nanoparticles could reach 15 $emu.g^{-1}$ and the nanoparticles showed the characteristics of superparamagnetism. The nanoparticles have a potential for use in assisted drug delivery systems, cell/enzyme immobilization and many other industrial processes.

References

[1] Li G -y *et al.* 2008 *J. Alloys Compd* **466** 451
[2] Denkbas E B *et al.* 2002 *Reac. & Func. Poly.* **50** 225
[3] Juang R S *et al.* 2001 *Biores. Tech.* **80** 187
[4] Juang R *et al.* 2002 *Adva. Envi. Re* **6** 171
[5] Jiang D S e*t al.* 2005 *Bio Eng* J. **25** 15
[6] Feride Cengelli *et al.* 2006 *JPET Fast Forward.* JPET#101915
[7] Denkbas E B *et al.* 2002 *Reac. & Func. Poly.* **50** 225
[8] Liu J Chen-Guang 2005 *Agric. Food Chem.* **53** 1729
[9] Korecka L *et al.* 2005 *JMMM* **293** 349

Journal of Physics: Conference Series **187** (2009) 012037 doi:10.1088/1742-6596/187/1/012037

Spin-dependent transport in double ferromagnetic-gate graphene structures

Nguyen Viet Hung[1,2], A. Bournel[1], P. Dollfus[1] and Nguyen Van Lien[2]

[1]Institut d'Electronique Fondamentale, UMR 8622, CNRS, Univ. Paris Sud, 91405 Orsay, France

[2]Theoretical Department, Institute of Physics, VAST, P.O. Box 429 Bo Ho, Hanoi, Vietnam

E-mail: viet-hung.nguyen@u-psud.fr

Abstract. Based on the non-equilibrium Green's function method recently developed to solve the quantum relativistic-like Dirac's equation, spin dependent transport in double ferromagnetic-gate graphene structures has been investigated. The study is focussed on the possibility of controlling the spin polarization and the tunnelling magneto-resistance by the gate voltages. Different from the case of ordinary semiconductors, obtained results show that the transport depends weakly on the well width, but strongly on the confined hole states in the gated regions.

Keyword: Spin polarized transport, Klein tunnelling, Green's functions method, graphene.

1. Introduction

Graphene is a single layer of carbon atoms filling in a honeycomb lattice with the lattice constant of ~0.14 nm. This novel material was discovered in 2004 [1] and has been attracting a lot of attention, from both experimental and theoretical viewpoints, in different fields of research, e.g., condensed matter physics [2], materials science [3] and device applications [4]. This intensive interest is due to the unique electronic properties of graphene such as the gapless and the conical-shape of the electronic structure around the Dirac points (or the K-points). Accordingly, the low energy excited states behave as those of massless relativistic-like particles with the typical charge-conjugation symmetry between the electron and the hole states. These lead to observations of specific and anomalous transport phenomena in this single carbon-atomic layer, e.g. the finite value of the conductance at zero-concentration [3], the spin valve effect [5], the unusual half-integer quantum Hall effect [6], etc. Such phenomena can be successfully described using the massless fermion Dirac's model [7].

Because of high mobility and gate-controllable carrier conduction, graphene appears to be very promising for designing high performance transistors. Additionally, due to very weak spin orbit interaction [8, 9] which leads to a long spin flip length ($\sim 1\,\mu m$) [5], it offers good potential for spintronics applications [10]. Some possibilities of inducing ferromagnetism in graphene have also been discussed. For instance, the ferromagnetism can be intrinsically induced in graphene by defects [11, 12], impurities [13] or by using an external electric field [14]. Haugen *et al.* [15] recently suggested that the ferromagnetic correlations can be externally induced in graphene by the proximity effect of a magnetic gated insulator deposited on the graphene sheet. Moreover, possibilities of

© 2009 IOP Publishing Ltd

controlling the spin current in single ferromagnetic-gate graphene structures have been discussed [16, 17]. However, such studies are usually based on the scattering matrix approach which is intuitive but difficult to implement in a general framework, for instance including some important effects as scattering by impurities and defects. Recently, we have developed a calculation method based on the nonequilibrium Green's functions (NEGF) to solve the massless fermion Dirac's equation [18]. The approach has been efficiently applied to study the electronic transport and the spin polarization effect in single gate graphene structures. In this work, we use that method to study the spin dependent transport in double ferromagnetic-gate graphene structures, as schematized in figure 1. Our study is focussed on the role of confined charge states in both the well and the gated regions and on the analysis of the two typical quantities, the spin polarization (P) and the tunnelling magnetoresistance (TMR).

2. Model and formulation

The considered double ferromagnetic-gate structure is modelled by the Dirac-like Hamiltonian

$$H = \mathrm{v}_\mathrm{F}\left(\sigma_x p_x + \sigma_y p_y\right) + Q \ , \tag{1}$$

where $\mathrm{v}_\mathrm{F} \approx 10^6 \mathrm{ms}^{-1}$ is the Fermi velocity; σ_x and σ_y are the Pauli matrices; p_x and p_y are the momentum operators; and

$$\begin{aligned} Q(x) = &[U_1 - \eta\Delta]\Theta(x)\Theta(L_1 - x) \\ &+ [U_2 \pm \eta\Delta]\Theta(x - L_1 - W)\Theta(L_1 + W + L_2 - x) \ , \end{aligned} \tag{2}$$

here Θ is the Heaviside step function used to define the two gated regions. The potential barrier $U_{1(2)} > 0$ is tuned by the gate voltage $V_{G1(G2)}$ all along the gate length $L_{1(2)}$. The two gated regions separating the source/drain leads from the well are covered by a ferromagnetic insulator, which will induce the exchange splitting Δ in graphene. $\eta = \pm 1$ stands for the spin orientation of conduction electrons in the structure. The magnetic moment of the gated region 1 is fixed while that of the gated region 2 may be reversed.

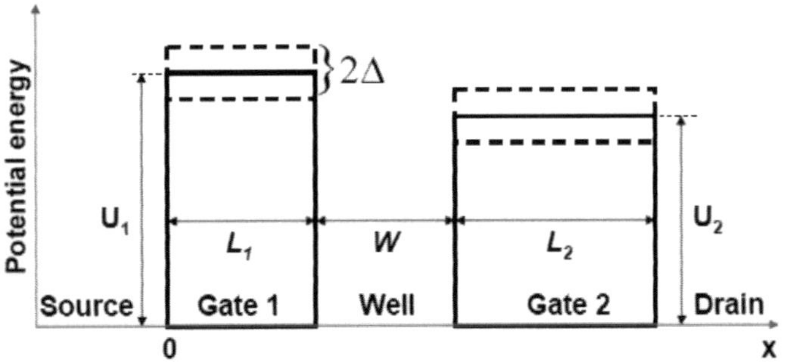

Figure 1. Sketch of the potential along the considered double ferromagnetic-gate graphene structure.

Within the model described by equation (1), the transport is considered to be ballistic according to the NEGF formalism. The transport quantities can be determined through the transmission coefficient which is expressed in terms of the retarded and advanced Green's functions $G^{r,a}$ as

$$T(E) = \sum\nolimits_{n,m}^{2} \Gamma_{1,n}^{L} G_{n,2N-1}^{r} \Gamma_{1,m}^{R} G_{2N-2+m,1}^{a} \ , \tag{3}$$

where $\Gamma^{L(R)} = i\left(\Sigma_{L(R)}^{r} - \Sigma_{L(R)}^{a}\right)$ is defined in terms of the retarded and advanced self-energies $\Sigma^{r,a}$ which describe the coupling between the graphene sheet and the source (drain) electrode. The conductance is then calculated using the Landauer formula [19]. Similar to Ref. [18], hereafter, the temperature is chosen to be zero for simplicity and the energy unit is $E_0 = \hbar v_F / 2a$, where the mesh-spacing a is 0.2nm.

On the basis of the above formalism, the spin polarization P and the tunnelling magnetoresistance TMR are determined by

$$\begin{aligned} P &= (G_\uparrow - G_\downarrow)/(G_\uparrow + G_\downarrow) \\ TMR &= (G_P - G_{AP})/(G_P + G_{AP}) \end{aligned} \quad , \tag{4}$$

where $G_{\uparrow(\downarrow)}$ corresponds to the conductance for up (down) spin channel and $G_{P(AP)}$ is the conductance for parallel (anti-parallel) alignment configuration of the magnetic moments.

3. Numerical results and discussions

As discussed in Ref. [18], due to the confined hole states inside the barrier (gated) regions, the conductance of each spin channel oscillates versus the barrier height. The exchange splitting, which shifts the conductance of the spin channel relatively to each other, can lead to the oscillation of the spin polarization P and of the TMR as well. We display in figure 2 the dependence of P and TMR on the barrier height U_1 while the barrier height U_2 is fixed in a symmetric structure with $L_1 = L_2 = 20$ nm. In this case, the spin polarization behaves similarly to that in the case of the single barrier (single gate) structures, i.e., it is an oscillatory function of U_1 or/and the exchange splitting Δ. But for appropriate values of Δ, the spin polarization P can be a positive/negative-definite function of the barrier height [e.g., $\Delta = 0.015$ or/and $\Delta = 0.045$]. The period of the oscillation is then estimated to be $U_P = 2\pi a / L_1$. For instance, for $L_1 = 20$nm the period is $U_P \approx 0.062$. When tuning Δ, this period does not change but the amplitude of the oscillations does.

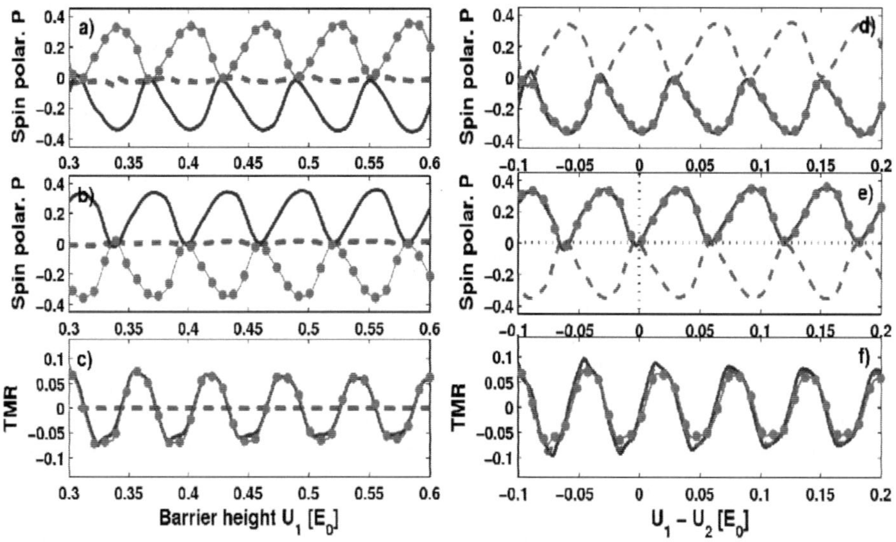

Figure 2. (Colour online) Oscillations of the spin polarization and the correlated TMR as a function of the barrier height U_1 in a structure: $L_1 = W = L_2 = 20$ nm. The Fermi energy E_F in the source-drain leads and the well is 0.1. The figures (a) and (d) [(b) and (c)] correspond to the parallel (anti-parallel) alignment of the magnetic moments. In the left column: $\Delta = 0.015$ (solid), 0.03 (dashed) and 0.045 (solid-circle) with $U_2 = 0.4$. In the right column: $U_2 = 0.34$ (solid), 0.37 (dashed) and 0.4 (solid-circle) with $\Delta = 0.015$.

For the *TMR*, the period of the oscillation is about $\pi a/L_1$ which is one half of that of the spin polarization (see the left column of figure 2). This agrees well with the estimation in Ref. [15] where $TMR \propto P^2$ in symmetric structures. Moreover, the maximum (minimum) of the spin polarization P for the parallel configuration corresponds to the minimum (maximum) of that for the anti-parallel one. Because the two barriers are equivalent, P and *TMR* also oscillate when changing U_2 with the period $2\pi a/L_2$ (see the right column of figure 2). However, the spin polarization is completely suppressed for the anti-parallel configuration if $L_1 = L_2$ and $U_1 = U_2$ when the pictures of the confined hole states are the same in the two barrier regions.

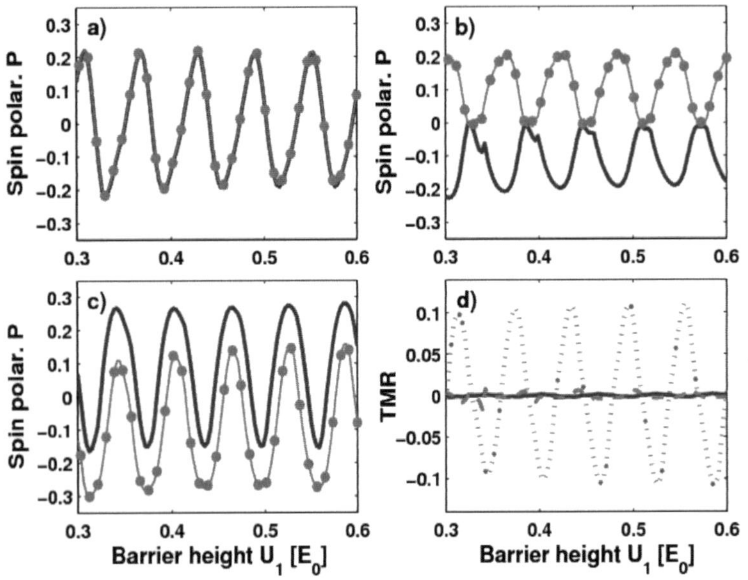

Figure 3. (Colour online) Oscillations of the spin polarization as a function of the barrier height U_1 in an asymmetric structure: $L_1 = W = 20$ nm, $L_2 = 30$ nm, $U_2 = 0.4$, $E_F = 0.1$ and for different values of Δ: 0.02 (a), 0.03 (b) and 0.045 (c). The solid (solid-circle) curve corresponds to the parallel (anti-parallel) alignment. The correlated TMR (d) is also plotted for: $\Delta = 0.02$ (solid), 0.03 (dotted) and 0.045 (dashed).

To investigate the effect of asymmetry of the gate lengths, we present in figure 3 the obtained results for a structure with $L_1 = 20$ nm and $L_2 = 30$ nm. It is interesting to note that by tuning Δ one can find some values which satisfy the condition $\Delta \approx n\pi a/L_1$ or $m\pi a/L_2$ (e.g. $\Delta = 0.02$ or 0.03) for which the hole states in the gated region 1 or the gated region 2 are identical for the two spin channels. Hence the structure behaves similarly to the case of a ferromagnetic barrier coupled with a normal (nonmagnetic) barrier. This leads to identical spin polarization for the parallel and the anti-parallel configurations and therefore a complete suppression of the *TMR*. If the above condition is not satisfied, e.g., $\Delta = 0.045$, the spin polarizations for the parallel and for anti-parallel configurations are different and the *TMR* is an oscillatory function of the barrier height. Moreover, in this case, the spin polarization for the anti-parallel configurations does not vanish when $U_1 = U_2$.

Since the conductance does not only oscillate versus the barrier height, but also as a function of the exchange splitting, one can expect a modulation of these oscillations when changing another parameter such as the gate lengths [17, 18]. Indeed, figures 4 and 5 show the beating behaviour of the spin polarization and the *TMR* versus the gate length L_1 for different values of Δ and U_2. The period of this modulation can be estimated to be $L_P \approx 2\pi a/\Delta$ [for example, with $L_P \approx 61.5$ nm and $L_P \approx 30.6$ nm for $\Delta = 0.02$ and $\Delta = 0.04$, respectively, see figure 4]. For $\Delta = 0.03$ ($\approx \pi a/L_2$), as discussed above, the modulation effect of the spin polarization is also similar to that of the single ferromagnetic gate structure and the *TMR* is suppressed. In comparison with the results presented in figure 2, we additionally remark that when tuning U_2 (figure 5), the mean value and the amplitude of the spin polarization and TMR oscillate with the period $2\pi a/L_2$ but the phase of modulated oscillations seems to be unchanged.

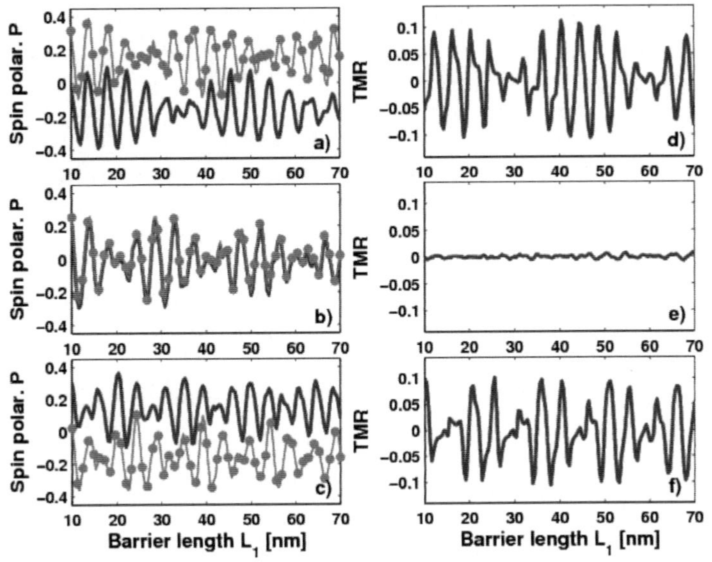

Figure 4. (Colour online) Modulation of the spin polarization P and the correlated TMR as a function of the barrier width L_1 for $U_1 = U_2 = 0.4$, $W = L_2 = 20$ nm, $E_F = 0.1$ and different values of Δ: 0.02 (a,d), 0.03 (b,e) and 0.04 (c,d). In the plots of P, the solid (solid-circle) curves correspond to the parallel (anti-parallel) alignment.

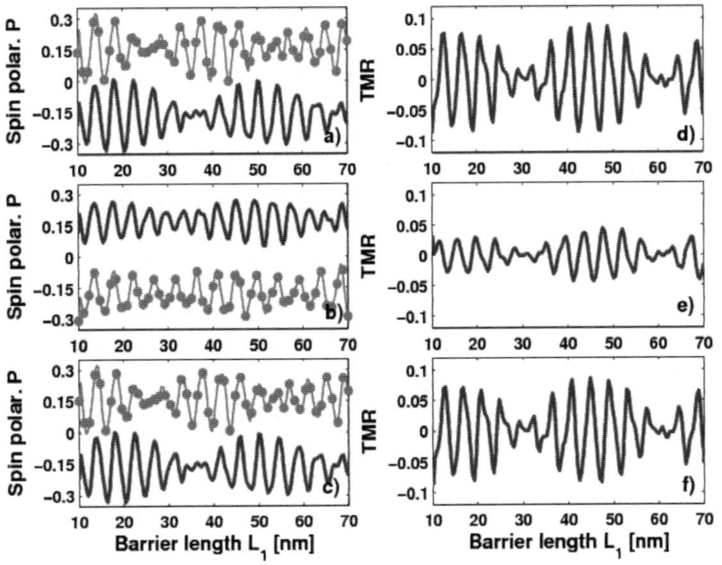

Figure 5. (Colour online) Modulation of the spin polarization P and the correlated TMR as a function of the gate length L_1 for $U_1 = 0.4$, $W = L_2 = 20$ nm, $\Delta = 0.02$, $E_F = 0.1$ and different values of U_2: 0.34 (a,d), 0.37 (b,e) and 0.4 (c,f). In the plots of P, the solid (solid-circle) curves correspond to the parallel (anti-parallel) alignment.

Another point is the possible role of the well width W (separating the two gates). It is well-known that in double barrier structures based on normal semiconductors, this layer is very important since states of charge are quantized and govern the transport properties of the system. However, for the structures under study here, the numerical results show that the conductance and also the spin current depend weakly on the width W. In contrast, the alignment of the confined hole states inside the two barrier regions is important, which is nothing, but the resonant transport condition.

4. Conclusion

We have studied the spin current in double ferromagnetic-gate graphene structures by using *NEGF* formalism. We find that, due to the exchange splitting in the gated regions, the spin polarization and *TMR* have an oscillatory behaviour with respect to the barrier heights, which can be tuned by the gate voltages, and exhibit modulating behaviour related to the gate lengths. The study also shows that not the confined charge states in the well, but the alignment of the hole states inside the gated regions, responds for the resonant transport condition which leads to the oscillation of the spin current. The simulation technique developed and these results may provide an additional understanding of the spin dependent transport in graphene-based spintronic devices.

References

[1] Novoselov K S, Geim A K, Morozov S V, Jiang D, Zhang Y, Dubonos S V, Grigorieva I V, Firsov A A 2004 *Science* **306** 666
[2] Geim A K and Novoselov K S 2007 *Nature Materials* **6** 183
[3] Li D and Kaner R B 2008 *Science* **320** 1170
[4] Lemme M C, Echtermeyer T J, Baus M and Kurz H 2007 *IEEE Electron Dev. Lett.* **28** 282
[5] Tombros N, Jozsa C, Popinciuc M, Jonkman H T, van Wees B J 2007 *Nature* **448** 571
[6] Zhang Y, Tan Y W, Stormer H L and Kim P 2005 *Nature* **438** 201
[7] Novoselov K S, Geim A K, Morozov S V, Jiang D, Katsnelson M I, Grigorieva I V, Dubonos S V and Firsov A A 2005 *Nature* **438** 197
[8] Kane C L and Mele E J 2005 *Phys. Rev. Lett.* **95** 226801
[9] Huertas-Hernando D, Guinea F and Brataas A 2006 *Phys. Rev. B* **74** 155426
[10] Semenov Y G, Kim K W and Zavada J M 2007 *Appl. Phys. Lett.* **91** 153105
[11] Yazyev O V and Helm L 2007 *Phys. Rev. B* **75** 125408
[12] Wehling T O, Novoselov K S, Morozov S V, Vdovin E E, Katsnelson M I, Geim A K and Lichtenstein A I 2008 *Nano Lett.* **8** 173
[13] Peres N M R, Guinea F and Castro Neto A H 2005 *Phys. Rev. B* **72** 174406
[14] Son Y -W, Cohen M L and Louie S G 2006 *Nature* **444** 347
[15] Haugen H, Huertas-Hernando D and Brataas A 2008 *Phys. Rev. B* **77** 115406
[16] Yokoyama T 2008 *Phys. Rev. B* **77** 73413
[17] Soodchomshom B, Tang I M and Hoonsawat R 2008 *Phys. Lett. A* **372** 5054
[18] Nam D V, Hung N V, Dollfus P and Bournel A 2008 *J. Appl. Phys.* **104** 63708
[19] Milton Pereira J Jr, Vasilopoulos P and Peeters F M 2007 *Appl. Phys. Lett.* **90** 132122

Acknowledgments

One of the authors (Nguyen Viet Hung) thanks Do V. Nam for sharing the computer code. This work was partially supported by the European Community through Network of Excellence NANOSIL.

CIP spin torque effect in the spin valve pinned with an oxide antiferromagnetic layer

Nguyen Chi Thuan[1], Nguyen Van Dai[1], Dao Nguyen Hoai Nam[1,2], Nguyen Xuan Phuc[1] and Le Van Hong[1]

[1] Institute of Materials Science, Vietnam Academy of Science and Technology, 18 Hoang Quoc Viet Road, Cau Giay Distr., Hanoi, Vietnam
[2] Department of Materials Science and Engineering, University of Virginia, Virginia 22903, USA

E-mail: thuannc@ims.vast.ac.vn

Abstract. Spin valve $Ni_{0.85}Co_{0.15}O/Co_{85}Fe_{15}/Cu/Co_{85}Fe_{15}$ was manufactured by a RF magnetron sputtering system. M(H) and R(H) characteristics of the spin valve were measured in CIP configuration at room temperature has a magnetoresistance ratio of about 8% and a high exchange bias at room temperature. The current density and angle between the applied magnetic field and injection current were changed in an aim to observe their effects on MR and exchange bias of the spin valve. The current density and angle strongly affect MR and exchange bias. Both MR and exchange bias clearly decrease in dependence of the current density and direction of the magnetic field. It is supposed to be related with a current-induced spin torque in device.

Keywords: GMR, exchange bias, spin torque, effect.

1. Introduction

The NiFe and CoFe based spin-valves pinned with an antiferromagnetic layer are of interest in the development of the high density magnetoresistance (MR) heads and sensors. Various antiferromagnetic film layers such as FeMn, NiMn, and oxidic layers such as NiO have been used as a biasing layer, which is exchange coupled with the NiFe and CoFe sensing films [1, 2]. The spin valve effect is associated with relative orientation of magnetization vectors in adjacent magnetic layers, which are very sensitive to crystallographic orientations and interface roughness of layers [3]. A theoretical calculation by Inoue *et al.* [4] supposed that the spin-dependent scattering factor of CoFe is larger than that of NiFe. Therefore, it is expected that the MR ratio could be improved in the spin valve $Ni_{1-x}Co_xO$/FeCo/Cu/FeCo. Shirota *et al.* [5] obtained MR ratio of 7.9% for NiO/2F/2.6Cu/1.6F/5Si-N spin-valve device, where F is $Co_{90}Fe_{10}$. Egelhoff *et al.* [6] has improved the MR ratio up to over 20% at room temperature for this kind of spin-valve device with two pinning layers in the symmetrical configuration. Using a ferromagnetic $Fe_{15}Co_{85}$ buffer layer with an appropriated thickness we improved the MR ratio [7], which was explained as a result of improving the plane magnetic anisotropy of the antiferromagnetic $Ni_{1-x}Co_xO$ layer.

Generally, the ferromagnetic (FM) material exhibits symmetric hysteresis loop centered at zero magnetic field (H = 0). As the first, Meiklejohn and Bean observed a displacement of the hysteresis loop of slightly oxidized Co particles from H = 0. It was explained as a result of the exchange

© 2009 IOP Publishing Ltd

coupling between the surface antiferromagnetic (AFM) Co oxides and the FM Co core [8, 9]. This exchange coupling can be better revealed in the FM/AFM bilayers such as NiFe/FeMn, NiFe/NiO [10], or $Ni_{1-x}Co_xO$/FeCo bilayer presented in this paper. When the FM/AFM bilayer was cooled in a magnetic field from temperatures above Neél temperature T_N of the AFM to lower temperatures, exchange coupling has been locked in. Effectively the hysteresis loop is shifted from H = 0 by an amount called the exchanged field H_{EX}, which can reach as high as of 1 KOe. The shift of hysteresis loop is substantially accompanied by coercivity H_C which is much larger than the intrinsic value of the FM layer. Besides, it depends on anisotropy axis, in which the exchange coupling in FM/AFM layer was locked. According to theoretical calculations the exchange coupling at the FM/AFM interface depends on anisotropic axis of the constituent magnetic layers, consequently the exchange coupling reveals the angular dependence.

In this work we present the angular dependence of the exchange coupling of the spin valve $Fe_{15}Co_{85}$(1.6 nm)/$Ni_{0.85}Co_{0.15}O$(40 nm)/$Fe_{15}Co_{85}$(3 nm)/Cu(3.8 nm)/$Fe_{15}Co_{85}$(4.5 nm). In general, the spin transfer or spin torque in a spin-valve (or MTJ) structure is expected to be strong when spin valve works in the CPP (current perpendicular to plane) configuration [11]. Some recent results on spin torque in spin valve pinned by a metal antiferromagnetic layer, working in CPP configuration were reported by [12, 13]. Recently, Tang *et al.* [14] reported a new result for the case of spin valve working in CIP configuration. It showed that the spin torque effect occurs only when the sweeping magnetic field induced by the injection current is antiparallel to the exchange-bias field. Exchange bias of spin valve systematically decreases when increasing the flowing current. Their observation confirmed that the critical spin torque induced current in an antiferromagnet is smaller than the typical value for switching in a ferromagnet.

In this report, we present an observation of the current induced spin torque of $Fe_{15}Co_{85}$(1.6 nm)/$Ni_{0.85}Co_{0.15}O$(40 nm)/$Fe_{15}Co_{85}$(3 nm)/Cu(3.8 nm)/$Fe_{15}Co_{85}$(4.5 nm) valve spin working in CIP configuration. We also present a new result of study on the dependence of exchange bias upon the angle between the external applied field and the exchange field.

2. Experimental

$Fe_{15}Co_{85}$(1.6 nm)/$Ni_{0.85}Co_{0.15}O$(40 nm)/$Fe_{15}Co_{85}$(3 nm)/Cu(3.8 nm)/$Fe_{15}Co_{85}$(4.5 nm) samples were fabricated by the successive deposition of layers in a magnetron sputtering system (EDWARDS AUTO 306) equipped with 3 targets, a DC and RF source in a chamber with the base vacuum of 8×10^{-6} mBar. The metallic alloy layer $Fe_{15}Co_{85}$ and non-magnetic Cu have been deposited in Ar gas of pressure about 5×10^{-3} mBar and 3×10^{-3} mBar, respectively. $Ni_{0.85}Co_{0.15}O$ was deposited in Ar gas of pressure of 3×10^{-3} mBar, containing 20% of oxygen. A quartz vibrating thickness-meter has controlled thickness of all layers during the deposition processing. An external magnetic field of 500 Oe was applied in-plane during the deposition processing to create unidirectional magnetic exchange field. We found that inserting a CoFe seed layer with a pertinent thickness would help stabilize the antiferromagnetic layer and considerably improve the magnetoresistance signal, probably due to an improvement of the in-plane anisotropy of the magnetic layers deposited afterward. Quality of the films was examined by X-ray diffraction, energy-dispersive X-ray spectroscopy, magnetization, and conductance techniques. Resistance (R) and magnetoresistance $\{MR_H=(R_H-R_{\uparrow\uparrow})x100\%/R_{\uparrow\uparrow}\}$ of the spin valves were measured by the standard four-probe techniques in an external magnetic field aligned both in-plane and out-of-plane under various angles at room temperature by a Quantum Design PPMS.

3. Result and discussion

In this investigation two experiments were done, the first observed the angular dependence of exchange coupling in FM/AFM layer by measuring MR and exchange field H_{EX} of the spin valve in dependence on angle between the magnetic measuring field and the magnetic field that is applied in aim of creating an initial exchange anisotropy in the spin valves, the second observed the spin torque induced by sweeping current in CIP configuration. According to the aims we performed two measurement configurations as presented in figure 1, the in-plane measurement configuration (figure

1b) and the out-of-plane measurement configuration (figure 1c).

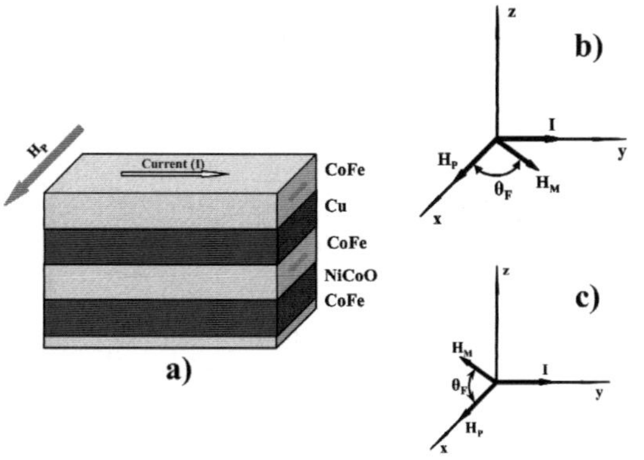

Figure 1. Spin valve structure (a), in-plane measurement configuration (b) and out-of-plane measurement configuration (c).

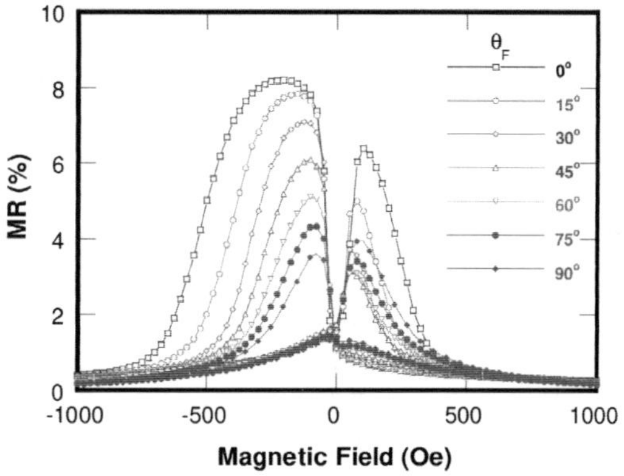

Figure 2. The MR ratio versus angle θ_F measured in the in-plane configuration.

As seen at this figure, in the case of the in-plane configuration the injection current is aligned along y-axis, perpendicular to the pinning magnetic field (H_P). Conversely, the measuring magnetic field (H_M) is applied in-plane, respecting the pinning magnetic field under an angle (θ_F), which changed in values of 0, 15, 30, 45, 60, 75, and 90 degrees. The measured results are presented in figure 2. It shows that MR and H_{EX} of the spin valve significantly decreases when angle θ_F was increased. According to the two models proposed by Mauri [15] and Malozemoff [16] the exchange field is proportional to $(A_{AFM}K_{AFM})^{0.5}$, where A_{AFM} and K_{AFM} are the exchange constant and the anisotropy of the AFM layer, respectively. The exchange field conclusively demonstrates an unidirectional behavior and then depends on angles between the measuring magnetic field and the anisotropy axis. The most general angular dependence of the exchange field is presented as follows [17]

$$H_{ex} \approx \sum_n b_n \cos n\theta \ , \qquad (1)$$

where b_n is a constant and n is a positive integer. According to equation (1) the exchange field decreases when θ is increased. It well agrees with the obtained experimental results H_{EX}, which has been evaluated as $H_{EX}= -(H_{\uparrow}^{+} + H_{\downarrow}^{-})/2$, where arrows ↑ and ↓ mark the direction of the measurement magnetic field, $H_{\downarrow}^{-}$ and $H_{\uparrow}^{+}$ the fields at which the MR of spin valve reduces to value of $MR_{max}/2$ in branch H<0 and conversely the MR of spin valve rises up to value of $MR_{max}/2$ in branch H>0, respectively.

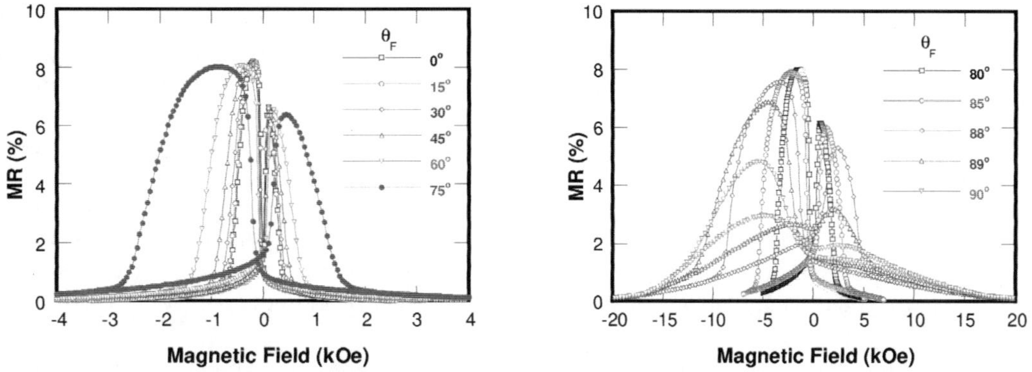

Figure 3. The MR ratio versus angle θ_F measured in the out-of-plane configuration.

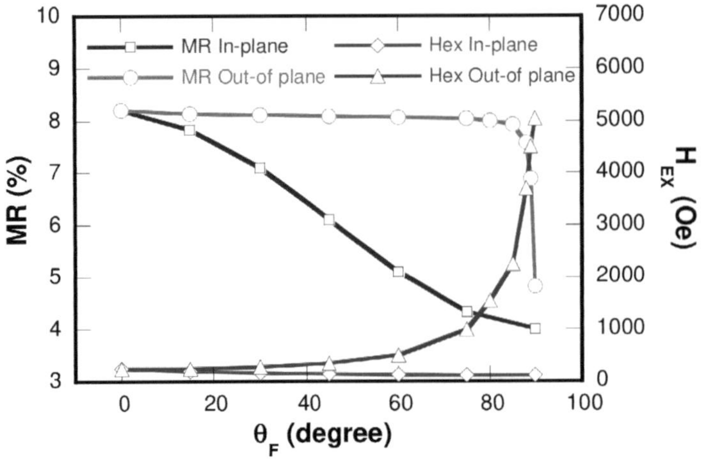

Figure 4. The dependence of MR and H_{EX} on angle θ_F in two configuration measurements.

Table 1. MR ratio and H_{EX} in two configuration measurements with varying angle θ_F.

θ_F (degree)	0	15	30	45	60	75	80	85	88	89	90
MR in-plane (%)	8.2	7.83	7.10	6.10	5.10	4.33					4.00
H_{EX} in-plane (Oe)	244	171	155	143	131	122					115
MR out-of-plane (%)	8.2	8.13	8.11	8.09	8.06	8.04	8.00	7.92	7.57	6.89	4.82
H_{EX} out-of-plane (Oe)	244	252	285	350	504	995	1536	2253	3705	4504	5036

By this way H_{EX} of the spin valve was evaluated and presented in table 1 as well as in figure 4. As shown in table 1 the obtained exchange field decreases from 170 to 115 Oe when θ_F was increased from 0 to 90 degree. According to Tang [14] at the FM/AFM interface the moments of electrons flowing from the FM into the AFM are not parallel to the AFM. Consequently they induce torques on moments in the antiferromagnetic layer. In addition we also observe a decrease of MR of spin valve when the angle θ_F increases. It may be related with a canting of magnetization moment in the FM layer due to the measuring field. In the case of the out-of-plane measurement configuration, the magnetic field measurement was applied in the z-x plane, perpendicular to the spin valve plane, respecting the anisotropy axis under angles from 0 to 90 degrees. It means that by this configuration we can observe the perpendicular exchange coupling in the spin valve. The obtained results are presented in figure 3. We receive absolutely different dependence of MR and H_{EX} on the applied magnetic field. In contrary to the case of in-plane configuration the H_{EX} in this case much increases in dependence of the angle θ_F beside a quite unchanged in MR. H_{EX} raises up to higher than 1 kOe at the θ_F equal to 75 degree. This very large value of the recorded perpendicular H_{EX} is probably due to the domination of the in-plane anisotropy of the FM/AFM bilayer.

Our second study is observing spin torque effect in CIP induced by pumping different currents through the spin valve. The obtained R(H) and MR(H) curves are presented in figure 5a and figure 5b, respectively. Since the current is flowing in the layer planes, it produces a magnetic field which is perpendicular to the sweeping field, and then the real magnetic field acting on the pinned layer, which is shown in figure 1b, rotates by an angle θ_F. The angle θ_F increases with increasing current. Hence we receive the same situation as in the case of angular dependence of the exchange bias, presented previously.

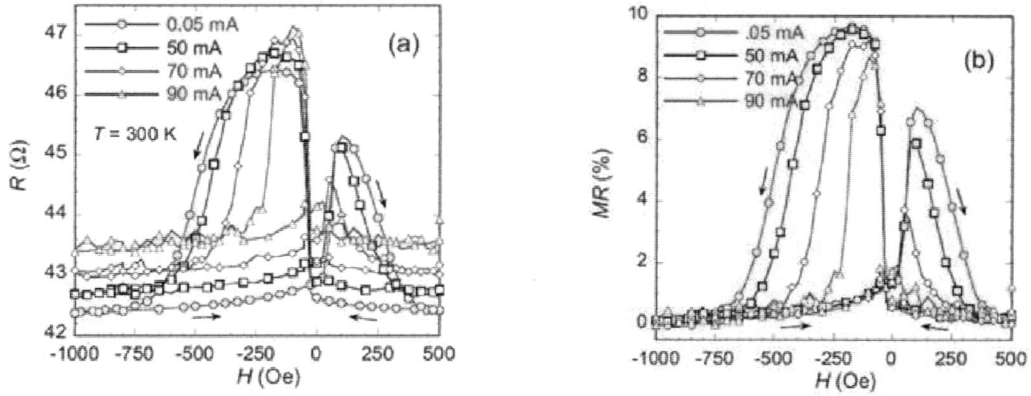

Figure 5. Variation of (a) the R(H) and (b) corresponding MR(H) curves with current amplitude.

As seen in figure 5, with increasing current the reversal field of the free layer seems to be unaffected, while that of the pinned layer (in both field sweeping directions H>0 and H<0) is substantially suppressed. It is remarkable that the results in figure 5 just resemble those previously observed in spin valves with a metallic antiferromagnetic layer, where electron flow induces a torque of its magnetic moment, therefore leading to a change in the exchange bias [12, 14]. In our case, since the in-plane current is confined to flow only in the CoFe(3.0 nm)/Cu(3.8 nm)/CoFe(4.5 nm) top layers, there would be no such current-induced spin torques in the antiferromagnetic layer. Our results in figure 5 show quite the same results obtained in our first experiment. In the same way as presented previously we evaluated H_{EX} values for all the measurement lines. It clearly shows a reduction of H_{EX} from 240 to about 85 Oe, induced by the injection current. The critical current density which noticeably reduced H_{EX} has been estimated at about 10^9 A/m^2. Note that this current density is more than about 3 orders higher than the switching current reported in Ref. 14 for an antiferromagnet (10^6 A/m^2) and still below the typical level for switching a ferromagnet ($10^9 \div 10^{11}$ A/m^2). Probably, the current flowing in the CoFe(3.0 nm)/Cu(3.8 nm)/CoFe(4.5 nm) top layers induces a magnetic field which rotates electrons with a force acting on the antiferromagnetic layer under angles in range from 0 to 90 degree. Consequently a torque of moments in the AFM was induced and exchange bias decreases with increasing of the injection current. As seen in figure 5a resistance of spin valve rises up a little when applying the current higher 50 mA. It is related to the Joule heating which was confirmed in Ref. [18] that the Joule heating by flowing current plays a lesser role in the depression of H_{EX} compared to the current induced effect.

4. Conclusions

Spin valve $Ni_{0.85}Co_{0.15}O/Co_{85}Fe_{15}/Cu/Co_{85}Fe_{15}$ was successfully manufactured by a RF magnetron sputtering. A depression of H_{EX} in dependence of the angle between applied magnetic field and injection current were observed. H_{EX} decreases with increasing the angle in the case of the in-plane configuration of CIP measurement. Conversely H_{EX} significantly increases with increasing the angle in case of the out-of-plane configuration. It is supposed to be related with the anisotropy of the antiferromagnetic layer. A depression of H_{EX} due to the spin torque effect induced by the injection current in CIP measurement configuration was also observed. It is supposed that the torque of moments in AFM is an effect induced by the flow of electrons which is not parallel to AFM layer.

References

[1] Tsang C, Fontana R E, Lin T, Heim D E, Speriosu V S, Gurney B A and Williams M L 1994 *IEEE Trans. Magn.* **30** 3801
[2] Lin T, Tsang C, Fontana R E and Howard J K 1995 *IEEE Trans. Magn.* **31** 2585
[3] Han D-H, Zhu J-G and Judy J H 1997 *J. Appl. Phys.* **81** 4996

[4] Inoue J, Itoh H and Maekawa S 1992 *J. Phys. Soc. Jpn.* **61** 1149
[5] Shirota Y, Kawawake Y, Satomi M, Sakakima H 1998 *Jpn. J. Appl. Phys.* **37** 5984
[6] Egelhoff W F, Ha T, Misra R D K, Kadmon Y, Nir J, Powell C J, Stiles M D, McMichael R D, Lin C L, Sivertsen J M, Judy J H, Takano K, Berkowitz A E, Anthony T C, Brug J A 1995 *J. Appl. Phys.* **78** 273
[7] Nguyen Chi Thuan, Nguyen Dang Thanh, Le Van Hong and Nguyen Xuan Phuc 2006 *Proc. of the 1st IWOFM-3rd IWONN Conf. (Halong, Vietnam)* p 325
[8] Meiklejon W H and Bean C P 1956 *Phys. Rev.* **102** 1413
[9] Meiklejon W H and Bean C P 1957 *Phys. Rev.* **105** 904
[10] Carey M J and Berkowitz 1992 *Appl. Phys. Lett.* **60** 3061
[11] Slonczewski J C 1996 *J. Magn. Magn. Mater.* **159** L1
[12] Núñez A S, Duine R A, Haney P and MacDonald A H 2006 *Phys. Rev. B* **73** 214426
[13] Wei Z, Sharma A, Nunez A S, Haney P M, Duine R A, Bass J, MacDonald A H and Tsoi M 2007 *Phys. Rev. Lett.* **98** 116603
[14] Tang X L, Zhang H W, Su H, Zhong Z Y and Jing Y L 2007 *Appl. Phys. Lett.* **91** 122504
[15] Mauri D, Siegmann H C, Pagus P S and Kay E 1987 *J. Appl. Phys.* **62** 3047
[16] Malozemoff A P 1987 *Phys. Rev.* B **35** 3679
[17] Ambrose T, Sommer R L and Chien C L 1997 *Phys. Rev.* B **56** 83
[18] Dai N V, Thuan N C, Hong L V, Phuc N X, Lee Y P, Wolf S A and Nam D N H 2008 *Phys. Rev. B* **77** 132406

Acknowledgments
This work was done with the support of Grant 4.076.06 in the Fundamental Research Program of Vietnam (2006-2008) and the Basic Research Project of IMS under grant number CS06.08. The technique support in part by the National Key Laboratory for Electronic Materials and Devices is acknowledged.

Growth and luminescent properties of $Li_2B_4O_7$ single crystal doped with Cu

Bui The Huy[1], Bui Minh Ly[1], Vu Xuan Quang[1], Huynh Ky Hanh[1], Doan Phan Thao Tien[1], Vinh Hao[1] and Tran Ngoc[2]

[1]Nha Trang Institute of Technology Research and Application,
2 Hung Vuong Street, Nha Trang, Vietnam
[2]Quang Binh University, Quang Binh, Vietnam

E-mail: buithehuy.nt@gmail.com

Abstract. The authors have primarily succeeded in the study of the technology for growing single crystal $Li_2B_4O_7$ doped with Cu ions by Bridgman technique. The TL-3D spectra show peaks at around 375 nm ($3d^94s \rightarrow 3d^{10}$ radiative excitation transition). This success opened up an opportunity in the radiotherapy to manufacture scintillators for neutron detection. The kinetic parameters of thermal stimulation luminescence were investigated by the three point method.

Keywords: Single crystal, kinetic parameter, three points, TL spectra.

1. Introduction

Lithium Tetraborate, $Li_2B_4O_7$ (LBO) single crystals have attracted much attention of researchers for use in surface acoustic wave (SAW), bulk acoustic wave (BAW) devices and dosimeter in radiotherapy. LBO is considered to be a promising material for piezoelectric materials due to its high electromechanical coupling coefficient (k^2=1) and very low temperature coefficient of delay (TCD, 0 ppm/^{o}C) [1]. LBO materials have found application in neutron detection and nonlinear harmonic generation [2]. Initially, Lithium Tetraborate single crystals were grown by the Czochralski method and subsequently both vertical and horizontal Bridgman methods were applied [3, 4]. The LBO single crystals of larger size were grown by Bridgman [5].

Thermoluminescence (TL) phenomenon is a result of the stimulated radiative recombination of released electrons, initially trapped by some centers and defects in the material lattice after being exposed to a source of ionizing radiation. The studies of TL have revealed useful information on the properties of the various types of defects within insulator or semiconductors. Many TL glow curve have been determined and a variety of methods for analyzing TL glow curve have been developed. Several methods are applied to obtain the trap parameters of the glow peaks [6-9].

LBO single crystals Cu-doped were grown in a graphite crucible by the Bridgman method at our laboratory. The three point method of M.S. Rasheedy [10] has been applied for the investigation of the kinetic parameters of thermal stimulated luminescence as the order kinetics, energy of trap, etc.

© 2009 IOP Publishing Ltd

2. Theoretical basis of TL kinetics

The calculations will be given here in outline form taken from the original paper by M.S. Rasheedy [10]. For general-order, the behavior of TL intensity of a phosphor is governed by the following equation:

$$I = -\frac{dn}{dt} = \frac{n^b}{N^{b-1}} S \exp\left(\frac{-E}{kT}\right) , \tag{1}$$

where I (in arbitrary unit) is the TL intensity, n (cm^{-3}) the electron concentration trapped at time t (s), N (cm^{-3}) the trap concentration and k (eV/K) is the Boltzmann's constant. Equation (1) is more general than the two equations describing the first and second-order kinetics. The solution of equation (1) for b # 1 is given as

$$I = \frac{n_0 S'' \exp(-E/kT)}{\left(1 + [((b-1)S'')/\beta] \int_{T_0}^{T} \exp(-E/kT')dT'\right)^{b/(b-1)}} , \tag{2}$$

where β (Ks^{-1}) is the linear heating rate and n_0 (cm^{-3}) is the concentration of traps populated at the starting heating temperature T_0 (K). The pre-exponential factor S" = $S(n_0/N)^{b-1}$ is constant for a given dose.

These expressions assuming the concentration of populated traps at a temperature T_i during the TL run is proportional to the area A_i, where A_i is the area under the glow peak between the T_i and T_f (T_f is the final temperature of the glow peak).

The kinetic parameters E, b, S" and n_0 was given from the below equations:

$$E = \left[(\ln z) - b\ln\left(A_x / A_y\right)\right]\left(\frac{kT_xT_y}{T_x - T_y}\right) , \tag{3}$$

$$b = \frac{T_y\left(T_x - T_z\right)\ln(y) - T_z\left(T_x - T_y\right)\ln(z)}{T_y\left(T_x - T_z\right)\ln\left(\frac{A_x}{A_y}\right) - T_z\left(T_x - T_y\right)\ln\left(\frac{A_x}{A_z}\right)} , \tag{4}$$

$$S'' = \frac{\beta E \exp(E/kT_m)}{\left(bkT_m^2\right) - (b-1)E\Phi \exp(E/kT_m)} , \tag{5}$$

with T_m(K) is the temperature corresponding to the maximum intensity I_m of the glow peak, y = Ix/Iy, z = Ix/Iz and

$$\Phi = \int_{T_0}^{T_m} \exp\left(\frac{-E}{kT'}\right) dT' , \tag{6}$$

$$n_0 = \frac{\mathrm{I_m} \exp(E/kT_m)}{S''}\left[\frac{bkT_m^2 S''}{\beta E \exp(E/kT_m)}\right]^{b/(b-1)} . \tag{7}$$

3. Experimental

Single crystal of LBO:Cu was grown by the modified vertical Bridgman method. The electric furnace has two independent heating zones. The raw material was a polycrystalline disc of 99.99% purity. The raw material was charged into the graphite crucible after being doped Cu_2O 0.03 wt.%. The inside diameter of the crucible is 20 mm and the length is 110 mm. The charge was melted in a nitrogen atmosphere. A seed crystal oriented to <110> direction with a dimension of 5 mm in diameter and 50 mm in length was placed at the bottom of the crucible.

An important parameter of the growth process is temperature gradient of the furnace system. It requires that the gradient should be from 15 to 20°C/cm. The furnace was heated to 930°C in both zones, the temperature gradient was reached at 17°C/ cm.

The raw materials were melted for 10 hours and then the crucible was moving down with the rate of movement of 0.4 mm/hour in 100 hours. The furnace was cooled with rate of 20°C/hour to room temperature. After the growth, the crucible was peeled off to get the crystal out.

4. Results and discussion

The crystal sample has the cylinder shape with about 35 mm in length. The crystal is transparent but there are some bubbles on the out side surface of sample. The reason is the difference between heat transfer coefficients of graphite and the crystal. The block crystal is shown in figure 1.

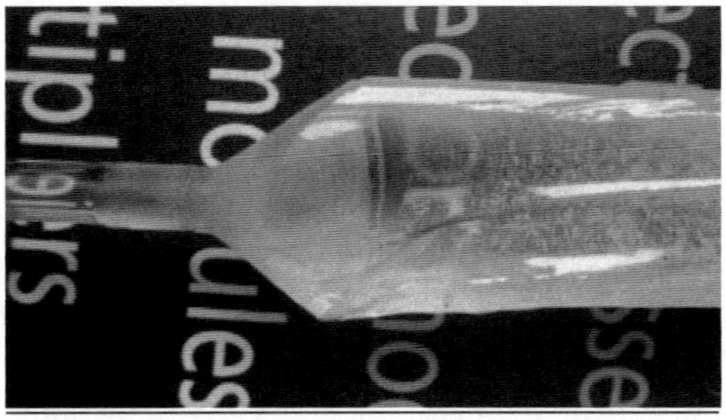

Figure 1. Block crystal $Li_2B_4O_7$ doped Cu.

The block crystal was cut perpendicularly to the growth direction and polished to mirror finish. The crystal sample was measured by slice surface measurement method at Shonan Institute of Technology-Japan. The measurements were carried out with the four sites 0, 90, 180, and 270 degree of cylinder slice-sample. At the each site there were 3 measurements with the positions 0, +7, and -7. The figure 2 shows X-ray diffraction pattern of crystal at site 0. It proves that the quality of the single crystal is good. At the others sites, the results are similar.

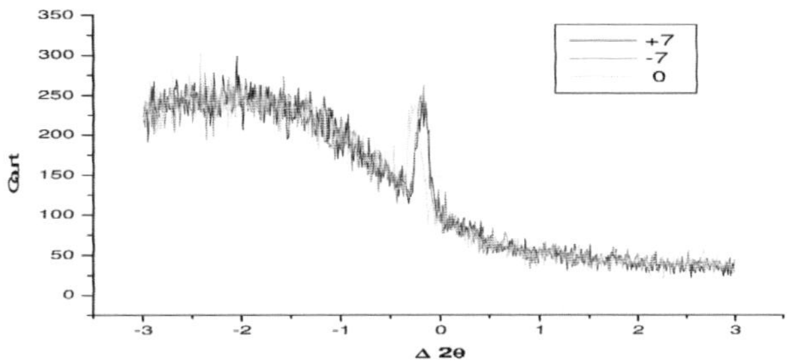

Figure 2. X-ray diffraction pattern of LBO:Cu crystal.

The TL glow curves were received from Harshaw Reader 3500 with the heating rate 1°C/sec as shown in figure 3. There are three individual peaks P1, P2 and P3 at 209, 254 and 345°C, respectively.

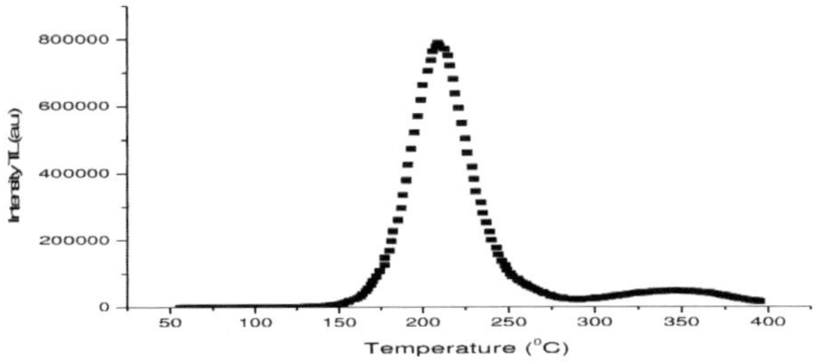

Figure 3. TL glow curve of LBO:Cu crystal.

The crystal samples were measured the TL- 3D and shown in figure 4. The spectra show that there is a peak at around 375 nm ($3d^9 4s \rightarrow 3d^{10}$ radiative ed-excitation transition) and 200°C.

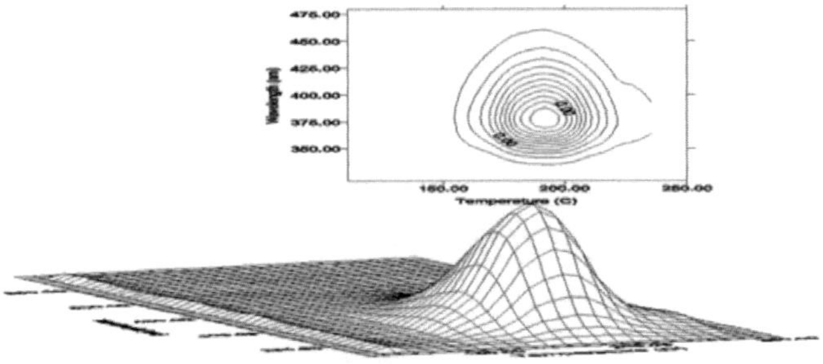

Figure 4. TL-3D spectra of LBO:Cu crystal.

In order to determine the kinetic parameters it was at first started with obtaining the kinetic parameters of the highest temperature peak 345°C - P3 by (3), (4), (5) and (7). Secondly, these values are used in equation (2) in order to fit the experimental glow curve into the theoretical glow curve. Five runs of the three points method were performed with random selections on the glow curve. The average values of the parameters in this peak are found equal to b = 1.80 ± 0.03, E = 1.42 ± 0.02 eV, $S'' = 2.011E \pm 10.s^{-1}$ and $n_0 = 4.089E \pm 6$. The results of two glow curves are illustrated in figure 5. The theoretical curve of peak 3 fits well with the experimental one.

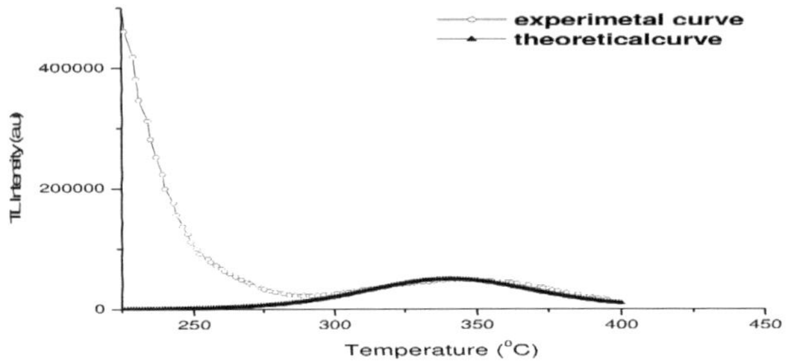

Figure 5. The experimental and theoretical curve of peak 3.

The experimental curve after subtraction of the peak P3 will be used for the determination of parameters P1, P2. The procedures are similarly repeated for P2 and P1. The results of all calculations are listed in table 1.

Table 1. The results of kinetic parameters of the P1, P2 and P3.

	P1 (209°C)	P2 (254°C)	P3 (345°C)
b	1.88 ± 0.02	1.43 ± 0.05	1.80 ± 0.03
E (eV)	1.73 ± 0.05	1.96 ± 0.08	1.42 ± 0.02
S'' (s^{-1})	0.9631E+17	6.7682E+17	2.011E+10
n_0	3.4719E+07	2.2726E+06	4.089E+6

The glow curve of P1, P2, P3 and experimental data are shown in the figure 6. The theoretical curves of the peaks are a good fitting to the experimental ones.

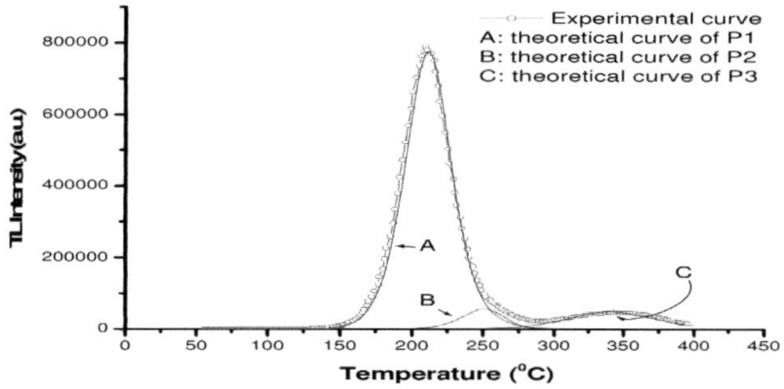

Figure 6. Experimental curve (circle), theoretical curve of peaks P1-P3 (line).

5. Conclusion

Single crystals of $Li_2B_4O_7$ have been grown to a size of 20 mm in diameter and 35 mm in length by the Bridgman method. These results open an opportunity in the radiotherapy to manufacture single crystal scintillators for neutron detection. The kinetic parameters were determined by the three point method. The results presented in table 1 show that each peak has individual values which agree with the experimental data. Therefore this method is reliable and accurate in evaluation of kinetic parameters of thermoluminescence curves.

References

[1] Ishii M 2004 *Piezoelectric Single crystal and Their Application* eds. Trolier-McKinstry S, Cross L E, Yamashita Y (Pennsylvania: University Park) p. 90
[2] Komatsu R, Sugawara T, Sassa K, Sarukura N, Liu Z, Izumida S, Segawa Y, Uda S, Fukuda T, Yamanouchi K 1997 *Appl. Phys. Lett.* **70** 3492
[3] Garrett J D, Natarajan Iyer M, Greedan J E 1977 *J. Crystal growth* **41** 225
[4] Katsumataa T, Konouraa H, Imagawaa H, Konnob A, Takeib K, Shinoharab M, Kanoc K and Takahashi K 1993 *J. Crystal growth* **128** 924
[5] Tsutsui N, Ino Y, Imai K, Senguttuvan N, Ishii M 2000 *J.Crystal growth* **211** 271
[6] Rasheedy M S 1996 *Physica B.* **228** 342
[7] Manam J, Sharma S K 2005 *Radia. Phys. And Chem.* **7** 423
[8] Ogundare F O, Balogun F A, Hussain L A 2006 *Radia. Measu.* **41** 892
[9] Rasheedy M S, Abd-Elmageed A I 2007 *J. of Phys. and. Solids* **68** 243
[10] Rasheedy M S, El-Sherif M A, Hefni M A 2007 *Nuclear Instrument and Method in Physics Research B* **258** 440

Acknowledgments

The authors wish to thank Prof. M. Ishii, Shonan Institute of Technology- Japan for his assistance in providing our laboratory with the Bridgman furnace as a gift. This work is supported in part by the Vietnam Academy of Science and Technology.

Low-temperature synthesis of nano-TiO_2 anatase on nafion membrane for using on DMFC

Nguyen Manh Tuan[1], **Ngo Thanh Nha**[2] **and Nguyen Hoang Tuyen**[1]

[1] Ho Chi Minh City Institute of Physics, Vietnam Academy of Science and Technology (VAST), 01 Mac Dinh Chi Street, 1 District, Ho Chi Minh City, Vietnam
[2] Can Tho University, Can Tho, Vietnam

E-mail: manhtuan2k@yahoo.com

Abstract. Low-temperature synthesis of 60-70°C of anatase nanocrystalline titanium dioxide TiO_2 using sol-gel technique on Nafion membrane is investigated and characterized. Titan tetraisopropoxide (TTIP) is used as precursor and ethanol as the solvent. The best precursor to solvent weight ratio has been used for the synthesis of nano-TiO_2 particles. The X-ray diffractograms and TEM images show the formation of anatase structure of nanocrystalline TiO_2 at low temperatures as shown with average particle size below 12 nm. The films deposited by spin coating technique using these nanoparticles show the crystalline and porous nature of the films. The nano-TiO_2 film as shown can be used to reduce the cross-over permeation of methanol through the PEM and increase electric power of the DMFC.

Keyword: TiO_2 anatase, sol-gel technique, nanoparticles.

1. Introduction

The fabrication of nanocrystalline titanium dioxide (TiO_2) has recently attracted much attention in terms of their versatile applications in solar cells, fuel cells, electrical and photocatalytic systems because it is highly stable, non-toxic and has a suitable redox potential for photodegrading pollutants. Several studies have focused on the use of TiO_2 nanoparticles for the purposes of improving the catalytic efficiency with respect to the high surface-to-volume ratio. The development of the immobilization technique that can maintain high surface area and excellent physicochemical properties of TiO_2 for photocatalysis is thus required.

In recent years, a great interest in nanocrystalline TiO_2 films has been growing, especially for use in fuel cell (FC) and also direct methanol fuel cell (DMFC) as well. Many works have been published on the preparation of titania thin films using sol-gel technique, direct deposition from aqueous solution, sputtering technology, ultrasonic spray pyrolysis and hydrothermal crystallization. The sol-gel technique of preparing TiO_2 thin film on the substrates has many advantages over other methods such as chemical vapour deposition (CVD), plasma spraying, anodization and thermal oxidation of Ti metal mainly because of the following reasons: (1) no special apparatus are required, (2) uniform muticomponent films can easily be formed if a homogeneous solution is available, (3) phase structure of film can be controlled, in this case TiO_2 (anatase) can be obtained, (4) the resultant films may be characterized by structure of a gel with a large specific area characteristic. Recently, TiO_2 thin films of high specific area were also reported to be obtained from a TiO_2 sol prepared by hydrothermal treatment of peroxotitanic acid sol or alkoxide derived TiO_2 gel [3].

© 2009 IOP Publishing Ltd

In this work, we present results of investigations on processing and structural characteristics of TiO_2 thin films deposited on glass and Nafion membrane substrates at very low temperatures of 60°C by sol-gel technique using polyethylene glycol (PEG) as a template. An application of this thin film on Nafion as a special polymer substrate for using in DMFC shown above need such low temperature in processing due to properties of Nafion. The influences of concentration and type of PEG template, the amount of water added, the complexing agents and solvents were involved in this study because of the morphologies of thin films are sensitive to them. One of the main impediments to the practical realization of the DMFC as a power source is cross-over of methanol through the polymer electrolyte Nafion. The methanol permeation reduces the fraction of useful methanol since the methanol passed to the cathode is converted into carbon dioxide leads to additional losses so that the electric power decreases. In fact, TiO_2 nanocrystalline thin film on Nafion membrane can reduce the methanol permeation through the polymer electrolyte and then increase significantly the electric power of DMFC. We will show in this work very good results of low-temperature synthesis at ~60°C of nanocrystalline TiO_2 anatase with some characterizations included.

2. Experimental

Low-temperature synthesis of nanocrystalline TiO_2 anatase is really a very big problem when using sputtering technology, CVD method... Temperatures used in those synthesis methods usually are known at about 300°C and higher for getting crystalline anatase. In this work sol-gel technique was used with glass and Nafion substrates and we got a very good result of nanocrystallline at low-temperature of 60°C. Titanium tetraisopropoxide (TTIP) (98% Merck), ethanol (99.5%), HCl (37%), water and PEG 600 were used as the starting materials. Water and 3% of PEG 600 (average molecular weight of 600) were then introduced to the mixture. All additions were accompanied by vigorous stirring at 80°C to allow the formation of a stable, homogenous and transparent sol solution.

Following the process, titania sol formed by the procedure shown above was deposited on the substrates of glass and Nafion N117 membrane by spin coating. The substrates coated by titania sol were taken out and dried at 60°C in a vacuum dryer machine for 8 hours with flow chart of preparation as shown in detail in figure 1 below. The sol solutions and PEG prepared at 3 different pH of 1.5, 2.0 and 1.0 before spin coating were shown in figure 2, respectively.

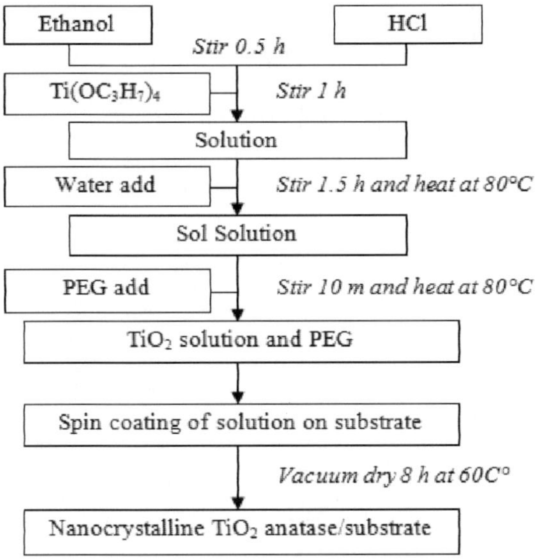

Figure 1. Flow chart of preparation of nanocrystalline TiO_2 anatase by sol-gel technique.

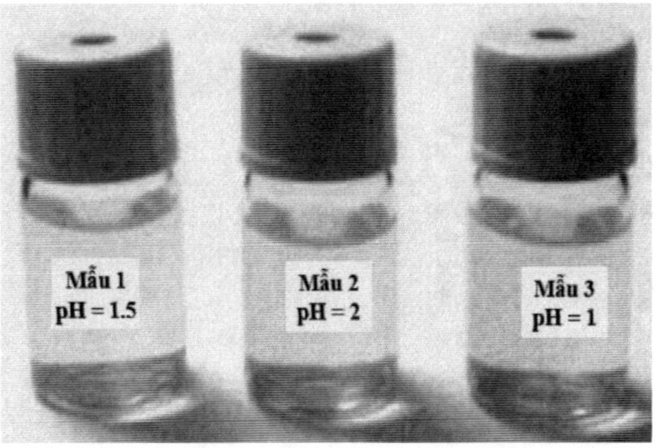

Figure 2. Sol solution and PEG prepared at different pH before spin coating.

Journal of Physics: Conference Series **187** (2009) 012040 doi:10.1088/1742-6596/187/1/012040

For application of TiO_2 coated Nafion membrane shown above to direct methanol fuel cell (DMFC), protonic conductivity measurements were performed on the membrane in a typical two electrode cell at different temperatures. A sample of the membrane (1.5 cm in diameter) was placed between two platinum electrodes and measured with AC potentiostatic condition. Methanol permeability was measured by electrochemical technique. A 1M C_2H_5OH solution was pumped into the anode side of the cell. The reaction occurring at the cathode is the oxidation of methanol that crosses through the membrane. When the applied voltage is sufficient enough to oxidize rapidly all the methanol diffusing to the cathode side, and a limiting current is achieved. The diffusion coefficient of methanol can be easily estimated from the limiting current.

3. Results and discussion

Nanocrystalline TiO_2 anatase was well prepared at low temperatures by sol-gel technique on Nafion membrane substrate. Simple hydrolysis with water was conducted at an ageing temperature of 70°C. This temperature was used as it has been reported elsewhere [4] to be sufficient for the formation of highly crystalline anatase nanoparticles. The transmission electron microscopy (TEM) images of the TiO_2 particles of our samples on Nafion membrane were shown in figure 3 with the average particle diameter is bellow 20 nm according to different pH of 1.0, 1.5 and 2.0 sol solution as described, respectively. And that the above PEG aided sol-gel process produce a very stable colloidal solution. As water is added together with PEG in the hydrolysis mixture, some water molecules as known will associate with the PEG through hydrogen bonding. The sol particles formed by the sol-gel reaction are covered by the PEG chains to form particle-PEG complex [5]. Physical cross-linking and agglomeration of these particulate complexes leads to an inorganic polymer composite.

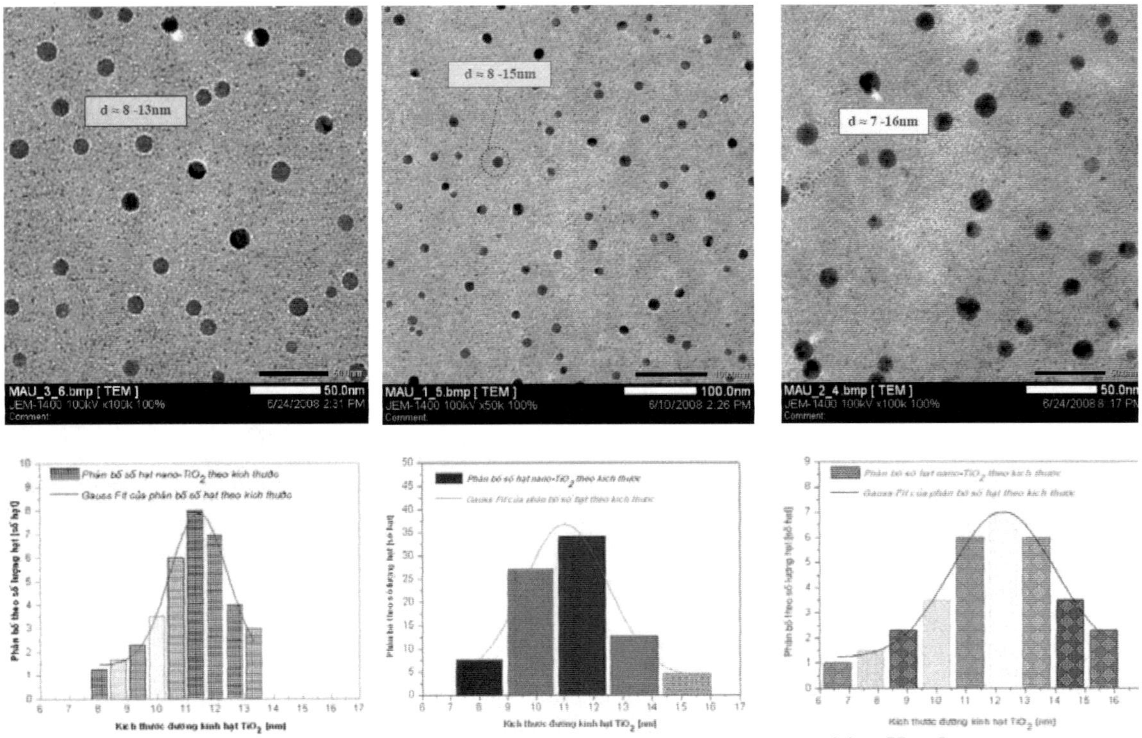

(a) pH = 1 (b) pH = 1.5 (c) pH = 2

Figure 3. TEM images of nanoparticles of TiO_2 anatase prepared at (a) pH = 1 (b) pH = 1.5 and (c) pH = 2 of sol solutions show best result at pH = 1 with homogeneous nanoparticle size as shown.

Thin films of TiO_2 were formed on the Nafion membrane by spin coating. The titania content was controlled by the TiO_2 concentration in sol and the coated times. The results show that the titania content increases with the TiO_2 concentration in the sol. When the TiO_2 concentration in the sol is more than 1.5 wt.%, the titania content increase rapidly. This may be caused by an increase in viscosity with titania content.

X-ray diffraction pattern of TiO_2 nanoparticles was performed as shown in figure 4 with peaks for the anatase phase and those for the rutile phase are not observed naturally. This result indicates that the powder consists of nanocrystalline TiO_2 anatase with all anatase peaks as shown. In figure 4, the upper is reference of XRD of anatase and rutile phase and the lower is our powder sample of TiO_2 with corresponding anatase peaks. By contrast, a clear peak at $2\theta = 25.8$ (101 anatase) is subjected to the ageing treatment at very low temperature of 60°C instead of 300°C and higher with other method preparation. This is clearly the characteristic peak of the anatase phase. All the clear peaks in figure 4 could be indexed as TiO_2 anatase as they basically all in agreement with those of the standard anatase. Those results show that thin solid film of TiO_2 on Nafion membrane prepared by sol-gel technique as described above is single anatase phase of nanosize TiO_2 particles. The nano TiO_2 thin films is dense and appears to be well attached to the membrane are no cracks when observed in SEM and then can reduce the methanol permeation through the polymer Nafion electrolyte and then significantly increase the electric power of DMFC as we expected. As a result, the protonic conductivity decreases with increasing titania content. The methanol permeability as a function of the TiO_2 content was investigated and characterized with the TiO_2 films on the substrates.

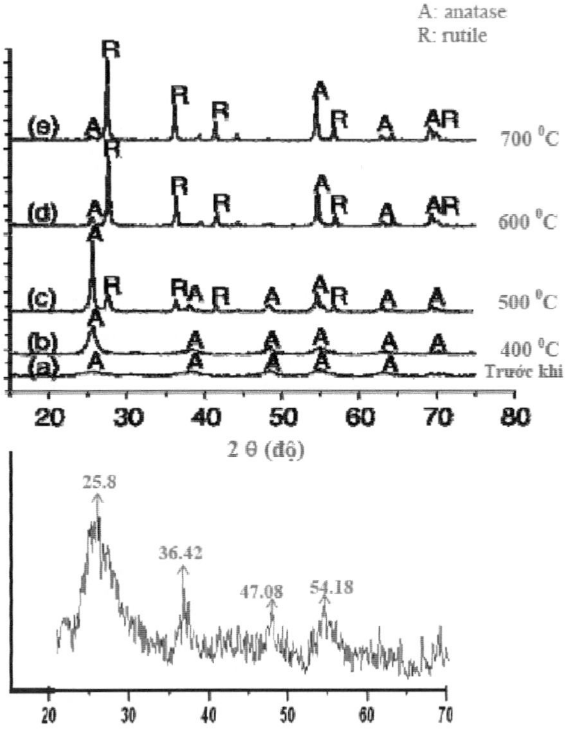

Figure 4. X-ray diffraction pattern of powder of nanocrystalline TiO_2 anatase in comparison with sample (also with rutile peaks).

The results were shown as formed on the Nafion N117 membrane and also on glass substrates. The permeability of the coated membranes decreases with increasing TiO_2 content in our samples with different contents and temperatures. Thus, rise in temperature leads to a strong increase in permeation by factor about 3. The methanol permeability depends on temperature and about 4×10^{-6} cm^2s^{-1} at 26°C. One reason for this behavior is that the nanoscale titania particles cover the cracks in Nafion membrane. A nano TiO_2 thin film with high content in the membrane display a so called mesoporous network texture, which hinders the permeation of methanol. With the methanol permeation data, the poor resistance to methanol cross-over of the membrane with low titania content may be due to incomplete coverage of the cracks in the membrane.

The free, non associated water molecules are readily available to TTIP for hydrolysis in the early stages. As hydrolysis exhausts the supply of free water, water will be drawn from the associated water. The sol particles formed by the sol-gel reaction are covered by PEG chains to form particle PEG complex. Some water molecules associate with the PEG through hydrogen bonding [6]. This feature will actually loosen the PEG chains in due course.

Nanocrystalline anatase uniform sized TiO_2 with average particle size varied from 8 to 13 nm has been synthesized by sol-gel technique on Nafion membrane at a very low temperature of 60°C. Particle size, shape and crystallinity of the nanoparticles are found to be dependent on the reaction temperature, time and precursor to solvent ratio. The particle size and the crystallinity increased with reaction time and temperature. For reaction temperatures as described, meta-stable anatase phase product is obtained. With the increase in the reaction temperature higher and also solvent to precursor ratio, products with random shaped TiO_2 nanoparticles with mixed phase (anatase and rutile) are obtained.

4. Conclusions

Thin films of anatase nanocrystalline TiO_2 were formed at a very low temperature of 60°C on the Nafion membrane by a simple sol-gel and spin coating procedure. Highly crystalline TiO_2 anatase nanoparticles are formed in an ethanol-water solution. X-ray powder diffraction analysis shows that the TiO_2 particles have crystallinity that is further refined by ageing treatment at 60°C. The TiO_2 films on membrane that are coated by a thick titania sol is very dense and no cracks are found. The protonic conductivity of the membranes decreases with increasing titania content, but their methanol permeability is also reduced. In fact, TiO_2 nanocrystalline thin film on Nafion membrane described above can reduce the methanol permeation through the polymer electrolyte and then increase significantly the electric power of DMFC as an application. The titania content was controlled by the TiO_2 concentration in sol and the coated times.

References

[1] Wainright J S, Wang J T, Weng D, Savinell R F, Liu M 1995 *J.Electrochem. Soc.* **142** (7) L121
[2] Su C, Hong B Y, Tseng C M 2004 *Catalysis Today* **96** 119
[3] Bu S J, Jin Z G, Liu X X, Yang L R, Cheng Z J 2005 *J. of the European Ceramic Soc.* **25** 673
[4] Fujima A and Honda K 1972 *Nature* **37** 238
[5] Shimizu K, Imai H, Hirashima H and Tsukuma K 1999 *Thin Solid Films* **351** 220
[6] Uchida H, Mizuno Y, Watanabe M 2000 *Chem. Lett.* **29** 1268

Acknowledgments

This work is supported in part by the Grant-in-Aid for Basic Scientific Research through the Ministry of Science and Technology.

Electron field emission characteristics of carbon nanotube on tungsten tip

Phan Ngoc Hong[1], Bui Hung Thang[1], Nguyen Tuan Hong[1,2], Soonil Lee[2] and Phan Ngoc Minh[1]

[1]Institute of Materials Science, Vietnam Academy of Science and Technology, 18 Hoang Quoc Viet Road, Cau Giay Distr., Hanoi, Vietnam
[2]Division of Energy Systems Research, Ajou University, Suwon 443-749, Korea

E-mail: hongpn@ims.vast.ac.vn

Abstract. Electron field emission characteristic of carbon nanotubes on tungsten tip was investigated in $2x10^{-6}$ Torr vacuum. The measurement results showed that the CNTs/W tip could emit electron at 0.7 V/μm (nearly 10 times lower than that of the W tip itself) and reach up to 26 μA at the electric field of 1 V/μm. The emission characteristic follows the Fowler-Nordheim mechanism. Analysis of the emission characteristic showed that the CNTs/W tip has a very high value of field enhancement factor ($\beta = 4.1\times10^{4} cm^{-1}$) that is much higher than that of the tungsten tip itself. The results confirmed the excellent field emission behavior of the CNTs materials and the CNTs/W tip is a prospective candidate for advanced electron field emitter.

Keywords: Field emission, Schottky emission, CNT, electron sources.

1. Introduction

Carbon nanotubes (CNTs) have attracted much attention as an electron source because of their ultra-small radius, high aspect ratio and remarkably high electrical conductivity [1, 2]. They can be applied to the cathodes of a flat-panel display, cathode-ray-tube lighting elements, and as field-emission sources in modern electron microscopies [3-5]. In high-resolution electron microscopy applications, CNTs offer several advantages such as enhanced current stability, low threshold voltage and long lifetime. Continuous progress in the semiconductor industry, nanotechnology requires significant enhancement in the resolution and progressing speed of currently available electron sources [6, 7]. These needs can be met by improving the performance of cold electron emitters with the aid of CNTs.

Furthermore, because of high mechanical strength and chemical characteristic, CNT is considered to be the ideal tip for scanning probe microscopy (SPM) application such as atomic force microscopy (AFM) and scanning tunneling microscopy (STM) [8]. A CNT tip for SPM offers important advantages over the conventional metal tips by virtue of high-lateral-resolution imaging, tracing of rough surfaces with steep and deep feature, preventing the multi-tip effect and reducing damage to delicate samples. Moreover, in electrical resistance measurements, the CNT tips are expected to measure electrical resistance of nanostructure materials. Conventionally metal tips such as W or Pt/Ir are not suitable for measuring electrical resistance of nanostructure materials of sizes below 100 nm.

© 2009 IOP Publishing Ltd

This is because a radius of curvature of electrochemically etched metal tip is usually larger than 50 nm. In case of the AFM application, the CNT tip has already proven above-mentioned advantages [9, 10]. The CNT for AFM is well established and is now commercially available.

Various techniques to fabricate CNT tips for atomic force microscopy have developed recently. For example, CNT tips prepared by mechanical assembly under optical microscopy or scanning electron microscopy (SEM) were shown to enhance the spatial resolution of the AFM [11, 12]. Recently, Konishi *et al.* [13] have announced that they could synthesize conductive CNT tips in high-yield in the following way: a) attaching a CNT at the apex of a supporting W tip by a dielectrophoresis method, b) deposition of amorphous hydrocarbon around the junction portion between the CNT and the W tip by electron beam deposition (EBD), c) annealing of the tip at 500°C, and d) coating the surface of the tip with a Pt/Ir layer by pulsed laser deposition (PLD). This method not only reduces the contact resistance at the junction between the CNT and the supporting W tip but also enhances the sticky ability of the CNT on the W tip. However, this method is expensive and requires a complex experimental setup.

For fabricating CNT in high production yield, high density and low cost, CVD is the most suitable method [14]. CNT fabricated by CVD using support catalyst is ideal for field emission, sensors, and other applications requiring the growth on patterned substrate. In this work, we use CVD method to fabricate CNTs/W tip by directly growing the CNT on the apex of the sharpened tungsten tip. Detailed process of fabrication of the CNTs/W tip was presented in Ref. [15]. In this paper we present the result of electron field emission characteristics of the fabricated CNTs/W tip.

2. Experimental

CNTs were directly grown on a sharp tungsten tip using CVD method. At first, tungsten tip of sub-50 nm in diameter was formed by electrochemically etching in a KOH 5 wt.% solution. In order to grow CNT on the sharp tungsten tip, Fe nanograins were created at the apex of the W tip by using electroplating in one circle of 20 Hz pulse mode in $FeCl_3$ 0.01 wt.%. solution. CNT was then grown from the Fe nanograins on the sharp W tip using hot-filament CVD apparatus. During the growing process, the temperature in CVD chamber was 750°C, feedstock gas was a mixture of (C_2H_2 : H_2, 50 sccm : 600 sccm) gases and CVD process took place in 30 minutes. In order to enhance the alignment and support the individual CNT formation, a DC electric field was applied between the W tip and the electrode with an electric field of approximately 8.10^4 V/m. To measure the electron field emission characteristic of the CNTs/W tip we used the emission measurement system as shown in figure 1. The set-up consists of a vacuum chamber having cathode to hold emitter tip in vertical position. Anode was used to collect the electron emission and connected to the computerized Keithley-248 source and Keithley-2001 measurement units. The gap between anode and cathode can be manually controlled. The measurement was carried out at pressures of $2x10^{-6}$ Torr at room temperature.

3. Results and discussion

Figure 2 shows SEM image of the sharp tungsten tip after electrochemical etching from a commercial W wire of 0.3 mm diameter. The diameter at the apex of the tip was approximately 50 nm. We found that the diameter of the tip played a critical role for growing the CNTs. If the W tip is too sharp (less than 20 nm) it is not easy to create individual Fe nanograin catalyst by electroplating method. Therefore, probability of growth of CNT is difficult. Inversely, if the W tip is too obtuse (more than 1μm) many Fe nanograin catalysts are created. Consequently, several CNTs will be grown at the top of the tip. Besides, to evaluate the quality of the electrochemically etched W tip, we estimated the diameter of the commercial STM tip by observing SEM image. Figure 3 shows SEM image of the Pt/Ir STM tip of Agilent Technology Company. From figures 2 and 3, it is clearly seen that the diameter at the apex of our W tip is nearly 10 times smaller than the diameter at the apex of the commercial Pt/Ir STM tip.

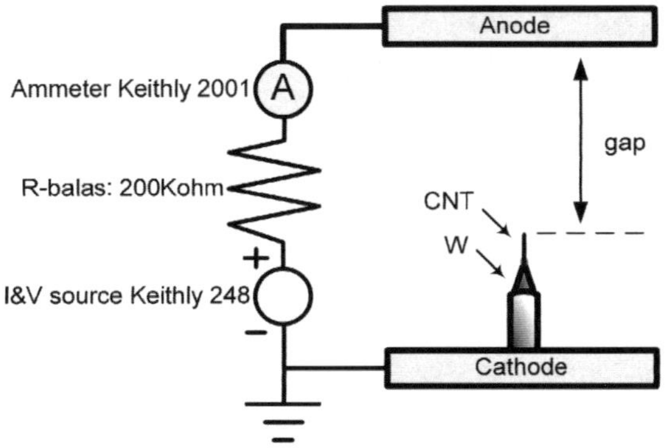

Figure 1. The schematic view of the experimental system for field emission measurement.

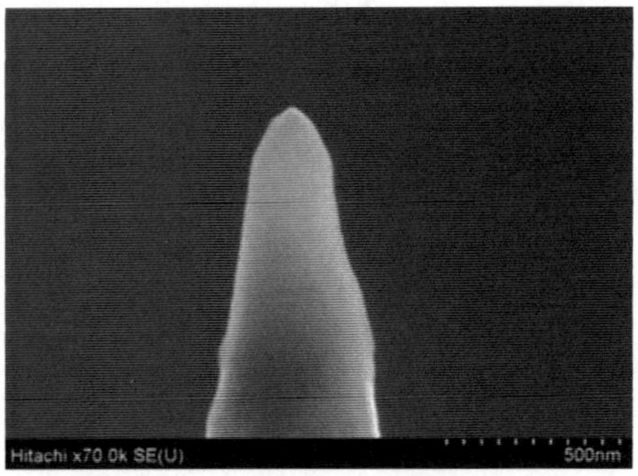

Figure 2. SEM image of the apex W tip after electrochemical etching.

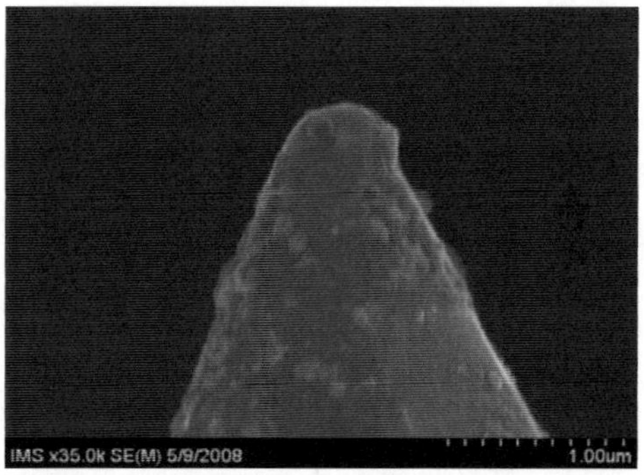

Figure 3. SEM image of the commercial Pt/Ir STM tip of Agilent Technology Company.

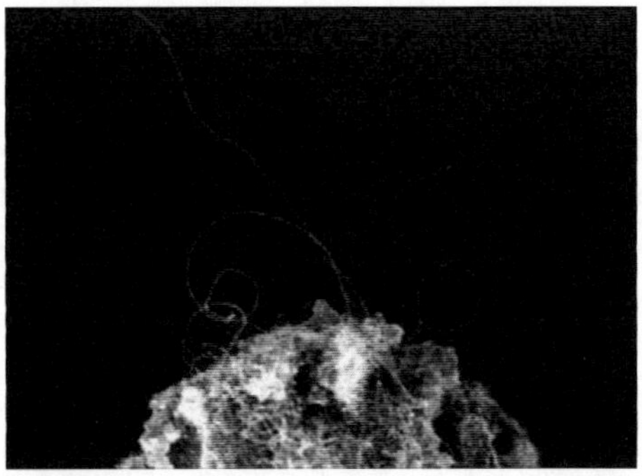

Figure 4. SEM image of the bundle CNT/W tip.

Figure 5. SEM image of the individual CNT/W tip.

By CVD method, CNTs were grown successfully on the apex of tungsten tip as shown in figures 4 and 5. Using measurement system as shown in figure 1, the electron field emission characteristics of the W and CNTs/W tips were characterized. This measurement system enables us to investigate the *I-V* characteristic of mentioned emitters. Figure 6 is the total electron emission current from the bundle CNTs/W tip and W tip when applying voltage to the anode. As shows in figure 6, the CNTs/W tip started emitting electron at 0.7 V/μm and rising up to 26 μA at the electric field of 1 V/μm. Whereas the W tip could emit electron at 5.8 V/μm and only reach to 13 μA at the electric field 6.3 V/μm. Thus, the CNTs/W tip could emit electron at threshold voltage nearly 10 times lower than that of the W tip itself. The threshold voltage is defined as the voltage at which the emission current starts increasing from the baseline. These results indicate that the CNTs/W tip shows a better performance as electron field emitter. From Fowler-Nordheim theory, the emission current I can be expressed in a simplified form [16, 17]

$$I = 1.54\times10^{-6}\frac{S\beta^2}{\phi}V^2\exp(-6.83\times10^7\frac{\phi^{3/2}}{\beta V}) \quad . \tag{1}$$

Here, the emission current I and the applied voltage V are expressed in A and V, respectively. The work function ϕ is expressed in eV and the emitting area S is expressed in cm^2. The field enhancement factor β corresponds to the geometry shape of the emitter, and is inversely proportional to the diameter of the emitting tip. By rearranging equation (1), we obtain:

$$\ln(I/V^2) = A(1/V) + B \quad , \tag{2}$$

where $A = -6.83\times10^7\phi^{1.5}/\beta$ and $B = \ln(1.54\times10^{-6}S\beta^2/\phi)$.

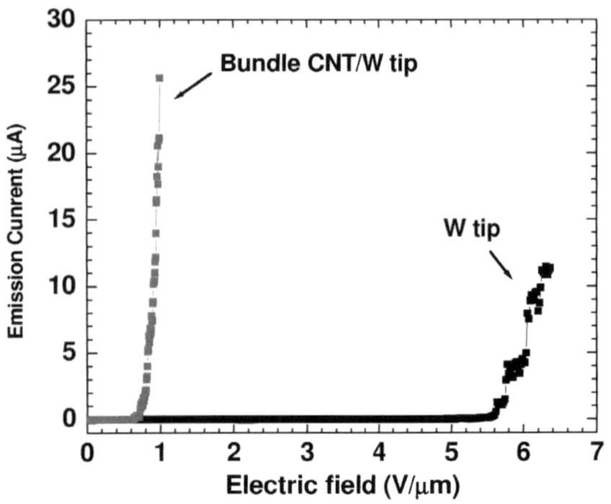

Figure 6. Electron emission current as a function of anode voltage from CNTs/W tip and W tip.

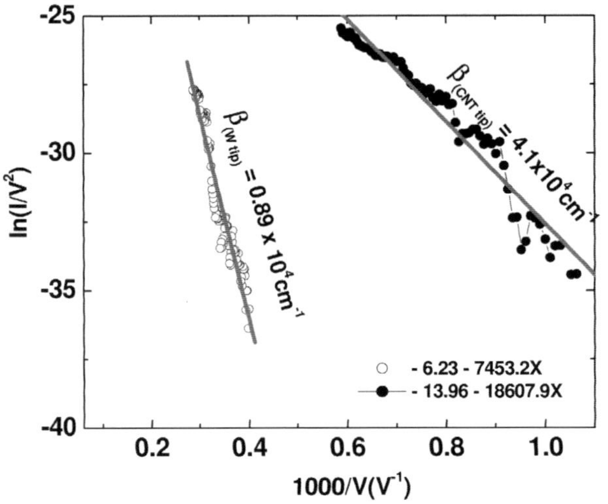

Figure 7. Fowler-Nordheim plots of I-V curves in figure 6.

The corresponding Fowler-Nordheim plots and the values of field enhancement factor β are shown in figure 7. By adopting work functions of tungsten and CNTs of 4.55 and 5.0 eV, respectively, the field enhancement factor were determined to be $0.89\times10^4 cm^{-1}$ and $4.1\times10^4 cm^{-1}$ for W tip and CNT/W tip, respectively. The field enhancement factor for CNT/W tip was 5-times larger than that of the W tip.

The results confirmed the excellent field emission behavior of the CNTs materials and the CNTs/W tip is a prospective candidate for the advanced electron field emitter. Although the electron emission characteristic of individual CNT/W tip was not done in this time, according to recent papers, individual CNT tip had lower threshold voltage, the field enhancement factor higher than the bundle CNT tip [18]. The emission on individual CNT/W tip will be done the next time for our further work. The CNTs/W tip was also successfully applied as an advanced STM tip. The result will be presented elsewhere.

References

[1] Teo K B K, Lacerda R G, Yang M H, Teh A S, Robinson L A W, Dalal S H, Rupesinghe N L, Chhowalla M, Lee S B, Jefferson D A, Hasko D G, Amaratunga G A J, Milne W L, Legangneux P, Gangloff L, Minoux E, Schnell J P, and Pribat D 2004 *IEEE Proc.-G: Circuits, Devices Syst.* **151** 43
[2] Zhang L, Balzano L, and Resasco D E 2005 *J. Phys. Chem. B* **109** 14375
[3] Bonard F-M, Stockli T, Noury O, and Chatelain A 2005 *Appl. Phys.* **44** 1648
[4] Charbonnie F 1996 *Appl. Surf. Sci.* **94-95** 26
[5] Ferrer D, Shinada T, Tanii T, Kurosawa J, Zhong G, Kubo Y, Okamoto S, Kawarada H and Ohdomari I 2004 *Appl. Surf. Sci.* **234** 72
[6] Jonge N D 2004 *J. Appl. Phys.* **95** 673
[7] Jonge D, Allioux M, Oostveen J T, Teo K B K and Milne W I 2005 *Appl. Phys. Lett.* **87** 133118
[8] Wong S S, Joselevich E, Wolley A T, Cheung C L, Lieber C M 1998 *Nature* **52** 394
[9] Wong S S, Woolley A T, Odom T W, Huang J L, Kim P, Vezenov D V and Lieber C M 1998 *Appl. Phys. Lett.* **73** 3465
[10] Hafner J H, Cheung C L and Lieber C M 1999 *Nature* **398** 761
[11] Dai H, Hafner J H, Rinzler A G, Colbert D T, Smalley R E 1996 *Nature* **384** 147
[12] Nishijima H, Kamo S, Akita S, Nakayama Y 1999 *Appl. Phys. Lett.* **74** 4061
[13] Koshini H, Murata Y, Wongwiriyapan W, Kishida M, Tomita K, Motoyoshi K, Honda S, Katayama M, Yoshimoto S, Cubo K, Hobara R, Matsuda I, Hasegawa S, Yoshimura M, Lee J-G and Mori H 2007 *Rev. of Sci. Ins.* **78** 013703
[14] Chuc N V, Dung N D, Hong P N, Quang L D, Khoi P H and Minh P N 2008 *J. Korean Phys. Soc.* **52** 1368
[15] Hong P N, Cuong V V, Thang B H, Khoi P H and Minh P N 2008 *J. Korean Phys. Soc.* **52** 1386
[16] Fowler R H and Nordheim L M 1928 *Proc. R. Soc. London, Ser.* A **119** 173
[17] Gomer R 1961 *Field Emission and Field Ionization* (Cambridge:Harvard University Press, MA)
[18] Koizumi T, Ikeda M, Takagi D, Chiashi S, Rokuta E, Oshima C and Homa Y 2007 *The Japan Soc. of Appl. Phys.* 6087

Acknowledgements

This work was supported by VAST and National Basic Research Project on Nanostructured Carbon Materials. We would like to thank the support of the Vietnam-Korea and Vietnam-France Bilateral Science Collaboration Programs. A part of this work was done with the help of the National Key Laboratory for Electronic Materials and Devices.

Electrochemical direct immobilization of DNA sequences for label-free herpes virus detection

Phuong Dinh Tam[1], Tran Trung[2], Mai Anh Tuan[1] and Nguyen Duc Chien[3]

[1] International Training Institute for Materials Science
[2] Department of Electrochemistry, Hung-Yen University of Technology and Education
[3] Institute of Engineering Physics, Hanoi University of Technology, 1 Dai Co Viet Road, Hanoi, Vietnam

E-mail: tr_trunghut@yahoo.com

Abstract. DNA sequences/bio-macromolecules of herpes virus (5'–AT CAC CGA CCC GGA GAG GGA C–3') were directly immobilized into polypyrrole matrix by using the cyclic voltammetry method, and grafted onto arrays of interdigitated platinum microelectrodes. The morphology surface of the obtained PPy/DNA of herpes virus composite films was investigated by a FESEM Hitachi-S 4800. Fourier transform infrared spectroscopy (FTIR) was used to characterize the PPy/DNA film and to study the specific interactions that may exist between DNA biomacromolecules and PPy chains. Attempts are made to use these PPy/DNA composite films for label-free herpes virus detection revealed a response time of 60 s in solutions containing as low as 2 nM DNA concentration, and self life of six months when immerged in double distilled water and kept refrigerated.

Keyword: Electrochemically, bio-macromolecules, herpes virus.

1. Introduction

Label-free, electrochemical DNA detection have become an attractive, and a major focus in biotechnology and clinical diagnostics [1-3]. This technique offers high sensitivity, low cost, and fast analysis and can be easily integrated into small devices [4]. Typically, a single-strand DNA is firstly immobilized onto a conducting polymer coated on an electrode, then the microsensor converts the chemical recognition event into a measurable electronic signal. The immobilization of the DNA, normally by means of covalent attachment between the DNA oligos and polymeric membrane, plays a major role in determining the overall performance of an electrochemical DNA sensor [5, 6].

The conducting polymers offers an excellent interface for immobilized DNA probes/microsensor's surface [7-10] among which, thanks to large surface area and binding ability with DNA, bio-compatible, high hydrophilic, well stable and facile incorporation of many counter ions. Polypyrrole (PPy) is the most commonly used [11]. In addition, PPy has attracted much interest since both its good thermal and mechanical stability and maintains its conductivity for several months or a year [12, 13].

DNA is a large and highly structured biomacromolecule of which single-strand backbone is composed by alternatively a deoxyribose and a phosphate group. The latter confer to DNA its polyanionic character. The rationale hypothesis for the study on interaction of DNA biomacromolecules of herpes virus (HSV-1) and PPy chains is that PPy in their oxidized state

© 2009 IOP Publishing Ltd

provides positively charge density for electrostatically interaction with the highly negative charged DNA. In general, the hybridization reaction of a single strand of DNA (target) and an immobilized DNA probe is of extreme importance in the achievement of suitable sensitivity. Shimidzu [14] and Livache [15, 16] have proposed an electrochemically directed polymerization which allows preparing PPy/DNA composite films directly or in one-step for immobilization of DNA onto the electrodes. During the preparation of PPy/DNA composite films, deoxyribonucleic acid is well suited to act as a natural polyelectrolyte, meaning as dopant for conducting polymers such as polypyrrole that carry a net positive charge in their oxidized state, as illustrated in figure 1.

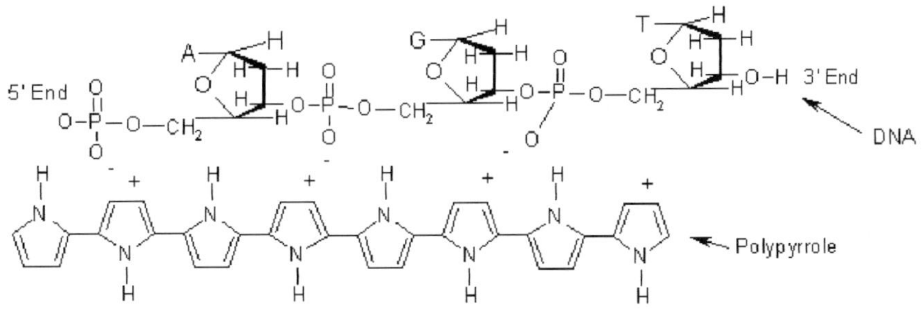

Figure 1. Illustration of the covalent interaction of a polypyrrole chain and DNA biomacromolecules of herpes virus.

Cyclic voltammetry (CV), is well known as an excellent analytical technique with CV, one can control the programmable delivery of electrons through the electrode and the potential applied to the electrode to couple the soluble oligomers of pyrrole and the deposited PPy chains during the direct electrosynthesis of PPy/DNA composite films. Also, the DNA probes are reoriented in better order in the field caused by applied potential. It is noted that the DNA uptake is proportional to increasing ionic strength resulting in the observed adsorption of polyelectrolyte onto hydrophobic surfaces [17]. The steady increasing state of potential during anodic potential sweep, usually positive sign, may decrease in the electrostatic repulsion between the DNA probes. Meanwhile cathodic potential sweep, the steady decreasing state of potential appears to make it advantage for the reorientation of the DNA probes. Therefore a more compact structure of the couple of DNA and pyrrole ligomer is obtained, resulting in more favorable adsorption on the adsorbent surface. In this work, the PPy/DNA (herpes virus) films were electrochemically synthesized directly by cyclic voltammetry and grafted onto arrays of interdigitated platinum microelectrodes which are used for the determination of the herpes.

2. Experimental

2.1. Chemicals and reagents

Pyrrole was purchased from Merck. Doubly distilled water in a quartz apparatus was used to prepare aqueous solutions. To remove oxygen from stirred aqueous solutions 0.1 M $LiClO_4$ (Merck) were bubbled with nitrogen gas for 30 minutes before adding pyrrole monomer (without further purification) and during electrosynthesis of PPy/DNA (herpes virus) composites.

Custom oligonucleotides, specific to herpes virus (5'–AT CAC CGA CCC GGA GAG GGA C–3'), were synthesized by Invitrogen Corp.

2.2. Preparation of PPy/DNA electrodes

0.3 M pyrrole and 0.1 μM single strand DNA of herpes virus was placed in a 5 ml three-electrode cell, at room temperature. The potential applied on the working electrode was relative to an Ag/AgCl (3 M NaCl) reference electrode; and a Pt sheet (1 cm^2) serves as auxiliary electrode. The process was controlled by EG&E 362 (Princeton Applied Research) incorporating the Ecunive-HH5 program. The microelectrodes were fabricated at International training Institute for Materials Science (ITIMS), Hanoi University of Technology, by using standard microelectronic technology.

Firstly, the microelectrodes were immerged in a mixture of 1 M KCr_2O_7 and 1 M H_2SO_4, and washing for several times by double distilled water; followed by cyclic voltammetry in a solution of 0.5 M H_2SO_4 with potentials ranging of -1.5 and +1.5 V, at a scan rate of 25 mV/s for cleaning and activating the surface. Thereafter the direct immobilization of DNA of herpes virus was done by cyclic voltammetry (potential sweep of 100 mV/s, potential range: -0.3 to +0.9 V).

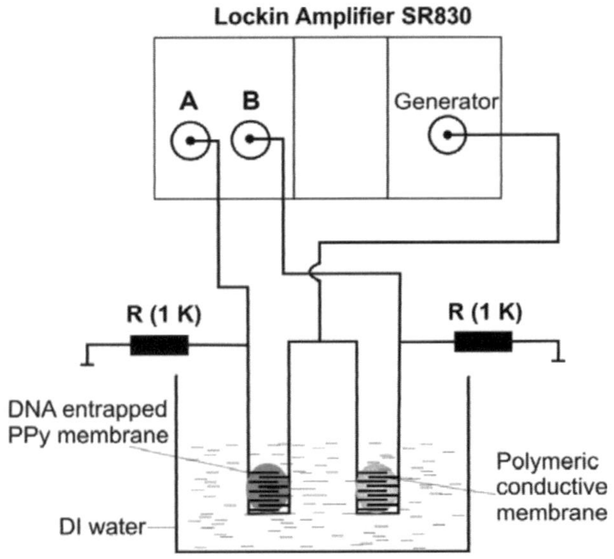

Figure 2. Diagram illustrates measurements of the sensibility of PPy/DNA electrodes. A signal of alternative current that had frequency of 10 KHz and amplitude of 100 mV taken out from generator of the Lock-in Amplifier SR830 was applied on PPy/DNA electrodes (working electrode) and other PPy electrodes (reference electrode).

The morphology of the surface of obtained PPy/DNA films was investigated by a field emission scanning electron microscopy Hitachi-S 4800. The presence of DNA in the PPy matrix was confirmed with FTIR measurements using the Thermo Nicolet 6700. The sensibility of the PPy/DNA electrodes was investigated by a Lock-in Amplifier SR830 (figure 2). A signal of alternative current (10 KHz, 100 mV) was applied on PPy/DNA electrodes serving as working electrodes, another one is the reference electrode. The output signal was acquired by measuring the voltage drop on two 1 KΩ resistances by the channels A and B of the Lock-in Amplifier, which was set on automatically differential measurement mode.

3. Results and discussion

A typical voltammogram of a one-step immobilization of DNA of herpes virus in copolymerization with pyrrole was accomplished by repeated potential cycling over the range of -0.3 and 0.9 V. As seen in figure 3, the irreversible oxidation of polymerization of pyrrole in the presence of DNA appears to be more easily, at 0.2 V (figure 3b) than that of polypyrrole, at 0.48 V (figure 3a). It indicates that the existence of covalent/hydrogen bonds between the DNA biomacromolecules and pyrrole monomers makes its advantage for the polymerization process, and plays an important role in decreasing the oxidation potential of pyrrole and expanding the oxidation potential range.

As the results, the oxidation of a conducting polymer increases the density of charge carriers in the polymer film. This oxidation process also creates deep Coulomb traps due to the presence of the dopants. In parallel, the mentioned covalent/hydrogen bonds limit the diffusion of species of soluble olygomers of pyrrole to the electrode surface. Thus, the value of peak current density of the later, 0.08 $mA.cm^{-2}$, is smaller in comparison with the one of the first, 0.1 $mA.cm^{-2}$. To reduce this, a supporting 0.1 M $LiClO_4$ electrolyte was added. In this case an oxidative current density is steadily increasing up to 0.8 $mA.cm^{-2}$, several times higher more than the one of the first two (figure 3c).

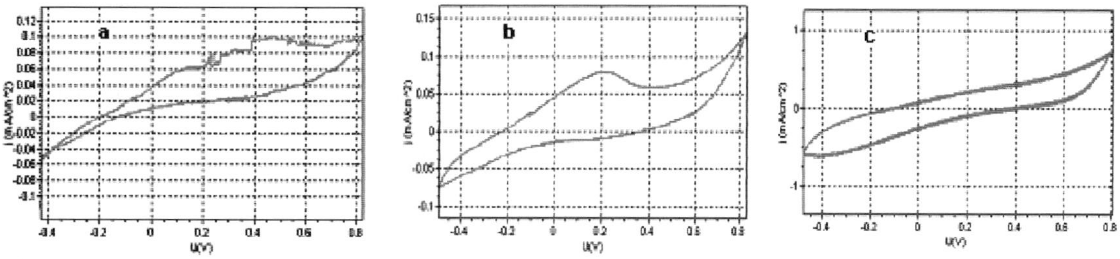

Figure 3. Cyclic voltammograms of the electropolymerizations of (a) Pyrrole were conducted in aqueous solutions of 0.3 M pyrrole, at scan rate of 100 mV.s^{-1}; (b) with 0.1 μM single strand DNA biomacromolecules of herpes virus; and (c) with 0.1 μM single strand DNA biomacromolecules of herpes virus and 0.1 M $LiClO_4$.

Figure 4 shows the scanning electron micrographs of the PPy and PPy/DNA films. In case of PPy film, an irregular and heterogeneous membrane illustrated by numerous cauliflowers like projections and channels between them was observed. Meanwhile, since DNA bio-macromolecules act as template reagents, the PPy/DNA membrane showed an ordered and stacked structure, a continuous film consisting of globular granules that coalesced together into a more smoothing and compact surface covered entirely the surface electrode.

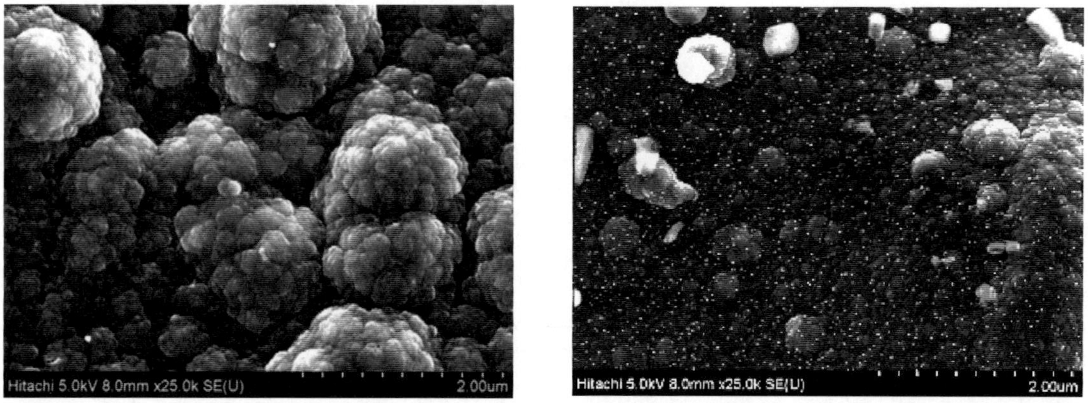

Figure 4. FESEM imagines of the PPy film (a) and the PPy/DNA membrane (b). Both were electrosynthesized in 0.1 M $LiClO_4$ and 0.3 M pyrrole solutions. The left-hand side is without and the right-hand side is with 0.1 μM DNA of HSV-1.

The channels appear to be disappeared, but there are still several aggregates sized in submicron. This can be explained by the linkage of DNA biomacromolecules and soluble species of pyrrole oligomers during copolymerization to form the bi-dimensional growth of the PPy/DNA composite film. In consequences, DNA biomacromolecules may be incorporated into the whole of a PPy matrix for easier hybridization with the DNA target in solutions.

As obtained the PPy/DNA composite films possess the oxidative current density (figure 3c) up to 0.8 mA.cm^{-2} and the ordered and stacked structure (figure 4, rhs.). It may be considered as a result of an improved interchain hopping of polarons and higher mobility of charge carriers/polarons in the PPy/DNA composite films.

In figure 5, the spectroscopic characteristics for PPy are observed at the vibration bands in a range of 500-1650 cm^{-1} for both PPy and PPy/DNA samples [18]. The analysis of the FTIR spectra also presents the increase and extension of the absorption band centered at around 3400 cm^{-1} belonging to asymmetric stretching of NH groups (figure 5, lower) which resulted from the covalent attachment of the DNA biomacromolecules with NH groups of a moiety of PPy matrix (figure 1). Unlike PPy (figure

5, upper), the spectrum of a PPy/DNA sample showed a continuous increase tail in absorption from 3650 to 4000 cm^{-1} (figure 5, lower). It is the tail of the ~1 eV (~8066 cm^{-1}) bipolaron absorption band that is the signature for electrical conductivity of conducting polymers [19, 20]. In addition, for the PPy sample, a rippled obtuse peak having featureless shape that was observed in the 1350-1760 cm^{-1} region, is constructed with several component vibrations owing to the imine/amine species.

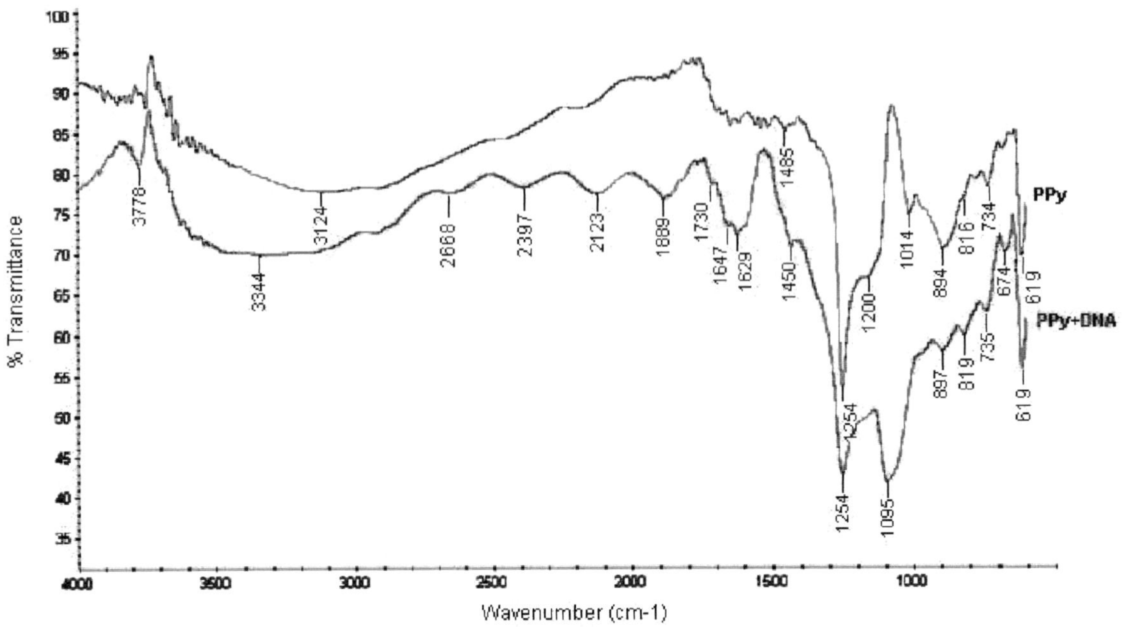

Figure 5. FTIR spectra in transmission mode of the PPy film (upper) and the PPy/DNA of herpes virus film (lower).

Among these the wave-numbers above 1600 cm^{-1} may be caused by the conjugated molecular structure of PPy and the free charge carriers in the polymer. The bands assigned to ν(C=C) stretch are observed in the region of 1400-1620 cm^{-1}. The presence of DNA biomacromolecules of HSV-1 causes the split of spectroscopic characteristics (figure 5, lower). Indeed, the 1350-1760 cm^{-1} region of PPy/DNA consists of several features sharply, such as the bands in the regions 1450 cm^{-1} are assigned to the C–N bond stretching. Besides, an intensive band centered at 1629 cm^{-1} that was constructed with the 1647 cm^{-1} vibration band due to C=N stretching and the 1600 cm^{-1} vibration band due to C=C bonds associated with C–N stretching and bending vibration [21, 22]. Furthermore, due to the presence of DNA biomacromolecules, the adsorption band expanding from 1800 cm^{-1} upward that is attributed to free charge carriers [23], is not just split and grouped into smaller vibration bands centered at 1889, 2123, 2397 and 2668 cm^{-1}, but with higher intensity. It reveals that the imine/amine charge carriers and the conjugated structure of PPy chains may be disturbed by the interactions between them and biomacromolecules of HSV-1 and thus limited the extent of charge delocalization along the polymer chains, leading to increase of spectral features.

In the range 1000 to 1300 cm^{-1}, the increase of spectral feature of absorption bands was also observed. These bands present the relationship in population of charge carriers, amine and imine species of PPy. Without DNA strands, a sharp peak at 1254 cm^{-1} and a shoulder at ~1200 cm^{-1} characterize C–C and C–N stretching vibrations of amine species (aromatic structure), and a peak at 1014 cm^{-1} is attributed to C–H in plane vibration [24]. Oppositely, as seen in the spectrum of PPy/DNA composite (figure 5, lower), a new sharp peak was observed at 1095 cm^{-1}, corresponding to the in-plane stretching vibration of NH^+ of imine species (quinonoid structure) that was performed by protonation of PPy chains [25].

It implies that a part of aromatic structure of PPy chains converts into quinonoid structure, and the formation of quinonoid structure was extended during electropolymerization of pyrrole in the presence of DNA of HSV-1. Furthermore it might be resulting from the Coulomb interaction of π-conjugated electron in C=C of pyrrole and the lone pairs of electrons in nitrogen atoms with phosphate groups of the DNA biomacromolecules.

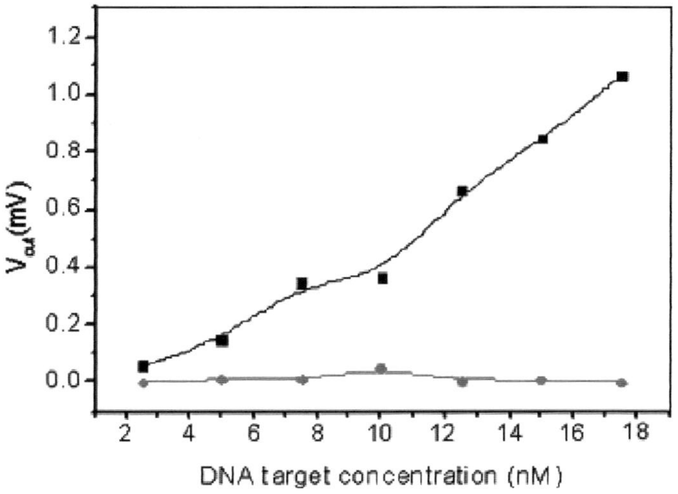

Figure 6. The change in voltage drop on a resistance connected to a PPy/DNA composite film as a function of DNA target concentration (upper) that is its sensing characteristic in comparing with the constant in the case of PPy film (lower).

The increase of charge carriers in PPy chains and higher mobility of charge carriers/polarons in the PPy/DNA composite films led to the increase in conductivity as seen in their sensing measurements. The voltage drop on a resistance connected to the PPy/DNA film reaches a stable value after 60 seconds and recorded. A quasi linear increase of the voltage drops with DNA target concentration was observed and shown in figure 6. Meanwhile the voltage drop on another film connected to the PPy film appears to be constant. These results are still repeated after the PPy/DNA films immerged in doubly distilled water, and stored in refrigerator for six months.

4. Conclusions

The PPy/DNA composite films having an ordered and stacked structure, as confirmed by FE-SEM studies, were electrosynthesized directly by cyclic voltammetry at 0.2 V onto the surface of the microelectrodes. The analysis of FTIR data mentioned above revealed that DNA bio-macromolecules of herpes virus act as template reagents. The intensive interactions of the DNA bio-macromolecules and PPy chains may disturb imine/amine charge carriers and the conjugated structure of PPy chains and thus limited the extent of charge delocalization along the polymer chains. A part of aromatic structure of PPy chains converts into quinonoid structure and causes an increase of population of charge carriers in PPy chains of the PPy/DNA composite films.

The PPy/DNA based sensors can detect the HSV-1 DNA from 1–18 nM. The response time was 60 seconds. These PPy/DNA sensors can be used for rapid screening in food and medical applications.

References

[1] Drummond T G, Hill M G, Barton J K 2003 *Nature Biotech.* **21** 1192
[2] Fojta M 2002 *Electroanalysis* **14** 1449
[3] Obayashi M K, Kusakawa T, Saito M et al. 2004 *Electrochem. Commun.* **6** 337
[4] Palecek E, Fojta M 2001 *Anal. Chem.* **73** 253
[5] Hang Tin C, Guiseppi- Elie A 2004 *Biosen. Bioelectron.* **19** 1537
[6] Blohm D H, Guiseppi-Elie A 2001*Curr. Opin. Biotechnol* **12** 41
[7] Christine E S, Venkatram R S, Joseph P V, Robert L 1997 *Proc. Natl. Acad. Sci. USA* **94** 8948
[8] Ramanavičius A, Ramanavičiene A, Malinauskas A 2006 *Electrochim. Acta.* **51** 6025

[9] Ionescu R E, Herrmann S, Cosnier S, Marks R S 2006 *Electrochem. Commun.* **8** 1741
[10] Gajovic-Eichelmann N, Ehrentreich-Föster E, Bier F F 2003 *Biosen. Bioelectron.* **19** 417
[11] Pethrick R A 1990 *Conducting Organic Polymers* (London, New-York: Elsevier Applied Science) p 149
[12] Kathirgamanathan P and Fowcett A H 1991 *High Value Polymers* (London: Royal Society of Chemistry, Cambridge)
[13] Trung T, Tri N V 2000 *Functional Materials* **13** 309
[14] Shimidzu T 1987 *React. Polym* **6** 221
[15] Livache T, Roget A, Dejean E, Barthet C, Bidan G, Teoule R 1994 *Nucleic Acids Res.* **22** 2915
[16] Livache T, Roget A, Dejean E, Barthet C, Bidan G, Teoule R, Mathis G 1998 *Anal. Biochem.* **225** 188
[17] Hesselink F T 1984 *Adsorption from Solution at Solid-Liquid Interface,* ed D Parfitt and C H Rochester (London: Academic Press)
[18] Can M, Ozaslan H, Pekmez N O, Yildiz A 2003 *Acta Chim. Slov.* **50** 741
[19] Allen N S, Murray K S, Flaming R J, Saunders B R 1997 *Synthetic Met.* **87** 237
[20] Cui X, Martin D C 2003, *Sensors and Actuators A Physical* **103** 384
[21] Prabhakar N, Arora K, Singh S P, Singh H, Malhotra B D 2007 *Anal. Biochem.* **366** 71
[22] Arora K, Chaubey A, Singhal R, Singh R P, Key M, Pandey, Samanta S B, Malhotra B D, Chand S 2006 *Biosen. Bioelectron.* **21** 1777
[23] Saunders B R, Murray K S, Fleming R J, McCulloch D G, Brawn L J, Cashion J D 1994 *Chem. Mater.* **6** 697
[24] Hsu C F, Zhang L, Peng H, Travas-Sejdic J, Kilmartin P A 2008 *Current Applied Phys.* **8** 316
[25] Omastová M, Pionteck J, Trchová M 2003 *Synth. Met.* **135-136** 437

Acknowledgment
This work is financially supported by the Vietnam Ministry of Education and Training (under code B2008-21-09. We are grateful to Vietnam National Center for Hygiene and Epidemiology for HSV-1 sequence samples.

Atomic layer deposition of TiO_2 photonic crystal waveguide biosensors

E Jardinier[1], G Pandraud[2], M H Pham[2], P J French[1] and P M Sarro[2]
[1]Electronic Instrumentation Laboratory, Delft University of Technology, 4 Mekelweg, 2628 CD Delft, The Netherlands
[2] Electronic Components, Technology and Materials, Delft University of Technology, 17 Feldmannweg, 2628 CT Delft, The Netherlands

E-mail: g.pandraud@tudelft.nl

Abstract. A photonic crystal waveguide biosensor in the visible is presented for biosensing. The sensor is applied to Refractive Index (RI) measurements. The sensitivity at different wavelength is presented for both air holes and air core configurations of photonic crystal waveguide (PCW) made of TiO_2. It is shown that by using Atomic Layer Deposition (ALD) the expected sensitivity of the air core configuration outperforms the previously reported results.

Keyword: Photonic crystal waveguide, biosensor, ALD.

1. Introduction

Photonic crystals (PC) are artificial optical materials with a periodic modulation of the refractive index. Depending on the exact periodic modulation PCs may posses a photonic bandgap. Thus, a given band width of light cannot be transmitted through such a material. A wide range of devices based on bulk PCs have been presented for biosensing where the incident light perpendicular to the PC is reflected and measured. Such a method has been used for protein [1] or antibodies [2] detection.

Planar PCs in which the light is guided along defects, such as missing rows of holes or rods can be designed to obtain a very high and spatially selective sensitivity to changes in RI superior to bulk devices. The basic property of a PCW is that a given bandwidth of light can be guided in the waveguide. As in a standard waveguide, the light is confined by the film thickness vertically and by the PC horizontally. The sensing properties of PCW have already been exploited for different simple parameters. Applications of PCW include nanofluidic tuning [3] and RI measurements [4], however they have not been used for biosensing in the visible and the ultra violet. In comparison with bulk PC biosensors, PCW sensors are ultra-compact and can be integrated with both additional optical and electronic components onto one single chip. We present in this paper a 2D Finite-Difference-Time-Domain (FDTD) modelling of a PCW made of TiO_2. We describe the fabrication of air core TiO_2 PCWs by ALD and compare its modelled sensitivity to the one of an air core device.

2. Photonic-crystal waveguide consideration

Among the great variety of deposition processes and technologies, ALD has been receiving increased attention due to its ability to deposit very thin layers at low temperatures with precise thickness and stress control and exceptional uniformity and conformality. For example, TiO_2 can be deposited by

© 2009 IOP Publishing Ltd

ALD and presents the characteristics of being transparent in the visible as well as having a high refractive (n = 2.47). Further ALD has proven to be the appropriate technique to produce air core PCs has shown in figure 1 [5]. At first an Anodic Aluminum Oxide (AAO) template is prepared on a substrate figure 1a. It consists of a periodic array of holes with a hexagonal lattice. The TiO_2 is then deposited by ALD, layer by layer, into the AAO pores figure 1b. The deposition rate is of about 0.5 Angstrom per cycle. Later the upper TiO_2 is removed by polishing figure 1c and the AAO template is selectively etched to obtain a periodic array of TiO_2 rods figure 1d. The used etchant can be phosphoric acid or KOH, both being selective for AAO against TiO_2. The line defect to produce a PCW is made by photolithography in the AAO template. To unsure the single-mode condition for a TE polarized incident light, the waveguide thickness, where the line defect is introduced, is chosen to be smaller then 100 nm.

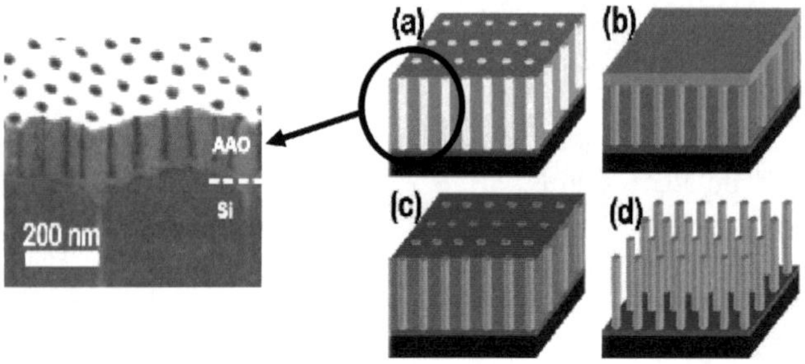

Figure 1. Different fabrication steps to produce TiO_2 pillars from an AAO template by ALD.

3. Modelling

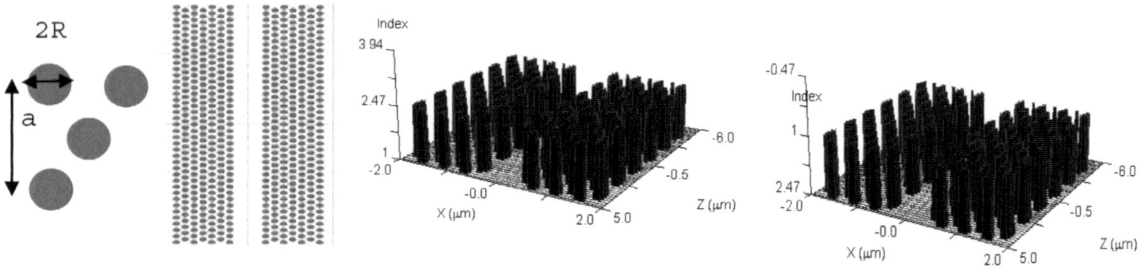

a) **b)** **c)**

Figure 2. a) Photonic Crystal Waveguide configuration, b) index profile of an air core PCW and c) the index profile of an air holes PCW.

Figure 2 shows the PCW simulated using a 2D FDTD modeling and the corresponding index profiles for both air holes and air core structures. It consists of a hexagonal lattice in which the three central rows are removed. At first, simulations are performed to optimize the lattice constant and the rods radius R (figure 2). Indeed the interesting property of PCW is the sudden drop in transmission when the transmission of the fundamental mode is no longer possible. A sharp change in transmission is a clear advantage for the actual sensor operation as it is simple to detect and gives the possibility for precise detection even for spectrum with a high degree of noise. Four lattice constant were investigated, a = 130, 260, 390 and 520 nm and for each of them the radius varies as r = 0.2a, 0.3a and 0.4a. Those dimensions are compatible with the dimensions of the AAO templates fabricated at Delft University of Technology [6]. They are also close to the ones usually used in PCW designs made of silicon.

The launched light is a continuous wave (CW), of Gaussian shape, of transverse electric polarization (TE), with a power normalized at 1. The output signal is collected in X = 0 at the end of the device (figure 2). Both structures presented in figure 2b and 2c were studied and the results are presented in the following.

3.1. Air core PCW

Figure 3 shows the main working points of such a device when the core refractive index is varying. Two PCW strutures b) and c) allow sensing in the visible range and in the red where sources are easy to find and cheap. The struture a) allows sensing in the extreme UV range.

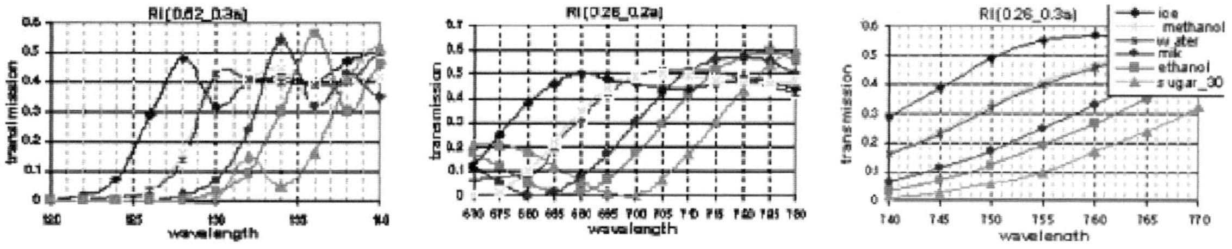

a) b) c)

Figure 3. Simulated transmission spectra of the air core PCW for various liquids for a) a = 0.52, r = 0.3a, b) a = 0.26, r = 0.2a and c) a = 0.26, r = 0.3a.

Figure 4 shows the sensitivity of the above structures. The change in wavelength is plotted against the change in refractive index. The points are all taken from the curves of figure 3 at a transmission level of 0.3. It shows that the most sensitive region of the spectrum is the region around 700 nm.

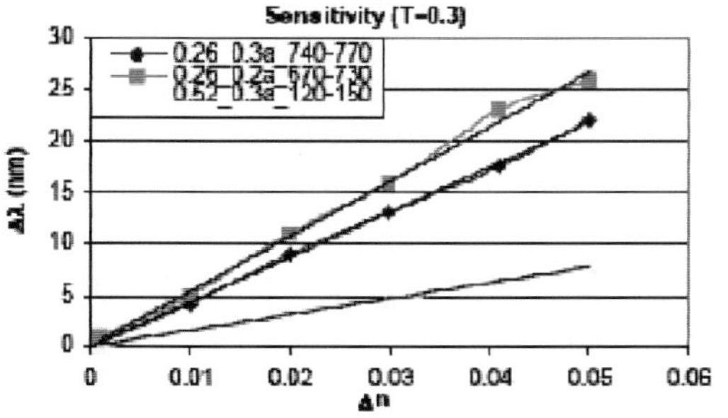

Figure 4. Sensitivity of an air core TiO_2 PCW for different lattice size and radius.

3.2. Air holes PCW

Figure 5 shows the transmission spectrum of a air holes PCW for two sets of lattice parameters and hole radius. It presents a sharp drop at the lower band edge with potential high sensitivity in the UV region. However, a deeper study showed that there is no shift in wavelength while the refractive index of the cover is changing. This assumption has to be verified by experiments as it is very likely that we are reaching the limits of the FDTD software package.

The upper band edge highlights a second possible working point of the system. It is situated in the near infrared range where PCW biosensors have already been reported. However, to get all the working points of PCW made of TiO_2, we performed the same study as for the air core PCW. This time the refractive indexes chosen were larger (figure 6) than in the previous case showing that the sensitivity is lower for this PCW configuration.

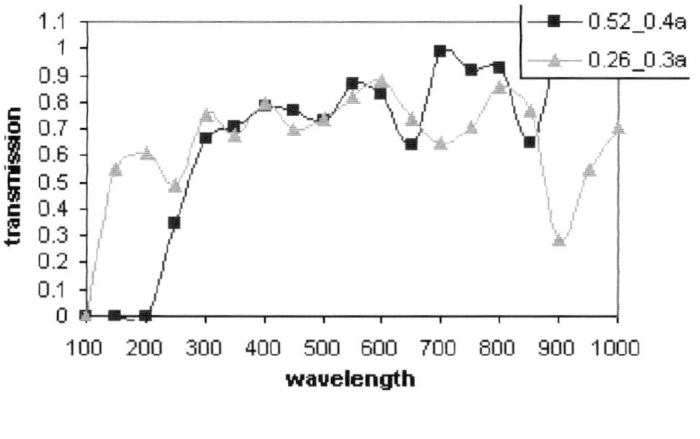

Figure 5. Calculated transmission spectra for TE-like polarization vs. wavelength from a TiO_2 PCW in water with a = 0.52 and R = 0.4a and a = 0.26 and R = 0.3a.

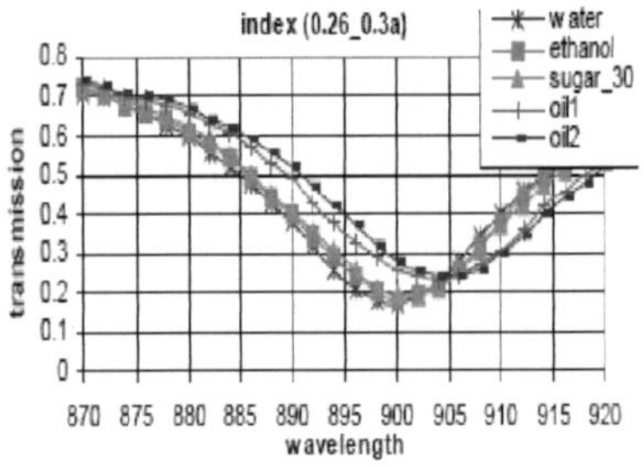

Solution	Refractive index
Water	1.33
Ethanol	1.36
Sugar 30%	1.38
Oil1	1.48
Oil2	1.518

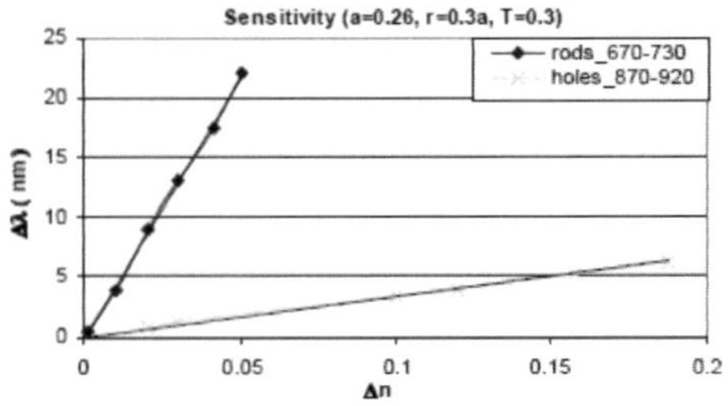

Figure 6. a) Calculated transmission spectra for TE-like polarization vs wavelength from a TiO_2 PCW for different covers with the different refractive indexes used; b) Sensitivity comparison between air holes and air core TiO_2 PCW with a = 0.26 and R = 0.3a at a transmission value of 0.3.

4. Conclusion

We have presented a photonic crystal waveguide used for RI measurements fabricated using ALD TiO_2. We have shown that it is a good candidate for sensing applications from the UV to the near infrared. By comparing an air holes structure to an air core structure we found out that by far the more sensitive one is the air core configuration with sensitivity as high as 0.002 nm^{-1}. Such structure presents also sensitivity higher than air holes biosensors made of silicon that exhibits a sensitivity of 0.0157 nm^{-1} [7].

References

[1] Chan L L, Cunningham B T, Li P Y and Puff D 2007 *Sens. Actuators* B**120** 392
[2] Levi O, Suh W, Lee M M, Zhang J, Brueck S R J, Fan S and Harris J S 2006 *Proc. CLEO/QELS Technical Conference*, Long Beach, USA
[3] Erickson D, Rockwood T, Emery T, Scherer A and Psaltis D 2006 *Opt. Lett.* **31** 59
[4] Topolanick J, Bhattacharya P, Sabarinathan J and Hu P C 2003 *Appl. Phys. Lett.* **82** 1143
[5] Yang C J, Wang S M, Liang S W, Chang Y H and Chen C 2007 *Appl. Phys. Lett.* **90** 033104
[6] Yan B, Pham H T M, Ma Y, Zhuang Y and Sarro P M 2007 *Appl. Phys. Lett.* **91** 053117
[7] Skivesen N, Têtu A, Kristensen M, Kjems J, Frandsen L H and Borel P I 2007 *Opt. Exp.* **15** 3169

Giant magnetoimpedance in layered composite micro-wires for high-sensitivity magnetic sensor applications

Le Anh Tuan, Nguyen Thanh Huy and Pham Thanh Huy

Department of Nano Science and Technology, Hanoi Advanced School of Science and Technology (HAST), Hanoi University of Technology (HUT),
F Building, 40 Ta Quang Buu Street, Hai Ba Trung District, Hanoi, Vietnam

E-mail: tuanla-hast@mail.hut.edu.vn

Abstract. Currently, the design and development of novel magnetic materials with optimized properties is of relevance to study new phenomena and to use them for multifunctional applications of sensor technologies and magnetoelectronic devices. In this letter, we present a detailed investigation of giant magnetoimpedance (GMI) effect in layered magnetic microwires. A new family of layered microwire composite materials was fabricated by using rapidly quenching and coating techniques. The experimental measurements were performed using an impedance analyzer (HP 4191A) in the frequency range of 100 – 1000 MHz and a varying applied dc magnetic field within ± 300 Oe. Interestingly, the GMI ratio and magnetic-field sensitivity reached the largest values of 3200 and 2400%/Oe at a resonance frequency of 890.15 MHz for layered Co-rich microwire sample, respectively. These results are very ideal for the development of a new family of ultra-sensitive and high-frequency magnetic sensors. Finally, the layered composite magnetic micro-wire developed with enhanced sensitivity performance can be ideally used as sensing elements in magnetic sensor technologies.

Keywords: Giant magnetoimpedance, layered magnetic microwires, magnetic sensor.

1. Introduction

Nowadays, high-performance sensors and transducers have an increasing interest in the society, industry and academia, because of their importance in many technological applications [1]. All modern vehicles and transport means use a vast variety of sensors and transducers. The operation of all medical instruments is also based on sensors and transducers. Industry is also employing more and more transducers for the monitoring and control of production lines [2].

A recent discovery of giant magneto-impedance (GMI) effect in amorphous magnetic wires [3] has given rise to a new direction in magnetic sensor technology. These soft magnetic wires exhibited outstanding properties such as: (i) relatively high resistivity, (ii) tiny dimensions, (iii) high permeability (or low coercivity), and (v) increased mechanical strength. With these excellent properties, these wires are to be promising candidate for micro-technological applications, and present an attractive option for high-sensitivity magnetic sensors, advanced measurement and control systems, and high-density magnetic recording applications [4, 5].

In the present work, a new class of layered composite micro-wires was designed and fabricated by using rapidly quenching and coating techniques. A detailed investigation of the high-frequency GMI behaviours in these materials was performed. It has been revealed that the layered composite magnetic

© 2009 IOP Publishing Ltd

micro-wire developed with enhanced sensitivity performance ideally represents a new class of sensing materials with great technological potential in high-sensitivity sensor devices.

2. Experimental works

Figure 1 shows the schematic view of fabrication process and composite micro-wire sample. The inner ferromagnetic core is a Co-rich microwire of nominal compositions $Co_{83.2}B_{3.3}Si_{5.9}Mn_{7.6}$ fabricated by Taylor-Ulitovsky method [6], in which the alloy pieces were previously introduced in a glass tube having a sealed end, and then the alloy was melted by the high frequency field of an inductive coil and the end of the glass tube was softened. Hence, around the molten metal drop, there was a softened glass cover which allows the drawing of the capillary. The glass tube displaced with a uniform speed ranging between 0.5 and 7 mm/min in order to ensure the continuity of the process. The as-formed wire was cooled by a water jet at approximately 1 cm under the high frequency induction coil. The diameter of the ferromagnetic core was about 16 μm and the thickness of the insulating glass coating was about 5 μm. The prepared microwire sample of 15 mm in length was fixed on a glass substrate. A conductive layer with 0.5 μm in thickness was deposited on the glass-coated micro-wire at two ends of the microwire by coating technique. And then, two copper contact terminals of 10 mm in length were connected with the sample.

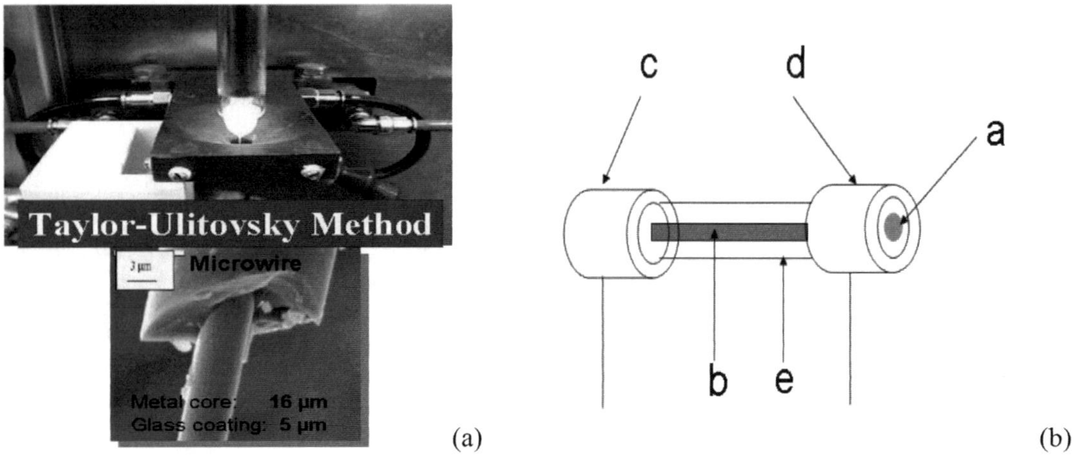

Figure 1. a) A schematic view of Taylor-Ulitovsky method (above, left side) for fabrication of amorphous composite micro-wire (below, left side); b) A layered composite micro-wire (right side) consists of a,b: Co-rich ferromagnetic core; c,d: conductive layers; e: glass coating.

The impedance values of all investigated samples were directly measured by an impedance analyzer (HP4191A, 1 MHz – 1 GHz) in a wide frequency range from 100 - 1000 MHz. During the measurement, the power of the output terminal in HP4191A was fixed as 1 mW. Therefore, the current amplitude was varied as a function of frequency and also abruptly changed at near the resonance point. Impedance measurements were conducted along the sample axis under a longitudinal applied dc magnetic field. The external dc magnetic field, created by a solenoid, was swept through the entire cycle equally divided by 800 intervals from -300 to +300 Oe.

The percentage change of magnetoimpedance, e.g. GMI ratio, with applied dc magnetic field has been expressed as [7]

$$GMI\ (\%) = \Delta Z/Z\ (\%) = 100\ \%\ \text{x}\ [Z(H)-Z(H_{max})]/Z(H_{max}), \quad (1)$$

and the dc magnetic-field sensitivity of GMI as

$$\xi = [\Delta Z/Z\ (\%)]_{max}/\Delta H, \quad (2)$$

where H_{max} is an applied maximum dc magnetic field and H_{max} = 300 Oe in the present work. ΔH stands for the full width at half maximum (FWHM).

3. Results and discussion

From the fundamental point of view, the GMI can be understood as a large change in the alternating current (ac) impedance of a magnetic conductor with an ac when subjected to a dc magnetic field. It originates from the skin effect as a consequence of the changes in the penetration depth (δ_m) induced by the applied dc magnetic field through modification of the effective permeability of a magnetic material [8]. This change of the penetration depth, for example, for a magnetic wire, can be described by formula

$$\delta_m = \sqrt{\frac{\rho}{\pi f \mu_\phi}} \, , \tag{3}$$

where f is the frequency of the ac current flowing along the sample, ρ is the electrical resistivity, and μ_Φ is the circumferential magnetic permeability [9]. Since application of a dc magnetic field increases the skin depth and thereby decreases the impedance, the penetration depth should be as small as possible in the absence of an applied magnetic field, in order to obtain a GMI effect. It reveals that the GMI effect would exist in materials having (i) low resistivity, (ii) high effective permeability, and (iii) low relaxation parameter [10]. In this context, Co-rich amorphous wires are good candidates for GMI sensor applications, because they possess extremely high circumferential permeability arising from their circumferential domain structure [9, 10]. The GMI characterizations are strongly dependent on the frequency of the ac current flowing along the sample and the applied magnetic field. We will analyze these dependences in detail.

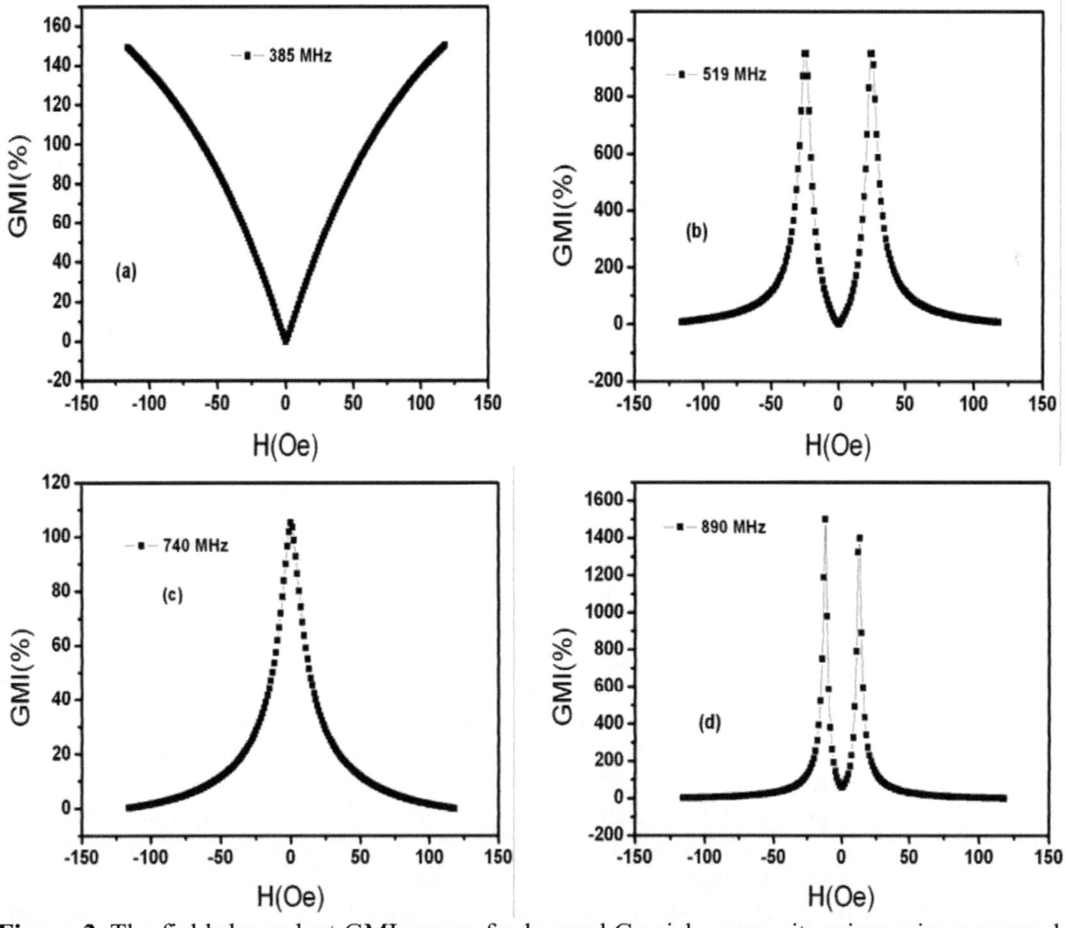

Figure 2. The field-dependent GMI curves for layered Co-rich composite micro-wire measured at different frequencies of a) 385 MHz; b) 519 MHz; c) 740 MHz; and d) 890 MHz.

3.1. Driving frequency dependence of GMI

Figure 2 displays the field-dependent GMI curves measured at different frequencies. It can be easily seen that the shape of GMI curves varies dramatically as the frequency increases. GMI profiles show a single-peak feature, or a double-peak one, depending upon a certain frequency region. According to the GMI theory [8], at high frequencies (1 MHz $< f <$ 1 GHz), the GMI effect was ascribed to variations of the magnetic penetration depth due to strong changes of the effective permeability caused by an applied dc magnetic field. In this case, both domain walls motion and magnetization rotation contribute to the circular permeability and consequently to the GMI effect. In our present case, because the measuring frequency varies from 100 – 1000 MHz, consequently, the observed GMI behavior was mainly related to the rotation of magnetic moments arising from the circular magnetization process. In the investigated frequency range, large GMI ratio values of 900% and 1500% were found at measuring frequencies of 519 and 890 MHz, respectively, as shown in figure 2b and 2d. These results make the layered wire-shaped material very ideally for developing high-sensitivity magnetic sensor applications.

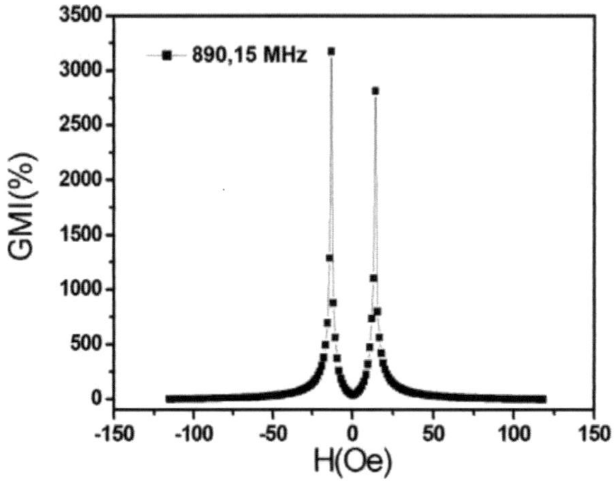

Figure 3. The field-dependent GMI curve measured at 890.15 MHz.

3.2. Magnetic field dependence of GMI

To measure GMI effect, a dc magnetic field is usually applied parallel to the ac current along the longitudinal direction of the sample [8, 9]. At a given frequency, the application of a dc magnetic field (H_{dc}) changes the circumferential permeability μ_Φ and hence the penetration depth δ which in turn alters the magnetoimpedance until the value of δ reaches the radius of the sample (a). To achieve a large GMI effect the penetration depth should be as small as possible in the absence of an applied magnetic field. Large circumferential permeability along with a low value of the resistivity gives rise to a small penetration depth at high frequency range. A large increase of the circumferential permeability can be achieved by applying an ac current of the frequency sufficiently high to excite the resonance of the sample. This large circumferential permeability at the resonance strongly decreases the penetration depth and, therefore increases the impedance of the sample. As observed in figure 2, it is worth noting that, at the resonance frequency of 890 MHz, the maximum GMI value is much larger than that obtained at 519 MHz. This implies that the resonance at 890 MHz produced a larger circumferential permeability when compared to that at 519 MHz. This hypothesis can be acceptable because the ferromagnetic resonance actually occurs around 890 MHz (~ 1 GHz) resulting in an increase of the permeability with respect to frequency. Therefore, a smaller dc magnetic field is required to excite the sample at 890 MHz and a larger GMI effect is consequently observed. This probably explains a shift of the position of the resonance peaks towards lower values of the applied field, when the resonance frequency increases from 519 to 890 MHz. As one can see from figure 2d and 2d, at the frequency of 519 MHz, the resonance peaks appear at $H_{dc} = \pm$ 24.6 Oe, whereas it occurs at $H_{dc} = \pm$ 12.3 Oe at the frequency of 890 MHz.

In order to further improve the GMI sensitivity, we investigate the GMI behavior around the resonance frequency of 890 MHz. It is noted that from the magnetic field dependence of GMI we can calculate the field sensitivity of GMI ξ (%/Oe) as illustrated by equation (2). In practice, based upon the magnitude of ξ, the operating regime and potential applications of a magnetic material can be determined. Interestingly, the largest GMI value of 3200 % was observed at a measuring frequency of 890.15 MHz (figure 3). The field sensitivity of GMI calculated is about 2400 %/Oe. This is one of the highest values of field sensitivity of GMI reported till now in the research of GMI. These results are very ideal for developing ultra-sensitive magnetic sensors operating at the high-frequency region.

4. Conclusion

In the present work, a new class of layered Co-rich composite micro-wire was fabricated by the two-step process including rapid quenching and coating. A study of high-frequency GMI behavior in newly layered composite micro-wire was performed. Noticeably, the GMI effect was observed at two frequencies of 519 MHz and 890 MHz. Very interestingly, the largest field sensitivity of GMI reached a large value of 2400 %/Oe at a measuring frequency of 890.15 MHz. These results revealed that the layered material can be used as a sensing element for high-performance sensor devices.

References

[1] Ripka P and Acuna M H 2001 *Magnetic sensors and magnetometers* (Norwood, MA: Artech House)
[2] Jiles D C and Lo C C H 2003 *Sens. Actuators A.* **106** 3
[3] Pannia L V and Mohri K 1994 *Appl. Phys. Lett.* **65** 1189
[4] Vazquez M and Hernando A 1996 *J. Phys. D: Appl. Phys.* **29** 939
[5] Vazquez M *Physica B.* **299** 302
[6] Chiriac H and Ovari T A 1996 *Prog. Mater Sci.* **40** 333
[7] Tuan L A et al. 2007 *J. Phys. D: Appl. Phys.* **40** 4582
[8] Knobel M and Pirota K R 2002 *J. Magn. Magn. Mater.* **242-245** 33
[9] Mandal K, Puerta S, Vazquez M and Hernando A 2000 *Phys. Rev. B.* **62** 6598
[10] Phan M H, Peng H X, Wisnom M R and Yu S C 2005 *J. Appl. Phys.* **98** 014316

Acknowledgments

This work was acknowledged to an internationally scientific collaboration program between HAST-Vietnam, CNU-Korea, and CSIC-Spain, BK 21 and KOSEF program in Korea.

Preparation and investigation of ferroelectric $Pb(Zr_{0.53}Ti_{0.47})O_3$ by modified Pechini method

Truong Van Chuong, Huynh Duy Nhan, Le Quang Tien Dung and Nguyen Duy Anh Tuan

Department of Physics, College of Science, Hue University, Vietnam

E-mail: truongvanchuong@yahoo.com

Abstract. This paper describes a new method, namely modified Pechini for one, for the preparation of nano-sized ferroelectric $Pb(Zr_{0.53}Ti_{0.47})O_3$ (PZT (53/47)) from low cost precursors. With this method, PZT (53/47) was prepared by microwave-assisted dissolving a pre-calcinated mixture of component oxides in dilute HNO_3 for a short time and precipitating in NH_4OH solution (pH = 9-10). On the other hand, the ferroelectric thin films PZT (53/47) on Al substrates were also prepared. The films and the ceramic powders of PZT (53/47) were obtained with high quality in terms of structure and microstructure. The ferroelectric properties of thin films have also been investigated.

Keywords: Ferroelectric materials, PZT (53/47), microwave-assisted, Pechini method.

1. Introduction

The ferroelectric $Pb(Zr_xTi_{1-x})O_3$ (PZT), La-doped PZT (PLZT), and $PbTiO_3$ materials are attractive research subject not only in sense of basic research but also of applications such as in fabricating sensors and actuators. There are two common preparation methods for these compounds: a traditional solid state reaction and a sol-gel method. By using a solid state reaction, the title compounds are prepared from low cost starting materials but the prepared powders possess microscale average grain size. In contrast, the sol-gel method provides nanoscale average grain size but requires expensive precursors such as $TiCl_4$ or $Ti[O(CH_2)CH_3]_4$ [1]. The advantages (nano-sized grains and low cost precussors) of these two method can be combined by applying the modified Pechini method [2]. With this method, from precursors (nitrate salts, ethylene glycol and citric acid), a fixed network enclosing the metal ions is formed and, as a result, the particles size is inhibited to grow. In addition, this method has the capability to prepare multicomponent compounds with high homogeneity.

For the preparation of ferroelectric PZT and PLZT materials by wet chemical route using TiO_2 as one of starting materials, the most difficult problem is the low solubility of TiO_2 in dilute HNO_3. J. A. Eiras et al. proposed to make the PZT solutions from pre-calcinated PZT powders. However, due to the use of normal heating calcination methods it could not be dissolved totally PZT in dilute HNO_3 [3-5]. In this paper, we prepare and investigate the PZT (53/47) by dissolving calcined PZT (53/47) in dilute HNO_3 using microwave-assisted power. In addition, the Pechini method was also used to prepare the ferroelectric thin films.

© 2009 IOP Publishing Ltd

2. Experimental

2.1. Experimental setup

The microwave-assisted heating is a heat supplying technique by creating molecular oscillation at high speed. Its ability is to supply a homogeneous heat field similar to that of the hydrothermal process. This is a combination of normal calcination process and the heat due to rub of molecules. The heating process is performed immediate inside the sample. With microwave oven having the frequency of 2.45 GHz, the water molecules spin according to field and to be in contact with together 2.45 billion times per second, as a result, a very large and uniform heat field was created. The NE-5670 microwave oven was operated with the power of 500 W and the frequency of 2.45GHz. The low temperature microwave process device was described on figure 1.

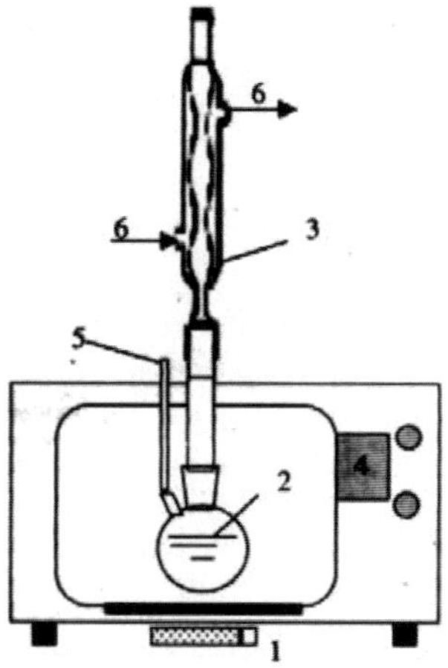

1. The magnetic stirrer
2. Reactor
3. Recover pipe
4. Microwave tube
5. Thermometer
6. The water cooler

Figure 1. The low temperature microwave-assisted processing system.

2.2. Preparation sample

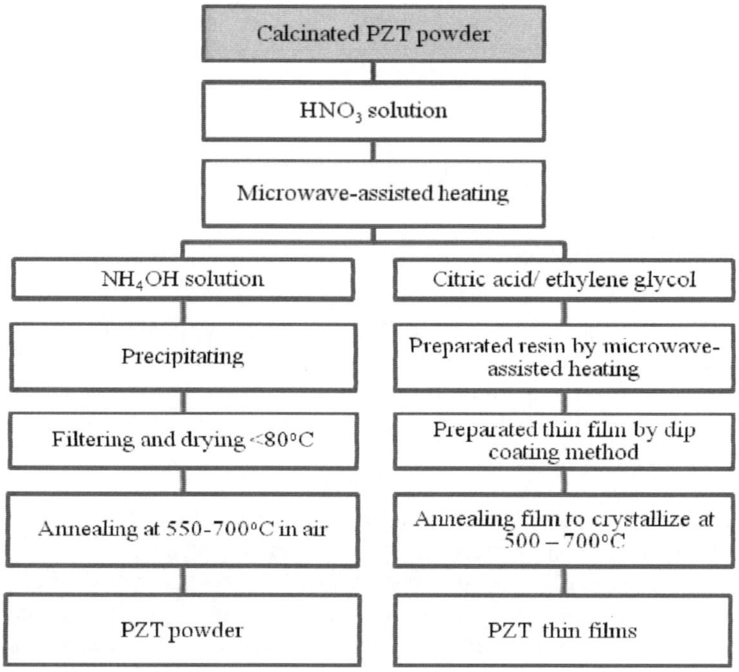

Figure 2. The diagram prepared PZT (53/47) materials by modified with the Pechini method.

The investigated sample under a general chemical formula of $Pb(Zr_{0.53}Ti_{0.47})O_3$ (PZT (53/47) was obtained following the preparation procedure described in figure 2. The starting materials (PbO, ZrO_2, TiO_2), with an excess amount of 10 wt.% PbO in order to compensate the lost in weight during the firing, were used. The oxides were mixed and milled in 12 h and were then calcined in air at 850°C for 4 h. The calcined powder was given into a tank of dilute 10 wt.% HNO_3 solution in a 250 ml glass jar, and was placed in the microwave oven and set at medium mode for 15 minutes. The PZT (53/47) material dissolved completely in dilute HNO_3 solution, the obtained solution is transparent, forming the sol solution contained (Pb^{2+}, Zr^{4+}, Ti^{4+})$(NO_3)_x$.

2.2.1. Preparation PZT (53/47) ceramic powders

The solution contained (Pb^{2+}, Zr^{4+}, Ti^{4+})$(NO_3)_x$ was cooled to room temperature and then under continuous agitation an aqueous NH_4OH solution was added, to promote the precipitation. The constituents were precipitated as $Pb(OH)_2$, $TiO(OH)_2$ and $ZrO(OH)_2$. The precipitated slurry was obtained when the pH value of the solution was adjusted from 9 to 10.

2.2.2. Preparation PZT thin film

For preparation of the resin by the Pechini method, citric acid and ethylene glycol (citric acid/ethylene glycol = 45/55 in mol. %) were mixed with 100 ml from solution containing (Pb^{2+}, Zr^{4+}, Ti^{4+})$(NO_3)_x$. The mixture was processed by microware-assisted setting at medium mode for 2 minutes after that the solution compound is agitated for 12 h. A transparent resin of light yellow was obtained. Films of the resins PZT were deposited on hot substrates of Al at a speed of 0.5 mm/minute by dip coating method. The deposited films were previously heated on a hot plate before they were heat treated for crystallization.

3. Results and discussions

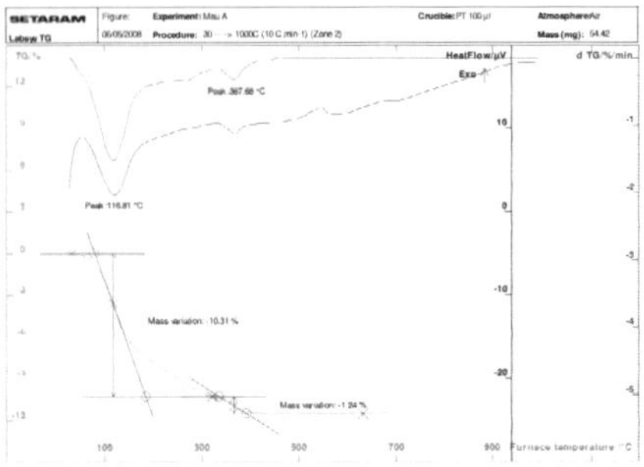

Figure 3. DTA/TGA curves of the coprecipitated PZT powders.

The endothermic peak on DTA curve (figure 3) at 368°C was related to the formation of the PZT and this formation ended up at 700°C. The PZT was formed without intermediate reactions and the mass lost of only 1.24 wt.% on TGA curve. This is affirmed by X-ray diffraction diagram (XRD) in figures 4 and 5. Figure 4a showed the XRD pattern of powders annealed at temperature of 600°C for 2 h. At this temperature the crystallization of PZT is not completely and pyrochlore phase is observed. The powders were crystallized completely at 700°C for 2 h and forming tetragonal phase at angles 44.5, 50, and 55° and rhombohedral at 21.7, 31.2, 38.2° (figure 4b).

Figure 5a shows the XRD pattern of a PZT film on Al substrates annealed at 550°C for 30 min. The tetragonal PZT phase is found with (100), (101), and (210) peaks at corresponding 2θ-angles of 21.9°, 31°, and 50.2°. Besides, the rhombohedral phase is also detected with peaks (101), (200) at 31°, 44.5°, respectively. However, the peak intensities are still low and the thin film may be only partially crystallized. The thin film annealed at 600°C for 30 min. (figure 5b) exhibits better crystallinity than film on figure 5a. The PZT film is completely crystallized with the coexistence of tetragonal and rhombohedral PZT phases.

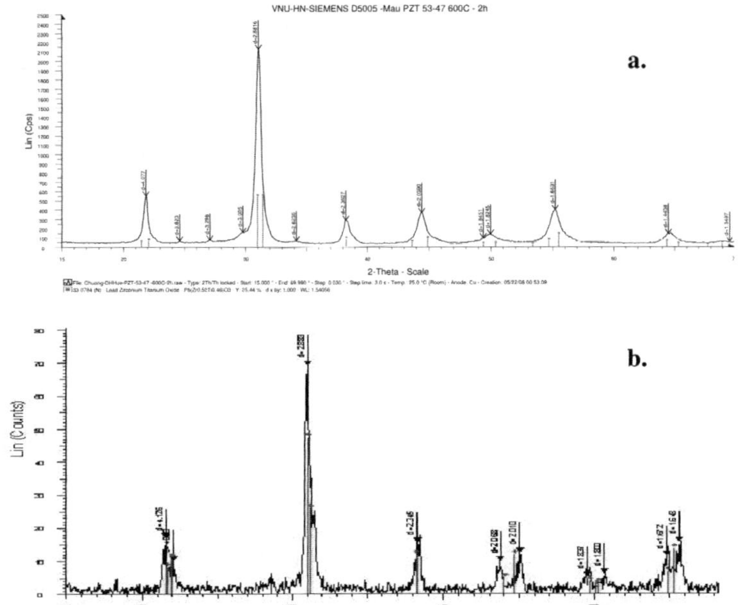

Figure 4. X-ray diffraction diagram of PZT (53/47) ceramic powders prepared by modified Pechini method: (a) annealed at 600°C for 2 h and (b) annealed at 700°C for 2 h.

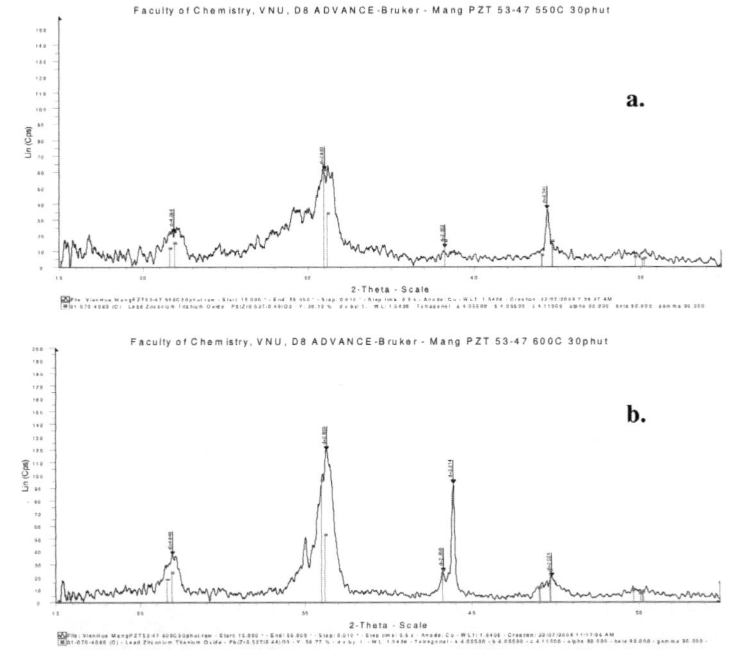

Figure 5. X-ray diffraction diagram of PZT (53/47) thin film on Al substractes prepared by modified Pechini method: (a) annealed at 550°C for 30 minutes and (b) annealed at 600°C for 30 minutes.

Figure 6. FESEM images of PZT (53/47) thin film on Al substrates prepared by modified Pechini method annealing at: (a) 600°C for 30 minutes, (b) 650°C for 30 minutes, (c) 700°C for 30 minutes.

Figures 6 a, b and c showed FESEM images of the PZT (53/47) thin films on Al substrates annealed at 600, 650, and 700°C for 30 minutes, respectively.

The surface morphology of a PZT film deposited on Al substrates exhibits a dense microstructure. These films were prepared with 7 layers of coating and its average grain size of about 80 nm.

The EDS spectrum of PZT (53/47) powder annealed at 700°C for 2 h, showed the presence of all component chemical elements according to ratio of composition and no strange elements are detected (figure 7).

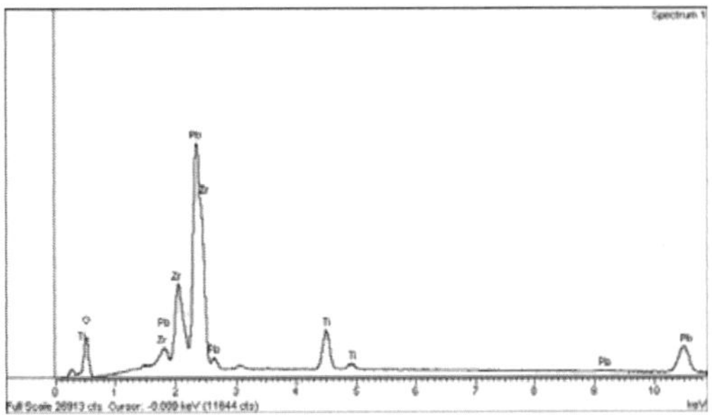

Element	Weight (%)	Atomic (%)
O K	21.38	69.83
Ti K	6.58	7.18
Zr L	15.02	8.60
Pb M	57.03	14.38
Totals	100.00	

Figure 7. The EDS spectrum of PZT (53/47) powder prepared by modified Pechini method annealing 700°C for 2 h.

Figure 8 shows the hysteresis loops of the PZT (53/47) ceramics prepared from the coprecipitated powders and sintered at 950, 1000, 1050, and 1100°C for 3 h. The measured ferroelectric parameters of PZT powders are given in table 1. The results show that when the sintered temperature increased, the remanent polarization (P_r) also increased and the coercive field (E_c) decreased. The coercive field (E_c) decreased from 24 to 11 kV/cm and the remanent polarization (P_r) increased from 17 to 47 μC/cm^2.

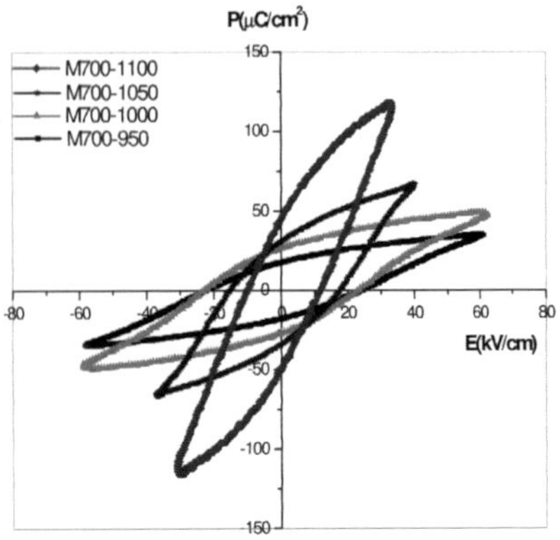

Figure 8. The ferroelectric hysteresis loops of ceramic samples.

Table 1. The ferroelectric parameters of ceramic samples prepared from the coprecipitated PZT powders.

Sample	The sintered temperatures and period	P_r (μC/cm^2)	E_c (kV/cm)
M700 - 950	950°C – 3 h	17	24
M700 - 1000	1000°C – 3 h	26	22
M700 - 1050	1050°C – 3 h	32	17
M700 - 1100	1100°C – 3 h	47	11

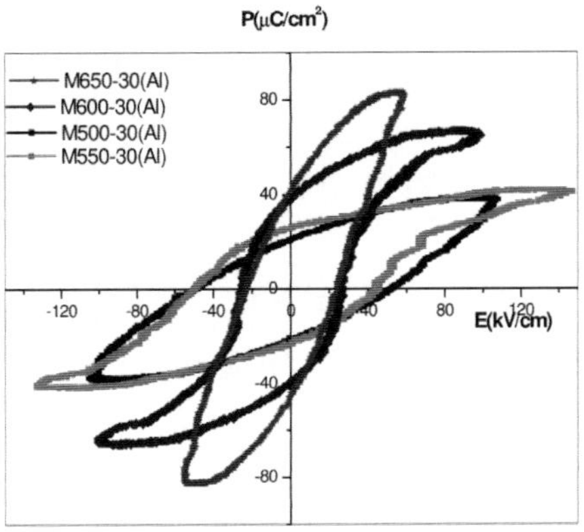

Figure 9. Hysteresis loops of PZT films on Al substrates annealed at 500, 550, 600, 650°C for 30 minutes.

Table 2. The ferroelectric parameters of the PZT (53/47) thin films on Al substrates

Thin films	The annealead temperatures and period	P_r (μC/cm^2)	E_c (kV/cm)
M500 – 30(Al)	500^{0}C – 30 minutes	21	50
M550 – 30(Al)	550^{0}C – 30 minutes	25	47
M600 – 30(Al)	600^{0}C – 30 minutes	40	25
M650 – 30(Al)	650^{0}C – 30 minutes	43	23

From the table 2, one can see the remanent polarization (P_r) of the thin films on Al substrates is increased from 21 to 43 μC/cm^2 and the coercive field (E_r) decreased from 50 to 23 kV/cm. Generally, the PZT (53/47) thin films on Al substrates prepared by dip coating method exhibit good ferroelectric.

4. Conclusion

A new method, the modified Pechini method, was proposed and developed for preparation of ferroelectric ceramic powders and thin films. This method allows to make use of low cost precursors, like oxides or carbonates, and provides nano-sized prepared PZT powders or thin films. Another advantage of this method is the possibility of obtaining a source solution containing all constituents, with low crystallization temperature.

This method caused a full reaction of PZT (53/47) powders in diluted HNO_3, had no immediate reactions during the crystallization process from 367°C and completely crystallized to form the tetragonal and rhombohedral phases at 700°C. The prepared samples exhibit homogenous structure and microstructure and having the average particles size of about 80 nm for the ceramics as well as the thin films. These samples also show good ferroelectricity with the remanent polarization P_r from 17 to 47 μC/cm^2 and the coercive field E_c from 11 to 24 kV/cm for the ceramics, the remanent polarization P_r from 21 to 43μC/cm^2 and coercive field E_c from 23 to 50 kV/cm for the films on Al substrates.

References

[1] Dung L Q T, Chuong T V and Dan V D 2004 *Proc. of the Second International Workshop on Nanophysics and Nanotechnology* p.187
[2] Pechini M P, Adams N, Mass 1963 *Method of preparing lead and alkaline earth titanates and niobates and coating method using the same to form a capacitor* Ser. **No. 304**434 7 Claims. (Cl, 117-215)
[3] Rodirigues J B, Eiras J A 2002 *J. Eur. Cer. Soc.* **22** 2927
[4] Araujo E B, Eiras J A 1999 *J. Mater. Res. Vol. 2.* **No. 1** 17
[5] Menegazzo B A and Eiras J A 1993 *J. Am. Cer. Soc.* **76** 2734

Surface modification of iron oxide nanoparticles and their conjuntion with water soluble polymers for biomedical application

Nguyen Thanh Huong[1], Lam Thi Kieu Giang[1], Nguyen Thanh Binh[1] and Le Quoc Minh[1, 2]
[1] Institute of Materials Science, Vietnam Academy of Science and Technology
18 Hoang Quoc Viet Road, Cau Giay District, Hanoi, Vietnam
[2] College of Technology, National University, Hanoi

E-mail: nthuong@ims.vast.ac.vn

Abstract. Superparamagnetic iron oxide nanoparticles (SPION) coated with suitable bio-compatible substances have been used in biomedicine, particularly in magnetic resonance imaging (MRI), tissue engineering, and hyperthermia and drug delivery. In this study, we describe the synthesis of SPION and its surface modification for in-vitro experiments. The particle diameter and structure were estimated by FESEM, TEM, XRD analyses. The saturation magnetization was characterized. SPION with a mean size of 12 nm have been prepared under N_2 atmosphere, with support of natural polymeric starch, by controlling chemical coprecipitation of magnetite phase from aqueous solutions containing suitable salts ratios of Fe^{2+} and Fe^{3+}. The surface of SPION-nanoparticles was treated with a coordinatable agent for higher dispersion ability in water and remaining the superparamagnetic behavior. The prepared iron oxide nanoparticles were coated with starch, dextran, PEG or MPEG to extend the application potential in the quite different engineering field of nano biomedicine.

Keywords: Nanoparticles, superparamagnetism, colloid, surface engineering, biomedical application.

1. Introduction

Superparamagnetic iron oxide nanoparticles (SPION) combined with MRI is under clinical evaluation to enhance the detection of neurodegenerative diseases. A major improvement would be to link therapeutic drugs to the SPION to achieve targeted drug delivery, either at the cell surface or intracellular, together with active disease detection, without inducing cell reaction [1, 2].

Dextran was used as a coating agent for SPION to form a core of iron oxide and a coating of biocompatible dextran. The size of the core affects the sensitivity of detection in MRI and the overall size can be modified by varying the thickness of the dextran layer giving the nanoparticle possibilities of extracellular transport or restricting it to the intravascular compartment. This type of nanoparticles have been successfully used in cancer diagnostics in liver tumors and pathologies where SPION are selectively taken up by healthy liver cells while sclerotic parts of the liver cannot. Consequently, in MRI the healthy regions are darkened and the diseased regions remain bright. The general tendency is to develop MRI devices with higher magnetic field strength for in-vivo applications, which give the opportunity for faster recordings at higher spatial resolution.

© 2009 IOP Publishing Ltd

Colloidal processing of magnetic nanoparticles has been extensively studied, emphasizing the characteristics for dispersing SPION as a ferrofluid. To maintain SPION in a colloidal solution the attractive forces from the SPION must be considered, such as Van der Waals' forces and magnetic dipole-dipole interactions generated from remanent magnetic forces, which tend to agglomerate the particles. These forces must be stronger than the attractive forces in order to stabilize the ferrofluid in a colloidal solution. Surfactants are typically introduced to modify the surface properties, while Coulombic repulsion accounts for the stability in ionic ferrofluids. Potential candidates of biomaterials like drug carriers and coating agents for MR purposes can be natural polymers such as carbohydrates and proteins. Starch, a long-chain polymer of D-glucose, is one of the most abundant naturally occurring polysaccharides. It has been used experimentally as a drug carrier because of its biocompatibility, biodegradability, and non-toxicity [3]. Two advantages of SPION are their low toxicity to humans and the possibility to exploit their superparamagnetic properties [4, 5]. The aim of the present investigation is to design target specific markers that can be used for detecting morphological and physiological changes in vivo using magnetic resonance imaging (MRI) technique.

In this paper we report the results of synthesizing of iron oxides by co-precipitating of Fe^{2+} and Fe^{3+} in presence of a natural water-soluble polymer under oxygen-free environment by passing N_2 gas. We have treated the ironnanoparticles with some polymers such as starch, dextran, polehthylenglycol (PEG) or methylpolyehtylene glycol (MPEG). This paper is focused on coating magnetic cores with biocompatible outer shells.

2. Experimental

All the chemicals were of reagent grade used without further purifications. Iron (III) chloride hexahydrate ($FeCl_3.6H_2O$ > 99%), Iron (II) chloride tetrahydrate ($FeCl_2.4H_2O$ > 99%) and soluble starch (Zukowsky GR), sodium hydroxide (NaOH > 99%) and hydrochloric acid (HCl > 37%) were obtained from Merck. Water was de-ionized (specific conductance < 10^{-8} S) and deoxygenated by bubbling N_2 gas for 1 h prior to use.

A dispersion of SPION was prepared by alkalinizing an aqueous mixture of ferric and ferrous salts with NaOH at room temperature. N_2 gas was bubbled through the reaction medium during synthesis in a closed system. 0.2 g of starch was dissolved in 160 ml of distilled and de-ionized water in a bottle of 250 ml at the end the solution was heated up to 80°C. 1.7 g $FeCl_2.4H_2O$, 4.7 g $FeCl_3.6H_2O$ and some drops of HCl 37% were prepared as a source of iron by dissolving the respective chemicals in de-ionized water under vigorous stirring. In the same way, stock solutions of 5.0 g NaOH in 80 ml of de-ionized water were prepared as alkali source. A dilute HCl was prepared for surface neutralization, while a solution of dextran, PEG, MPEG and starch at pH from 10.0 to 11.0 was prepared for coating.

The magnetic nanoparticles were prepared by controlled coprecipitation method [5]. After that the reaction suspension was precipitated and washed by de-ionized water and supernatant was removed from the precipitate by magnetic decantation. After 3 times of washing the powder, 0.01M HCl was mixed to neutralize the anionic charge on the particle surface. The cationic colloidal particles have modified their surface by interacting with the polymer dextran, PEG, MPEG and starch.

The resulting particle suspension was then washed several times with analytical grade methanol with the help of external magnets. Then methanol solution was mixed with toluene and heated at 95°C under nitrogen until half of the solution evaporated, so that the water was thoroughly removed. A solution of 3-aminopropyltrimethoxy silane (APTMS) was added to the solution. The ferrofluid was stirred and heated under N_2 atmosphere at 110°C for 10 hours.

All the main synthesis and surface treatment steps were carried out under N_2 gas passing through the solution media to avoid possible oxygen contamination during the operations.

Journal of Physics: Conference Series **187** (2009) 012046 doi:10.1088/1742-6596/187/1/012046

3. Results and discussion

Figure 1. FESEM micrograph of Fe_3O_4 nanoparticles.

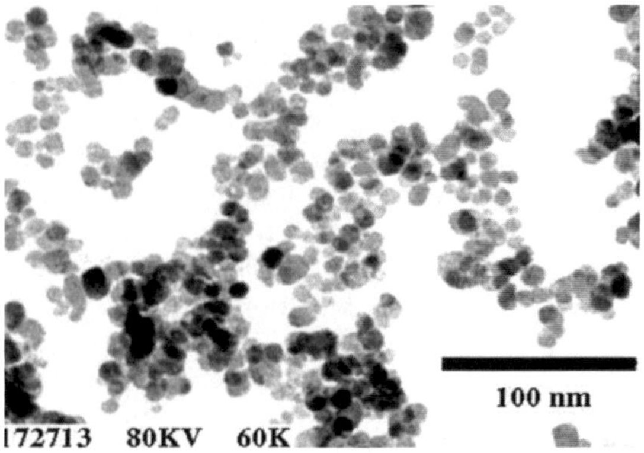

Figure 2. TEM image of Fe_3O_4 nanoparticles.

During the precipitation process, in order to prevent the possible oxidation in air as well as the agglomeration of Fe_3O_4 nanoparticles produced by reaction, they were usually coated with organic molecules such as diethylenglycol (DEG). The FESEM (figure 1) and the TEM images of the typical SPION (figure 2) confirm that the fabricated Fe_3O_4 nanoparticles have a average size below 12 nm and relatively narrow particle size distribution. This result is close to the crystal size calculated from XRD data.

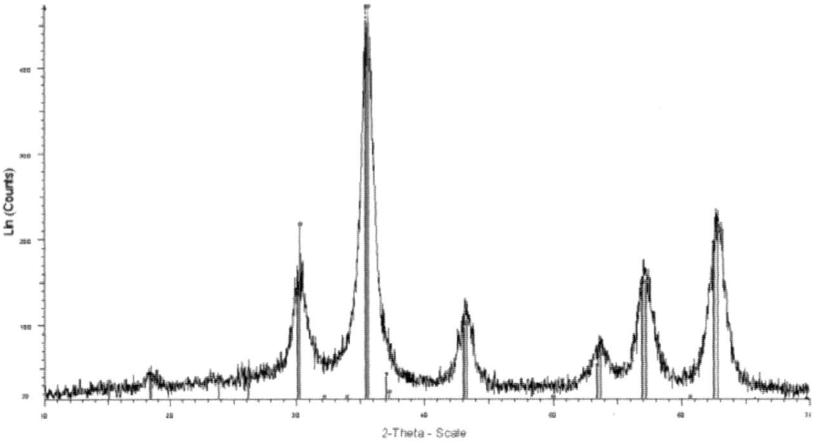

Figure 3. X-ray diffraction pattern of the iron oxide nanoparticles.

To receive structural information on the obtained magnetite powders, the samples were analyzed by X-ray diffraction measurements (figure 3). XRD measurements were performed using a Siemens D5005 diffractometer. The XRD spectrum shows a high degree of crystallinity of the iron oxide nanoparticles and the presence of diffraction peaks related to magnetite and maghemite crystallographic phase. The Scherrer's formula [7] was used for the estimation of the average crystal size. The mean diameter was calculated and estimated 6-9 nm in size.

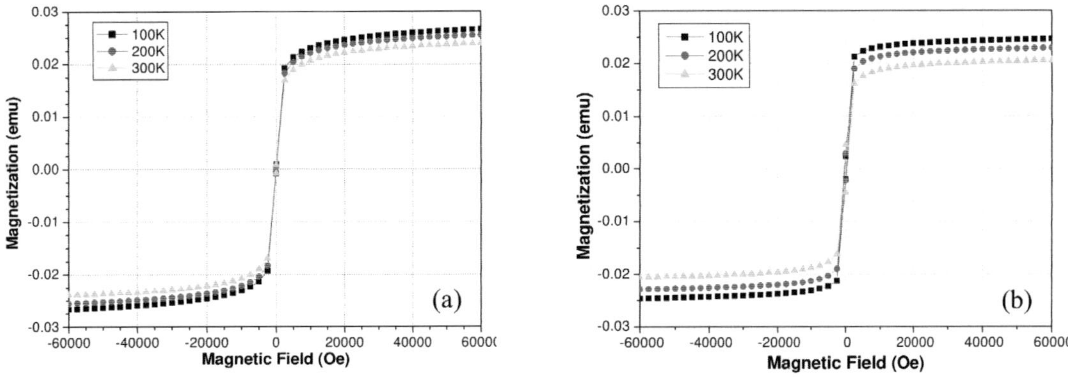

Figure 4. Magnetizations vs applied magnetic field of fabricated nanoparticles (a) and a commercial sample of SPION (b).

The magnetization of fabricated nanoparticles is shown in figure 4a, the typical characteristics of superparamagnetic behavior are observed by showing almost immeasurable coercivity from 100-300K and comparable to a commercial sample (figure 4b). The SPION are ferri- or ferromagnetic particles. For an assembly of particles, magnetic moment directions are redistributed by thermal fluctuations and the particles exhibit no remanence without an applied magnetic field. To produce a stable superparamagnetic core, the SPION particles have to be kept at a distance to reduce the influence of their interactions from the surface of each magnetic nanoparticle. The specific magnetisation with a value of about 60 emu/g was obtained for the DEG coated Fe_3O_4 nanoparticles. These values are lower for other coating polymers such as dextran, PEG, MPEG and starch, because the surfactant-coated layer can be considered as a magnetically dead layer at the surface, thus affecting the uniformity or magnitude of magnetization due to quenching of surface moments.

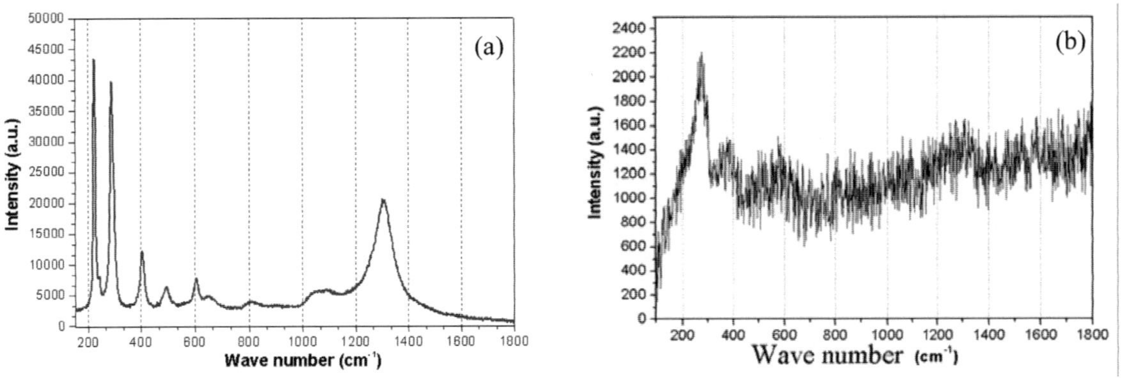

Figure 5. Raman spectrum of DEG- (a) and amin - (b) coated Fe_3O_4 nanoparticles.

Figure 5 shows the Raman spectrum of DEG- (a) and amin- coated Fe_3O_4 nanoparticles. As seen in figure 5b, the Raman bands of magnetite close to 300, 410, 600 and 1300 cm^{-1} [8] are very clearly shown. With a further increasing the thickness of surfactant-coated layer, the intensities of the Raman bands of magnetite gradually vanish (figure 5a).

Magnetic nanoparticles obtained by coprecipitation method were covered with successive layers of different water-soluble polymer such as dextran, PEG, MPEG and starch. The polymer matrix hinders the growth of SPION and in addition, the polymer layer on the surface of the SPION prevents further oxidation, which affects the phase transition during the coprecipitation process. These polymeric networks cover a large number of continuous magnetic monodomains and keep them apart against surface tension forces. Also, to transport this hydrophilic substance into biological membrane systems, surface modifications of the nanoparticles are necessary. In aqueous media at room temperature, the SPION shows extremely stable behavior for months.

4. Conclusion

We have prepared a controlled coprecipitation method upon support of watersoluble polymer, which provides the feasibility of synthesizing crystalline nanoparticles of Fe_3O_4 with a size below 12 nm. The fabricated Fe_3O_4 nanoparticles have a narrow particle size distribution. It creates a precursor for direct incorporation with appropriate coating materials. These nanoparticles have superparamagnetic behavior and can be used for relevant physical and biomedical applications.

References

[1] Franger S, Berthet P, Dragos O, Baddour-Hadjean R, Bonville P and Berthon J 2007 *J. Nanoparticles Research*, DOI **10**.1007 389
[2] Qiang Y, Anthony J, Sharma A, Nutting J, Sikes D and Meyer D 2005 *J. Nanoparticles Research* DOI **10**.1007
[3] Kim D K, Mikhaylova M, Wang F H, Kehr J, Bjelke B, Zhang Y, Tsakalakos T and Muhammed M 2003 *Chem. Mater.* **15** 4343
[4] Gupta A K, Wells S 2004 *IEEE, Trans Nanobiosci* **3** (1) 66
[5] Levy L, Sahoo Y, Kim K S, Bergey E J and Prasad P N 2002 *Chem. Mater.* **14** 3715
[6] Huong N T, Tinh T X, Giang L T K, Tuyen L D and Minh L Q 2006 *1st IWOFM-3rd IWONN Conference* Vietnam p 21
[7] Massart R, Cabuil V 1987 *J. Phys. Chem.* **84** 967
[8] Shebanova O N and Lazor P 2003 *J. Raman Spectrosc.* **34** 845

Acknowledgments

This work has been supported by the project for Nanostructured Hybrid Materials and Application and partially financed by the Institute of Materials Science, Vietnam Academy of Science and Technology. A part of the work was done in the National Key Laboratory for Electronic Materials and Devices, Institute of Materials Science, VAST. The authors want to thank all the members of Raman, FESEM and Magnetic groups for their technical assistance.

Preparation of drug nanoparticles by emulsion evaporation method

Le Thi Mai Hoa[1], Nguyen Tai Chi[2], Nguyen Minh Triet[2], Le Ngoc Thanh Nhan[2] and Dang Mau Chien[1]

[1] Laboratory for Nanotechnology, Vietnam National University of Ho Chi Minh City, 6 Community, Linh Trung Ward, Thu Duc District, Ho Chi Minh City, Vietnam
[2] Faculty of Pharmacy, The University of Medicine and Pharmacy at Ho Chi Minh City, 41 Dinh Tien Hoang, Ben Nghe Ward, 1 District, Ho Chi Minh City, Vietnam

E-mail: ltmhoa@vnuhcm.edu.vn

Abstract. Polymeric drug nanoparticles were prepared by emulsion solvent evaporation method. In this study, prepared the polymeric drug nanoparticles consist of ketoprofen and Eudragit E 100. The morphology structure was investigated by scanning electron microscopy (SEM). The interactions between the drug and polymer were investigated by Fourier transform infrared spectroscopy (FTIR). The size distribution was measured by means of Dynamic Light Scattering. The nanoparticles have an average size of about 150 nm. The incorporation ability of drugs in the polymeric nanoparticles depended on the integration between polymer and drug as well as the glass transition temperature of the polymer.

Keywords: Nanoparticles, Eudragit, emulsion evaporation method, polymeric nanoparticle, drug.

1. Introduction

Drug nanoparticles can be defined as drug-containing particles having size measured by µm. They have potential applications for tissue targeting in cancer therapy, controlled release, carrier action for the delivery of peptides, and increase of the solubility of drug [1, 2]. Drug nanoparticles consist of the drug material and, usually, a stabilising polymer. First polymer nanoparticles for pharmaceutical application were prepared in the late 1960's and early 1970's [3, 4].

Various techniques have been used to manufacture nanosized drug particles with size down to hundreds and even tens of nanometres. One of the popular methods for the encapsulation of drug within water - insoluble polymer is the emulsion solvent evaporation method. The emulsion evaporation techniques were developed at the end of the 1970s and have been used successfully in the preparation of microspheres made from several biocompatible polymers.

Eudragits are biocompatible co-polymers synthesized from acrylic and methacrylic acid esters. These polymers have been widely used in pharmaceutical industry for various coating applications as well as for microencapsulation and nanoparticle synthesis purpose [5]. Eudragit E 100 is a co-polymer consisting of 1:2:1 molar ratio of methol methacrylate, N, N – dimethylaminoethyl methacrylate, and butyl methacrylate monomers.

The aim of the study was to prepare drug nanoparticles using emulsion solvent evaporation method. Eudragit E 100 was used as stabilisation materials in this study. The model drug used was ketoprofen.

© 2009 IOP Publishing Ltd

Journal of Physics: Conference Series **187** (2009) 012047 doi:10.1088/1742-6596/187/1/012047

2. Experimental

2.1. Materials
The following chemicals were obtained from commercial sources and used as received: Ketoprofen (USP-30) (Röhm Pharma, Darmstadt, Germany), Eudragit E 100, chloroform (Merck, Germany), Lactose (Merck, Germany), Sodium dodecyl sulphat (SDS) (Merck, Germany), Aceton (Merck, German).

2.2. Preparation of the drug nanoparticles
Eudragit E100 was used as stabilisation materials in this study. The model drug used was ketoprofen, which is an example of a poor water soluble corticosteroid. Ketoprofen and Eudragit were prepared by separately dissolving in chloroform at room temperature using a magnetic stirrer. After that, the solution of Eudragit and ketoprofen was combined at respective amounts. Sodium dodecyl sulphat was then added into solution containing ketoprofen and Eudragit to achieve a final solution. The mixture was agitated with a magnetic stirring until the complete evaporation of organic solvent was accomplished. After that, the protective excipients glucose (50 - 400 mg) and lactose (100 - 300 mg), as aqueous solutions, were added to the nanoparticle dispersions. The nanoparticle dispersions were frozen at -32°C for a minimum of 12h and freeze-dried at -55°C and 0.5 kPa for 24 h.

2.3. Characterization of particles
The morphology of the particles was analyzed using a scanning electron microscopy (SEM) (JSM 6480LV-JEOL- JAPAN). The size distribution was measured by means of Dynamic Light Scattering (HORIBA Lb-902-JAPAN).

The interactions between the drug and the polymers were investigated using a Fourier transform spectrometer (FTIR) (TENSOR™ 37-BRUNER-USA) in the wavelength region from 4000 to 400 cm^{-1}.

The crystal phase identification of the studied samples was carried out using X-ray diffraction (XRD) (Philips PW 1710 difractometer, Eindhoven and Almelo, The Netherlands) with Cu-Kα radiation at 40 kV and 100 mA, the scan range (2θ) was from 5° to 40°.

3. Results and discussion
Figure 1 shows scanning electron microscopic photographs of ketoprofen (a) and nanoparticle containing ketoprofen and Eudragit E100 (b and c). In figure 1c, the morphology of particles was spherical and not aggregated. The spheres have mean diameters around 150 nm. The nanoparticles dry powder consists of individual nanoparticles, which touch each other, but retain their original size and shape.

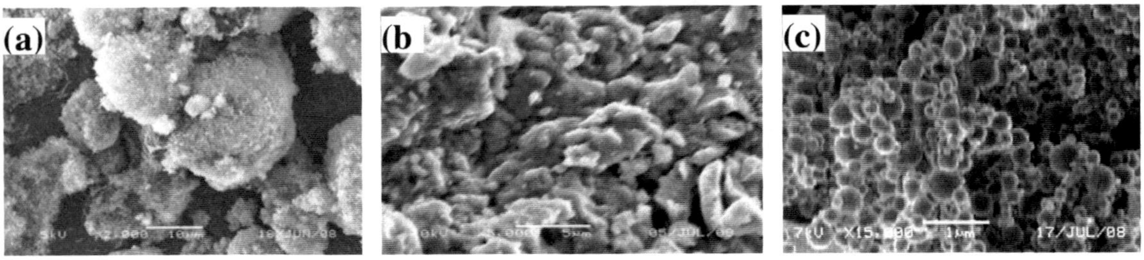

Figure 1. SEM images of Ketoprofen (a) nanoparticles containing ketoprofen and Eudragit E 100; (b) magnification 500 x; and (c) magnification 15000 x.

Infrared spectroscopy was used to study the interactions between the drug and the polymers. Figure 2 shows IR spectra at a wavenumber range of 4000 - 400 cm^{-1}. In figure 2a, the IR spectra of the pure Eudragit E 100 shows the strong stretching vibration of the carbonyl moiety of ester groups at 1732

cm^{-1}. The peak corresponding to the amino groups of the Eudragit E 100 have been identified at 2958.35 and 2773.87 cm^{-1}. However, any change in the position of these peaks was not observed when ketoprofen was incorporated in the nanoparticles (figure 2c). It can be concluded that ketoprofen drug does not protonate the amino groups, but mainly interacts with the ester groups of the Eudragit E 100. Figure 2b shows the IR spectra of ketoprofen. The peak corresponding to the carboxylic acid group of ketoprofen at 1697.03 cm^{-1} was observed in the spectra. However, this peak was not observed in the spectra of the nanoparticle (figure 2c). The carboxylic group of the ketoprofen molecule interacts with the polymers, leading to disruption of the carboxylic acid.

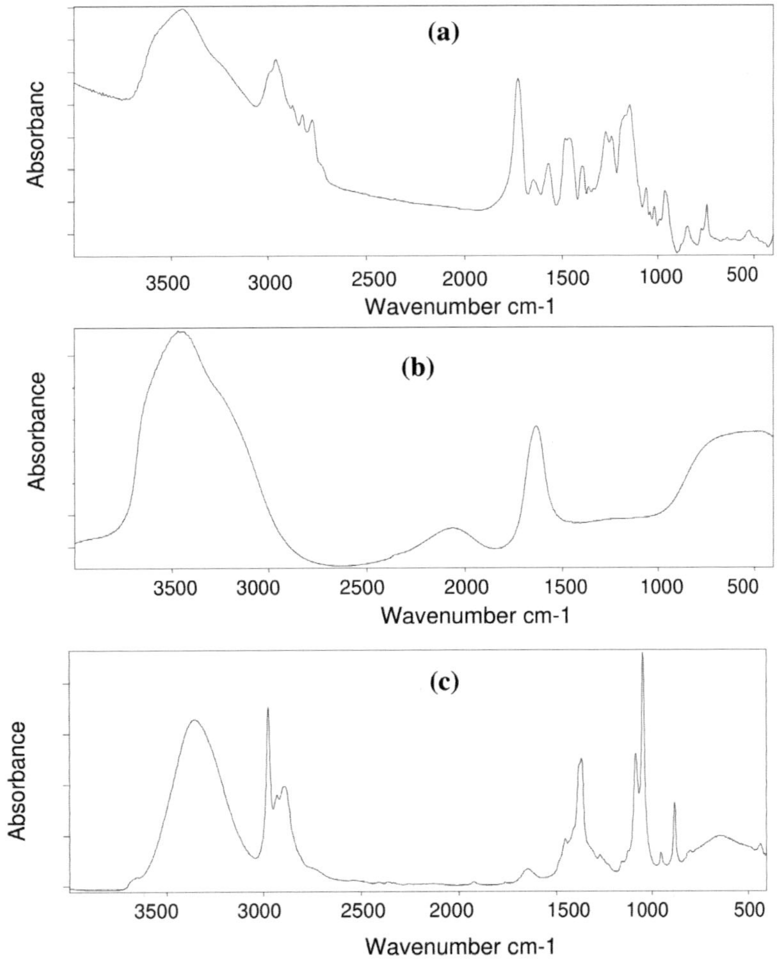

Figure 2. Infrared spectra at a wavenumber range of 4000 to 400 cm^{-1} of the following: (a) pure Eudragit E 100; (b) pure ketoprofen; (c) nanoparticles containing Eudragit E 100 and ketoprofen.

Figure 3 shows the particle size distribution of nanoparticles containing ketoprofen and Eudragit E 100. Size distributions were determined by Dynamic Light Scattering in aqueous solution, the size of the particles was in the range from 50 to 200 nm, the mean diameter was 150 nm.

Journal of Physics: Conference Series **187** (2009) 012047 doi:10.1088/1742-6596/187/1/012047

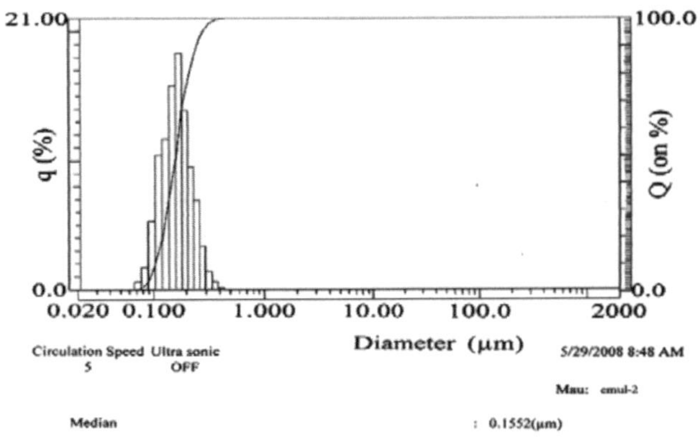

Figure 3. Particle size distributions of nanoparticle containing ketoprofen and Eudragit E 100.

Figure 4 shows X-ray diffraction patterns of (a) ketoprofen and (b) nanoparticles containing ketoprofen and Eudragit. In figure 4b, XRD shows diffraction pattern of the amorphous structure and no crystallinity. The strongest diffraction peaks of crystalline ketoprofen at 6.5° (2θ) could not see in the diffraction curve of the nanoparticles containing ketoprofen and Eudragit E 100. We can conclude that the nanoparticles powder as-prepared was amorphous, as peaks corresponding to diffraction from drug crystal lattice were not detected.

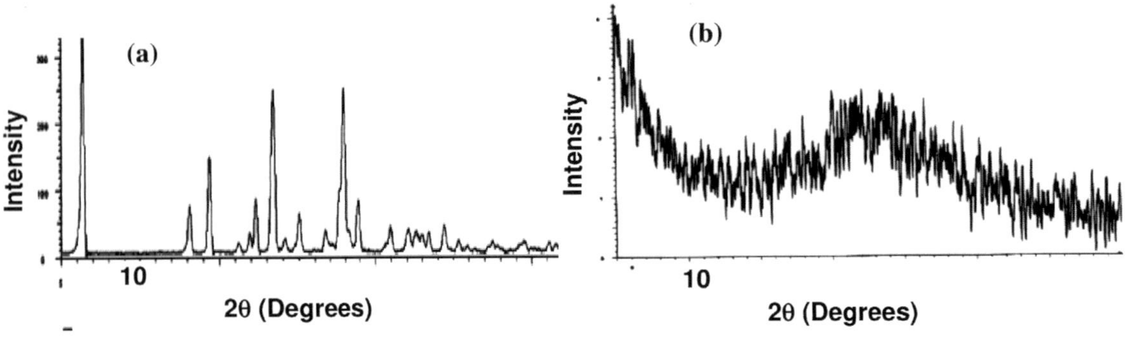

Figure 4. X-ray diffraction patterns: (a) Ketoprofen; (b) nanoparticle containing ketoprofen and Eudragit E 100.

4. Conclusion

Polymeric drug nanoparticles were prepared by emulsion solvent evaporation method. By using a mixture of protective excipients glucose and lactose as the dispersed organic phase, the more stable particles could be prepared. The SEM observations give the surface morphological features, morphology of particles was spherical and homogeneous. The size distribution of the nanoparticles prepared was found in the range from 50 to 200 nm, the mean diameter was 150 nm. The interaction between the drug and the polymer was determined by FTIR. The carboxylic group of the ketoprofen molecule interacts with the Eudragit.

References

[1] Ravi Kumar M N 2000 *Pharm Pharmaceut. Sci.* **3** 234
[2] Edelstein A S and Cammarata R C 2000 *Pro. App. JW Arrowsmith Ltd, Bristol*
[3] Soppimath K S, Aminabhavi T M, Kulkarni A R and Rudzinski W E 2001 *J. Cont. Release* **70** 1
[4] Panyam J and Labhasetwar V 2003 *Adv. Drug Del. Rev.* **55** 329
[5] Dittgen M, Durrani M and Lehmann K 1997 *Acr. Rev. Phar. Appl.* **7** 403

Preparation and characterization of magnetic nanoparticles coated with polyethylene glycol

Le Thi Mai Hoa[1], Tran Thi Dung[2], Tran Mau Danh[2], Nguyen Huu Duc[2] and Dang Mau Chien[1]

[1]Laboratory for Nanotechnology, Vietnam National University of Ho Chi Minh City, Community 6, Linh Trung Ward, Thu Duc District, Ho Chi Minh City, Vietnam
[2]Laboratory for Nano Magnetic Matrials and Devices, Faculty of Engineering Physics and Nanotechnology, College of Technology, Vietnam National University, Hanoi, E3 Building, 144 Xuan Thuy Road, Cau Giay District, Hanoi, Vietnam

E-mail: ltmhoa@vnuhcm.edu.vn

Abstract. Magnetic nanoparticles Fe_3O_4 were prepared in air environment by the coprecipitation method using molar ratios of Fe^{2+}: Fe^{3+} = 1: 2. The surface of magnetic nanoparticles was coated with sodium oleate as the primary layer and polyethylene glycol 6000 (PEG-6000) as the second layer. The morphology of the particles was investigated by scanning electronic microscopy (SEM). X-ray diffraction (XRD) indicated the sole existence of inverse cubic spinel phase of Fe_3O_4 and an average size of about 25 nm. Fourier transform infrared spectroscopy (FTIR) analysis indicated existence of two distinct surfactants on the particle surface. In addition, the results of FT-IR indicated that the coated Fe_3O_4 particles improved the thermal stability due to the interaction between the Fe_3O_4 particles and protective layers.

Keywords: Magnetic nanoparticles, polyethylene glycol, nanomaterials, bilayer, synthesis.

1. Introduction

Magnetic nanoparticles Fe_3O_4 have attracted researchers in various fields such as physics, medicine, and biology due to their multifunctional properties such as small size, superparamagnetism and low toxicity, etc. [1-3]. Magnetic nanoparticles are used for many applications such as drug delivery, cell tracking, hyperthermia. However, Fe_3O_4 nanoparticles tend to aggregate due to strong magnetic dipole-dipole attractions between particles. So, modification of the surface of the Fe_3O_4 nanoparticles with various biocompatible and biodegradable polymer have been widely investigated [4]. These particles consist of iron oxide magnetic cores coated with a protective layer. Stability of the coated particles depends on the strength of the bond between core and coating.

Polyethylene glycol (PEG) has been coated on nanoparticle surfaces to disperse the nanoparticles, reduce nonspecific protein adsorption and clearance by macrophages, and render the nanoparticles capable of crossing the cell membrane. Moreover, there is growing interest in attaching PEG conjugate to the surfaces of nanoparticles through chemical bonding. PEG is attached to the nanoparticles surface to

© 2009 IOP Publishing Ltd

prevent the nanoparticles from agglomeration, to make them more biocompatiticles [5, 6]. Because PEG plays different roles in intracellular uptake of nanoparticles, it is expected that the uptake efficiency will be further improved after the PEG conjugates are attached to nanoparticle surfaces.

In this work, we prepared Magnetic nanopaticles with an inner layer of sodium oleate and outer layer of polyethylene glycol (PEG-6000). The obtained products were investigated by diffused reflectance Fourier transform infrared spectroscopy (FTIR), X-ray diffraction (XRD) and scanning electron microscopy (SEM).

2. Experimental

Magnetic nanoparticles Fe_3O_4 were prepared using co-precipitation method. Aqueous solutions of 0.1 M $FeCl_3$. $6H_2O$ (50 ml) and 0.05 M $FeCl_2$. $4H_2O$ (50 ml) were mixed by stirring on a magnetic stir plate. 20 ml ammonia aqueous solution (NH_4OH) with a concentration of 25% was then quickly added into the solution with vigorous stirring, followed by more ammonia aqueous solution (25%) being dropped into the mixture slowly stirring until the pH value of the solution reached 9, and then the solution was stirred for additional 30 min. Finally, the Fe_3O_4 nanoparticles was collected by magnetic field separation and washed several times with deionized water and dried at 40°C for 24 h.

The polyethylene glycol solutions with different concentration were prepared by dissolving PEG-6000 powder in distilled water. The mixture of water and Fe_3O_4 particles was vigorously stirred for 15 min and then a solution of 10 ml sodium oleate was added to the mixture with slow agitation. The PEG-6000 solution with different solutions was then added to the suspension. After stirring for 30 min, the mixture was centrifugated 6000 rpm to obtain a stable ferrofluid. All magnetic nanoparticles, both before and after modification, were investigated.

The surface morphology of prepared powders was observed by the scanning electron microscopy (SEM-JSM-6480 LV, Japan). The binding of sodium oleate and polyethylene glycol to the magnetic nanoparticles was confirmed by Fourier transform infrared spectroscopy (FTIR) TENSOR™ 37, Bruker in the range of 4000-400 cm^{-1}. Pellets for FTIR analysis were prepared by mixing a small of nanoparticles powder with KBr powder. X-ray diffraction (XRD) pattern was recorded by step scan in an Philips PW 1830 difractometer, using Cu-Kα radiation at 40 kV and 100 mA, the scan range (2θ) was from 25° to 65°.

3. Results and discussion

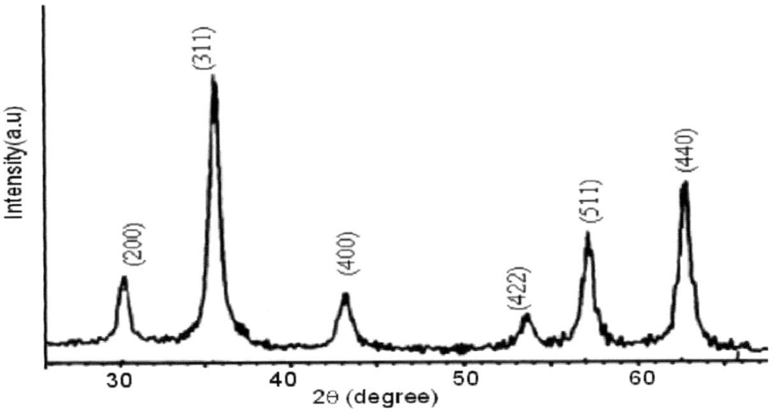

Figure 1. XRD patterns of the as-synthesized Fe_3O_4 nanoparticles.

Figure 1 shows X-ray diffraction patterns of the bare Fe_3O_4 nanopaticles. The XRD peaks of the Fe_3O_4 are compared with those of standard one in JCPDS file (PDF No. 65-3107). A series of characteristic peaks at $2\theta = 30.17°$, $35.46°$, $43.38°$, $53.69°$, $57.23°$, and $62.77°$, which corresponds to (220), (311), (400), (422), (511), and (440) Bragg reflection, respectively, in figure 1 agree with standard magnetite (Fe_3O_4) XRD patterns, identify that the Fe_3O_4 nanoparticles are cubic spinel structure [7].

The average diameter of Fe_3O_4 nanoparticles was determined according to Sherrer's equation: $D = k\lambda/ \beta \cos\theta$ [8]. Here the X-ray wavelength of CuKα radiation λ is 1.54 A°, k is the shape factor which can be assigned a value of 0.89 if the shape is unknown, θ is Bragg angle and β is the full width at half maximum in radians. The reflecting peak at $2\theta = 35.46°$ is chosen to calculate the average diameter, the estimated average size of the Fe_3O_4 nanoparticles are about 25 nm.

Figure 2 shows the SEM image of the bare Fe_3O_4 nanoparticles (a), the Fe_3O_4 nanoparticles coated with sodium oleate (b), the Fe_3O_4 nanoparticles coated with sodium oleate and PEG-6000 (c). According to the SEM images, the agglomeration is very strong in the bare Fe_3O_4 nanoparticles. The agglomeration is due to the Van der Waals force between the particles. After surface modification, there is a little aggregation.

Figure 2. The SEM images of Fe_3O_4 nanoparticles: a) the bare Fe_3O_4 particles, b) the Fe_3O_4 nanoparticles coated with oleate sodium, c) the Fe_3O_4 nanoparticles coated with sodium oleate and PEG-6000.

Figure 3 shows FTIR spectrum of bare Fe_3O_4 nanoparticles (a), the Fe_3O_4 nanoparticles coated with sodium oleate (b), the Fe_3O_4 nanoparticles coated with sodium oleate and PEG-6000 (c). The characteristic absorption of Fe_3O_4 was at 582.85 cm^{-1} relates to Fe–O bond. The peak at 3438.06 cm^{-1} observed in curve (a), (b) and (c) relates to the OH group. The cure b and c show strong CH_2 bands at 2924.72 cm^{-1} and 2846.03 cm^{-1}. These bands are known to be the characteristic bands of CH_2 groups that are presented in oleate sodium. This can be explained as the carboxyl groups of sodium oleate combined with the Fe atoms on the surface of Fe_3O_4 nanoparticles. The peak at 1455.90 cm^{-1} observed in curve (b) and (c) indicate a complex reaction between hydroxyl groups on the surface of mangetic nanoparticles Fe_3O_4 and carboxylate groups of the oleate sodium. The peak at 1740.67 cm^{-1} relates to free sodium oleate (C=O). The –C=C– group of sodium oleate was characterized by band at 1099.94 cm^{-1}. Absorption at 1212.35 cm^{-1} (–C–O–C–) (the associated hydroxy groups) indicated the existence of PEG-6000 (see figure 3c). This result suggested the the Fe_3O_4 successfully coated by sodium oleate and PEG-6000. Figure 4 shows the FTIR spectra of Fe_3O_4 nanoparticles coated with different concentration of PEG-6000 ((a) 3 wt.% and (b) 10 wt.%), the intensity of all peak increased with higher PEG-6000 concentration (figure 4b - 10 wt.%). The FTIR spectra developed strong and well-defined bands at 1212.35 cm^{-1}. This can explained the hydroxy groups (–C–O–C–) bonding to Fe_3O_4 nanoparitcles increased with increasing PEG-6000 concentration.

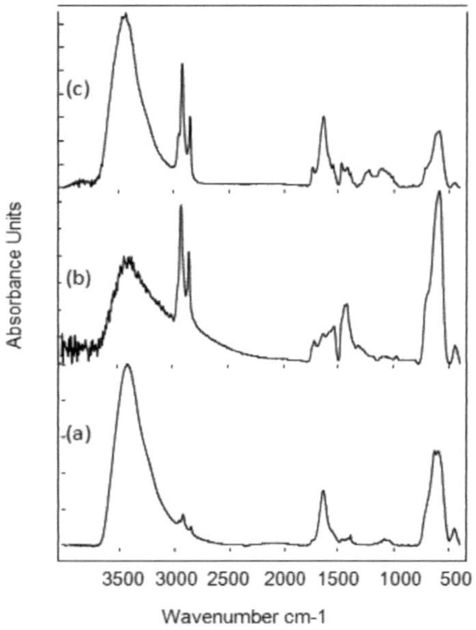

Figure 3. The FTIR spectra of the bare Fe_3O_4 nanoparticles (a), the Fe_3O_4 nanoparticles coated with sodium oleate (b), the Fe_3O_4 nanoparticles coated with sodium oleate and PEG-6000 (c).

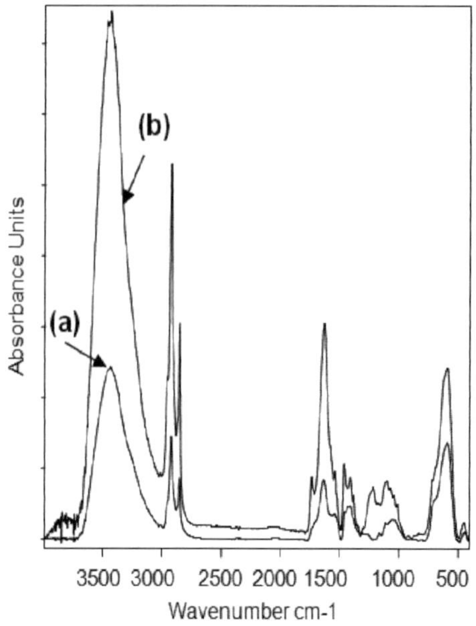

Figure 4. The FTIR spectra of the Fe_3O_4 nanoparticles coated with different concentration of PEG-6000, (a) 3 wt.%, (b) 10 wt.%.

4. Conclusions

The inverse cubic spinel phase of Fe_3O_4 (magnetite) was prepared by coprecipitation method, the average size of the Fe_3O_4 nanoparticles is about 25 nm. The measurements of FTIR indicated that the Fe_3O_4 nanoparticles are successfully coated by sodium oleate and PEG-6000.

References

[1] Gu H W, Xu K M, Xu C J and Xu B 2006 *Chem. Comm.* **6** 941
[2] Roger J, Pons J, Massart R, Halbreich A and Bacri J 1999 *Eur. Phys. J. Appl. Phys.* **5** 321
[3] Wunderbaldinger P, Josephson L and Weissleder R 2002 *Bioconjugate Chem* **13** 264
[4] Kim D H, Lee S H, Im K H, Kim K N, Kim K M, Shim I B, Lee M H, and Lee Y K 2006 *Curr. Appl. Phys.* **6** 242
[5] Zhang Y, Kohler N and Zhang M Q, 2002 *Mat. Res. Soc. Symposium Proceeding* MRS San Francisco 981
[6] Zhang Y, Kohler N and Zhang M Q 2002 *Biomaterials* **23** 1553
[7] Cheng F Y, Su C H, Yang Y S, Yeh C S, Tsai C Y, Wu C L, Wu M T and Shieh D B 2005 *Biomaterials* **26** 729
[8] Hammond C 1997 *The Basics of Crystallography and Diffraction* (Oxford:Oxford University Press)

Acknowledgments

This work was carried out at the Laboratory for Nanotechnology – Vietnam National University of Ho Chi Minh City. The authors are grateful for financial support from the Fundamental Research Program.

Modeling of planar carbon nanotube field effect transistor and three dimensional simulation of current-voltage characteristics

Dinh Sy Hien[1], Nguyen Thi Luong[1], Thi Tran Anh Tuan [1] and Dinh Viet Nga[1]

[1] HCM University of Natural Sciences, 227 Nguyen Van Cu Street, 5 District, Ho Chi Minh City, Vietnam

E-mail: vietnga704@yahoo.com

Abstract. We provide a CNTFET model with planar geometry. Planar CNTFETs constitute the majority of devices fabricated to date, mostly due to their relative simplicity and moderate compatibility with existing manufacturing technologies. We explore the possibilities of using non-equilibrium Green function method to get I-V characteristics for CNTFETs. This simulator also includes a graphic user interface (GUI) of Matlab that enables parameter entry, calculation control, intuitive display of calculation results, and in-situ data analysis methods. In this paper, we review the capabilities of simulator, and give examples of typical CNTFET 3D simulations. The I-V characteristics of CNTFET are also presented.

Keywords: Planar CNTFET, GUI, NEGF, simulation of I-V characteristic of CNTFET.

1. Introduction

The carbon nanotube field-effect transistor (CNTFET) is a promising candidate for future electron devices. Rapid progress in the field has recently made it possible to fabricate CNTFET based circuits, such as logic gates, static memory cells, and ring oscillators.

This three-terminal device consists of a semiconducting nanotube bridging two contacts (source and drain) and acting as a carrier channel, which is turned on or off electrostatically via the third contact (gate). Presently, there are various groups pursuing the fabrication of such devices in several variations, achieving increasing success in pushing performance limits, while encountering myriad problems, as expected for any technology in its infancy. Although progress in CNTFETs has been rapid, there are still many issues to address.

In this work, we summarize experimental works on planar CNTFETs and introduce a CNTFET model with planar geometry. We also review the capabilities of the simulator, summarize the theoretical approach and experimental results, and give examples of typical planar CNTFET's 3D simulations (current-voltage characteristics which are functions of parameters such as the length of CNTFET, the gate thickness and temperature). The obtained I-V characteristics of the CNTFET are also presented by analytical equations.

© 2009 IOP Publishing Ltd

2. Simulation results

2.1. Planar CNTFET geometry

We now provide a brief description of typical CNTFET geometries, which are grouped in two major categories, planar and coaxial. A CNTFET, whether planar or coaxial, relies on similar principles, while being governed by additional phenomena such as 1D density of states (DOS) and ballistic transport.

Planar CNTFETs constitute the majority of devices fabricated to date, mostly due to their relative simplicity and moderate compatibility with existing manufacturing technologies. The nanotube and the metallic source-drain contacts are arranged on an insulating substrate, with either the nanotube being draped over the pre-patterned contacts, or with the contacts being patterned over the nanotube. In the latter case, the nanotubes are usually dispersed in a solution and transferred to a substrate containing pre-arranged electrodes; transistors are formed by trial and error. Manipulation of an individual nanotube has also been achieved by using the tip of an atomic force microscope (AFM) to nudge it around the substrate; due to its strong, covalent bonds, this is possible to do without damaging the molecule. In the case where the electrodes are placed over the tube, manipulation of the CNT is not required and alignment markers, pre-arranged on the substrate, allow accurate positioning of the contacts once the nanotube is located via examination by a scanning tunneling microscope (STM). The gate electrode is almost always on the back side of the insulating substrate, or alternatively is patterned on top of an oxide-covered nanotube.

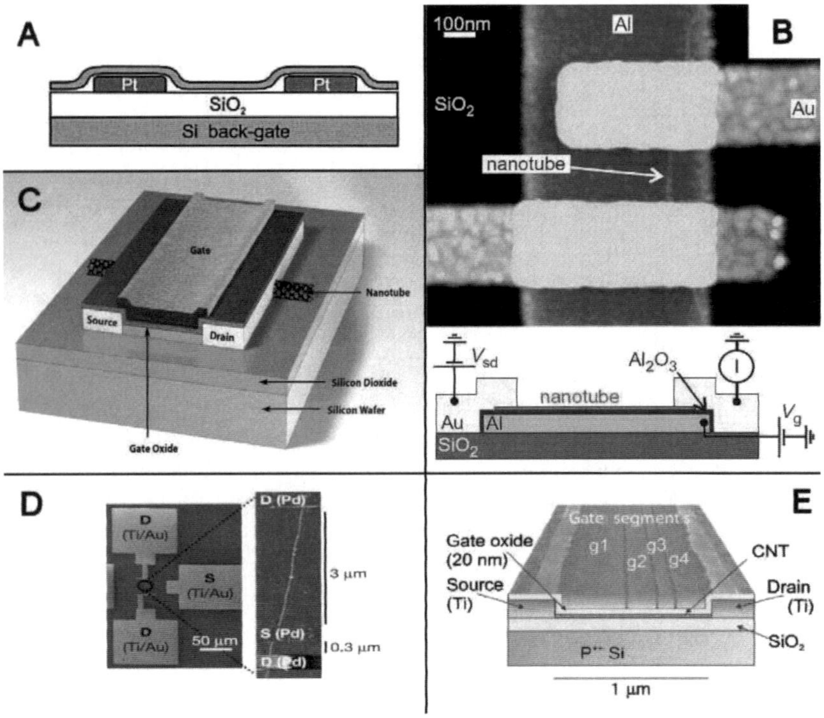

Figure 1. Examples of planar CNTFETs: (A) Ref. [1], (B) Ref. [2], (C) Ref. [3], (D) Ref. [4], and (E) Ref. [5].

The first CNTFET devices were reported in 1998, and involved the simplest possible fabrication. They consisted of highly-doped Si back gates, coated with thick SiO_2, and patterned source-drain metal contacts, either using Au or Pt, as shown in figure 1A [1]. Experimentations with different metals such as Ti, Ni, Al, and Pd have been carried out by several groups, primarily to manipulate the work function difference between the end contacts and the nanotube. Subsequent work also produced a device that replaced the back gate with an electrode placed over the substrate, perpendicular to the source and drain contacts, as illustrated in figure 1B [2]. Here, the nanotube was separated from this gate electrode by a thin insulating layer of Al_2O_3, with the source-drain electrode strips placed over the tube ends. Recent theoretical studies comparing the interfaces between different bulk metals and metallic nanotubes, studying both end- and side-contacted tubes, concluded that Ti contacts yield superior conductance over their Au and Al counterpart's contact resistance. Figure 1C shows a further improvement in CNTFETs through the placement of the gate electrode over the semiconducting nanotube, thus improving the channel electrostatics via the thin gate oxide [3]. Moreover, the Ti source-drain metallization in this device form titanium carbide abrupt junctions with the nanotube, yielding increased conductance. Another attempt to obtain better gate electrostatics involved materials with high dielectric constants, such as zirconia (ZrO_2) and hafnia (HfO_2), being used as gate insulators. Figure 1D illustrates a device built with Pd source-drain contacts in order to exploit the sensitivity of this material's work function to hydrogen [4]. A multi-gate device, as shown in figure 1E, has recently been reported, whereby parallel top gates are used to independently control the electrostatics of different sections of the channel, thus facilitating a study of the transport characteristics of the nanotube channel [5]. Most recently, a device with excellent DC characteristics was fabricated with Pd end contacts, Al gate, and hafnia for the insulator. The coaxial geometry maximizes the capacitive coupling between the gate electrode and the nanotube surface, thereby inducing more channel charge at a given bias than other geometries. This improved coupling is desirable in mitigating the short-channel effects that plague technologies like CMOS as they downsize device features. It is also of importance to low-voltage applications, a dominating trend in the semiconductor industry and to allow, potentially, for easier integration with modern implementations of existing technologies such as CMOS. Besides the wraparound gate, special attention must also be paid to the geometry of the end contacts, since these play a role in determining the dimensions of the Schottky barriers that are present in the channel near the device ends and have a direct effect on current modulation. We hereinafter deal specifically with the planar geometry of the CNTFET.

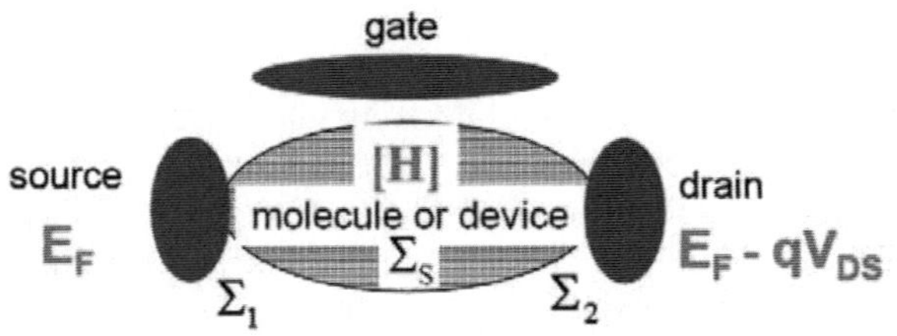

Figure 2. The generic transistor with a molecule or device channel connected to the source and drain contacts. The source-drain current is modulated by a third electrode, the gate. The quantities in the NEGF calculation are also shown.

2.2. Review of the NEGF formalism

A carbon nanotube can be viewed as a rolled-up sheet of graphene with a diameter typically between one and two nanometers. The nanotube can be either metallic or semiconducting, depending on how it is rolled up from the graphene sheet (i.e. depending on its chirality). Semiconducting nanotubes are suitable for transistors. In order to correctly treat carbon nanotube transistors, strong quantum confinement around the tube circumferential direction, quantum tunneling through Schottky barriers at the metal nanotube contacts, and quantum tunneling and reflection at barriers in nanotube channel need to be considered. The non-equilibrium Green's function (NEGF) formalism, which solves Schrödinger equation under non-equilibrium conditions and can treat coupling to contacts and dissipative scattering process, provides a sound basis for quantum device simulations. The NEGF simulation approach has demonstrated its usefulness for simulating nanoscale transistors from conventional Si MOSFETs, MOSFETs with novel channel materials, to CNTFETs [6], and molecular transistors [7]. In this section, we give brief summary of the NEGF simulation procedure. For a more thorough description of the technique, one can refer to work [8]. Figure 2 shows a generic transistor and defines some terms for the NEGF simulation. The first step is to identify a suitable basis set and Hamiltonian matrix for an isolated channel. The self-consistent potential, which is a part of the Hamiltonian matrix, is included in this step. The second step is to compute the self-energy matrices, Σ_1, Σ_2 and Σ_S, which describe how the ballistic channel couples to the source-drain contacts and to the scattering process. For simplicity, only ballistic transport is treated in this work. After identifying the Hamiltonian matrix and the self-energies, the third step is to compute the retarded Green's function,

$$\mathbf{G}(E) = [(E + i0+)\mathbf{I} - \mathbf{H} - \Sigma_1 - \Sigma_2]^{-1} \tag{1}$$

The fourth step is to determine the physical quantities of interest from the Green's function. In the ballistic limit, states within the device can be divided into two parts: 1) states filled by carriers from the source according to the source Fermi level, and 2) states filled by the drain according to the drain Fermi level. Within the device, the source (drain) local-density-of-states (LDOS) is $\mathbf{D}_{S(D)} = \mathbf{G}\Gamma_{S(D)}\mathbf{G}^+$, where $\Gamma_{S(D)} = i(\Sigma_{1(2)} - \Sigma^+_{1(2)})$ is the energy level broadening due to the source (drain) contact. The charge density within the device is computed by integrating the LDOS, weighted by the appropriate Fermi level over energy. The charge contributed by the source contact is

$$Q_S(z) = (-e)\int_{E_N}^{+\infty} D_S(E,z) f(E - E_{FS})dE + e\int_{-\infty}^{E_N} D_S(E,z)\{1 - f(E - E_{FS})\}dE, \tag{2}$$

where e is the electronic charge, and E_N is the charge neutrality level. The total charge is

$$Q(z) = Q_S(z) + Q_D(z) = (-e)\int_{-\infty}^{+\infty} dE.\text{sgn}[E - E_N(z)]\{D_S(E,z) f(\text{sgn}[E - E_N(z)](E - E_{FS})) + D_D(E,z) f(\text{sgn}[E - E_N(z)](E - E_{FD}))\}, \tag{3}$$

where sgn (E) is the sign function, and $E_{FS,D}$ is the source (drain) Fermi level. For a self-consistent solution, the NEGF transport equation is solved with iteratively the Poisson equation until self-consistency is achieved after which the source-drain current is computed from:

$$I = \frac{4e}{h}\int T(E)[f_S(E) - f_D(E)]dE, \tag{4}$$

where $T(E) = \text{Trace}(G_1 G^+_2)$ is the source-drain transmission and the extra factor of two comes from the valley degeneracy in the carbon nanotube energy band structure. The computationally expensive part of the NEGF simulation is finding the retarded Green's function, according to equation (1), which requires the inversion of a matrix for each energy grid point. The straightforward way is to explicitly invert the matrix, whose size is the size of the basis set. This, however, is impractical for an atomistic simulation of a nanotube transistor. In the ballistic limit, the problem is simplified because only a few columns of the Greens's function are needed. Still, reducing the size of the Hamiltonian matrix and developing computationally efficient approaches are of great importance for an atomistic simulation.

2.3. Main screen of planar CNTFET

Purpose of the project was to make a user-friendly simulator that provides as much control as possible over every aspect of the simulation. Consequently, graphic user interface (GUI) development in Matlab was a major part of the program. Here we present an overview of the most important GUI features.

The main screen shown in figure 3 is the central location where one can control the simulation programme. From the main screen, one can quickly enter simulation parameters with a minimum of typing. Clicking the left mouse pointer on each item of the main screen to enter the parameters such as types of source-drain materials (Au, Pt, Pd); types of gate materials (Al_2O_3 (k = 5), HfO_2 (k = 20), ZrO_2 (k = 26), TiO_2 (k = 65), $SiTiO_3$ (k = 175)); length of CNT (nm); temperature (K); the gate thickness (nm); colours (blue, green, pink, yellow, black); voltage V_g (V); V_d (V). Clicking the right mouse pointer on the item which is used to calculate the current voltage characteristics in 2D (plot I_d-V_{ds} or plot I_d-V_{gs}), reset used to clear current data or in 3D (plot I_d-V_{ds}-T, plot I_d-V_{ds}-L and plot

I_d-V_{ds}-t_{ox}). During calculation, symbol "running" will be appeared. Source-drain material (Au), Gate material (Al_2O_3, k = 3), Colour (red) are default settings. Simulations in 2D take less time than in 3D. Thanks to simulations in 3D, one can get more detail of tendency and effects of parameters such as temperature, the length of CNT, the gate thickness on I-V characteristics.

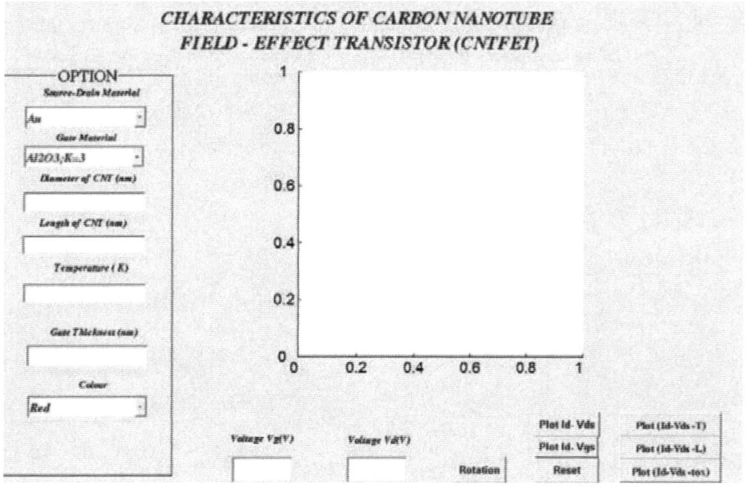

Figure 3. The main screen of simulation program of planar CNTFET. Source-drain material (Au), Gate material (Al_2O_3, k = 3), Colour (red) are default settings.

2.4. Simulations of current-voltage characteristics of planar CNTFET

The drain I-V characteristics in 2D are shown in figure 4. The saturation current at V_{GS} = 0.5 V is around 6 μA, which is not inconsistent with values emerging from recent experimental work [9].

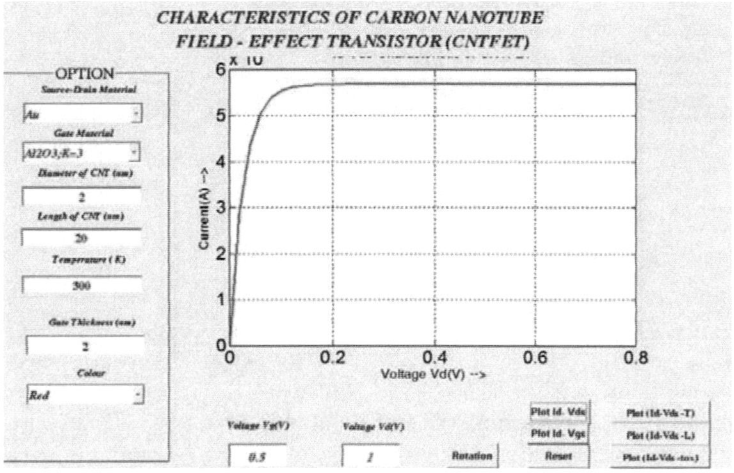

Figure 4. Drain current-voltage characteristics of planar CNTFET.

The drain I-V characteristics in 3D are shown in figure 5. The parameters used in 3D simulation are drain current-voltage characteristics and temperature. Drain I-V characteristics exhibited dependence of saturation drain current on temperature. When CNTFET is cooled, drain saturation currents were lightly decreased.

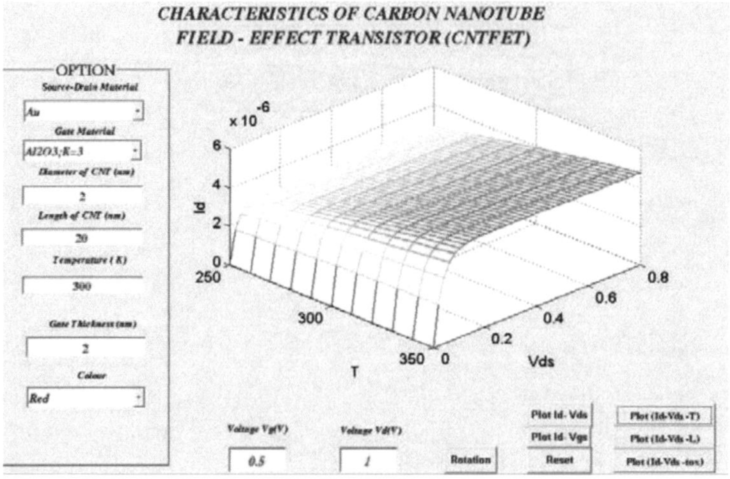

Figure 5. Drain current-voltage characteristics in 3D. When CNTFET is cooled, its saturation drain currents lightly decrease.

Drain current-voltage characteristics in 3D exhibited dependence of saturation currents on CNTFET length, L are shown in figure 6. Tendency of saturation currents is decreased, when CNTFET length is increased.

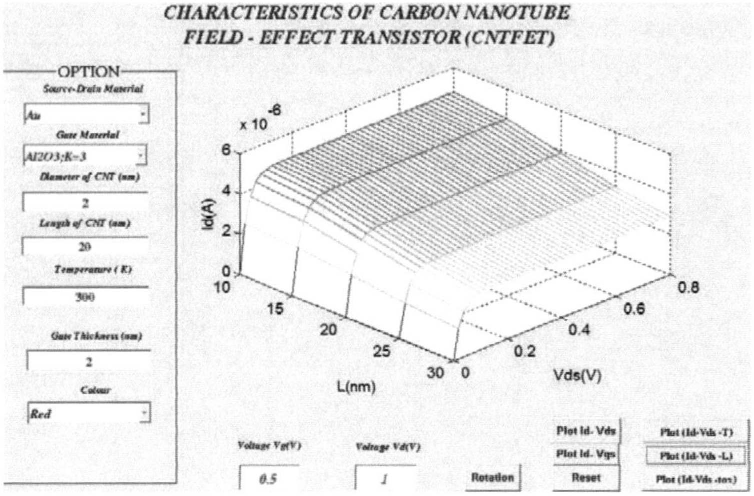

Figure 6. Drain current-voltage characteristics in 3D exhibited dependence of saturation drain currents on CNTFET length.

Drain current-voltage characteristics in 3D exhibited dependence of saturation current on the gate thickness of CNTFET, t_{ox} are shown in figure 7.

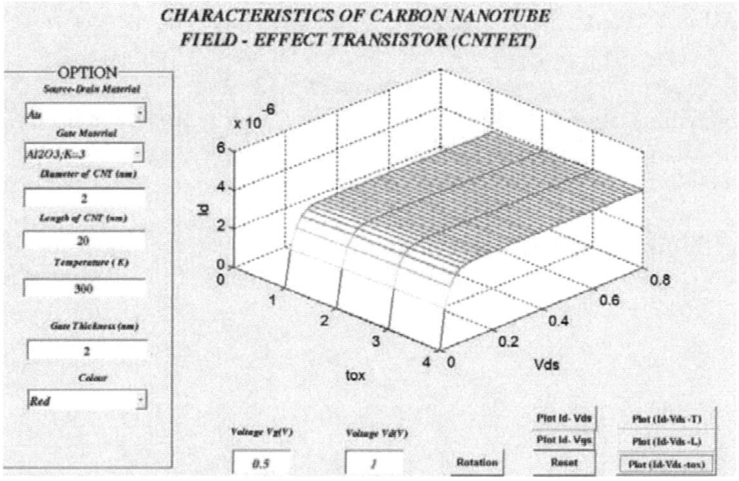

Figure 7. Drain current-voltage characteristics in 3D exhibited dependence of saturation drain currents on the gate thickness of CNTFET.

The current voltage curve can be divided into two regions: linear and saturation. Drain current in the linear region of CNTFET can be described as follows:

$$I_d = \frac{W}{L}\mu C_{ox}[(V_{gs} - V_T)V_{ds} - \frac{V_{ds}^2}{2}] \tag{5}$$

or

$$I_d = K_n\left[2(V_{gs} - V_T)V_{ds} - V_{ds}^2\right] , \tag{6}$$

where K_n is conductance of CNTFET, W is the width of CNTFET, L is the length of CNTFET, μ is mobility of carriers, C_{ox} is gate capacitance.

We can also obtain saturation current of CNTFET by replacing $V_{ds(sat)} = V_{gs} - V_T$. Then the expression of saturation current of CNTFET can be written:

$$I_{d(sat)} = K_n(V_{gs} - V_T)^2 . \tag{7}$$

3. Conclusions

A planar model for CNTFET has been reported, which is specially intended for 3D simulations of drain current-voltage characteristics. The proposed model has been verified and a good agreement with recent experimental data [9] is found. A set of 3D simulations is then successfully performed for different parameters of materials, the length, the gate thickness of CNTFET, and temperature. The tendency and effects of these parameters on I-V characteristics have been predicted. Analytical equations of drain current-voltage curve have been also presented.

References

[1] Tans S J, Alwin V R M and Dekker C 1998 *Nature* **393** 49
[2] Bachtold A, Hadley P, Nakanishi T and Dekker C 2001 *Science* **294** 1317
[3] Wind S J, Appenzeller J, Martel R, Derycke V and Avouris P 2002 *Appl. Phys. Lett.* **80**(20) 3817
[4] Javey A, Guo J, Wang Q, Lundstrom M and Dai H 2003 *Nature* **424** 654
[5] Wind S J, Appenzeller J and Avouris P 2003 *Phys. Rev. Lett.* **91**(5) 058301-1–058301-4
[6] Guo J 2004 Ph.D. Thesis, *Carbon Nanotube Electronics: Modeling, Physics, and Applications* (Purdue University)

[7] Damle P, Rakshit T, Paulsson M and Datta S 2002 *IEEE Transactions on Nanotechnology* **1** 145
[8] Datta S 2000 *Superlattices and Microstructures* **28** 253–278
[9] Javey Ali, Tu Ryan, Damron B, Farmer, Guo Jing, Gordon Roy G and Dai H *Nano Letters* **5** (No.2) 345

Acknowledgments
The authors are grateful to Prof. Dang Luong Mo of Hosei University, Japan for helpful discussions. This work was supported in part by Vietnam National University Ho Chi Minh City, under research contract No. 2005-18-06.

Preparation and characterization of hybrid materials based on polypyrrole and silver nanoparticles

Nguyen Duc Nghia and Ngo Trinh Tung

Institute of Chemistry, Vietnam Academy of Science and Technology,
18 Hoang Quoc Viet Road, Cau Giay District, Hanoi, Vietnam

E-mail: ducnghia264@fpt.vn

Abstract. Hybrid material is one of the most promising materials classed in the 21st century because of its unique properties and its advanced applications. In this work, hybrid materials based on polypyrrole (Ppy) and silver nanoparicles were prepared and characterized. The preparation of the hybrid material was performed by the chemical polymerization method. The structure , electrical and thermal properties of Ppy/Ag hybrid materials were characterized by XRD, SEM, and TGA and the conventional four probe method. The results showed that the Ag particles of 4-8 nm were agglomerated during the in-situ polymerization of PPy and formed some clusters with the diameter of 25 -150 nm. By the addition of Ag particles, the electrical conductivity of Ppy increased with increasing Ag concentration. The thermal stability of Ppy was significantly improved by modification with Ag particles.

Keyword: Hybrid material, polypyrrole, silver nanoparticle.

1. Introduction

Over the past few years, nanoparticles of noble metals have been studied with growing interest. The size-dependent properties of small metal particles are known to yield particular optical, electrochemical and electronic properties [1, 2]. However, the instability of metal nanoparticles suspended in solutions inhibits their application because of their irreversible aggregation over time. On the other hand, it is possible to confine nanoparticles into a conducting polymer matrix. Among various conducting polymers, polypyrrole (PPy) has attracted considerable attention because it offers reasonably high conductivity and has fairly good environmental stability, and it can be widely used in batteries, supercapacity, sensor and microwave absorption materials [3-6]. Meanwhile, it was found that the conductivity and sensing behavior of conductive polymers could be further improved by imbedding metal particles into the polymer matrix to form metal – organic hybrid materials [6].

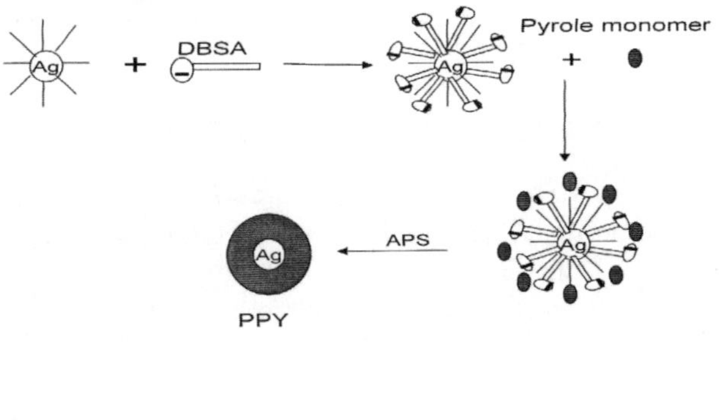

Figure 1. Schematic preparation route of PPy/Ag "core-shell" hybrid materials.

© 2009 IOP Publishing Ltd

In this work, PPy/Ag hybrid material was prepared and its structure properties were investigated. Figure 1 shows the schematic preparation route of PPy/Ag hybrid polymer as described earlier in our work [7]. The effect of the Ag concentration on the electrical conductivity and thermal stability of the hybrid materials have been investigated and discussed.

2. Experimental

2.1. Materials

Pyrole, ammoniumpersulfate (APS) and dodecylbenzulfonic acid (DBSA) were purchased from the Merck Company. Silver nitrate ($AgNO_3$), potassium hydroxyl (KOH), Glucose, NH_4OH, acid myristic and other solvents were purchased from Sigma-Aldrich. All materials were used as received. Silver nanoparticles was prepared according to [8] in our laboratory.

2.2. Preparation and characterization methods

Pyrrole of 0.30 mol, DBSA of 0.15 mol and various amounts of Ag nanoparticles were emulsified in distilled water of 500 ml with vigorous stirring. The above emulsion, APS of 0.2 mol was dissolved in distilled water and slowly added, respectively. The reaction was carried out for 14 hours at $0^{\circ}C$. Polymerization was terminated by pouring excess methanol and precipitated PPy powder was filtered. The resulting PPy powder was washed sequentially with distilled water, methanol and acetone, followed by filtering and dried in a vacuum oven at $25^{\circ}C$ for 24 hours.

The structure of the hybrid polymer was characterized by XRD (X-ray diffraction) and SEM (Scanning electron microscopy) methods. The XRD investigation was performed at the PCM-Bruker Instrument. SEM images of the materials was recorded on Hitachi S-4800 instrument. The electrical properties of the hybrid polymer were characterized by the conventional four probes method. The materials were pressed to form a pellet with a thickness of approx 200 μm. Silver paste was used to connect a copper probe with a sample, and four probes were arranged by the Van der Pauw method.

3. Results and discussion

Figure 2 shows the X-ray diffraction pattern (XRD) of Ag nanoparticle and PPy/Ag hybrid material.

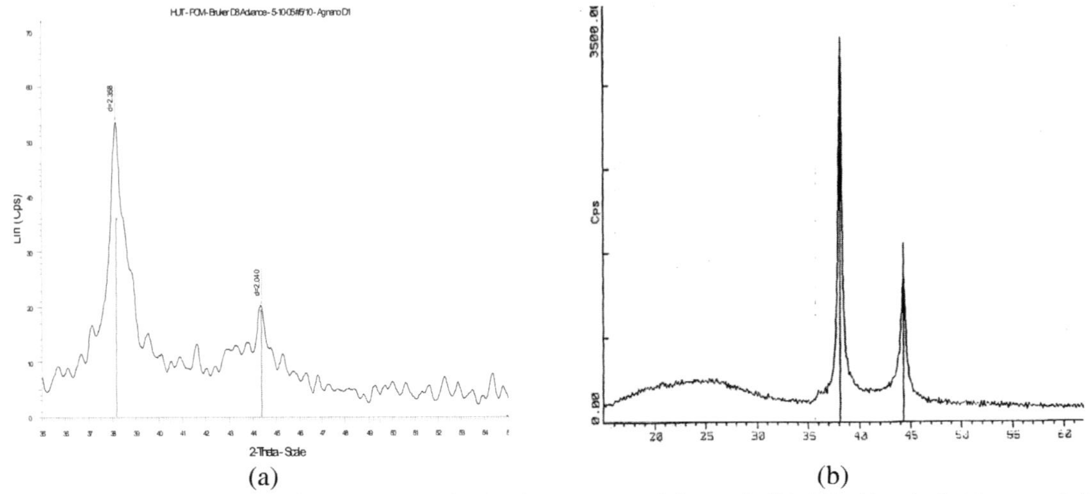

(a) (b)

Figure 2. X-ray diffraction pattern of (a) Ag nanoparticle and (b) PPy/Ag hybrid material (Ag content: 5%).

From the XRD diagram of Ag it is clear to see that the formation of silver metallic particles was obtained. It showed broad peaks assigned to the 111, 200 planes of a face centered cubic lattice of silver. The average particle size of the metal nuclei calculated from the peak broadening of XRD patterns by the Scherrer equation was about 9 nm.

Since PPy/Ag is a hybrid material, there exists a broad halo peak in the 2θ range from 1 to 30° besides the crystalline XRD pattern of Ag. This result indicates that PPy has an amorphous structure. The formation of Ag nanoparticle could be confirmed by SEM image showed in figure 3a.

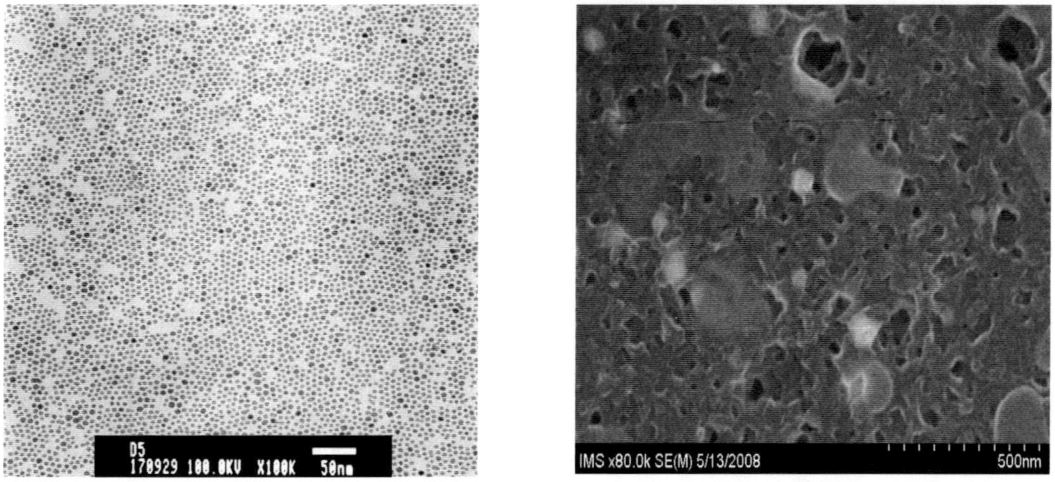

(a) (b)

Figure 3. SEM – Images of (a) Ag nanoparticle and (b) PPy/Ag hybrid materials (Ag content: 5%).

The Ag nanoparticles are identical and have the size of 4 - 8 nm. This result is in good agreement with the particle size analysis by XRD measurement. By the preparation of PPy/Ag hybrid material, the Ag phase was not well dispersed in the PPy matrix (figure 3b). The Ag cluster size varies from 25 to 150 nm. This result means that the agglomeration of Ag nanoparticle occurred during the in-situ polymerization process.

Figure 4 shows the dependence of the electrical conductivity of PPy/Ag hybrid material on the Ag-concentration.

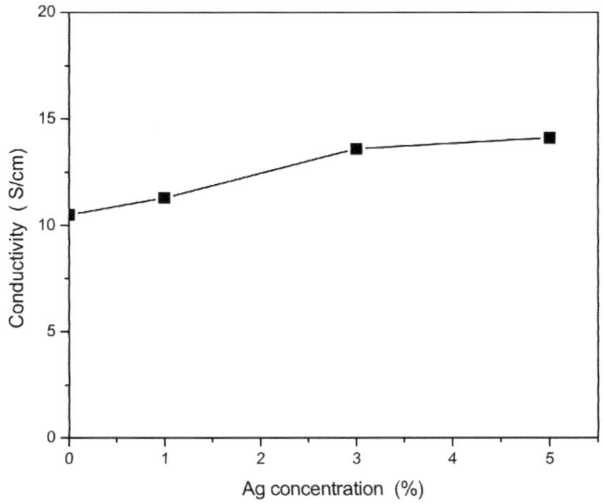

Figure 4. Dependence of the conductivity on the Ag concentration in PPy/Ag hybrid material.

With increasing Ag concentration, the conductivity of PPy/Ag hybrid material significantly increases from 10.5 S/cm of pure PPy to 13.2 S/cm by the Ag concentration of 3%. For further increasing Ag concentration, the conductivity of PPy/Ag hybrid material slightly increases. Figure 5 compares the thermal stability of PPy and PPy/Ag hybrid materials.

Journal of Physics: Conference Series **187** (2009) 012050 doi:10.1088/1742-6596/187/1/012050

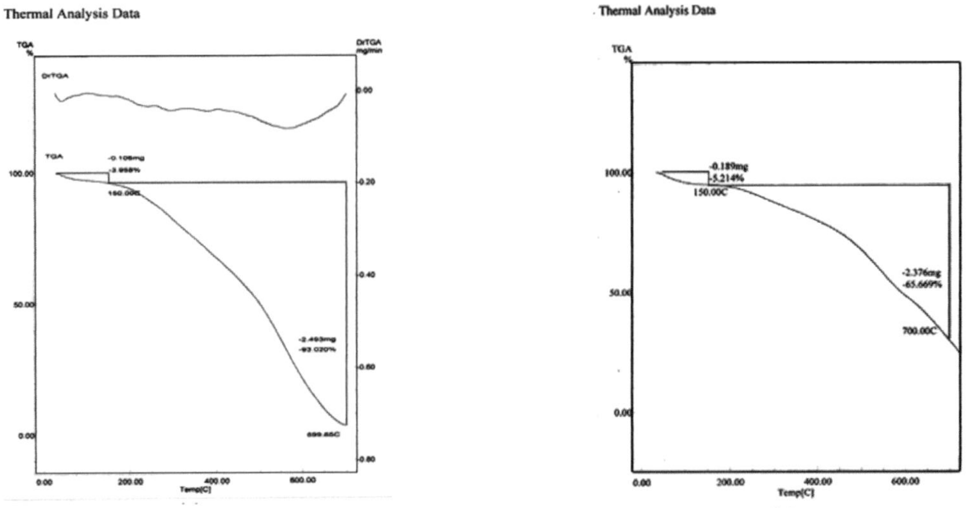

(a) (b)

Figure 5. TGA diagrams of (a): PPy and (b): PPy/Ag hybrid materials (Ag content: 5%).

In the two cases, the weight loss in the temperature until approximately 100°C is assigned to the evaporation of water contained in the sample. The weight loss of PPy in the temperature range from 150 to 700°C was 93.02%. In comparison, the weight loss of PPy/Ag hybrid materials in the same temperature range was only 65.67%. This result clearly indicates that the thermal stability of PPy is significantly improved by modification with Ag nanoparticles. This thermal behavior of PPy/Ag hybrid materials can be explained by high thermal stability and high thermal conductivity of the metal Ag.

4. Conclusion

PPy/Ag hybrid materials could be prepared by the in-situ chemical polymerization method. The Ag particles as prepared have a quite uniform size from 4 to 8 nm. By the polymerization process, the Ag nanoparticles formed some clusters with different diameter varying from 25 to 150 nm. By the addition of Ag nanoparticle, the electrical conductivity of PPy increased with increasing Ag concentration. The thermal stability of PPy was also improved by modification with Ag. For the fine dispersion of Ag nanoparticles in the PPy matrix, further investigations are required.

References

[1] Liu Y C, Lee H T, Yang S J 2006 *Electrochimica Acta* **51** 3441
[2] Chen W, Li C M, Chen P, Sun C Q. 2007 *Electrochimica Acta* **52** 2845
[3] Wanekaya A K, Lei Y, Bekyarova E, Chen W, Haddon R, Mulchandani A, Myung N V 2006 *Electrianalysis* **18** 1047
[4] Shi S W G 2007 *Materials Chemistry and Physics* **102** 255
[5] Pinter E, Patakfalvi R, Fulei T, Gingl Z, Dekany I and Visy C 2005 *J. Phys. Chem.* **109** 7474
[6] Strike D J, Rooij N F D 1992 *J Appl. Electrochem.* **22** 922
[7] Luyen N T, Nghia N D, Tung N T 2007 *Vietnamese Journal of Chemistry* **45** 123
[8] Ha H M 2005 Master Thesis *Preparation and Investigation of the Phasemorphology, Structure and Properties of Silver Nanoparticle* Hanoi National Uni. – University of Technology

Application of multiwall carbon nanotubes for thermal dissipation in a micro-processor

Bui Hung Thang, Phan Ngoc Hong, Phan Hong Khoi and Phan Ngoc Minh

Institute of Materials Science, Vietnam Academy of Science and Technology, 18 Hoang Quoc Viet Road, Cau Giay District, Hanoi, Vietnam

E-mail: minhpn@ims.vast.ac.vn

Abstract. One of the most valuable properties of the carbon nanotubes materials is its high thermal conductivity with 2000 W/m.K (compared to thermal conductivity of Ag 419 W/m.K). It suggested an approach in applying the CNTs in thermal dissipation media to improve the performance of computer processors and other high power electronic devices. In this research, the multiwall carbon nanotubes (MWCNTs) made by thermal chemical vapour deposition (CVD) at our laboratory was employed as the heat dissipation media in a microprocessor a Personal Computer with configuration: Intel Pentium IV 3.066 GHz, 512Mb of RAM and Windows XP Service Pack 2 Operating System. We directly measured the temperature of the microprocessor during the operation of the computer in two modes: 100% usage CPU mode and over-clocking mode. The measured results showed that when using our thermal dissipation media (a mixture of the mentioned commercial thermal compound and 2 wt.%. MWCNTs), the temperature of the microprocessor decreased 5°C, and the time for increasing the temperature of the microprocessor was three times longer than that when using commercial thermal compound. In over-clocking mode, the processor speed reached 3.8 GHz with 165 MHz of system bus clock speed; it was 1.24 times higher than that in non over-clocking mode. The results confirmed a promising way of using MWCNTs as the thermal dissipation media for microprocessor and high power electronic devices.

Keywords: MWCNTs, thermal compound, microprocessor, thermal dispersion, thermal dissipation.

1. Introduction

Currently the density and the speed of electronic and optoelectronic devices are increasing very quickly. For long-term operation, such devices consume power and liberate thermal energy that seriously decline the performance of the devices or even destroy them. Finding an effective way for thermal dissipation therefore becomes a very important issue. The future of high density, high power electronic and optoelectronic devices strongly rely on the thermal dissipation management.

Carbon nanotubes (CNTs) was known as the highest thermal conductivity materials compare to diamond or other metallic materials (K_{CNTs} = 3000 W/m.K compared to K_{Ag} = 419 W/m.K) [1]. This valuable property of the CNTs opened up an obvious way to utilize the CNTs as an ideal thermal dissipation media in high power, high-density electronic devices. In this paper, we present the initial results on thermal dissipation efficiency in a microprocessor of personal computer (P.C.) using multiwall CNTs as thermal matching media. A simple and effective model was developed for

© 2009 IOP Publishing Ltd

evaluation of the thermal dissipation efficiency by directly measuring the temperature of the microprocessor.

2. Experimental

A commercial silver base thermal grease for PC (named commercial thermal compound) and its mixture with MWCNTs (named CNTs-thermal compound) were utilized as the thermal matching media on the PC (Intel Pentium IV, 3.066 GHz, 512Mb RAM, Windows XP Service Pack 2 Operating System). The thermal dispersive efficiency is evaluated by directly measuring the temperature and temperature response of the CPU using a dedicate software and a temperature sensor built-in inside the microprocessor. Since the surface of the CPU and the heat-sink is not entirely flat, there are many tiny gaps between the two components that make a negative effect on the heat transfer. Material with high thermal conductivity is needed to fill these gaps and improve heat transfer between the CPU and the heat-sink. The testing thermal dissipation compound is located between the surface of the CPU and the CPU fan (heat-sink) as shown in figure 1.

Multiwall carbon nanotubes (MWCNTs) were produced by thermal CVD technique on iron mesh catalyst in a gas mixture of acetylene, hydrogen and nitrogen [2]. The diameter and length of the grown MWCNTs used in this experiment was 20-80 nm, several μm, respectively. Typical Scanning Electron Microscopy (SEM) image of the grown MWCNTs is shown in figure 2. Such MWCNTs have been successfully applied for field emission and advanced plating applications [3-5]. This paper is our first report on utilizing the grown MWCNTs for thermal dissipation application. Functionalized MWCNTs was mixed with the commercial thermal compound mentioned above at different concentrations by mechanical method. The thermal matching media in a volume of 0.12 ml was carefully applied on an area of 3x3 cm^2 of the microprocessor. The thickness of the thermal matching media was approximately 130 μm. Figure 3 shows a SEM image of the 2 wt.%. CNTs thermal compound. It showed that the CNTs were well dispersed in the compound. The testing computer is kept in an environment of 26°C for all measurements. Figure 4 shows the experiment configuration on the actual microprocessor of the PC.

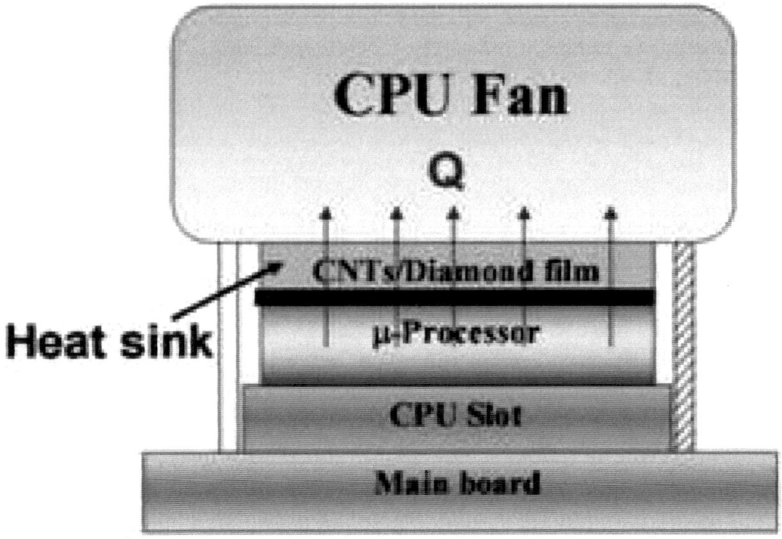

Figure 1. Schematics of the thermal dispersive system of the CPU [2].

Figure 2. The MWCNTs used in this work.

Figure 3. SEM image of the CNTs thermal compound.

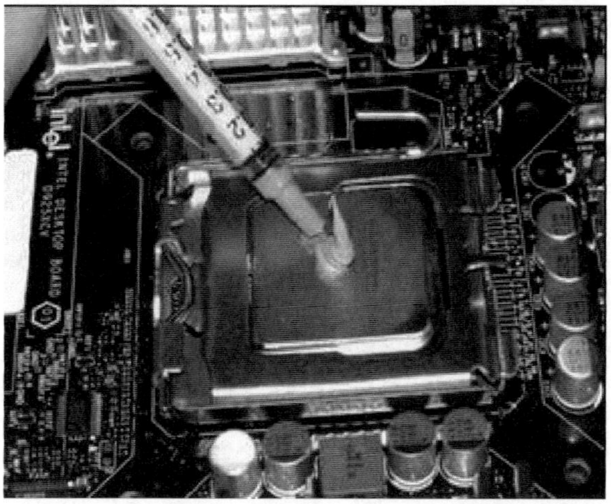

Figure 4. Experiment configuration on the actual microprocessor of the computer.

3. Results and discussion

We measured directly the temperature of the microprocessor during the operation of the computer in 100% usage CPU mode. Figure 5 is the measured temperature of the microprocessor as a function of working time in three cases: (i) no thermal matching media, (ii) utilizing commercial thermal compound and (iii) CNT-thermal compound. It is clearly seen from figure 5 that without thermal matching media, the temperature of the microprocessor reached 85°C within 20 seconds and the computer was automatically shutdown. This obviously confirmed the essential of the thermal matching media for the device. By using 2 wt.% CNTs thermal compound, the temperature increasing time and the maximum temperature of the microprocessor were 200 seconds and 65°C, respectively. Whereas these values are 75 seconds and 70°C for the commercial thermal compound, respectively. The measurement was repeated many times and remained stable for long periods of operation. The improvement of the thermal dissipation is absolutely attributed by the added MWCNTs component. Noted that 2 wt.% CNTs was optimized for keeping the adhesive and sticking property of the media.

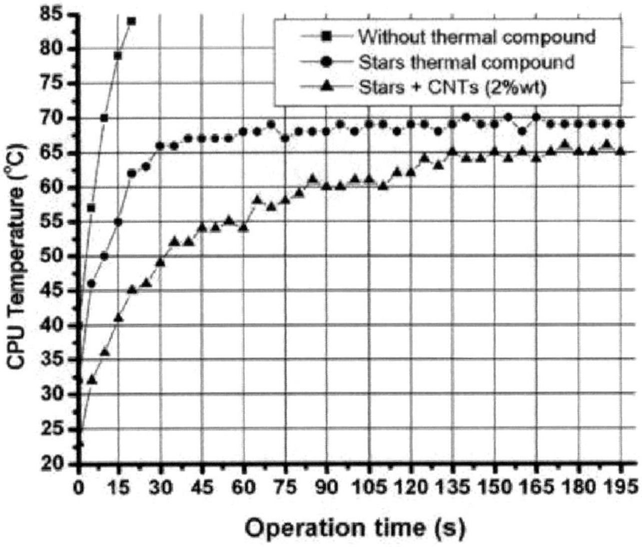

Figure 5. The temperature of the microprocessor as a function of working time.

The temperature increase of the microprocessor can be expressed by differential equation

$$\frac{dT}{dt} = \frac{(P-J)}{C} \quad , \tag{1}$$

where T is temperature of the microprocessor, P is the radiate thermal power of the microprocessor, J is the rate of radiate heat flow and C is heat capacity of the system

$$J = \frac{(T-T_0)}{R} \quad , \tag{2}$$

where R is heat resistance of thermal dispersive system, and T_0 is environmental temperature (T=26°C). From equations (1) and (2), we have:

$$\frac{dT}{dt} = (P - \frac{T-T_0}{R})\frac{1}{C} \quad , \tag{3}$$

$$\frac{dT}{dt} = \frac{PR + T_0 - T}{RC} \quad , \tag{4}$$

$$\frac{dT}{PR + T_0 - T} = \frac{dt}{RC} \quad . \tag{5}$$

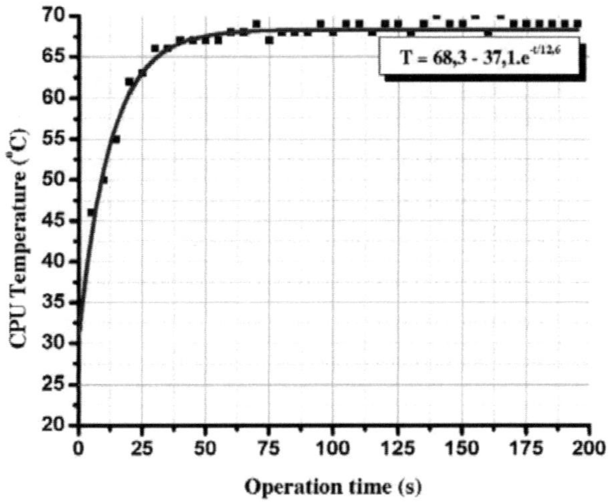

Figure 6. The graph of fitting exponential function of microprocessor temperature when using commercial thermal compound.

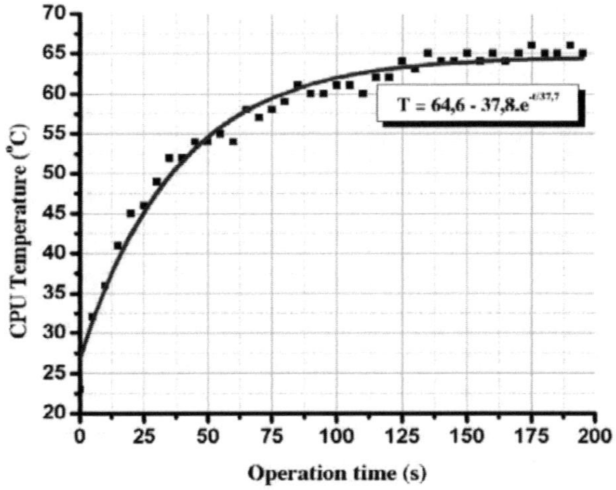

Figure 7. Graph of fitting exponential function of microprocessor temperature when using CNTs thermal compound.

Since the temperature of the microprocessor at the time t = 0 is slightly different from T_0 due to the running of the computer before starting to count the time so from equation (5) we have:

$$\int_{T_1}^{T} \frac{dT}{PR+T_0-T} = \int_0^t \frac{dt}{RC} \quad , \tag{6}$$

$$-\int_{T_1}^{T} \frac{d(PR+T_0-T)}{PR+T_0-T} = \frac{1}{RC}\int_0^t dt \quad , \tag{7}$$

$$-\ln(PR+T_0-T)\Big|_{T_1}^{T} = \frac{t}{RC} \quad , \tag{8}$$

$$\ln\left(\frac{PR+T_0-T}{PR+T_0-T_1}\right) = -\frac{t}{RC} \quad , \tag{9}$$

$$\frac{PR+T_0-T}{PR+T_0-T_1} = e^{-\frac{t}{RC}} \quad , \tag{10}$$

$$PR+T_0-T = (PR+T_0-T_1)e^{-\frac{t}{RC}} \quad , \tag{11}$$

$$T = (PR+T_0)-(PR+T_0-T_1)e^{-\frac{t}{RC}} \quad . \tag{12}$$

Thus, temperature T of the microprocessor increases as an exponential function of working time t:

$$T = a - be^{-\frac{t}{c}} \quad , \tag{13}$$

where:

$$a = PR + T_0; b = PR + T_0 - T_1; c = RC \quad . \tag{14}$$

Figures 6 and 7 showed the measured and fitting results according to equation 13 of the microprocessor's temperature as a function of working time for the commercial and CNTs-thermal compounds, respectively. The fitting results showed a function $T = 68.3 - 37.1e^{-\frac{t}{12.6}}$ (°C) for commercial compound and $T = 64.6 - 37.8e^{-\frac{t}{37.7}}$ (°C) for CNTs-thermal compound. The fitting curves excellently matched with the measured values. Noted that the values of P (radiate thermal power of the microprocessor) and T_0 do not depend on the nature of thermal dissipation media. The value of T_1, R and C depend on the type of thermal matching media. For the case of using commercial thermal compound matching media, we have:

$$PR_1 + T_0 = 68.3 \quad , \tag{15}$$

$$PR_1 + T_0 - T_1 = 37.1 \quad , \tag{16}$$

$$R_1C_1 = 12.6 \quad . \tag{17}$$

For the case of using CNT thermal compound matching media, we have:

$$PR_2 + T_0 = 64.4 \quad , \tag{18}$$

$$PR_2 + T_0 - T_2 = 37.8 \quad , \tag{19}$$

$$R_2C_2 = 37.7 \quad . \tag{20}$$

From equations (15)-(20) and T_0 = 26°C, we received: T_1 = 31.2°C; T_2 = 26.8°C; PR_1 = 42.3; R_1C_1 = 12.6; PR_2 = 38.6; R_2C_2 = 37.7; R_1/R_2 = 42.3/38/6 = 1.0958 and $C_1/C_2 = (R_1/R_2)$.(12.6/37.7 = 0.305 or $C_2 = 3.279C_1$. These values confirmed that the heat capacity of the system with CNTs thermal compound (C_2) was more than three times larger than that with the commercial thermal compound (C_1).

4. Conclusion

The results confirmed the advantage of the MWCNTs for the thermal dissipation media for microprocessor of the PC or high power electronic devices in general. The measured and calculated results showed that temperature of CPU increases as an exponential function of working time. When using CNTs thermal compound, the temperature of the microprocessor decreased 5°C, and the heat capacity and the time for increasing the temperature of the microprocessor was more than three times larger than that when using commercial thermal compound.

References

[1] *Handbook of Chemistry and Physics* 2006-2007 (CRC BocaRaton)

[2] Chuc N V, Dung N D, Hong P N, Quang L D, Khoi P H and Minh P N 2008 *J. Korean Phys. Soc.* **52** (5) 1368

[3] Tuyen L T T, Minh P N, Ruduner E, Chi P T D, Ono T, Miyashita H, Khoi P H and Esashi M 2005 *Chemi. Phys. Lett.* **415** 333

[4] Hong P N, Cuong V V, Thang B H, Khoi P H and Minh P N 2008 *J. Korean Phys. Soc.* **52** (5) 1386

[5] Minh P N, Khoi P H 2006 *Proc. of JAIST Int. Symp. on Nanotech.* (Ishikawa, Japan)

Acknowledgements

We would like to thank the AOARD 084056 project for the financial support. This work is also supported by the National Basic Research and Vietnamese-France bilateral Science Collaboration Programs. A part of this work was done with the help of the National Key Laboratory for Electronic Materials and Devices, Institute of Materials Science.

On growth mechanisms and dynamic simulation of growth process based on the experimental results of nanowire growth by VLS method on semiconductor substrates

Dao Khac An[1], Nguyen Xuan Chung[2], Pham Hong Trang[3], Hoang Van Vuong[3], Phan Viet Phong[1] and Phan Anh Tuan[1, 4]

[1] Institute of Materials Science, Vietnam Academy of Science and Technology, 18 Hoang Quoc Viet Road, Cau Giay District, Hanoi, Vietnam
[2] Physics Department, Geo and Mining University, Tu Liem, Ha Noi, Vietnam
[3] Students of Faculty of Materials Science and Technology, Hanoi University of Technology, 1 Dai Co Viet Road, Hai Ba Trung District, Hanoi, Vietnam
[4] Tohoku University, Japan

E-mail: andk@ims.vast.ac.vn

Abstract. Recently the production of nanowires is attracting many scientists but it also holds many challenges. Many problems including growth mechanisms are still not understood clearly. This paper will show briefly some our experiment results of nanowires (GeO_2 and Ga_2O_3) growth by VLS, the rest of the paper will discuss nanowire growth mechanisms and then showing a developed programme on PC for dynamic simulation of nanowire growth process. This running simulation software for nanowires growth process has contained main mechanisms and we can see directly some physical phenomena concerning growth process by VLS method.

Keywords: GeO_2 and Ga_2O_3 nanowires, VLS method, growth mechanisms, dynamic simulation of growing nanowires.

1. Introduction

Nanowires, nanorods and nanotubes play a very important role in nanoelectronics [1-4]. They have many applications in many fields such as for making nanowires connecting between nano devices in modern nanoelectronic circuits, for tips or probes in modern microscopes, for making various kinds of sensors and pn junction diodes, swishings, memory nodes, nanowire FET devices [5-8]. It is not like as fabrication process of nano particles (0D) as well as nano thin films (2D), where technologies are somewhat developed, the nanowire growing process along one axis (1D) gets more and more difficult, for example, how to control the uniform morphology with fitable sizes of nanowires (diameter and length) during nanowire growing, how to grow nanowires in array forms and how we can treat nanowire easily and measurable parameters to make contact with them…[1, 3-5, 8].

So far the fabrication of nanowires is achieved by two approaches, top-down and bottom-up [1-5, 14-17]. Concretely, nanowires could be produced by different methods such as by VLS method (Vapor Liquid Solid), CVD (Chemical Vapor Deposition), LCG (Laser Ablation Catalytic Growth), using template Assisted Synthesis, FLS (Fluid Liquid Solid), SLS (Solution Liquid Solid), OAG (Oxide Assisted Growth), MBE (molecular Beam Epitaxy) ... Among them the VLS method is seem to be a

© 2009 IOP Publishing Ltd

simple, popular, cheap method. Depending on applications of nanowires and growing method so far one could produce many kinds of nanowires on single semiconductor substrates (Si, Ge,...) or on compound semiconductor substrates such as GaAs, GaP, InAs, InP,GaAs/P, InAs/P, ZnS, ZnSe, CdS, CdSe...[4]. The common requirements for growth nanowires, depending on each application, could be in a fitable range of resistivity, diameters of nanowires could be distributed from several tens of nm to several hundreds nm, their lengths are in a range of micrometers. Besides, the treatment and, measurement process could be done easily on the growth nanowires. Unfortunately, these requirements are not easily to satisfy. These facts are also challenges for technologists [1-8]. In order to see clearly and get better understanding of the growth process, this paper shows briefly some our experiment results of GeO_2 and Ga_2O_3 nanowires growing on Ge and GaAs substrates by VLS method with an Au thin catalyst layer on the front side [9-14], the rest of the parts of the paper discuss growth mechanisms and showed the results of dynamic simulation of a whole nanowire growing process containing main mechanisms in 2D.

2. Experimental procedures

Starting materials were the n-type Ge slices with (111) orientation and both p-type and n-type GaAs slices with (100) orientation. For the first step we have cut Ge slices into many small square shape samples with dimensions of 5×5×1 mm. The small samples were mechanically lapped, polished, cleaned by ultra-sound vibration in three solutions alternatively: solution of acetone (CH_3COCH_3 99.5%); solution of alcohol (CH_3CH_2OH 95%) and ionized water, then chemically treated in HF: H_2O with a volume ratio of 1:15. The last step is vibrating in ion water and drying. A nano thin layer of Au was evaporated onto the surface of semiconductor samples in vacuum 10^{-4}-10^{-5} torr. The thickness of Au thin layer was measured using the SEM cross-section method. Its thickness is about 50 nm. The same way of chemical treatment is done for the ready GaAs slices.

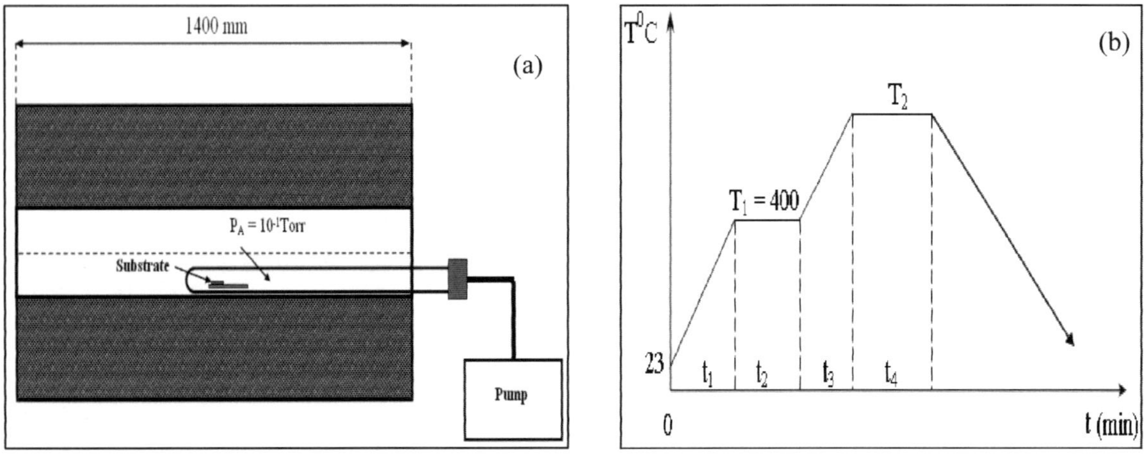

Figure 1. Blocks of experiment equipment furnace (a); Profile of experiment temperatures for GeO_2, in period 1: heating up to T_1, around the eutectic temperature with t_1 and t_2 duration times, then in period 2 heating up to T_2 with t_3 and t_4 duration times where the nanowires growing process are mainly occurring (b).

The furnace used for nanowires growing was Lindberg/ Blue M Three-zone tube Furnace. The block of experiment equipment can be seen in figure 1 (a). We have heated the furnace for different temperature profiles for different experiments of growing nanowires. The furnace's tube temperatures were automatically measured by Keithley 2000 Multimeter connecting with PC. The temperature profile of the furnace for GeO_2 and Ga_2O_3 nanowires growing as well as the growing times can be seen in table 1.

Table 1. Technological conditions for the experiment cases of GeO_2 and Ga_2O_3 nanowires growing on Ge, GaAs substrates.

Experiment Conditions		Our works	Period 1: heating to around eutectic temperature, T_1			Period 2: heating up to the growing temperature, T_2		
Experiment Cases			T_1 (^{0}C)	t_1 (min)	t_2 (min)	T_2 (^{0}C)	t_3 (min)	t_4 (min)
GaAs Substrate	figure 3	[9,10,19]	450	35	35	550 -700	30	30-50
Ge Substrate	figure 4	[10]	400	55	75	520- 720	34	55 -75

Here we would like to emphasize that the temperature T_1 in period 1 must be the temperature around (lower or upper) the eutectic temperature, and we have chosen the temperatures T_2 in the range of 520 to 720°C. After growing nanowires, the samples have been investigated by SEM, EDX, and TEM, AFM... The experimental main steps can be seen briefly in figure 2. The experimental procedures can be seen in details for both nanowire growing of GeO_2 and Ga_2O_3 in our works [9, 10, 19].

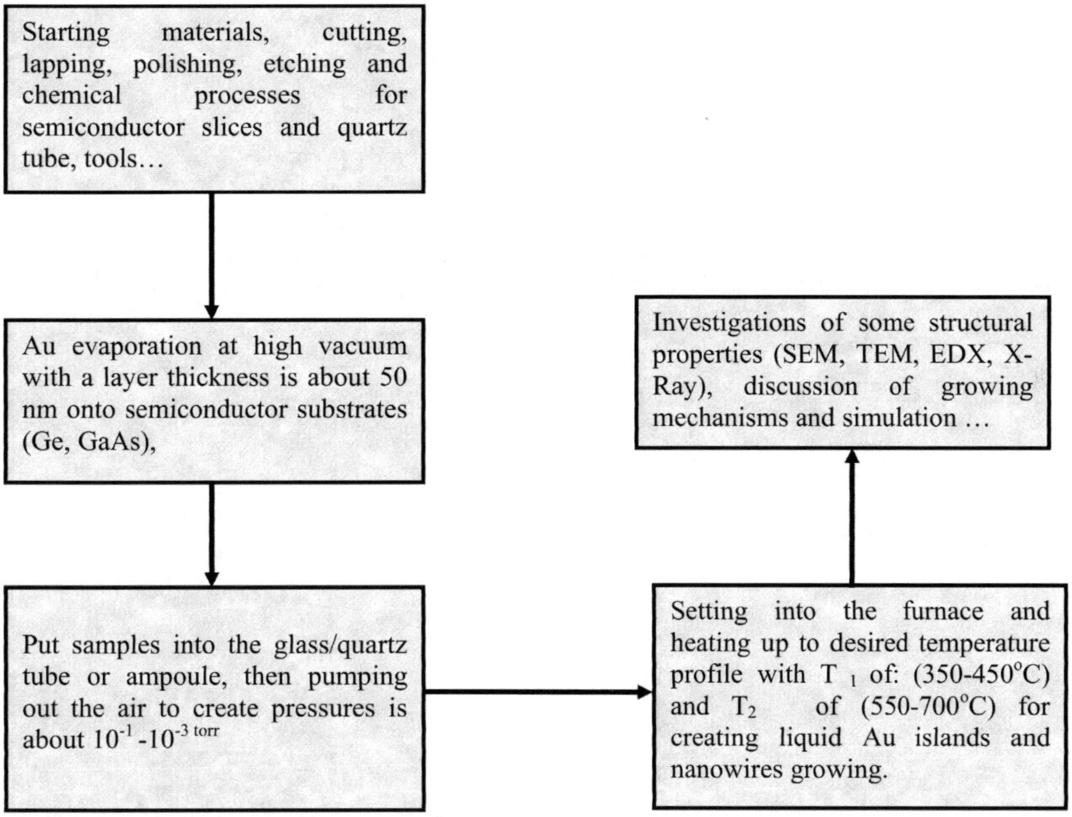

Figure 2. Schematic diagram of experiment procedures of nanowire growing by VLS method.

3. Experiment results

Figures 3, 4 show some different forms or morphologies of Ga_2O_3, GeO_2 nanowires on GaAs, Ge substrates which depend strongly on the different technological conditions. The results can be seen in details of our works [9, 10, 19].

Journal of Physics: Conference Series **187** (2009) 012052 doi:10.1088/1742-6596/187/1/012052

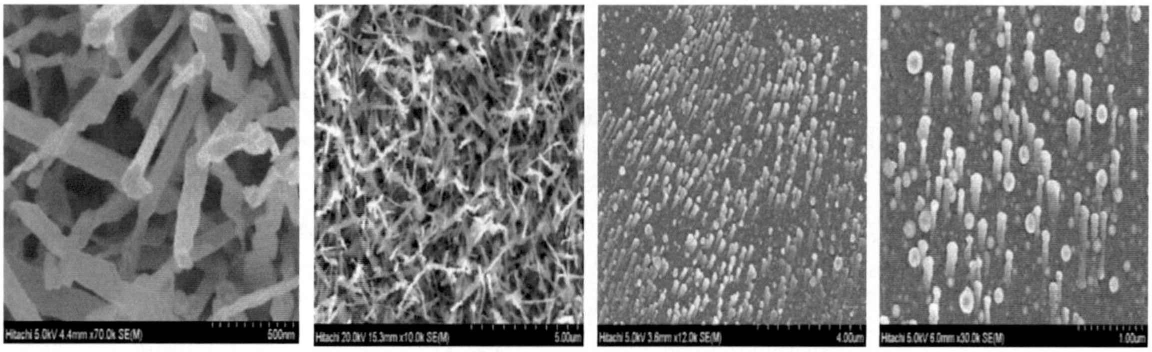

Figure 3. SEM images of Ga_2O_3 nanowires formed on GaAs substrate at the different growing conditions. The morphologies of nano wires vary from random forms to rather vertical orientation with more regular (from left to right).

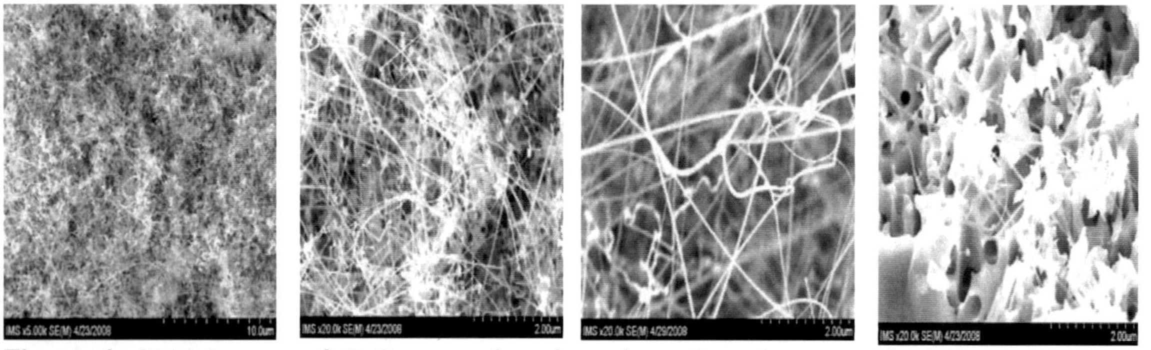

Figure 4. SEM Images of GeO_2 nanowires formed on Ge substrate with different growing conditions The morphologies of nano wires are tangly whiskers as a porous surface.

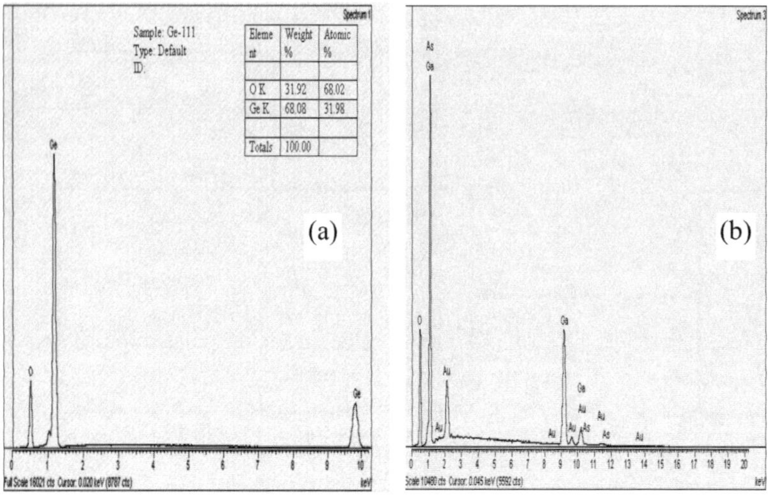

Element	Weight%	Atomic%
O K	34.17	69.66
Ga K	51.94	24.30
As L	13.89	6.05
Totals	100.00	

(c)

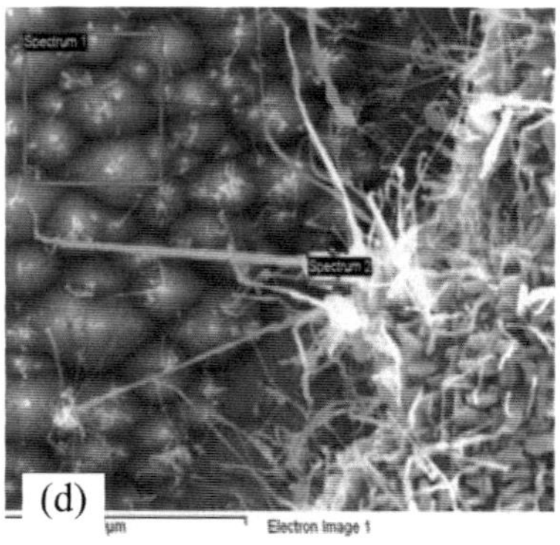

Figure 5. Energy-dispersive x-ray spectroscopy (EDX) characterizing nanowires grown on Ge substrate containing oxygen (68.2% atomic, Ge (31.98% atomic) (a); elements in Ga_2O_3 nanowires containing oxygen (69.662% atomic), Ga (24.30% atomic) and As (6.05% atomic %) (b,c); The place of measurement is in the centre of the sample (d).

Figures 5 shows Energy-dispersive X-ray spectroscopy (EDX) of GeO_2, FESEM and EDX of Ga_2O_3 nanowires at the marked place. From our experiment results in figures 3, 4, 5 as well as our other results we observed that the GeO_2 and Ga_2O_3 strongly depend on technological conditions: heating profiles T_1, T_2 and the growing time t_4 [9, 10, 19].

4. Discussions of growth mechanisms of VLS method

The nanowires growth mechanisms on semiconductor substrates with Au thin catalyst film layer have been discussed very intensively [1-5, 13-17] but so far some points concerning mechanisms are not yet clearly understood. It is a fact that the growth mechanisms can be established on the base of the phase diagram. Figure 6 shows two phase diagrams of two systems used in growing nanowires of GeO_2 and Ga_2O_3. The eutectic temperatures of Au-Ge and Au-GaAs are 361°C and 630°C, respectively. The chosen temperature T_1 is vitally important for formation of Au droplets and making alloy between Au and substrate material. The size of Au droplets on the semiconductor surface as well as the uniform of liquid droplets will determine the size of nanowires. This process depends on many factors such as the surface morphology, surface and bulk defect situation of semiconductor substrate, the heat amount or T_1 for formation of Au droplet, the diffusion of Au into the semiconductor, the diffusion of the host atoms to the liquid-solid interface, the formation of the seed from that nanowire will be grown…Unfortunately we do not yet understood these processes clearly.

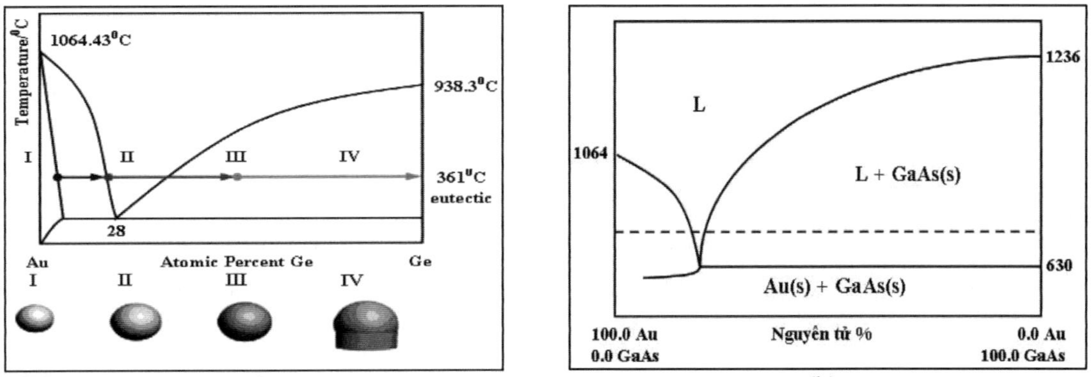

(a) (b)

Figure 6. Phase diagram of the Au - Ge binary system concerning four main stages of GeO_2 nanowire growth (a); and phase diagram of the Au – GaAs system (b).

When we heat up the sample in a furnace system containing certain low air pressure to temperature T_2 which is much higher than the eutectic temperature, many physical-chemical processes occurred. The growth mechanisms also strongly depend on these processes.

4.1. Some requirements and some main factors affecting growth mechanisms

- Semiconductors substrate must be completely clean and flat, it may contain a few defects. The fact shows that Au droplets as well as nucleation forms strongly depend on semiconductor surface and sample's defects.
- Au catalyst thin film's thickness is very important. This film should not be too thin and not too thick, its value is approximately 40-80 nm. Depending on temperature T_1 (lower or upper eutectic temperature) Au droplets will be formed and make an alloy with substrate material. The forms of Au droplets and alloy will strongly determine the sizes of nanowires in later stages. Experiment results showed that Au liquid droplets on the top of nanowire also play a very important role during nanowire growth, especially chemical potential (μ_{LV}) at the interface between liquid and solid phase determining the growth direction.

- Vapour ambient in the ampoule or in the low pressure furnace is a very important factor of the VLS method. The presence of certain air, containing some oxygen atoms or some H_2O vapor for oxide formation inside or on the surface of nanowire, could also create an oxide liquid phase on the surface of nanowire in order to assist the growth process, so called, oxide assisted growth (in several cases). This process is very complex. In some cases ones also uses some semiconductor materials in gaseous phase. Depending on the material of substrate the air pressure in the ampoule is in range of 10^{-1} to 10^{-5} torr. For GeO_2 and Ga_2O_3 nanowire growth pressure is usually in range of 10^{-1} to 10^{-2} torr. This pressure has to choose an optimum value together with other factors.
- Temperature profile of the growth process is most important and it depends on what solid materials (catalyst metals and semiconductor substrates) have been chosen. The temperature profile could strongly affect the formation of the Au droplet sizes, diffusion of Au atoms, oxygen atoms into substrate material as well as into liquid Au droplets and allowing regions. Some of these effects are also not yet clearly understood, and so we can not yet control them. The speed of temperature increase (slope of temperature profile) also strongly affects the formation of Au droplet and alloying. In principle, when the growth temperature is higher, Au droplets are smaller, and then the nanowire diameters are also smaller.
- Growth times could be understood to be the time for whole temperature profiles. But the duration time t_2 is long enough for Au droplets formation, alloying, and duration time t_4 is also chosen so long depending on requirements of length of nanowire.

4.2. Some phenomena of growth process and main growth mechanisms

We have investigated some main phenomena of the growth processes concerning mechanisms by SEM images at different temperatures and different growth times. These can be seen in figure 7 [9, 10, 19]. The phenomena concerning the mechanisms of each nanowires growing system by VLS may be different from each other but, for example, we discuss Au-Ge and Au-GaAs systems, in essence, their growing processes have similarities. The main growth mechanisms are followings:

- A growing system consists of semiconductor substrate (Ge, GaAs) in ampoule or sealed tube with low pressure air as in figure 7 a)
- The furnace is heated up to temperature near T_1 lower than the eutectic temperature (as in figure 6), then Au thin film catalyst layer is disturbed; the alloy of Au-Ge could be beginning to form on semiconductor slice surface (figure 7 b, c).
- Temperature increases to T_1 after duration time t_2, Au droplets (or islands) as well as the alloy regions under Au droplets are formed more clearly, sharply (figure 7 d). The semiconductor atoms (Ge, Ga...) more and more diffused into the interface between Au droplet and alloy. When semiconductor atom concentration in alloy becomes saturated, semiconductor atoms precipitate (or cluster) at the liquid-solid interface. This process is very important for growing nanowires in the next steps.
- Growth process is developed when temperature continuously increases from T_1 toward T_2, from phase diagram we can see that semiconductor atomic % (Ge, Ga) will increase. At the increased semiconductor atom (Ge, Ga) weight percentage of about 50-60% one big precipitate or more small precipitates will be formed. Because the sticking coefficient on liquid Au droplet is higher then that on solid surfaces, semiconductor crystal growth is occurred only where the liquid metallic catalyst is present (figure 7e). Depending on the number and sizes of semiconductor precipitates in an Au droplet, one big nanowire or more small whiskers could be grown up from one Au droplet. Due to the surface tension the liquid catalyst forms a ball situated on the top of the growing nanowire (figure 7 f).

Journal of Physics: Conference Series **187** (2009) 012052 doi:10.1088/1742-6596/187/1/012052

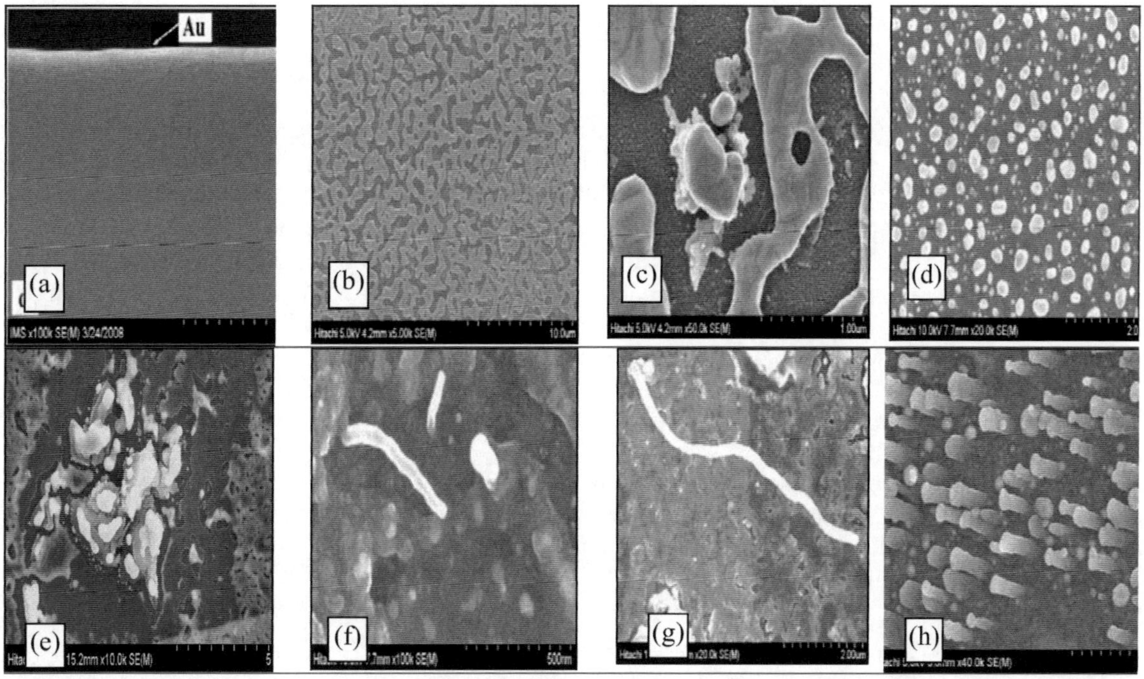

Figure 7. Some main results showing our main phenomena concerning Ga_2O_3 nanowires growing process on GaAs substrate concerning growing mechanisms: Au thin film layer on semiconductor substrate with low pressure a); Au layer is disturbed on the surface b); Au droplets and alloy formed at near temperature T_1 d); the nucleation's are formed and semiconductor crystals began to grow e); nanowires diameters depend on the forms of Au droplets on the surface f); a long Ga_2O_3 nanowire of about 8 micrometers length and 180 nm diameter g); and Ga_2O_3 nanowires formed in vertical orientation with rather uniform of array h).

4.3. Further discussions

The diameter of the nanowire grown by VLS method is determined by the form or diameter of Au droplet. Its size remains essentially unchanged during the entire process of wire growth (see figure 7 f, g and h). The speed of growth, depending on technological conditions and substrate materials, is about several hundreds nm per minutes. The orientation of growing nanowire is related to the locations of the diffused semiconductor atoms in solution of alloy near the interface of Au droplet. Preferred growth directions lie perpendicular to planes along which semiconductor atoms diffused readily allowing for completion of mentioned planes to the diameter of the catalyst, presumably by incorporation of added atoms at steps edges. The chemical reactions occurred in the low vacuum tube /ampoule containing low oxygen or vapour water can be written for three kinds of growing systems (on Ge, Si and GaAs substrates) as the following, respectively:

$$Ge + O_2 \rightarrow GeO_2 \quad , \qquad (1)$$

$$4\,Si + O_2 \rightarrow SiO_2 \quad , \qquad (2)$$

$$Ga + 3\,O_2 \rightarrow 2\,Ga_2O_3 \quad , \qquad (3)$$

$$2Ga + 3H_2O \rightarrow Ga_2O_3 + 3H_2 \quad . \qquad (4)$$

Here we would like to note that semiconductor atoms (Ge, Si or Ga) from gaseous phase or from bulk substrate could be diffused out into gas ambient or diffused into the alloy region. The source of oxygen (O_2) from gaseous phase is in the growing tube.

The diameter of a grown nanowire d can be determined by the following way:

$$d_c = \frac{4\alpha\Omega}{RTLn\dfrac{C}{C_\infty}} \qquad (5)$$

where α is specific free energy of the nanowire at the liquid-vapour interface, Ω is gram molecular volume, R is real gas constant, T is absolute temperature, C is concentration of semiconductor atom (Ge, Si, Ga) in the liquid phase, C_∞ is equilibrium concentration of semiconductor element in liquid phase. From Eq. (5) we also see that the diameter increases linearly with α and Ω and inversely with T and in(C/C_∞). Our experimental results agreed with these facts. From obtained results on figures 3, 4, 5, 7 we also observed that GeO_2, Ga_2O_3 nanowires sometimes grow spontaneously with tangly whiskers forms, and grow orderly depending on chosen technological conditions. The diameters of GeO_2 nanowires are in the range of about 10 to about 40 nm, for Ga_2O_3 nanowires the diameters are in the range of about 80 - 200 nm with length of several μm. We observed that the higher substrate temperatures T_2 is, the smaller nanowire's diameter and more nanowires density becomes.

5. Dynamic 2D simulation of nanowire growth process

Based on our experiment results obtained and the understanding of nanowires growth mechanisms we have developed a simulation program for dynamic simulation of nanowires growth in 2D.

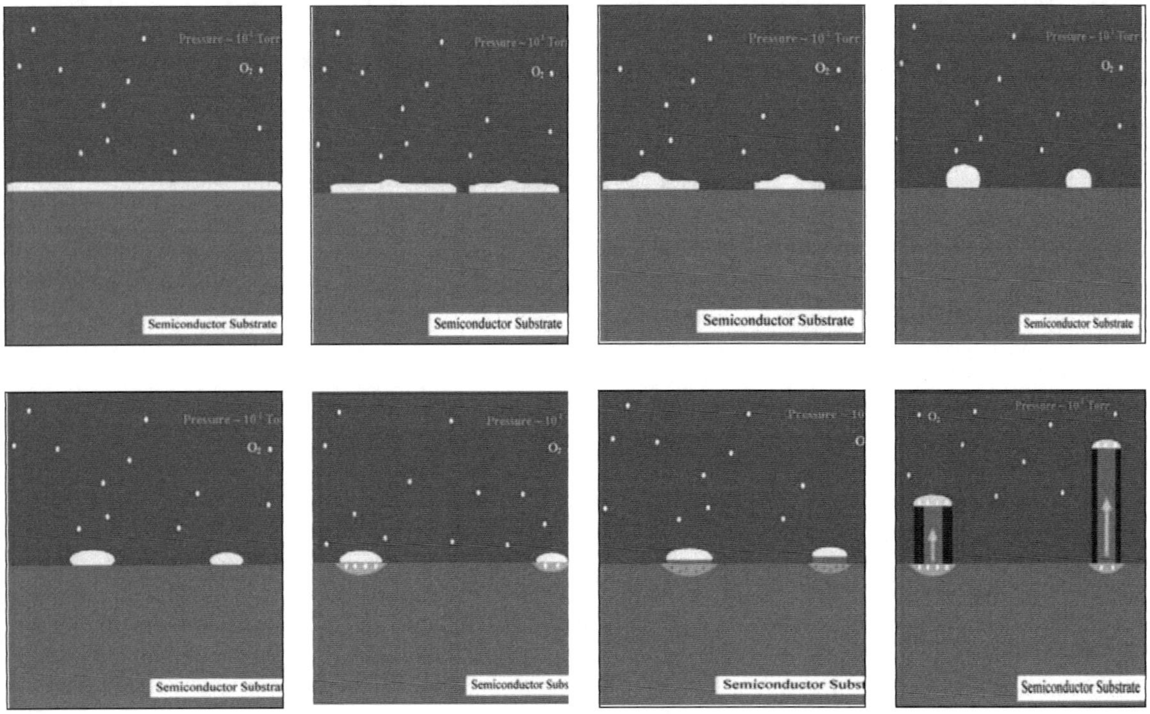

Figure 8. Eight standing simulation pictures of dynamic simulation program at typical growth temperatures and growth times corresponding to the real growth phenomena and growth mechanisms in figure 7. *Upper row*: Au layer is on the surface then it is disturbed and becomes Au droplets. *Under row*: Au droplets is alloying to form liquid-solid phase; the nucleation's seed are formed and semiconductor crystals began to grow in the forms of Au droplets on the surface

Figure 8 shows some standing simulation pictures attracted from the dynamic simulation program of nanowire growth process. The program can run with different velocities, it can also zoom to small or larger sizes. The programme should be running in a rather strong configuration PC. The program has transformed up into Avi or Flash format in order to run in multimedia. It has been inserted in our website together with other simulation programmes. Software programme consists of some blocks using many commands of Matlab 7.0 to create figures linking together to form a dynamic simulation programme of nanowires growth on semiconductor substrate in air pressures 10^{-1} - 10^{-2} torr. The flow chart of simulation program can be seen in figure 9.

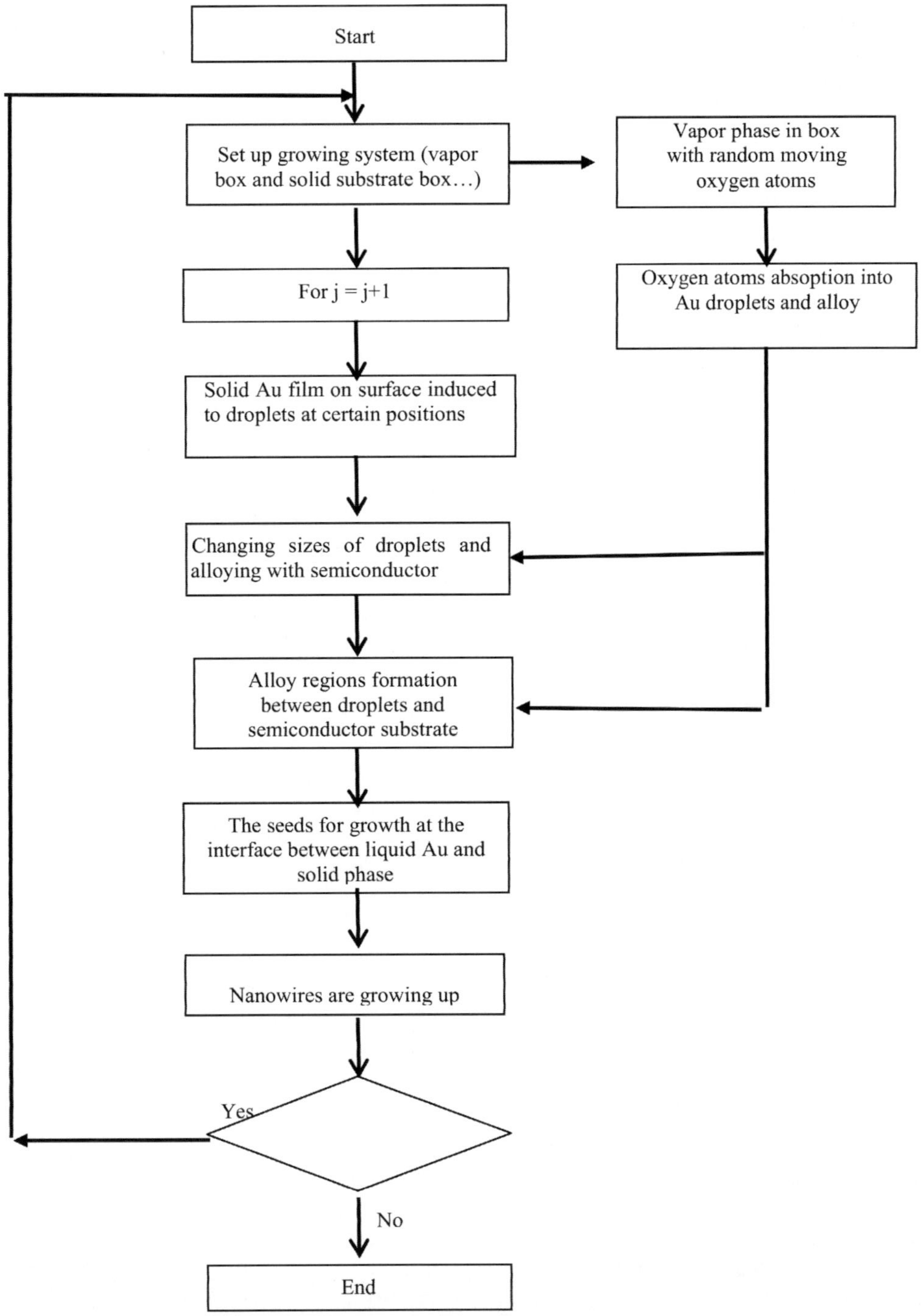

Figure 9. Flow chart of dynamic simulation programmes for nanowires growth on semiconductor substrate.

6. Conclusions

We have synthesis of GeO_2, Ga_2O_3 nanowires using a thermal VLS method in an ampoule or a sealed glass/quartz tube at a pressure maintaining of 10^{-1} -10^{-2} torr with different temperatures and different growth times. The nanowires were investigated by SEM, EDX, AFM and SEM… The nanowires have diameters distributed from 15 to 40 nm for GeO_2 and 20-180 nm for Ga_2O_3. Their lengths are ranging from 20 nm to several μm. We observed also that the higher growth temperature is, the diameter smaller becomes, its density is also increased. A dynamic simulation program of nanowires growth process on semiconductor substrate has been developed. This simulation contains some main phenomena concerning growth mechanisms, and from this simulation we could see directly and lively growth process in PC screen. We would like to note that the technology of VLS method for growing nanowires is simple, low cost but in order to grow good quality, uniform nanowires with fitable sizes for special applications at desired places on substrate we still faced with some technological difficult problems concerning controlling growth mechanisms.

References

[1] Patently M J 2004 Nano materials – the driving force *Nanotoday* **7**(12) 20
[2] Roco 2002 Small Wonders *Nanoscience, Engineering and Technology (NSET)* 1
[3] Bharat Bhushan 2004 *Handbook of Nanotechnology* **Part A** (Springer) 99
[4] Yang P, Wu Y, Fan R 2002 *International Journal of Nanoscience* **1**(1) 1
[5] Lauhon L J, Gudiksen M S and Lieber C M 2004 *Phil. Trans. R. Soc. Lond. A* **362** 1247
[6] Huang Y, Xiangfeng, Cui Y and Mlieber C 2002 *Nano Letters* **2**(2) 101
[7] Greytak A B, Lauhon L J, Gudiksen M S and Lieber C M 2004 *Applied Physics Letters* **84** 24
[8] Atashbar M Z, Yu M-F and Chen X 2002 Investigation and characterization of Ga_2O_3 nanowire for gas sensing applications *IEEE Sensors*
[9] Tuan P A, An D K, Manh D H, Phong P V and Hoa L T 2007 *Proc. of the 5th National Solid State conference, Vietnam* 359
[10] Trang P H, Vuong H V, Phong P V, Manh N H and An D K 2008 *Proc. of the APCTP–ASEAN Workshop on Advanced Materials Science and Nanotechnology (Nha Trang, Vietnam)* p. 975
[11] Mojzes I, Kokényesi S, Szabó I A, Ivan I and Pécz B 2006 *Nanopages* **1** 85
[12] Hemant A, Ann M F, Irene A G, Christopher E D C and Paul C 2007 *ACS Nano.* **1**(5) 415
[13] Civale Y, Nanver L K, Hadley P 2004 *Proc. of 7th Annual Workshop on Semiconductor Advances for Future Electronics and Sensors (Veldhoven, The Netherlands*) 692
[14] Noborisaka J, Motohisa J, Hara S and Fukui T 2005 *Applied Physics Letters* **87** 093109
[15] Wen J G, Lao J Y, Wang D Z, Kyaw T M, Foo Y L and Ren Z F 2003 *Chem. Phys. Lett.* **372** 717
[16] Kalache B, Cabarraocas P R, Morral A F 2006 *Jpn. J. Appl. Phys.* **45** L190
[17] Verheijen M A, Immink G, Smet T de, Borgstrom M T and Bakkers E P 2006 Growth kinetics of heterostructured GaP-GaAs nanowires *J. Am. Chem. Soc.* **128** 1353
[18] Chiem C V, Ha N T T, Tam N T T, Chuc N V, Li H and Muyng S J 2008 *Proc. of the APCTP–ASEAN Workshop on Advanced Materials Science and Nanotechnology (Nha Trang, Vietnam)* p. 837
[19] An D K, Chung N X and Tuan P A 2008 *Proc. of the APCTP–ASEAN Workshop on Advanced Materials Science and Nanotechnology (Nha Trang, Vietnam)* p. 518

Acknowledgements
Authors would like to express their thanks for Mr. Do Hung Manh for his measurement of SEM, EDX of samples. Authors also would like to express their thanks to the Institute of Materials Science VAST as well as NAFOSTED for financial funding to the basic research project in 2007, 2008 and 2009 to carry out these topics.

Some physical properties of $ZnAl_2O_4$: Cr^{3+} (Co^{2+}) powders prepared by hydrothermal method

Le Hong Ha, Phung Thi Lanh, Nguyen Ngoc Long and Trinh Thi Loan

Hanoi University of Science, 334 Nguyen Trai Street, Thanh Xuan District, Hanoi, Vietnam

E-mail: phungthilanh@gmail.com

Abstract. $ZnAl_2O_4$: Cr^{3+} (Co^{2+}) nanopowders were prepared by hydrothermal method and annealed at various temperatures. The samples were characterized by X - ray diffraction, transmission electron microscopy, photoluminescence and photoluminescence excitation spectra. The results show that crystal structure and optical properties of the samples were affected by heat treatment regime. When annealing temperature increased, crystal structure of the samples became better. Optical spectra of the ions Cr^{3+} are sensitive to crystal structure of the host lattice. Electron transitions difference in absorption and radiation processes between the samples $ZnAl_2O_4$: Co^{2+} unannealed and annealed at high temperature was investigated.

Keywords: $ZnAl_2O_4$: Cr^{3+} (Co^{2+}), nanopowders, hydrothermal.

1. Introduction

Spinels have attracted interest because of their properties such as high chemical and thermal stability, high mechanical resistance, low sintering temperature and high quantum yields [1]. They are widely used as a phosphor host material for applications in mechano - optical stress sensors, optical coating, thin film electroluminescence displays, stress imaging devices, catalysts and catalysts supports [2, 3]. There are many methods to prepare spinels with nanostructures, such as solid state reaction [4], wet chemical routes: co - precipitation [5], sol - gel [6], hydrothermal method [7].

In this work, we have succeeded in synthesizing nanopowders $ZnAl_2O_4$: Cr^{3+} (Co^{2+}) with single phase and controlled particle size and morphology by the hydrothermal techniques. We also studied structural, optical properties of Cr^{3+} (Co^{2+}) ions doped samples $ZnAl_2O_4$ unannealed and annealed at various temperature in the range of 200 - 1200°C.

2. Experiment

In this experiment, $AlCl_3.6H_2O$, $ZnCl_2$ and $Co(NH_2)_2$ were used as source materials. At first, the source materials were dissolved in distilled water and the products were stirred at room temperature. Chromium (cobalt) nitrate was added to above solution with the molar ratio Cr^{3+} (Co^{2+})/$ZnAl_2O_4$ varying up 5:1000. After that, an amount of NaOH and then C_2H_5OH were supplemented to the solution. Next the solution was put into Teflon - lined stainless steel autoclave and kept at the reaction temperature of 200°C for 96 hours. Thereafter, the solid material was filtered off, washed with distilled water and then air - dried at 100°C.

All the synthesized samples were characterized for their crystallinity phase by power X - ray diffraction (XRD) on a SIMENS D5005 diffractometer using Cu - Kα radiation (λ = 1.54056 Å). The

© 2009 IOP Publishing Ltd

morphology and nanostructures of particles were examined by TEM using a JEOL JEM 1010 transmission electron microscope. The optical properties were investigated by photoluminescence (PL) and photoluminescence excitation (PLE) spectra at room temperature on a spectroflourometer Fluorolog FL3 -22 Jobin Yvon Spex USA.

3. Results and discussion

3.1. Cr^{3+} doped $ZnAl_2O_4$ spinel nanopowders

The XRD patterns of nanopowders Cr^{3+} (Co^{2+}) doped $ZnAl_2O_4$ annealed at various temperatures: 700, 900, 1200°C are shown in figure 1. As seen from the XRD patterns, all the samples are single - spinel phase and no characteristic peaks associated with impurity phase are observed. The peaks can be indexed as (220), (311), (400), (331), (422), (511) and (440) diffraction planes of spinel. With the increase of heat treatment temperature, the intensity of the diffraction peaks increases and their full - width at half- maximum (FWHM) decreases. All of these are associated with the increase of crystallinity and grain sizes. The average grain sizes of the powders sintered at different temperatures were determined by means of the Scherrer's formula and shown in the table 1. It can be noticed that the grain sizes of powders increase with the annealing temperature. The unannealed spinel $ZnAl_2O_4$: Cr^{3+} grains have an average size of 5 nm and those calcined at 1200°C have an average size of 18.3 nm. The unit - cell parameters and the grain sizes of all samples were also calculated from the XRD patterns and are shown in table 1.

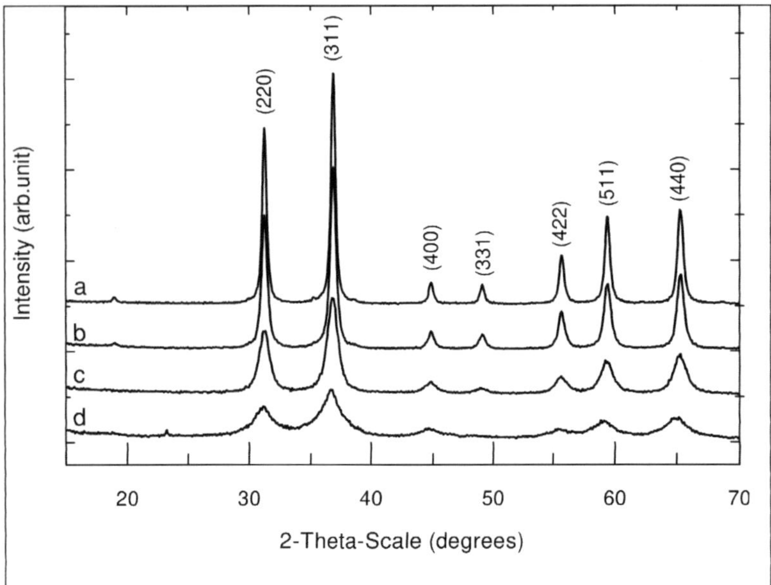

Figure 1. XRD patterns of samples annealed at various temperatures: a) 1200°C, b) 900°C, c) 700°C, d) unannealed.

Figure 2 shows the TEM micrographs of the powder sample formed at 700°C and the unnealed sample. The spinel particles consist of uniform quasi - spherical crystallites with the diameters of 7 nm (annealed at 700°C) and 5 nm (unnealed), which is consistent with calculated results from the XRD patterns.

To confirm the presence of the octahedral Cr^{3+} ions in the $ZnAl_2O_4$ nanocrystals, the room - temperature PLE and PL spectra of the powder samples were studied and the results are displayed in figure 3. It shows that the PL spectra of nanoparticles $ZnAl_2O_4$: Cr^{3+} is similar to those of bulk crystals doped with octahedral Cr^{3+} ions [8]. This fact indicates that doped ions Cr^{3+} were in the places of octahedral symmetry of synthesized $ZnAl_2O_4$ nanocrystals and the emission peaks were assigned to the $^2E(^2G) \rightarrow {}^4A_2(^4F)$ transition of octahedral Cr^{3+} ions. It is found from figure 3a that the

PL spectra of all samples are similar, but the FWHM of emission lines corresponding to zero-phonon $^2E(^2G) \rightarrow {}^4A_2(^4F)$ transition in the wavelength range of 683 - 693 nm are different.

Table 1. The dependence of the unit - cell parameters and the grain sizes on annealing temperatures.

Annealing temperatures (°C)	Unit - cell parameters (Å)	Average grain sizes (nm)
1200	8.080 ± 0.002	18.3
900	8.082 ± 0.004	14.5
700	8.087 ± 0.003	7.5
Unannealing	8.121 ± 0.004	5.1

In the unannealed sample the broad emission line resulted from an overlapping of N_1 line (689 nm) associated with disordering surround of ions Cr^{3+} and R line (686 nm) related to ordering surround. In this case the ratio of the intensity of these lines is approximately equal to 1. So that, with the increase of annealing temperature this ratio is significantly reduced.

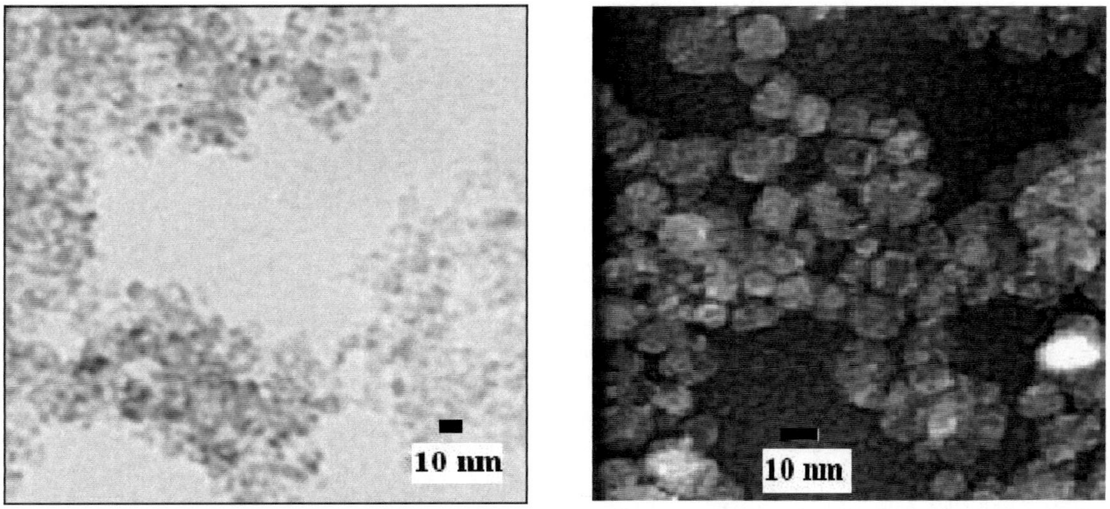

Figure 2. TEM micrographs of powder sample unannealed (left) and annealed at 700°C (right).

The excitation spectra recorded by monitoring the emission of transition $^2E(^2G) \rightarrow {}^4A_2(^4F)$ corresponding to zero - phonon R line were displayed in the figure 3b. It is seen that the PLE spectrum of Cr^{3+} doped $ZnAl_2O_4$ nanoparticles consists of two broad bands. The first belongs to the wavelength range of 360 - 450 nm and the second of 450 - 600 nm. Both of them were associated with absorption transitions of octahedral Cr^{3+} ions from the ground level $^4A_2$ to the excited levels $^4T_1$ and $^4T_2$, respectively. Because of the spin - orbit interaction, the level $^4T_1$ is separated into levels $^4A_2(^4T_1)$ and $^4E(^4T_1)$, so in the first band it is found two absorption peaks at 390 and 417 nm. Moreover, the difference of peak positions in the second band between the unannealed and annealed samples ($\Delta\lambda = 5$ nm) is related to a change of the unit - cell parameters. It is known, in opposite of the first transition $^4A_2 \rightarrow {}^4T_1$, a peak position of the second absorption transitions $^4A_2 \rightarrow {}^4T_2$ depends on the distances of Zn^{3+} - O^{2-}, Al^{3+} - O^{2-}. It is seen from table 1 that the unit - cell parameters were the highest in case of unannealed sample $ZnAl_2O_4$: Cr^{3+}, therefore the distances were large too. And the crystal - field is smaller in comparison with annealing cases. That is why the peak positions corresponding to above mentioned transitions $^4A_2 \rightarrow {}^4T_2$ shifted to longer wavelength bands in the unannealed sample.

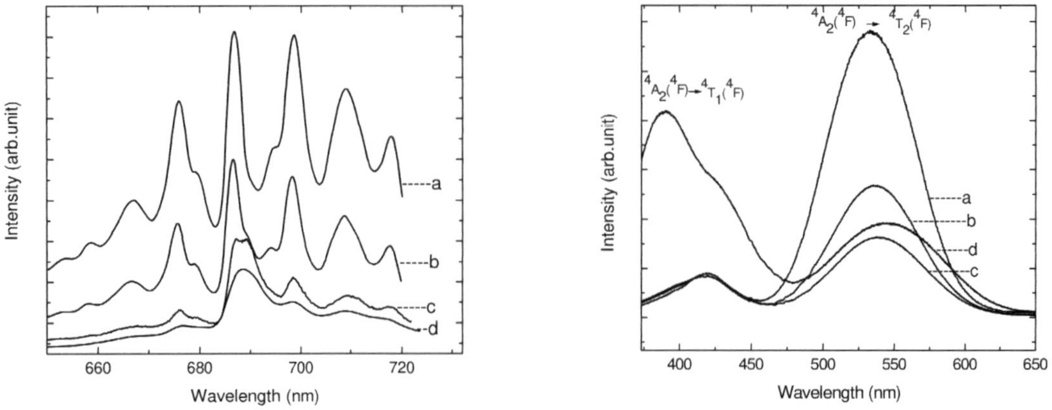

Emission spectra, $\lambda_{ex} = 540$ nm. Excitation spectra, $\lambda_{em} = 686$ nm.

Figure 3. Emission and excitation spectra of $ZnAl_2O_4$: Cr^{3+} sample annealed at various temperatures. a) 1200°C, b) 900°C, c) 700°C, d) unannealed.

Both emission and excitation spectra of the Cr^{3+} doped $ZnAl_2O_4$ may give detailed information about the surrounding of Cr^{3+} ions in a lattice. That is why the structural and optical probe is very useful.

3.2. Co^{2+} doped $ZnAl_2O_4$ spinel nanopowders

According to the computation of Tanabe - Sugano, the tetrahedral symmetry ions $3d^7$ (Co^{2+}) have the same energy level structure as the octahedral symmetry ions $3d^3$ (Cr^{3+}). Due to the investigation of the influence of heat - treatment regime on crystal structure and optical properties of $ZnAl_2O_4$: Cr^{3+} samples in the previous section, in this section we study the change of optical properties of $ZnAl_2O_4$: Co^{2+}.

Up to now there is not unified explanation of the nature of the broad band centered at 660 nm in the emission spectrum. The authors [9, 10] claim that the broad band centered at 660 nm is associated with the $^2E(^2G) \rightarrow {}^4A_2(^4F)$ transition. Meanwhile, Denisov and his partners [11] assert that this radiation band corresponding to the $^4T_1(^4P) \rightarrow {}^4A_2(^4F)$ transition. In this work, the experiment results show that the broad band centered at 660 nm isn't associated with $^2E(^2G) \rightarrow {}^4A_2(^4F)$ transition, but it may relate to the $^4T_1(^4P) \rightarrow {}^4A_2(^4F)$ transition. The reasons are as follows:

- Firstly, with the selective excitation wavelength of 390 nm, one can observe a series of narrow lines which is similar to the PL spectrum of Cr^{3+} ion due to the $^2E(^2G) \rightarrow {}^4A_2(^4F)$ transition in the perfect lattice of spinel.
- Secondly, from the Tanabe - Sugano's diagram [12], the luminescence peaks which are due to the spin - allowed transition $^2E(^2G) \rightarrow {}^4A_2(^4F)$ are narrow, while those due to the $^4T_1(^4P) \rightarrow {}^4A_2(^4F)$ transition depend strongly on host lattice and are usually broad.

Besides, the excitation spectrum of $ZnAl_2O_4$: Co^{2+} sample unannealed and annealed at 1200°C were presented in figure 4b. It is found that the excitation spectrum of Co^{2+} doped unannealed sample is similar to the PLE spectrum of Cr^{3+} doped $ZnAl_2O_4$ with two broad bands at 393 and 535 nm corresponding to the $^4A_2(^4F) \rightarrow {}^4T_1(^4F)$ and $^4A_2(^4F) \rightarrow {}^4T_2(^4F)$ transitions, respectively. Meanwhile, the PLE spectrum of sample sintered at 1200°C is different to unanealed case. The PLE spectrum of sample annealed at 1200°C displays three peaks around 565, 592 and 620 nm which cause the fluorescence $^4T_1(^4P) \rightarrow {}^4A_2(^4F)$.

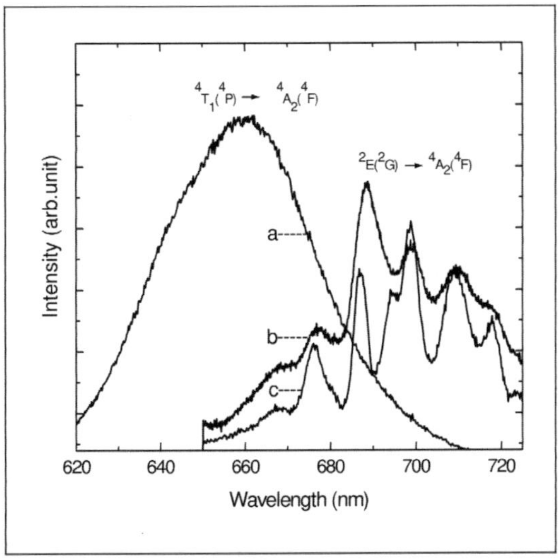

PL spectrum of $ZnAl_2O_4$:Co^{2+} sample.
a- annealed at 1200°C, λ_{ex} = 592nm,
b- annealed at 1200°C, λ_{ex}=390nm,
c- unannealed λ_{ex} = 390nm.

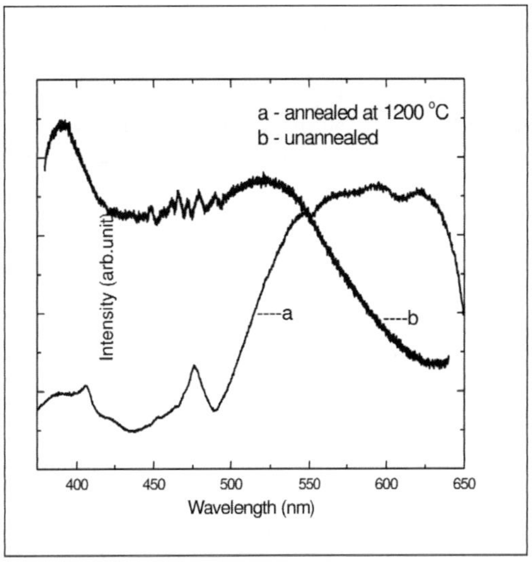

PLE spectrum of $ZnAl_2O_4$:Co^{2+} sample unannealed and annealed at 1200°C, λ_{em} = 686 nm.

Figure 4. Emission and excitation spectra of $ZnAl_2O_4$: Co^{2+} sample annealed at various temperatures.

4. Conclusion

$ZnAl_2O_4$: Cr^{3+} (Co^{2+}) nanopowders were formed by hydrothermal method and subsequently calcined at various temperatures. The $ZnAl_2O_4$: Cr^{3+} nanopowders annealed in the temperature range of 200 - 1200°C exhibit an increase in the particle sizes from 5 nm (for unanealed sample) to 18 nm (for sample annealed at 1200°C). The PL spectrum of the $ZnAl_2O_4$: Co^{2+} unannealed nanopowders is similar to the $ZnAl_2O_4$: Cr^{3+} nanopowders. The PL spectrum of the sample annealed at 1200°C includes one broad band centered at 660 nm when the excitation wavelength is 592 nm. The series of narrow lines in the range of 670 - 710 nm are due to $^2E(^2G) \rightarrow {}^4A_2(^4F)$ transition and broad band at 660 nm is due to $^4T_1(^2P) \rightarrow {}^4A_2(^4F)$ transition.

References

[1] De Camargo A S S, Nunes L A O, Silva J F, Costa A C F M, Barros B S, Silva J E C, De Sas G F and Alves Jr S 2007 *J. Phys. Condens. Matter.* **19** 2462099
[2] Vijay, Masuqul H M D and Dong K K 2007 *Bull. Korean Chem. Soc.* **28** 2477
[3] Cheng B, Qu S, Zhou H and Wang Z 2006 *Nanotechnology* **17** 2982
[4] Sampath K and Cordano K 1998 *J. Amer. Ceram. Soc.* **81** 649
[5] Hetting G F, Worl H and Weiter H H 1956 *Z. Anorg. Allg. Chem.* **283** 207
[6] Arean C O, B.S. Sintes B S, Palomino G T, Carbonell C M, Platero E E and Soto J B P 1997 *Microporous Mater.* **8** (3-4) 187
[7] Zawadazki M and Wrzyszcz J 2000 *Mater. Res. Bull.* **35** 109
[8] Mikendar W and Preisinger A. 1980 *J. Lumin.* **26** 53
[9] Tanaka M K, Ishihara T, Hirao K, Sogo N, Ashida M and Kato R 1993 *J. Amer. Ceram. Soc.* **769** 2839
[10] Xiulan D, Duorong Y, Zhihong S, Caina L, Dongying P, Dong X and Mengkai L 2005 *J. Alloys and Compounds* **386** 311
[11] Denisov A, Volk Y V, Malyarevich A M, Yumashev K V, Dymshits O S, Zhilin A A, Kang U and Lee K 2003 *J. Appl. Phys.* **93** 3827

[12] Tanabe Y and Sugano 1954 *J. Phys. Soc. Jpn.* **9** 753

Acknowledgments

This work was completed with financial support from Viet Nam Fundamental Research Program (Project 4 055 06) and Vietnam National University (No QG 07 - 05).

Synthesis of nanosilver particles by reverse micelle method and study of their bactericidal properties

Tran Thi Ngoc Dung[1], Ngo Quoc Buu[1], Dang Viet Quang[1], Huynh Thi Ha[2], Le Anh Bang[1], Nguyen Hoai Chau[1], Nguyen Thi Ly[1] and Nguyen Vu Trung[3]

[1] Institute of Environmental Technology, Vietnam Academy of Science and Technology, 18 Hoang Quoc Viet Road, Cau Giay Distr., Hanoi, Vietnam
[2] Institute of Materials Science, Vietnam Academy of Science and Technology, 18 Hoang Quoc Viet Road, Cau Giay Distr., Hanoi, Vietnam
[3] National Institute for Infectious and Tropical Diseases, 1 Ton That Tung, Dong Da Distr., Hanoi, Vietnam

E-mail: ttndzung@yahoo.com, buu_nq@yahoo.com

Abstract. Nanosilver particles have been synthesized by the reverse micelle method, where $AgNO_3$ was used as a silver ions source, $NaBH_4$ and quercetin - as reducing agents, CTAB, SDOSS and AOT- as surfactants, while the stabilizer was Vietnamese chitosan. Studying the factors influencing the process of nanosilver particle formation, it was shown that the particle size of the nanosilver products depends on the concentration of the reaction components and their stoichiometric ratio. It was also shown that the reaction system using AOT surfactant is capable of producing nanosilver particles with smallest nanoparticles (ϕ_{av} ~ 5 nm) and good particle-size distribution. The study on bactericidal activity of the nanosilver products indicated that the disinfecting solution with a nanosilver concentration of 3 ppm was able to inhibit all *E.coli* and *Coliforms*, TPC and fungi at 15 ppm, while *Vibrio cholerae* cells were inactivated completely with 0.5 ppm of nanosilver after 30 minutes exposition.

Keywords: Reverse micelle, nanosilver, AOT, antibacterial agent.

1. Introduction

Among inorganic antibacterial agents, silver has been employed most extensively since ancient times to fight infections and control spoilage. Catalytic oxidation by metallic silver and reaction with dissolved monovalent silver ion probably contribute to its microbicidal effect [1]. Microbes are unlikely to develop resistance against silver, as they do against narrow-target antibiotics, because the metal attacks many targets in the organisms, which means that they would have to develop a host of mutations simultaneously to protect themselves [2, 3]. Thus, at present silver ion is being widely used for disinfection, especially due to the advances in nanotechnology, which make possible the delivery of ionic silver during disinfection process [3-6].

For these reasons, we studied the synthesis of nanosilver for disinfection purposes using reverse micelle systems as one of the simplest methods for nanosilver production and its antibacterial activity.

Water/carbohydrate reverse microemulsions in the presence of a surfactant are used as a microreactor for synthesizing nanoparticles from different metals such as Au, Ag, Cu, Zn, and Fe. In a solvent, different micelle can be formed under different conditions [7-10] as shown in figures 1 and 2. From figure 2 it can be seen that the formation of micelles depends on the geometrical ratio P. So, to obtain a reverse micelle emulsion it is necessary to have P >1 [8, 9].

© 2009 IOP Publishing Ltd

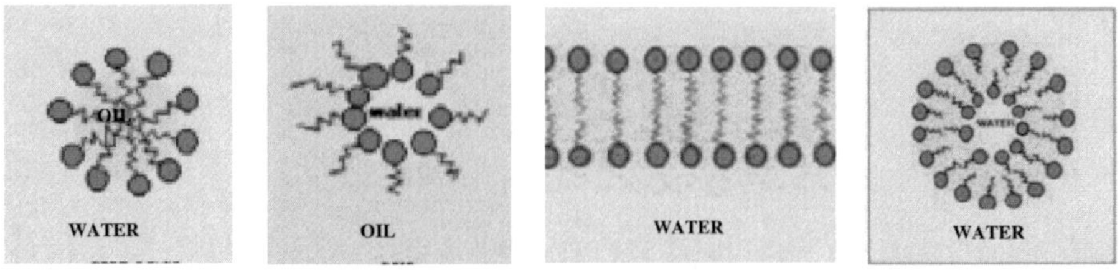

Normal micelle | Reverse micelle | Double-layer membrane | Micellar vesicle

Figure 1. Different forms of micellae in a water/surfactant/carbohydrate system [7, 8].

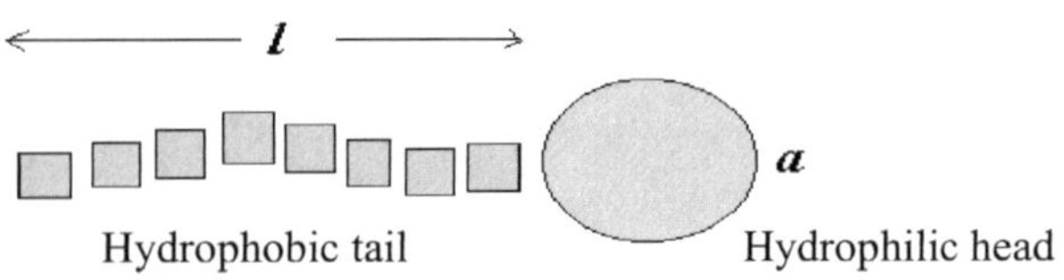

$$P = \frac{v}{a.\ell}$$

ℓ, v – lengh and volume of the hydrophobic tail
a – cross-section of the hydrophilic head
P < 1 normal micelle
P > 1 reverse micelle
P ~ 1 double-layer membrane or vesicle

Figure 2. Structure of a surfactant and influence of its geometrical parameters on the micellar formation [9].

The procedure of nanosilver preparation can be done by using reverse micelle method as follows. Micellar solutions are produced by successively mixing silver nitrate water solution and water solution of a reductant with surfactant in a solvent. Then the reductant-containing microemulsion is added to the silver nitrate -containing microemulsion while stirring vigorously for two hours.

The exchange of the solubles ($AgNO_3$ and $NaBH_4$) between the micelles takes place according to the following stages: 1) diffusion process of the micelles resulting in their collision; 2) destruction of certain parts of surfactant layer (CTAB or AOT) around the micelles; 3) diffusion exchange of the solubles in the micelles; 4) formation of new micelles with appearance of nanosilver particle therein [9-11].

These stages are illustrated in figure 3, where the third one was approved to be the slowest and, thus, the limiting stage, and considerably depending on the speed of stirring. In a triple component system "carbohydrate – water – surfactant" the solubility ratio ω ($[H_2O]_{mol}$ / $[Surfactant]_{mol}$) is a crucial factor for the formation of nano-sized silver particles [9].

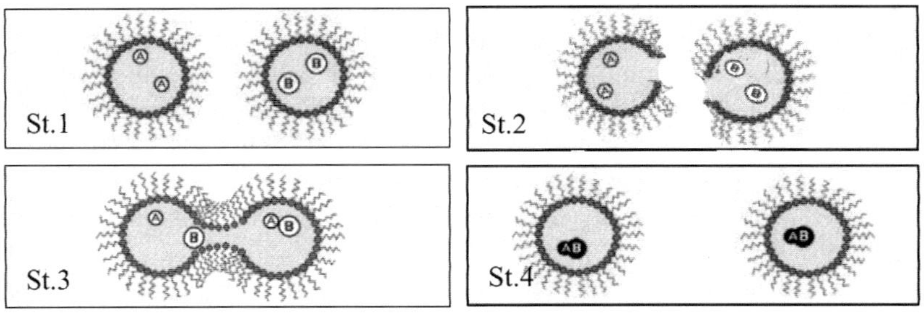

Figure 3. Stages of the nanosilver particles formation in reverse micelles [10].

Figure 4 illustrates the influence of the solubility ratio on the size of water micelle in a triple system "Hexan – H_2O – AOT", where one can see that diameter of a water micelle increases with increasing of solubility ratio.

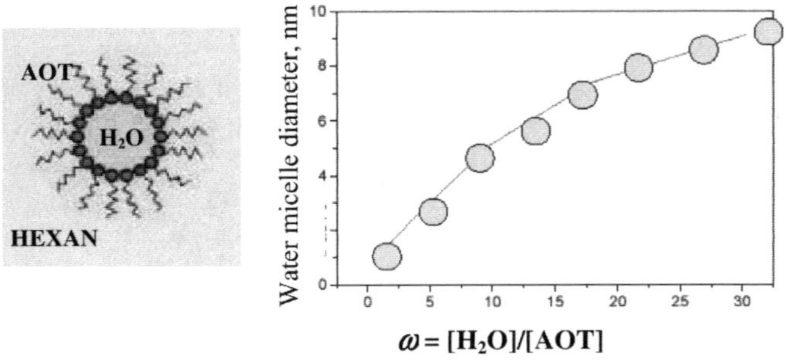

Figure 4. Diameter of water micelles depends upon the solubility ratio in system "Hexan- H_2O- AOT" [7].

Although mechanism of inhibitory action of nanosilver on microbes is not fully clear, there is a lot of proof on the bactericidal action of silver [4, 5, 12-14]. Nanosilver can provide a control delivery of ionic silver and works in a number of ways to disrupt critical functions in an organism. It has a high affinity for negatively charged side groups on biological molecules such as sulfohydryl, carboxyl, phosphate and other charged groups distributed throughout microbial cells. This binding reaction alters the molecular structure of the macromolecule, rending it worthless to the cell. Silver simultaneously attacks many sites within the cell to inactivate critical physiological functions such as cell-wall synthesis, membrane transport, nucleic acid synthesis and translation, electron transport, which is important in generating energy for the cell. Without these functions, the microorganism is inhibited or killed. Therefore at present, with more and more bacteria developing resistance to antibiotic drugs, the healthcare researchers began to consider nanosilver as one of the most potent antimicrobial agents.

2. Experimental

2.1. Materials

Chemicals such as silver nitrate, sodium borohydride, quercetin, chloroform, isooctan, cetyltrimethyl ammonium bromide (CTAB), sodium bis(2-ethylhexyl) sulfosuccinate (AOT) and sodium dioctyl sulfosuccinate (SDOSS) were of high purity (Merk, Aldrich, Canto Chemical). Vietnamese β-chitozan (10% deacetylated) was provided by Institute of Chemistry, VAST.

E.coli, Coliforms and *Vibrio cholerae* were isolated from hospital pathogenous waste water, while total aerobic bacteria (TPC) and fungi were isolated from air. Nutrient broth Chromocult and PCA in agar medium was used to grow and maintain the bacterial cultures.

2.2. Methods

Silver nanoparticles were synthesized by reverse micelle method using two representative reverse micelle systems $AgNO_3/NaBH_4$/CTAB/Chloroform and $AgNO_3$/quercetin/SDOSS(or AOT)/isooctan. In these systems isooctan and chloroform were used as solvent, quercetin and sodium borohydride as reductant, while CTAB, AOT and SDOSS were used as surfactant which met the requirement of the reverse micelle formation (P>1).

For experiment, the following solutions were prepared:

1. Water silver nitrate solutions, 1 M and 3 M
2. Water solutions of the reducing agents: $NaBH_4$ (0.1 M, 1.0 M, 1.5 M) and quercetin (0.06 M) dissolved in 1 M NaOH solution (20 mg/ml)
3. 0.1 M CTAB solution in chloroform
4. 0.1 M SDOSS and 0.1 M AOT solutions in isooctan
5. Water solution of β-chitosan 0.5%.

In the first system, to restrict the nanoparticle aggregation a stabilizing agent should be introduced into solution during the nanoparticle formation. Figure 5 illustrated the formation of an anti-agglomeration layer around a nanosilver cluster in a system $AgNO_3/NaBH_4/CTAB/CHCl_3$ using thiolglycerin as a stabilizing agent [7].

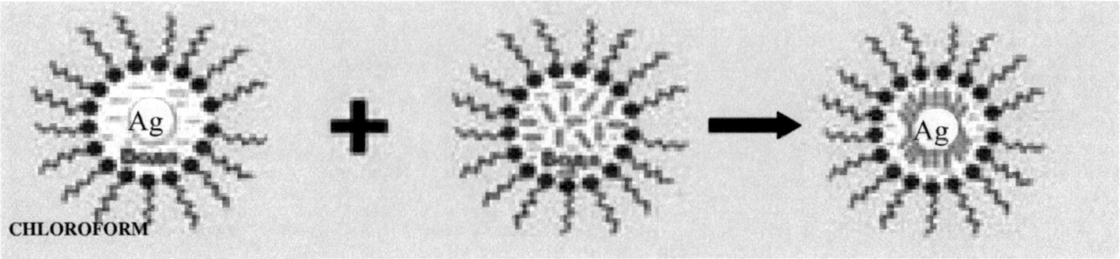

Figure 5. Thiolglycerin stabilizes nanosilver particle against aggregation in a reverse micelle system $AgNO_3/NaBH_4/CTAB/CHCl_3$.

As mentioned above, another important factor which controls the parameters of nanosilver particles in reverse micelles is to keep the molar solubility ratio (ω) as small as possible [15].

To make nanosilver of better quality, the two reaction systems with different constituents have been studied. A procedure for preparation of a nanosilver solution according to the reverse micelle system $AgNO_3/NaBH_4$/CTAB/chloroform could be as follows:

- 0.5 ml of 1 M silver nitrate water solution was added to 30 ml of 0.1 M CTAB in chloroform
- 0.4 ml of 1 M $NaBH_4$ water solution was added to 30 ml of 0.1 M CTAB in chloroform
- The two solutions were vigorously stirred for one hour to form reverse micelle emulsions (RMEs)
- Then the emulsions were mixed together and ultrasonic stirring was continued for 2 hours, and during this time 0.2 ml of 0.5% chitosan solution was added in order to stabilize the nanoparticles obtained.

For the reverse micelle system $AgNO_3$/quercetin/AOT(or SDOSS)/isooctan the synthesis could be performed in a similar manner:

- 0.1 ml of 3 M $AgNO_3$ solution was added to 25 ml of 0.1 M AOT or 0.1 SDOSS in isooctan
- 0.2 ml of 0.06 M quercetin solution was added to 25 ml of 0.1 M sodium dioctyl sulfosuccinate in isooctan
- The two solutions were vigorously stirred for two hours to form RME, then mixed together and ultrasonic stirring was continued for 2-3 hours to obtain nanosilver particles.

For the second reaction system, according to the Russian researchers [12], due to the peculiar reducing and stabilizing properties of quercetin, the use of stabilizers becomes unnecessary. The same researchers confirmed that quercetin makes it possible to prepare only one quercetin-containing RME solution, whereas silver nitrate water solution could be poured directly into the quercetin RME. This peculiarity allows to considerably decrease the molar solubility ratio, giving rise to the decrease of nanosilver particle size.

The nanosilver products then underwent antibacterial activity tests. To examine the bactericidal effect of nanosilver particles on *Escherichia coli, Coliforms,* TPC and fungi, bacteria was incubated with nanosilver particles at concentration of 3, 5, 10, 15 and 30 ppm for 30 minutes. Then, bacteria were cultured on Chromocult agar plates and colony-forming units were counted.

3. Results and discussion

Reaction parameters of the nanosilver formation in reverse micelles were presented in table 1. In the first system with borohydride reductant, CTAB in chloroform was used as a surfactant and chitosan – as a stabilizer, while in the second one with quercetin reductant, AOT or SDOSS in isooctan was used instead of CTAB. In this system a stabilizer was not required because quercetin possesses stabilizing property to keep silver nanoparticles from oxidizing and agglomeration.

Table 1. Experimental parameters and results of preparation of nanosilver using different reverse micelle systems.

a) $AgNO_3$/ $NaBH_4$/ CTAB/ chloroform/ stabilizer

N^0	$[AgNO_3]$ (M/ml)	Rednt $NaBH_4$ (M/ml)	Stabilizer (ml)		Surfactant CTAB (M)	$CHCl_3$ (ml)	Water phase (ml)	$[H_2O]$ / [CTAB]	[Ag]* (ppm)	Remarks and particles features
1	1.0/ 0.4	1.5/ 0.4	0	0	0.1	80.8	0.8	5.5	531	Precipitate after 48 h staying; ϕ_{av} ~ 15 nm
2	1.0/ 0.3	1.5/ 0.3	Chts**	0.2	0.1	81.1	0.8	7.5	400	ϕ_{av} ~10 nm; good particle size distribution; stable

[Ag]* - maximally available nanosilver concentration; **Chts - chitosan, concentration 0.5%

b) $AgNO_3$/ quercetin/ SDOSS/ isooctan

N^0	$[AgNO_3]$ (M/ml)	Rednt Qr** 0.06M (ml)	Stabi lizer (ml)		Surfact-ant (M)	Iso-octan (ml)	Water phase (ml)	$[H_2O]$ / [DOSS]	Ag* (ppm)	Remarks and particles features
3C	1.0/0.15	0.08	0	0	SDOSS 0.1	35	0.23	3.62	460	Slight precipitation after 24 h staying; less uniform particle size distribution; ϕ_{av} ~ 10 nm,
4F	1.0/0.15	0.2	0	0	SDOSS 0.1	50	0.35	3.86	320	Stable particles, more or less uniform particle size distribution; ϕ_{av} ~10 nm

*[Ag] - maximally available nanosilver concentration; ** Qr - quercetin (0.06 M)

c) $AgNO_3$/ quercetin/ AOT/ isooctan

N^0	$[AgNO_3]$ (M/ml)	Rednt Qr** 0.06M (ml)	Stabi lizer (ml)		Surfac tant (M)	Iso-octan (ml)	Water phase (ml)	$[H_2O]$ / [AOT]	Ag* (ppm)	Remarks and particles features
5K	3.0/0.06	0.3	0	0	AOT 0.1M	50	0.36	3.97	380	Stable particles after staying; more uniform particle size distribution; ϕ_{av} ~ 5-7 nm
6I	3.0/0.10	0.3	0	0	AOT 0.1M	50	0.40	4.44	640	Water silver nitrate solution was poured directly into quercetin RME and then ultrasonically stirred. Silver partly precipitated; less uniform particle size distribution; ϕ_{av} ~ 10 nm

Experimental data depicted in table 1a confirmed the stabilizing role of chitosan for the reaction system using sodium borohydride as a reductant. In experiment $N^0$1 without stabilizer a black precipitate appeared in the finished solutions after 48 hours, meanwhile for the experiment using chitosan stabilizer ($N^0$2) nanosilver solutions remained transparent after staying. For the system $AgNO_3/NaBH_4$/ CTAB/chloroform, nanosilver particles of the best quality have been obtained by using chitosan as a stabilizer and solubility ratio $\omega = 7.5$ (exp. $N^0$2, table 1a). TEM image of this nanosilver product was shown in figure 6, where one can see a rather good particle-size distribution with an average size of 10 nm.

Tables 1b and 1c represent the experimental data for the system $AgNO_3$/quercetin/AOT (or SDOSS)/isooctan. The results show that using quercetin reductant it is possible to obtain nanosilver particles stable without stabilizer, but in condition that the quercetin concentration should be enough, comparable with that of silver nitrate (exp. $N^0$3,4). In the case of low quercetin concentration, nanoparticles can not be avoided from oxidation (exp. $N^0$3); consequently a precipitate appeared in the nanosilver reverse micelle emulsion after 24 hours staying.

TEM image of nanosilver particles prepared following the system $AgNO_3$/quercetin/ SDOSS/isooctan and presented in figure 6b shows small average particle size (<10 nm), but not uniform (exp. $N^0$3). In exp.

$N^0$4 quercetin concentration was considerably increased, comparable with that of silver nitrate, resulting in stable and more uniform nanoparticles.

Nanosilver particles produced according to the system $AgNO_3$/quercetin/AOT/isooctan presented in table 1c demonstrate the higher quality of the particles with an average particle diameter 5 nm (figure 1c) in comparison with a system, where reducing agent was SDOSS (table 1b). It is due to the fact that, although both AOT and SDOSS have the same molecular weight (444Da), in SDOSS molecule there are two linear octyl chains, but in AOT there are two branched 2-ethylhexyl groups which better fit the reverse micelle formation requirement (P>1).

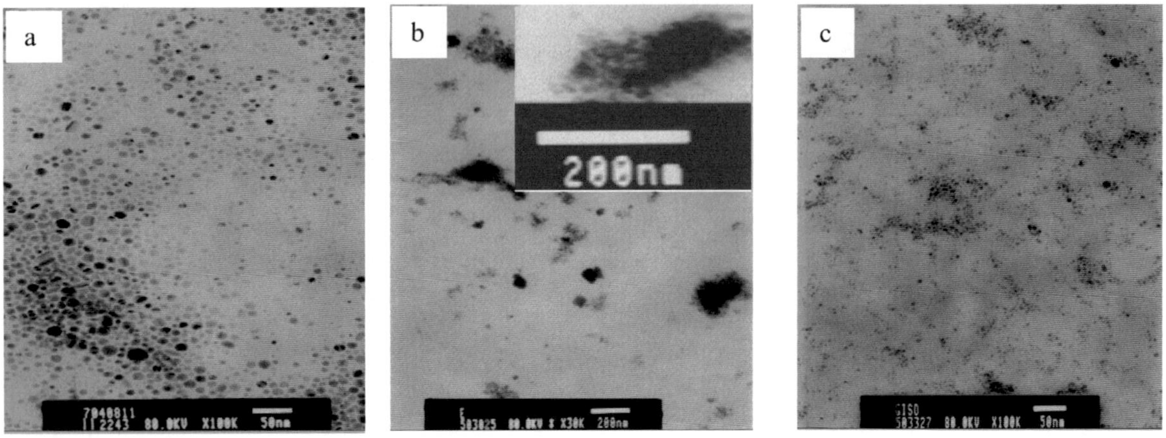

a)$AgNO_3$/$NaBH_4$/CTAB/$CHCl_3$/ chitosan b)$AgNO_3$/quercetin/SDOSS/ isooctan c) $AgNO_3$/quercetin/AOT/isooctan

Figure 6. Nanosilver particles obtained by reverse micelle method in the systems.

Table 2. Bactericidal activity of the nanosilver produced by using reverse micelle method. Exposition time 30 minutes.

Nanosilver concentration	TPC		*E.coli*		*Coliforms*		*Fungi*	
	(cfu/ml)	Inhibite (%)	(cfu/ml)	Inhibited (%)	(cfu/ml)	Inhibited (%)	(cfu/ml)	Inhibited (%)
Control	3.5×10^7		2.2×10^4		3.1×10^4		2.9×10^6	
3 ppm	7.3×10^3	99.98	0	100	0	100		
5 ppm	6.8×10^3	99.98	0	100	0	100		
7 ppm	7×10^2	99.99	0	100	0	100	8×10^2	99.97
10 ppm	6×10^2	99.99	0	100	0	100	1.1×10^2	99.99
15 ppm	4×10^2	99.999	0	100	0	100	50	99.99

The results of bactericidal activity tests were illustrated in table 2. The experimental data indicated that *E.coli* and *Coliforms* were totally killed in a solution with nanosilver concentration of 3 ppm after 30 minutes of exposition, whereas TPC bacteria and fungi being more resistant were inactivated 99.99% only at 10 ppm of nanosilver. Figure 7 illustrates the antimicrobial activity of the reverse micelle nanosilver against TPC bacteria and fungi on the tile and wood surfaces respectively after being exposed 30 minutes to 10 ppm of nanosilver.

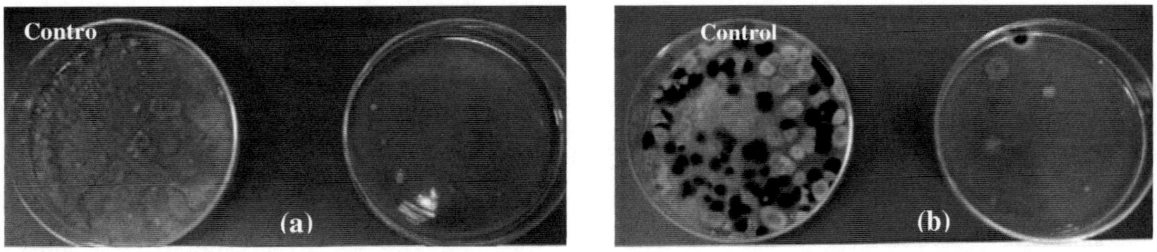

Figure 7. TPC bacteria (a) and fungi (b) were inactivated on the tile and wood surfaces respectively after being exposed 30 minutes to 10 ppm of nanosilver.

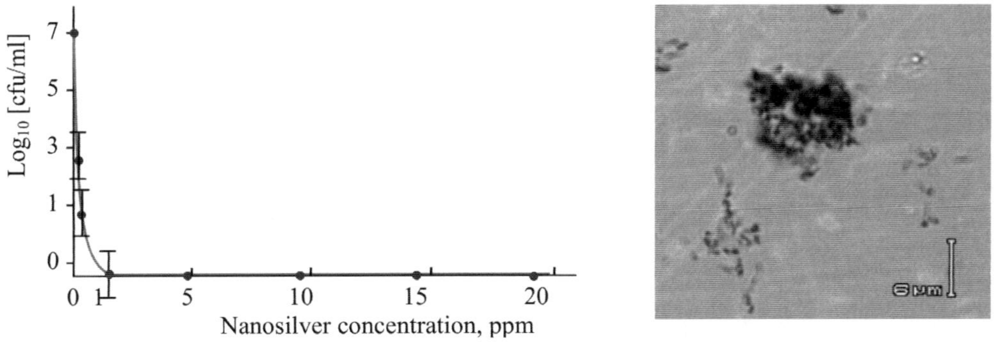

Figure 8. Influence of nanosilver concentration on the inactivation rate of *Vibrio cholerae*. Exposition time 30 minutes.

Reverse micelle nanosilver solutions of different concentration have also been tested against *Vibrio cholerae* bacterium. Figure 8 depicted the inactivation rate of *Vibrio cholerae* cells depending on the silver concentration. The data proved that this bacterium is very sensitive to the destructive action of nanosilver: 0.5 ppm of nanosilver was able to inactivate completely *Vibrio cholerae* during 30 minutes of exposition. Microscopic image on the right shows that almost *Vibrio cholerae* cells were destroyed in the presence of 5 ppm of nanosilver after 30 minutes of exposition.

4. Conclusion

Nanosilver particles have been synthesized by reverse micelle method, where $AgNO_3$ was used as a silver ions source, $NaBH_4$ and quercetin - as reducing agents, CTAB, SDOSS and AOT- as surfactants, while for the system using sodium borohydrite as a reducing agent, a stabilizer was Vietnamese chitosan. Studying the factors influencing the process of nanosilver particle formation, it was shown that the particle size of the nanosilver products depends on the concentration of the reaction components and their stoichiometric ratio, as well as the way of their introduction into reaction mixture. It was also shown that the reaction system using AOT surfactant is capable of producing nanosilver particles with smaller particles (ϕ_{av} ~ 5 nm) and good particle-size distribution.

The study on bactericidal activity of the nanosilver products indicated that a disinfecting solution with a nanosilver concentration of 3 ppm was able to inhibit all *E.coli* and *Coliforms*, TPC and fungi at 10 ppm, while *Vibrio cholerae* cells were inactivated completely with 0.5 ppm of nanosilver after 30 minutes exposition.

References

[1] James G V 1971 *Water treatment* (Cleveland OH: CRC Press) p 38
[2] Elechiguerra J L, Burt J L, Morones J R *et al.* 2005 Interaction of nanosilver particles with HIV- 1 *J. Nanobiotechnol.* **3** (6) 41
[3] Sucdeb P, Yu K T, Joon M S 2007 Does the antibacterial activity of silver nanoparticles depend on theshape of the nanoparticle *Appl. & Env. Microbiol.* **73** (6) 1712

[4] Oka M, Tomioka T, Tomita K *et al.* 1994 Inactivation of enveloped viruses by a silver thiosulfate complex *Metal-based druds* **1** 511

[5] Sondi I, Salopek-Sondi 2004 Silver nanoparticles as antimicrobial agent: a case study on E. coli as a model for gram-negative bacteria *J. Colloid Interface Sci.* **275** 177

[6] Wiley B, Sun Y, Mayers B and Yxia 2005 Shape controlled synthesis of metal nanoostructures: the case of silver *Chem. Eur. J.* **11** 454

[7] Razumov V F 2003 Nanoparticles and chemical reactions in micellar systems *Report on the scientific session of the Section of Chemical Sciences (Department of Chemistry and Material Sciences Russian Academy of Sciences 9-11 April)*

[8] Robinson B H, Khan-Lodhi A N, Towey T 1989 *Microparticle synthesis and characterization in reverse micelles* ed Pileni M P (Amsterdam: Elsevier) p 199

[9] Pileni M-P 1989 *Structure and reactivity in reverse micelles* ed Pileni M P (Amsterdam: Elsevier)

[10] Petit C, Lixon P, Pileni M P 1993 In situ synthesis of silver nanocluster in AOT reverse micelles *J. Phys. Chem.* **97** 12974

[11] Ershov B G 1997 Ions of metals in unusual and unstable oxidation states in aqueous solutions: the receipt and properties *Successes Chemistry* **66** (2) 103

[12] Fung M C, Bowen D L 1996 Silver products for medical indications: risk-benefit assessment *Clinical Toxicol.* **34** 119

[13] Yakabe Y, Sano T, Ushiho H, Yasumaga T 1980 Kinetic studies of the interaction between silver ion and deoxyribonucleic acid *Chem. Lett.* **4** 373

[14] Gibbins B 2003 The antimicrobial benefits of silver and the relevance of microlattice Technology *Ostomy Wound Management* **49** (6) 5

[15] Egorova E M, Revina A A 2002 Optical properties and size of nanoparticles of silver in mitselyarnyh solutions *Colloid Journal* **64** (3) 334

Acknowledgement

This research was partially sponsored by the Basic Research Program from Ministry of Science & Technology of Vietnam.

On theory of single-molecule transistor

Tran Tien Phuc

Nha Trang University, 02 Nguyen Dinh Chieu, Nha Trang, Khanh Hoa, Vietnam

E-mail: trantienphuc@gmail.com

Abstract. The results of the study on single-molecule transistor are mainly investigated in this paper. The structure of constructed single-molecule transistor is similar to a conventional MOSFET. The conductive channel of the transistors is a single-molecule of halogenated benzene derivatives. The chemical simulation software CAChe was used to design and implement for the essential parameter of the molecules utilized as the conductive channel. The GUI of Matlab has been built to design its graphical interface, calculate and plot the output I-V characteristic curves for the transistor. The influence of temperature, length and width of the conductive channel, and gate voltage is considered. As a result, the simulated curves are similar to the traditional MOSFET's. The operating temperature range of the transistors is wider compared with silicon semiconductors. The supply voltage for transistors is only about 1 V. The size of transistors in this research is several nanometers.

Keywords: Single-molecule transistor, benzene ring, nanotransistor.

1. Introduction

There are many studies on molecular transistors which exhibit structure and properties similar to conventional MOSFET but much smaller in size [1-5]. In several research papers the U_{DS} voltage on I-V curve of molecular transistor is reported to be as high as 40 V [6] and even 100 V [7]. This paper presents the results of research on molecular transistors which have the structures like traditional MOSFET's in shape, but the conductive channel is made from commercial chemicals, such as 1,4-dihalo-benzenes (1,4-$C_6H_4X_2$, with X: F, Cl, Br, and I).

Regarding the structure of molecules, every disubstituted benzene ring falls into three isomers, namely 1,2-; 1,3-; and 1,4-isomer). Only 1,4-isomer is chosen due to the suitable symmetrical contact to drain (D) and source (S) electrode of MOSFET structure.

The chemical simulation software CAChe is applied to design molecules used for conductive channel and to determine the parameters for I_{DS} current computation and draw I-V curve. To evaluate the applicability of transistors in practical purposes, investigation of the effect of temperature to I-V curve is needed. The dimension of molecules used for conductive channel is computed via limitation of Van der Waals force. The effect of dimension quantization to I-V curve is also considered. Every molecule which has one benzene ring of 1,4 associated with two halogen atoms is completely the same. However, when using molecules as a conductive channel, the electrical characteristic may be different due to the non-identical size. The above modification in parameters of molecule is under investigation via a Matlab GUI interface.

2. Theory

Figure 1 shows the configuration outline of a transistor, which has the electrode symbols like electrodes of an insulated-gate field-effect transistor (MOSFET).

© 2009 IOP Publishing Ltd

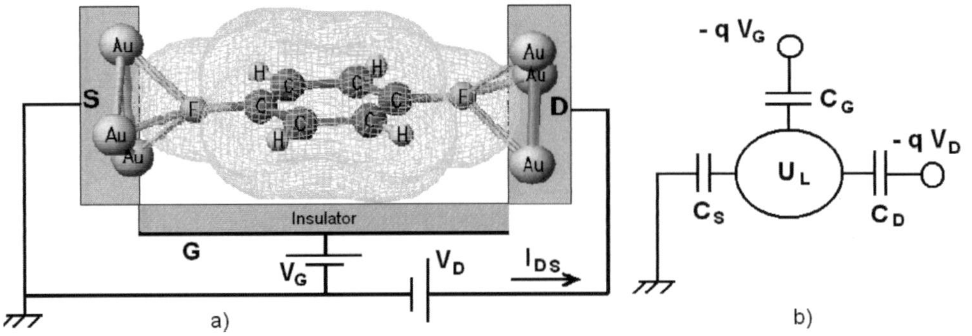

Figure 1. a) Configuration outline of a molecular transistor, b) The equivalent capacitive circuit.

The conductive channel is the molecules of 1,4-dihalo-benzenes (1,4-$C_6H_4X_2$, with X: F, Cl, Br, and I). The simulation results in CAChe show that all atoms of above molecules are coplanar. Therefore, the selected G electrode has a rectangular area and was opposite to the plane of the molecules to increase the control effect of the voltage V_G. The length L and the width W of the molecule calculated on the surface of the Van der Waals force were considered to be the dimension of the electrode G. Every halogen atom has a single bond with carbon atom in the benzene ring and three double bonds with three Gold atoms at the electrodes.

A molecular channel must have an energy spectrum which is separated into three bands at the beginning state. The valence band of an inorganic semiconductor corresponds to the highest occupied molecular orbital (HOMO), whilst the lowest unoccupied molecular orbital (LUMO) represents the conduction band (the difference between the energy levels of the HOMO and LUMO). The band gap energy is wide enough so that no electron receives thermal effect.

Considering the case of the transistor with the structure as in figure 1a, the beginning energy level is distributed so that Fermi level is in the middle of the band gap.

The supply voltage $V_D \neq 0$ leads to the electrochemical potential difference $\mu_1 - \mu_2 = qV_D$ between two electrodes. The gate voltage $U_G \neq 0$ causes the conduction band or the valence band of a molecular channel, and then the current I_{DS} of the external circuit is created.

Contact 1 (S-Channel) would like to see $f_1(\varepsilon)$ electrons, while contact 2 (D-Channel) would like to see $f_2(\varepsilon)$ electrons occupying the state, where f_1 and f_2 are the source and drain Fermi functions, respectively. The average number of electrons N at steady state will be some number intermediate between $f_1(\varepsilon)$ and $f_2(\varepsilon)$. The steady-state current per spin is determined by formula

$$I = \frac{q}{\hbar}\frac{\gamma_1\gamma_2}{\gamma_1+\gamma_2}\left[f_1(\varepsilon) - f_2(\varepsilon)\right]. \tag{1}$$

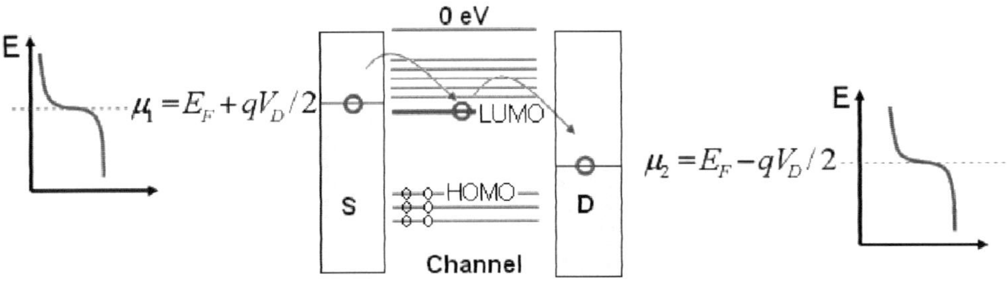

Figure 2. The current creation via a transistor.

We can interpret the rate constants $\gamma_1/\hbar$ and $\gamma_2/\hbar$ as the rates at which an electron placed initially in the level ε will be escaped into the S and D contacts, respectively. In principle, these quantities, γ_1 and γ_2, which have a measurement unit of Joule energy, could be measured experimentally [9]. Hence, only energy levels between the electrochemical potentials of μ_1 and μ_2, which have $f_1(\varepsilon) \neq f_2(\varepsilon)$, are responsible for the creation process of I_{DS}.

In the conductive channel, the energy spectrum is expanded to a band with density of state (DOS) *D(E)*. For both spins over the energy level of ε we have

$$D_\varepsilon(E) = 2\times\frac{\gamma/2\pi}{(E-\varepsilon)^2+(\gamma/2)^2}, \tag{2}$$

where $\gamma=\gamma_1+\gamma_2$. From studying a molecular channel with a defined size, we have: $D(E)=m_c WL/\pi\hbar^2\,\upsilon(E-E_C)$ and $\gamma_1=\gamma_2=\hbar\upsilon_R/L$ [9]. With respect to the electrical aspect, we can see the molecular channel like a point and the capacitive circuit is showed in figure 1b. So its potential energy is given by the sum of the Laplace potential U_L and an additional term proportional to the change in the number of electrons

$$U = U_L + \frac{q^2}{C_E}\Delta N. \tag{3}$$

The number of electrons *N* and the current *I* show the energy level with extended effect [9].

$$N = \int_{-\infty}^{+\infty} \mathrm{d}E D_\varepsilon(E-U)\frac{\gamma_1 f_1(E)+\gamma_2 f_2(E)}{\gamma_1+\gamma_2}, \tag{4}$$

$$I = \frac{q}{\hbar}\int_{-\infty}^{+\infty} \mathrm{d}E D_\varepsilon(E-U)\frac{\gamma_1\gamma_2}{\gamma_1+\gamma_2}\left[f_1(E)-f_2(E)\right]. \tag{5}$$

The self-consistent field (SCF) method of two expressions in a loop is applied for the caculation. If we know the number of energy levels in the regions, we can calculate the total current through the molecular conductive channel.

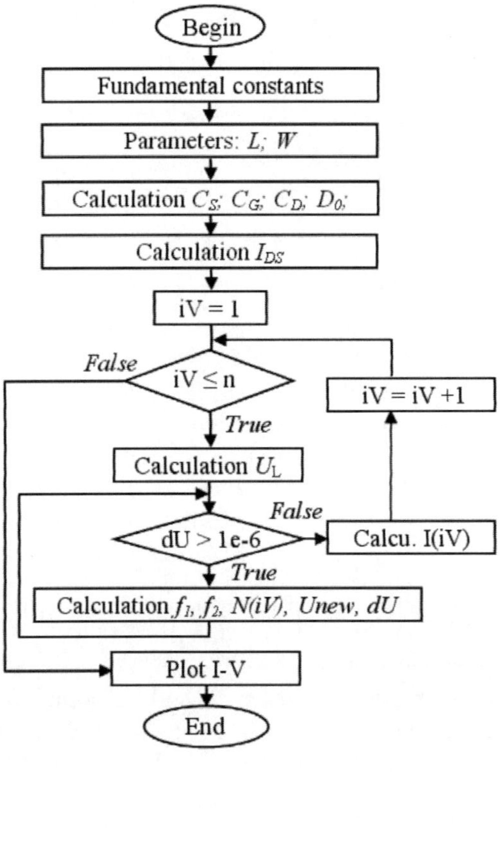

Figure 3. Algorithm diagram flow for plotting I_{DS}.

Figure 3 represents the diagram algorithm with a self-compatible method for plotting I_{DS} in Matlab [10]. The conditions can be applied for this algorithm is:

- The size of the molecules (according to Van der Waals force) has to be able to be defined, and this value is almost considered as the channel size of transistors; it is also the size of the capacitive side C_G in an equivalent circuit shown in figure 1b.
- The Van der Waals surface of molecules is almost smooth (like the surface of a dish). It can be combined with opposite capacitive side G to form a capacitor C_G controlling the molecules in the channel.
- The number of energy levels in each band and the level density of conductive band which is a factor of generative activities of current I_{DS} must be able to calculate.
- The energy level density of conduction band is small enough to satisfy conditions for this application of self-consistent field (SCF) method.

3. Results and discussion

3.1. The parameters of molecules used to form conductive channel of transistors

The CAChe software was used to simulate and define the parameters of molecules. The CAChe software can be used to design the molecules and optimize the structure, examine the electron bonds, measure the size of the molecule according to Van der Waals force, calculate the energy level and also the density of energy levels in the band. These are necessary parameters to apply to a loop algorithm with a self-compatible method for calculating and plotting output characteristic curve of I_{DS}.

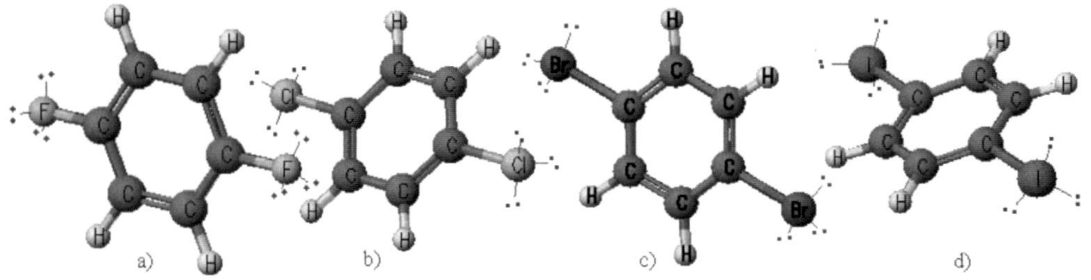

Figure 4. Molecules used for forming conductive channel of transistor in this research.

Table 1. Energy levels with responding orbits.

Orbitals	Eigenvalues (eV)					
	1,4_$C_6H_4S_2$	1,4_$C_6H_4O_2$	1,4_$C_6H_4Cl_2$	1,4_$C_6H_4F_2$	1,4_$C_6H_4Br_2$	1,4_$C_6H_4I_2$
26	1.282	2.086	2.292	2.113	2.216	2.361
25	0.157	1.890	1.329	2.060	-0.182	-0.064
24	0.135	1.718	1.294	2.042	-0.202	-0.140
23	-0.178	0.873	-0.052	-0.295	-0.295	-0.484
22	-1.088	0.027	LUMO -0.243	LUMO -0.332	LUMO -0.318	LUMO -0.653
21	LUMO -3.291	LUMO -1.707	HOMO -9.235	HOMO -9.870	HOMO -9.869	HOMO -8.947
20	HOMO -9.489	HOMO -10.922	-10.236	-10.492	-10.314	-9.470
19	-9.522	-11.162	-10.661	-12.876	-11.319	-9.500
18	-9.862	-11.453	-10.680	-13.276	-11.322	-9.544
17	-11.418	-11.684	-10.716	-14.837	-11.376	-10.166
10	-12.754	-14.194	-12.155	-15.007	-12.303	-11.105
Band gap	6.198	9.215	8.992	9.538	9.551	8.294

Figure 4 shows the results for designing the molecules, optimizing the structure, and examining the double bonds. Atoms forming the molecules are on the surface. Each atom in the halogen group at two ends of the molecule remains six free electrons. When putting the molecule on the conductive channel, three double bondings with three Gold atoms at electrodes will be created. Mechanically, three couples of bonds at the same side, together with symmetric characteristics of molecules will be excellent conditions for good location ability in the channel.

The levels of LUMO, HOMO and band gaps were calculated by using chemical simulation software. The calculated results of ten basic energy levels of molecules are shown in table 1 and figure 5.

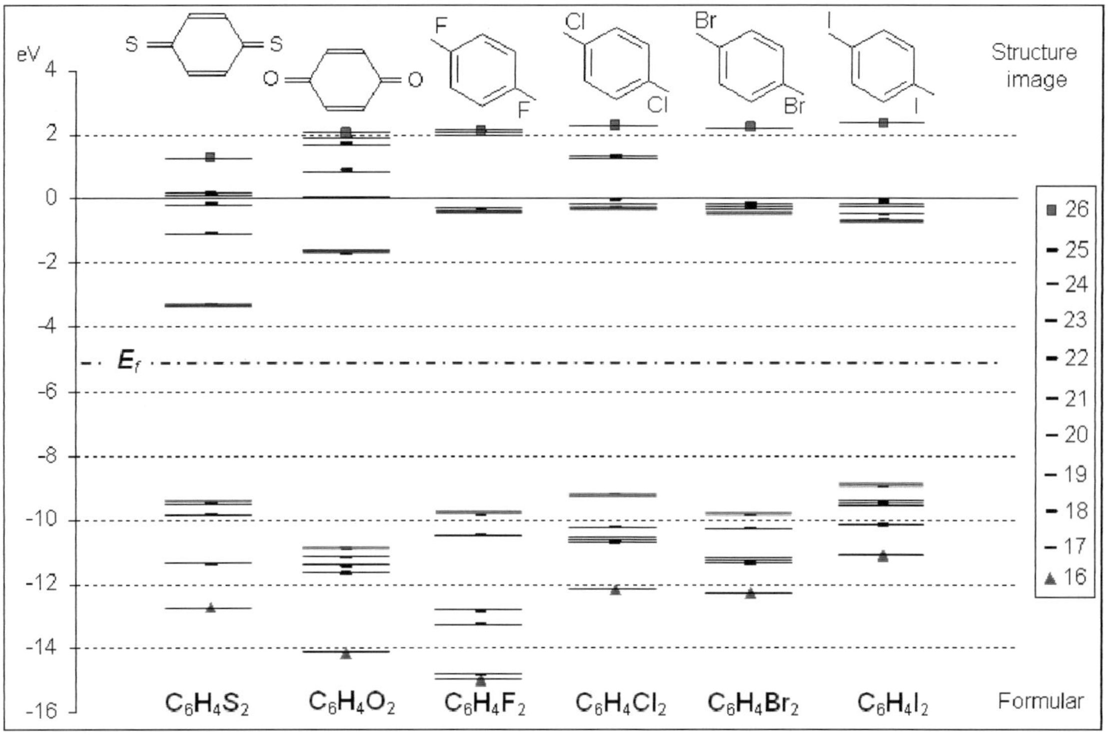

Figure 5. Structure image, energy level graph and formula of molecules.

The first band must have enough numbers of energy levels to obtain the saturation value of current responding to supply voltage less than 1 V. If the band contains many energy levels the saturation current at U_{DS} will be large. However, according to the theoretical research [6, 7], nanotransistors with required large U_{DS} are not feasible due the penetrating electric field depletes. In order that the curve at saturation point lasts, the energy levels in the next conduction band must be far away from the present one. If this condition is not satisfied, the characteristic curve will be in stair shape. Table 1 also introduces the data of molecules of Dithio-p-benzoquinone ($C_6H_4S_2$) and p-Benzoquinone ($C_6H_4O_2$). The simulated and calculated results for these molecules show that levels of HOMO energy band on valence band satisfies the conditions to become conductive channels of transistors with negative polarization ($U_G < 0$). Calculation and sketch of characteristic curves of transistors with molecular channel of 1,4_$C_6H_4S_2$ by NEGF method are published in reference [1].

The Van der Waals surface of molecules in this topic has the same shape, almost as smooth as the surface of a dish. Observing the following perpendicular direction to the surface containing the atoms in the molecule, limitation surface of Van der Waals force is shaped like a six-petal flower. The area of the dish is a rectangle shape with the length of L and width of W. The sizes shown in table 3 are average values. Following the CAChe software, the size of molecules is able to be presented at seven different values and equally distances. We can consider that the size of molecules is also quantized and their effects on I-V curve should also be evaluated. Figure 6 shows the measured results of 1,4-difloro-benzene, and table 2 also presents the average size value of molecules in this research.

The molecules of 1,4-dihalo-benzenes (1,4-$C_6H_4X_2$, with X: F, Cl, Br, and I) have three couples of electron for double bonding at each side. CAChe software is used to simulate this bonding and to explore the effect of Gold atom on energy levels used to form conductive channels (figure 7). As the results shown in figure 7, after bonding with Gold atoms and being optimized, the structure of molecules which are used to form the conductive channel is not deformed. Therefore, the measured results in table 3 can be used for calculation programming for I_{DS} of transistors.

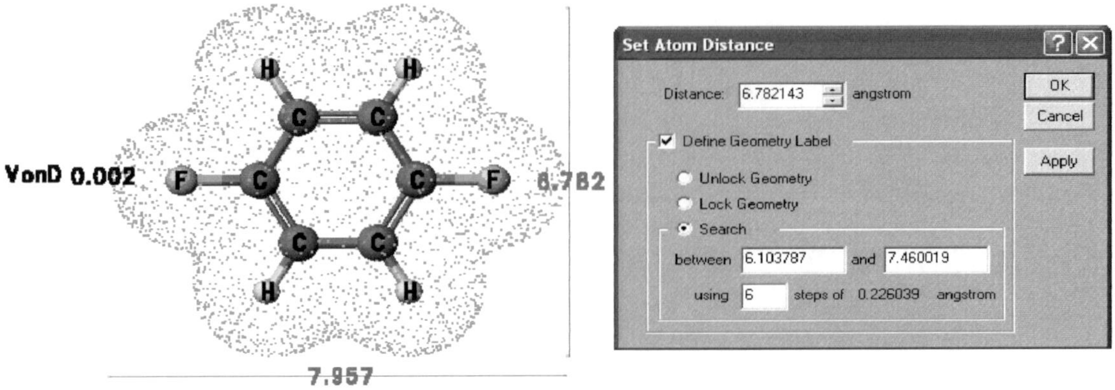

Figure 6. The size of molecules calculated by the Van der Waals force.

Table 1. The size of molecules consisting of 1,4 isomer of benzene ring with some atoms of S, O, F, Cl, Br, I.

Molecule	1,4_$C_6H_4S_2$	1,4_$C_6H_4O_2$	1,4_$C_6H_4F_2$	1,4_$C_6H_4Cl_2$	1,4_$C_6H_4Br_2$	1,4_$C_6H_4I_2$
High *H* nm	0.3797	0.3309	0.3839	0.3366	0.3803	0.3530
Long *L* nm	0.9584	0.7699	0.7957	0.9340	1.0025	1.0321
Wide *W* nm	0.6574	0.6770	0.6782	0.6696	0.6714	0.6637

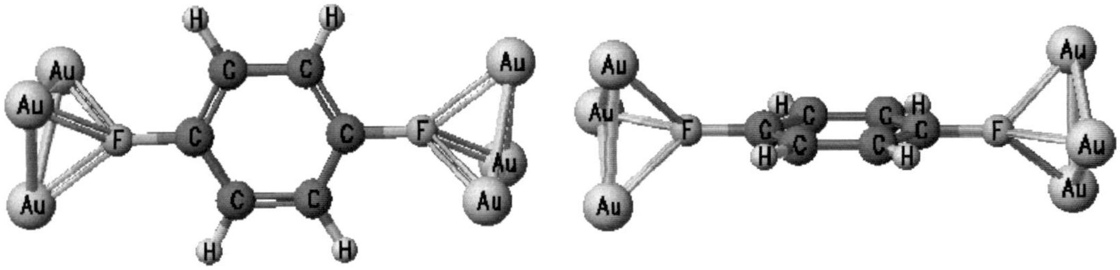

Figure 7. The molecule of 1,4_$C_6H_4F_2$ bonding with Gold in two different observation directions.

3.2. Transistor simulation with GUI in Matlab

From the main interface, in popup menu, we can click the mouse to choose one of the four molecules used to make the transistor's channel, then, click mouse on "Transistor characterize" to move to the next interface (figure 8). The interface includes image of transistor's construction, basic information of channel material's molecule and four slides to modify parameters of: length L, width W, gate voltage, temperature and I-V characteristic.

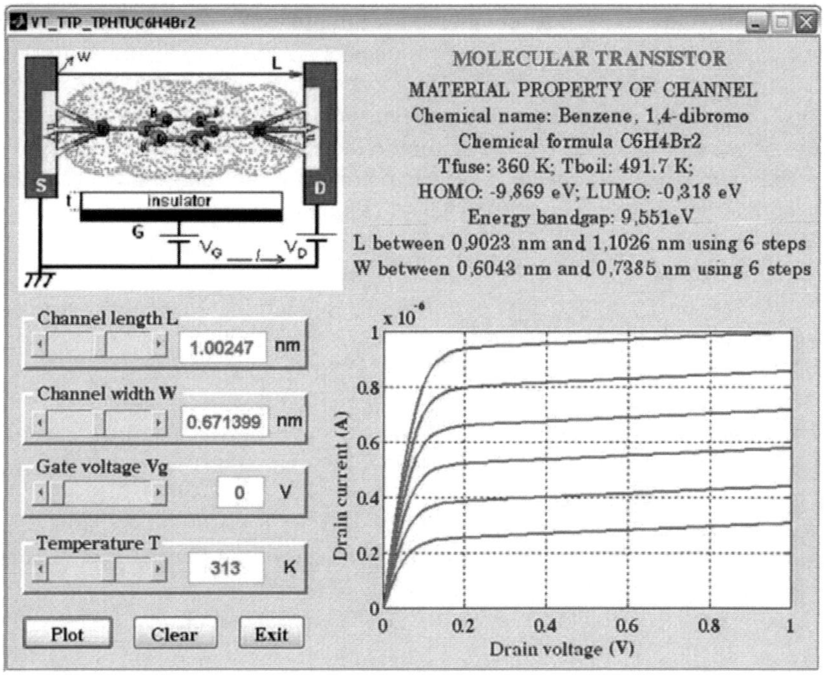

Figure 8. Survey interface of transistor with conductive channel of molecule of 1,4-dichloro_benzene.

3.3. Transistor's characteristic curves family

The result of transistor characterizing four kinds of transistors which are made of a 1,4-dihalo-benzene molecule varying U_G from 0 to 0.5 V, shows that its characteristic has the shape of traditional MOSFET's. The family of characteristic curves is divided into two separate regions: the steep region (I_{DS} current increases rapidly) when $U_{DS} < 0.2$ V and saturation region (I_{DS} current increases quite slowly) when $U_{DS} > 0.2$ V. Supply voltage for these kinds of transistor is merely about 1 V.

The result of testing impact of temperature from -80 to +120°C demonstrates that output characteristics keep unchanged for all four channel material molecules. Impact levels of temperature on the value of current I_{DS} are different with these four molecules. The range of working temperature with them is also larger than semi-conductor devices such as Si and Ge.

The result of impact of quantizing length and width of molecules on output characteristic shows that current I_{DS} increases when the width increases. In contrast, value of length does not give impact on current I_{DS}. This is compatible with quantum theory: conductivity at a level of energy is a constant.

In the same condition of temperature, voltages V_D and V_G, current I_{DS} of the transistors using molecular channel of 1,4-dihalo-benzenes (1,4_$C_6H_4X_2$, with X: F, Cl, Br, and I) has different values. This is related to the ability of power supply at the output of specific schematic when using molecular transistor. Among four kinds of testing, the largest power supply ability is the transistor with molecular channel of 1,4-diiodo-benzene, the second largest is the transistor with molecular channel of 1,4-dibromo-benzene, and the smallest is the transistor with molecular channel of 1,4-dichloro-benzene.

4. Conclusion

Dithio-p-benzoquinone [1] and p-Benzoquinone can be used as the p-type conductive channel for transistors but with the voltage $V < 0$. Chemically, each of oxygen and sulfur atoms has six valence electrons. Based on chemical structure, either oxygen or sulfur atom has one double bond with carbon atom in the benzene ring. Therefore, it is necessary to have four single bonds or two double bonds between atoms of oxygen and sulfur and electrodes of D and S. On the other hand, the transistor requires the negative bias to increase supply voltage. The higher value of supply voltage leads to the less density of devices in nanoelectronic circuits.

Based on the molecular transistor theory and the chemical simulation software of CAChe and Matlab, the author has outlined a nanotransistor. The molecule transistor's structure is similar to that of the conventional MOSFET, and the conductive channel is a 1,4-substituted benzene ring. The size of this transistor is about one nanometer. In comparison to a conventional MOSFET, these transistors are more advantageous with the simpler circuit and the lower required value of supply voltage due to the positive bias. The effect of other relating factors for each transistor using the conductive channel from molecules mentioned above has been investigated in order to evaluate its application and verify experimental results in future.

References

[1] Damle P, Rakshit T, Paulsson M, and Datta S 2002 *IEEE* **1** 145
[2] Samanta M P, Tian W, Datta S, Henderson J I and Kubiak C P 1996 *Phys. Rev. B* **53**
[3] Damle P 2003, PhD thesis, *Nanoscale device modeling: from mosfets to molecules* (Purdue University)
[4] Lyshevski S E 2007 ed *Nano and Molecular Electronics Handbook* (Boca Raton, London, New York: CRC Press) p 6-66
[5] http://www.nature.com/nnano/journal/v1/n3/fig_tab/nano.2006.130_F5.html
[6] Hengjun Z, Ying W, Kuizhan S, Yunqi L, Shiyan C, Wenfeng Q, Xiaobo S, Ting Q, Yongqiang M, Gui Y, Zhongmin S and Daoben Z 2006 *Chem. Comm.* **7** 755
[7] Xike G, Weiping W, Yunqi L, Wenfeng Q, Xiaobo S, Gui Y and Daoben Z 2006 *Chem. Comm.* **1** 2750
[8] Datta S 2003 *Lecture 2: What Makes Electrons Flow* (Purdue University)
[9] Datta S 2005 *Quantum Transport: Atom to Transistor* (Cambridge University Press)
[10] Phuc Tran Tien 2007 *Proceedings of 5th Nationwide Conference on Solid State Physics* (Vung Tau, Vietnam) p 768
[11] http://www.mcc.uiuc.edu/summerschool/2002/Supriyo_Datta/Datta_files/COURSE02.pdf

Acknowledgements

The author wishes to thank Prof. Dr. Dinh Sy Hien (HCMC University of Natural Sciences) - advisor on this topic, as well as Prof. Dr. Phan Hong Khoi (Institute of Materials Science, VAST), Prof. Dr. Le Khac Binh (HCMC University of Natural Sciences) for their help in the research and writing of this paper. I would also like to thank Prof. Dr. Supriyo Datta regarding online courses.

Optimization of planar Hall effect sensor for magnetic bead detection using spin-valve NiFe/Cu/NiFe/IrMn structures

Bui Dinh Tu[1*], Le Viet Cuong[2], Do Thi Huong Giang[1], Tran Mau Danh[1] and Nguyen Huu Duc[1,2]

[1] Department of Nano Magnetic Materials and Devices, Faculty of Engineering Physics, College of Technology, Vietnam National University, Hanoi
[2] Laboratory for Micro-Nano Technology, College of Technology, Vietnam National University, Hanoi

E-mail: buidinhtu@vnu.edu.vn

Abstract. Present paper deals with the planar Hall effect (PHE) of Ta(5 nm)/NiFe(t_f)/Cu(1.2 nm)/NiFe(t_p)/IrMn(15 nm)/Ta(5 nm) spin-valve structures. Experimental investigations are performed for 50 × 50 μm^2 junctions with various thicknesses of free and pinned layer t_f = 4, 8, 10, 15, 20 nm and t_p = 2, 3, 6, 8, 9, 12 nm. The results show that the thicker free layers, the higher PHE signal is obtained. In addition, the thicker pinned layers, the lower PHE signal. The highest PHE sensitivity S of 15.6 mΩ/Oe is obtained in the spin-valve configuration with t_f = 20 nm and t_p = 2 nm. This optimum structure is rather promising for micro magnetic bead detections.

Keywords: Biosensors, Hall effect, magnetization reversal, magnetoresistance.

1. Introduction

Magnetic label detection is usually accompanied by using giant magnetoresistance effect, PHE, as well as magnetic tunneling junctions. Among them, PHE has recently been receiving great attention for spintronic biosensor design thanks to its nano-Tesla sensitivity and high signal-to-noise ratio [1-6]. PHE is based on the anisotropic magnetoresistance (AMR) of ferromagnetic (FM) materials. The transverse voltage on a planar Hall cross depends on the orientation of the magnetization of the free ferromagnetic (FFM) layer with respect to the longitudinal current running through the material. Thus, the large PHE is expected to be observed in exchange coupling based structures because they can ensure a sufficient uniaxial anisotropy with well defined single domain state to introduce an unidirectional anisotropy. For this purpose, Ejsing *et al.* [3, 4] have reported a single PHE sensor of NiFe/IrMn/NiFe. Furthermore, a PHE magnetic bead array counter microchip integrating 24 single sensors based on a simple NiFe/IrMn bilayer structure was successfully prepared [5]. Recently, Thanh *et al.* [6] have found that the sensor signal can be further improved by using spin-valve structure of Ta(5 nm)/NiFe (6 nm)/Cu (3.5 nm)/NiFe (3 nm)/IrMn (10 nm)/Ta (5 nm) with the size of 3×3 μm^2 when detecting the 2.8 μm magnetic beads.

The present paper deals with studies of the sensitivity dependence on the thickness of the individual ferromagnetic layers in patterned 50×50 μm^2 PHE sensor based on Ta/NiFe/Cu/NiFe/IrMn/Ta spin-valve structure. This PHE sensor has been proposed to apply to magnetic bead detection.

© 2009 IOP Publishing Ltd

2. Experimental

Spin-valve Ta(5 nm)/NiFe(t_f)/Cu(1.2 nm)/NiFe(t_p)/IrMn(15 nm)/Ta(5 nm) structures (with t_f = 4, 8, 10, 15, 20 nm and t_p = 2, 3, 6, 8, 9, 12 nm) are fabricated by DC magnetron sputtering system with the base pressure less than 7×10^{-7} mTorr. The spin-valve structures are sputtered on SiO_2 wafer at room temperature with Argon working pressure of 3×10^{-3} mTorr. During the sputtering process, a uniform magnetic field of H_x = 400 Oe is applied in plane of the films, parallel to the *Ox* direction. This magnetic field induces a magnetic anisotropy in the free and pinned ferromagnetic (PFM) layers and then aligns the pinning direction of the antiferromagnetic (AFM) IrMn layer. The PHE sensors are structured by using lithography techniques into four-electrode bars with the patterned size of 50 × 50 μm^2 (figure 1a) The sensors were passivated by a sputtered 150 nm thick Si_3N_4 layer to protect against the fluid used during experimentation. The bead array counter (BARC) microchip was fabricated by integrating 10 sensor patterns as shown in figure 1b.

The PHE characteristics of sensors was measured by using a nanovoltmeter in external magnetic fields H_y up to 30 Oe applied along *Oy* direction and sensing currents I_x of 1 mA. Magnetization is measured by means of a Lakeshore 7400 vibrating sample magnetometer (VSM) on defined 12×12 mm^2 films.

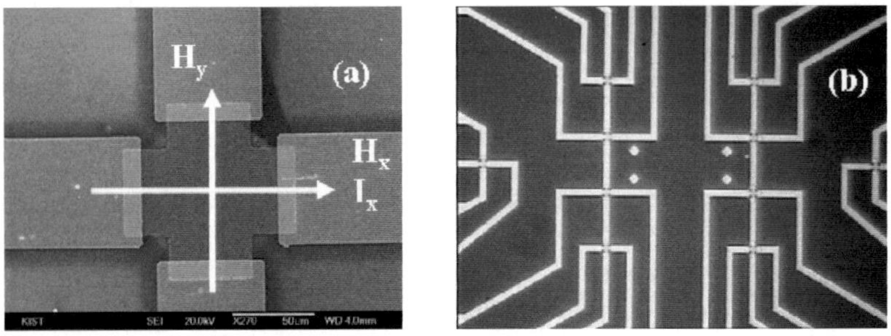

Figure 1. (a) - Top view micrograph of the single 50×50 μm2 PHR cross. The pinning direction Hx as well as the direction of the bias field Hy and sensing current Ix are indicated. (b) – the bead array counter microchip including 10 of single PHE sensors (with 8 single sensors in the two middle lines and 1 single sensor in each edge line).

3. Results and discussion

3.1. Fixed pinned layer thickness system

Figure 2a presents magnetization data of spin-valve Ta(5 nm)/NiFe(t_f)/Cu(1.2 nm)/NiFe(3 nm)/IrMn(15 nm)/Ta(5 nm) structures with different thickness (t_f) from 4 to 20 nm. It is clearly seen that all samples exhibit two hysteresis loops. The magnetization accounting from the first loop increases with increasing x while that from the second loops is almost constant. These two hysteresis loops are attributed to the free and pinned layers, respectively. The free layer is expected to dominate the sensor response at low magnetic fields. The values of the anisotropy (H_c or H_K) and exchange coupling (H_{ex}) fields determined from these magnetization data are collected and listed in table 1. Note that H_K and H_{ex} are almost constant for $t_f > 10$ nm.

Shown in figure 2b is the sensor voltage as a function of the applied fields. It can be seen from this figure that the PHE voltage initially develops rather fast at low fields, reaching a maximal value at $H \sim$ 10 Oe and finally decreases with further increase of the magnetic fields. For this fixed pinned layer spin-valve system, however, the maximal value of the PHE voltage increases with increasing free layer thickness. It increases from the value of 15 μV for sample with t_f = 4 nm to the value of 48 μV for t_f = 20 nm. Consequently, the sensor sensitivity S (= dV/dH, see below) is enhanced from the value of 1.7 mΩ/Oe to 7.6 mΩ/Oe, respectively.

It is well known that when the magnetization vector M makes an angle θ with easy axis along the Ox direction (and/or with I_x), the transverse induced PHE voltage V_{PHE} (or V_y) parallel to Oy direction is given as follows:

$$V_y = I_x \Delta R \sin\theta \cos\theta \ , \quad (1)$$

where $\Delta R = (\rho_{//} - \rho_{\perp})/t$ with $\rho_{//}$ and $\rho_{\perp}$ are the resistivity measured with the current parallel and perpendicular to the magnetization, respectively; t_f is the free ferromagnetic layer thickness.

Typically, these $V_{PHE}(H)$ curves are fitted well by using the single domain model [7] with the magnetic energy per unit of the magnetic layer. In this case, the Stoner-Wohlfarth energy can be expressed as:

$$E_{FM} = K_u t \sin^2\theta_f - M_s t H \cos(\alpha - \theta_f) - J\cos(\theta_f - \theta_P) \ . \quad (2)$$

Here, the θ_f and θ_p are the angles between magnetization of the free and pinned layers and easy axis direction, respectively; K_u (= $H_K/2M_s$) is the effective anisotropy constant, M_s is the saturation magnetization of the free layer and J is the interlayer coupling constant that can be extracted from the relationship with the exchange coupling field between two FM layers (H_{ex}) ($J = tH_{ex}M_s$).

If the exchange biased field between PFM and AFM layers is strong enough, the angle between magnetization and easy axis direction of the pinned layer will be fixed at low applied magnetic fields, *i.e.* θ_p equals to zero. This can be applied for the present case, where the magnetization reversal of the free and pinned layers occurred separately (see figure 2a).

For small angles, cos $\theta \approx 1$, the PHE voltage exhibits linear characteristics as well as high sensitivity in low fields (H < 10 Oe). In this case, the sensitivity of the sensor is given as:

$$S = \frac{V_y}{IH_y} = \frac{\Delta R}{H_K + H_{ex}} \ . \quad (3)$$

Applying this theoretical approach to experimental data, we can determine the values for H_K, H_{ex} fields and the sensor sensitivity S. The obtained results are summarized in table 1 (in brackets). Note that the values of S and H_{ex} obtained for the fit of the PHE data are in excellent agreement with those derived from experimental data. The values of H_K, however, are systematically larger than those obtained from the VSM measurements.

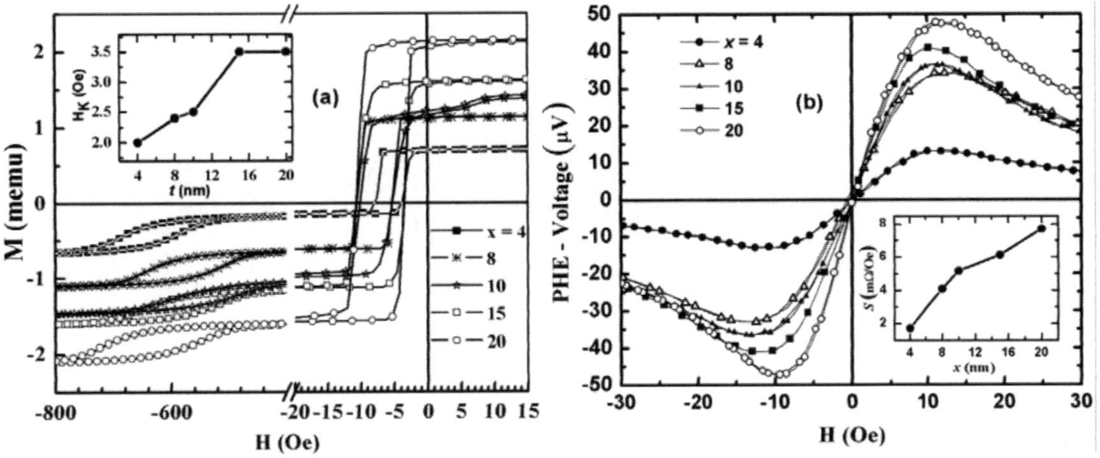

Figure 2. Hysteresis loops (a) and low field $V_{PHE}(H)$ characteristics (b) of spin-valve structures with the fixed pinned layer thickness t_p = 3 nm and free layer thickness (t_f) varying from 4 to 20 nm. The inserted figures are anisotropy field and sensor sensitivity extracted from experimental data, respectively.

Table 1. Experimental and calculated (in brackets) anisotropy (H_K), exchange coupling field (H_{ex}) and sensor sensitivity (S) for spin-valve system with different free layer thicknesses.

t_f (nm)	S (mΩ/Oe)	H_K (Oe)	H_{ex} (Oe)
4	1.7 (3.5)	2 (2.1)	5.4 (5.4)
8	4.1 (4.6)	2.4 (6.7)	7.9 (7.9)
10	5.2 (5.5)	2.5 (5.5)	7.7 (7.7)
15	6.1 (6.5)	3.5 (5.3)	7.1 (7.1)
20	7.6 (7.6)	3.5 (5.3)	7.1 (7.1)

3.2. Fixed free layer thickness system

Figure 3a presents magnetization data of Ta(5 nm)/NiFe(10 nm)/Cu(1.2 nm)/NiFe(t_p nm)/IrMn(15 nm)/Ta(5 nm) spin-valve structures with different pinned layer thickness (t_p) from 2 to 12 nm. Here, all samples exhibit two hysteresis loops too. However, contrary to the fixed pinned layer thickness system, the magnetization accounting from the first loop is almost constant while that from the second loop increases with increasing t_p. The values of the anisotropy (H_K) and exchange coupling (H_{ex}) fields determined from these magnetization data are collected and listed in table 2. Note that H_K and H_{ex} increase with increasing t_p.

Shown in figure 3b is the sensor voltage as a function of the applied fields. For this fixed pinned layer spin-valve system, it is clear that with increasing t_p, the maximal value of the PHE voltage decreases. In addition, this peak shifts to higher magnetic fields. Consequently, the sensor sensitivity S is reduced strongly from 8.8 mΩ/Oe to 3.4 mΩ/Oe when t_p increases from 2 nm to 12 nm (figure 3a, insert).

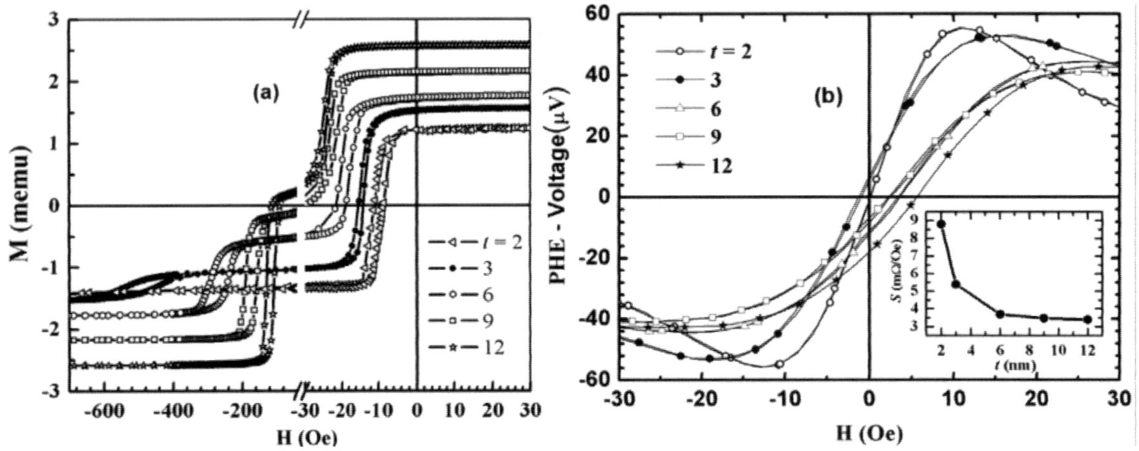

Figure 3. Hysteresis loops (a) and low-field $V_{PHE}(H)$ characteristics (b) measured in spin-valve structures with the fixed free layer thickness x = 10 nm and different pinned layer thickness (x) from 2 to 12 nm.

Calculated values of anisotropy (H_K), exchange coupling field (H_{ex}) and sensor sensitivity (S) are listed in table 2 for the fixed pinned layer thickness spin-valve system. Again, an excellent consistence with experimental results is obtained for H_{ex} and S only.

Table 2. Experimental and calculated (in brackets) anisotropy (H_K), exchange coupling field (H_{ex}) and sensor sensitivity (S) for spin-valve system with different pinned layer thicknesses.

t_p (nm)	S (mΩ/Oe)	H_K (Oe)	H_{ex} (Oe)
2	8.8 (8.9)	1.2 (2.1)	10.3 (10.3)
3	5.4 (5.9)	0.9 (2.9)	14.4 (14.4)
6	3.7 (3.8)	1.3 (4.4)	19.0 (19.0)
9	3.3 (3.3)	0.6 (2.5)	22.9 (22.9)
12	3.4 (3.3)	1.3 (1.2)	25 (25.0)

3.3. Optimal spin-valve configuration for PHE sensor sensitivity

It was provided from above mentioned investigation that the large PHE sensor sensitivity can be reached in spin-valve structures with thin PFM and thick FFM layers. This may relate to the spin collinear structure in the ferromagnetic layers. In the PFM layer, the well-aligned spin part is usually formed near PFM/AFM interface. Further increasing the pinned layer thickness will lead to an enlarging the twisted structure where the magnetization is pinned in different directions from the easy axis (*i.e.* $\theta_p \neq 0$) [7-10]. In this context, the twisted part can be assumed to be eliminated in the structure with thin pinned layer t_p = 2 nm. Practically, the maximal PHE voltage and the highest sensitivity of sensor were observed in this configuration. For the free layers, the magnetic influence and then the twisted part can be established near NM/FFM interface only. The thick free layers thus dominate the collinear ferromagnetic part and enhance the PHE voltage. Combining these two optimal tendencies, we prepared Ta(5 nm)/NiFe(20 nm)/Cu(1.2 nm)/NiFe(2 nm)/IrMn(15 nm)/Ta(5 nm) spin-valve structure, *i.e.* with t_f = 20 nm and t_p = 2 nm. Its magnetization and PHE data are presented in figure 4. It is interesting to note that this spin-valve configuration shows sensor sensitivity as large as 15.6 mΩ/Oe.

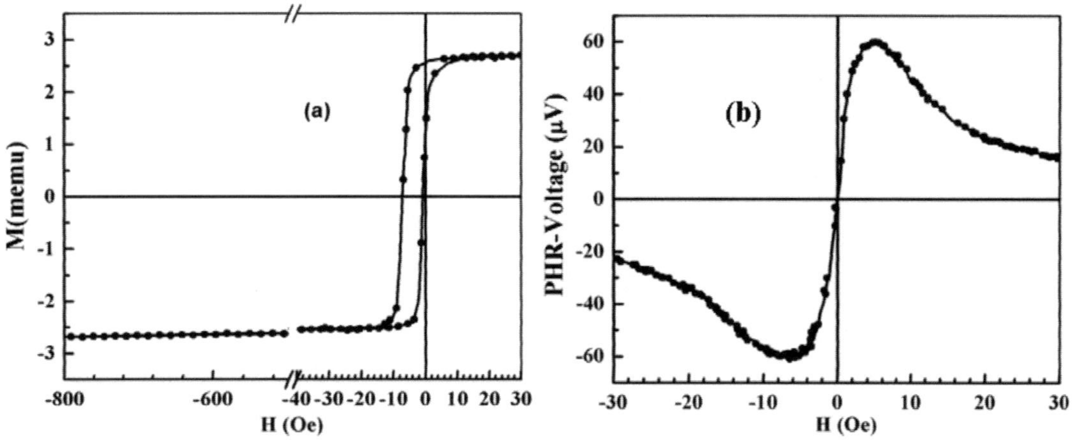

Figure 4. Hysteresis loops (a) and low-field $V_{PHE}(H)$ characteristics (b) measured in Ta(5 nm)/NiFe(20 nm)/Cu(1.2 nm)/NiFe(2 nm)/IrMn(15 nm)/Ta(5 nm) spin-valve structure, *i.e.* with t_f = 20 nm and t_p = 2 nm.

Conclusion remarks

The influence of the individual free and pinned layer thickness to the sensitivity of PHE sensor based on the spin-valve structure of NiFe(t_f)/Cu(1.2nm)/NiFe(t_p)/IrMn(15nm) with the size of 50×50 μm^2 has been studied. The results show that the thicker free ferromagnetic layers enhance the PHE signal, whereas the thicker pinned ferromagnetic layers lower PHE one. For a good combination, the highest PHE sensitivity of 15.6 mΩ/Oe is obtained in the spin-valve configuration with t_f = 20 nm and t_p = 2 nm. The result is rather promising for application of micro magnetic bead detections in the biology field.

References

[1] Schuhl A, Dau F N V, Childress J R 1995 *Appl. Phys. Lett.* **66** 2751
[2] Dau F N V, Schuhl A, Childress J R, Sussiau M 1996 *Sensors and Actuators A* **53** 256
[3] Ejsing L, Hansen M F, Menon A K, Ferreira H A, Graham D L, Freitas P P 2004 *Appl. Phys. Lett.* **84** 4729
[4] Ejsing L, Hansen M F, Menon A K, Ferreira H A, Graham D L, Freitas P P 2005 *J. Magn. Magn. Mater.* **293** 677
[5] Tu B D, Thanh N T, Danh T M, Duc N H, Kim C G 2008 *J. Appl. Phys.* **104** 74701
[6] Thanh N T, Parvatheeswara R B, Duc N H and Kim C G 2007 *Physica Status Solidi (a)* **204** 4053
[7] Robert C. O'Handley 2000 *Modern Magnetic Materials* (John Wiley & Son)
[8] Slonczewski J C 2002 *J. Magn. Magn. Mater.* **247** 324
[9] Ralph D C, Stiles M D 2008 *J. Magn. Magn. Mater.* **320** 1190
[10] Shuai W, Yuan X and Ke X 2008 *Physical Review B* **77** 184430

Acknowledgments

This work was supported by the Vietnam National University of Hanoi under the project QG.TD 07.10 and QC.09.13.

Magnetoelectric effects in piezoelectric/soft magnetic amorphous FeCo-based ribbon composites

Do Thi Huong Giang[1], Le Khac Quynh[2], Nguyen Van Dung[3] and Nguyen Hoang Nghi[3]

[1] Laboratory for Nano Magnetic Materials and Devices, Faculty of Engineering, Physics and Nanotechnology, 144 Xuan Thuy Road, Cau Giay, Hanoi, Vietnam
[2] Hanoi University of Education 2, Xuan Hoa Community, Phuc Yen District, Vinh Phuc, Vietnam
[3] Laboratory of Amorphous and Nanocrystalline Materials, Hanoi University of Technology, Vietnam

E-mail: giangdth@vnu.edu.vn

Abstract. Magnetoelectric effect has been investigated in several multiferroic laminates generated by inserting a lead-zirconate titanate PZT plate between sheets of highly soft magnetic amorphous FeCo-based alloy ribbon. The results show that the magnetoelectric coefficient α_E (= dE/dH) depends on the rest glass-forming elements as well as the concentration of Co substitution. Giant magnetoelectric is found for the composite of $Fe_{73.5}Cu_1Nb_3B_9Si_{13.5}$ ribbon, where α_E as high as 1000 mV/cmOe at low field of 50 Oe. α_E can be further improved up to 1170 mV/cmOe at 20% Fe substituted by Co. The higher the Co concentration, the lower magnetoelectric effect is observed. High magnetoelectricty at low field makes these composites suitable for special applications such as smart sensors for detecting microtesla magnetic fields and especially biosensor for magnetic label detection.

Keywords: Magnetoelectric, magnetostrictive, soft magnetic ribbons, piezoelectric.

1. Introduction

Over the past several decades, melt-spun Co- and Fe-based alloys have been extensively investigated for applications in magnetic devices requiring magnetically soft materials [1]. Because there is no crystalline magnetic anisotropy, these alloys exhibit very good soft magnetic properties in the amorphous state by a high saturation magnetization for Fe-based alloys and high permeability for Co-based alloys [2]. Up to date, most researches have mainly focused on improving their magnetic properties and exploiting their application ability in components of devices [3]. Amorphous magnetic ribbons containing about 70-85 at % of the transition metals Fe and Co and the rest glass-forming elements (*e.g.* B, Si, Nb, Cu, Cr, *etc…*) are well known as materials with excellent soft magnetic properties. Although these alloys possess an extremely low magnetostriction ($\lambda_s \sim 10^{-6}$) compared with the conventional rare earth - transition metal intermetallic compound ($\lambda_s \sim 10^{-3}$), their superior soft magnetostrictive properties at low-field is rather promising for magnetostriction-based applications. However, this advantage has not been taken into consideration.

Regarding the magnetostriction-based applications, multiferroic composites using magnetostrictive ferrites have been studied intensively from the beginning of this century [4 and refs. therein]. In these

© 2009 IOP Publishing Ltd

Journal of Physics: Conference Series **187** (2009) 012057 doi:10.1088/1742-6596/187/1/012057

materials, an electric polarization can respond to an applied magnetic field. In applied *dc* magnetic fields (H), the sample undergoes poling that creates an electric field $E = \alpha_E \cdot H$, where α_E denotes the magnetoelectric (ME) voltage coefficient. As a results, a ME-voltage $V_{ME} = t \cdot E$ ($= \alpha_E \cdot t \cdot H$) appears between the surface of the sample with thickness t. Large magnetoelectric voltage coefficients α_E ($= dE/dH = V_{ME}/h_{ac}.t$) offer potential device applications as highly sensitive magnetic field sensors, microwave filters, transformers, gyrators, *etc*. [4]. It has been already reported that to obtain a large ME voltage at low magnetic fields in these composites, the magnetostrictive phase must have a large magnetostriction (λ) and large magnetostrictive susceptibility ($d\lambda/dH$) [5]. In the light of these requirements, the amorphous FeCo-based ribbons are extremely suitable for use in ME composites. Besides their magnetic and magnetostrictive softness advantages, it must be mentioned the cheap materials constituted, low cost and simple preparation techniques. In the frame of this paper, we focused to study the magnetoelectric effect in the piezoelectric/soft magnetic amorphous FeCo-based alloy ribbon composites and their application abitily in biosensor for magnetic label detection. The dependence of ME effect on the rest glass-forming elements as well as concentration of the Co substitution will be considered and studied.

2. Experimental

Amorphous FeCo-based ribbons with different glass-forming elements (*e.g.* B, Si, Nb, Cu, Cr, *etc*...) and Co concentration have been prepared by the melt-spinning technique. The ME composites are formed by bonding an out-of-plane poled piezoelectric PZT plate (APCC-855) with thickness t_{PZT} = 200 μm (supplied by the American Piezoceramics Inc) between ribbon sheets in size 5x5 mm. The optimum configuration for this ME composite uses two ribbon sheets bonded on both sides of PZT plate (figure 1) [6]. Thanks to mechanical coupling between these components, when the magnetostrictive sheets are strained under applied in-plane magnetic fields, the PZT plate will undergo a forced strain. In this case, the ME-voltage V_{ME} is induced across the thickness of the piezoelectric plate. Practically, the V_{ME} is measured directly as a response of the ME composite to a weak *ac* magnetic field h ($= h_0\sin(2\pi f_o t)$) oscillating at resonant frequency of 2.5 kHz in a *dc* bias field H provided by an electromagnet. A lock-in amplifier (7265 DSP) is used to generate a controllable input current to the *ac* field coil and to measure the output voltage (V_{ME}) induced across the PZT layer. The magnetization was recorded by a Lake Shore 7400 series VSM.

3. Results and discussion

3.1. Dependence of ME effect on glass-forming elements

Shown in figure 2 are hysteresis loops and ME coefficient respects to *dc* magnetic field of three composites of three typical ribbons with composition $Fe_{76.5-x}Cr_xB_{10}Si_{13.5}$ with x = 0, 2 and 4 at%. As seen clearly in this figure that all these ribbons exhibit extremely low magnetic coercivity (less than 0.5 Oe) but very different magnetically softness depending on the Cr concentration. The more Cr concentration, the higher slope angle of the loop will be achieved. The optimum magnetic softness with highest slope is observed in the ribbon with 4 at% Cr. Consequently, the high magnetoelectric softness characterized by the low magnetic field H_o, where the ME coefficient is expected to be optimum at 4% Cr. Indeed, H_o is decreased from 205 Oe to 100 Oe with increasing x from 0 to 4 at%. Although having better low-field ME effect, the ribbon with higher Cr (x = 4) give lower maximum value α_E of 680 mV/cmOe rather than that of 800 mV/cmOe in the case without Cr (x = 0). That can be well understood by the higher rest glass-forming elements concentration lowers the saturation magnetostriction due to their non magnetostriction contribution, thereby, lower maximum ME coefficient. For sensor application to detect *dc* magnetic field, the sensitivity of output voltage S_H with respects to magnetic field is most important. S_H reaches maximum of 0.75 mV/Oe in the ribbon with 4 at% Cr.

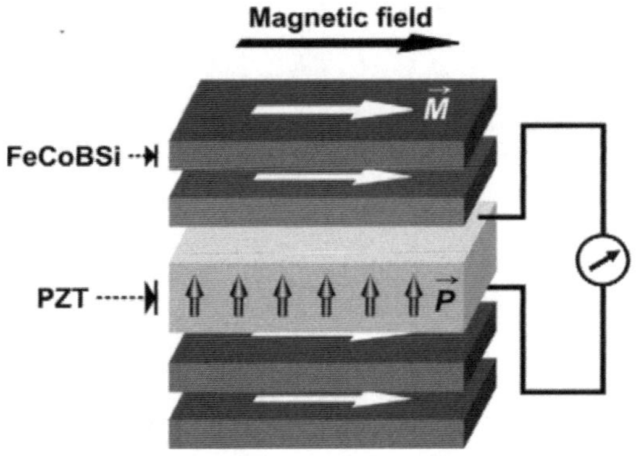

Figure 1. Schematic illustrations of ME composite based on amorphous magnetic FeCo-based ribbons (with double ribbon sheets on each outer side).

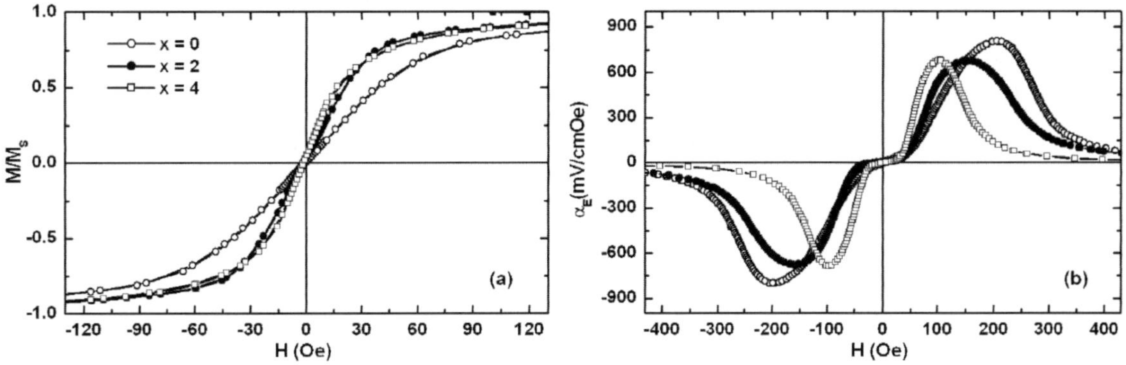

Figure 2. Relative magnetization (a) and ME coefficient (b) *vs* magnetic dc field for composites of Fe-based ribbons with different glass-forming Cr concentration (*x*).

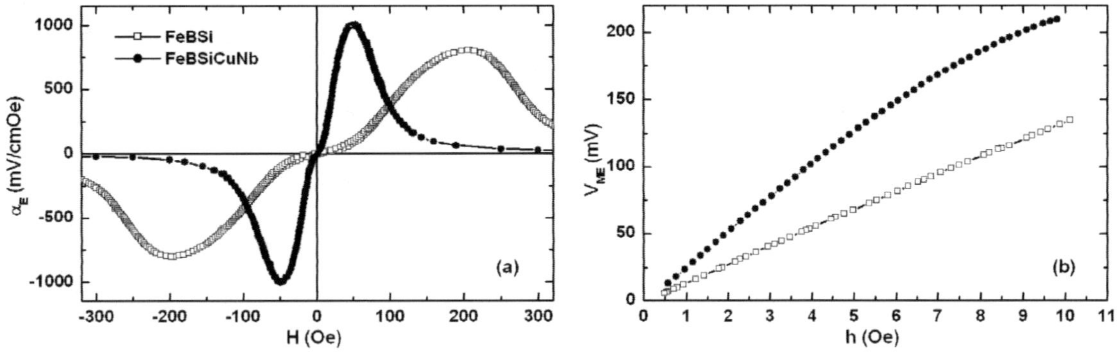

Figure 3. ME coefficient (a) and ME voltage (b) as a function of *dc* and *ac* magnetic fields, respectively, for the composites of Fe-based ribbons with and without small amounts of glass-forming Cu and Nb.

The ME effect can be significantly improved in the ribbon with small amounts of Fe substituted by Cu and Nb. In this case, the optimum composition has found in the composite of $Fe_{73.5}Cu_1Nb_3B_9Si_{13.5}$ (Finemet) ribbon, where α_E as high as 1000 mV/cmOe obtained at very low magnetic fields of 50 Oe (see in figure 3a). In addition, the slope of ME coefficient respect to the external field close to zero is much higher compared with that without Cu and Nb giving a sensitivity S_H of 2.8 mV/Oe, four times higher than that without Cu and Nb. It shows application ability to use this composite in sensor highly sensible to low magnetic field. This magnetoelectric sensor can sense to not only *dc* magnetic field but also especially with minus *ac* magnetic field. Shown in figure 3b is the variation of ME-voltage as a function of *ac* field measured at resonant frequency. The composite using $Fe_{73.5}Cu_1Nb_3B_9Si_{13.5}$ ribbon reveals a higher sensitivity, whereas the linear range respects to *ac* field is larger in the composite of $Fe_{76.5}B_{10}Si_{13.5}$ ribbons.

3.2. Dependence of ME effect on Co addition

Various composites using $Fe_{100-y}Co_y$-based ribbons with different composition of Co (y) have been prepared and investigated. The hysteresis loops and ME coefficient are shown in figure 5. The higher Fe concentration substituted by Co exhibits better magnetic softness with higher low-field slope and lower saturation field (figure 5a). As a consequent, the magnetic field H_o at which the α_E reaches maximum is decreased from 200 to 60 Oe with the increase of y from 0 to 100 %. Whereas, the highest value of α_E of about 1170 mV/cmOe at 120 Oe was observed in the ribbon with 20 at% Fe substituted by Co. High Co concentration leads to lower α_E and almost vanished (α_E = 6 mV/cmOe) at concentration more than 90%. These findings can be attributed to the strong relationship between ME voltage and the magnetostrictive susceptibility ($d\lambda/dH$) of the magnetic ribbon, $V_{ME} \sim d\lambda/dH$ [7]. As we known that the magnetic elements Fe and Co in their pure crystalline forms have positive and negative magnetostriction, respectively. The amorphous Fe-based alloys have high positive magnetostriction λ of about 30×10^{-6} [8] compared with negative or nearly zero magnetostriction ranging between -6×10^{-6} and -0.1×10^{-6} [9] of Co-based alloys. As a result, the ME effect is null at more than 90% Fe substituted by Co due to zero magnetostriction caused by Co atom cancelled out part of the positive magnetostriction. The optimum concentration for ME effect is observed at 20 at% Co because of both their high λ and $d\lambda/dH$ [6]. Further decreasing Co concentration will lower the ME effect due to the decline of $d\lambda/dH$.

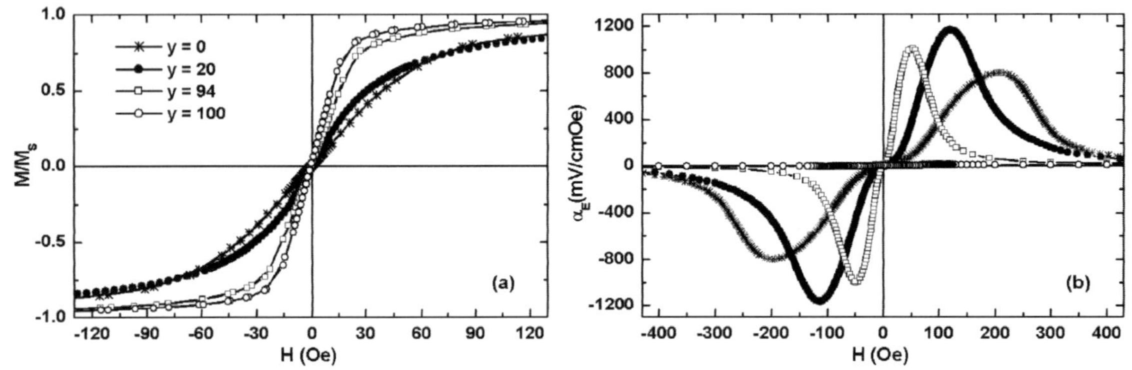

Figure 4. Relative magnetization (a) and ME coefficient (b) *vs* magnetic dc field for composites based on amorphous magnetic $Fe_{100-y}Co_y$-based ribbons with different Fe concentration (y) substituted by Co.

For magnetoelectric sensor applications, the $Fe_{80}Co_{20}$-based ribbon will be the best candidate. That gives both high magnetoelectric effect and sensitivity ME-voltage at low magnetic field [6]. This sensor is promising not only for microtesla magnetic field sensors but also for biosensor applications. Here, for biosensor fields to detect magnetic labels, we also start studying the potentiality of using this composite for magnetic nanoparticles detection. Shown in figure 5 is real-time voltage change for the detection of nickel ferrite ($NiFe_2O_4$) nanoparticles using composite of $Fe_{80}Co_{20}$-based ribbon. There are two cycles where the high state shows the signals after dropping 2μl of the $NiFe_2O_4$ solution (5 wt%) on the surface of the composite and the low state shows the signals after washing (figure 5, insert). As seen from this figure, the total output signal changes were found of about 30 μV. After further process with another droplet/washing cycle, the similar signal response was repeated. This voltage response can be much higher by increasing the *ac* magnetic field.

This high value of nanoparticles response may recommend our future work in developing biochip for biosensor based on our composites. For a real biosensor application to detect magnetic labels, this sensor is required to diminish in micrometer-size. The project is still in progress.

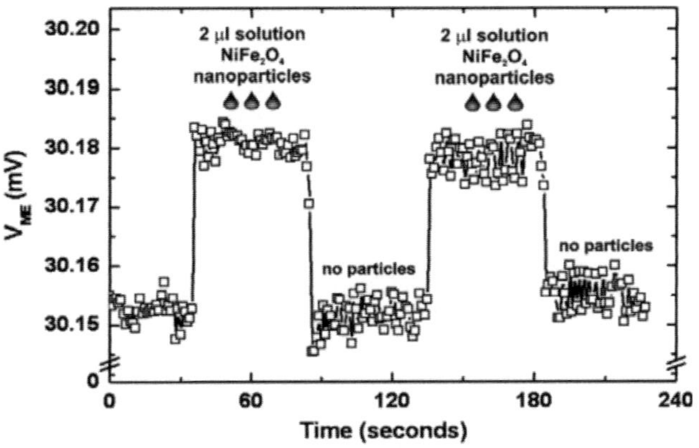

Figure 5. Real-time voltage changes fornickel ferrite nanoparticles detection using composite of $Fe_{80}Co_{20}$-based ribbon. This profile was carried out in an applied magnetic field of 5 mT.

4. Conclusion remarks

Although amorphous FeCo-based ribbons possess an extremely low magnetostriction ($\lambda_s \sim 10^{-6}$), their superior soft magnetic and magnetostrictive properties is promising for low-field magnetoelectric applications. The magnetoelectric investigations in the composites generated by inserting a PZT plate between melt-spinning ribbon sheets show that the magnetoelectric effect depends on the concentration of the Co substitution as well as the rest glass-forming elements. The giant magnetoelectric α_E as high as 1000 mV/cmOe at very low field of 50 Oe is found for the composite of $Fe_{73.5}Cu_1Nb_3B_9Si_{13.5}$ ribbon. This coefficient can be further increased up to 1170 mV/cmOe by substituting 20% Fe by Co. The highly magnetoelectricity at low field makes these composites promising for applications in magnetic sensor and biosensor fields.

References

[1] McHenry M E, Willard M A and Laughlin D E 1999 *Progress in Materials Science* **44** 291
[2] Makino A, Inoue A and Masumoto T 1995 *Mater Trans. JIM* **36** 924
[3] Ryusuke Hasegawa 2005 *Applications of amorphous magnetic alloys* (Kluwer Academic Publishers)
[4] O'Dell T H 1965 *Electron Power* **11** 266
[5] Duc N H and Huong Giang D T 2008 *J. Alloys Compounds* **449** 214
[6] Giang D T Huong and Duc N H 2009 *Sensors and Actuator A* **149** 129
[7] Fetisov Y, Bush A, Kamentsev K, Ostashchenko A, Srinivasan G 2004 *Proc. of IEEE Sensors* **3** 1106
[8] Ryusuke Hasegawa 2003 *J. Non-Crystalline Solids* **329** 1
[9] Takaomi Itoi and Akihisa Inoue 2000 *Mater. Trans.* **41** 1256

Acknowledgments

This work was supported by the Fundamental Research Program of Vietnam under the Project 410.406 and the Vietnam National University, Hanoi under the project QC. 07.07.

Journal of Physics: Conference Series **187** (2009) 012058 doi:10.1088/1742-6596/187/1/012058

Synthesis, properties and application of polyindole/TiO_2 nanocomposites

Vu Quoc Trung[1] and Duong Ngoc Huyen[2]

[1] Faculty of Chemistry, Hanoi National University of Education, 136 Xuan Thuy Road, Cau Giay District, Hanoi, Vietnam

[2] Institute of Engineering Physics, Hanoi University of Technology, 1 Dai Co Viet, Hanoi, Vietnam

E-mail: vuquoctrung@netnam.vn

Abstract. Nanocomposites of polyindole (PIn) and TiO_2 were prepared by chemical polymerization. The morphology of nanocomposite particles was studied by scanning electron microscopy and transmission electron microscopy. The chemical structure of conducting polymers in nanocomposites was characterized by FTIR and Raman spectra. The thermal analysis showed that all conducting polymers in the nanocomposites were stable at more than 600°C. The corrosion protection of the nanocomposites was investigated.

Keywords: Nanocomposite, polyindole, titanium dioxide, corrosion protection.

1. Introduction

In recent years, several conducting polymers have received increasing attention as material coatings for the protection of common metals against corrosion. In particular, polyaniline and polypyrrole have been found to sufficiently reduce the corrosion of steel and other oxidizable metallic materials [1–3]. Several reports have been published on polyindole and its derivatives although its close structural similarities with polyaniline and polypyrrole [4–6].

Polyindole is an electroactive polymer, which can be usually obtained after anodic oxidation of indole in various electrolytes; electro-chemical oxidation of indole in $LiClO_4$ containing acetonitrile electrolyte medium gives an electrochromic polymer film with good air stability. In its doped state, polyindole has a green colour and an electrical conductivity in the range of 10–10 Scm, depending on the nature of the counter ion [7, 8]. It is also reported that polyindole films have the advantages of fairly good thermal stability, high redox activity and slow degradation rate in comparison with polypyrrole and polyaniline [9, 10]. However, there was a lack of reports on the chemical electrosynthesis of PIn or PIn composites.

In this paper, nanocomposites of polyidole and TiO_2 were prepared by chemical polymerization. The morphology and properties of nanocomposite and were studied by scanning electron microscopy (SEM), transmission electron microscopy (TEM), FTIR, Raman spectra and thermal gravimetric analysis (TGA). The corrosion protection of the nanocomposites was investigated.

© 2009 IOP Publishing Ltd

Experimental

1.1. Preparation of nanocomposites
All nanocomposites were chemically prepared as the procedure described in Ref. [11]. Firstly, a dispersion was prepared by mixing (for 30 minutes) of 10.0 g TiO_2 (Merk) and 1.0 mL indole monomers (Guangzhou Jihuada Chemical, China) in 50.0 ml distilled water. Then 4.5 g $FeCl_3$ (water-free, Fluka Chemie) was added to the oxide particle dispersion during stirring. After 2 hours of stirring, the particles were cleaned by distilled water, filtered, extracted for 10 hours and dried at 50°C for several days under low pressure.

1.2. Characterization of nanocomposites
TGA of prepared samples was measured by Ghimashu-50 H with scan rate of 10°C/min in ambient conditions. The X-ray diffraction patterns of TiO_2 and TiO_2 nanocomposites were measured by Simens D-5005. The chemical structure of the nanocomposites was characterized by Fourier transform infrared spectroscopy (FT-IR) and Raman spectroscopy. FTIR spectra were performed by GBC Cintra 40-Nicolet Nexus 670 FT-IR. Raman spectra were measured by a Laser Raman spectrophotometer (Ramalog 9I, USA). The Van der Pauw method was used to measure the dependencies of dark conductivity on reciprocal temperature, as acquired on pressed pellets. The electromagnetic shielding features of the nanocomposites were performed by HP8720D Network Analyzer (USA).

1.3. Investigation of metal corrosion inhibition [12]
The corrosion inhibition of the PIn/TiO_2 nanocomposite was performed by electrochemical methods using potentiostatic technique. The electrochemical equipment was AUTOLAP 30 composed from a potentiostat in order to make a polarized potential or current. A computer with suitable software (GPES) controls the measurement system and analyses the obtained data.

The solution containing typical eroded ion Cl^- (using HCl 3%) was used. The material for corrosion investigation was a cylinder (horizontal section: 1 cm), made by the construction steel CT3 (TISCO company, produced in Thainguyen, Vietnam) with standard: GOST 001-2001 - TCVN 1656-75 (C: 0.16 %; Mn: 0.62 %; Si: 0.15 %; P: 0.010 % and S: 0.042 %). Before using, it was polished by rough paper 1000, and then cleaned by distilled water and ethanol.

The sample of the PIn/TiO_2 nanocomposite (marked DTC) was prepared as a mixture of 0.1g nanocomposite and 1.0 g polyvinyl alcohol (PVA). Then this mixture was covered to be a thin film on the surface of the steel. Three other samples were prepared for comparison. The first one was made by only 1.0 g PVA (marked D_0). The second one was made by a mixture of 0.1 g TiO_2 and 1.0 g PVA (marked DT). The third one was made by a mixture of 0.1 g PIn and 1.0 g PVA (marked by DP). Then these mixtures were covered on the surface of the steel as in the case of the PIn/TiO_2 nanocomposite.

A three-electrode cell was used. A saturated calomen electrode (SCE) was used as reference electrode, Pt was the counter electrode and steel bar was the working electrode. All chemicals were pure (PA, China).

2. Results and discussion

2.1. Morphology
Figure 1 shows the SEM image of the PIn/TiO_2 nanocomposite particles. The average size of the nanocomposite particles was around 60 nm while the average size of TiO_2 particles was 50 nm. Figure 2 shows the TEM image of the PIn/TiO_2 nanocomposite particles. The black is TiO_2 material and the light is the PIn polymer. Generally, SEM and TEM images show that the PIn/TiO_2 nanocomposite particles presented with core-shell structure.

Figure 1. SEM image of the PIn/TiO_2 nanocomposite particles.

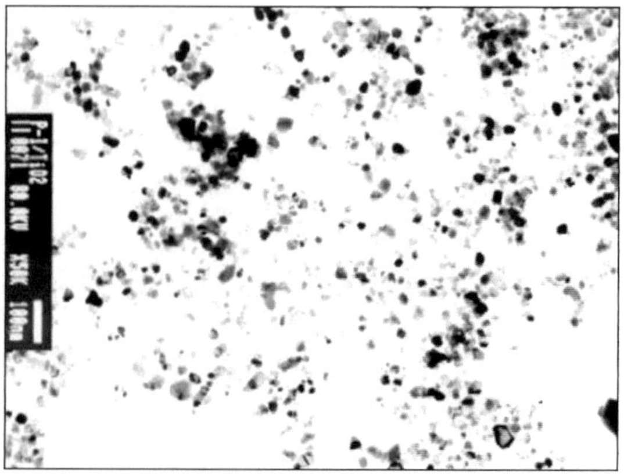

Figure 2. TEM image of the PIn/TiO_2 nanocomposite particles.

2.2. Thermal analysis and conductivity of the PIn/TiO_2 nanocomposite

Thermal analysis of the PIn/TiO_2 nanocomposite was studied. Below 120°C, the weight reduction is originated from water inside samples (1.4%). In the range of 120-330°C, the weight reduction (3.2%) is very small, corresponding to the decomposition of redundant oligomers. At higher temperatures (300-600°C), the change of weight (10%) is attributed to the decomposition of the oxidized PIn.

Conductivity of the PIn/TiO_2 nanocomposite was performed by two-probe methods with the pressed pellet from material powder. It showed that the conductivity of the PIn/TiO_2 nanocomposite is 1.75 $S.cm^{-1}$. Low conductivity could be explained by the presence of TiO_2.

2.3. FTIR and Raman spectra

Chemial structure of PIn nanocomposites was characterized by FTIR and Raman scattering spectroscopy. The FTIR spectrum is of good quality and the infra-red bands are well defined (figure 3). The spectrum associated with the oxidized PIn are characterized by a very large adsorption band located in the spectral domain between 3700 and 3100 cm^{-1}, which is characteristic of the OH groups belonging to residual water molecules trapped in the polymer matrix as well as water molecules absorbed in PIn. The peaks at 3401, 2924, 1593, and 1312 cm^{-1} could be attributed to N-H stretching, C-H (aromatic) stretching, C=C stretching and C-N stretching (between two idole units), respectively.

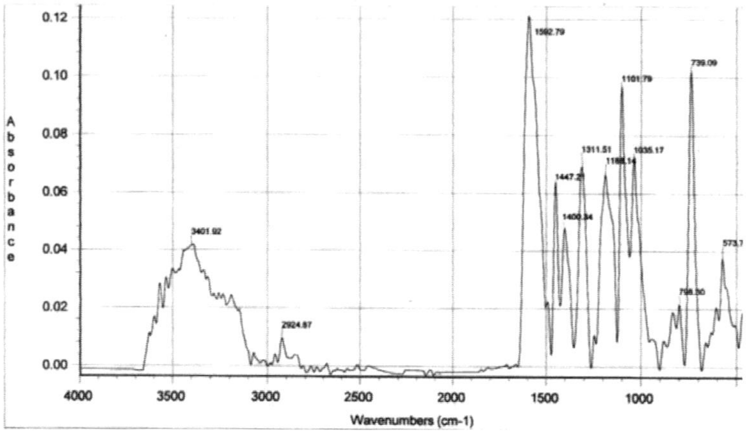

Figure 3. FTIR spectrum of PIn/TiO_2 nanocomposite.

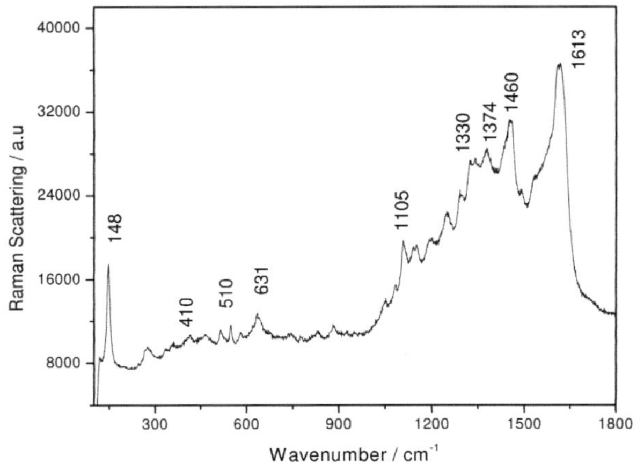

Figure 4. Raman spectrum of PIn/TiO_2 nanocomposite.

Figure 4 presents Raman spectrum of the PIn/TiO_2 nanocomposite measured at 514 nm with 1 mW laser power. The peaks at 148, 410, 510, and 631 cm^{-1} correspond to the characteristic signals of TiO_2. The peaks at 1620, 1372, and 1107 cm^{-1} were attributed to C=C stretching, C-N stretching and C-H in plane deformation.

2.4. Study of the corrosion inhibition

The potentiostatic method was applied to study the corrosion inhibition of the eugenol derivatives. The curves describing the dependent of the open circuit potential on time (E_0 *vs.* t) are showed in figure 5. Value E_0 is a parameter to appreciate the corrosion inhibition of samples. It shows that all samples DT, DP and DTC had higher E_0 than that of D_0. It means that these samples could inhibit steel against corrosion better than PVA. Among these samples, DTC had highest E_0 and, thermodynamically, E_0 presented the best corrosion protection.

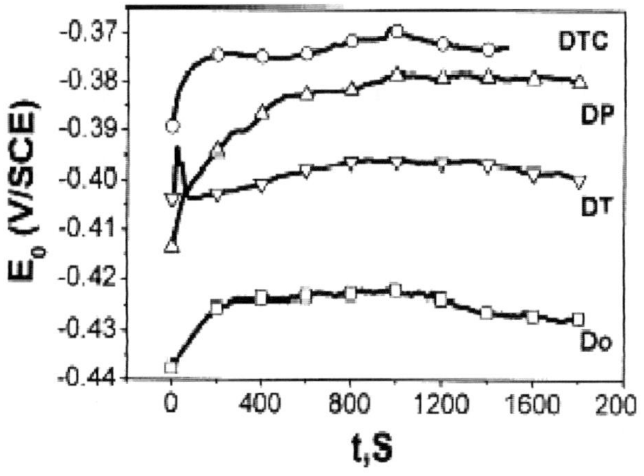

Figure 5. The plots of open circuit potential of the test samples vs. time (in solution of HCl 3%).

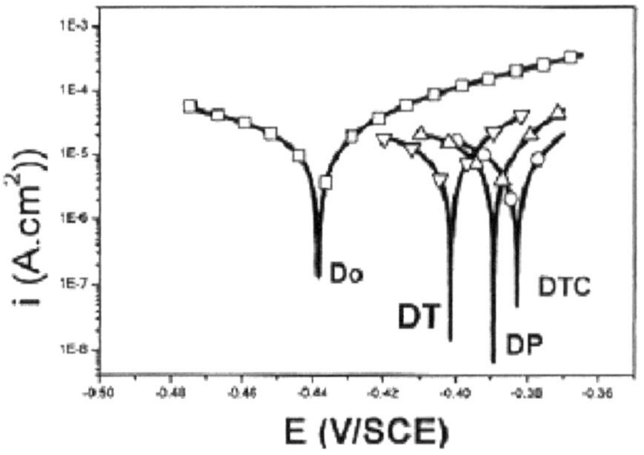

Figure **6.** Tefel curves of the test samples (in solution of HCl 3%).

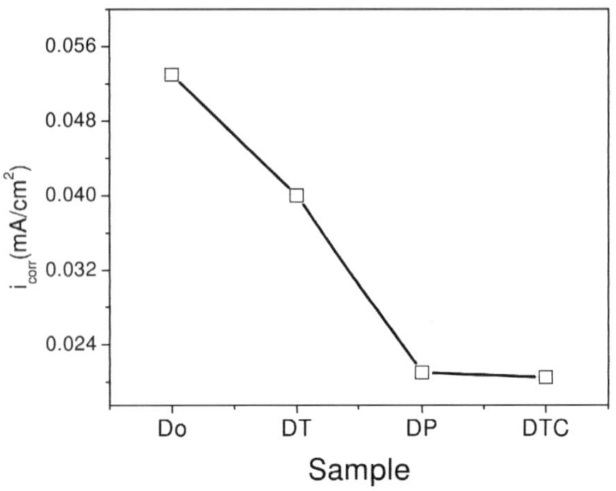

Figure 7. Corrosion current density of the test samples.

In a 3% solution of HCl, the corrosion inhibition of DTC presented remarkably well by transferring positively the Tafel plots (figure 5). Therefore, the corrosion potential was also more positive. From Tefel plots and GPES software, the corrosion current density was determined (figure 6). All samples containing TiO_2, PIn and PIn/TiO_2 nanocomposite have lower corrosion current densities, in which sample DTC performed the lowest current density.

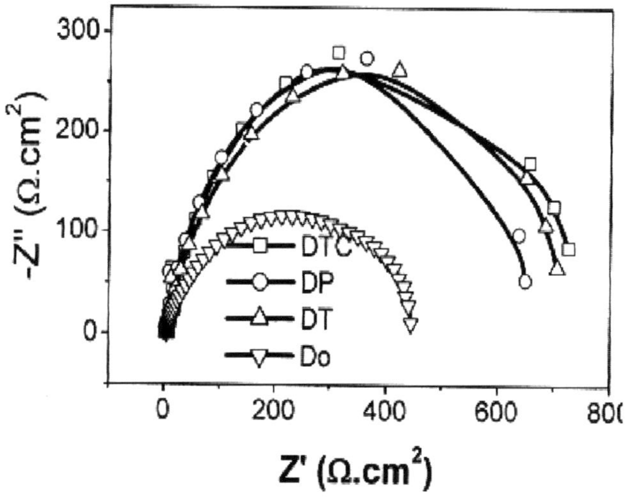

Figure 8. Electrochemical impedance spectroscopy of the test samples (Z" is the imaginary part and Z' is the real part).

Figure 8 shows the electrochemical impedance spectroscopy (EIS) of the test samples. The order of the impedance is DTC >DP > DT >> D_0. By using FRAS software, resistances of the charge transfer (R_{ct}) and of the polarization (R_p) were determined (figure 9). It shows both the resistances of the charge transfer and of the polarization of samples.

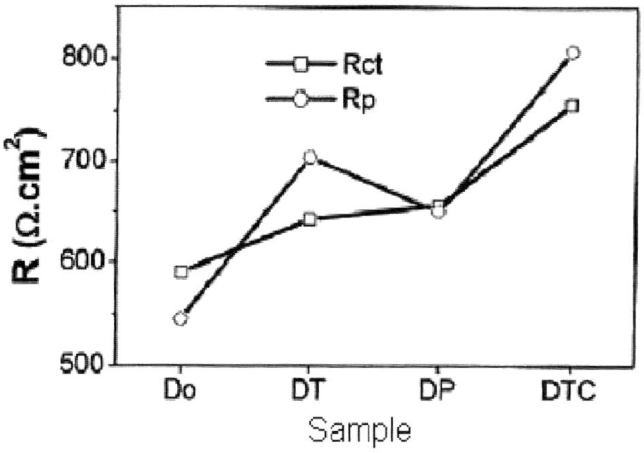

Figure 9. Resistance of charge transfer (Rct) and of the polarization (Rp).

Generally, from the results of EIS measurements were in the agreement with open circuit potential and corrosion current measurements. These results show that the corrosion inhibition of the PIn/TiO_2 was the best.

3. Conclusions

In this paper, the PIn/TiO_2 nanocomposites were prepared by in situ polymerization. SEM and TEM images show that the composites are obtained with nanostructure. The chemical structure of PPy in nanocomposites was characterized by FTIR and Raman spectra. The thermal analysis TGA showed that PPy in the nanocomposites was stable at around 600°C. The corrosion inhibition of the PIn/TiO_2 nanocomposite was investigated by the polarization and electrochemical impedance methods. It showed that the PIn/TiO_2 nanocomposite can protect the mild steel against corrosion. It is the foundation for producing the environment-friendly paints.

References

[1] Tan C K, Blackwood D J 2003 *Corros. Sci.* **45** 545
[2] Herrasti P, Ocon P 2001 *Appl. Surf. Sci.* **172** 276
[3] Ahmad N, Mac Diarmid A G 1996 *Synth. Met.* **78** 103
[4] Sazou D 2002 *Synth. Met.* **130** 45
[5] Dudukcu M, Yazici B, Erbil M 2004 *Mat. Chem. Phys.* **130** 138
[6] Ismail A A, Sanad S H, El-Meligi A A 2000 *J. Mat. Sci. Tech.* **16**(4) 397
[7] Billaud D, Maarouf E B, Hannecart E 1994 *Polym. Commun.* **35** 2010
[8] Talbi H, Humbert B, Billaud D 1997 *Synth. Met.* **84** 875
[9] Billaud D, Maarouf E B, Hannecart E 1995 *Synth. Met.* **69** 571
[10] Abthagir P S, Dhanalakshmi K, Saraswathi R 1998 *Synth. Met.* **93** 1
[11] Q.-T. Vu, Pavlik M, Hebestreit N, Pfleger J, Rammelt U, Plieth W 2005 *Electrochim. Acta.* **51** 1117
[12] *Corrosion Handbook* 2000 (John Wiley and Sons)

DNA enrichment by functionalized magnetic nanoparticles for on-site and fast detection of virus in biomedical application

Mai Anh Tuan[1] and Nguyen Hoang Hai[2]

[1] International Training Institute for Materials Science (ITIMS), Hanoi University of Technology, 1 Dai Co Viet, Hanoi, Vietnam
[2] Center for Materials Science, Faculty of Physics, Hanoi University of Science, 334 Nguyen Trai Road, Hanoi, Vietnam

E-mail: tuanma-itims@mail.hut.edu.vn

Abstract. Label-free electrochemical DNA sensor is a promising technique for simple, fast, on-site virus detection. However, the low sensitivity is still a challenge for this method. This report shows a way to improve the sensitivity by magnetic enrichment of the DNA concentration before measuring DNA concentration by the DNA sensor. The enrichment was performed by conjugating superparamagnetic nanoparticles with the DNA probe single strand. Then, the system of nanoparticles-DNA probe was matched with the target DNA followed by magnetic decantation and compulsorily unfolding the DNA strands at a high temperature. The obtained solution was determined by the electrochemical DNA sensor. The results showed that the DNA enrichment process by magnetic nanoparticles improved the sensitivity of the electrochemical DNA sensor to about 200 times, which can be used for on site virus detection.

Keywords: DNA sensors, DNA amplification, label free detection.

1. Introduction

In recent years, many studies have been devoted to develop DNA sensors due to the simplicity, specificity, exceptional sensitivity and selectivity for the detection of specific genes [1–3]. Biosensors have opened a perspective to overcome most of disadvantages of the conventional detection methods for the applications in food, agriculture, environment, clinics and drugs... [3–12]. With the ability of amplification from small amounts of DNA into readable quantities, the conventional methods such as polymerase chain reaction (PCR) and real-time PCR have been applied mostly in the fields of biomedicine. However, these techniques require time, sample preparation, well-equipped apparatus and well-educated operators [3, 9].

DNA sensor is considered as a promising tool in pre-diagnostics, prevention and control of the infectious diseases which supposed to be able to be used for real-time and on site analysis [2, 3, 9]. High sensitive and selective DNA sensors are of great importance in diagnosis of genetic diseases, detection of infectious agents, and identification in forensic and environmental cases [1–12]. There are various types of DNA sensors which have been developed. Methods used for DNA detection in those sensors have been reported to be based on radiochemical, enzymatic, fluorescent, electrochemical, optical, and acoustic wave techniques [13]. Optical DNA sensors are used predominantly and promising results have been reported. The disadvantage of the optical sensors, however, is the requirement of a separate labeling process, equipment to stimulate the transducer, high complexity,

© 2009 IOP Publishing Ltd

and thus, a higher cost in order to conduct an analysis [14].

Electrochemical method for hybridization detection is an alternative way compared to well-developed fluorescent detection. Over the past decade, a significant progress has been made towards the development of electrochemical DNA sensors. Considerable advantages of these devices are ascribed to their potential for obtaining specific information in a faster, simpler and inexpensive way. In addition, the sensors have a high potential for automation and miniaturization since only basic electrochemical equipment is required [2, 12, 15–17]. However, in practice the concentration of the DNA trace is too small to be detected. For in-lab analysis, samples can be pretreated before the final measurement. The pretreatment must be simple and useful. One of the pretreatments is magnetic enrichment. In this work, we reported the sample enrichment by using magnetic nanoparticles to detect low concentration Herpes virus sample.

Magnetic nanoparticles (MNPs) are the subject of many recent studies, especially in biological applications such as cell separation and sorting, magnetic drug delivery, magnetic resonance imaging [19] due to the fact that MNPs have following advantages: - small particle size of order of nm allow MNPs attach to biological objects without changing their functions; - large surface areas permit strong bond between MNPs and biological entities as well as drug molecules which connect to biological objects through MNPs; - strong magnetic nature of nanoparticles (NPs) allow control system of biological objects/MNP by an external magnetic field, which leads to changes in some physical or chemical properties of the system and the environment in an expected way. In addition, at the scale smaller than a critical length, ferromagnetic behavior disappears and changes to a superparamagnetic (SP) state, which shows strong magnetic properties under an applied field and nonmagnetic behavior when the field is removed.

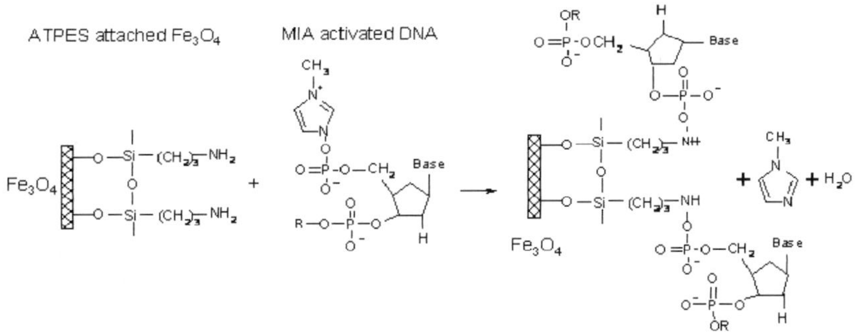

Figure 1. Schematic illustration of the functionalization of magnetic nanoparticles.

The critical length in this case is of the order of a few tens of nm at which the thermal agitation dominates the ferromagnetic order of the particle and, as a result, the particle does not show magnetic properties under zero fields. SP state is very important in many biological applications because it prevents the aggregation in the bloodstream. Magnetite nanoparticles are the most commonly used magnetic materials in the field of biophysics due to its high saturation magnetization M_s (90 Am^2/kg), biological compatibility, environmental stability, and technical simplicity and inexpensiveness of preparation process. Therefore, we used magnetite nanoparticles for study in this work.

2. Experiments

Magnetite nanoparticles have been prepared by the conventional coprecipitation of Fe^{3+} and Fe^{2+} salts by NH_4OH at room temperature. In a typical synthesis, 4.17 g of $FeCl_3.6H_2O$ and 1.52 g of $FeCl_2.4H_2O$ (such that $Fe^{3+}/Fe^{2+} = 2$) were dissolved in 80 ml water (concentration of Fe^{2+} was 0.1 M) with vigorous stirring. A solution of 6 ml NH_4OH 35% was added with the rate of 1 drop per second at room temperature during constant stirring. Black precipitates of magnetite were formed and isolated from the solvent by magnetic decantation. Water washing and decantation processes were repeated four times to remove excess solution. Procedures are conducted under N_2 atmosphere. Nanoparticles with a diameter of about 15 nm have been obtained for further treatments.

The surface functionalization of MNPs was based on the reaction between amino group of the 3-aminopropyl triethoxysilane (APTES) and the phosphate group of DNA sequence. The functionalization procedure of MNPs is the following. Nanoparticles were mixed with a solution containing methanol and chloric acid (vol. ratio: 1/1) for 30 min. Then MNPs were immersed with another solution consisting of 30 vol. % of APTES and 70 vol. % of ethanol for one hour to obtain APTES attached to the Fe_3O_4 MNPs (APTES-Fe_3O_4). The APTES-Fe_3O_4 MNPs were reacted with 1-Methyllmidazole (MIA) activated DNA (MIA-DNA). The MIA-DNA was a product of the reaction between 5'-phosphorylated DNA and [3-(Dimethylamino) propyl]ethylcarbodiimide (EDC). The role of MIA was to stabilize the activated EDC molecules which were labile in aqueous solution. The oxygen atom in phosphate group of the DNA (5' terminal) in MIA-DNA was used as interface media to bind DNA probe with amino group of APTES (figure 1, left side). As a result, the surface of the Fe_3O_4 MNPs was coated with the DNA probe (DNA-Fe_3O_4). Schematic illustration of amino modified magnetite nanoparticles is presented in figure 1. The DNA-Fe_3O_4 was heated in de-ionized water at 37°C for 18 hours. DNA probe, that was the Herpes virus in this work, with a specific sequence for HSV-1 of 5'-AT CAC CGA CCC GGA GAG GGA C-3' and complementary DNA target sequence for HSV-1 of 3'-TA GTG GCT GGG CCT CTC CCT G-5' was supplied by Invitrogen Life Technologies Company.

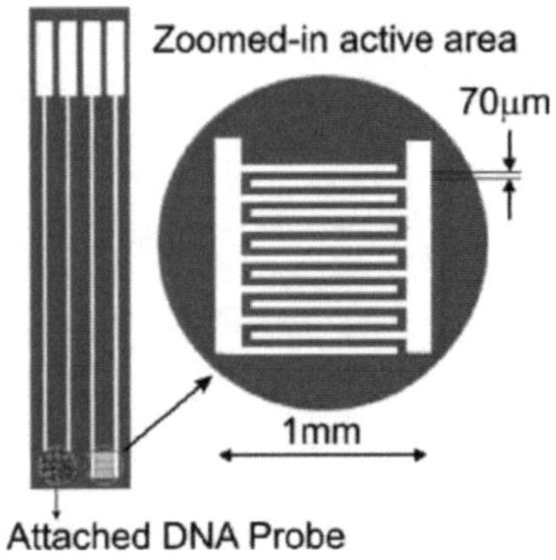

Figure 2. Electrochemical DNA sensor to determine DNA concentration by measuring the resistance change due to the hybridization of probe target DNA strands. The resistance is proportional to the concentration of DNA. Sensivity of the sensor is limited by 10 nM/L.

As described in the previous work [21], the DNA concentration in real sample was very small. With the current sensor configuration, real concentration is sometimes out of detection range. This is why the samples need to be enriched before the analysis. After that, the probe DNA attached to MNPs hybridized with the target DNA by stirring. In a typical experiment, 1 ml of probe DNA attached to the Fe_3O_4 nanoparticles was mixed with 1 ml HSV1-DNA sequence for 15 minutes at room temperature. For faster hybridization process, the reaction cell was stabilized at 37°C. Magnetic decantation was applied to obtain the condensed MNP/DNA system. The condensed sample was heated up to 98°C for DNA unfolding in 15 minutes and rapidly cooled down to room temperature. The probe DNA attached to the Fe_3O_4 nanoparticles were removed from the sample by applying a magnetic field from outside. The enriched sample was finally analyzed by electrochemical DNA sensor.

The electrochemical sensor to measure concentration of DNA consisted of a pair of microelectrodes placed on the surface of a silicon substrate. One microelectrode served as the working electrode, another as the reference. The dimensions of the electrode were 30×70 μm^2 (figure 2). The signal measurement relied on the hybridizing of two DNA sequences which changes electric density on the surface of the DNA sensor. The reference signal of an AC current with frequency of 10 KHz and amplitude of 100 mV was given by a Lock-in Amplifier (SR830). The output signal was acquired by measuring the voltage change between two 1 KΩ resistances. The change of the voltage resulted from the hybridization possibility of the probe DNA attached on the surface and the target DNA strands between the two electrode. If there was no hybridization, there would be no change of the output voltage.

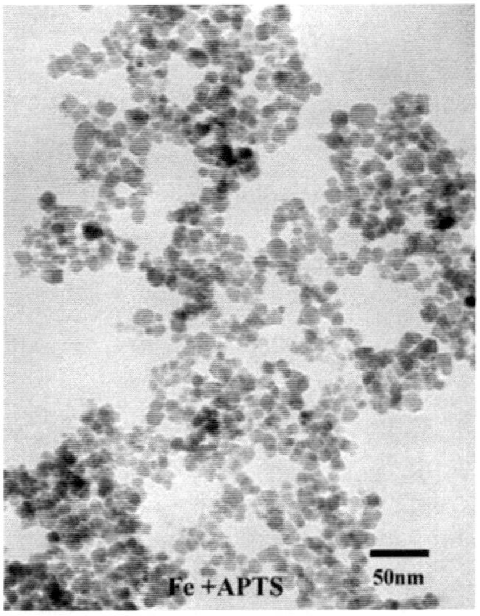

Figure 3. TEM image of image of the ATPES coated Fe_3O_4 nanoparticles.

The possibility of hybridization depended on the concentration of the target DNA. Therefore, the output signal was a function of the concentration of the target DNA strands. Our study revealed a linnear dependence of the output signal and the concentration in the range from 10 nM/l to 10 μM/l. We used the DNA sensor to determine the concentration of DNA of Herpes virus before and after the DNA magnetic enrichment. Some available tools allow calculating the molecular weight of DNA sequence just simply by the number of bases of the DNA. In our case, the molecular weight of 21 base HSV1 sequence is 6.4 K.Daltons. Limit of the sensivity of the sensor is 10 nM/l. For samples with concentration lower than the limit, other techniques must be applied. To overcome this difficulty, we combined magnetic enrichment with the electrochemical sensor. A certain volume of solution of 0.1 nM/l of the Herpes DNA was magnetically concentrated from 20 - 200 times using MNPs and an external magnetic field of a strong commercial magnet. After magnetic decantation, the concentration of the DNA increased to the value higher than the sensitivity of the sensor. The concentration of the concentrated solution obtained from measurement by electrochemical sensor was compared to the value obtained from the reduction of the solution volume before and after enrichment process.
Transmission electron microscope measurements were carried out using a Transmission Electron Microscope (TEM) JEM-1010 working at an accelerating voltage of 100 kV. Samples for TEM were prepared by ultrasonically dispersing the products into absolute ethanol, placing a drop of this suspension onto a copper grid coated with an amorphous carbon film or onto a copper plate, respectively, and then drying in vacuum. Fourier Transformed Infrared (FTIR) Spectra were recorded in the transmission mode on a Nicolet Impact 410 spectrometer. Magnetic properties were measured by Vibrating Sample Magnetometer DMS 880-CTS.

3. Results and discussion

The transmission electron microscope (TEM) image of the ATPES coated Fe_3O_4 nanoparticles was given in figure 3. Particles were spherical with diameters from 13 - 18 nm. The particles were separated due to a layer of ATPES coating on the surface of the particles. As illustrated in figure 1, APTES was bound to the surface of nanoparticles leaving the NH_2 group outwards which can conjugate with phosphate groups of DNA probe. Magnetic measurement showed a SP behavior with the saturation magnetization of 65 emu/g.

Fourier Transformed Infrared Spectrum of the DNA-APTS complexes (figure 4) revealed absorption bands at 2972 and 2925 cm^{-1} of the stretching vibration of the C-H bond, 1091 cm^{-1} of the stretching vibration of the C-N bond, 1051 cm^{-1} of the stretching vibration of the Si-O bond, 885 cm^{-1} of the bending vibration of the NH_2 group, 1750-1600 cm^{-1} of the vibration plane of the G-C pairs and A-T base pairs, 1085 cm^{-1} of the backbone phosphate.

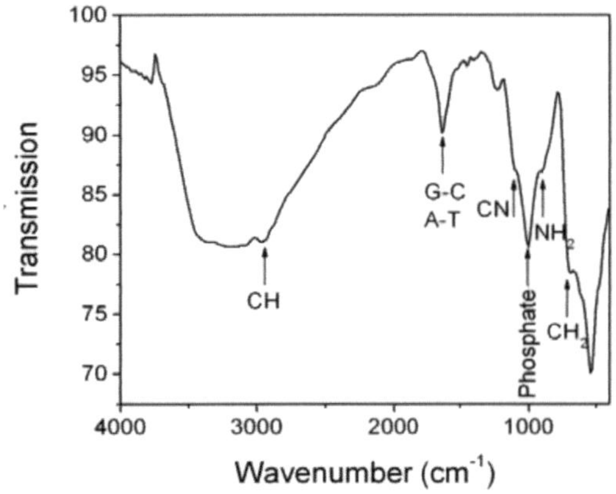

Figure 4. FTIR spectra of the probe DNA attached to the nanoparticles.

AC reference signal (10 KHz, 100 mV sine wave) generated by the Lock-in Amplifier SR830 was applied on the two microelectrodes of the DNA sensor. The output signal was acquired by measuring the voltage change between the two 1 K resistances. The probe DNA with concentration of 100 mM/l was immobilized on APTES coated electrodes by a way which was similar to the way the probe DNA attached to the MNPs. All measurements were performed at room temperature. In this experiment, five DNA sensors were used to test the hybridization of DNA sequences.

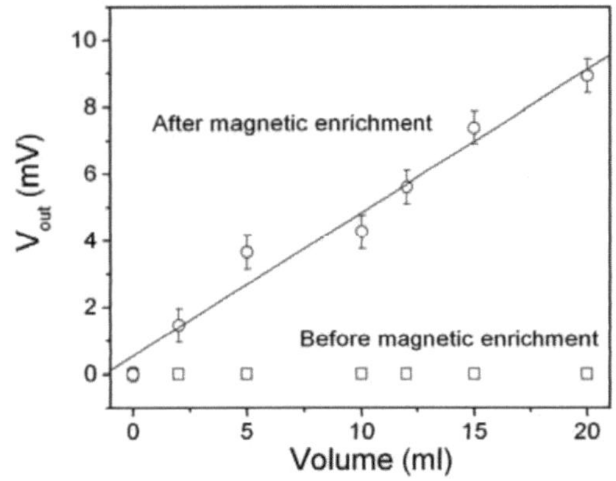

Figure 5. Dependence of the output signal on the volume of the solution containing 0.1 nM/l of the Herpes DNA before the magnetic enrichment.

Figure 4 presents the dependence of the output signal on the volume of the solution containing 0.1 nM/l of the Herpes DNA before the magnetic enrichment. The initial solution contained 0.1 nM/l of the DNA which was much smaller than the sensitivity of the sensor. Therefore, the measurement of the solution before magnetic enrichment was almost zero (figure 5, opened square). After magnetic enrichment, depending on the initial volume of the solution, the output signals linearly increased with the increasing of the solution volume: the higher the volume, the higher the concentration. As a result, the higher output signals were obtained. This means that the concentration of HSV-1 DNA was much condensed after the enrichment. The linearity of the output signals is very important for the calibration. The UV-VIS spectrometer (Ultrospec 2000) was also used in this work to determine the concentration of DNA target after the enrichment. The concentration obtained by comparing the initial volume and the final volume was almost consistent with the concentration obtained from the electrochemical sensor. With the highest initial volume that was used in our studies, the concentration after magnetic enrichment was 200 times higher than the initial concentration. With the magnetic enrichment process, we can measure solution with the low concentration of 0.1 nM/l, which is 10 times smaller than the limit of the electrochemical sensor.

4. Conclusion

This work reported the application of Fe_3O_4 nanoparticles for DNA enrichment for onsite and fast label-free electrochemical DNA sensor. The magnetic enrichment was simple which can be applied easily in under equipped laboratories. It is promising for on-site analysis of viruses in remote areas.

References

[1] Wang J 2002 *Anal. Chim. Acta* **469** 63
[2] Drummond G, Hill M G and Barton J K 2003 *Nature Biotech.* **21** 1192
[3] Malhotra B D, Singhal R, Chaubey A, Sharma S K and Kumar A 2005 *Current Appl. Phys.* **5** 92
[4] Lermo A, Campoy S, Barbe J, Hernndez S, Alegret S and Pividori M I 2007 *Biosensors and Bioelectronics* **22** 2010
[5] Bystricka D, Lenz O, Mraz I, Piherova L, Kmoch S and Sip M 2005 *J. Virological Methods* **128** 176
[6] Rodriguez-Mozaz S, Lopez de Alda M J, Marco M P and Barcelo D 2005 *Talanta* **65** 291
[7] Gooding J J 2006 *Anal. Chim. Acta.* **559** 137
[8] Lazckaa O, Campo F J D and Munoz F X 2007 *Biosensors and Bioelectronics* **22** 1205
[9] Riccardi C S, Dahmouche K, Santilli C V, Inacio da Costa P and Yamanaka H 2006 *Talanta* **70** 637
[10] Ye Y K, Zhao J H, Yan F, Zhu Y L and Ju H X 2003 *Biosensors and Bioelectronics* **18** 1501
[11] Gonzalez R, Masquelier B, Fleury H, Lacroix B, Troesch A, Vernet G and Telles J N 2004 *J. Clin. Microbio.* **42** 2907
[12] Kara P, Meric B, Zeytinoglu Aysin and Ozsoz M 2004 *Anal. Chim. Acta.* **518** 69
[13] Pearson J E, Gill A and Vadgama P 2000 *Ann. Clin. Biochem.* **37** 119
[14] Gabl R, Feucht H D, Zeininger H, Eckstein G, Schreiter M, Primig R, Pitzer D and Wersing W 2004 *Biosensors and Bioelectronics* **19** 615
[15] Fan C, Plaxco K W and Heeger A J 2003 *Proc. Natl. Acad. Sci. U.S.A.* **100** 9134
[16] Wang J, Liu G and Merkoci A 2003 *J. Am. Chem. Soc.* **125** 3214
[17] Wang J, Polsky R, Merkoci A and Turner K 2003 *Langmuir* **19** 989
[18] Costello Michael, Sabatini L and Yungbluth P 2006 *Clinical Microbiology Newsletter* **28** 185
[19] Leslie-Pelecky D L, Labhasetwar V and Kraus J 2005 *Nanobiomagnetics in Advanced Magnetic Nanostructures* (New York: Kluwer)
[20] Thach C V, Hai N H and Chau N 2008 *J. Korean Phys. Soc.* **5** 1332
[21] Tuan M A et al. 2007 *Proc. 2nd Int. Conf. Biomed. Engineer (Hanoi, Vietnam)* p 273
[22] Yamaura M, Camilo R L, Sampaio L C, Macedo M A, Nakamura M and Toma H E 2004 *J. Magn. Magn. Mater.* **279** 210

Acknowledgments

The authors would like to thank the Asia Research Center, Vietnam National University, Hanoi (DT 34/2007/HD-DT) and the Ministry of Science and Technology (Protocol for Scientific Collaboration with Israel) for finance support.

Modelling transport in single electron transistor

Dinh Sy Hien, Huynh Lam Thu Thao and Le Hoang Minh

HCM University of Natural Sciences, 227 Nguyen Van Cu Street, 5 District, Ho Chi Minh City, Vietnam

E-mail: huynhlamthao1980@yahoo.com

Abstract. We introduce a model of single electron transistor (SET). Simulation programme of SET is used as the exploratory tool in order to gain better understanding of process and device physics. This simulator includes a graphic user interface (GUI) in Matlab. The SET was simulated using GUI in Matlab to get current-voltage (I-V) characteristics. In addition, effects of device capacitance, bias, temperature on the I-V characteristics were obtained. In this work, we review the capabilities of the simulator of the SET. Typical simulations of the obtained I-V characteristics of the SET are presented.

Keywords: Coulomb blockable oscillation, single electron transistor, graphic user interface, I-V characteristics.

1. Introduction

Incredible progress in microelectronics has pushed the MOSFET dimension toward the nanoscale limits (10 nm). In the future it is probable that MOSFETs will be replaced by fundamental new devices, such as single electron transistors (SET). SETs have recently attracted much attention because of their nano feature size, ultralow power dissipation, new functionalities and CMOS compatible fabrication process. SET shows unique advantages in terms of low power consumption and of new characteristics related to its Coulomb blockage oscillations.

Until now, only two compact analytical models (Uchida *et al*, for single gate resistive symmetric SET [1] and Mahapatra *et al*, for asymmetric SET [2]) have been reported. We propose an improved model, which is more flexible and can be adapted for symmetrical and asymmetrical device geometries.

The aim of this work is to propose a model of SET, to build a simulator, and to simulate current-voltage characteristics of SET using graphic user interface (GUI) in Matlab.

In this work, we also review the capabilities of the simulator, summarize the theoretical approach and experimental results, and give examples of typical simulations of SET's current-voltage characteristics.

2. Simulation results

A model of SET is shown in figure 1a. A SET is made from two tunnel junctions that share a common electrode. A tunnel junction consists of two pieces of metal supported by a very thin (about 1 nm) insulator. The only way for electrons is one of the metal electrodes to travel to the other electrode to tunnel through the insulator. Since tunnelling is a discrete process, the electric charge flows through the tunnel junction in multiples of e, the charge of electrons.

© 2009 IOP Publishing Ltd

A quantum dot (QD) is usually formed in two dimensional electron gas (2DEG) in GaAs/AlGaAs using standard electron beam lithography. The QD is connected to the source and drain electrodes through tunnel barriers. The potential in the dot can be controlled by the gate electrode which is capacitive coupled to the QD (figure 1b). The current through the dot can be periodically modulated by the gate voltage (Coulomb oscillations). When the current is zero (Coulomb blockade), the number of electrons is fixed.

A proper operation of a SET device requires: 1) the tunnel junction resistances (drain resistance, R_D and source resistance, R_S) to be greater than the quantum resistance (25.8 kΩ) to confine the electrons in the QD, 2) the charging energy of the QD capacitance to be larger than the available thermal energy to avoid electron tunnelling due to the thermal emission, and the total capacitance of QD is equal to the sum of all device capacitances, i.e. $C_T = C_G + C_D + C_S$.

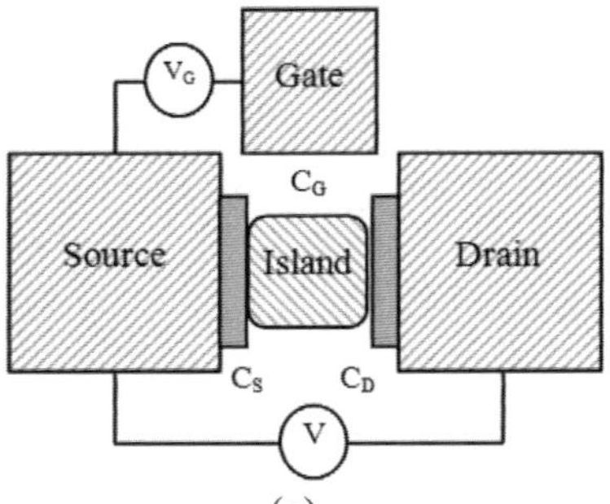

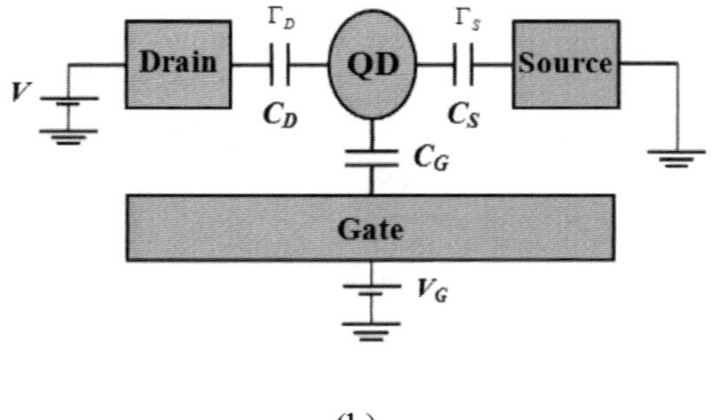

Figure 1. a) Structure of single electron transistor, b) equivalent diagram of SET.

The QD is connected to the source and drain electrodes through small tunnel barriers. The potential in the QD can be modified by the gate electrode which is capacitive coupled to the QD. The DC bias (V_{DS}) is applied and the current is measured as a function of V_{DS} and V_{GS}. The SET parameters are: C_S, C_D, C_G, Γ_S, Γ_D.

The energy diagram of the SET is presented in figure 2 and it shows quantization in QD.

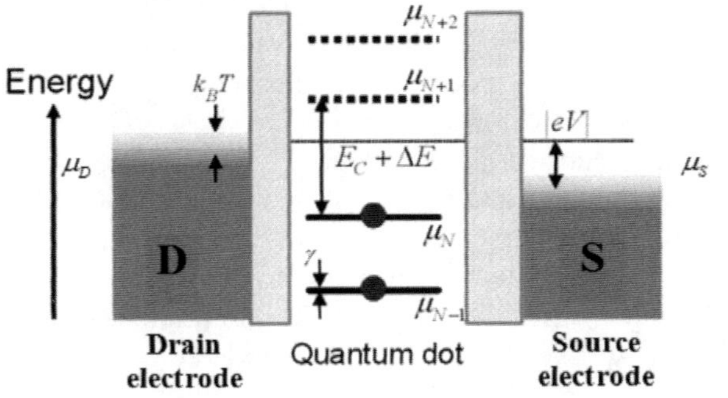

Figure 2. Energy level diagram for SET and quantization in QD.

We have developed the model by solving the state equation for single electron tunnelling [3]. The drain current can be calculated by using the tunnelling rate of an electron through the tunnelling junction. The drain current (I) of single electron transistor is obtained from

$$I = |e|[P_0\Gamma_D f_D - P_1\Gamma_D(1-f_D)] = |e|\left(\frac{\Gamma_S\Gamma_D}{\Gamma_S+\Gamma_D}\right)(f_D - f_S) = |e|\Gamma(f_D - f_S) \ , \tag{1}$$

where

$$P_1 = 1 - P_0, \ P_0 = \frac{\Gamma_S(1-f_S)+\Gamma_D(1-f_D)}{\Gamma_S+\Gamma_D} \ , \tag{2}$$

$$f_D = \frac{1}{1+e^{\frac{\mu_{N+1}-\mu_D}{k_B T}}} = \frac{1}{1+e^{\frac{\mu_{N+1}+|e|V_D}{k_B T}}}\ , \tag{3}$$

$$f_S = \frac{1}{1+e^{\frac{\mu_{N+1}-\mu_S}{k_B T}}} = \frac{1}{1+e^{\frac{\mu_{N+1}}{k_B T}}}\ , \tag{4}$$

and

$$\mu_{N+1} = E_0 - |e|\frac{C_G V_G + C_D V_D}{C_T} = -|e|\frac{C_G(V_G - V_C) + C_D V_D}{C_T}\ . \tag{5}$$

Here, Γ_S and Γ_D are the electron tunnelling rate from source to island and from island to drain, respectively. The number of electrons in the island could be 0 or 1, P_n is the probability of finding n electrons in the island. f_S, f_D are probable distribution functions in source and drain, respectively. μ is Fermi energy (electrochemical potential).

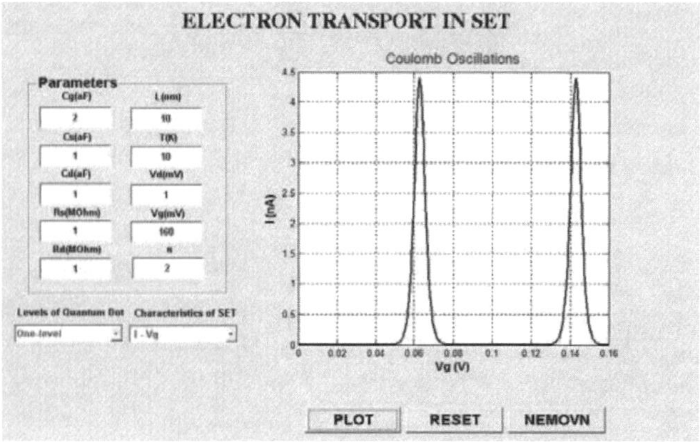

Figure 3. Typical I_{DS}-V_{GS} characteristics simulated by the simulator with value of V_{DS} = 1 mV. The symmetrical SET device parameters are: C_G = 2 aF, C_S = C_D= 1 aF, R_D = R_S = 1MΩ.

The calculations of drain current (I) are awkward. Therefore, the characteristics of SET devices are usually calculated with numerical simulators. By utilizing the model, the I_{DS} – V_{GS} characteristics of SET having the parameters of C_S = C_D = 1 aF, C_G = 2 aF, and R_D = R_S = 1 MΩ are calculated as shown in figure 3. The results numerically calculated using the simulator are also illustrated. There are two factors undermining characteristics of SET: temperature and the drain voltage. Figure 3 demonstrates the typical Coulomb blockade oscillation behaviour in symmetrical SET I_{DS}-V_{GS} characteristics (R_D = R_S = 1 MΩ). It shows that the SET Coulomb blockade oscillation period (e/C_G, e is the electronic charge) is dictated by SET gate capacitance. Here, it should be emphasized that the peak and the valley currents of Coulomb oscillations are perfectly represented by the model despite its simplicity.

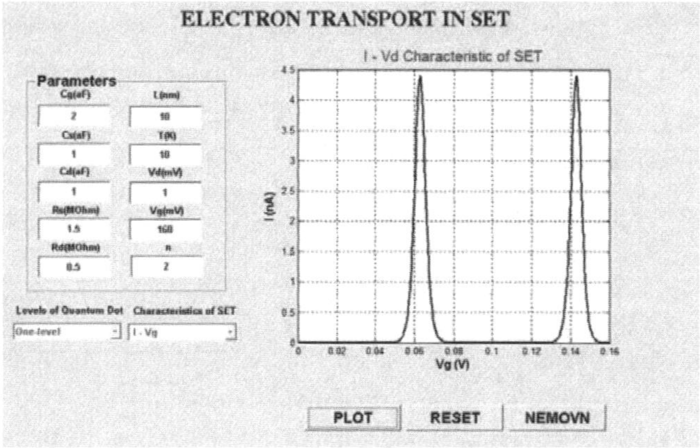

Figure 4. Typical I_{DS}-V_{GS} characteristics simulated by the simulator with value of V_{DS} = 1 mV. The asymmetrical SET device parameters are: C_G = 2 aF, C_S = C_D= 1 aF, R_D = 0.5 MΩ, R_S = 1.5 MΩ.

Figure 4 demonstrates the typical Coulomb blockade oscillation behaviour in asymmetrical SET I_{DS}-V_{GS} characteristics ($R_D = 0.5$ MΩ, $R_S = 1.5$ MΩ) with the same voltage, V_{DS}.

Figure 5 represents I_{DS}-V_{GS} characteristics with value of $V_{DS} = 1$ mV at different temperatures. One can note that the effects of temperature on Coulomb oscillations. The Coulomb oscillations of SET are clear at low temperature (10K). Current-voltage (I_{DS}-V_{GS}) characteristics showing the suppression of the Coulomb blockade by broadening at higher temperature (100 K). It also reveals the fact that it is not possible to obtain the Coulomb blockade in the device characteristics at high temperature.

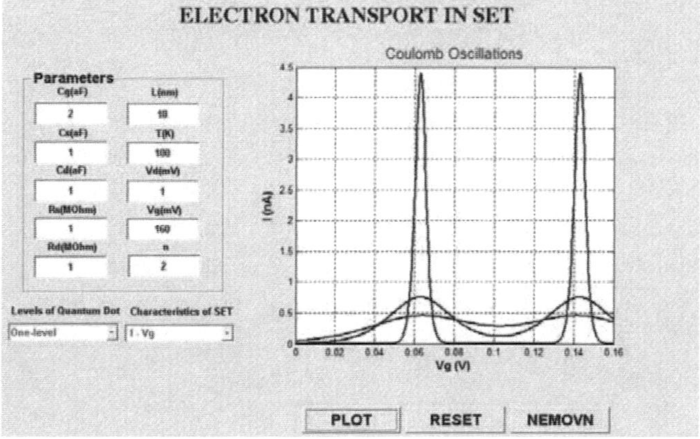

Figure 5. Typical I_{DS}-V_{GS} characteristics with value of $V_{DS} = 1$ mV at different temperatures. Temperature range is from 10 K (top curve) to 100K (bottom curve) and the step is 50K. The symmetrical SET device parameters are: $C_G = 2$ aF, $C_S = C_D = 1$ aF, $R_D = R_S = 1$ MΩ.

Figure 6 demonstrates I_{DS}-V_{DS} characteristics of symmetrical SET at low temperature (10 K).

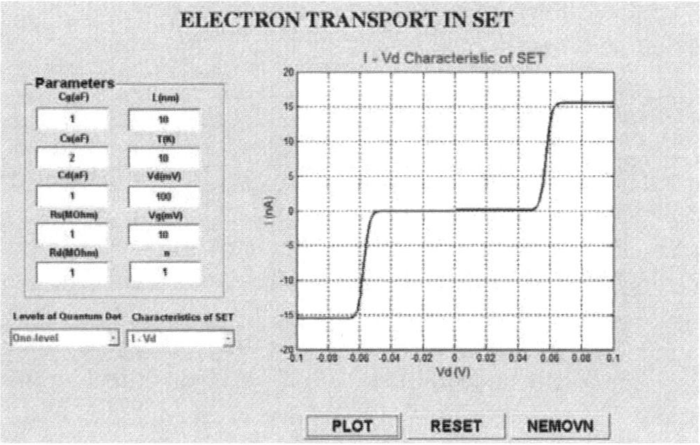

Figure 6. Typical I_{DS}-V_{DS} characteristics of SET. The SET device parameters are: $C_S = 2$ aF, $C_G = C_D = 1$ aF, $R_D = R_S = 1$ MΩ.

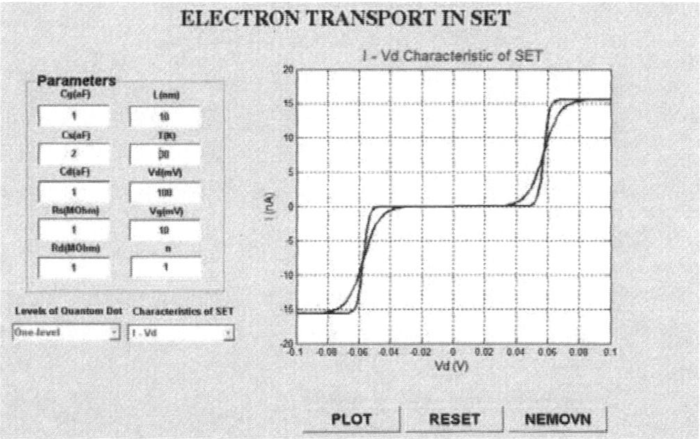

Figure 7. The effect of temperature (T) on I_{DS}-V_{DS} characteristics. The Coulomb blockade region at low temperature (10 K) becomes thinner at higher temperature (100 K).

The effects of temperature on I_{DS}-V_{DS} characteristics of SET are demonstrated in figure 7 and it shows that the Coulomb blockade region becomes thinner at higher temperature. Therefore, our accurate analytical model can be able to capture both the effect of temperature and the effect of high V_{DS} on the device characteristics.

Figure 8 demonstrates conductance of SET device and it shows that the simulator can be able to compute another important characteristic of SET (I_D/V_{DS}).

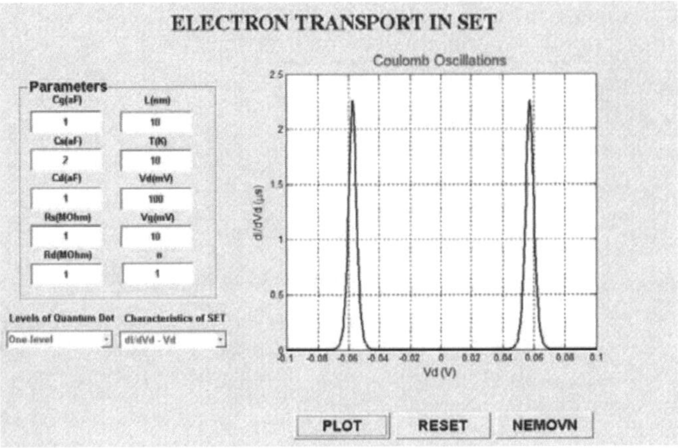

Figure 8. Conductance of the SET device.

3. Conclusions

A model for SET devices has been reported. The proposed model has been verified at device level for both symmetrical and asymmetrical devices. A set of simulations is then successfully performed for different parameters of SET devices. The model is able to accurately describe the I_{DS}-V_{DS} and I_{DS}-V_{GS} SET characteristics at different temperature.

References

[1] Uchida K, Matsuzawa R, Koga J, Ohba R, Takagi S and Toriumi A 2000 *Jpn. J. Appl. Phys.* **3** (4B) 2321

[2] Mahapatra S, Ionescu A M and Banerjee K 2002 *IEEE Electron Device Lett* **23** 366

[3] Wasshuber C 2002 *Computational Electronics* (New York: Springer-Verlag)

Acknowledgments

The authors are grateful to professor Dang Luong Mo of Hosei University, Japan for helpful discussions. This work was supported in part by Vietnam National University Ho Chi Minh City, under research contract No. 2005-18-06.

3D Simulation of coaxial carbon nanotube field effect transistor

Dinh Sy Hien, Nguyen Thi Luong, Thi Tran Anh Tuan and Dinh Viet Nga

HCM University of Natural Sciences, 227 Nguyen Van Cu Street, 5 District, Ho Chi Minh City, Vietnam

E-mail: ngthiluong@yahoo.com

Abstract. We provide a model of coaxial CNTFET geometry. Coaxial devices are of special interest because their geometry allows for better electrostatics. We explore the possibilities of using non-equilibrium Green's function method to get I-V characteristics for CNTFETs. This simulator also includes a graphic user interface (GUI) of Matlab. We review the capabilities of the simulator, and give examples of typical CNTFET's 3D simulations (current-voltage characteristics are a function of parameters such as the length of CNTFET, gate thickness and temperature). The obtained I-V characteristics of the CNTFET are also presented by analytical equations.

Keywords: CNT, carbon nanotube, CNTFET, NEGF, I-V characteristics.

1. Introduction

The carbon nanotube field-effect transistor (CNTFET) is a promising candidate for future electron devices. Rapid progress in the field has recently made it possible to fabricate CNTFET based circuits, such as logic gates, static memory cells, and ring oscillators.

This three-terminal device consists of a semiconducting nanotube bridging two contacts (source and drain) and acting as a carrier channel, which is turned on or off electrostatically via the third contact (gate). Presently, there are various groups pursuing the fabrication of such devices in several variations, achieving increasing success in pushing performance limits, while encountering myriad problems, as expected for any technology in its infancy.

Manufacturing issues will ultimately play a decisive role in any future CNT electronic technology. It is still too early to tell what role CNTFETs will play in electronic systems of the future, but they provide us with a specific context in which to develop technology and understand transport, contacts, interfaces, etc., which are likely to be important for CNT electronics in general. It is appropriate, however, to say a few words about where CNTFET technology stands today. Early CNTFETs were fabricated using nanotubes synthesized by a laser ablation process using nickel-cobalt catalysts. The nanotubes were then suspended in a solvent and dispersed on an oxidized silicon wafer with predefined metal contact pads. The result was a random distribution of CNTs some of which bridged the contacts. Subsequently, catalytic chemical vapor deposition (CVD) methods were developed to grow CNTs on predefined catalyst islands [1]. The nanotubes thus fabricated are rooted in the catalyst islands and grow in random directions on the wafer. Some CNTs terminate on another island and create the bridges. CVD techniques provide more control over device fabrication and have led to rapid progress in device performance.

© 2009 IOP Publishing Ltd

CNTFETs are typically p-type devices that operate as so-called Schottky barrier (SB) transistors. The p-type characteristics have been attributed to the alignment of source-drain metal Fermi level near the valance band of the CNTs rather than to background doping or charges. The holes in the channel are electrostatically induced by applying a negative gate voltage. Transistor action occurs because the gate modulates the SB width for hole tunnelling near metal-CNT contact. While early transistors used gold (Au) as contact metals, significant performance improvements were obtained by using palladium contacts instead, which seem to eliminate the Schottky barrier for holes and produce near-ballistic operation [2].

Early CNTFETs were fabricated on oxidized silicon substrates with a back-gated geometry and a thick SiO_2 layer that resulted in poor gate control of drain current.

The use of a top-gated geometry produced immediate performance improvements [3]. Wind *et al.* deposited a thin dielectric layer (15-20 nm) on top of CNTs, and lithographically defined metal electrodes for gating and contacts. A transconductance ($g_m = dI_{ds}/dV_{gs}|_{Vds}$) of 3.25 µS and subthreshold swing ($S = \ln(10)\,[dV_{gs}/d(\ln(I_{ds}))]$) of 130 mV/decade were obtained, which was a significant improvement in device performance. Later, the incorporation of high-κ dielectrics in a top-gated structure produced even better device characteristics. Javey *et al.* employed a high-κ ZrO_2 (κ ~ 25) gate dielectric with a thickness of ~ 8 nm and obtained $g_m \approx 12$ µS and $S \approx 70$ mV/decade [4]. Although not always with top-gated geometry, other groups have also reported the use of high-κ dielectrics such as HfO_2 (κ ~ 11), TiO_2 (κ ~ 40-90), $SrTiO_3$ (κ ~ 175), and even electrolyte gating mechanisms for attaining improved performances. In the case of electrolyte gating, Siddon *et al.* have reported $S \approx 62$ mV/decade which is very close to the theoretical limit of 60 mV/decade. All of these devices appear to operate as CNT MSDFETs, some with essentially no barrier and others with rather large (half band gap) barriers.

Several recent improvements to CNTFET design collectively incorporate various techniques that have been developed during the past few years. For instance, Javey *et al.* reported a self-aligned top gate structure that uses the catalytic CVD method for CNT growth, a thin. HfO_2 top gate dielectric ~50 nm in length and self-aligned palladium source/drain contacts. A transconductance of 30 µS, subthreshold swing of 110 mV/decade, and a saturation current of ~25 µA at a power supply of V_{DS}~1 V were obtained. Novel CNTFET device structures that enable high current operation and high integration densities have also been reported. These devices are all of the metal source-drain (MSDFET) variety, but it is recognized that the use of a metal source will limit the drain current (unless the SB is sufficiently negative). Very recently, CNT MOSFETs with doped source/drain regions have been reported [5, 6]. Substitutional doping is not practical because of the strong carbon-carbon bond, so charge transfer approaches analogous to modulation doping in III-V heterostructures are used.

Although progress in CNTFETs has been rapid, there are still many issues to discuss. The potential for digital logic was demonstrated early on. Techniques that modify the behaviour of the nanotube from p-type to n-type have been implemented, which allowed their use in complementary CMOS logic. Following this, Derycke *et al.* demonstrated an inverter structure based on nanotubes. Other nanotube based elementary digital logic gates with high gain and high I_{on}/I_{off} ratios, such as a NOR gate, a ring oscillator and an SRAM cell, have also been implemented [7, 8].

In this work, we start with a brief summary of progress in CNTFET technology since the first reports in 1998. This discussion is intended to provide background for whom not familiar with CNTFET technology. As our understanding of CNTFET device physics has evolved, an ability to model and simulate them has also been developed. We also briefly describe the device simulation approach that we use. We provide a model of coaxial CNTFET geometry. Coaxial devices are of special interest because their geometry allows for better electrostatics than their planar counterparts. Capitalizing on the inherent cylindrical shape of nanotubes, these devices would exhibit wrap-around gates that maximize capacitive coupling between the gate electrode and the nanotube channel.

In this work, we also review the capabilities of the simulator, summarize the theoretical approach and experimental results, and give examples of typical CNTFET's 3D simulations (current-voltage characteristics are a function of parameters such as the length of CNTFET, the gate thickness and temperature). The obtained I-V characteristics of the CNTFET are also presented by analytical equations.

2. Simulation results

2.1 Coaxial CNTFET geometry

We now provide a brief description of typical CNTFET geometries, which are grouped in two major categories, planar and coaxial. A CNTFET, whether planar or coaxial, relies on similar principles, while being governed by additional phenomena such as 1D density of states (DOS) and ballistic transport. The coaxial geometry maximizes the capacitive coupling between the gate electrode and the nanotube surface, thereby inducing more channel charge at a given bias than other geometries. This improved coupling is desirable in mitigating the short-channel effects that plague technologies like CMOS as they downsize device features. It is also of importance to low-voltage applications a dominating trend in the semiconductor industry and to allow, potentially, for easier integration with modern implementations of existing technologies such as CMOS. Besides the wrap around gate, special attention must also be paid to the geometry of the end contacts, since these play a role in determining the dimensions of the Schottky barriers that are present in the channel near the device ends and have a direct effect on current modulation. We hereinafter deal specifically with the coaxial geometry of the CNTFET shown in figure 1, but we note that most concepts and results discussed in this work are still transferable to planar devices, at least in a qualitative sense. The key device dimensions are: the gate inner radius, R_g, and thickness, t_g; the nanotube radius, R_t, and length L_t; the insulator thickness $t_{ins} = R_g / R_t$; the end-contact radius, t_c (the source and drain may sometimes be of different sizes), and length, L_c; and the gate-underlap L_u. Note that a closed-cylinder structure was employed for simplicity in treating the electrostatics, wherein $L_c = t_g = L_u = 0$, $t_c = R_g$ and $L_g = L_t$.

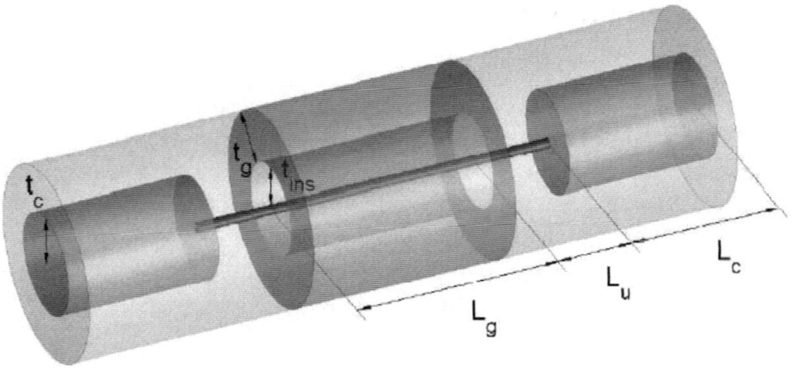

Figure 1. Coaxial CNTFET structure. The insulator fills the entire simulation space not occupied by metal or the nanotube [9].

2.2 Review of the NEGF formalism

A carbon nanotube can be viewed as a rolled-up sheet of graphene with a diameter typically between one and two nanometers. The nanotube can be either metallic or semiconducting, depending on how it is rolled up from the graphene sheet (i.e. depending on its chirality). Semiconducting nanotubes are suitable for transistors. In order to correctly treat carbon nanotube transistors, strong quantum confinement around the tube circumferential direction, quantum tunnelling through Schottky barriers at the metal nanotube contacts, and quantum tunnelling and reflection at barriers in nanotube channel need to be considered. The non-equilibrium Green's function (NEGF) formalism, which solves Schrödinger equation under non-equilibrium conditions and can treat coupling to contacts and dissipative scattering process, provides a sound basis for quantum device simulations. The NEGF simulation approach has demonstrated its usefulness for simulating nanoscale transistors from conventional Si MOSFETs, MOSFETs with novel channel materials, to CNTFETs [10], and molecular transistors [11]. In this section, we give brief summary of the NEGF simulation procedure. For a more thorough description of the technique one can refer to reference [12]. Figure 2 shows a generic transistor and defines some terms for the NEGF simulation.

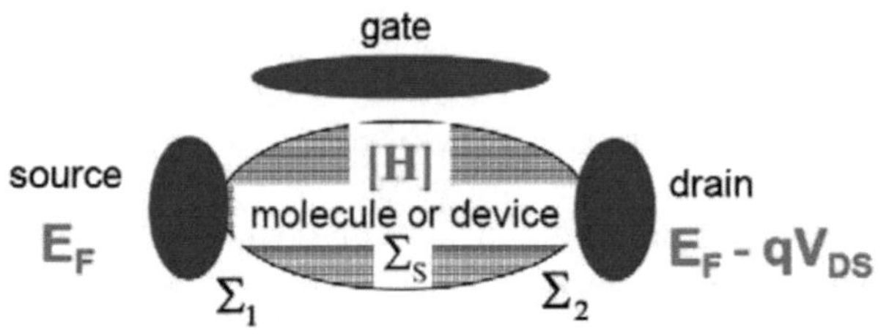

Figure 2. The generic transistor with a molecule or device channel connected to the source and drain contacts. The source-drain current is modulated by a third electrode, the gate. The quantities in the NEGF calculation are also shown.

The first step is to identify a suitable basis set and Hamiltonian matrix for an isolated channel. The self-consistent potential, which is a part of the Hamiltonian matrix, is included in this step. The second step is to compute the self-energy matrices, Σ_1, Σ_2 and Σ_S, which describe how the ballistic channel couples to the source/drain contacts and to the scattering process. For simplicity, only ballistic transport is treated in this work. After identifying the Hamiltonian matrix and the self-energies, the third step is to compute the retarded Green's function,

$$\mathbf{G}\,(E) = [(E + i0+)\,\mathrm{I} - \mathbf{H} - \Sigma_1 - \Sigma_2]^{-1} \qquad (1)$$

The fourth step is to determine the physical quantities of interest from the Green's function. In the ballistic limit, states within the device can be divided into two parts: 1) states filled by carriers from the source according to the source Fermi level, and 2) states filled by the drain according to the drain Fermi level. Within the device, the source (drain) local-density-of-states (LDOS) is $\mathbf{D}_{S(D)} = \mathbf{G}\Gamma_{S(D)}\mathbf{G}^{+}$, where $\Gamma_{S(D)} = I(\Sigma_{1(2)} - \Sigma^{+}_{1(2)})$ is the energy level broadening due to the source (drain) contact. The charge density within the device is computed by integrating the LDOS, weighted by the appropriate Fermi level over energy. The charge contributed by the source contact is

$$Q_S(z) = (-e)\int_{E_N}^{+\infty} D_S(E,z) f(E - E_{FS})dE + e\int_{-\infty}^{E_N} D_S(E,z)\{1 - f(E - E_{FS})\}dE \;, \qquad (2)$$

where e is the electronic charge and E_N is the charge neutrality level. The total charge is:

$$Q(z) = Q_S(z) + Q_D(z) = (-e)\int_{-\infty}^{+\infty} dE \cdot \mathrm{sgn}[E - E_N(z)]\{D_S(E,z) f(\mathrm{sgn}[E - E_N(z)](E - E_{FS}))$$

$$+ D_D(E,z) f(\mathrm{sgn}[E - E_N(z)](E - E_{FD}))\}, \qquad (3)$$

where sgn (E) is the sign function, and $E_{FS,D}$ is the source (drain) Fermi level. For a self-consistent solution, the NEGF transport equation is solved with iteratively the Poisson equation until self-consistency is achieved after which the source-drain current is computed from formula

$$I = \frac{4e}{h}\int T(E)[f_S(E) - f_D(E)]dE \quad , \tag{4}$$

where T (E) = Trace ($G_1G^+_2$) is the source-drain transmission and the extra factor of two comes from the valley degeneracy in the carbon nanotube energy band structure. The computationally expensive part of the NEGF simulation is finding the retarded Green's function, according to equation (1), which requires the inversion of a matrix for each energy grid point. The straightforward way is to explicitly invert the matrix, whose size is the size of the basic set. This, however, is impractical for an atomistic simulation of a nanotube transistor. In the ballistic limit, the problem is simplified because only a few columns of the Green's function are needed. Still, reducing the size of the Hamiltonian matrix and developing computationally efficient approaches are of great importance for an atomistic simulation.

2.3 Main screen of coaxial CNTFET

Purpose of the project was to make a user-friendly simulator that provides as much control as possible over every aspect of the simulation. Consequently, graphic user interface (GUI) development in Matlab was a major part of the programme. Here we present an overview of the most important GUI features.

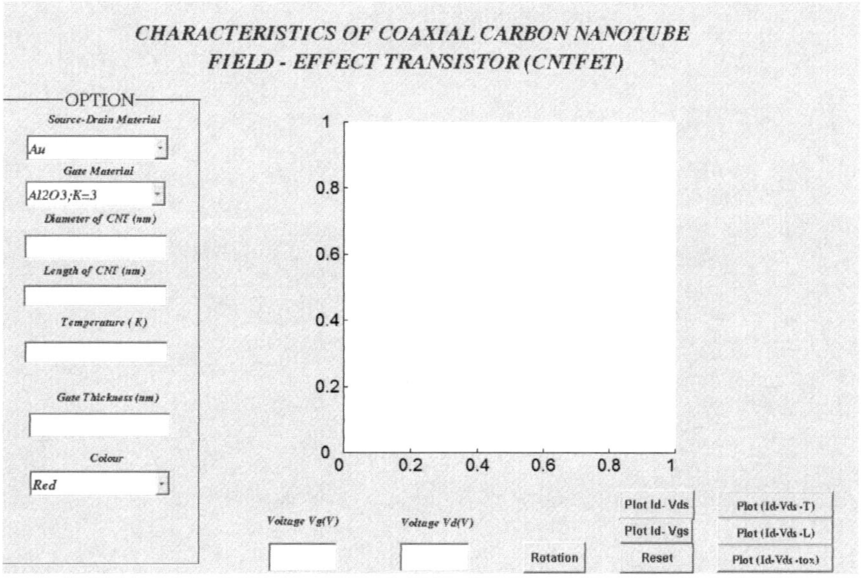

Figure 3. The main screen of simulation programme of coaxial CNTFET. Source-drain material (Au), Gate material (Al_2O_3, k = 3), Colour (Red) are default settings.

The main screen shown in figure 3 is the central location where one can control the simulation program. From main screen, one can quickly enter simulation parameters with minimum of typing. Clicking the left mouse pointer on each item of the main screen to enter the parameters such as types of source-drain materials (Au, Pt, Pd); types of gate materials (Al_2O_3 (k = 5), HfO_2 (k = 20), ZrO_2 (k = 26), TiO_2 (k = 65), $SiTiO_3$ (k = 175)); length of CNT (nm); temperature (K); gate thickness (nm); colours (blue, green, pink, yellow, black); voltage V_g (V); V_d (V). Clicking the right mouse pointer on the item which is used to calculate the current voltage characteristics in 2D (plot I_d-V_{ds} or plot I_d-V_{gs}), reset used to clear current data or in 3D (plot I_d-V_{ds}-T, Plot I_d-V_{ds}-L and plot I_d-V_{ds}-t_{ox}). During calculation, symbol "running" will be appeared. Source-drain material (Au), Gate material (Al_2O_3, k = 3), Colour (Red) are default settings. Simulations in 2D take less time than in 3D. Thanks to simulations in 3D, one can get more information of tendency and effects of parameters such as temperature, the length of CNT, the gate thickness on I-V characteristics.

Journal of Physics: Conference Series **187** (2009) 012061 doi:10.1088/1742-6596/187/1/012061

2.4 Simulations of current-voltage characteristics of coaxial CNTFET

The drain I-V characteristics in 2D are shown in figure 4. The saturation current at V_{GS} = 0.5 V is around 6 μA, which is not inconsistent with values emerging from recent experimental work [13].

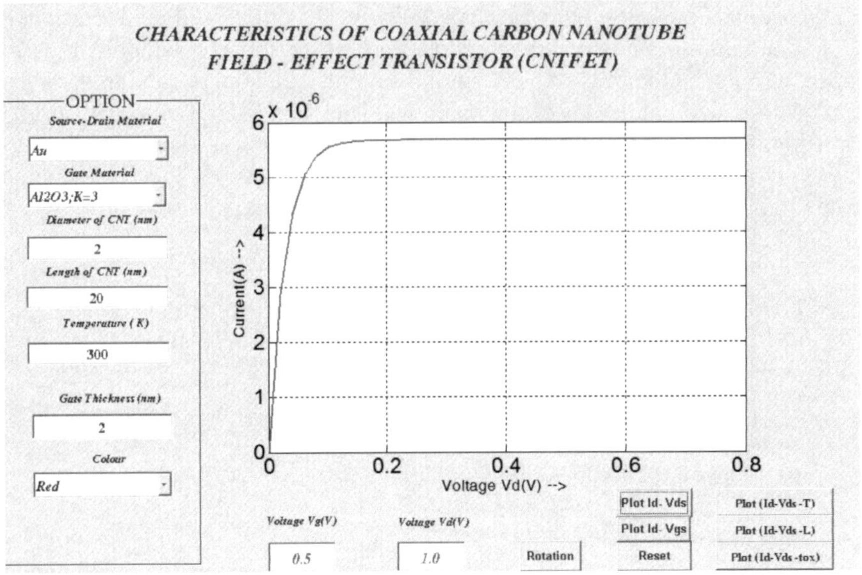

Figure 4. Drain current-voltage characteristics of coaxial CNTFET.

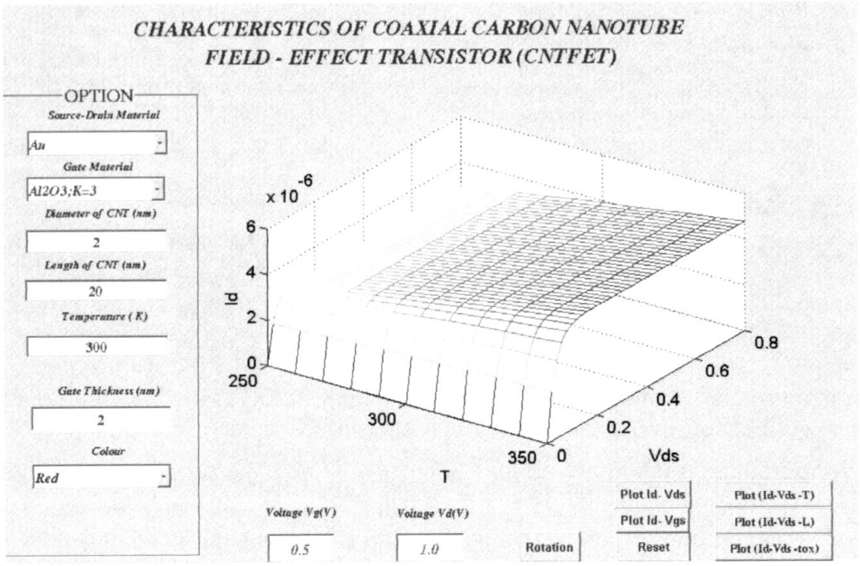

Figure 5. Drain current-voltage characteristics in 3D. When CNTFET is cooled, its saturation drain currents were lightly decreased.

The drain I-V characteristics in 3D are shown in figure 5. The parameters used in 3D simulation were drain current-voltage characteristics and temperature. Drain I-V characteristics exhibited dependence of saturation drain current on temperature. When CNTFET is cooled, saturation currents were lightly decreased.

Drain current-voltage characteristics in 3D exhibited dependence of saturation currents on CNTFET length L are shown in figure 6. Tendency of saturation currents is decreased, when CNTFET length is increased.

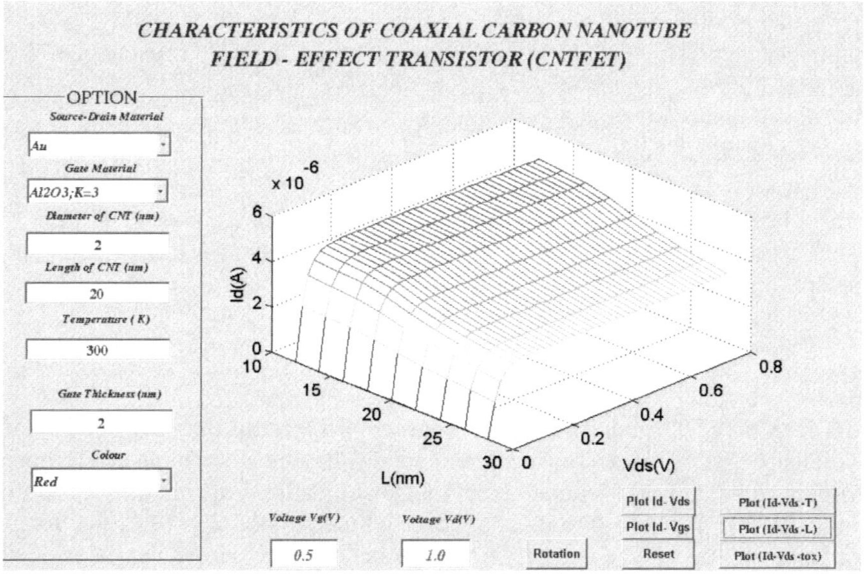

Figure 6. Drain current-voltage characteristics in 3D exhibited dependence of saturation drain currents on CNTFET length.

Drain current-voltage characteristics in 3D exhibited dependence of saturation current on the gate thickness of CNTFET, t_{ox} are shown in figure 7.

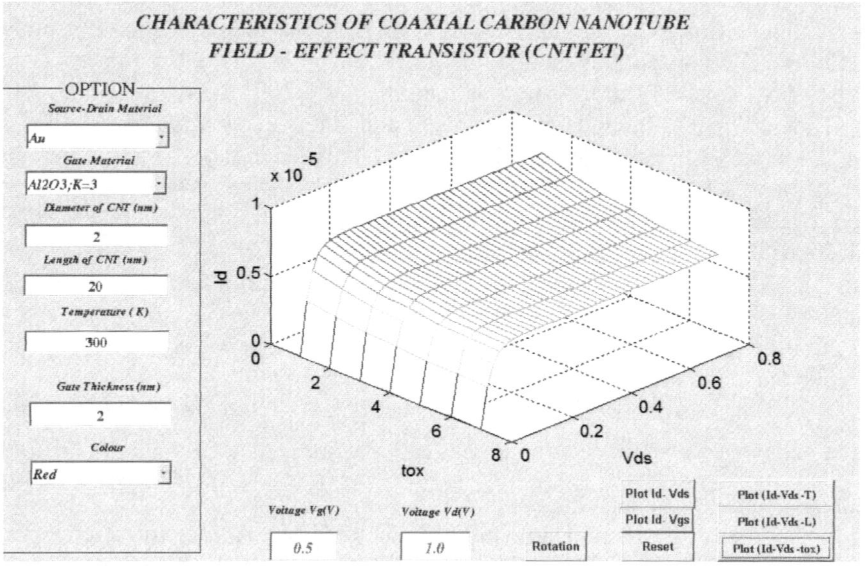

Figure 7. Drain current-voltage characteristics in 3D exhibited dependence of saturation drain currents on the gate thickness of CNTFET.

The current voltage curve can be divided into two regions: linear and saturation. Drain current in the linear region of CNTFET can be described as follows:

$$I_d = \mu C_{ox}[(V_{gs} - V_T)V_{ds} - \frac{V_{ds}^2}{2}] \text{ ,} \tag{5}$$

or

$$I_d = K_n\left[2\left(V_{gs} - V_T\right)V_{ds} - V_{ds}^2\right], \quad (6)$$

where K_n is conductance of CNTFET, μ is mobility of carriers, C_{ox} is gate capacitance.

In the coaxial CNTFET

$$C_{ox} = 2\pi R_g L_g k\varepsilon_0 \left(\lg\frac{R_g}{R_t}\right)^{-1}, \quad (7)$$

where k is relative dielectric constant, ε_0 is dielectric constant in vacuum, 8.85×10^{-14}Fcm^{-2}.

We can also obtain saturation current of CNTFET by replacing $V_{ds(sat)} = V_{gs} - V_T$. Then the expression of saturation current of CNTFET can be written as follows:

$$I_{d(sat)} = K_n\left(V_{gs} - V_T\right)^2. \quad (8)$$

3. Conclusions

A coaxial model for CNTFET has been proposed, which is specially intended for 3D simulations of drain current-voltage characteristics. The proposed model has been verified and a good agreement with recent experimental data [13] is found. A set of 3D simulations was then successfully performed for different parameters of materials, the length, the gate thickness of CNTFET, and temperature. The effects of these parameters on I-V characteristics have been predicted. Analytical equations of drain current-voltage curve are also presented.

References

[1] Kong J, Soh H T, Cassell A M, Quate C F and Dai H 1998 *Nature* **395** 878
[2] Javey A, Guo J, Wang Q, Lundstrom M and Dai H J 2003 *Nature* **424** 654
[3] Wind S J, Appenzeller J, Martel R, Derycke V and Avouris P 2002 *Appl. Phys. Lett.* **80** 3817
[4] Javey A, Kim H, Brink M, Wang Q, Ural A, Guo J, McIntyre P, McEuen P, Lundstrom M and Dai H J 2002 *Nature Materials* **1** 241
[5] Radosavljevic M, Appenzeller J, Avouris Ph and Knoch J 2004 *Appl. Phys. Lett*. **84** 3693
[6] Javey A, Tu R, Farmer D B, Guo J, Gordon R G and Dai H 2005 *Nano Lett.* **5** (2) 345
[7] Derycke V, Martel R, Appenzeller J and Avouris P 2001 *Nano Lett.* **1** 453
[8] Bachtold A, Hadley P, Nakanishi T, and Dekker C 2001 *Science* **294** 1317
[9] Leonardo de Camargo e Castro 2006 *Modeling of Carbon Nanotube Field-Effect Transistor* Ph.D Thesis (The University of British Colombia)
[10] Jing Guo 2004 *Carbon Nanotube Electronics: Modeling, Physics and Applications* Ph.D. Thesis (Purdue University)
[11] Damle P, Rakshit T, Paulsson M and Datta S 2002 *IEEE Transactions on Nanotech.* **1** 145
[12] Datta S 2000 *Superlattices and Microstructures* **28** 253
[13] Javey A, Ryan Tu, Damron B, Farmer, Guo J, Gordon R G and Dai H 2005 *Nano Lett.* **5** No.2 345

Acknowledgments

The authors are grateful to professor Dang Luong Mo of Hosei University, Japan for helpful and stimulating discussions. This work was supported in part by Vietnam National University Ho Chi Minh City, under research contract No. 2005-18-06.

Magnetic and sensitive magnetoelastic properties of Finemet nanostructured ribbon

Pham Duc Thang[1], Hoang Hai Duong[2] and Nguyen Hoang Nghi[2]

[1] Faculty of Engineering Physics & Nanotechnology and Laboratory of Micro & Nanotechnology, College of Technology, Vietnam National University, 144 Xuan Thuy, Cau Giay, Hanoi, Vietnam
[2] Laboratory of Amorphous and Nanocrystalline Materials, Hanoi University of Technology, 1 Dai Co Viet, Hanoi, Vietnam

E-mail: pdthang@vnu.edu.vn

Abstract. Soft-magnetic $Fe_{73.5}Cu_1Nb_3Si_{13.5}B_9$ (Finemet) ribbon has been fabricated by using melt-spinning techniques. After annealing at suitable temperature the ribbon changes from an amorphous to crystalline state which related to the formation of Fe nanocrystallites. Study on the magnetic and magnetoelastic properties of the ribbon is presented. Furthermore, based on the fabricated ribbon stress sensors are simply constructed. The sensors showed high sensitivity of 3.8 mV/MPa as well as a wide working range up to 17 MPa. These sensors are potential for practical applications such as detecting small stress and movement in civil structures.

Keywords: Magnetic and magnetoelastic properties, Finemet, nanostructured materials.

1. Introduction

Recently, interest in the development of magnetoelastic sensors has greatly increased because of their wide applications, such as for small magnetic field detection, stress, strain, pressure, displacement, and vibrations sensors [1-7]. It is well-known that a magnetostrictive material changes its size when placed in an external magnetic field. By applying a mechanical stress to the material, one observes an inverse magnetoelastic effect (namely the Villari effect) in which the magnetization direction varies leading to the variation of the induction and magnetic permeability. Consequently, an electromotive force is induced in a pick-up coil surrounding the material.

Fe-based amorphous alloys exhibit a good magnetic as well as magnetostrictive softness. In addition, they have also good mechanical properties. In the as-prepared alloys, however, the permeability easily reaches maximum at low stress and then the sensitivity is reduced at high stress. In order to extend the working range, suitable heat treatments or (Fe, Co)-based amorphous ribbons are usually used [8].

In this paper we present the study on crystallographic and micro-structure, magnetic and magnetoelastic properties of $Fe_{73.5}Cu_1Nb_3Si_{13.5}B_9$ (Finemet) ribbon. In addition, the dependences of the stress sensor sensitivity on frequency and annealing temperature are also investigated.

2. Experimental

$Fe_{73.5}Cu_1Nb_3Si_{13.5}B_9$ ribbon has been prepared by using melt-spinning technique. The purity of the starting elements was at least 99.9%, excepting Si and B (99.5%). In the melt-spinning process, the

© 2009 IOP Publishing Ltd

bulk alloy was melted at 1450°C in a quartz tube under an Ar pressure of 0.5 atm and produced in ribbon form by rapid quenching on a Cu roller rotating at the speed of 2100 turns/minute. The typical dimensions of the as-prepared ribbon were 10 mm in width and 20 μm in thickness. Different strips cut from the same ribbon with 5 mm in width and 15 mm in length were submitted to the magnetoelastic measurement. For other analysis, the size of the ribbons were around 5×5 mm^2. Conventional annealing treatments have been performed in vacuum at different annealing temperatures (T_a) up to 550°C for 1h.

The crystallographic structure of the samples was characterized by using an X-ray diffractometer (XRD) Siemens D-5005 with CuK_α wavelength. The microstructure was observed by means of a field emission scanning electron microscopy (FE-SEM) Hitachi S-4800. The magnetic hysteresis loops of the samples, measured at room temperature along and perpendicular to the ribbon surface, were carried out by using a vibrating sample magnetometer (VSM) Lake Shore 7404.

For magnetoelastic analysis, stress sensors have been constructed using the above 15 mm-length ribbon and a two-coil system, as shown in figure 1. At one end of the ribbon the excited coil produces an alternating magnetic field

$$H = H_0 \sin(2\pi f_o t) \ , \tag{1}$$

in which H_0 is the amplitude of the exciting magnetic field and the frequency is tuned to the fundamental resonance frequency $f_0 = v/2l$ (v is the sound velocity, l is the length of the ribbon) [9, 10]. Due to the magnetoelastic effect, the magnetic field induces an alternating strain ε that determines a local excitation propagating through the ribbon as a magnetoelastic wave. Then the magnetic component of the generated wave induces an electromotive force in the pick-up coil located at the opposite end of the ribbon. As a consequence of the inverse magnetoelastic effect, an alternating voltage can be detected as

$$V(t) = V_0 \sin(2\pi f_0 t + \varphi) \ , \tag{2}$$

where the voltage amplitude, i.e. the stress sensor's signal V_0, can be expressed as

$$V_0 = H_0 \mu k^2 \left(\frac{E}{\rho}\right)^{1/2} , \tag{3}$$

in which μ, E, ρ and k are the ribbon's magnetic permeability, Young's modulus, mass density and magnetomechanical coupling factor, respectively.

When a stress is applied along the ribbon length, one can measure the change in the sensor's signal V_0 as a function of external stress σ.

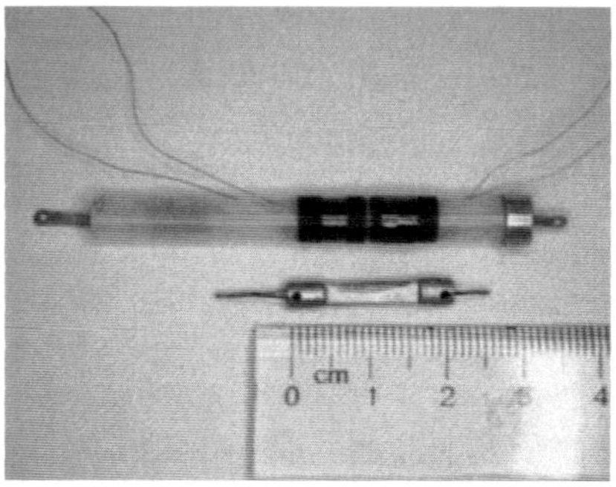

Figure 1. Image of a used typical stress sensor.

Journal of Physics: Conference Series **187** (2009) 012062 doi:10.1088/1742-6596/187/1/012062

3. Results and discussion

For as-prepared Finemet ribbon, a flat XRD pattern is observed indicating its amorphous state. After increasing annealing temperatures up to T_a = 350°C, the XRD patterns are still almost flat although nanocrystallites may have formed during the heat treatment. Only at T_a around 500°C, some peaks clearly appear which corresponds to Fe nanocrystallites. These samples, however, were too brittle for other measurements.

The FE-SEM images of the ribbons, presented in figure 2, also confirmed the XRD result. As can be seen from this figure, the as-prepared sample has a homogeneous structure of amorphous state. Meanwhile after annealing at T_a = 350°C for 1h the sample displays signs of crystallization. The influence of these different microstructures on the properties of the studied ribbons will be discussed later.

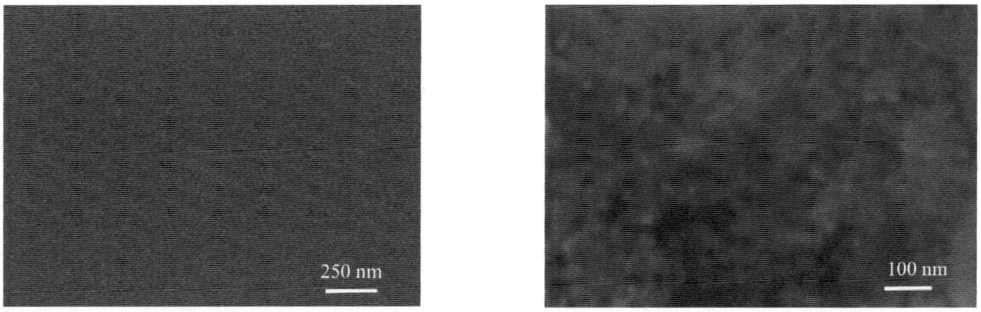

Figure 2. FE-SEM images of the as-prepared Finemet ribbon (left-hand side) and after annealing at 350°C for 1h (right-hand side).

The in-plane and perpendicular magnetic hysteresis loops of the as-prepared and annealed ribbons, measured at room temperature, are plotted in figure 3. All samples show soft magnetic behavior in the in-plane, with the coercivity around 1 Oe. With increasing the annealing temperature, the saturation magnetization M_s increases from 95 to 129 emu/g for the as-quenched ribbon and 350°C-annealed one, respectively (see also table 1). This observation can be explained in terms of the formation of Fe nanocrystallites in the annealed sample.

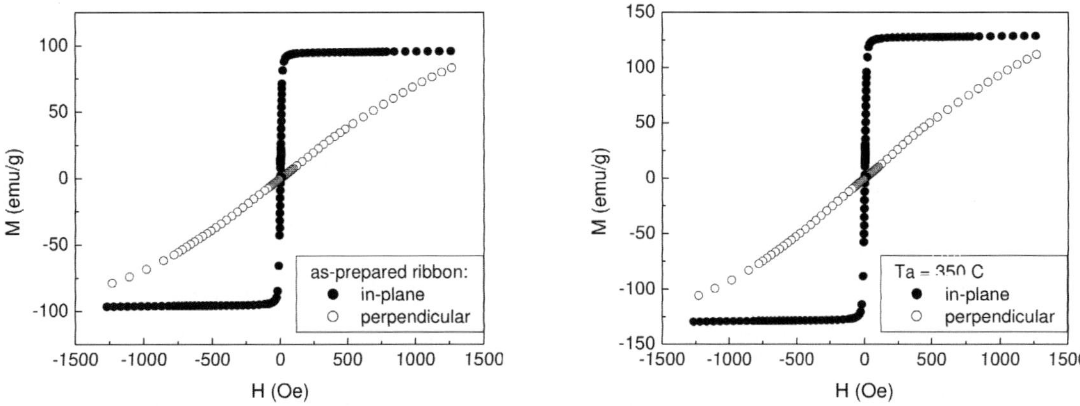

Figure 3. In-plane and perpendicular magnetic hysteresis loops of the as-prepared (left-hand side) and annealed ribbons (right-hand side).

Figure 4 represents the stress sensor's signal V_0 responding to longitudinal applied tensile stress σ of the ribbon annealed at 250°C, measured at different frequencies. With increasing the applied stress, firstly the sensor signal linearly increases and then reaches saturation state. Besides, the stress sensor's signal and therefore its change in the linear part, ΔV_0, exhibit higher values at high frequency. From the slope of the linear part of the V_0-σ graph, one can calculate the sensitivity of the stress sensor.

Table 1. Characterized parameters of the stress sensors based on Finemet ribbon.

Sample	M_s (emu/g)	Maximum sensitivity (mV/MPa)	ΔV_0 (mV)	Linear working range (MPa)
As-prepared	95	4.0 @ 150 kHz	17.8	4.5
T_a = 150°C	98	3.1 @ 120 kHz	54.5	17.5
T_a = 250°C	118	3.8 @ 180 kHz	82.0	21.5
T_a = 350°C	129	1.4 @ 50 kHz	29.2	21.5

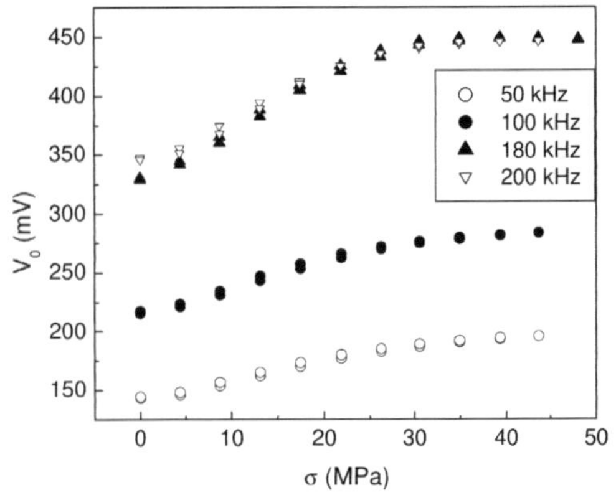

Figure 4. Stress dependence of sensor signal of 250°C-annealed ribbon at different frequencies.

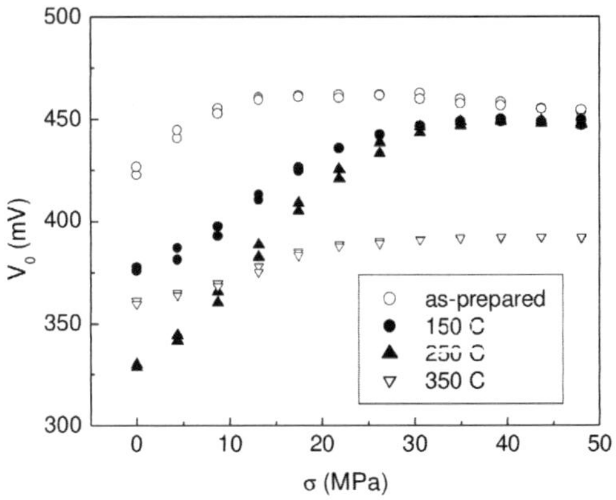

Figure 5. Stress sensor signal of as-prepared and annealed ribbons.

In figure 5 the stress dependences of the sensor signals of the as-prepared and annealed ribbons are plotted. These measurements are taken at frequencies around 150 kHz (excepting 350°C-annealed ribbon having a frequency of 50 kHz) at which the sensors exhibit maximum sensitivity. As indicated in table 1, the sensor using as-prepared ribbon displays highest sensitivity of 4.0 mV/MPa with a linear working range from 0 to 4.5 MPa. Meanwhile for the 250°C-annealed ribbon, a considerable change in the voltage amplitude ΔV_0 = 82 mV is obtained, resulted in a comparable sensitivity of 3.8 mV/MPa. Furthermore, the working range is also extended four-times, from 0 to 21.5 MPa. This sensitivity is in the same order as the previous result for Fe-Co-Si-B ribbon [8]. The results obtained on the Finemet ribbon meet the requirements for application as high-sensitivity stress sensors.

4. Conclusions

The stress sensor is constructed using the as-prepared and annealed Finemet ribbons. The sensor based on 250°C-annealed ribbon exhibits a high sensitivity of 3.8 mV/MPa as well as a linear stress dependence up to almost 22 MPa. These stress sensors are potential for practical applications such as detecting small stress and movement in civil structures.

References

[1] Kouzoudis D and Grimes C A 2000 *Smart Mater. Struct.* **8** 885
[2] Germano R, Ausanio G, Iannotti V, Lanotte L and Luponio C 2000 *Sensors and Actuators A* **81** 134
[3] Ausanio G, Iannotti V, Luponio C, Lanotte L, Germano R, D'Agostino A, Inverno M and Sorrentino R 2001 *Sensors and Actuators A* **91** 123
[4] Jain M K, Cai Y Q and Grimes C A 2001 *Smart Mater. Struct.* **10** 347
[5] Ishiyama K, Imamura K and Arai K I 2002 *J. Magn. Magn. Matter.* **242-245** 1163
[6] Gomez-Polo C, Barandiaran J M and Gutierez J 2003 *Sensors and Actuators A* **106** 155
[7] Bydzovsky J, Kraus L, Svec P, Pasquale M and Kollar M 2004 *Sensors and Actuators A* **110** 82
[8] Thang P D, Huong Giang D T, Tinh B C, Danh T M, Tuan N H and Duc N H 2007 *Phys. Stat. Sol.* (c) **4** 4585
[9] Lanotte L, Kaczkowski Z and Iannotti V 1998 *Mol. Quantum Acoust.* **19** 149
[10] Iannotti V and Lanotte L 2000 *Philos. Mag. B* **80** 1903

Acknowledgements

This work was partly supported by the research projects of the Asia Research Center and the College of Technology (project QC.08.09), Vietnam National University in Hanoi. The authors would like to thank D. H. Manh for his kind help in microstructure analysis.

Improvement of electrical property for $Pb(Zr_{0.53}Ti_{0.47})O_3$ ferroelectric thin film deposited by sol-gel method on SRO electrode

Vu Ngoc Hung[1], Le Van Minh[1], Bui Thi Huyen[1] and Nguyen Duc Minh[1,2]

[1]International Training Institute for Materials Science (ITIMS), Hanoi University of Technology, Hanoi, Vietnam
[2]Twente University, MESA+, Netherlands

E-mail: hungvungoc@itims.edu

Abstract. $Pb(Zr,Ti)O_3$ thin films with Zr/Ti ratio of 53:47 were deposited on SRO/YSZ/Si(100) and Pt/Ti/SiO_2/Si(100) substrates using the sol-gel method. The thermal process of the PZT thin films included treating two-step pyrolysis, at 300°C for 30 minutes and 400°C for 10 minutes, and annealing crystallization, at 600°C for 10 minutes in air ambient. The remnant polarization and the coercive field of the PZT/SRO/YSZ/Si were 28 μC/cm^2 and 65 kV/cm^2, and those of the PZT/Pt/Ti/SiO_2/Si were 12 μC/cm^2 and 79 kV/cm^2. This indicates that the use of the metallic SRO-electrodes significantly improves electrical property of PZT piezoelectric thin film as compared with a metal-electrode Pt.

Keywords: Ferroelectric thin film, SRO-electrode, sol-gel method.

1. Introduction

Ferroelectric lead zirconate titanate (PZT) thin film capacitors have been studied with great interest in the last decade because of their potential applications in piezoelectric sensors, actuators, power generators and non-volatile memory devices [1, 2]. Electrode materials can play a significant role in determining the quality of ferroelectric thin films. Among metal electrode materials, platinum presents very attractive properties as bottom electrode material for ferroelectric thin films due to its high electrical conductivity, good stability in oxidation at high electrical conductivity, good stability in oxidation at high temperatures and a high Schottky barrier, which lowers the leakage currents. However, when using Pt film as a ferroelectric device electrode, the Pt electrode cannot be used for more than 10^6 switching cycles because of the fatigue. On the contrast, conductive metal oxide such as $SrRuO_3$ (SRO) [3], YBa_2CuO_{7-x}, IrO_2, RuO_2, $LaNiO_3$ (LNO) [4], and $La_{0.5}Sr_{0.5}CoO_3$ (LSCO) were found to be effective candidates to improve the fatigue property for PZT thin films. All these materials possess relatively good electrical conductivity. The lattice-matched hetero-structures of PZT/oxide electrode ferroelectric thin films can be expected to improve the electrical properties, especially for the fatigue property. Perovskite-type oxide electrode such as LNO and SRO are promising candidates because of their similar lattice parameters (the lattice parameters for a PZT of a morphotropic phase boundary: a=4.036 Å, c=4.146 Å [5] and the pseudocubic lattice parameter of a SRO: a_t=3.928 Å [6]). Recently, great efforts have been taken to research in the interfaces between the ferroelectric film and electrode for optimizing the properties of ferroelectric devices.

© 2009 IOP Publishing Ltd

Journal of Physics: Conference Series **187** (2009) 012063 doi:10.1088/1742-6596/187/1/012063

In this study, the results on the fabrication of PZT thin films with SRO and Pt electrodes by sol-gel spin coating method are presented. It is shown that the microstructure and ferroelectric properties of the fabricated PZT thin films are significantly affected by SRO conducting oxide electrode.

2. Experimental procedure

Our sol-gel experiment was prepared to make $Pb(Zr,Ti)O_3$ (Zr:Ti=53:47) thin films with 15% added lead. The raw materials for the solution synthesis were lead acetate $[Pb(CH_3COO)_2 3H_2O]$, Titan iso-propoxit $[Ti(i\text{-}C_3H_7O)_4]$ and Zirconi n-propoxit$[Zr\ (n\text{-}C_3H_7O)_4]$. Acetic acid ($CH_3COOH$) and 2-methoxyethanol $[CH_3OCH_2CH_2OH]$ were a stabilizer and solvent, respectively. The PZT solutions were deposited on both SRO/YSZ/Si and Pt/Ti/SiO_2/Si substrates by the spin coating at 4000 rpm for 30 sec. After that, the PZT thin films were treated by thermal process shown in figure 1. Every PZT-coating layer was pyrolyzed at two-step heat treatment - the first step at 300°C for 30 minutes and the second at 400°C for 10 minutes. The crystallization of every two PZT-coating layers was performed by annealing at 650°C in air ambient environment. In our work, the PZT thin films with four-coating layers were fabricated.

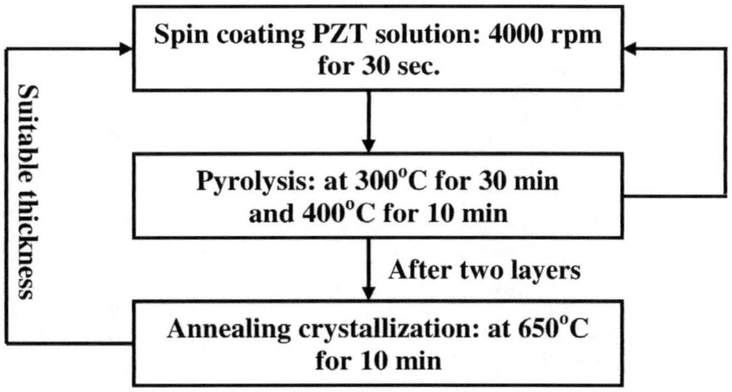

Figure 1. Diagram for the thermal process of PZT thin films.

The orientation of the PZT thin film was determined by X-ray diffraction (XRD) method, Philips X'Pert SR5056. The PZT thin film thickness was measured by using cross-section scanning electron microscopy (SEM), FE-SEM Hitachi S-4800. The atomic force microscopy (AFM) was used to examine the surface morphology and roughness of thin films. The electrical properties of PZT thin films were measured by TF2000 ferroelectric test system.

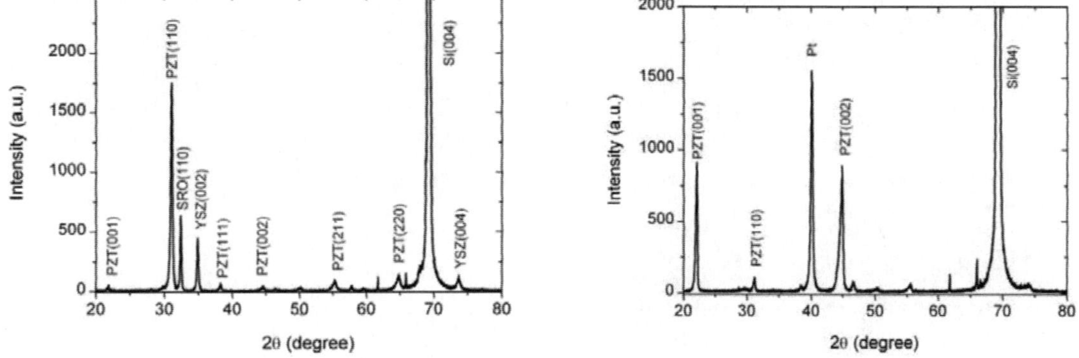

(a) (b)

Figure 2. XRD patterns for PZT/SRO/YSZ/Si (a) and PZT/Pt/Ti/SiO_2/Si (b).

Journal of Physics: Conference Series **187** (2009) 012063 doi:10.1088/1742-6596/187/1/012063

3. Results and discussion

Figures 2a and 2b illustrate the XRD patterns for the PZT/SRO/YSZ/Si film and the PZT/Pt/Ti/SiO_2/Si film, respectively. It is shown that the SRO film exhibited the (110) preferred orientation. The PZT phase in PZT/SRO/YSZ/Si showed preferential (110) orientation. The PZT lattice constant and the SRO lattice constant were very close, therefore, it is assumed that the (110)-oriented PZT film was grown on the pseudo-cubic SRO (110) plane. Meanwhile, the PZT phase in the PZT/Pt/SiO_2/Si film showed a high (001) orientation, thus the resultant PZT film on Pt was not greatly affected by the Pt substrate.

Microstructure of the film affects its electrical properties, therefore, the surface morphology of the resultant films was observed by AFM. Figures 3 and 4 show the AFM images of the PZT/SRO/YSZ/Si film and the PZT/Pt/Ti/SiO_2/Si film, respectively. The average roughness values of the PZT/SRO/YSZ/Si film and the PZT/Pt/Ti/SiO_2/Si film were 1.1 nm and 7.5 nm, respectively. It was confirmed that the surface of PZT/SRO/YSZ/Si is smoother than that of PZT/Pt/Ti/SiO_2/Si. The average particle size of the PZT on SRO/YSZ/Si (40-60 nm) was smaller than that on Pt/Ti/SiO_2/Si (80-100 nm). On the other hand, the numbers of PZT/SRO/YSZ/Si particles are larger than that of PZT/Pt/Ti/SiO_2/Si.

Figure 3. The AFM images of PZT/SRO/YSZ/Si thin film.

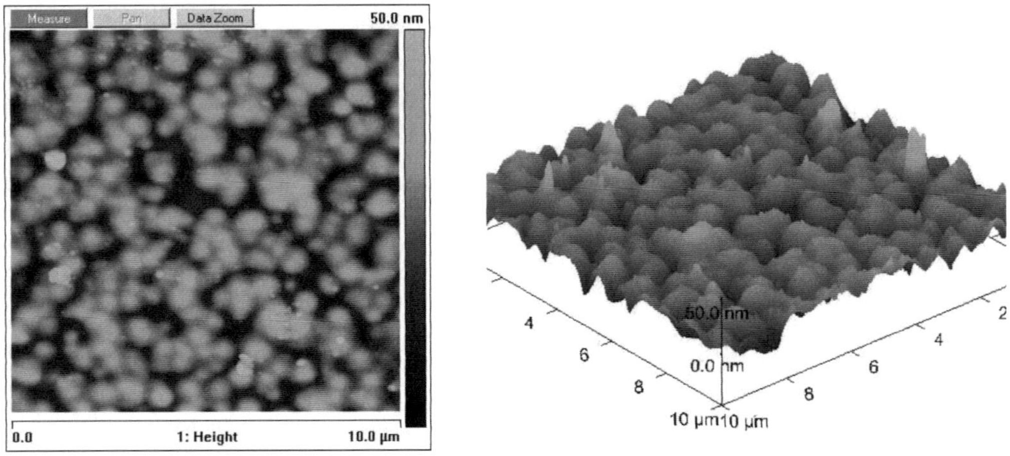

Figure 4. The AFM images of PZT/Pt/Ti/SiO_2/Si thin film.

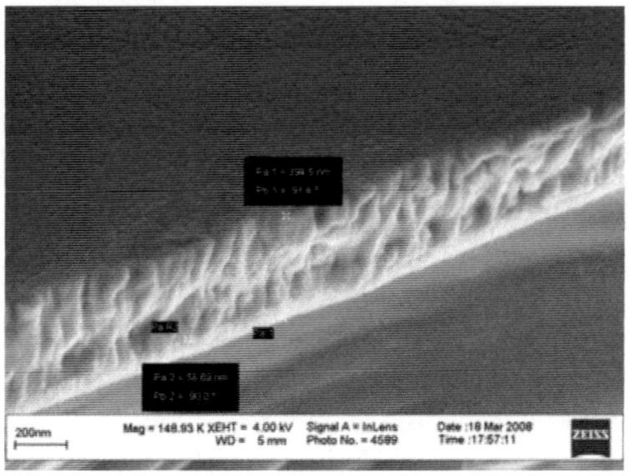

(a)

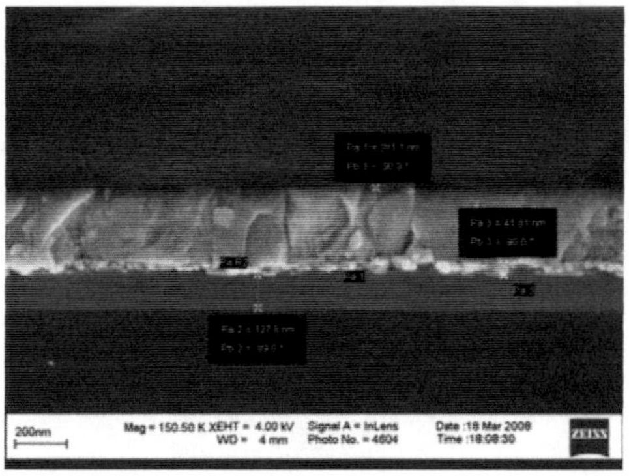

(b)

Figure 5. The cross-section SEM of PZT/SRO/YSZ/Si (a) and PZT/Pt/Ti/SiO2/Si (b) thin film.

Figures 5a and 5b show the cross-section SEM of PZT thin film on SRO/YSZ/Si and Pt/Ti/SiO_2/Si substrate, respectively. In figure 5b, the microstructure of PZT thin film on Pt/Ti/SiO_2/Si is the dense structure. In contract, the microstructure of PZT thin film on SRO/YSZ/Si (figure 5a) is grown in columnar structure. Therefore, the PZT column became anisotropic geometry along the c-axis on the SRO/YSZ/Si substrate film. It is assumed that the particle size on the film surface depends on the column width, and this result agrees with the SEM observation.

These results of orientations and microstructures reveal the effect of materials electrode on growing PZT thin film. As a consequence, they affect the electrical properties of PZT thin films.

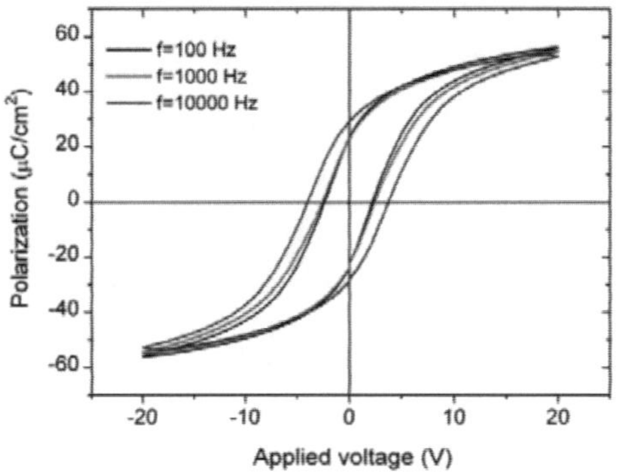

(a)

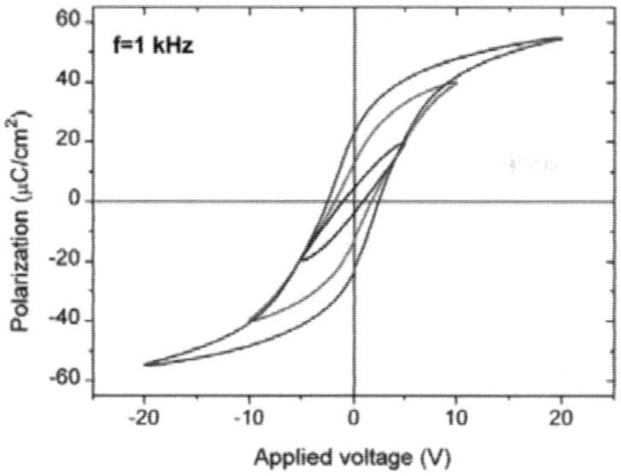

(b)

Figure 6. Ferroelectric hysteresis loops of PZT/SRO/YSZ/Si thin films: a) at different frequencies; b) at different applied voltages 5 V, 10 V and 20 V with a frequency, 1 kHz.

A gold top electrode was deposited on the PZT films by sputtering to evaluate the electrical properties of the PZT films. Figures 6 and 7 show the hysteresis loops for the PZT/SRO/YSZ/Si film and the PZT/Pt/Ti/SiO_2/Si film, respectively. It displays the typical hysteresis of the ferroelectric thin films. These P-E loops were measured at different frequencies, 100 Hz, 1 kHz and 10 kHz. In addition, these P-E loops of PZT thin films were examined at three different voltages, 5 V, 10 V and 20 V. Both the loops indicated good symmetries and imprint was not observed. These figures indicate that the SRO electrode has improved the ferroelectric property of the resultant PZT thin films on the SRO/YSZ/Si substrate. The remanent and saturated polarization of the resulting PZT films were larger than that of (001)-oriented PZT thin film on the commonly used substrate of a Pt/Ti/SiO_2/Si. Thus, the

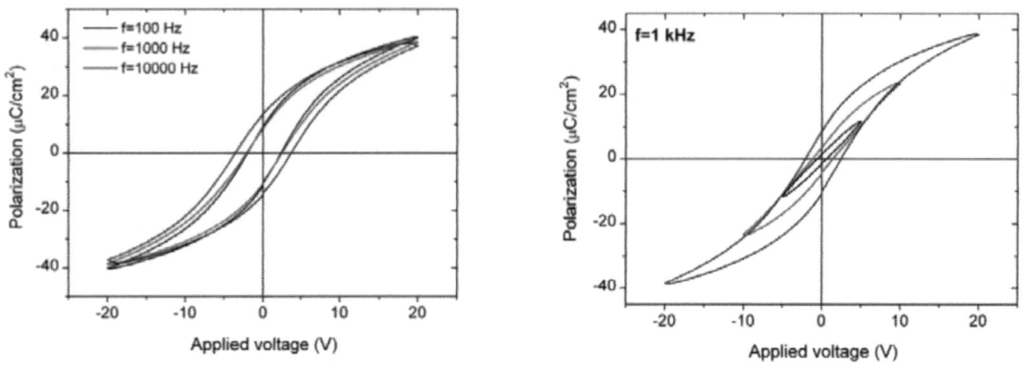

(a) (b)

Figure 7. Ferroelectric hysteresis loops of $PZT/Pt/Ti/SiO_2/Si$ thin films: a) at different frequencies; b) at different applied voltages 5 V, 10 V and 20 V with a frequency, 1 kHz.

remanent polarization and the saturation polarization of the PZT/SRO/YSZ/Si film were $28\mu C/cm^2$ and 58 $\mu C/cm^2$, and those of the $PZT/Pt/Ti/SiO_2/Si$ film were 12 $\mu C/cm^2$ and 38 $\mu C/cm^2$, respectively. In addition, coercive field of the resulting PZT films was also reduced by using the SRO electrode because of the better contact of the interface between electrode and the ferroelectric PZT thin film. The coercive field for the PZT/SRO and the PZT/Pt capacitor was 60 kV/cm and 75 kV/cm, respectively.

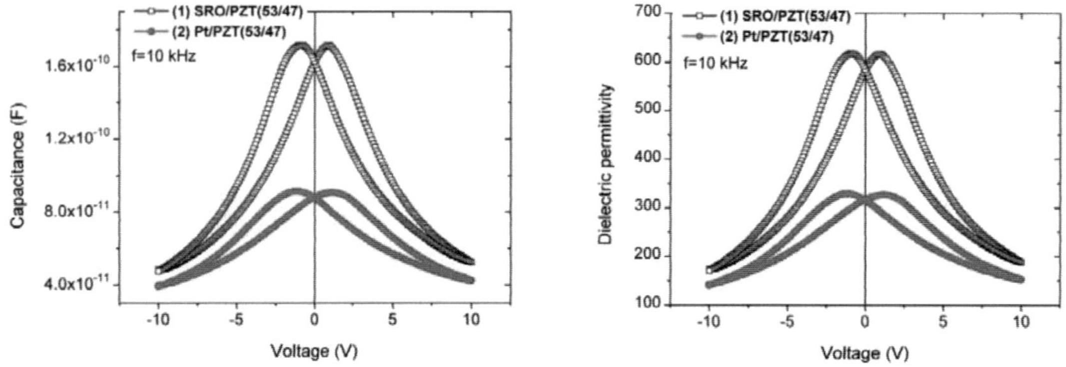

(a) (b)

Figure 8. The capacitance (a) and dielectric permittivity (b) of PZT thin films on SRO and Pt bottom electrodes depend on applied voltages.

The C-V curves (figure 8a) and D-V curves (figure 8b) reveal the dielectric non-linearity and hysteresis in ferroelectric thin films. The expected butterfly-type shapes are the natural result of switching behavior where the maximum occurs near the coercive voltage. This is because the domain wall contribution is maximized due to the large number of coexisting domain walls at these voltages.

Dielectric constant for these PZT/SRO thin film-capacitors with (110) preferred at 10 kHz were about 580 and 310, respectively (figure 8). On the other hand, dielectric constant for the (001)-preferred PZT thin film on a $Pt/Ti/SiO_2/Si$ substrate at 10 kHz was about 310, which was smaller than that of the PZT film on the SRO/YSZ/Si substrate, suggesting the higher c-axis orientation of the PZT thin film on the (110)-oriented SRO thin film.

4. Conclusion

We have successfully deposited PZT thin films on the SRO/YSZ/Si substrates with preferred orientation by sol-gel process. The orientation and microstructure of the PZT thin films are affected strongly by the SRO bottom electrode. These PZT/SRO thin films exhibited better ferroelectric properties compared with those of the (001)-textured PZT/Pt thin films on the Si substrates. These results indicate that the thin-film oxide electrode is essential for the led-based ferroelectric thin films.

References

[1] Xie J, Hu M, Ling S F and Du H 2006 *Sensors and Actuators* **A 126** 182
[2] Jeon Y B, Sood R, Jeong J H and Kim S G 2005 *Sensors and Actuators A* **122** 16
[3] Wang Y K, Tseng T Y and Lin P 2002 *Appl. Phys. Lett.* **80** 3790
[4] Sama N, Herdier R, Jenkins D, Soyer C, Remiens D, Detalle M and Bouregba R 2008 *J. Cryst. Growth* **310** 3299
[5] Suzuki H, Miwa Y, Naoe T, Miyazaki H, Ota T, Fuji M and Takahashi M 2006 *J. of the European Ceramic Society* **26** 1953
[6] Lee H N, Senz S, Zakharov N D, Harnagea C, Pignolet A, Hesse D and Gosele U 2000 *Appl. Phys. Lett.* **77** 3260

Fabrication and properties of highly luminescent materials from $Tb(OH)_3@SiO_2$ and $Tb(OH)_3@SiO_2:Eu^{3+}$ nanotubes

Tran Thu Huong[1], Tran Kim Anh[1] and Le Quoc Minh[1, 2]

[1] Institute of Materials Science, Vietnam Academy of Science and Technology, 18 Hoang Quoc Viet Road, Cau Giay Distr., Hanoi, Vietnam
[2] College of Technology, National University Hanoi, Vietnam

E-mail: huongtt@ims.vast.ac.vn

Abstract. Luminescent nanomaterials with one-dimensional (1D) structures have attracted much attention due to their unique properties and potential applications in nanophotonics and nanobiophotonics. In this paper, we report a synthesis of terbium - hydroxide - at - silica $Tb(OH)_3@SiO_2$ and $Tb(OH)_3@SiO_2:Eu^{3+}$ nanotubes. Terbium - hydroxide tubes were synthesized by soft template method. The size of the tubes can be controlled precisely and have outer diameters ranging from 80 to 120 nm, wall thickness of about 30 nm, and lengths ranging from 300 to 800 nm. To fabricate core/shell materials, the seed growth method is used. FESEM, X-ray diffraction, Raman spectra of $Tb(OH)_3$ and $Tb(OH)_3@SiO_2$ nanotubes were investigated. The photoluminescence (PL) spectrum of $Tb(OH)_3$ under 325 nm excitation consists of four main peaks at 488, 542, 582, and 618 nm. Furthermore, a preliminary suggestion for the mechanism of growth of the $Tb(OH)_3$ nanotubes using the soft – template synthesis technique has been proposed. The PL intensity from $Tb(OH)_3@SiO_2$ or $Tb(OH)_3@SiO_2:Eu^{3+}$ nanotubes is much stronger than that of $Tb(OH)_3$.

Keywords: Nanophotonics, nanobiophotonics, soft-template synthesis, seed growth method.

1. Introduction

One-dimensional (1D) nanomaterials have attracted much attention since the last decade due to their unique properties and potential applications in nanodevices. Many attempts have been made to fabricate 1D nanomaterials using template techniques so far [1, 2]. Recently, tubular nanostructures have stimulated intensive and world-wide research activities because their morphology is associated with an intrinsic multifunctional that arises from different contact regions; these properties make nanotubes very promising candidates for highly functional and more effective nanomaterials [3, 4]. Rare earth oxides are a kind of advanced material and have been widely used as high performance luminescent devices, magnets, catalysts and other functional materials [5]. Most of them show a collective function correlated with the bonding states of rare earth atoms or ions, so if rare earth oxides can be obtained in a nanotubular form, they might exhibit more effective collective functions resulting from shape-specific and size effects. To date some rare earth (Er, Tm, Yb, Lu, Dy, Ho, and Y) oxide nanotubes have been successfully fabricated by various methods [6, 7]. Inspired by the all routes to rare earth oxides, we designed a template-assisted hydrothermal route to obtained $Tb(OH)_3$. An organic reagent, which was considered to be a template, was adopted to induce $Tb(OH)_3$ growth into nanotubes in the course of hydrothermal crystallization. In this paper, we report some results about

© 2009 IOP Publishing Ltd

synthesis of Terbium - hydroxide $Tb(OH)_3$, Terbium - hydroxide-at-silica $Tb(OH)_3@SiO_2$ and $Tb(OH)_3@SiO_2:Eu^{3+}$ nanotubes. Terbium- hydroxide tubes were synthesized by soft template method. The core/shell materials were fabricated by the seed growth method.

2. Experiment

2.1. Sample Preparation

The core-shell-like Silica (Silica:Eu^{3+}) active - Terbium hydroxide nanotubes, where the core is the Terbium hydroxide tubes and the shell is a SiO_2 or $SiO_2:Eu^{3+}$, were prepared using the following protocol: 1) - The core Terbium hydroxide $Tb(OH)_3$ nanotubes were prepared from $TbCl_3.H_2O$ by soft template method; 2) - The shell was obtained by a seeded growth method. Briefly, 300 mg of Terbium hydroxide $Tb(OH)_3$ nanotubes have outer diameters ranging from 80 to 100 nm, wall thickness of about 30 nm, and lengths ranging from 300 to 500 nm, obtained by the soft template method were added to a solution with the molar ratio $TEOS:CH_3COOH:H_2O$ of 1:3:3 for $Tb(OH)_3@SiO_2$ sample. The preparation procedure of $Tb(OH)_3@SiO_2:Eu^{3+}$ is similar to that of $Tb(OH)_3@SiO_2$ sample with an exception of adding 1at.% $EuCl_3$. The mixture was stirred for 24 h with a magnetic stirrer. After stirring the mixture for 24 h, the samples were separated from the solution by centrifuging at 5900 rpm and washed at least twice with pure water. Subsequently, the Terbium – hydroxide, core-shell-like Terbium - hydroxide-at-Silica $Tb(OH)_3@SiO_2$ and $Tb(OH)_3@SiO_2:Eu^{3+}$ nanotubes were treated at 60°C for 6 h.

2.2. Characterization

The morphology of the as-synthesized samples was observed using a field emission scanning electron microscopy (FESEM, Hitachi). X-ray diffraction (XRD) measurements of the products were performed on an X-ray diffractometer (Siemens D5000 with $\lambda = 1.5406$ Å in the range of $15^\circ \leq 2\theta \leq 75^\circ$). The PL spectra of the $Tb(OH)_3$, $Tb(OH)_3@SiO_2$ and $Tb(OH)_3@SiO_2:Eu^{3+}$ nanotubes were measured at room temperature using a visible-ultraviolet spectrophotometer with a xenon lamp as the excitation light source. Raman spectra were also investigated. The PL spectra of $Tb(OH)_3$, $Tb(OH)_3@SiO_2$ and $Tb(OH)_3@SiO_2:Eu^{3+}$ nanotubes were also measured.

3. Result and discussion

The morphology of the obtained $Tb(OH)_3$ products was measured by field emission scanning electron microscopy (FESEM). An FESEM image taken of a large area (figure 1) reveals that the $Tb(OH)_3$ product consists of nanotubes with outer diameters ranging from 80 to 100 nm, wall thickness of about 30 nm, and lengths ranging from 300 to 500 nm in contrast to the irregularly shaped large particles of the starting material. It is clearly seen that the tips of these tubes are open, and larger tubes have hexagonal prism morphology.

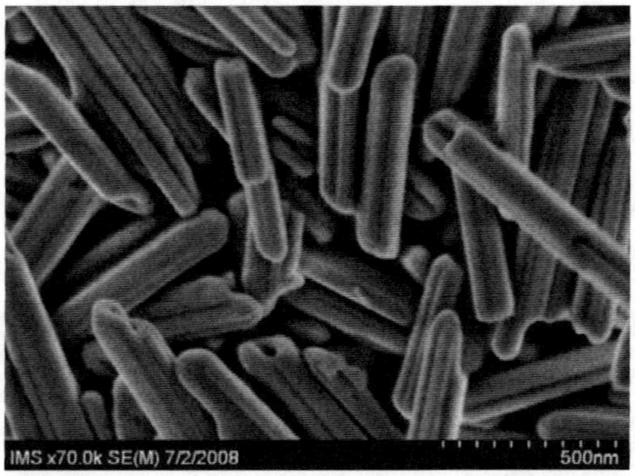

Figure 1. FESEM images of the $Tb(OH)_3$, nanotubes at 200°C for 24 h.

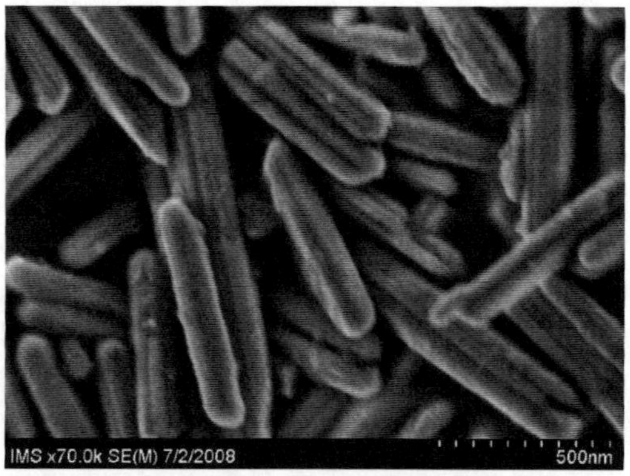

Figure 2. FESEM images of the $Tb(OH)_3@SiO_2$ tubes at 200°C for 24 h.

Figures 2 and 3 present FESEM images of the $Tb(OH)_3@SiO_2$ and $Tb(OH)_3@SiO_2:Eu^{3+}$ nanotubes. When core/shell materials of $Tb(OH)_3$ nanotubes are fabricated the diameter of the $Tb(OH)_3$ nanotubes is more than about 90–120 nm, wall thickness of about 30 nm, and lengths ranging from 300 to 700 nm.

Notably, after fabricating $Tb(OH)_3$, $Tb(OH)_3@SiO_2$ and $Tb(OH)_3@SiO_2:Eu^{3+}$ nanotubes, we can separate nanotubes to individual nanotubes. A FESEM image of the $Tb(OH)_3@SiO_2$ prepared at 200°C for 24 h is shown in figure 4. We can select a single nanotube. It has a variety of promising applications, for example, in nanoluminescence, magnets, catalysts, and other functional devices.

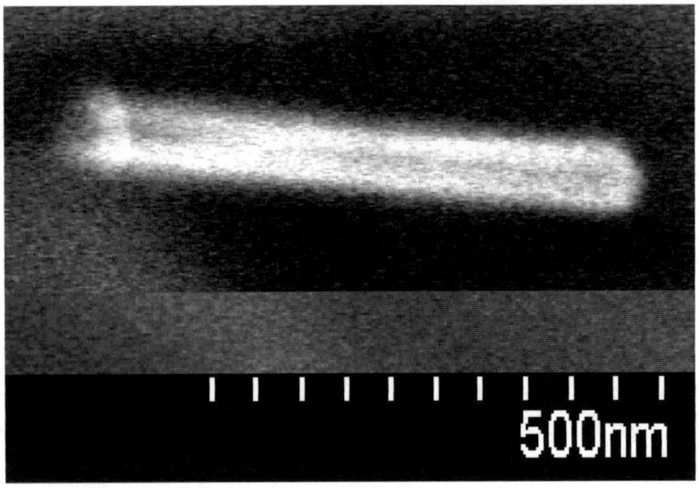

Figure 4. FESEM image of the $Tb(OH)_3@SiO_2$ nanotube at 200°C for 24 h.

The X-ray diffraction pattern of the as-prepared $Tb(OH)_3$ tubes is shown in figure 5. It can be seen from figure 5 that all of the diffraction peaks can be indicated a hexagonal phase of $Tb(OH)_3$ which is well-known to exhibit anisotropic growth and no by-products can be found. The XRD result confirmed also that the crystalline phase of the as-obtained $Tb(OH)_3$ nanotubes is very pure.

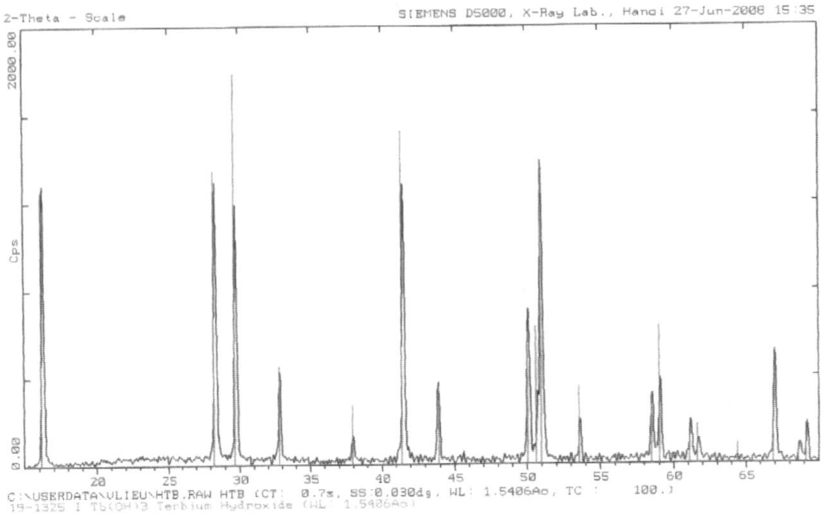

Figure 5. XRD pattern of $Tb(OH)_3$ nanotubes at 200°C for 24 h.

Figure 6 shows Raman scattering of $Tb(OH)_3$ nanotubes at 200°C for 24 h. Peaks at 136, 179, 307, 390, 496 and 692 cm^{-1} show vibration modes of Tb–O, O–H bonds.

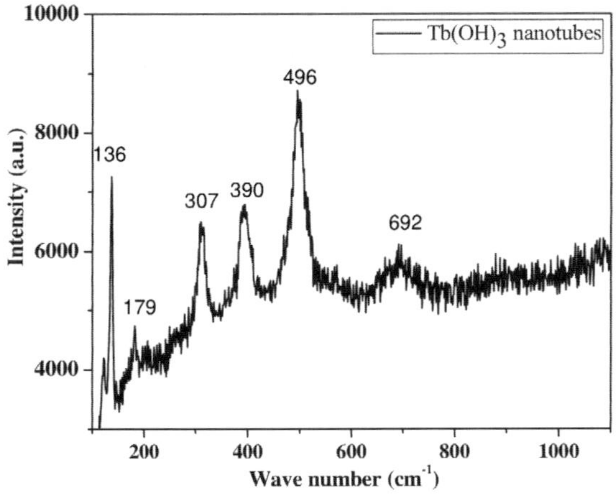

Figure 6. Raman scattering of $Tb(OH)_3$ nanotubes at 200°C for 24 h.

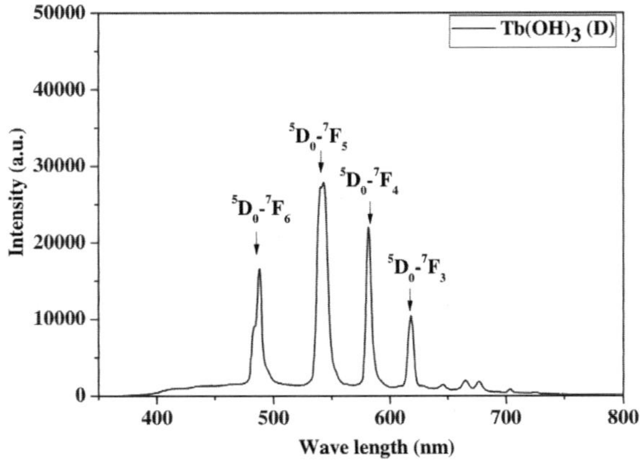

Figure 7. PL spectrum of $Tb(OH)_3$ nanotubes 200°C/24 h under an excitation wavelength of 365 nm at room temperature.

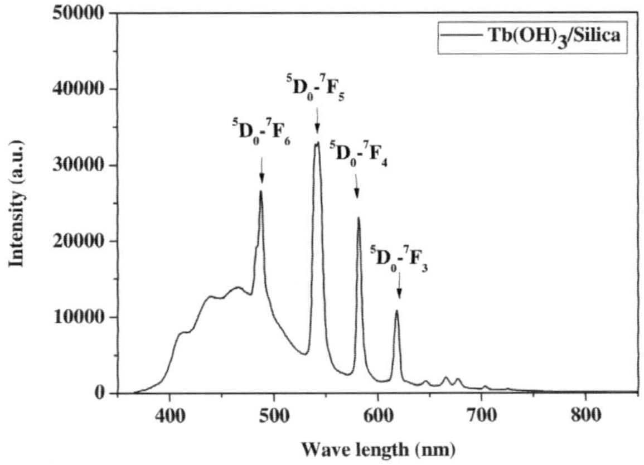

Figure 8. PL spectrum of $Tb(OH)_3@SiO_2$ nanotubes 200°C/24 h under an excitation wavelength of 365 nm at room temperature.

The PL spectrum of the $Tb(OH)_3$ nanotubes under an excitation wavelength of 365 nm at room temperature is displayed in figures 7, 8 and 9. The $Tb(OH)_3@SiO_2$ nanotubes displayed green emission with narrow bands corresponding to the intra-4f transitions $^5D_4 \rightarrow {}^7F_n$ (n = 6, 5, 4, 3). The bands were at 488 nm ($^5D_4 \rightarrow {}^7F_6$), 542 nm ($^5D_4 \rightarrow {}^7F_5$), 582 nm ($^5D_4 \rightarrow {}^7F_4$), and 618 nm ($^5D_4 \rightarrow {}^7F_3$), with the strongest emission at 542 nm. Figures 7, 8, and 9 show the PL spectrum $Tb(OH)_3$, $Tb(OH)_3@SiO_2$, $Tb(OH)_3@SiO_2:Eu^{3+}$ tubes 200°C/24 h, respectively. We found that the PL intensity from $Tb(OH)_3@SiO_2$ or $Tb(OH)_3@SiO_2:Eu^{3+}$ nanotubes is much stronger than that of $Tb(OH)_3$ one (about 30%).

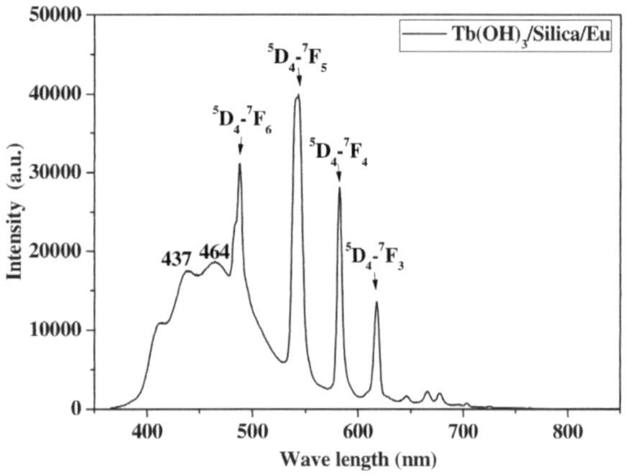

Figure 9. Photoluminescence spectrum $Tb(OH)_3@SiO_2:Eu^{3+}$ nanotubes 200°C/24 h under an excitation wavelength of 365 nm at room temperature.

4. Conclusion

In summary, the $Tb(OH)_3$, $Tb(OH)_3@SiO_2$ and $Tb(OH)_3@SiO_2:Eu^{3+}$ nanotubes have been in high yield prepared by soft- template method and were coated with Silica/Silica:Eu^{3+}. The size of $Tb(OH)_3$ is well built in diameter from 80 to 100nm, in length from 300 to 500nm with wall thickness of 30 nm. The coated shell could be adjusted from several ten nano to hundred one by using the seed growth technique. The XRD result indicates that the phase of the as-obtained $Tb(OH)_3$ nanotubes is very pure. The growth mechanism of rare earth hydroxide nanotubes can be explained well by their highly anisotropic crystal structure. These new types of rare earth compound nanotubes with open ends have a variety of promising applications, for example, in luminescent devices, magnets, catalysts, and other functional materials. The PL spectrum of the $Tb(OH)_3$ nanotubes displayed green emission with narrow bands corresponding to the intra-4f transitions $^5D_4 \rightarrow {}^7F_n$ (n = 6, 5, 4, 3) with the strongest emission at 542 nm ($^5D_4 \rightarrow {}^7F_5$). Special, the PL intensity from $Tb(OH)_3@SiO_2$ or $Tb(OH)_3@SiO_2:Eu^{3+}$ nanotubes is much stronger than that of $Tb(OH)_3$ one (about 30%).

References

[1] Xia B Y, Yang P, Sun Y, Wu Y, Mayers B, Gates B, Yin Y, Kim F and Yan H 2003 *Adv. Mater.* **15** 353
[2] Tao F, Wang Z, Yao L, Cai W and Xiaoguang Li 2006 *Nanotech.* **17** 1079
[3] Tang Q, Shen J, Zhou W, Zhang W, Yu W and Qian Y 2003 *J. Mater. Chem.* **13** 3103
[4] Wakefield G, Keron H A, Dobson P J and Hutchison J L 1999 *J. Phys. and Chem. of Solids* **60** 503
[5] Wang X and Li Y 2003 *Chem. Eur. J.* **9** 5627
[6] Fang B Y-P, Xu A-W, You L-P, Song R-Q, Yu J C, Zhang H-X, Li Q and Liu H-Q 2003 *Adv. Funct. Mater.* **13** 955
[7] Lin B Y-S, Hung Y, Lin H-Y, Tseng Y-H, Chen Y-F and Mou C-Y 2007 *Adv. Mater.* **19** 577

Acknowledgments

This work was supported by Vietnam Basic Research Programming Physic Science, project 408806 and National Key Lab of Electronic Materials and Devices.

Control of growth mode of multiwalled carbon nanotubes

Nguyen Hong Quang[1] and Do-Hyung Kim[2]

[1] Department of Physics, Vinh University, 182 Le Duan Street, Vinh City, Vietnam
[2] Department of Physics, Kyungpook National University, 1370 Sangyeok-dong, Bukgu, Daegu 702-701, Korea

E-mail: quangnh2007@yahoo.com

Abstract. We have conducted an experimental study to investigate the synthesis of multi-walled carbon nanotubes (CNTs) by a dc plasma-enhanced chemical vapour deposition (PECVD) technique. The synthesis of base and tip-type of CNTs was selectively controlled by changing the catalyst size, catalyst film thickness correlated with altering the NH_3 pretreatment plasma current. These types of CNT showed distinctive properties in nanotube structure, growth rate and vertical alignment, which were confirmed by scanning electron microscopy (SEM), transmission electron microscopy (TEM), and in situ optical interference measurement. The vertically aligned behaviour of CNT was systematically studied by using a fine-patterned catalyst layer with diverse critical dimensions. Freestanding single CNT was successfully realized by optimum tip-type CNT growth, conventional photolithography and wet-etch process.

Keywords: Carbon nanotube (CNT), multiwalled, growth, synthesis.

1. Introduction

Since their discovery in 1991, carbon nanotubes (CNTs) have attracted much attention due to their unique electronic and outstanding mechanical properties [1]. A wide range of potential applications including electron field emitters [2-6], gas sensors [7-11], nanoelectronic devices [12-14], supercapacitors [15], have been well demonstrated. Among various applications, the use of CNTs as electron field emitters is one of the most fascinating areas. Up to now, CNT can be mainly synthesized by arc discharge, laser ablation and chemical vapor deposition (CVD) [16]. Among these methods, the arc-discharge method can produce CNT on a large scale but the production usually contains several impurities [17]. Laser ablation process enables to synthesize CNT with higher purity but it is not amenable for scale up [18]. In recent years, CVD techniques have been widely used to produce CNTs with large scale and high quality [19, 20]. In this approach, a feedstock such as CO or hydrocarbon is heated to 800-1000°C with a transition metal catalyst to promote nanotube growth [21].

In comparison with conventional CVD, plasma-enhanced chemical vapour deposition (PECVD) is advantageous for direct growth of CNTs on substrates at a relatively low temperature due to the presence of uniform plasma [22-24]. In addition, PECVD has been considered as a good controllable method for growing vertically aligned and highly selective-positioned CNTs for field emitter applications [25-28], which are currently being considered for use in flat panel displays [29, 30]. Vertically aligned CNTs have also been proposed as nano-electrodes for sensitive biosensors [31, 32].

© 2009 IOP Publishing Ltd

Here we report on a simple method of control the growth mode between base and tip type CNTs as well as properties of grown CNTs. The dynamics of the CNT growing in each mode are also discussed in this work.

2. Experimental procedures

CNTs were grown by a dc glow discharge PECVD technique. The substrates used were n-type Si wafers and Ni films were sputtered as catalyst with various thicknesses. Acetylene (C_2H_2) and ammonia (NH_3) was used as a carbon source and etchant, respectively.

Firstly, the NH_3 pretreatment with a flow rate of 180 sccm was maintained for 5 minutes in order to etch the continuous catalyst Ni thin film. The initial smooth Ni catalyst film was transformed into nanometer-scale islands after the NH_3 pretreatment and reserved as seeds for the CNT nucleation. The NH_3 flow rate of 180 sccm and treating time of 5 minutes were kept for all catalyst film samples. A -450 V bias was applied to the substrate in order to create a dc-plasma. The plasma current with different thicknesses was adjusted from 20 mA to 60 mA in order to alter the morphology of the films. In the next step, C_2H_2 gas with the flow rate of 60 sccm was additionally introduced for CNT growth. To eliminate sample-to-sample variations and maintain a growth environment, the substrates experienced to various NH_3 pretreatments were processed under identical CNT growth condition.

The CNTs growth was performed at 600°C and monitored using an in situ optical interference system [33, 34]. It consists of a focused 650 nm laser diode with 1 mm beam diameter and a photo-detector. The growth experiments and measurements were automatically controlled by a personal computer with an in-house prepared software [34]. The SEM and TEM images of the grown CNT samples were performed using a FE-SEM 4300 and Hitachi 9000 system, respectively.

3. Results and discussions

In the PECVD techniques for CNT growth, it is believed that the catalyst particles play a crucial role. The thickness of catalyst film and its pre-treatment were also essential to control the size and shape of the subsequent products. This belief was verified by the experimental results as illustrated in figure 1.

(a) (b) (c)

Figure 1. SEM images of Ni islands from the catalytic film thickness of 10 nm, 30 nm and 20 nm, respectively (top panel) and CNT arrays grew on the corresponding substrates (bottom panel).

Figure 1 (top panel) shows the typical SEM images of catalytic Ni islands which were formed during the NH_3 pretreatment and served as the CNT nucleation sites. The uniformity and size of the Ni islands were affected and controlled by altering the thickness of Ni film and NH_3-pretreatment. The initial thin 10 nm-Ni film was changed to uniform Ni islands with an average 20 nm diameter after 60 mA-NH_3-pretreatment as shown in figure 1 (a, top panel). The etched Ni islands using thick 30 nm-Ni film showed a much larger average diameter of about 100 nm as illustrated in figure 1 (b, top panel). After 40 mA-NH_3-pretreatment using 20 nm-Ni film, the Ni islands have broad distribution in diameter (ranged from 20~70 nm) as depicted in figure 1 (c, top panel). The thinner Ni film or/and stronger plasma current of pretreatment were used, the smaller average size of catalyst islands were created and the thinner diameter of grown CNT could be produced. The initial thickness of the catalyst film did not only affect the diameter of grown CNTs, but also influenced their growth mode. As depicted in the bottom panels of figure 1 (a, b, c) which is corresponding to the thin film (10 nm), thick film (30 nm), and medium film (20 nm), respectively. It is observed that the CNTs which grew on the thin Ni film are predominantly vertical with an average diameter of 20 nm and height of 10 μm. Some tubes are curved due to its weight and aspect (the height-to-diameter ratio). The growth rates of these CNTs have values ranged from 100 nm/sec to 120 nm/sec in the initial stage with no catalyst particles were observed at the tubes top. The dependence of CNT length on growth time in base and tip mode is depicted in figure 2.

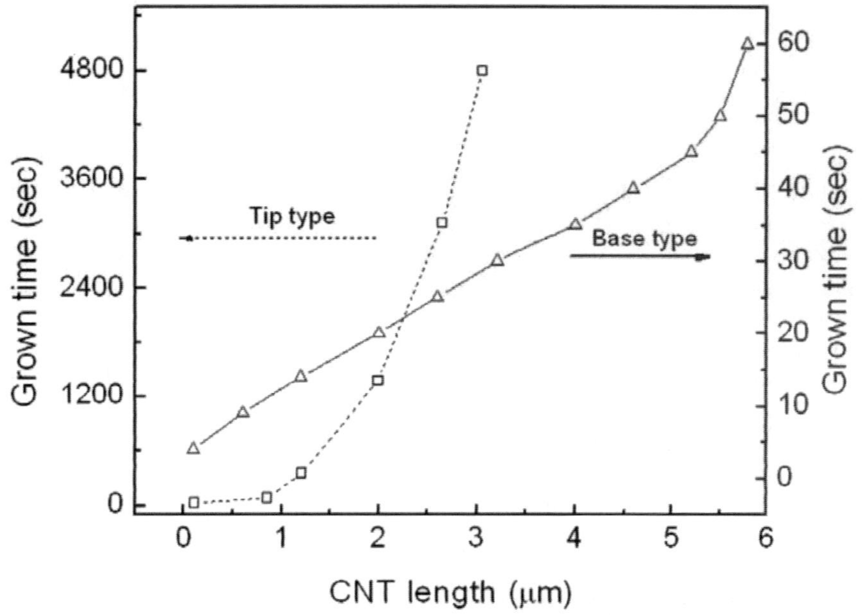

Figure 2. The dependence between CNT length and growth time in the base and tip mode growth.

As one can see from the figure that the tip-type CNT grew very slowly compared to the base type CNTs. Meanwhile the base type CNT rapidly grew and could achieve to 6 μm in length within one minute, the tip CNTs took more than one hour (80 minutes) to get 3-μm length. The length of tip CNTs almost no longer developed after the synthesis time lasted for more than two hours.

The magnified TEM images of the CNTs clearly showed that the catalyst particles located at the base of the tubes in case a thin film of Ni catalyst was used (figure 3a) or at the tip tube when a thicker size of Ni catalyst was employed (figure 3b). If medium size of Ni catalyst particles were used, a mixture of base and tip CNTs were formed (figure 3c). The similar result could be taken place when plasma current of NH_3 pretreatment was raised to 40 mA. No catalyst particles were observed at the tip of the longest CNT tube, inferring that the catalyst particles, if they have, must locate on the base of the substrate. In this case, the CNT has a small diameter (about 5 nm, see the inset in figure 3c). The fraction of base mode CNTs increased when the plasma current increased from 40 mA to 60 mA.

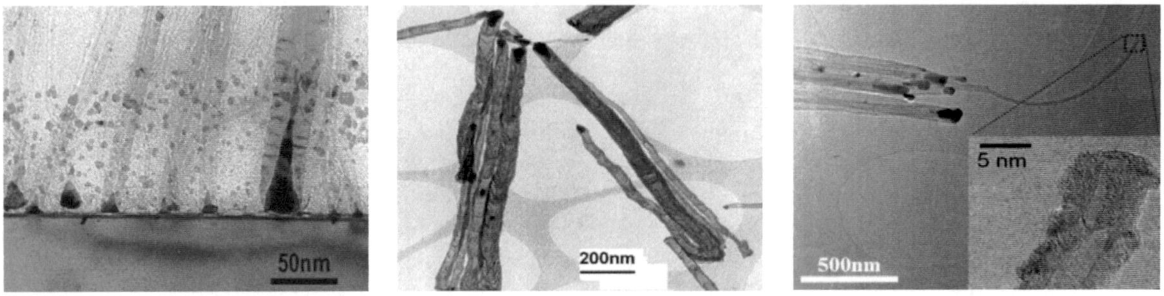

(a) (b) (c)

Figure 3. TEM images of CNT in base, tip and mixed growth mode, respectively.

From the experimental results, we assure that the stronger plasma current had formed smaller catalytic islands on the substrates. In case of thin Ni catalyst film (10 nm), the pretreatment with relatively weak plasma current had transformed the initial smooth film into Ni particles with uniform shape and size. The annealing temperature during CNT growth might melt the catalyst particles, changing them into wetting islands. The adhesion force between the wetting islands and the substrate which must be larger than that between solid state particles and the substrate is probably responsible to stick the catalyst particles on the substrate. That is the reason why the base growth of CNTs was promoted when a thin Ni film was used as catalyst. Moreover, when plasma current of the pretreatment was stronger than 60 mA, most of etched catalyst Ni islands had gone with no CNTs observed on the substrate. For thicker catalyst Ni films, other effects might take place during the plasma etching pretreatment. Due to thermal influence, some etched Ni particles could be agglomerated into coarsened islands. In this case, the splitting is inhomogeneous, giving a broad size distribution of catalyst islands, including very larger clusters [35] as seen from figure 1b, c (top panels). The small catalyst islands usually support the base mode of CNT growth, meanwhile the larger ones promote for the tip growth mode. However, it is possible for large CNTs to grow in base mode if they were supported by large-wetting catalyst islands because of large adhesion force between substrate and catalyst islands in the wetting phase. Subsequently, the CNTs grown on the thick films have a wide distribution of diameter with a mixture of base and tip growth modes.

4. Conclusions

In summary, we have discussed the experimental results on the CNT synthesis via a dc PECVD technique with Ni used as catalyst in different film thicknesses. The C_2H_2 and NH_3 were used as feedstock and etching gas, respectively. The diameter and growth mode of CNT were controlled by changing the film thickness and the plasma current of the NH_3 pretreatment. For thin catalyst film, the CNT grew in base mode for all of the available plasma currents. For thicker catalyst film, a tip mode of CNT was observed only when a 20 mA plasma current of pretreatment had been applied. In the other films, a mixture of base and tip mode CNT coexisted on the substrate. The mechanism of growth mode change has been discussed in this work.

References

[1] Saito R and Dresselhaus M S 1998 *Physical Properties of Carbon Nanotubes* (Imperial College Pres) p 35

[2] Gulyaev Y V, Chernozatonskii L A, Kosakovskaja Z J, Sinitsyn N I, Torgashov G V and Zakharchenko Y F 1995 *J. Vacu. Sci. Technol. B* **13** 435

[3] deHeer W A, Bonard J M, Stockli T, Chatelain A, Forro L and Ugarte D 1997 *Zeitschrift Fur Physik D-Atoms Molecules and Clusters* **40** 418
[4] Manohara H M, Bronikowski M J, Hoenk M, Hunt B D and Siegel P H 2005 *J. Vacuum Sci. Technol. B* **23** 157
[5] Sakai Y, Haga A, Sugita S, Kita S, Tanaka S I, Okuyama F and Kobayashi N 2007 *Rev. Sci. Instrum.* **78** 013305
[6] Rakhi R B, Reddy A L M, Shaijumon M M, Sethupathi K and Ramaprabhu S 2008 *J. Nanopart. Res.* **10** 179
[7] Ahn K S, Kim J H, Lee K N, Kim C O and Hong J P 2004 *J. Korean Phys. Soc.* **45** 158
[8] Liang Y X, Chen Y J and Wang T H 2004 *Appl. Phys. Lett.* **85** 666
[9] Quang N H, Van Trinh M and Huh J S 2005 *Eco-Materials Processing & Design Vi* **486-487** 485
[10] Hoa N D, Van Quy N, Cho Y and Kim D 2007 *Sens. Actu. B-Chem.* **127** 447
[11] Maklin J, Mustonen T, Kordas K, Saukko S, Toth G and Vahakangas J 2007 *Phys. Stat. Soli.B* **244** 4298
[12] Avouris P, Hertel T, Martel R, Schmidt T, Shea H R and Walkup R E 1999 *Appl. Surf. Sci.* **141** 201
[13] Terrones M 2004 *Int. Mater. Rev.* **49** 325
[14] Johnston D E, Islam M F, Yodh A G and Johnson A T 2005 *Nat. Mater.* **4** 589
[15] Wu F and Xu B 2006 *New Carbon Materials* **21** 176
[16] Dresselhaus M S, Dresselhaus G and Avouris P 2001 *Carbon nanotubes: synthesis, structure, properties, and applications*, in *Topics in applied physics* (Berlin - Heiselberg: Springer-Verlag) p 1
[17] Ando Y and Zhao X L 2006 *New Diamond Front. Carbon Technol.* **16** 123
[18] Zhang M F, Yudasaka M and Iijima S 2001 *Chem. Phys. Lett.* **336** 196
[19] Kasumov Y A, Shailos A, Khodos I I, Volkov V T, Levashov V I, Matveev V N, Gueron S, Kobylko M, Kociak M, Bouchiat H, Agache V, Rollier A S, Buchaillot L, Bonnot A M and Kasumov A Y 2007 *Appl. Phys. A-Mate. Sci. & Proc.* **88** 687
[20] Lee W Y, Lin H, Gu L, Leou K C and Tsai C H 2008 *Diam. Relat. Mater.* **17** 66
[21] Meyyappan M, Delzeit L, Cassell A and Hash D 2003 *Plas. Sour. Sci. Technol* **12** 205
[22] Park D, Kim Y H and Lee J K 2003 *Carbon* **41** 1025
[23] Hesamadeh H, Ganjipour B, Mohajerzadeh S, Khodadadi A, Mortazavi Y and Kiani S 2004 *Carbon* **42** 1043
[24] Yuji T and Sung Y M 2007 *Ieee Trans. Plas. Sci.* **35** 1027
[25] Zhang W D, Thong J T L, Tjiu W C and Gan L M 2002 *Diam. Relat. Mater.* **11** 1638
[26] Lee S B, Teh A S, Teo K B K, Chhowalla M, Hasko D G, Milne W I, Amaratunga G A J, and Ahmed H 2003 *Nanotechnol.* **14** 192
[27] Yeh C M, Chen M Y, Hwang J, Gan J Y and Kou C S 2006 *Nanotechnology* **17** 5930
[28] Kim D, Lim S H, Guilley A J, Cojocaru C S, Bouree J E, Vila L, Ryu J H, Park K C and Jang J 2008 *Thin Solid Films* **516** 706
[29] Wei S, Kang W P, Davidson J L, Choi B K and Huang J H 2006 *J. Vacu. Sci. Technol. B* **24** 1190
[30] Ma Y P, Shang X F, Gu Z Q, Li Z H and Xu Y B 2007 *Acta Phys. Sinica* **56** 6701
[31] Nguyen C V, Delzeit L, Cassell A M, Li J, Han J and Meyyappan M 2002 *Nano Lett.* **2** 1079
[32] Yu X, Kim S N, Papadimitrakopoulos F, and Rusling J F 2005 *Mol. BioSyst.* **1** 70
[33] Kim D H, Jang H S, Kim C D, Cho D S, Yang H S, Kang H D, Min B K and Lee H R 2003 *Nano Lett.* **3** 863
[34] Kim D H, Jang H S, Kim C D, Cho D S, Kang H D, Jee J G and Lee H R 2003 *Carbon* **41** 583
[35] Hofmann S, Cantoro M, Kleinsorge B, Casiraghi C, Parvez A, Robertson J and Ducati C 2005 *J. Appl. Phys.* **98** 034308

Density matrix of strongly coupled quantum dot - microcavity system

Nguyen Van Hop

Hanoi National University of Education,
136 Xuan Thuy Road, Cau Giay Distr., Hanoi, Vietnam

E-mail: hopnvdhsp@yahoo.com

Abstract. Any two-level quantum system can be used as a quantum bit (qubit)-the basic element of all devices and systems for quantum information and quantum computation. Recently it was proposed to study the strongly coupled system consisting of a two-level quantum dot and a monoenergetic photon gas in a microcavity-the strongly coupled quantum dot-microcavity (QD-MC) system for short, with the Jaynes-Cumming total Hamiltonian, for the application in the quantum information processing. Different approximations were applied in the theoretical study of this system. In this work, on the basis of the exact solution of the Schrodinger equation for this system without dissipation we derive the exact formulae for its density matrix. The realization of a qubit in this system is discussed. The solution of the system of rate equation for the strongly coupled QD-MC system in the presence of the interaction with the environment was also established in the first order approximation with respect to this interaction.

Keywords: Quantum bit, quantum information, quantum dot, rate equations.

1. Introduction

Cavity quantum electrodynamics (CQED) studies the properties of atoms coupled to discrete photon modes in high-Q cavities. Such systems are of great interest in the study of the fundamental quantum mechanics of open systems, the engineering of quantum states and the study of measurement-induced decoherence, and have also been proposed as possible candidates for use in quantum information processing and transmission [1-10]. This work is devoted to the study of the density matrix of the QD-MC system with the Jaynes-Cumming Hamiltonian. For the system without the dissipative interaction of the environment we derive exact expressions of the density matrix. In the presence of the weak dissipative interaction of the system with the environment the system of rate equations can be solved by means of the perturbation theory. We derive the expressions of their solution in the first order approximation.

2. System without decoherence

Consider the simplest model of strongly coupled system of a two-level quantum dot (QD) and monoenergetic photons in a microcavity (MC) with the Jaynes-Cumming total Hamiltonian

$$H_S = E_g c_g^+ c_g + E_e c_e^+ c_e + \omega\gamma^+\gamma + f\left(\gamma c_e^+ c_g + \gamma^+ c_g^+ c_e\right), \qquad (1)$$

where c_e and c_e^+ are the destruction and creation operators for the electrons at the energy level

© 2009 IOP Publishing Ltd

E_e ($E_e - E_g = E > 0$), and similarly for c_g and c_g^+, γ and γ^+ are the photon destruction and creation operators, ω is the photon energy and $f > 0$ is the effective constant of the electron-photon interaction. In the Hilbert space of state vectors there are following vector subspaces invariant under the action of the Hamiltonian:

1-d subspace V_1 with unit vector

$$|\varphi\rangle = |0\gamma, g\rangle, \tag{2}$$

2-d subspaces $V_2^{(n)}, n \geq 1$, with two basis vectors in each subspace:

$$\left|\varphi_1^{(n)}\right\rangle = \left|(n-1)\gamma, e\right\rangle = \frac{\left(\gamma^+\right)^{n-1}}{\sqrt{n!}} c_e^+ \left|vacuum\right\rangle, \tag{3}$$

$$\left|\varphi_2^{(n)}\right\rangle = \left|n\gamma, g\right\rangle = \frac{\left(\gamma^+\right)^{n}}{\sqrt{n!}} c_g^+ \left|vacuum\right\rangle. \tag{4}$$

Each pair of two states in any 2-d invariant subspace $V_2^{(n)}$ can be used as a physical realization of a qubit. In the absence of the damping, the eigenstates and eigenvalues of the Jaynes-Cumming Hamiltonian are

$$\psi_\alpha^{(n)} = A_\alpha^{(n)} \left|\varphi_1^{(n)}\right\rangle + B_\alpha^{(n)} \left|\varphi_2^{(n)}\right\rangle, \alpha = +,-,\ \psi^{(0)} = |\varphi\rangle, \tag{5}$$

$$E_\pm^{(n)} = \frac{\Delta}{2} + E_g + n\omega \pm \frac{1}{2}\Omega_n,\ E^{(0)} = E_g, \tag{6}$$

where

$$A_\pm^{(n)} = \left[\frac{1}{2}\left(1 \mp \frac{\Delta}{\Omega_n}\right)\right]^{1/2}, \tag{7}$$

$$B_\pm^{(n)} = \pm\left[\frac{1}{2}\left(1 \pm \frac{\Delta}{\Omega_n}\right)\right]^{1/2}, \tag{8}$$

$$\Omega_n = \left(\Delta^2 + 4f_n^2\right)^{1/2},\ f_n = \sqrt{n}f,\ \Delta = E - \omega. \tag{9}$$

In the 2-d subspaces $V_2^{(n)}, n \geq 1$, we have following 4 the elements of the density matrix

$$\begin{aligned}
\rho_{(g,n)(g,n)} &= \langle g,n|\rho|g,n\rangle \equiv \rho_{22}^{(n)}, \\
\rho_{(g,n)(e,n-1)} &= \langle g,n|\rho|e,(n-1)\rangle \equiv \rho_{21}^{(n)}, \\
\rho_{(e,n-1)(g,n)} &= \langle e,(n-1)|\rho|g,n\rangle \equiv \rho_{12}^{(n)}, \\
\rho_{(e,n-1)(e,n-1)} &= \langle e,(n-1)|\rho|e,(n-1)\rangle \equiv \rho_{11}^{(n)}.
\end{aligned} \tag{10}$$

Using exact solution (5) of the Schrodinger equation for this system without dissipation, we derive exact formulae for its density matrix

$$\begin{aligned}
\rho_{11}^{(n)}(t) = &\left\{1 - \frac{2f_n^2}{\Omega_n^2}\left[1-\cos(\Omega_n t)\right]\right\}\rho_{11}^{(n)}(0) + \frac{f_n}{\Omega_n}\left\{\frac{\Delta}{\Omega_n}\left[1-\cos(\Omega_n t)\right] + i\sin(\Omega_n t)\right\}\rho_{12}^{(n)}(0) \\
&+ \frac{f_n}{\Omega_n}\left\{\frac{\Delta}{\Omega_n}\left[1-\cos(\Omega_n t)\right] - i\sin(\Omega_n t)\right\}\rho_{21}^{(n)}(0) + \frac{2f_n^2}{\Omega_n^2}\left[1-\cos(\Omega_n t)\right]\rho_{22}^{(n)}(0),
\end{aligned} \tag{11}$$

$$\begin{aligned}\rho_{22}^{(n)}(t)=&\frac{2f_n^2}{\Omega_n^2}\left[1-\cos(\Omega_n t)\right]\rho_{11}^{(n)}(0)-\frac{f_n}{\Omega_n}\left\{\frac{\Delta}{\Omega_n}\left[1-\cos(\Omega_n t)\right]+i\sin(\Omega_n t)\right\}\rho_{12}^{(n)}(0)\\&-\frac{f_n}{\Omega_n}\left\{\frac{\Delta}{\Omega_n}\left[1-\cos(\Omega_n t)\right]-i\sin(\Omega_n t)\right\}\rho_{21}^{(n)}(0)+\left\{1-\frac{2f_n^2}{\Omega_n^2}\left[1-\cos(\Omega_n t)\right]\right\}\rho_{22}^{(n)}(0),\end{aligned} \tag{12}$$

$$\begin{aligned}\rho_{12}^{(n)}(t)=&\frac{f_n}{\Omega_n}\left\{\frac{\Delta}{\Omega_n}\left[1-\cos(\Omega_n t)\right]+i\sin(\Omega_n t)\right\}\left[\rho_{11}^{(n)}(0)-\rho_{22}^{(n)}(0)\right]+\frac{2f_n^2}{\Omega_n^2}\left[1-\cos(\Omega_n t)\right]\rho_{21}^{(n)}(0)\\&+\left[\frac{2f_n^2}{\Omega_n^2}+\left(1-\frac{2f_n^2}{\Omega_n^2}\right)\cos(\Omega_n t)-i\frac{\Delta}{\Omega_n}\sin(\Omega_n t)\right]\rho_{12}^{(n)}(0),\end{aligned} \tag{13}$$

$$\begin{aligned}\rho_{21}^{(n)}(t)=&\frac{f_n}{\Omega_n}\left\{\frac{\Delta}{\Omega_n}\left[1-\cos(\Omega_n t)\right]-i\sin(\Omega_n t)\right\}\left[\rho_{11}^{(n)}(0)-\rho_{22}^{(n)}(0)\right]+\frac{2f_n^2}{\Omega_n^2}\left[1-\cos(\Omega_n t)\right]\rho_{12}^{(n)}(0)\\&+\left[\frac{2f_n^2}{\Omega_n^2}+\left(1-\frac{2f_n^2}{\Omega_n^2}\right)\cos(\Omega_n t)+i\frac{\Delta}{\Omega_n}\sin(\Omega_n t)\right]\rho_{21}^{(n)}(0).\end{aligned} \tag{14}$$

Let us represent the density matrix in the form

$$\rho=\begin{pmatrix}\rho_{(g,n)(g,n)} & \rho_{(g,n)(e,n-1)}\\ \rho_{(e,n-1)(g,n)} & \rho_{(e,n-1)(e,n-1)}\end{pmatrix}=\sigma_1\rho_x^{(n)}+\sigma_2\rho_y^{(n)}+\sigma_3\rho_z^{(n)}+I\rho_0^{(n)}, \tag{15}$$

where $\sigma_1, \sigma_2, \sigma_3$ and I are the Pauli matrices and the identity matrix. From equations (11)-(15) it follows that

$$\rho_x^{(n)}=\left[\frac{4f_n^2}{\Omega_n^2}+\left(1-\frac{4f_n^2}{\Omega_n^2}\right)\cos(\Omega_n t)\right]\rho_x^{(n)}(0)-\frac{\Delta}{\Omega_n}\sin(\Omega_n t)\rho_y^{(n)}(0)-\frac{2f_n\Delta}{\Omega_n^2}\left[1-\cos(\Omega_n t)\right]\rho_z^{(n)}(0), \tag{16}$$

$$\rho_y^{(n)}=-\frac{\Delta}{\Omega_n}\sin(\Omega_n t)\rho_x^{(n)}(0)+\cos(\Omega_n t)\rho_y^{(n)}(0)-\frac{2f_n}{\Omega_n}\sin(\Omega_n t)\rho_z^{(n)}(0), \tag{17}$$

$$\rho_z^{(n)}=-\frac{2\Delta f_n}{\Omega_n^2}\left[1-\cos(\Omega_n t)\right]\rho_x^{(n)}(0)+\frac{2f_n}{\Omega_n}\sin(\Omega_n t)\rho_y^{(n)}(0)+\left\{1-\frac{4f_n^2}{\Omega_n^2}\left[1-\cos(\Omega_n t)\right]\right\}\rho_z^{(n)}(0). \tag{18}$$

Formulae (16)-(18) show that the density matrix at any $t>0$ is completely determined by the initial values $\rho_\alpha^{(n)}(0)$, $\alpha=x,y,z$. At the resonance $\omega=E$, we have

$$\rho_x^{(n)}=\rho_x^{(n)}(0), \tag{19}$$

$$\rho_y^{(n)}=\cos(\Omega_n t)\rho_y^{(n)}(0)-\sin(\Omega_n t)\rho_z^{(n)}(0), \tag{20}$$

$$\rho_z^{(n)}=\cos(\Omega_n t)\rho_z^{(n)}(0)+\sin(\Omega_n t)\rho_y^{(n)}(0). \tag{21}$$

3. System with decoherence

Coupling to additional uncontrollable degrees of freedom leads to energy relaxation and dephasing in the system. In the Born-Markov approximation, the density matrix is determined by the quantum Liouville equation [3,4,5]

$$\frac{d\rho}{dt}=-i\left[H_S,\rho\right]+L\rho, \tag{22}$$

where $L\rho$ describes the effect of the environment on the system

$$L\rho = \lambda_\gamma\left(L^\gamma\rho\right) + \lambda_\sigma\left(L^\sigma\rho\right) + \lambda_z\left(L^z\rho\right), \tag{23}$$

with

$$L^\gamma\rho = \gamma\rho\gamma^+ - \frac{1}{2}\gamma^+\gamma\rho - \frac{1}{2}\rho\gamma^+\gamma, \tag{24}$$

$$L^\sigma\rho = \sigma_-\rho\sigma_+ - \frac{1}{2}\sigma_+\sigma_-\rho - \frac{1}{2}\rho\sigma_+\sigma_-, \tag{25}$$

$$L^z\rho = \sigma_z\rho\sigma_z^+ - \frac{1}{2}\sigma_z^+\sigma_z\rho - \frac{1}{2}\rho\sigma_z^+\sigma_z. \tag{26}$$

In equation (23), λ_γ is the resonator rate of photon loss, λ_σ is the qubit energy dissipation rate and λ_z is the qubit pure dephasing rate. Equation (22) is obtained in the Markov approximation. For high-quality factor systems, like high-Q transmission line resonators or (most) superconducting qubit, this approximation is accurate as the system is probing the environment in a very small frequency bandwidth [3-5].

Suppose that at the initial time moment the system is excited only up to the states in the subspace $V_2^{(N)}$. In this case the matrix elements of ρ in subspaces with higher energies vanish. Substituting the expressions (23)-(26) into equation (22), we derive the following system of rate equations

$$i\frac{d\rho_{11}^{(N)}}{dt} = -i\gamma_{1N}\rho_{11}^{(N)} - f_N\left[\rho_{12}^{(N)} - \rho_{21}^{(N)}\right], \tag{27}$$

$$i\frac{d\rho_{22}^{(N)}}{dt} = -i\gamma_{2N}\rho_{22}^{(N)} + f_N\left[\rho_{12}^{(N)} - \rho_{21}^{(N)}\right], \tag{28}$$

$$i\frac{d\rho_{12}^{(N)}}{dt} = \left(\Delta - i\gamma_N\right)\rho_{12}^{(N)} - f_N\left[\rho_{11}^{(N)} - \rho_{22}^{(N)}\right], \tag{29}$$

$$i\frac{d\rho_{21}^{(N)}}{dt} = -\left(\Delta + i\gamma_N\right)\rho_{21}^{(N)} + f_N\left[\rho_{11}^{(N)} - \rho_{22}^{(N)}\right], \tag{30}$$

where

$$\gamma_N = \frac{(2N-1)\lambda_\gamma + \lambda_\sigma}{2} + 2\lambda_z,\ \gamma_{1N} = \left(N-1\right)\lambda_\gamma + \lambda_\sigma,\ \gamma_{2N} = N\lambda_\gamma. \tag{31}$$

The decoherence constants $\lambda_\gamma, \lambda_\sigma$ and λ_z are very small in comparison with the energy differences between the energy levels. In the first order approximation we obtain the following solution of the rate equations

$$\rho_{ij}^{(N)}(t) = \sum_{kl} G_{ij,kl}^{(N)}(t)\rho_{kl}^{(N)}(0), \text{ with } ij,kl = 11,12,21,22, \tag{32}$$

with

$$G_{ij,kl}^{(N)}(t) = A_{ij,kl}^{(N)}\exp\left(-\Gamma_{1N}t\right) + B_{ij,kl}^{(N)}\exp\left(-\Gamma_{2N}t\right) + C_{ij,kl}^{(N)}\exp\left(-\Gamma'_N t\right)\cos\left(\Omega_N t\right) + iD_{ij,kl}^{(N)}\exp\left(-\Gamma'_N t\right)\sin\left(\Omega_N t\right), \tag{33}$$

where

$$A_{11,11}^{(N)} = \frac{\left(\Gamma_{1N} - \gamma_{2N}\right)\Delta^2 + 2\left(\Gamma_{1N} - \gamma_N\right)f_N^2}{\left(\Gamma_{1N} - \Gamma_{2N}\right)\Omega_N^2},\quad A_{11,12}^{(N)} = A_{11,21}^{(N)} = A_{12,11}^{(N)} = A_{21,11}^{(N)} = \frac{\left(\Gamma_{1N} - \gamma_{2N}\right)f_N\Delta}{\left(\Gamma_{1N} - \Gamma_{2N}\right)\Omega_N^2},$$

$$A_{11,22}^{(N)} = A_{22,11}^{(N)} = \frac{2\left(\Gamma_{1N} - \gamma_N\right)f_N^2}{\left(\Gamma_{1N} - \Gamma_{2N}\right)\Omega_N^2},\quad A_{12,12}^{(N)} = A_{12,21}^{(N)} = A_{21,12}^{(N)} = A_{21,21}^{(N)} = \frac{\left[2\Gamma_{1N} - \left(\gamma_{1N} + \gamma_{2N}\right)\right] f_N^2}{\left(\Gamma_{1N} - \Gamma_{2N}\right)\Omega_N^2}, \tag{34}$$

$$A^{(N)}_{12,22} = A^{(N)}_{21,22} = A^{(N)}_{22,12} = A^{(N)}_{22,21} = -\frac{(\Gamma_{1N}-\gamma_{1N})f_N\Delta}{(\Gamma_{1N}-\Gamma_{2N})\Omega_N^2},\quad A^{(N)}_{22,22} = \frac{(\Gamma_{1N}-\gamma_{1N})\Delta^2+2(\Gamma_{1N}-\gamma_N)f_N^2}{(\Gamma_{1N}-\Gamma_{2N})\Omega_N^2}.$$

$$B^{(N)}_{11,11} = \frac{(\Gamma_{2N}-\gamma_{2N})\Delta^2+2(\Gamma_{2N}-\gamma_N)f_N^2}{(\Gamma_{2N}-\Gamma_{1N})\Omega_N^2},\quad B^{(N)}_{11,12} = B^{(N)}_{11,21} = B^{(N)}_{12,11} = B^{(N)}_{21,11} = \frac{(\Gamma_{2N}-\gamma_{2N})f_N\Delta}{(\Gamma_{2N}-\Gamma_{1N})\Omega_N^2},$$

$$B^{(N)}_{11,22} = B^{(N)}_{22,11} = \frac{2(\Gamma_{2N}-\gamma_N)f_N^2}{(\Gamma_{2N}-\Gamma_{1N})\Omega_N^2},\quad B^{(N)}_{12,12} = B^{(N)}_{12,21} = B^{(N)}_{21,12} = B^{(N)}_{21,21} = \frac{\left[2\Gamma_{2N}-(\gamma_{1N}+\gamma_{2N})\right] f_N^2}{(\Gamma_{2N}-\Gamma_{1N})\Omega_N^2}, \tag{35}$$

$$B^{(N)}_{12,22} = B^{(N)}_{21,22} = B^{(N)}_{22,12} = B^{(N)}_{22,21} = -\frac{(\Gamma_{2N}-\gamma_{1N})f_N\Delta}{(\Gamma_{2N}-\Gamma_{1N})\Omega_N^2},\quad B^{(N)}_{22,22} = \frac{(\Gamma_{2N}-\gamma_{1N})\Delta^2+2(\Gamma_{2N}-\gamma_N)f_N^2}{(\Gamma_{2N}-\Gamma_{1N})\Omega_N^2}.$$

$$C^{(N)}_{11,11} = -C^{(N)}_{11,22} = -C^{(N)}_{12,21} = -C^{(N)}_{21,12} = -C^{(N)}_{22,11} = C^{(N)}_{22,22} = \frac{2f_N^2}{\Omega_N^2},\quad C^{(N)}_{12,12} = C^{(N)}_{21,21} = \frac{\Omega_N^2-2f_N^2}{\Omega_N^2},$$

$$C^{(N)}_{11,12} = C^{(N)}_{11,21} = -C^{(N)}_{12,22} = -C^{(N)}_{21,22} = C^{(N)}_{12,11} = C^{(N)}_{21,11} = -C^{(N)}_{22,12} = -C^{(N)}_{22,21} = -\frac{f_N\Delta}{\Omega_N^2}. \tag{36}$$

$$D^{(N)}_{11,11} = D^{(N)}_{11,22} = D^{(N)}_{12,21} = D^{(N)}_{21,12} = D^{(N)}_{22,11} = D^{(N)}_{22,22} = 0,\quad D^{(N)}_{12,12} = -D^{(N)}_{21,21} = -\frac{\Delta}{\Omega_N},$$

$$D^{(N)}_{11,12} = -D^{(N)}_{11,21} = D^{(N)}_{12,11} = -D^{(N)}_{12,22} = -D^{(N)}_{21,11} = D^{(N)}_{21,22} = -D^{(N)}_{22,12} = D^{(N)}_{22,21} = \frac{f_N}{\Omega_N}. \tag{37}$$

$$\Gamma_{1N} = \frac{1}{2}\left[1+\sqrt{1-k_N}\right]\Gamma_N,\quad \Gamma_{2N} = \frac{1}{2}\left[1-\sqrt{1-k_N}\right]\Gamma_N. \tag{38}$$

$$\Gamma_N = \frac{\Delta^2[\gamma_{1N}+\gamma_{2N}]+2f_N^2[\gamma_{1N}+\gamma_{2N}+2\gamma_N]}{\Omega_N^2},\quad \Gamma'_N = \frac{f_N^2}{\Omega_N^2}(\gamma_{1N}+\gamma_{2N})+\left(1-2\frac{f_N^2}{\Omega_N^2}\right)\gamma_N. \tag{39}$$

$$k_N = 4\frac{(\Delta^2+4f_N^2)\left[\Delta^2\gamma_{1N}\gamma_{2N}+2f_N^2\gamma_N(\gamma_{1N}+\gamma_{2N})\right]}{\left[\Delta^2(\gamma_{1N}+\gamma_{2N})+2f_N^2(\gamma_{1N}+\gamma_{2N}+2\gamma_N)\right]^2}. \tag{40}$$

Due to the term $L\rho$ in equation (22) in the system of the rate equations for the matrix elements in the subspace $V_2^{(N-1)}$ there are the contributions from the known matrix elements in the subspace $V_2^{(N)}$. In general, for any set of matrix elements (10) with $n < N$ we have following rate equations

$$i\frac{d\rho_{11}^{(n)}}{dt} = -i\gamma_{1n}\rho_{11}^{(n)} - f_n\left[\rho_{12}^{(n)}-\rho_{21}^{(n)}\right] + in\lambda_\gamma\rho_{11}^{(n+1)}, \tag{41}$$

$$i\frac{d\rho_{22}^{(n)}}{dt} = -i\gamma_{2n}\rho_{22}^{(n)} + f_n\left[\rho_{12}^{(n)}-\rho_{21}^{(n)}\right] + i\left[\lambda_\sigma\rho_{11}^{(n+1)}+(n+1)\lambda_\gamma\rho_{22}^{(n+1)}\right], \tag{42}$$

$$i\frac{d\rho_{12}^{(n)}}{dt} = (\Delta-i\gamma_n)\rho_{12}^{(n)} - f_n\left[\rho_{11}^{(n)}-\rho_{22}^{(n)}\right] + i\lambda_\gamma\sqrt{n(n+1)}\rho_{12}^{(n+1)}, \tag{43}$$

$$i\frac{d\rho_{21}^{(n)}}{dt} = -(\Delta+i\gamma_n)\rho_{21}^{(n)} + f_n\left[\rho_{11}^{(n)}-\rho_{22}^{(n)}\right] + i\lambda_\gamma\sqrt{n(n+1)}\rho_{21}^{(n+1)}. \tag{44}$$

Beside of the matrix elements of the form (10) there are also matrix elements between two states belonging to different subspaces $V_2^{(n)}$ and $V_2^{(m)}$, $n \neq m$ as well as between the states in $V_2^{(n)}$ and V_1. The solutions of rate equations for all above-mentioned matrix elements were derived and will be published elsewhere.

Journal of Physics: Conference Series **187** (2009) 012066 doi:10.1088/1742-6596/187/1/012066

4. Conclusion

The rate equations for the density matrix of the strongly coupled QD-MC system were studied. It was shown that if the interaction of this system with the environment is neglected, then in each 2-dimensional invariant subspace $V_2^{(n)}, n > 1$, 4 matrix elements of density matrix ρ satisfy a closed exactly solvable system of 4 rate equations. In this case the pair of two states in each subspace $V_2^{(n)}$ can be considered as a qubit. In the presence of the interaction of the system with the environment the rate equations for the matrix elements in $V_2^{(n-1)}$ always contain the contribution of those in $V_2^{(n)}$. The solution of the rate equation in the first order approximation with respect to the system-environment interaction was also established.

References

[1] Pellizzari T 1997 *Phys. Rev. Lett.* **79** 5242

[2] Alexandre Blais, Ren-Shou Huang, Andreas Wallraff, Girvin S M and Schoelkopf R J 2004 *Phys. Rev.* A **69** 062320

[3] Jay Gambetta, Alexandre Blais, Schuster D I, Wallraff A, Frunzio L, Majer J, Devoret M H, Girvin S M and Schoelkopf R J 2006 *Phys. Rev.* A **74** 042318

[4] Alexandre Blais, Jay Gambetta, Wallraff A, Schuster D I, Girvin S M, Devoret M H and Schoelkopf R J 2007 *Phys. Rev.* A. **75** 032329

[5] Jay Gambetta, Alexandre Blais, Boissonneault M, Houck A A, Schuster D I and Girvin S M 2008 *Phys. Rev.* A **77** 012112

[6] Jun-Tao Chang and Suhail Zubairy M 2008 *Phys. Rev.* A **77** 012329

[7] Jun-Qi Li, Gang Chen and Liang J Q 2008 *Phys. Rev.* A **77** 014304

[8] Jaehak Lee, Jiyong Park, Sang Min Lee, Hai-Woong Lee and Ashfaq H Khosa 2008 *Phys. Rev.* A **77** 032327

[9] Wei-Tao Liu, Wei Wu, Ping-Xing Chen, Cheng-Zu Li and Jian-Min Yuan 2008 *Phys. Rev.* A **77** 032328

[10] Kaniber M, Laucht A, Neumann A, Villas-Bôas J M, Bichler M, Amann M-C and Finley J J 2008 *Phys. Rev.* B **77** 161303RA

Acknowledgments

The author expresses one's deep gratitude to Academician Nguyen Van Hieu for the suggestion of the problem and for the discussions.

The existence of large magnetocaloric effect at low field variation and the anti-corrosion ability of Fe-rich alloy with Cr substituted for Fe

Duong Thi Hong Gam, Nguyen Hoang Hai, Le Van Vu, Nguyen Hoang Luong and Nguyen Chau

Center for Materials Science, University of Science, Vietnam National University, 334 Nguyen Trai, Hanoi, Vietnam

E-mail: gamcms@gmail.com

Abstract. The influence of Cr substituted for Fe in $Fe_{78-x}Cr_xSi_4Nb_5B_{12}Ag_1$ ribbons on the crystallization and magnetic properties including the magnetocaloric effect have been investigated. DSC measurements showed that the crystallization temperature and the crystallization activation energy increased with Cr content. Thermomagnetic curves measured in low applied field indicated that there is a sharp ferromagnetic-paramagnetic phase transition at Curie temperature, T_C, of the amorphous phase. In addition, substitution of Cr for Fe led to approximate linear decrease of T_C with Cr content, namely from 450K to 280K for x = 0 and x = 8, respectively. From a series of the isothermal magnetization curves M(H) measured at different temperatures, magnetic entropy change ΔS_m was determined at magnetic field variations of 13.5, 10.0 and 5.0 kOe. Maximum value of ΔS_m (occurred near T_C) decreased with increasing Cr content and these values established are quite large at low field variation. Namely, at moderate low magnetic field variation of 5.0 kOe, $|\Delta S_m|_{max}$ is of 4.4 and 1.9 J/kg.K for x = 0 and x = 8, respectively. Our studied alloys system could be considered as the best magnetocaloric material candidates for magnetic refrigeration because of colossal magnetic entropy change at low field variation and working temperature could be controlled in large region by substitution effect. The anti-corrosion ability of alloys was also examined.

Keywords: Magnetocaloric effect, anti-corrosion, amorphous ribbons.

1. Introduction

Magnetic refrigeration recently is a current topic in solid state physics/materials science and is of interest to physicists, materials scientists, and electrical and/or electric engineers. Magnetic cooling using magnetocaloric effect (MCE) has much interest of commercial applications in magnetic refrigeration as well as of fundamental research. Magnetic refrigeration using solid substances offers a lot of advantages over conventional gas compression refrigeration, namely:

(i) high thermodynamic efficiency due to reversibility of the MCE (e.g. the efficiency of magnetic refrigeration can be reached up to 60% of a Carnot cycle, whereas the efficiency is only 5–10% for conventional vapour compression refrigeration),

(ii) high energy densities and more compact devices because the working medium is a solid rather than a gas,

(iii) natural recovery of magnetic work during the cycle,

© 2009 IOP Publishing Ltd

(iv) usage of common liquids as heat transfer fluids (e.g. water, water-alcohol solution, air, or helium gas) with no ozone - depleting and/or global - warning effects,
(v) lower cost as using inexpensive magnet systems (e.g. permanent magnets),
(vi) environmental protection because it eliminate the chloroflouro-carbons present in high-temperature gas-cycle systems.

Magnetic refrigeration has many potential applications in energy intensive industrial and commercial refrigerator systems, such as large–scale air conditioners, heat pumps, supermarket refrigeration units, waste separation, chemical processing, gas liquefaction, liquor distilling, sugar refining, grain drying etc... MCE, or adiabatic temperature change which is detected as the heating or cooling of magnetic materials due to a varying magnetic field, was originally discovered in iron by Warburg [1]. When an external magnetic field is applied to a material, the magnetic spins in the material attempt to align with the magnetic field, thereby reducing the magnetic entropy of the spin system. If this process is performed adiabatically, the reduction of the spin entropy is offset by an increase in lattice entropy, and the temperature of the sample will rise. If the magnetic field is subsequently removed, the temperature of sample will reduce. This warning and cooling is response to the application and removal of an external magnetic field is called MCE. Therefore the MCE is intrinsic to all magnetic materials.

As we well know, the magnetic entropy change, ΔS_m, and the adiabatic temperature change, ΔT_{ad}, are correlated with the magnetization, the magnetic field strength, the heat capacity and the working temperature by the fundamental Maxwell's relations [2]:

$$\Delta S_m(T,H) = \int_0^{H_{\max}} \left(\frac{\partial M(T,H)}{\partial T} \right)_H dH \ , \tag{1}$$

and

$$\Delta T_{ad}(T,\Delta H) = -\int_0^{H_{\max}} \left(\frac{T}{C(T,H)} \right) \left(\frac{\partial M(T,H)}{\partial T} \right)_H dH \ , \tag{2}$$

where H_{max} is the final applied magnetic field. Equations (1) and (2) have a fundamental importance on the understanding of the behaviour of the MCE in solid and serve as a guide for the search of new materials with a large MCE [3]. Conventional magnetic refrigerants fall into two categories: first, the paramagnetic salts used to obtain mK, and second, ferromagnetic materials working in the vicinity of their ordering temperature. Recently, there has been particular interest in extending the magnetic refrigeration technique to room temperature region because of the desire to eliminate the chlorofluoro-carbons presented in high-temperature gas-cycle systems (environmental concerns) as well as potential impact of energy savings. Therefore materials with large MCE are needed to improve energy efficiency.

The prototype material for the room temperature range is the lanthanide metal Gd which orders ferromagnetically at 294K. Its MCE has been studied independently by many authors and the ΔT_{ad} values at T_C are 6, 12, 16 and 20K for ΔH = 2, 5, 7.5 and 10 T, respectively (see, for example [4]). A series of $Gd_5(Si_xGe_{1-x})_4$ alloys was reported [5, 6] to display MCE at least two times larger than that of Gd near room temperature and between 2 to 10 times larger than that of the best (at that time) magnetocaloric materials in the low and intermediate temperature regions. Many other materials have been studied with respect to their magnetocaloric properties [7]. The compounds $La(Fe,Si)_{13}$ and $La(Fe,Al)_{13}$ [8], $MnFeP_{1-x}As_x$ [9], $MnAs_{1-x}Sb_x$ [10] are suggested to be the potential candidates for room temperature magnetic refrigerators. Another class of materials also displaying the large MCE (comparable to Gd) is based on perovskite [11-15]. Especially, for the first time we are examined the positive magnetic entropy change in manganite with charge-ordering [16].

Recently, the research group at Center for Materials of Science, Vietnam National University, Hanoi have examined the influence of P substituted for B [17], Ag, Zn and Au for Cu [18-20], Co, Mn, Cr and Nb for Fe [21-24] in Finemet on the crystallization and properties of materials. We have discovered giant magnetocaloric effect in a series of amorphous alloys typical Finemet ($Fe_{73.5}Si_{13.5}B_9Nb_3Cu_1$) and like-Finemet [25, 26]. The maximum magnetic entropy change reached quite large value comparing within world studies. Namely, this value is of 13.9 J/kg.K, 11.2 J/kg.K in reference [25] and 9.8 J/kg.K in reference [27] at moderate low field change (1.35 T) and occurs at a temperature that is higher than room temperature. Also, the working temperature can be easily controlled by substitution effect. However, this temperature is rather high and the value of magnetic entropy change strongly decreases when T_C reaches around room temperature.

In this report, we present the results of large MCE at a low field variation and anti-corrosion ability in Fe-rich alloys: $Fe_{78-x}Cr_xSi_4Nb_5B_{12}Ag_1$.

2. Experimental

The soft magnetic ribbons $Fe_{78-x}Cr_xSi_4Nb_5B_{12}Ag_1$ (x = 0 - 8) have been prepared by rapid quenching technique on a single copper wheel. The linear speed of the wheel for all ribbons was 30 m/s. The structure of the samples was examined by X-ray diffractometer D5005 Bruker. The thermal transition analysis was studied by a differential scanning calorimeter SDT 2960 TA Instruments. The ribbons were annealed in vacuum. The thermomagnetic curves of as-cast ribbons were measured by vibrating sample magnetometer (VSM) DMS 880 Digital Measurement System and the magnetic properties of studied samples were determined by using Permagraph AMH-401 A, Walker. In order to examine the chemical anti-corrosion, we chose the samples with different Cr content having the same thickness and width (the same area). These samples are soaked in a solution of diluted brine (concentration c = 5‰) and after 4 days, they are drawn out and dried. The mass depression is determined by the balance whose accuracy is 10^{-5} g.

3. Results and discussion

The XRD patterns of as-cast samples showed one broad peak centered at approximately $2\theta = 45^o$ which appears to be typical for an amorphous phase.

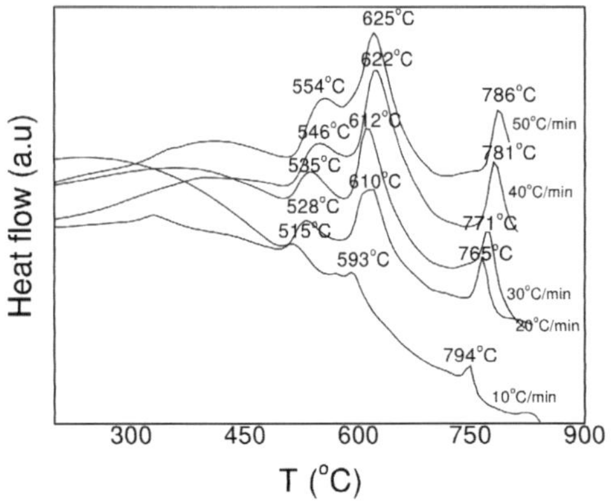

Figure 1. DSC curves of as-cast ribbon with x =0 measured with heating rate from 10 to 50°C/min.

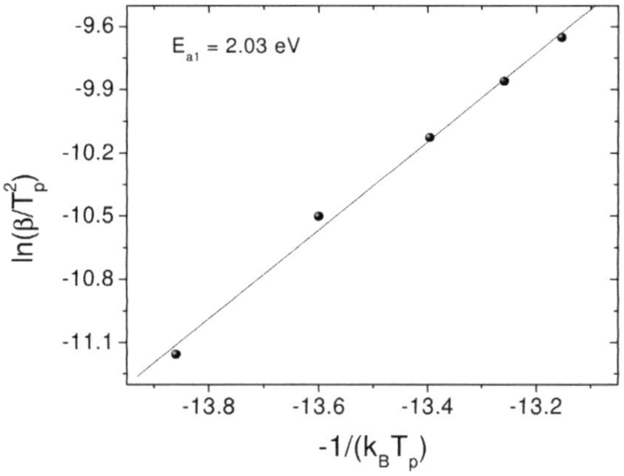

Figure 2. Kissinger plot for determination of crystallization activation energy of the first peak for the sample with x =0.

DSC measurement on as-cast amorphous ribbons was performed with heating rate from 10 to 50°C/min in Ar atmosphere. Figure 1 shows DSC curves for the sample with x = 0 as an example. There are three separated exothermal peaks T_{p1}, T_{p2} and T_{p3} on DSC curves. The first peak corresponds the formation of α-Fe(Si) phase and partial FeCr phase (T_{p1}). On a crystallographic level, α-Fe(Si) and FeCr phases are the same structure of bcc, so on the XRD patterns and DSC curves, three phases are close. Because of the resolution of device is not high, the phase splitting is not observed. The second and third peaks on DSC curves ascribe to the precipitation of boride phases (T_{p2} and T_{p3}) depending on the Cr content substituted for Fe.

Based on DSC patterns measured at different heating rates and Kissinger plots (figures 1 and 2), we have evaluated the crystallization activation energy E_{a1} of the first peak [28]. The formation of FeCr phase with laminar structure [27] makes the exothermal peaks (T_{p1}, T_{p2} and T_{p3}) and the crystallization activation energy increase with the Cr content due to the laminar structure have the shape magnetic anisotropy.

For the first time, Leu and Chin [29] have pointed out the use of DTA apparatus for estimating the crystallization volume fraction. The as-cast ribbons have been annealed from 505 to 545°C (depending on Cr content) for 90 minutes. Using Leu and Chin expression, we derive the crystallization volume fraction of the α-Fe(Si) phase and the results indicate to be 82%.

From the X-ray analysis, we can determine the average grain size of α-Fe(Si) particles formed after annealing according to the Sherrer expression [30] and the lattice constant using Bragg's diffraction. The results are shown in figure 3 and table 1. The grain size decreases while the lattice constant slightly increases with increasing Cr content. There is due to that the higher Cr content, the higher crystallization activation energy that leads to hard crystallization. Moreover, the radius of Fe ion is less than that of Cr ion, therefore the higher Cr content leads to the lattice deformation and make increasing the lattice constant. We know that the grain size and the crystallization phase depend on the annealing regime (annealing temperature and keeping time). The grain size depends majorly on annealing temperature, so the samples with higher Cr content annealed at the higher temperature, and the average grain size increases.

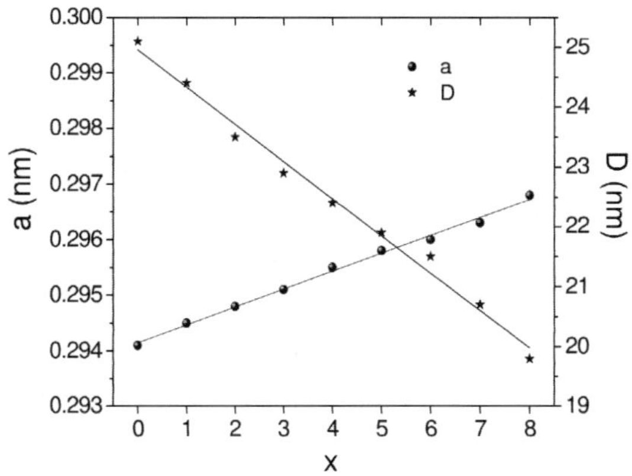

Figure 3. Dependence of the average grain size (D) and the lattice constant (a) on Cr concentration.

Because of nanosize and appropriate volume fraction of α-Fe(Si) phase, the soft magnetic properties of annealed sample have been remarkably improved compared to those of as-cast samples (figure 4 and table 1). However, the soft magnetic properties of the sample with x = 7 is worse than those of the sample with x = 0 due to ferromagnetic dilution and the formation of FeCr phase.

Table 1. Several characteristics of the ribbons with x = 0 and 7.

Sample	T_{p1} (°C)	E_{a1} (eV)	a (nm)	D (nm)	H_C (Oe)		μ_{max}	
					as-spun	annealed	as-spun	annealed
x = 0	528	2.03	0.2941	25.1	0.13	0.01	3,500	28,000
x = 7	580	2.70	0.2963	20.7	0.25	0.08	2,100	20,100

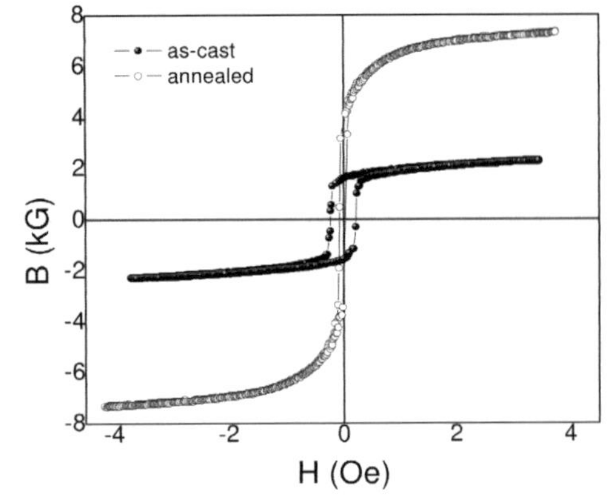

Figure 4. Hysteresis loops of the sample with x = 0.

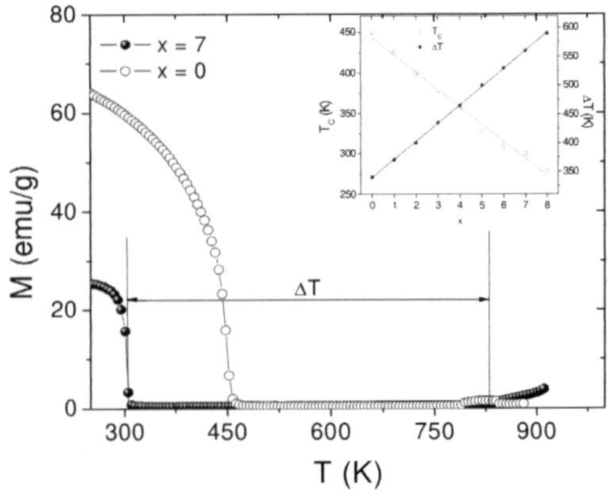

Figure 5. Thermomagnetic curves of the samples with x = 0 and 7 (Insert is the dependence of T_C and ΔT on Cr content).

Figure 5 shows the thermomagnetic curves M(T) of the ribbon with x = 0 and x = 7 measured in low field of 20 Oe. We can see that when the temperature increases, the magnetization is abruptly reduced at Curie temperature T_C of amorphous phase. With further increasing temperature up to region of onset of crystallization, the α-Fe(Si) crystallites start to be formed leading to the increase in magnetization. From M(T) curves, we define $\Delta T = T_{onset} - T_C$ as the stabilizing temperature region of amorphous phase, in which no phase transition is detected (figure 5). T_C and ΔT were determined for all samples and collected in table 2. Insert in figure 5 is the dependence of T_C and ΔT of the samples versus Cr content substituted for Fe. The results show that T_C significantly decreased while ΔT increased with increasing Cr content. It is clearly seen that we can control the working temperature of the ribbons by substitution effect and Cr plays the role of stabilizing amorphous structure.

The calculation of the magnetic entropy change, ΔS_m, associated with second-order phase transition was performed using the isothermal magnetization curves. Figure 6 shows the magnetization versus applied field up to 13.5 kOe at various temperatures of samples with x = 0 (figure 6a) and x = 7 (figure 6b). In our experiments, the changing rate of applied field is slow enough to get isothermal M-H curves.

In fact, ΔS_m is often evaluated by some numerical approximation methods. One way of approximation is to directly use the measurements of the M(T) curve under different magnetic field. In the case of small discrete field intervals, ΔS_m can be approximately determined from equation (1) as

$$\Delta S_m = \sum \left[\left(\frac{\partial M}{\partial T} \right)_{H_i} + \left(\frac{\partial M}{\partial T} \right)_{H_{i+1}} \right] \times \frac{1}{2} \times \Delta H_i \qquad (3)$$

where $\left(\frac{\partial M}{\partial T} \right)_{H_i}$ is the experimental value obtained from M(T) curve in magnetic field H_i.

Another method is to use isothermal magnetization measurements. In the case of magnetization measurements at small discrete field and temperature interval, ΔS_m can be approximately determined from equation (1) by

$$\Delta S_m = \sum \frac{M_i - M_{i+1}}{T_{i+1} - T_i} \Delta H_i \qquad (4)$$

where M_i and M_{i+1} are the experimental values of the magnetization at T_i and T_{i+1}, respectively, under an applied magnetic field of H_i.

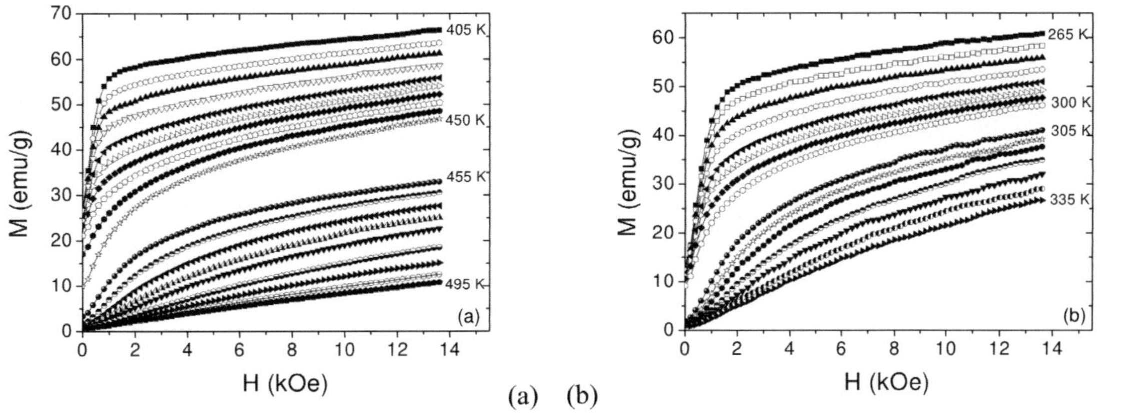

(a) (b)

Figure 6. Magnetization as a function of applied field at different temperatures for the samples with x=0 (a) and x=7 (b).

Table 2. Curie temperature (T_C), stabilizing temperature region (ΔT) and the maximum value of magnetic entropy change ($|\Delta S_m|_{max}$) of the studied samples.

Sample		x = 0	x = 1	x = 2	x = 3	x = 4	x = 5	x = 6	x = 7	x = 8
T_C (K)		450	425	400	375	355	330	310	300	280
ΔT (K)		340	370	400	435	465	500	530	560	590
$\|\Delta S_m\|_{max}$ (J/kg.K)	ΔH = 13.5	5.3	4.9	4.4	4.0	3.7	3.5	3.2	3.1	2.8
	ΔH = 10	4.9	4.6	4.2	3.8	3.4	3.2	2.6	2.5	2.1
	ΔH = 5	4.4	4.1	3.9	3.3	3.0	2.8	2.2	2.1	1.9

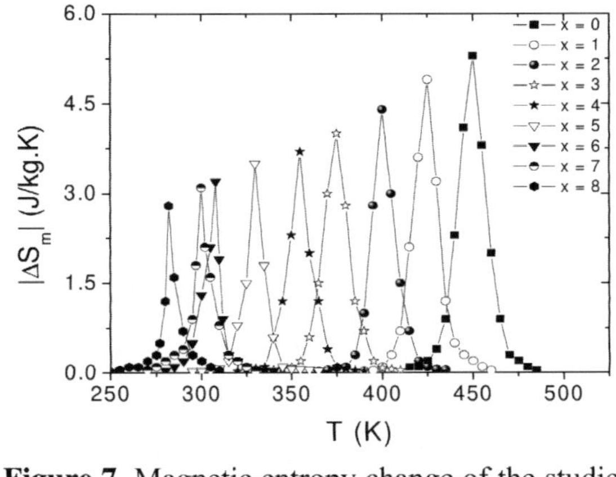

Figure 7. Magnetic entropy change of the studied samples with magnetic field variation of 13.5 kOe.

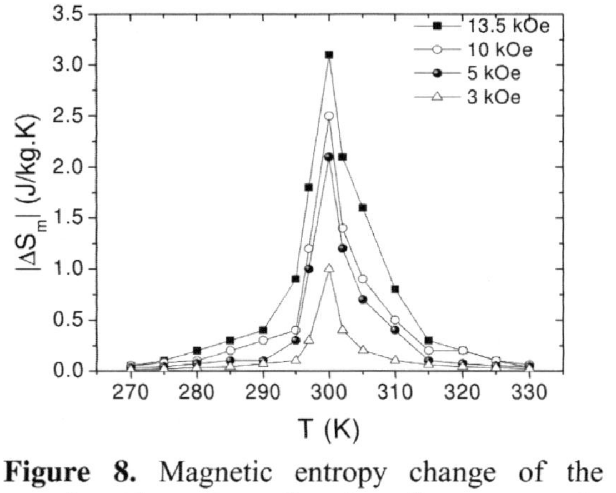

Figure 8. Magnetic entropy change of the sample with x =7 as a function of temperature in different field variations.

Figure 7 presents the temperature dependence of ΔS_m of the studied samples. It can be seen that the maximum value of $|\Delta S_m|$ occurs around T_C of amorphous phase and decreases with increasing Cr content. Namely, $|\Delta S_m|$ is of 5.3 J/kg.K and 2.8 J/kg.K measured in 13.5 kOe magnetic field variation for x = 0 and 8, respectively. ΔS_m has been also determined at lower field variation of 10 and 5 kOe and figure 8 shows the results for the ribbon with x = 7 as an example. It is interesting that the value of

$|\Delta S_m|_{max}$ reached the large value at the low field variation (see table 2). This can be only be explained in that our materials are ultrasoft magnetic materials. To our knowledge, these values of $|\Delta S_m|_{max}$ at magnetic field variation of 10kOe and 5kOe of the studied samples are the largest values that have not been reported so far. Especially, at magnetic field variation of 3 kOe, the value of $|\Delta S_m|_{max}$ of the samples is larger than 1 J/kg.K. Namely, $|\Delta S_m|_{max}$ is of 1.4 J/kg.K and 1.0 J/kg.K for x = 0 and 7, respectively. This has an important meaning for application that with the low magnetic field, the magnetic refrigerators only use a permanent magnet (AlNiCo, $Nd_2Fe_{14}B$) for a source of magnetic field.

From above mentioned results we could say that our studied materials have giant magnetocaloric effect working at a large temperature region (from under room temperature (280K) to high temperature (450K) by moderate low magnetic field variation, thereby they are the best refrigeration materials at the present time.

The chemical anti-corrosion of Cr was also studied. The results show that the mass of the samples without Cr or with less Cr content (x = 0÷2), strongly decreases with time when they are soaked in the solution of diluted brine (concentration c = 5‰), while the mass of the samples with higher Cr content is almost constant. The results are shown in figure 9. This can be only explained due to the anti-corrosion ability of Cr. So, the sample containing Cr can work in the breaking environments (seawater, acid, alkali…).

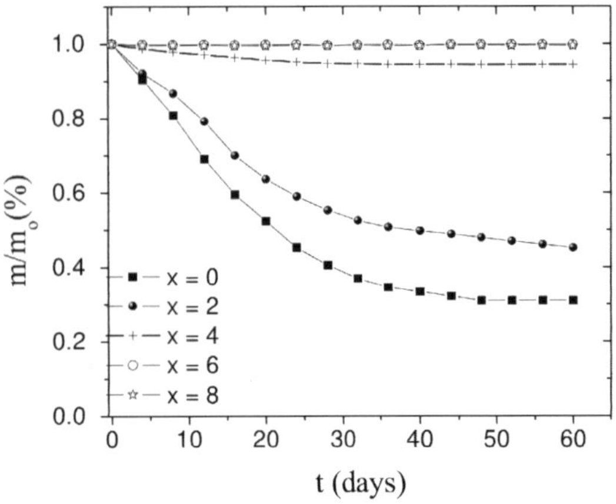

Figure 9. Mass of the studied samples after time of soak. (m_o is primary mass).

4. Conclusions

Magnetic ribbons of composition $Fe_{78-x}Cr_xSi_4Nb_5B_{12}Ag_1$ are prepared by rapid quenching technology with amorphous structure. Cr substitution makes increasing crystallization temperature of α-Fe(Si) phase and crystallization activation energy, shifting the Curie temperature of amorphous phase to room temperature region as well as stabilizing amorphous structure of samples. The large magnetocaloric effect (GMCE) has been established in all studied samples at low field variation. This has an important meaning for application with the low magnetic field, for example when the magnetic refrigerators only use a permanent magnet (AlNiCo, $Nd_2Fe_{14}B$) for a source of magnetic field. It is convinced that the studied samples have combined a number of advantages of magnetocaloric materials. Therefore they are the best materials working for active magnetic refrigeration at large temperature range from room to high temperature region. After appropriate annealing, materials become nanocomposites with ultrasoft magnetic properties. The existence of Cr increases the chemical anti-corrosion ability of the samples and these samples can be used in the breaking environments (seawater, acid, alkali…).

References

[1] Warburg E 1981 *Ann. Phys.* **13** 141

[2] Morrish A H 1963 *The principles of magnetic* (New York: Winley) chapter 3
[3] Pecharsky V K, Gschneidner J K A 1999 *J. Magn. Magn. Mater.* **200** 44
[4] Dankov S Y, Tishin A M, Pecharsky V K, Gschneidner Jr K A 1998 *Phys. Rev. B* **57** 3478
[5] Pecharsky V K, Gschneidner Jr K A 1997 *J. Magn. Mag. Mater.* **167** L179
[6] Gschneidner Jr K A, Pecharsky V K 1999 *J Appl. Phys.* **85** 5365
[7] Tishin A M, Spickin Y I, Magnetocaloric effect and its application 2003 *IOP Bristol Philadelphia Series in Condensed Matter Physcis*
[8] Chen Y F, Wang F, Shen B G, Sun J R, Wang G J, Hu F X, Cheng Z H, Zhu T 2003 *J. Appl. Phys.* **93** 6981
[9] Tegus O, Bruck E, Buschow K H J, de Boer F R 2002 *Nature* **415** 6868
[10] Wada H, Asano T 2005 *J. Magn. Magn. Mater.* **290-291** 703
[11] Phan M H, Phan T L, Yu S C, Tho N D, Chau N 2004 *Phys. Stat. Sol (b)* **241** 1744
[12] Phan M H, Peng H X, Yu S C, Tho N D, Chau N 2005 *J. Magn. Magn. Mater.* **290** 199
[13] Luong N H, Hanh D T, Chau N, Tho N D, Hiep T D 2005 *J. Magn. Magn. Mater.* **290** 690.
[14] Chau N, Hanh D T, Tho N D, Luong N H 2006 *J. Magn. Magn. Mater.* **303** e335
[15] Chau N, Tho N D, Luong N H, Giang B H, Cong B T 2006 *J. Magn. Magn. Mater.* **303** e402
[16] Chau N, Cuong D H, Tho N D, Nhat H N, Luong N H, Cong B T 2004 *J. Magn. Magn. Mater.* **272** 1292
[17] Chau N, Luong N H, Chien N X, Thanh P Q, Vu L V 2003 *Physica B* **327** 241
[18] Chau N, Hoa N Q, Luong N H 2005 *J. Magn. Magn. Mater.* **290-294** 1547
[19] Chau N, Hoa N Q, The N D, Vu L V 2006 *J. Magn. Magn. Mater.* **303** e415
[20] Hoa N Q, Gam D T H, Chau N, The N D, Yu S C 2007 *J. Magn. Magn. Mater.* **310** 2483
[21] Chau N, Chien N X, Hoa N Q, Niem P Q, Luong N H, Tho N D, Hiep V V 2004 *J. Magn. Magn. Mater.* **282** 174
[22] Tho N D, Chau N, Yu S C, Lee H B, Tuan L A, Hoa N Q 2006 *J. Magn. Magn. Mater.* **304** e868
[23] Gam D T H, Anh H D, Luong N H, Hai N H, Chau N, Vu L V 2008 *International Coference on Advanced Technologies for Communications ATC 2008 and REV'08*
[24] Hoa N Q, Gam D T H, The N D, Chau N, Son D V, Yu S C 2007 *Phys. Stat. Sol. (a)* **204** 4113
[25] Chau N, The N D, Huu C X 2004 *Proceeding of the 2nd International Workshop on Nanophysics and Nanotechnology (IWONN'04) (Hanoi Vietnam October 22-23)* p. 51
[26] Chau N, Yu S C, Huu C X, Hoa N Q, The N D, Tho N D 2007 *Mater. Sci. Eng. A*.**449-451** 360
[27] Chau N, Thanh P Q, Hoa N Q, The N D 2006 *J. Magn. Magn. Mater.* **304** 36
[28] Gam D T H, The N D, Hai N H, Chau N, Hoa N Q, Mahmud Md S 2008 *J. Korean Phys. Soc.* **52** 1423
[29] Leu M S, Chin T S 1999 *MRS. Symp. Proc.* **577** 557
[30] Cullity B D 2nd Edition 1978 *Element of X-ray Diffraction* Addison – Wesley Reading MA 10

Acknowledgements
The authors would like to thank Vietnam National Fundamental Research Program (Project 406506) for financial support.

Magnetic memory effect in a strong phase competition system $La_{0.7}Ca_{0.3}Mn_{0.925}Ti_{0.075}O_3$

Nguyen Chi Thuan[1], Dao Nguyen Hoai Nam[1], Nguyen Xuan Phuc[1], Nguyen Thi Yen Mai[1], Le Van Hong[1], Nguyen Van Dai[1,2], Vu Dinh Lam[1,2] and Young Pak Lee[2]

[1] Institute of Materials Science, Vietnam Academy of Science and Technology, 18 Hoang Quoc Viet Road, Cau Giay Distr., Hanoi, Vietnam
[2] Quantum Photonic Science Research Center and Dept. of Physics, Hanyang University, Seoul Korea

E-mail: honglv@ims.vast.ac.vn

Abstract. A magnetic memory effect was observed by means of zero-field-cooled magnetization measurements in a temperature range below T_C. The magnitude of magnetic memory effect, determined by the difference between the magnetization curve with a pause of temperature during the cooling process and its reference curve without the pause, showed that it is independent of the number of pauses. The memory effect is supposed to be related to the isolated states with an "infinite" degeneration, which is associated with the presence of magnetic frustration in a system with a strong magnetic phase competition. Besides, a resistance memory effect was concomitantly observed at the same pause temperatures.

Keywords: CMR, magnetic phase separation, magnetic memory effect.

1. Introduction

Slow magnetic and resistance relaxation, aging and memory phenomena are known as characteristic behaviour of glassy systems which were observed in several mixed-valence manganites, $R_{1-x}A_xMnO_3$ (R: trivalent rare earth; A: alkaline element) [1-4]. Although the dynamics of electrons is generally known as a very rapid process, glassy transport has been observed in various systems [5, 6] and commonly attributed to a slow evolution of competing phases coexisting in the material as a result of phase separation [7]. According to the molecular-field approximation, Nam *et al.* [8, 9] recently confirmed the existence of strong competition between double-exchange ferromagnetic and super-exchange antiferromagnetic interactions among Mn ions in $La_{0.7}Ca_{0.3}Mn_{1-x}Ti_xO_3$. The magnetic phase separation and strong competition between them were also considered as factors contributing to the establishment of a glassy state in manganite materials. It is known that one of the conditions for a magnetic system to become glassy is the appearance of magnetic frustration that causes a situation where no spin configuration can satisfy all the couplings in the system. For a loop of couplings, the "frustration function" Φ is defined as

$$\Phi = \prod_L \frac{J_{ij}}{|J_{ij}|} , \tag{1}$$

where J_{ij} is the exchange constant of the interaction between two magnetic moments. An interacting magnetic loop becomes "frustrated" if its frustration function is negative, implying that it contains an

© 2009 IOP Publishing Ltd

odd number of antiferromagnetic interactions. The magnetic frustration can cause an infinite degeneration of the ground state of a spin system, consequently, creating an infinite number of metastable states at the energy minima. In a system containing magnetic disorders caused by randomness, there always exist energy barriers separating the metastable states of the system. It has been commonly believed that the necessary and sufficient condition for a spin glass state of a system is a combination of magnetic frustration and disorder.

In this report we present our recent results on the magnetic memory effect in the polycrystalline $La_{0.7}Ca_{0.3}Mn_{0.925}Ti_{0.075}O_3$ compound, done by means of magnetization measurement in a temperature range below T_C with pauses at one or more temperatures during cooling from above T_C. We explain the obtained results in terms of magnetic frustration in a system of separated magnetic phases with a strong phase competition.

2. Experimental

The polycrystalline compound was synthesized by the conventional solid state reaction method. Pure (> 99.99%) raw powders with appropriate amounts of La_2O_3, $CaCO_3$, MnO_2 and TiO_2 were thoroughly ground, mixed, pressed in pellet and then calcined at several thermal processing steps with increasing temperature from 900 to 1200°C, sandwiched by intermediate regrinding and repressing. The product was then sintered at 1300°C for 48 h in the ambient atmosphere. The final samples were obtained after a slow cooling process from the sintering to room temperature for 5 h. Room-temperature X-ray diffraction showed that the samples were single phase of perovskite orthorhombic structure. Zero-field-cooled (ZFC) magnetization was measured in a magnetic field of 10 Oe with the same cooling and heating rate of 1 K/min by using a Quantum Design physical property measurement system (PPMS 6000). For the memory measurements, the cooling process was paused for various times at one, two or more temperatures. The measure of magnetic memory effect was evaluated as the variation of the recorded magnetization, compared to the reference one, which was done without pausing in the cooling process.

3. Result and discussion

Figure 1 presents ZFC magnetization curves measured in a magnetic field of 10 Oe with a pause at 65 K for 2, 4, 6, 8 and 10 h. The magnetization curve measured at the same conditions but without pausing is considered as the reference curve, which is used in the evaluation of magnetic memory effect of the compound. We can recognize that the ZFC magnetization measured with a pause repeats exactly the reference one for the whole temperature range except around the temperature where the pause was made.

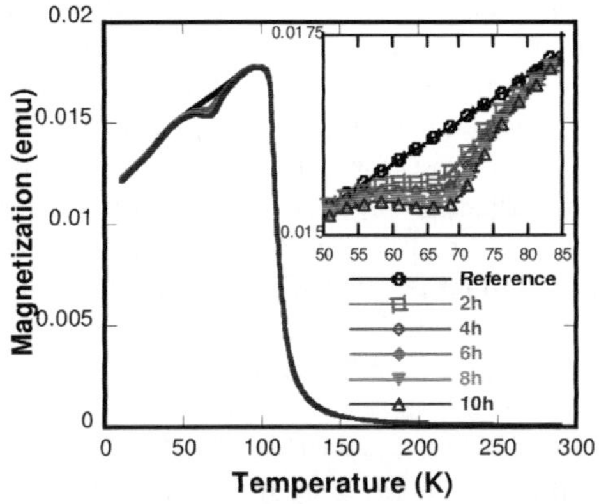

Figure 1. The ZFC magnetizations measured in a field of 10 Oe with a pause during cooling process at 65 K for 2, 4, 6, 8 and 10 h.

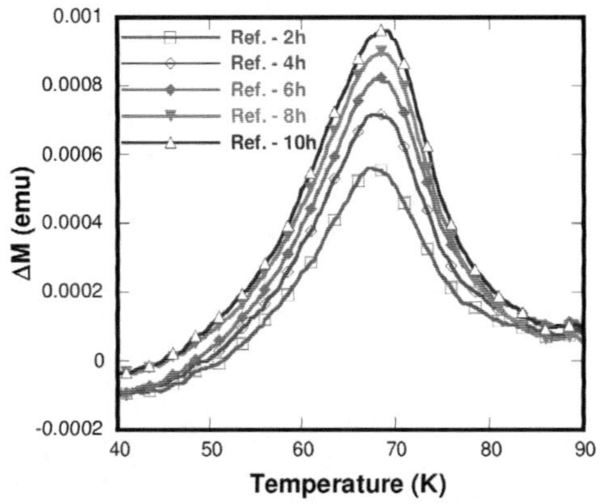

Figure 2. Magnitude of magnetic memory effect for the ZFC magnetization measured in a field of 10 Oe with a pause during cooling process at 65 K for 2, 4, 6, 8 and 10h.

As clearly seen in figure 2 the difference in magnetization appears at temperatures around the temperature where the cooling process was paused for different times. It is interesting that the maximum difference is always observed at temperatures close to the pause temperature, independent of the pause time. This observation confirms the existence of magnetic memory effect in the compound. However, the magnitude of memory effect slowly increases with the pause time. In a previous article [7], the magnetic and transport memory effects were observed in this compound. This was explained as a consequence of the magnetic glassiness with a strong competition between magnetic phases. In order to further understand the nonequilibrium dynamics of the glassy system, we established two magnetization measurements with a single pause at temperatures of 65 and 35 K, separately, and another measurement with double pause sequentially made at 65 and 35 K. Differences of these magnetization measurements are presented in figure 3. It is shown that the magnitudes of magnetization difference for the single and the double pause are equivalent, independent of the number of pauses.

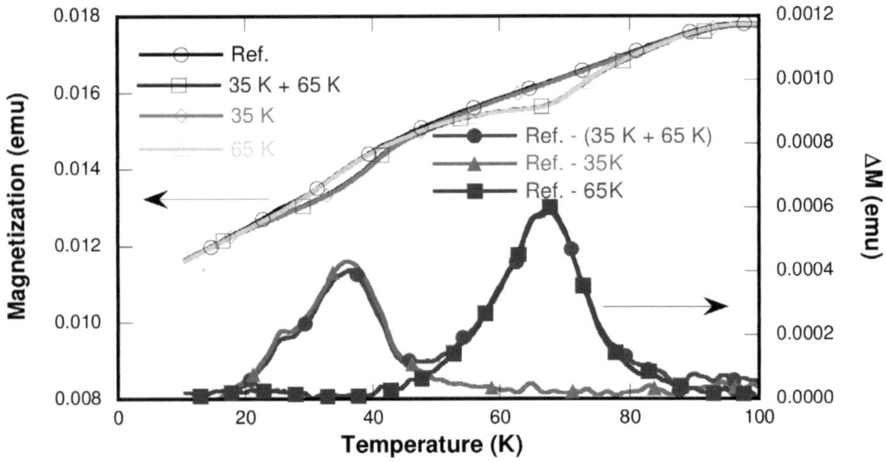

Figure 3. ZFC magnetization curves and their differences measured in a field of 10 Oe with a single pause at 65 K for 3 h, 35 K for 6 h and a sequent double pause at the same conditions in temperature and time.

This result suggests two important concepts: i) the observed effect is evidence for the existence of magnetic states separated by significantly high energy barriers that prevent the transition processes from one to another, and ii) the magnitude of the memory effect is independent of the number of initial states. It could be satisfied only when the amount of initial states is infinite. This is supposed to be related with the existence of a magnetic frustration in our system. Therefore, it is believed that the obtained result is reasonably an experimental evidence for the magnetic frustration. Generally such a situation has been observed only in a strongly phase separated system like $La_{0.7}Ca_{0.3}Mn_{0.925}Ti_{0.075}O_3$ [8]. It turns out that not only two separated states exist, but the number of separated metastable states could be infinite. This means we can make the memory effect at unlimited numbers of pause temperatures. As a test, a ZFC magnetization measurement has been done with pauses at temperatures of 30, 40, 50, 60, 70, 80, 90 and 100 K, below T_C during cooling process. The obtained result is presented in figure 4. It shows that the magnitude of memory effect depends on the temperature position of pause. A pause at 50 K exhibits a maximal magnitude, and the magnitude decreases on both sides of this temperature point. As explained in a previous article [7], the magnetic memory in a magnetic glassy system with a strong phase separation is a consequence of the competition between magnetic freezing and magnetic domain growth processes. Therefore, the position of temperature pauses where the maximal magnitude of memory effect is observed depends on the cooling and the heating rate. According to the same logic as in case of evaluation of the glass temperature of a spin glass, slow cooling and heating rates are required in identifying the essential position of the maximal magnetic memory.

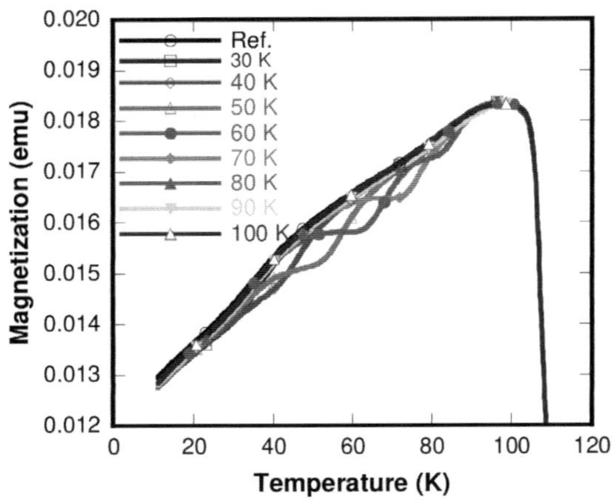

Figure 4. ZFC magnetizations measured in a field of 10 Oe with a pause at temperatures of 30, 40, 50, 60, 70, 80, 90, and 100 K for 2 h.

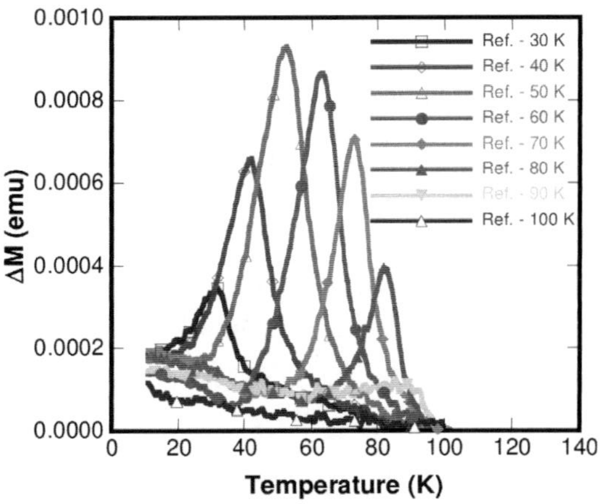

Figure 5. Magnitude of magnetic memory effect for the ZFC magnetizations measured in a field of 10 Oe with a pause at temperatures of 30, 40, 50, 60, 70, 80, 90, and 100 K for 2 h.

4. Conclusions

A magnetic memory effect was observed by means of ZFC magnetization measurements at temperatures below T_C. The magnitude of magnetic memory effect does not depend on the number of pauses. It is supposed to be related to the "infinite" number of separated states, which is associated with the presence of magnetic frustration in the system with a strong magnetic phase competition. This is reasonably connected with the resistance memory effect concomitantly observed at the pause temperatures.

References

[1] Nam D N H, Jonason K, Nordblad P, Khiem N V and Phuc N X 1999 *Phys. Rev. B* **59** 4189
[2] Nam D N H, Mathieu R, Nordblad P, Khiem N V and Phuc N X 2000 *Phys. Rev. B* **62** 8989
[3] Mathieu R, Nordblad P, Nam D N H, Phuc N X and Khiem N V 2001 *Phys. Rev. B* **63** 17445
[4] Nam D N H, Mathieu R, Nordblad P, Khiem N V and Phuc N X 2001 *JMMM* **226-230** 1340
[5] Tao J, Niebieskikwiat D, Salamon M B and Zuo J M 2005 *Phys. Rev. Lett.* **94** 147206
[6] De K, Majumdar S and Giri S 2007 *J. Appl. Phys.* **102** 103909
[7] Nam D N H, Khien N V, Dai N V, Hong L V and Phuc N X 2008 *Phys. Rev. B* **77** 214406
[8] Nam D N H, Bau L V, Khiem N V, Dai N V, Hong L V, Phuc N X, Newrock R S and Nordblad P 2006 *Phys. Rev. B* **73** 184430
[9] Nam D N H, Dai N V, Thanh T D, Tuong L T C, Hong L V, Phuc N X, Hong H S and Khien N V 2008 *Phys. Rev. B* **77** 224420

Acknowledgments

This work was done under the support of Grant 4.076.06 in the Fundamental Research Program of Vietnam (2006-2008). The financial support in part of Institute of Materials Science and National Key Laboratory for Electronic Materials and Devices is acknowledged. Nguyen Van Dai, Vu Dinh Lam and Young Pak Lee were supported by MEST through q-Psi, Seoul, Korea.

Magnetic fluid based on Fe_3O_4 nanoparticles: Preparation and hyperthermia application

Pham Hoai Linh[1], **Pham Van Thach**[1], **Nguyen Anh Tuan**[1], **Nguyen Chi Thuan**[1], **Do Hung Manh**[1], **Nguyen Xuan Phuc**[1] **and Le Van Hong**[1]

Institute of Materials Science, Vietnam Academy of Science and Technology, 18 Hoang Quoc Viet Road, Cau Giay Distr., Hanoi, Vietnam

E-mail: linhph@ims.vast.ac.vn

Abstract. The paper presents results of research on preparing the magnetic fluid based on Fe_3O_4 nanoparticles and its potential hyperthermia application. Magnetic fluids were manufactured by dissolving superparamagnetic nanoparticles coated by suitable biocompatible starch layer during the co-precipitation processing. The coated particle size changes in range of 15÷17 nm were characterized by FESEM images. At room temperature, the samples exhibit super-paramagnetic behaviour with a saturation moment of 65 emu/g. The concentration of magnetic nanoparticles contained in carried liquid reach to 15 mg/ml. The magnetic fluid was used as a mediator for heating by an AC magnetic field with the frequency of 184 kHz and field strength of 12 kA/m. The dependence of heating on nanoparticle concentration was observed and it implied that the magnetic fluid is a suitable mediator for cancer treatment by hyperthermia application with appropriate controlling the heating temperature ranges from 45 to 50°C.

Keywords: Nanotechnology, Fe_3O_4 nanoparticles, hyperthermia application.

1. Introduction

In recent years, the magnetic fluids have attracted great interest in research due to their offer of many possibilities for numerous in-vivo applications such as separation cells [1], magnetic resonance imaging as contrast agents [2], magnetic drug targeting and delivery [3], and especially they can generate heat in an alternating magnetic field and become promising material for tumor hyperthermia. Magnetic Hyperthermia is a method of cancer treatment using magnetic nanoparticles. In the process, magnetic particles were injected locally into cancer tissues that can be heated up to the required temperature of about 45°C with the help of an external alternating magnetic field. The method was first introduced by Gilchrist *et al.* since 50 years ago [4]. By now, several works devoted to this subject were reported [5, 6, 7]. It was found that magnetic fluid based on superparamagnetic Fe_3O_4 nanoparticles coated by biocompatible layer is suitable for hyperthermia because the particles have a high magnetic moment, strong magnetic specific loss power (SLP) and non-toxicity. The Food and Drug Administration approved that superparamagnetic Fe_3O_4 nanoparticles is biocompatible with human body [8]. In addition, for superparamagnetic nanoparticles where the magnetic hysteresis is vanished, the Néel and the Brownian losses are dominant so the specific loss power of magnetic fluid strongly depends on the alternating magnetic field and the nature and structure of the particles such as particle size and size distribution, anisotropy constant, saturation magnetization and surface

© 2009 IOP Publishing Ltd

modification [9, 10, 11]. However, the challenging of work is the development of heating mediator with high specific loss power, which allows reduction of ferrofluid doses in-vivo. In the aim of improving SLP of Fe_3O_4 nanoparticle, several methods were established [12, 13]. The highest value of SLP of 960 W/g at the field of 410 kHz and 10 kA/m was found for magnetosomes produced by magnetotactic bacteria [14]. Besides, it is important in hyperthermia to control saturation temperature in the range of 42÷47°C to destroy the cancerous cells within minimal damage to healthy tissue and limit negative side effects.

The purpose of this research was preparing magnetic fluid based on Fe_3O_4 nanoparticles that have high saturation magnetization to enhance the SLP value and focus on investigating the dependent of heating on the concentration of magnetic particles at alternating magnetic field. We are also able to control the heating temperature in the range of 42 to 47°C with low concentration of nanoparticles.

2. Experimental

1.1. Synthesis of Fe_3O_4 nanoparticles
Fe_3O_4 nanoparticles were prepared by the chemical co-precipitation method. Aqueous solution of $FeCl_3$, $FeCl_2$ salt were mixed and kept at 40°C while maintaining a molar ratio of $Fe^{2+}/Fe^{3+} = 1/2$. The mixed solution was added by solution of NaOH under vigorous mechanical stirring for 30 minutes. The precipitation and formation of nanoferrites took place by a conversion of the metal salts into hydroxides and that, in turn, immediately transformed into ferrites. During the reactions, magnetic-stirring speed was kept at 1000 round/min.

1.2. Coating particles
Starch was dissolved in water with a concentration of 0.1 g/ml at 80°C. After the starch was thoroughly dissolved, the solution was immediately placed in a 60°C water bath and dropped into the obtained aqueous of dispersed Fe_3O_4 particles after a co-precipitation reaction at 80°C in 2 h. Around 50 wt.% of water was evaporated during the boiling. The remaining suspension was then cooled to 0°C within 12 h. The formed gels were washed with deionized water until the pH of the solution was lowered to less than 8.5. Excess salts and ions were then removed by washing. To cleave glycosides bond and reduce the polymer chain to an average molecular weigh, starch-coated iron oxide particles were oxidized by H_2O_2 solution.

1.3. Characterization
The structure and the morphological investigation of the studied sample were taken by X-ray diffraction (XRD) and field-emission scanning electron microscopy (FESEM), respectively. The magnetic properties were carried out by using vibrating sample magnetometer (VSM) and physical property measurement system (PPMS).

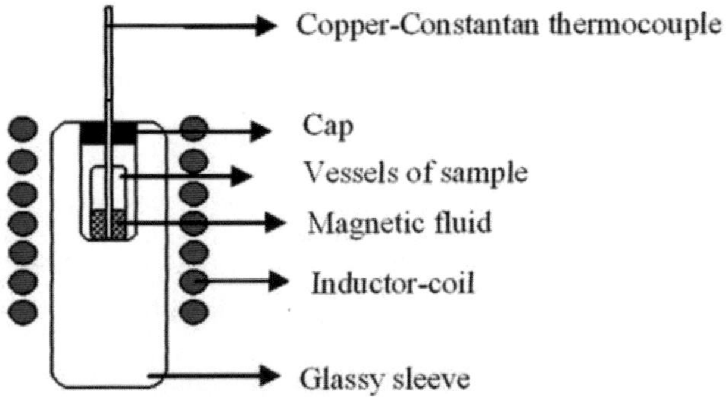

Figure 1. The experimental set up for SLP values measurements.

The magnetic heating experiments were performed under an alternating magnetic field with a frequency of 184 kHz and a field strength of 12 kA/m. The magnetic field is generated by a 5.5-loop induction coil with a 4.5 cm diameter and a 6.3 cm length. During the heating experiment, the temperature of the sample was observed by using a Copper-Constantan thermocouple. The magnetic heating experimental setup is shown in figure 1.

3. Result and discussion

Figures 2 and 3 present the FESEM image of the uncoated powder particles and the coated particles of the magnetic fluid sample, respectively. The magnetic nanoparticles have a mean size about 15÷17 nm with a narrow distribution. For the coated samples, there is good dispersion of nanoparticles in matrix polymers (starch) and particles were capsulated by the polymer layer. The FESEM images of uncoated and coated particles also show that the coating layer is probably very thin because the coated particles size is not much larger than that of uncoated particles. A typical XRD pattern of the nanoparticles is presented in figure 4. It indicates that the samples are single phase with the inversed cubic spinel structure of Fe_3O_4. This success in fabricating high quality Fe_3O_4 nanoparticles is essential for further research work.

Figure 2. FESEM image of uncoated particles.

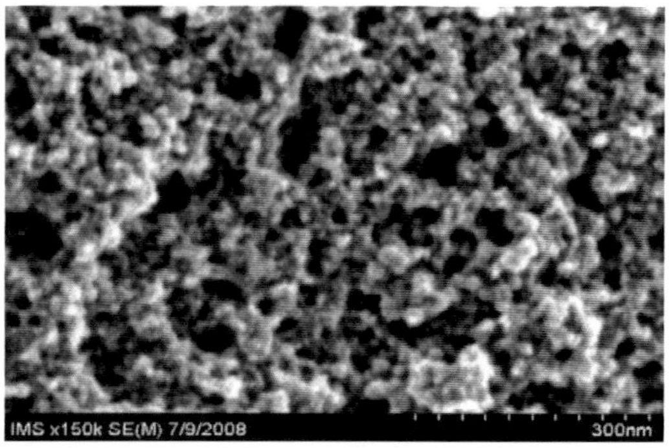

Figure 3. FESEM image of coated particles.

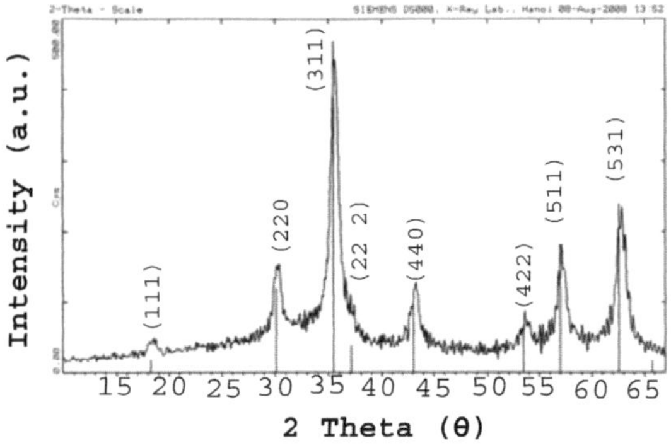

Figure 4. XRD pattern of Fe_3O_4 nanoparticles of magnetic fluid.

The magnetic hysteresis loop of the magnetic fluid measured at room temperature was shown in figure 5. At the field strength of 1.2 kOe, the saturation magnetization of magnetic liquid is measured about 65 emu/g and the coercive field closed to zero. It means that the sample exhibited superparamagnetic behaviour and has high magnetization that determines the heating power in magnetic heating experiments [15].

Figure 6 presents the time-dependence of heating temperature of the magnetic fluid samples with particles concentration varying from 3 to 15 mg/ml exposing an alternating magnetic field having a frequency of 184 kHz and its field strength of 12 kA/m. After about 15÷20 minutes of heating, the temperature of the samples comes to saturation when the energy loss to the surrounding medium equals to the energy generated by the alternating magnetic field. It is clear from this figure that the saturation temperature and rising temperature rate strongly depend on the concentration of particles in the liquid. The more concentrated samples have higher rising temperature rates and saturation temperatures (as shown in table 1). These dependences are almost linear as can be seen in figure 7.

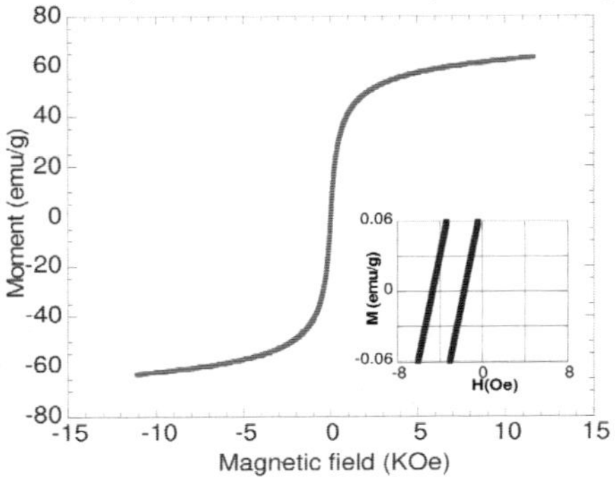

Figure 5. The hysteresis loop at 300 K of the magnetic fluid.

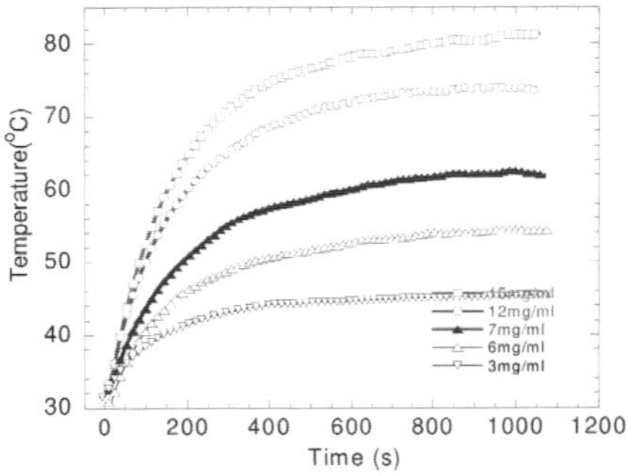

Figure 6. The heating curves of the magnetic fluid with various particles concentration.

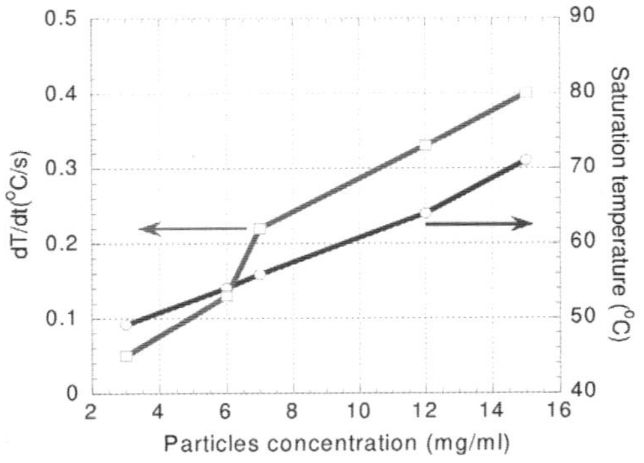

Figure 7. The dependence of the initial heating rate and saturation temperature on particles concentration.

The specific loss power (SLP) can be calculated by using the following formula [16]:

$$SLP = (C * m_s * \frac{dT}{dt}) / m_i ,$$

where C is the specific heat capacity of magnetic liquid, which was accepted equal to that of water, because of the mass mi of iron oxide in the liquid is much smaller than water contained in it, m_s is the total mass of magnetic liquid, dT/dt is the initial temperature rising rate, which was defined at the moment of a few seconds after starting the heat. The SLP value was estimated for the magnetic liquid with various concentrations, presented in table 1.

Table 1. The initial heating rate, SLP and saturation temperature values of different particles concentration of magnetic fluid.

Concentration of particles (mg/ml)	dT/dt (°C/s)	Saturation temperature (°C)	SLP values (W/g)
3	0.092	45	129
6	0.141	53	97
7	0.159	62	94
12	0.240	73	83
15	0.310	80	86

The highest SLP value is approximately 129 W/g, estimated for the liquid with the smallest concentration (3 mg/ml) and decreasing with increasing concentration of particles in liquid. As is known the response of magnetic nanoparticles is mainly derived by Néel and Brownian relaxation [15]. It is implied that the increasing concentration of magnetic particles would decrease distance between adjacent magnetic particles, and the interparticle interaction might affect the relaxation behaviour, that leads to reducing the SLP value. According to a previous report, magnetic particles with mean size of 15 nm and their SLP value of 100 W/g were effectively used in the tumour heating therapy [17]. Our results are particularly of interest for applications in the heating therapy because of the high SLP value at the examined magnetic field parameter (of 129 W/g) while allowing to achieve the saturation temperature of 45°C, which is ideally desired for cancer treatment application, with a minimal iron oxide dose.

4. Conclusion

The Fe_3O_4 nanoparticles and their starch-coated biocompatible ferrofluid have been successfully prepared by co-precipitation methods. The material exhibited the superparamagnetic behaviour with the magnetization is 65 emu/g and coated particles size are smaller than 20 nm. The heating ability investigation of the ferrofluid was performed under an alternating magnetic field of 184 kHz and field strength of 12 kA/m. The SLP of magnetic liquid linearly depends on nanoparticle concentration and achieve a high of 129 W/g at the low concentration of iron oxide. Ability of controlling the heating temperature of the magnetic liquid confirms its possibilities in hyperthermia applications

5. Reference

[1] Boyer M, Townsend L E, Vogel L M 2000 *J. Vasc. Surg.* **31** 181
[2] Kim Y S, Shin K S, Suh J S, Cheon J S 2005 *J. Am. Chem. Soc* **127** (16) 5732
[3] Schopf N T, Hofmann B, Hofmann H, Von Rechenberg M 2005 *JMMM* **293** 483
[4] Moroz P, Jones S K, Gray B N 2002 *Int. J. Hyperthermia* **18** 267
[5] Gordon R T, Hines J R, Gordon D 1979 *Medical Hypothesis* **5** 83
[6] Jordan A, Scholz R, Wust P, Fahling H, Krause J, Wlodarczyk W, Sander B, Vogl T, and Felix R 1997 *Int. J. Hyperthermia* **13** 587
[7] Yanase M, Shinkai M, Honda H, Wakabayashi T, Yoshida J, and Kobayashi T 1998 *Jpn. J. Cancer Res.* **89** 463
[8] Marco Lattuada and Alan Hatton T 2007 *Langmuir* **23** 2158
[9] Ma M, Wu Y, Zhou J, Sun Y K, Yu Z, Gu N 2004 *JMMM* **268** 33
[10] Chan D C F, Kirpokin D C, Bunn P A, Ha Feli U, Schutt W, Teller J, Zborowsk M 1997 *Scientific and Clinical Applications of Magnetic Carriers* (New York: Plenum Press)
[11] Brusentsov N A, Gogosov V V, Brusentsov T N 2001 *JMMM* **225** 113
[12] Muller R, Hergt R, Zeisberger M, Gawalek W 2005 *JMMM* **289** 13
[13] Zeng Q, Baker I, Loudis J A, Liao Y, Hoopes P J, and Weaver J B 2007 *Appl. Phys. Lett.* **90** 233112
[14] Hergt R, Hiergeist R, Zeisberger M, Schuler D, Heyen U, Hilger I and Kaiser W A 2005 *JMMM* **293** 80
[15] Hergt R, Andra W, Hilger I, et al 1998 *IEEE Trans. Magn.* **5** 3745
[16] Ingrid Hilger, Katrin Fruhauf, Wilfried Andra, Robert Hiergeist, Rudolf Hergt, Werner Kaiser A. 2002 *Academic Radiology* **9** 2
[17] Hergt R and Dutz S 2007 *JMMM* **311** 187

Acknowledgement

The work was done with the support of the Science and Technology Project VAST and the Basic Research Project of IMS under grant number CS08.08. The financial support in part by Institute of Materials Science and National Key Laboratory for Electronic Materials and Devices is acknowledged.

Tunnelling mafnetoresistance of (1 – x)$La_{0.7}Ca_{0.3}MnO_3$ +xAg composites

Pham Thanh Phong[1,2], Nguyen Van Khiem[3], Vu Van Hung[3], Do Hung Manh[2], Le Van Hong[2] and Nguyen Xuan Phuc[2]

[1]Nhatrang Pedagogic College, Khanh Hoa Province, Vietnam
[2]Institute of Material Science, Vietnam Academy of Science and Technology, 18 Hoang Quoc Viet Road, Cau Giay District, Hanoi, Vietnam
[3]Department of Natural Sciences, Hong Duc University, 307 Le Lai Street, Thanh Hoa City, Vietnam

E-mail: ptphong.nh@khanhhoa.edu.vn

Abstract. Magnetization and transport of (1 – x)$La_{0.7}Ca_{0.3}MnO_3$ +xAg (x = 0, 0.1, 0.2, 0.3, 0.4 and 0.5) composites has been reported in this paper. Low Field Magnetoresistance (LFMR), conductivity (σ) and metal – insulator transition temperature (T_p) increases with increasing content of Ag. Due to the presence of nomagnetic Ag in the ferromagnetic $La_{0.7}Ca_{0.3}MnO_3$ material, dc magnetization decreases but the corresponding T_p increases. Based on the spin polarized transport of conduction electrons at the grains boundaries, we find that our experimental data (*MR – H*) agree quite well with the dependence of temperature on tunneling magnetoresistance (*TMR*) described by an expression of Hwang *et al.*

Keywords: CMR manganite, low-field magnetoresistance, nanomanganite composites, tunnelling magnetoresistance, low-temperature resistivity.

1. Introduction

The discovery of the colossal magnetoresistance (*CMR*) effect in manganites [1, 2] has caused a renewed interest in these perovskites, which have been known for more than 50 years [3, 4]. Much research has been done on the manganites $Re_{1-x}A_xMnO_3$ (Re = rare earth, A = Ca, Sr, Ba). So far, two *CMR* effects have been found in these manganites, the intrinsic CMR and extrinsic CMR. The intrinsic *CMR* [5] is caused by the double exchange (*DE*) mechanism. The DE mechanism, which was proposed by Zener in 1951 [6], is useful to explain the *CMR* phenomena observed near Curie temperature (T_C) at a relatively high magnetic field (up to several kOe). The extrinsic *CMR*, which is related to the grain boundaries [7], can be explained by spin polarized tunneling [5] or spin dependent scattering [7]. Unfortunately, as temperature increases, the low field grain boundary magnetoresistance decreases rapidly, hindering, the utilization of this effect in sensor application.

Recently, experiments have indicated that the Ag-doped manganites can exhibit an improved T_C, metal – insulator transition temperature (T_p), temperature coefficient of resistance (*TCR*) and an enhanced extrinsic *MR* effect [8-12]. All the above mentioned results indicate that Ag-doped samples have improved MR effect and a significant low field magnetoresistance (*LFMR*) as compared with the undoped counterpart. In addition, these Ag-doped samples, in most cases, are two phase composites, in which the Ag metal phase occurs in the form of the second phase at grain boundaries and interfaces.

© 2009 IOP Publishing Ltd

In this paper we investigate the effects of Ag concentration on the *LFMR* of the bulk polycrystalline $(1\text{-}x)La_{0.7}Ca_{0.3}MnO_3 + x$Ag samples. Significant enhancements in T_p, dramatic reduction in resistivity, and also an improved *LFMR* are observed.

2. Experimental

$(1\text{-}x)La_{0.7}Ca_{0.3}MnO_3 + x$Ag (LCMO/Ag) with x = 0, 0.1, 0.2, 0.3, 0.4 and 0.5 samples were prepared by two steps. Firstly, the LMCO powder was prepared by the sol–gel process. Secondly, appropriate amounts of LMCO and Ag_2O powders were mixed according to the desired molar ratio. Then the mixtures were ground and pressed into pellets. The pellets were finally sintered at 900°C in air for 2 *h*, and then slowly furnace-cooled to room temperature. The crystalline structure of the samples was characterized by the X-ray diffraction. Magnetic properties were measured by a vibrating sample magnetometer (VSM) in the temperature range 100–300 K and in a field up to 5 kOe. The resistivity and the magnetoresistance (*MR* – *H*) of all the composites were examined by the PPMS (Physical Property Measurement Systems) in a magnetic field from 0 to 3kOe at a temperature range from 5 to 300 K.

3. Result and discussion

The phases of samples were characterized by X-ray diffraction (XRD) with Cu Kα radiation. Figure 1 shows the XRD patterns of LCMO/Ag composite samples with x = 0 and 0.5. The XRD patterns of the composites with x = 0.1 ÷ 0.5 show two different sets of diffraction peaks, corresponding to orthorhombic LCMO perovskite and cubic Ag phases, respectively, which clearly indicates the coexistence of LCMO and Ag phases. The pure LCMO sample has an orthorhombic unit cell with lattice parameters: a = 5.462 Å; b = 7.720 Å; c = 5.478 Å. The lattice parameters of LCMO in the LCMO/Ag composite do not change within the accuracy of diffractometer. This shows that LCMO maintains its identity and there is no reaction between LCMO and Ag is introduced as second phase and on the surfaces of the LCMO grains. The direct evidence of the coexistence of two phases also comes from SEM micrographs. The representative SEM micrographs of $(1\text{–}x)$LCMO/xAg composites are shown in figure 2. The interfaces between Ag and LCMO can be distinguished clearly. Moreover, energy dispersive X-ray (EDX) spectra of the doped composite for x = 0 and x = 0.5 shows the silver peak along with La, Ca, Mn and O peaks, which also supports the presence of Ag in the doped composites.

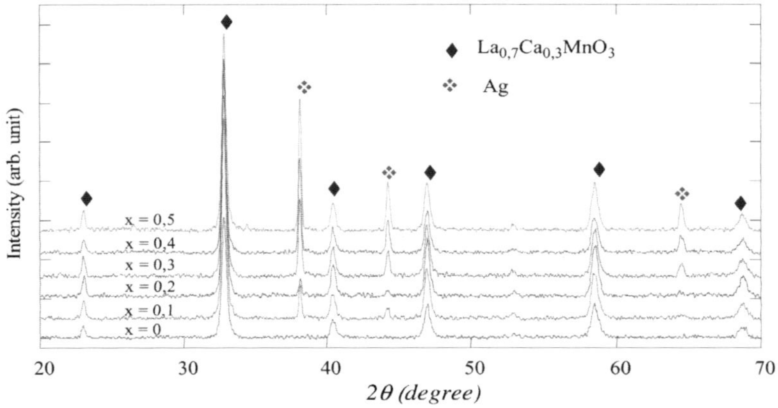

Figure 1. XRD patterns of $(1 - x)$LCMO + xAg composite.

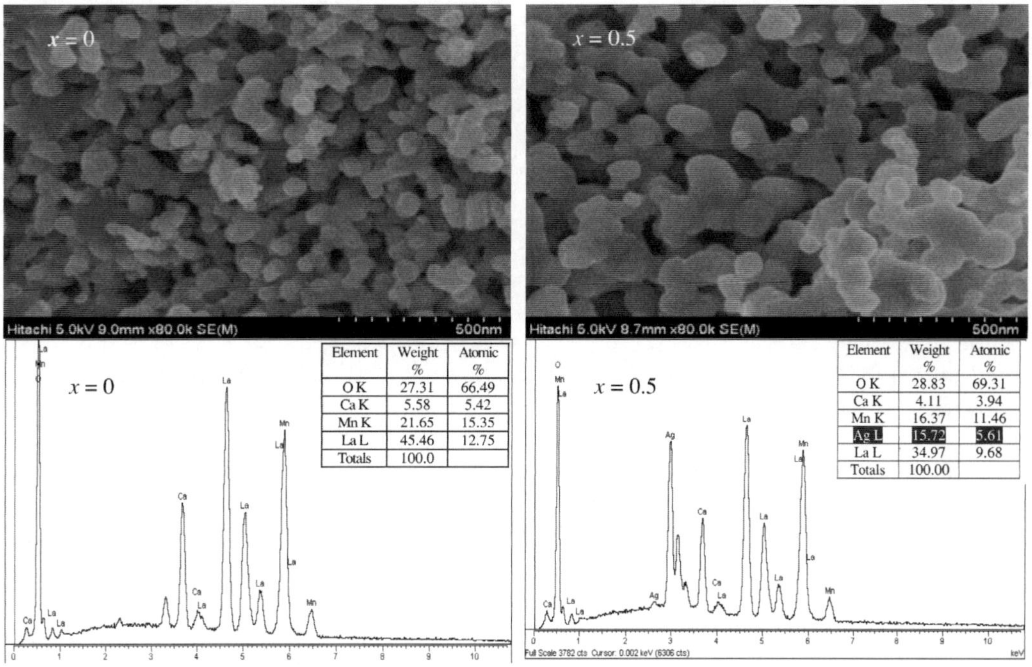

Figure 2. The SEM photographs for all samples and EDX of x = 0 and x = 0.5 samples.

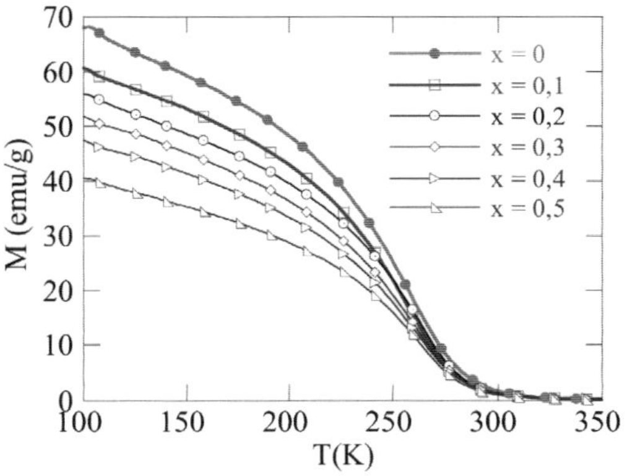

Figure 3. Temperature dependence of magnetization of LCMO/Ag in magnetic field 5 kOe.

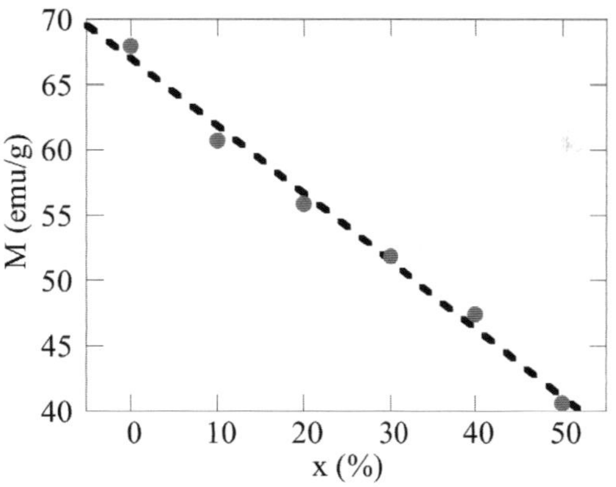

Figure 4. The value of magnetization at 100 K for LCMO/Ag in magnetic field 5 kOe. The line is linear fit to the data.

The temperature dependence of magnetization at 5 kOe for LCMO/Ag composites is shown in figure 3. The value of magnetization (*M*) at 100 K can be seen in figure 4, and it is clear that the decrease in *M* is due only to the silver percentage present in the sample, which leads us to take into account the extra magnetic disorder caused by Ag in the composites. This extra magnetic spin disorder is induced by grain boundaries in the composites and suggests the suppression of DE mechanism resulting suppression of the ferromagnetic alignment of Mn ions. Since the Ag is not incorporated into the LCMO lattice, and it segregates into the grain boundaries or interfacial regions, which blocks the Mn spins at grain boundaries and increases the anisotropy in the interfacial regions and misalignement of the magnetic moments of the neighboring FM domains [13]. Therefore, despite the nonmagnetic character of Ag, it is expected to increase the magnetic disorder by disrupting the Mn-O-Mn bonds in the interfacial regions and hence suppression of the long range FM order. Moreover, the paramagnetic (PM) to ferromagnetic (FM) phase transition temperature (T_C) determined from the peak of *dM/dT* is almost independent of Ag content and is ~ 260 K for all the samples. This is due to the fact that the PM-FM phase transition is an intrinsic and intragrain property. The observed constancy of T_C also indicates that stoichiometry of LCMO phase within the grains remains essentially unchanged as Ag is not accommodated within the perovskite structure and it sits only at the grain boundaries and on the surfaces of LCMO grains.

Figure 5 shows the temperature dependence of zero field resistivity (ρ) for the composites measured in the temperature region from 5 to 300 K. Results show that: (i) ρ decrease with Ag addition; (ii) adding Ag in LCMO sharps the ρ peak and shifts the peak temperature of ρ, T_p, to higher temperature.

Several authors [14-16] attempted to explain the thermal variation of electrical resistivity of Ag doped manganites. In an earlier attempt, Pi *et al.* [15] computed the total resistivity of $La_{0.67}(Ca_{0.65}Ba_{0.35})_{0.33}MnO_3/Ag_x$ composites assuming metal Ag segregated at the grain surfaces or boundaries improves the atomic disordered structure and magnetic state on the grain surfaces and increases the connectivity between the grains. The change of ρ peak can illuminate the improvement of grain boundaries. Later, Yuan *et al.* [14] proposed that in Ag doped manganites, there is the existence of high conductive metal Ag among the grains opens a new channel for the electron transport, which also decreases ρ of the samples. Li *et al* [16] reported that the increased oxygen content released from Ag_2O in the sintering process not only improves the magnetic inhomogenity of the samples and decreases the microstructure deficiency, but also enhances the concentration and hopping mobility of carriers through a little change of Mn^{3+}/Mn^{4+} ratio.

Figure 6 displays the magnetic field dependence at 30 K of the LFMR of LCMO/Ag composites in H = 3kOe. It can be observed from this figure that the MR ratio increases significantly with Ag addition, LFMR reaches the maximum when x = 0.2 and then decreases. The maximum LFMR for x = 0.2 samples is as large as 29%, larger than that of pure LCMO (about 25%).

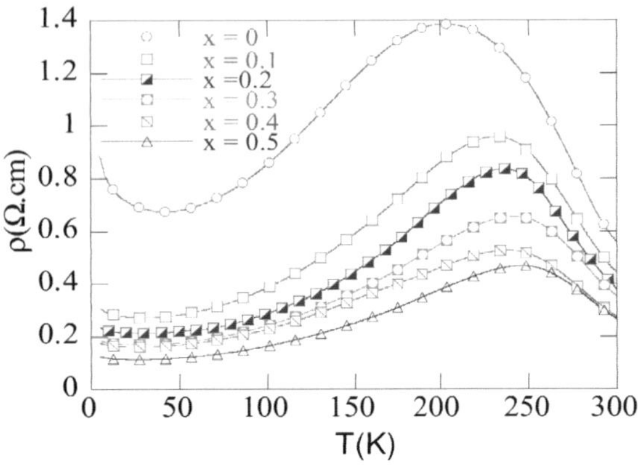

Figure 5. Temperature dependence of resistivity of LCMO/Ag composites.

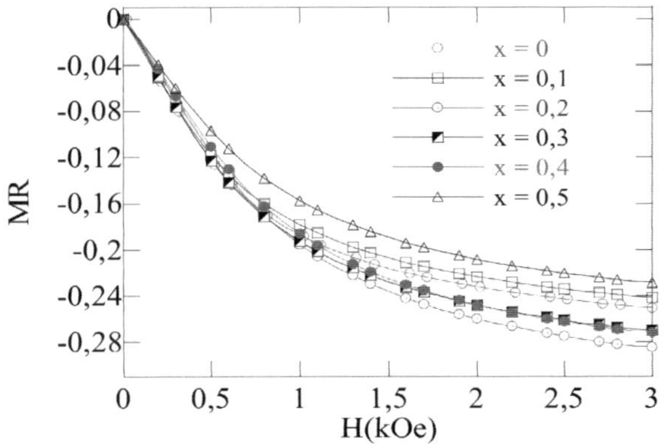

Figure 6. Magnetoresistance (MR) as a function of magnetic field at 30 K for the LCMO/Ag nanocomposites.

The great change of the resistivity of the Ag-added samples is not closely related to the grain size as mentioned above. So the enhancement of *MR* should not closely relate to the grain size. We attribute the large enhancement of *LFMR* to that the Ag addition improves the crystal structure and magnetic homogeneity of the grains and grain boundaries. These improvements are favorable to the increase in intrinsic properties of the compound, especially to the decrease in ρ dramatically. The ρ decrease is also an important factor inducing *MR* ratio to increase. The smaller is ρ of the sample, the larger is the relative change of LFMR.

In order to further illuminate the intrinsic property of the MR behavior observed in this Ag-added composite, we survey the field dependence of *MR* at various for x = 0, 0.2 and 0.5. Our primary approach is to separate out the part of the *MR* originating from SPT (MR_{spt}), from the part of the *MR* identified by the suppression of spin fluctuation (MR_{int}), and mainly to inspect their respective temperature dependencies. For this purpose, we have used the model as proposed by Raychaudhuri *et al.* [17] and Dey *et al.* [18], based on SPT transport of conduction electrons at the grain boundaries with attention paid to the magnetic domain wall motion at grain boundaries under the application of a magnetic field. According to this model we following expression for *MR*

$$MR = -A'\int_0^H f(k)dk - JH - KH^3 . \quad (1)$$

Within the approximation of the model, in zero field, the domain boundaries are pinned at the grain boundary pinning centers having pinning strengths k. The grain boundaries have a distribution of pinning strengths (defined as the minimum field needed to overcome a particular pinning barrier) given by $f(k)$, expressed as:

$$f(k) = A\exp(-Bk^2) + Ck^2\exp(-Dk^2) \ . \quad (2)$$

All adjustable fitting parameters, A, B, C, D are required to known from a nonlinear least square fitting to calculate MR_{spt}, which is defined as:

$$\mathrm{MR_{spt}} = -\int_0^H f(k)dk \ . \quad (3)$$

Differentiating equation (1) with respect to H and using equation (2), we obtain

$$\frac{d(MR)}{dH} = A\exp(-BH^2) + CH^2\exp(-DH^2) - J - 3kH^2 \ . \quad (4)$$

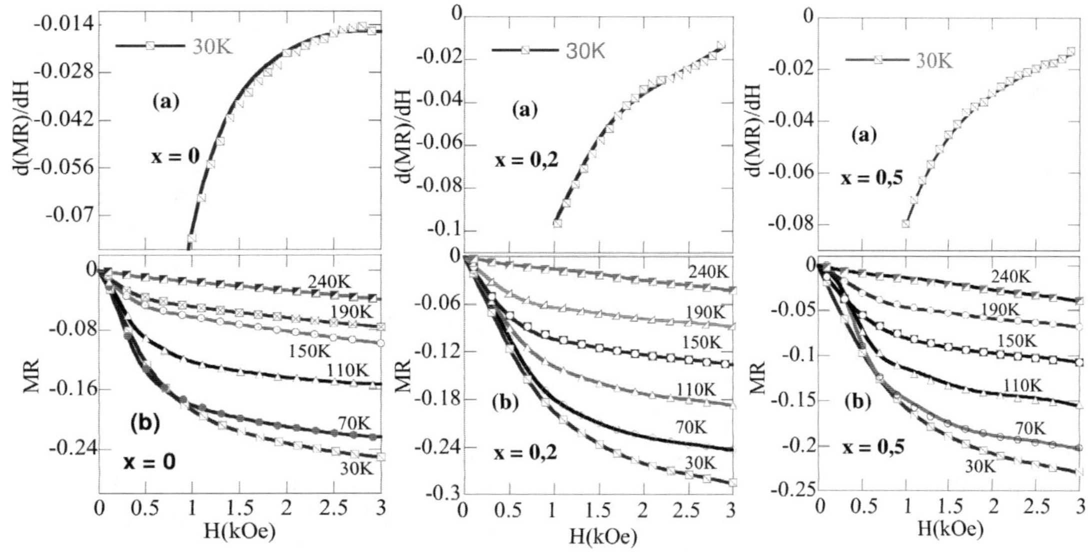

Figure 7. a) Derivative of the experimental (MR-H) curve (dot) and the fitted curve (line) using equation (5) at 30 K in the magnetic field range of (1 - 3 kOe) for sample with x = 0, 0.2 and 0.5. b) Experimental (MR-H) curve (dot) and the fitted curve (line) using equation (2) at various temperatures in the magnetic field range of (0 - 3 kOe) for sample with x =0, 0.2 and 0.5.

The experimental (*MR-H*) curves were differentiated and fitted to equation (4) to find the best-fit parameters at several temperatures. Figure 7a shows the differentiated curve and the best-fit function at T=30K for (1–x)LCMO/xAg composite samples. Using the best fit parameters we have fitted equation (1) to our experimental *MR* versus *H* curves at several temperatures. Figure 7b shows excellent fit for the experimental curves to equation (2) for samples at several temperatures below T_C. We observe that the total magnetoresistance is a nonmonotonic function of temperature with a slow decrease at low temperature followed by the increase as we approach T_C. The intrinsic contribution MR_{int}, however, follows the expected *DE* behavior with a steady increase of temperature. On the other hand MR_{spt} decreases steadily with temperature. In order to elucidate the basic physics behind temperature dependence of *MR*, Dey *et al.* [18] believed that the nature of the surface region of nanosize grains plays a very crucial role in electrical transport, magnetic and magneto-transport behavior of nanodimensional systems. When grain size of LCMO are 17 nm and 27 nm, MR_{SPT}(H) remains constant up to a high temperature (about $T \sim 200$ K) and then drops sharply with temperature. This effect gets enhanced with the decrease in particle size. This result for nanodimensional manganites is in contrast to the results reported by Hwang *et al.* [1] for $La_{0.67}Sr_{0.33}MnO_3$ polycrystalline sample prepared through conventional solid-state reaction process in air and thus have a large grain size (~μm). According to them the part of the *MR* most clearly identified with spin-polarized tunneling shows a gradual decrease with an increase of temperature. They had observed earlier that the temperature dependence of MR_{spt} is described quite well by an expression of the type $a + b/(c + T)$, which is a characteristic of spin polarized tunneling in granular ferromagnetic systems.

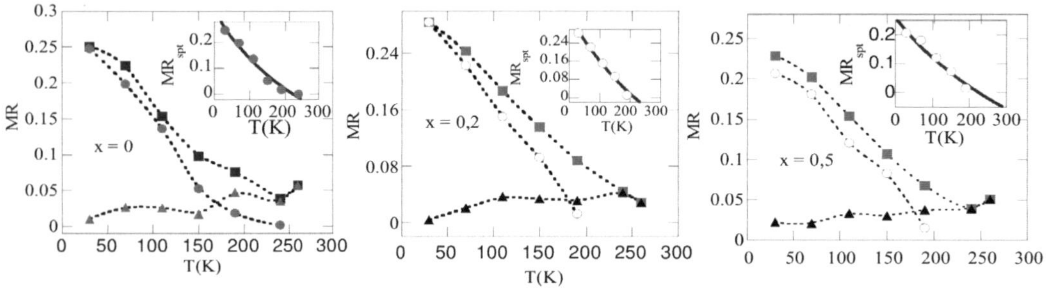

Figure 8. Temperature dependence of various components of magnetoresistance (MR) at 3kOe. (■) is the total MR, (●) is the spin polarized contribution to the magnetoresistance (MR_{spt}), (▲) is the intrinsic contribution MR_{int}. The inset shows the best fit of MR_{spt} to a function a + b/(c +T) with $x = 0$; $x = 0.2$ and $x = 0.5$.

Figure 8 shows the best fit of MR_{spt} with the expression $a + b/(c + T)$. The fitted curve matches well with the extracted values of MR_{spt} from model. However our values of *b* and *c* for the best fit are much higher compared to that observed by Hwang *et al.* although the T_C of our system is much smaller. The reason for this may be related to the structure of the composite. Ag distributed mostly at the grain boundaries can alter the grain behaviour like its size, magnetic and conducting properties. These procedures will significantly influence the tunnel of conduction electrons and hence enhanced *MR* will be expected. In this context we should note that the intergranular spin polarized tunneling have different temperature dependences for ferromagnetically and superparamagnetically coupled grains [19].

All the samples show the unusual low temperature resistivity minimum at some T_{min} (figure 5). The reresistivity upturn at $T < T_{min}$ in polycrystalline manganites is usually attributed to models such as Coulomb blockade (CB) [20], enhanced electron-electron interaction [21], or could be due to the Kondo effect [22, 23]. Here one encounters the phase coherence of two electrons, both becoming localized through elastic impurity scattering. The correction to the electrical conductivity is given by [24, 25]

$$\sigma(T) = \sigma(0) + BT^{1/2} \quad , \tag{5}$$

where $\sigma(0)$ is the residual conductivity contributed by the temperature independent scattering processes and *B* is proportional to the diffusion constant. However, inelastic scattering, which increases monotonically with temperature, competes with the e-e interaction [which decreases with temperature as given in equation (6)] and gives rise to the resistivity minimum. Assuming that all the temperature dependent scattering processes, like electron-phonon, electron-magnon, and electron-electron, are adequately described by a single power law (AT^n), one could write

$$\rho(T) = \frac{1}{\sigma(0) + BT^{1/2}} + AT^n \quad , \tag{6}$$

assuming Mathiessen's rule [24].

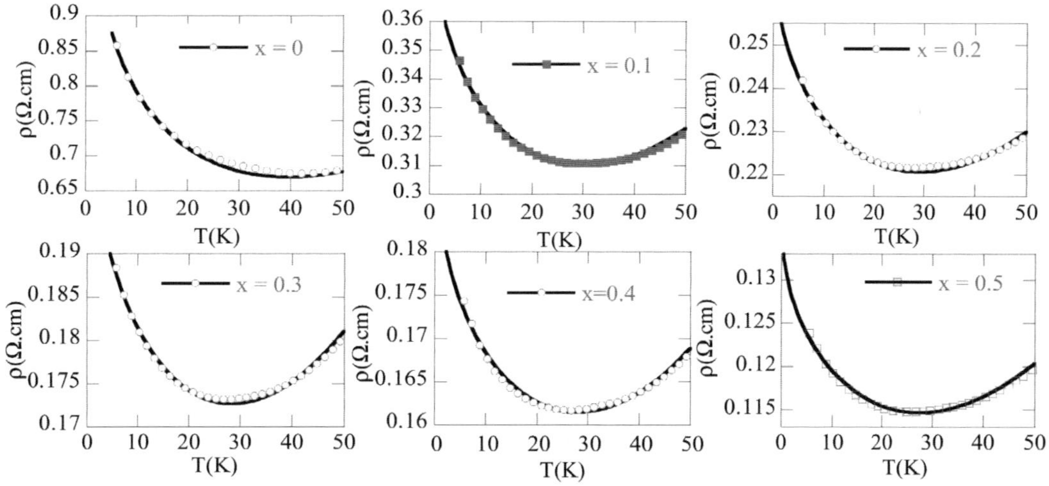

Figure 9. Resistivity (ρ) vs T plots for (1-x)LCMO/xAg. The solid lines are the best fits to equation (6).

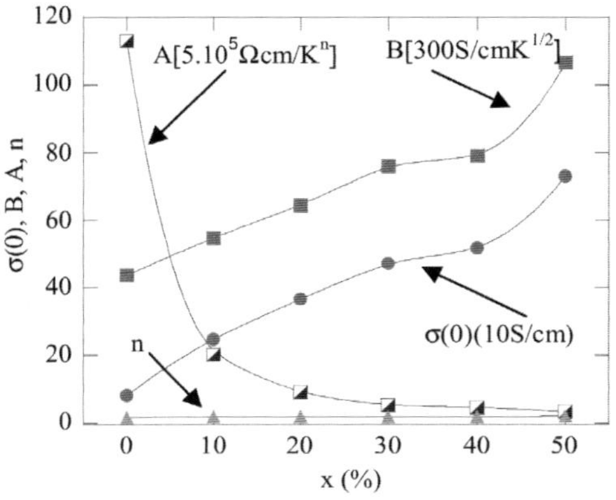

Figure 10. The best fitted parameters $\sigma(0)$, B, A and n [see equation (6) for definition] are plotted with x for LCMO/Ag. The solid lines are just guides to the eye.

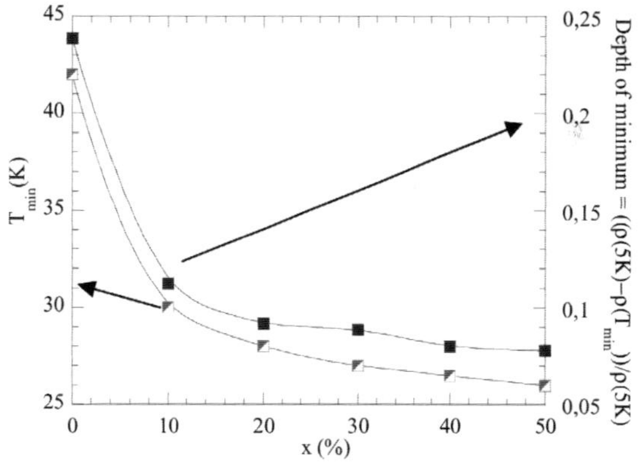

Figure 11. T_{min} and the depth of the minima are plotted against x for LCMO/Ag. The solid lines are just guides to the eye.

Table 1. Best fit parameters obtained from equation (6).

(1 — x)LCMO/x Ag	$\sigma(0)(\Omega cm)^{-1}$	$B(\Omega cmK^{1/2})^{-1}$	$A(10^{-5}\Omega cm/K^n)$	n
$x = 0$	0.82	0.15	22.58	1.64
$x = 0.1$	2.47	0.18	4.04	1.85
$x = 0.2$	3.65	0.21	1.83	1.94
$x = 0.3$	4.71	0.26	1.33	1.96
$x = 0.4$	5.17	0.26	1.17	1.97
$x = 0.5$	7.30	0.36	0.66	2.03

The $\rho(T)$ data for all samples shown in figure 9, are fitted to equation (6). Excellent fits are obtained as shown by the solid lines in figure 9. The coefficients of the fits are given in table 1. The best fitted coefficients $\sigma(0)$, B, A and n are plotted against x (x is Ag content) in figure 10. The coefficients B increase monotonically with x from 0.15 to 0.36 $(\Omega\text{cmK}^{1/2})^{-1}$. The decrease of A with x is understandable in view of the decrease in the depth of minima as shown in figure 11. The inelastic scattering temperature exponent n varies from 1.64 to 2.03 with x increases 0 to 0.5. Tiwari and Rajeev [26] found $n = 2.4$ at $H = 0$ in $La_{0.7}Ca_{0.3}MnO_3$ were prepared by the conventional solid sate method, while Kumar *et al.* [24] found $n = 1.9$ at $H = 0$ in $La_{0.7}Ca_{0.3}MnO_3$ thin film compared to $n =$ 1.64 of ours. The $\sigma(0)$ increases with x.

4. Conclusion

In summary, $(1 - x)$LCMO/xAg composites were prepared by a sol-gel process. The XRD and SEM results show that no reaction between Ag and LCMO takes place, and most Ag is distributed at the grain boundaries. We have studied the effect of Ag on microstructure and low field magneto transport properties of $(1 - x)$LCMO/xAg. As the Ag content increases, the ferromagenetic-paramagnetic (FM-PM) transition temperature remains almost constant, the metal-insulation transition temperature shift towards higher temperatures. The unusual low temperature upturn in resistivity minimum is seen for all the samples. The observed resistivity minimum is explained very well in terms of the electron-electron interaction. It has been found that *LFMR* increases as the Ag content increases. This enhanced *LFMR* is due to increased spin polarized tunneling behavior at lower temperature. We have analyzed our experimental *MR* data following a phenomenological model to separate out the *MR* arising from spin polarized transport, from the intrinsic contribution in our nanosize $(1 - x)$LCMO/xAg composite samples. It is found that the temperature dependence of the LFMR displays a Curie-Weiss law-like behavior, i.e. $LFMR(T) = a + b/(c + T)$, and the fitting parameter c increases monotonically when increasing the Ag content. A detailed study on magnetic behavior and magnetoresistance properties is in progress and results will be forthcoming.

References

[1] Kusters R M, Singleton J, Keen D A, McGreevy R, Hayes W 1989 *Physica B* **155** 362
[2] von Helmolt R, Wecker J, Holzapfel B, Schultz L, Samwer K 1993 *Phys. Rev. Lett.* **71** 2331
[3] Jonker G H, Van Santen J H 1950 *Physica* **16** 337
[4] Jonker G H 1956 *Physica* **22** 3707
[5] Hwang H Y, Cheong S W, Ong N P, Batlogg B 1996 *Phys. Rev. Lett.* **77** 2041
[6] Zener C 1951 *Phys. Rev.* **82** 403
[7] Gupta A, Sun J S 1999 *J. Magn. Magn. Mater.* **200** 24
[8] Huang Y H, Yan C H, Luo F, Song W, Wang Z M 2002 *Appl. Phys. Lett.* **81** 76
[9] Pal S, Banerjee A, Chatterjee S, Nigam A K, Chaudhuri B K, Yang H D 2003 *J. Appl. Phys.* **94** 3485
[10] Yuan X B, Liu Y H, Huang B X, Wang C J, Mei L M 2005 *Solid State Commun.* **135** 170
[11] Awana V P S, Tripathi R, Balamurugan S, Kishan H, Takayama-Muromachi E 2006 *Solid State Commun.* **140** 410
[12] Young S L, Chen H Z, Lin C C, Shi J B, Lance Horng, Shih Y T 2006 *J. Magn. Magn. Mater.* **303** e325
[13] Andres A de, Garcia-Hernadez M, Martinez J L, Prieto C 1999 *Appl. Phys. Lett.* **74** 3884
[14] Yuan X B, Liu Y H, Huang B X, Wang C J, Mei L M 2005 *Solid State Commun.* **135** 170
[15] Pi L, Hervieu M, Maignan A, Martin C, Raveau B 2003 *Solid State Commun.* **126** 229
[16] Li J, Huang Q, Li Z W, You L P, Xu S Y, Ong C K 2001 *J. Appl. Phys.* **89** 7428
[17] Raychadhuri P, Sheshadri K, Taneja P, Bandyopadhyay S, Ayyub P, Nigam A K, and Pinto R 1998 *J. Appl.Phys.* **84** 2048
[18] Dey P and Nath T K 2006 *Phys. Rev. B* **73** 214425
[19] Raychadhuri P, Nath T K, Nigam A K, and Pinto R 1999 *Phys. Rev. B* **59** 13919

[20] García Hernández M, Guinea F, De Andrés A, Martínez J L, Prieto C, Vázquez L 2000 *Phys. Rev. B* **61** 9549
[21] Bacells L I, Foncuberta J, Martínez B, Obradors X 1998 *Phys. Rev. B* **58** R14697
[22] Zhang J C, Xu Y, Cao S X, Cao G X, Zhang Y F, Jing C 2005 *Phys. Rev. B* **72** 054410
[23] E. Syskakis, G. Choudalakis, C Papastaikoudis *2003 J. Phys.D: Condens. Matter* **15** 7735
[24] Kumar D, Sankar J, Narayan J, Rajiv K. Singh, Majumdar A K 2002 *Phys. Rev. B* **65** 094407
[25] Goodenough J B, Zhou J S 1997 *Nature* **386** 229.
[26] Tiwari A, Rajeev K P 1999 *Solid State Commun.* **111** 33

Acknowledgments

The work was done with the support of the Science and Technology Project of VAST and the Basic Research Project of IMS under grant number 409306. The financial support in part of Institute of Materials Science and National Key Laboratory for Electronic Materials and Devices is acknowledged.

Preparation and magnetic properties of TiO_2 doped with V, Mn, Co, La

Nguyen Ngoc Hai[1], Nguyen The Khoi[2] and Pham Van Vinh[2]

[1] Hoang Quoc Viet High School, Dong Trieu District, Quang Ninh, Vietnam
[2] Department of Physics, Hanoi National University of Education,
136 Xuan Thuy Road, Cau Giay District, Hanoi, Vietnam

E-mail: haidhsphn@gmail.com

Abstract. Powder samples of TiO_2:A, where A is Mn, V, Co and La, with dopant concentrations of 0.2%, 0.6%, 1.0%, 5.0%, and 9.0%, were prepared by hydrothermal technique and annealed at temperatures ranging from 450°C to 850°C. The crystalline structure and phase of the samples were investigated by XRD, Raman scattering and SEM. The magnetic properties were studied by measuring the magnetization loops at room temperature. The structure and lattice parameters are found to depend on the dopant and annealing temperature. All doped samples exhibited ferromagnetism. A comparison between the doped samples with different elements was carried out in order to contribute to the understanding of the ferromagnetic mechanism.

Keywords: TiO_2, magnetic properties, ferromagnetism.

1. Introduction

In recent years, oxide diluted magnetic semiconductors (DMSs) with room temperature ferromagnetism have been studied for the potential spintronic applications. The investigations have been performed on the magnetic properties of TiO_2 doped with transition metals and some rare earth elements, such as Mn, Cr, Fe, N, V, La... [1-7]. Although ferromagnetism at room temperature has been observed in doped TiO_2 powder samples, some studies suggested the presence of second phases or precipitated ferromagnetic clusters as the origin of ferromagnetism, while other results seemed to indicate the existence of intrinsic ferromagnetism of transition metal substitution in the Ti lattice. In order to extend the study on the magnetic properties and ferromagnetism mechanism, many investigations on of Mn, Fe, Co... doped TiO_2 powders and bulk samples have been carried out [1-7].

In this paper, powder samples of TiO_2:A, where A is V, Mn, Co and La with dopant concentration of 0.2%, 0.6%, 1.0%, 3.0%, 5.0%, 7.0% and 9.0%, were prepared by hydrothermal technique and then annealed at different temperatures ranging from 450°C to 850°C. The effect of temperature and dopant concentration on magnetic properties was investigated in order to contribute to the understanding of the ferromagnetic mechanism.

2. Experimental

$TiCl_4$ was used as a starting material. The solution of $TiCl_4$ in water was stirred at 80°C. To have different Mn dopant concentrations, the predetermined amount of $Mn(NO_3)_2$ was put into the solution. Also, NH_4NO_3, NH_3 and HCl were also added to maintain pH constant at 7. The solution was further

© 2009 IOP Publishing Ltd

stirred for 30 minutes then the mixture was washed and dried to get the powder. Finally, the powder was annealed at temperatures ranging from 450 to 850°C for 1 hour. For TiO_2 powder doped with La, Co and V, the process was carried out similarly with $Mn(NO_3)_2$ replaced by $La(NO_3)_3$, $Co(NO_3)_2$ and V_2O_5/HCl. The dopant concentrations were calculated following the formula:

$$\%n_D = n_D/(n_D + n_{Ti}) ,$$

where n_D and n_{Ti} are the amounts (in moles) of dopant elements and Ti, respectively.

The crystalline structure and phase of the samples were characterized by X-ray Diffraction (XRD), Raman scattering and Scanning Electronic Microscopy (SEM). The magnetic properties were studied by a vibrating sample magnetometer (VSM).

3. Results and discussion

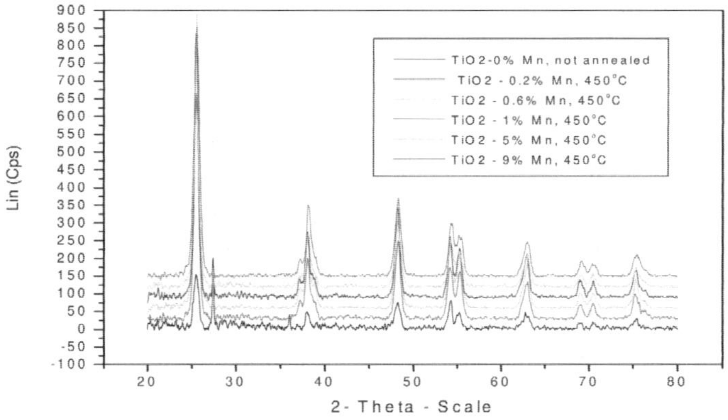

Figure 1a. XRD pattern of pure TiO_2 and TiO_2:Mn annealed at 450°C with Mn concentration: 0.2%, 0.6%, 1.0%, 5.0%, and 9.0%.

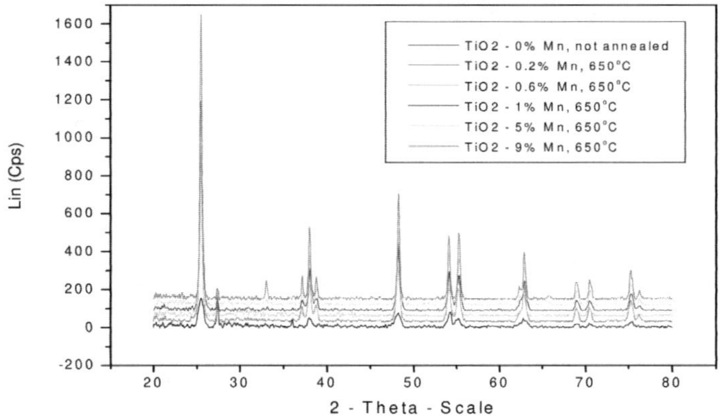

Figure 1b. XRD pattern of pure TiO_2 and TiO_2:Mn annealed at 650°C with Mn concentration: 0.2%, 0.6%, 1%, 5% and 9%.

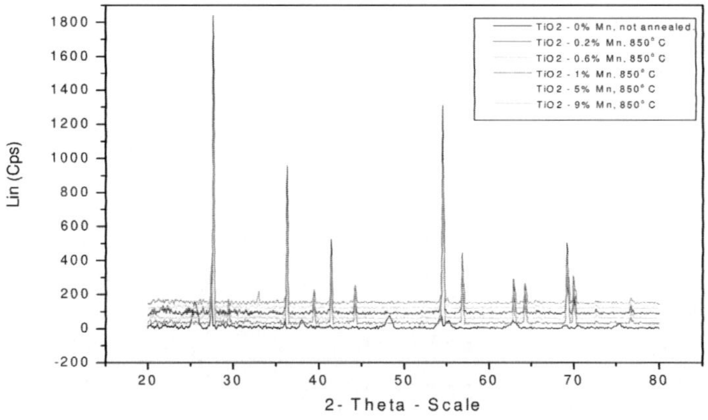

Figure 1c. XRD pattern of pure TiO_2 and TiO_2:Mn annealed at 850°C with Mn concentration: 0.2%, 0.6%, 1.0%, 5.0% and 9.0%.

Figures 1a, 1b and 1c show XRD diffractograms of TiO_2:Mn powder annealed at 450°C, 650°C, and 850°C with various Mn concentrations. The diffraction pattern shows the presence of crystalline TiO_2 anatase phase. The XRD peaks tend to shift towards lower 2θ values. The anatase phase is dominated for all samples annealed at 450°C. With the increase of annealing temperature, anatase phase tends to transform into rutile phase. The anatase phase was transformed into rutile phase completely when the annealing temperature reached 850°C.

The ionic radius of Mn^{2+}, Mn^{3+} and Mn^{4+} ions at octahedral sites are known to be 0.970Å, 0.785Å, and 0.670Å, respectively, while that of Ti^{4+} ion is 0.745Å. The shift of peaks was intended to the substitutions of Mn^{4+} to Ti^{4+} in the TiO_2 lattice. Besides, the XRD peaks are broadened with the increase of Mn concentrations from 0.2% to 1.0% and narrowed with the further increase of Mn concentrations. This indicates that the dopant concentrations has influenced on the TiO_2 crystalline size. At higher Mn concentrations (5.0 to 9.0%), some new peaks appear. They are related to the presence of secondary phases such as Mn_2O_3, MnO and MnO_2...

Figure 2 shows XRD diffractograms of TiO_2 powder doped with 1% Mn. The powder was annealed at temperatures ranging from 450 to 850°C. The diffraction pattern indicates that the annealing temperature has a significant effect on the crystalline size and phase of TiO_2:Mn. At temperatures of 450 and 650°C, the crystal exhibits the TiO_2 anatase phase. However, the crystalline phase is transformed to rutile phase at annealing temperature of 850°C. Beside the shift of peaks, the peaks are narrowed at higher annealing temperatures. This indicates that the crystalline size increases with the increase of annealing temperature. The recrystallization at high temperature is believed to be the reason for this event.

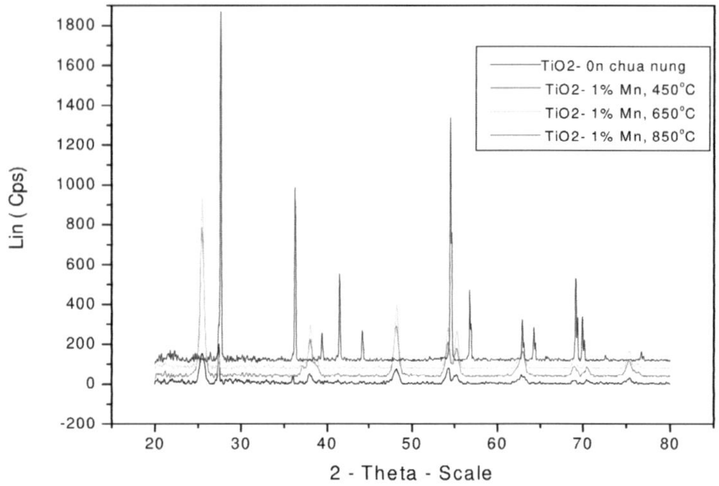

Figure 2. XRD pattern of pure TiO_2 and TiO_2:Mn, with Mn concentration: 1% annealed at temperatures ranging from 450°C to 850°C.

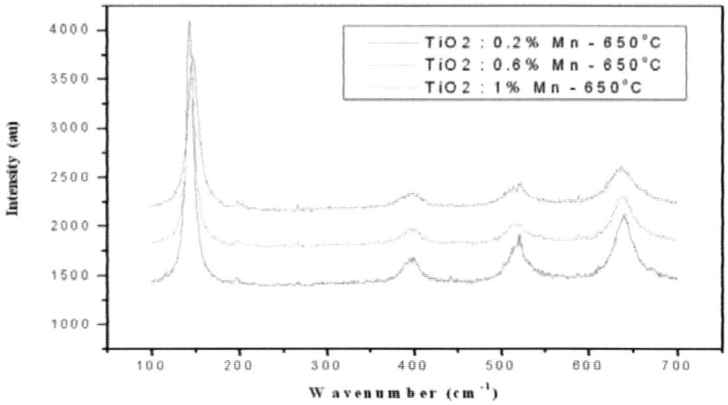

Figure 3. Raman spectra of TiO_2:Mn annealed at 650°C with Mn concentration: 0.2%, 0.6%, and 1.0%.

Journal of Physics: Conference Series **187** (2009) 012071 doi:10.1088/1742-6596/187/1/012071

Figure 3 presents the Raman spectra of TiO_2:Mn with Mn concentrations of 0.2, 0.6 and 1.0%. All peaks are shifted towards higher wavenumbers with the increase of Mn concentration. Also, the relative intensities of the high wavenumber peaks to the highest peak diminish with increase of Mn concentration.

The grain size calculated from XRD patterns by using Scherrer formula and observed from SEM images (figure 4 to 6) shows that the particle size increases with the increase of the annealing temperature and the concentration (figure 7).

Figure 4. SEM image of TiO_2:0.6% Mn annealed at 450°C.

Figure 5. SEM image of TiO_2:0.6% Mn annealed 650°C.

Figure 6. SEM image of TiO_2:0.6% Mn annealed at 850°C.

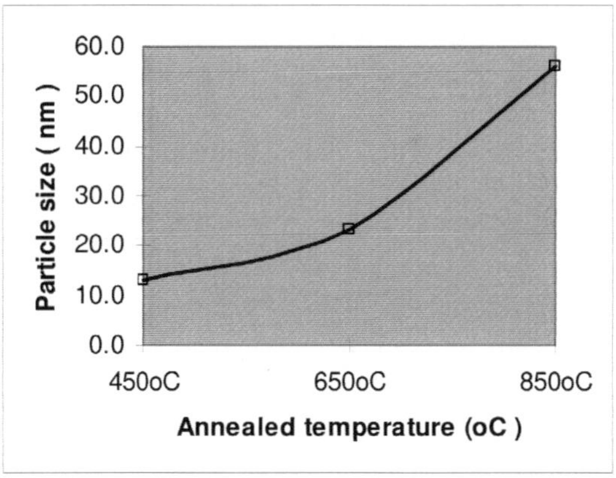

Figure 7. Particle size of TiO_2:0.6% Mn versus annealing temperature.

Figure 8 illustrates the magnetization hysteresis loops (M – H) at 300 K for TiO_2:0.6% Mn sample annealed at different temperatures 450°C, 650°C, and 850°C in air. As can be seen, all the samples exhibit a mixing behavior of paramagnetism and ferromagnetism. With the increase of annealing temperature, the ferromagnetic component decreases, while the paramagnetic component increases. As shown, the saturation magnetization of the samples annealed at 450°C is larger than that of the samples annealed at 650°C and 850°C. This sample also has the highest coercive field H_C and remanent magnetization M_r.

As shown from figure 9, the dependence of magnetization of TiO_2:Mn samples on Mn concentration 0.2%, 0.6%, 1.0%, 3.0%, 5.0%, 7.0% and 9.0%, annealed at the temperatures of 650°C. For the TiO_2 doped with 0.2% Mn, the coercive field H_C is about 37 Oe, the saturation magnetization

Journal of Physics: Conference Series **187** (2009) 012071 doi:10.1088/1742-6596/187/1/012071

approaches 0.038 emu/g and the remanent magnetization M_r is about 0.001 emu/g (figure 9). M_S tends to decrease when the Mn concentration increases to 3.0%. Further increasing Mn concentration, M_S increases. H_C and M_r in this case have the largest value. From this, it is seen that the increase of Mn concentration decreases the ferromagnetism and increases the paramagnetism of the samples. This also illustrates the ferromagnetism at room temperature can be attributed to the formation of Mn-dependant oxides. If the formation of the secondary phase was responsible for the ferromagnetic behaviour at room temperature, increasing Mn concentration, the volume fraction of the secondary phase and the corresponding magnetization would increase. In other words, a relatively large concentration of Mn in the samples does not contribute to the ferromagnetic ordering, which may be due to the formation of Mn clusters as a consequence of decreased Mn-Mn distance. Therefore, the ferromagnetism observed at room temperature can be due to the incorporation of Mn ions into the TiO_2 lattice. At present, although the origin of the ferromagnetism is not clear in oxide DMSs, many studies have argued that shallow defects, such as oxygen vacancies, play a critical role in activating ferromagnetism in these systems [1].

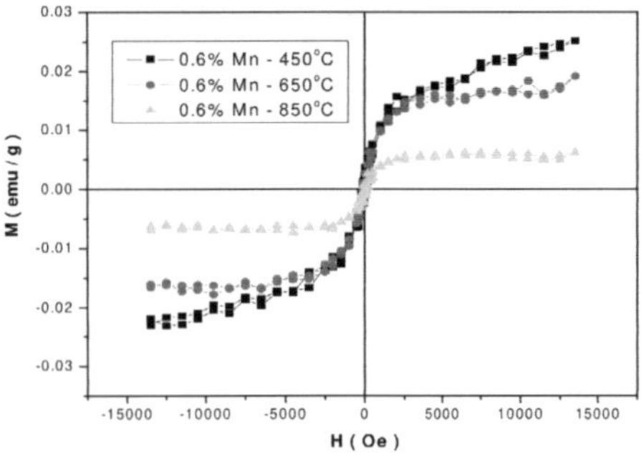

Figure 8. Magnetic hysteresis loops of TiO_2:Mn 0.6% at 300 K. Samples annealed at different temperatures 450°C, 650°C, and 850°C.

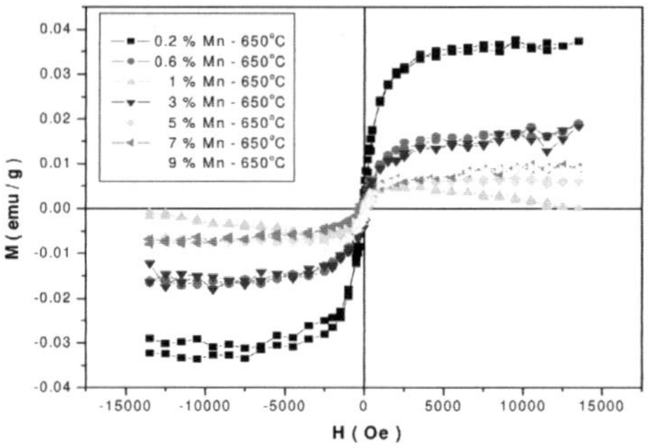

Figure 9. Magnetic hysteresis loops of TiO_2:Mn. The samples annealed at 650°C, with Mn concentration: 0.2%, 0.6%, 1.0%, 3.0%, 5.0%, 7.0% and 9.0%.

Table 1. M_S, H_C, M_r depend on the annealing temperatures.

Samples of TiO_2 : 0.6% Mn	M_S (emu/g)	H_C (Oe)	M_r (emu/g)
450°C	0.025	113	0.0014
650°C	0.019	82	0.0011
850°C	0.006	92	9.10^{-4}

Table 2. M_S, H_C, M_r depend on the concentration – TiO_2:Mn samples, annealed at 650°C - 1h.

TiO_2 : Mn (650°C-1h)	M_S (emu/g)	H_C (Oe)	M_r (emu/g)
0.2%	0.038	37	0.0010
0.6%	0.019	82	0.0011
1%	0.0049	54	$8x10^{-4}$
3%	0.0018	133	0.0025
5%	0.0068	53	$7.8x10^{-4}$
7%	0.0072	112	$6.6x10^{-4}$
9%	0.0093	84	$7.9x10^{-4}$

Thus, the reason for the decrease of the ferromagnetic properties should be the increase of the Mn concentration and the increase of the annealing temperature. The ferromagnetism at room temperature is attributed to the incorporation of Mn ions into the TiO_2 lattices. Although the XRD patterns shows the existence of Mn_3O_4 in the samples, the curie temperature of Mn_3O_4 (T_C) is about 43 K, it is not account for the ferromagnetism observed above [7].

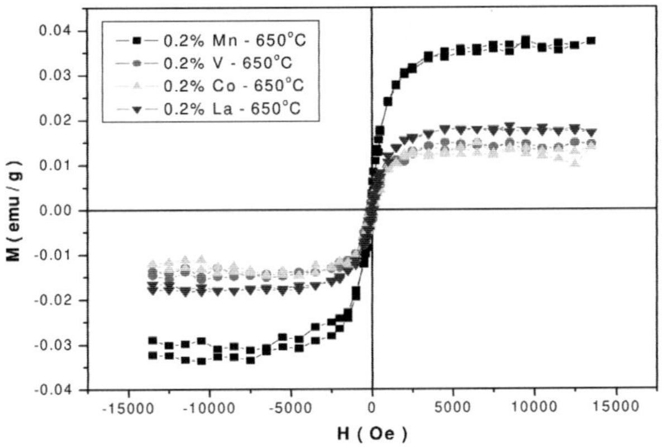

Figure 10. Magnetic hysteresis loops of TiO_2:Mn, TiO_2:V, TiO_2:Co and TiO_2:La at 300K. The samples annealed at 650°C, concentration: 0.2 %.

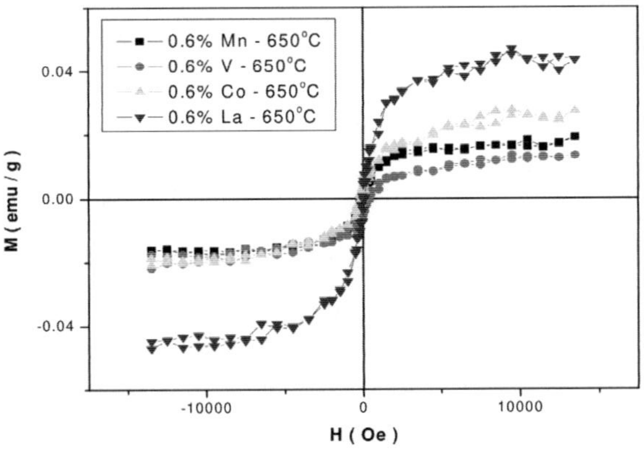

Figure 11. Magnetic hysteresis loops of TiO_2:Mn, TiO_2:V, TiO_2:Co and TiO_2:La at 300 K. The samples annealed at 650°C with concentration of 0.6 %.

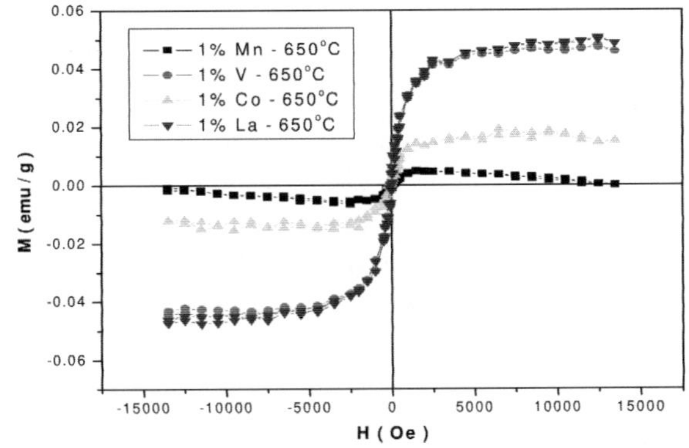

Figure 12. Magnetic hysteresis loops of TiO_2:Mn, TiO_2:V, TiO_2:Co and TiO_2:La at 300 K. The samples annealed at 650°C with concentration of 1%.

The magnetic hysteresis loops at 300 K of all the TiO_2:Mn, TiO_2:V, TiO_2:Co and TiO_2:La samples annealed at 650°C are shown in figures 10, 11, 12 and the range of experimental observation and theoretical computation based on data are also shown in tables 3, 4, 5. Although the results are at the first glance confusing, some general remarks can be extracted. It firstly strengthens the suggestion above that the ferromagnetism of the samples is superimposed by the paramagnetic components. It also shows that all the samples have ferromagnetic properties.

The saturation magnetization M_S of the TiO_2:La samples is highest in the cases of 0.2%, 0.6% and 1.0% dopant concentrations. From the figure 12, with the concentration of 1.0%, the coervive field H_C of the TiO_2:Mn sample is higher than the others while the highest remanent magnetism M_S is

0.0032 emu/g corresponding to the TiO_2:La sample. In figure 11, with lower concentration, the coervive field H_C of TiO_2:V sample is the largest while the smallest M_r belongs to the TiO_2:Mn sample. This conclusion is still right in the case of 0.2% concentration.

Table 3. M_S, H_C, M_r TiO_2 doped with Mn, V, Co and La samples, concentration: 1% annealed at 650°C/1h.

Samples annealed at 650°C, concentration : 1%	M_S (emu/g)	H_C (Oe)	M_r (emu/g)
TiO_2 : Mn	0.0049	54	8.10^{-4}
TiO_2 : V	0.0479	43	0.003
TiO_2 : Co	0.0190	46	$4\ .10^{-4}$
TiO_2 : La	0.0506	43	0.004

Table 4. M_S, H_C, M_r TiO_2 doped with Mn, V, Co, La samples, concentration: 0.6% annealed at 650°C/1h.

Samples annealed at 650°C, concentration : 0.6%	M_S (emu/g)	H_C (Oe)	M_r (emu/g)
TiO_2 : Mn	0.019	82	0.0011
TiO_2 : V	0.0014	140	0.0017
TiO_2 : Co	0.0278	69	0.0045
TiO_2 : La	0.0470	68	0.0047

Table 5. M_S, H_C, M_r TiO_2 doped with Mn, V, Co, La samples, concentration: 0.2% annealed at 650°C/1h.

Samples annealed at 650°C concentration : 0.2%	M_S (emu/g)	H_C (Oe)	M_r (emu/g)
TiO_2 :Mn	0.038	37	0.0010
TiO_2 :V	0.0152	74	0.0020
TiO_2 :Co	0.0149	91	0.0011
TiO_2 :La	0.0187	67	0.0017

The results above for transition metal dopants in TiO_2 are sparse. Although all of them contributed to the ferromagnetism of all the samples, we can see the differences among the samples doped with the elements in 3d group (V, Mn, Co) and 4f group (Lanthanide group).

4. Conclusion

Magnetic properties have been investigated in powder samples of TiO_2:A, where A is Mn, V, Co and La, with by the hydrothermal method. With the increase in the annealing temperature and concentration, the ferromagnetism of the samples decreases monotonically. For the samples annealed at 650°C with different dopant contents of Mn, V, Co and La, the ferromagnetism exhibits clearly. The main reason for the ferromagnetism at room temperature can be the substitution of the V^{5+}, Mn^{2+}, Co^{2+}, La^{3+} ions into the TiO_2 lattices and the formation of the dopant concentration 0.2%, 0.6%, 1.0%, 3.0%, 5.0%, 7.0% and 9.0% for Mn and 0.2%, 0.6% and 1 % for V, Co and La prepared clusters.

Mn, V, Co and La concentrations in TiO_2 play the key role in ferromagnetic mechanism but it still remaining the difference in the saturation magnetization M_S, the coercive field H_C and the remanent magnetism M_r in the respective cases.

References

[1] Tian Z M, Yuan S L 2007 *Effects of sintering temperature and atmosphere on magnetism in Mn-doped TiO_2 bulk samples* (Department of Physics, Huazhong University of Science and Technology, Wuhan 430074, PR China)

[2] Rebecca Janisch 2005 *Transition metal – doped TiO_2 and ZnO – present status of the field* (University of California Santa Barbara, CA, USA)

[3] Geng W T 2003 *Interplay of local structure and magnetism in Co – doped TiO_2 anatase* (Pohang University of Science and Technology, South Korea)

[4] Chambers S A 2003 *Clusters and magnetism in epitaxial Co – doped TiO_2 antase* (Pacific Northwest National Laboratory, Richland, Washington)

[5] Lee Yun- Hi 2004 *Magnetism in nanometer – thick TiOx/Co/TiOx/TiN/Si Multilayered structures* (Korea University, South Korea)

[6] Yuji Matsumoto 2001 *Room – Temperature Ferromagnetism in Transparent Transition Metal – doped Titanium Dioide* (Tokyo Institute of Technology, Japan)

[7] Cabrera A F, Errico L 2007 *Influence of thermal treatments on phase formation and magnetic behaviour in metal transition doped TiO_2* (Departamento de Physica, Facultad de Ciencias Exactas, UNLP, I.F.L.P, CONICET, Argentina)

Acknowledgments

This paper was supported by Basic Research Project of Hanoi National University of Education No.HD1-04. The authors would like to give the special thanks to the lecturers of Department of Solid State Physics, Hanoi National University of Education and the teachers of Hoang Quoc Viet High School, Dong Trieu, Quang Ninh.

Investigation of antibacterial activity of cotton fabric incorporating nano silver colloid

Ngo Vo Ke Thanh and Nguyen Thi Phuong Phong

Laboratory for Nanotechnology, Vietnam National University Ho Chi Minh City, Community 6, Linh Trung Award, Thu Duc District, Ho Chi Minh City, Vietnam

E-mail: ngovokethanh@yahoo.com

Abstract. In this work, silver nanoparticles were prepared by polyol process with microwave heating and incorporated on cotton fabric surfaces. The antibacterial performance of the antibacterial cotton fabric was tested for different concentration of nano-sized silver colloid, contact time germs, and washing times. It was found that antibacterial activity increased with the increasing concentration of nano-sized silver colloid. The antibacterial fabric with 758 mg/kg of silver nanoparticles on surface cotton was highly effective in killing test bacteria and had excellent water resisting property.

Keywords: Silver nanoparticles, antibacterial activity, cotton fabric, microwave radiation.

1. Introduction

The application of nanoparticles to textile materials has attracted considerable interest due to their novel physicochemical properties and their potential applications. However, some of these particles are toxic or poorly effective, which makes them not suitable for applications in medicine, filters, and textiles and for the exclusion of pollution. For example, nano-TiO_2 for seft-cleaning properties and applying in hygienic textiles [1-3] and ZnO nanoparticles for UV-blocking, antibacterial properties and applying in medical clothes, protective [4, 5]. For a long time, silver nanoparticles are non-toxic, no tolerant disinfectant which can remove more than 650 bacteria, virus, and fungi species. Therefore, silver nanoparticles attract an increasing amount of scientific and industrial interest from fields such as textile science [5, 6], medicine [7], agriculture [8], sensors and detectors [7], and catalysis [9]. Antibacterial activity of silver nanoparticles containing materials as diverse as polyurethane foam can be used for water treatment [10].

Cotton is a common material for the production of textiles for sport and leisure. It has excellent moisture absorption ability. However, the moist cotton can be easily attacked by bacteria. Decomposed products of body secretions have a characteristic odor [11].

In recent research, a good antibacterial effect of nano-sized silver colloidal solution on polymer and textile fabrics was shown [6, 12]. The objective of this study was not only the synthesis of silver nanoparticles used microwave radiation as a heating source but also investigated the surface morphologies of antibacterial cotton fabrics and the relationship between the antibacterial property and the content of nano-sized silver on cotton fabric was discussed. Moreover, the laundering durability of the bacteriostasis of the silver nanoparticles-treated cotton fabric was also examined.

© 2009 IOP Publishing Ltd

2. Experimental

2.1. Material

Plain weave cotton fabric (107 g/m^2) made by Phuoc Thinh Textile Company, Ho Chi Minh, Vietnam, was used. $AgNO_3$, Polyvinylpyrrolidone (PVA) and Ethylenlycol ($C_2H_4(OH)_2$) were obtained from MERCH-Germany was purchased from Hoa Nam Company-Vietnam. *Escherichia coli* (*E. coli*) (ATTC 25922) and Staphylococcus aureus (*S. aureus*) (ATCC 290408) were provided by Pasteur Institute-Ho Chi Minh, Vietnam. Nutrient agar was provided by Laboratory for Microbiology – Natural Science University, Ho Chi Minh city, Vietnam.

2.2. Synthesis of nano-sized silver colloid solution

An aliquot of 0.70 g of PVP and 50 ml $C_2H_4(OH)_2$ was mixed by magnetic stirrer at 80 – 90^{0}C for 1 h. After that, amount of 0.05 g $AgNO_3$ was added into PVP solution. This solution was placed into microwave oven chamber for 4 minutes at 160 W (figure 1). Finally, nano-sized silver colloid solution which has yellow color was obtained.

The UV-visible absorption behaviors for silver nanoparticles suspensions were recorded by using UV-VIS spectrophotometer (Varian, model 100, Australia). The size and shape of silver nanoparticles were measured using a Transmission Electron Microscopy (TEM) JEM model 1400, 100 kV. The size distribution of the particles was determined by using UTHSCSA Image Tool 3.00 software.

2.3. Preparation of antibacterial cotton fabric

The concentration of colloidal silver solutions was varied - 20, 50, 80 and 100 ppm - by diluting each nano–sized silver colloid solution with distilled water. The cotton fabric (d = 9 cm) were padded with each concentration of the colloid solutions about 5 minutes and squeezed to 100% wet pick-up using a laboratory pad at constant pressure. The samples were immediately dried at 80^{0}C for 15 minutes. After that, the samples were laundered in water for 15 minutes and dried at 80^{0}C.The antibacterial cotton fabrics were obtained.

The dispersibility of the silver nanoparticles on the surface of the cotton fabric were estimated by using a field-emission scanning electronic microscopy (FESEM, S-4800) was operated at 5kV and used at 20000 magnifications.

Inductively coupled plasma Atomic Absorption Spectroscopy (ICP-AAS) was used to measure the remained quantity of silver particles on cotton fabric surface. We compared the concentration of silver particles on fabrics before washing to the particles on them after 5, 10, and 15 times washing.

2.4. Antibacterial test

We used Colony Count Method to study the antibacterial properties of the fabrics finished with the nano-sized silver colloid solution. The effect of bacterial strains is determined by the relation

$$\eta = \frac{N1 - N2}{N1} \times 100\% \quad ,$$

where η is the percentage of bacterial reduction, N_1 is the number of surviving bacterial colonies from the control sample, and N_2 is the number of surviving colonies from test sample.

3. Result and discussion

3.1. Characterization of nano-sized silver colloid solution

The reduction of silver nanoparticles was visibly evident from the color changes associated with it (from white to yellow). Figure 2 shows the UV-VIS absorption spectra and image of nano-sized silver colloid after the process of reduction. As shown in figure 2, absorbance of nano silver synthesized by microwave heating was very high. From the Bee's law, a UV-visible absorbance is proportional to the path length and the concentration of the suspension.

Journal of Physics: Conference Series **187** (2009) 012072 doi:10.1088/1742-6596/187/1/012072

Figure 1. Solution of silver nanoparticles in microwave oven chamber.

Hence, it can be concluded that the suspension of microwave heating has a high concentration of Ag nanoparticles. The sharp peak at 407 nm can be proved a narrow size distribution of silver nanoparticles formed in the solution, because the maximum absorbance peak shifts to a longer wavelength when the particles becomes larger [13] and heating by microwave is more uniform [14]. As evident in figure 3, the UV-visible spectra of silver nanoparticles do not change after a month of ageing time. It can be concluded that the stable nature of nano-sized silver in solution is very high. TEM image and the size Ag nanoparticles distribution results showed that the silver nanoparticles were found to be spherical in shape, having a mean size about 7-11 nm (figure 4).

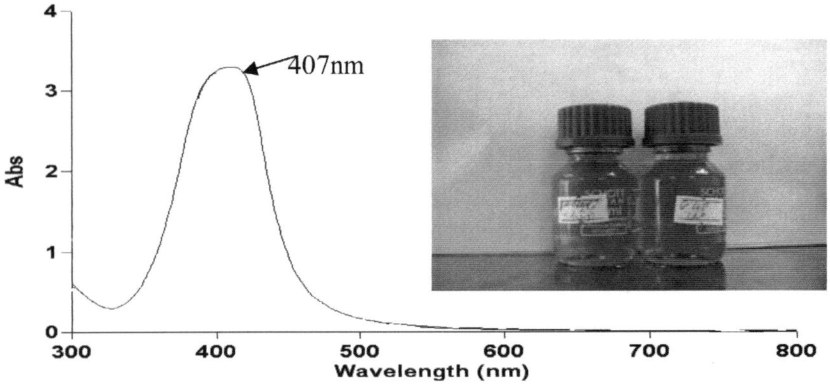

Figure 2. UV-VIS and image of nano silver colloid.

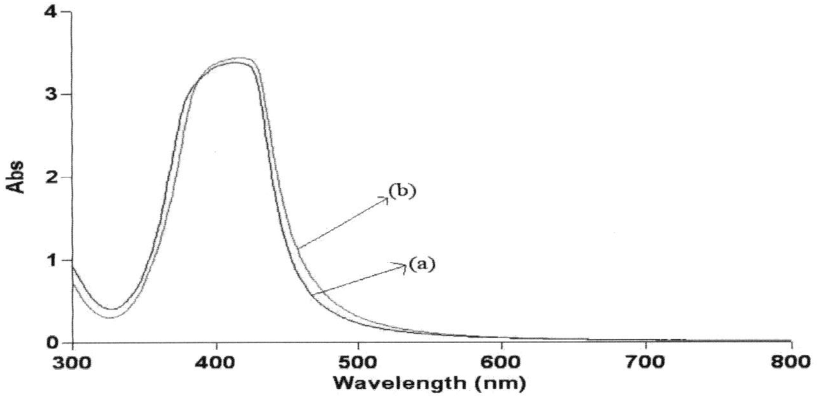

Figure 3. UV-visible absorption spectra of silver nanoparticles (a) after process of reduction; (b) after 1 month.

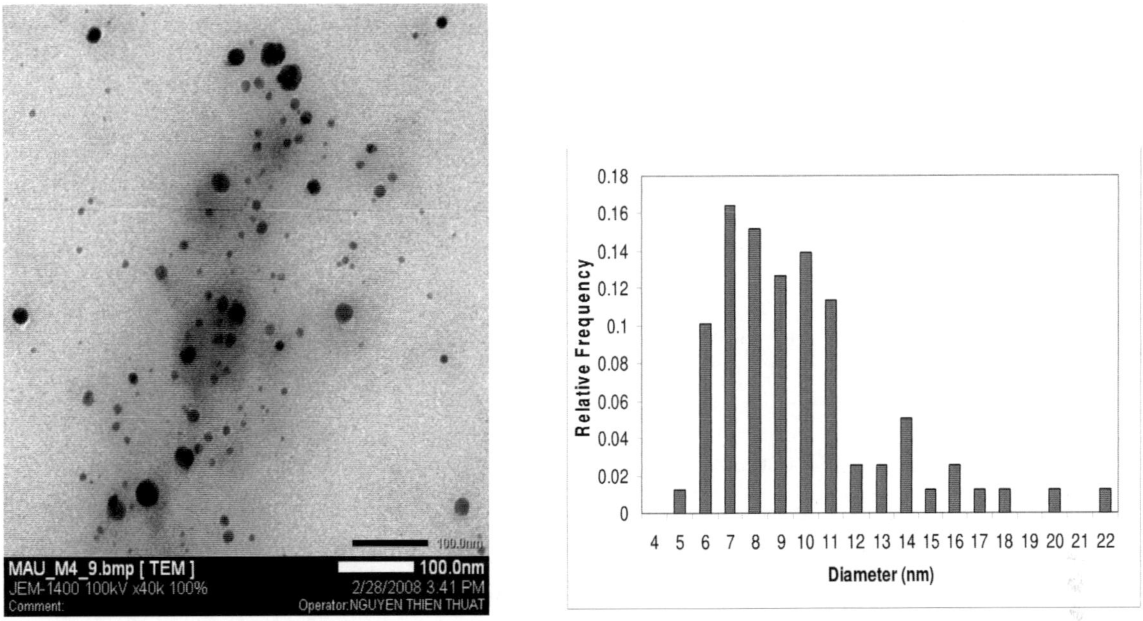

Figure 4. Transmission electron micrograph and particle size distribution of Ag nanoparticles.

3.2. Morphologies of antibacterial cotton fabrics

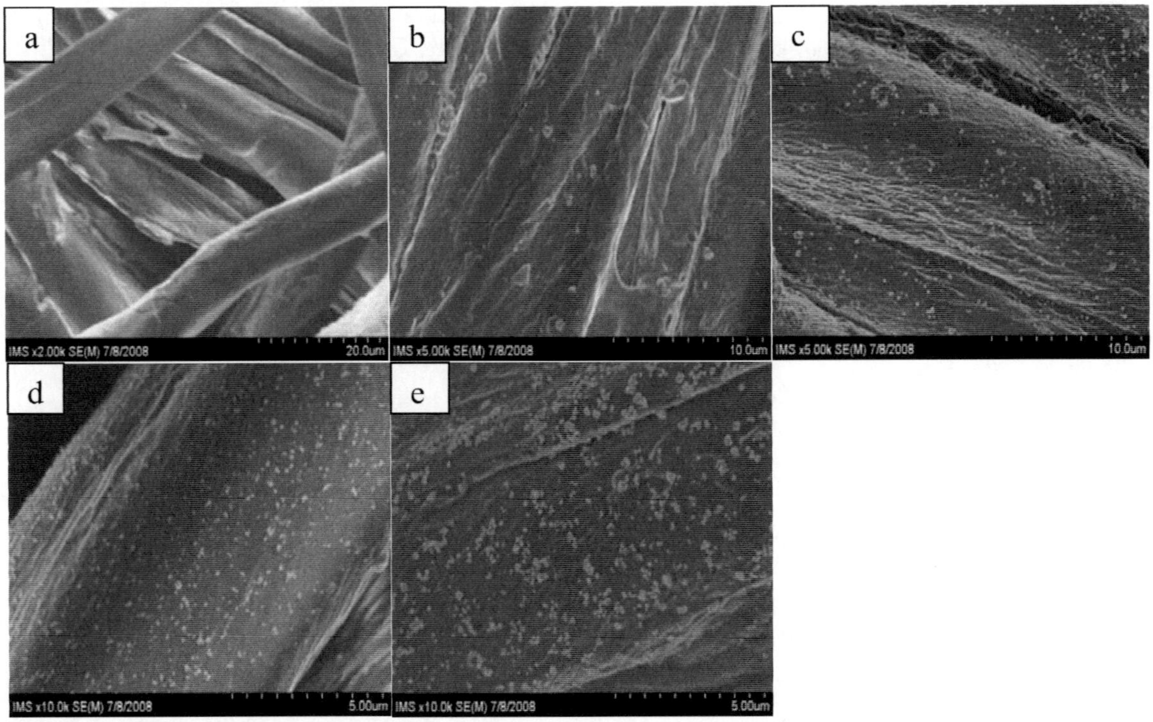

Figure 5. SEM images of antibacterial cotton fabrics treated by immersed in silver colloids: a) control; b) 20 ppm; c) 50 ppm; d) 80 ppm; e) 100 ppm.

Figure 5 is the FESEM images of the antibacterial cotton fabrics with different content of nano Ag colloid. As predicted, the silver nanoparticles were generally well-dispersed on fiber surfaces in each fabric. On the other hand, the surface morphology and the content of Ag nanoparticles on cotton fabrics (table 1) varied with the concentration of the nano-sized silver colloid solution. In figure 5b, the suspension concentration of nano silver colloid solution was low, silver nanoparticles were only observed at the cotton fiber surface, suggesting a low grafting content. As the concentration increased, Ag nanoparticles were well dispersed on cotton surface (figures 5c and 5d) with a few agglomerates, and the content of nano silver on cotton fabric was also increased. With a further increased suspension concentration, more Ag nanoparticles are aggregated to one another on the fibers because of their high surface free energy, so bigger agglomerates were observed on cotton fiber surface (figure 5e). However, Hoon Joo Lee and Sung Hoon Jeoong [6] concluded that the agglomerated silver particles do not seem to have any ill effect on the antibacterial activity of nano-sized silver colloidal on the cotton fabric samples.

Table 1. The content of silver nanoparticles on cotton fabrics.

Concentration of nano silver colloid solution	20 ppm	50 ppm	80 ppm	100 ppm
Content of nano silver on cotton fabrics (mg/kg)	276	698	702	758

3.3. Antibacterial property of antibacterial cotton fabrics

3.3.1. Effect of nano-sized silver colloidal concentration on the percentage reduction of bacteria

The antibacterial activity of cotton fabrics was resulted from the presence of nano silver particles grafted on their surface. The effect of the suspension concentration of nano silver colloidal on the antibacterial activity of cotton fabrics was shown in table 2. It has also been seen that the percentage reduction of bacteria increased with increasing concentration of nano-sized silver colloid and correspondingly with the content of nano silver on cotton fabrics. When the concentration was lower 50 ppm, the percentage reduction of *E. coli* bacteria was higher than and *S. aureus*. It has been known that the structures of gram-positive and negative cell walls have different compositions. The layer of peptidoglycan (about 20-30 nm) in gram-positive bacteria is thicker than gram-negative bacteria [7]. However, when the concentration was over 50 ppm, the percentage reduction of bacteria showed a slow increase. Above results can be found that silver nanoparticles on cotton surface are necessary for inhibition of the bacterial growth.

3.3.2. Effect of contact time on the reduction of bacteria

Table 2. The antibacterial activity of cotton fabrics treated in different nano-sized silver colloidal concentration.

Concentration of nano-sized silver colloid (ppm)	Content of nano-sized silver on cotton fabric (mg/kg)	*E. coli*		*S. aureus*	
		Bacterial reduction (%)	Standard deviation	Bacterial reduction (%)	Standard deviation
20	276	44.88	4.18	39,62	2.70
50	698	96.12	1.00	94,08	0,96
80	702	99.91	1.11	99,31	2.40
100	758	99.97	2.14	99,96	1.30

The effect of the contact time on percentage reduction of bacteria with antibacterial cotton fabrics (cotton fabrics were padded with concentration of nano-sized silver colloid 100 ppm) against *E. coli* and *S. aureus* is shown in figure 6. It can be found that the reduction of bacteria cotton fabric increased to 65% after 10 h in contact with *E. coli* or *S. aureus*. It can be seen clearly that the percentage reduction of *S. aureus* bacteria is lower than *E. coli* bacteria at 10 h, because the cell wall for gram-positive consists of linear polysaccharide chains cross-linked by short peptides to form a three dimensional rigid structure [15]. From figure 6, it can be seen that the percentage reduction of nano silver on cotton fabrics was to 99.99% and 99.96% after 15 h in contact with *E. coli* and *S. aureus*. Hence, it can suggest that the antibacterial activity of cotton fabrics immersed by nano silver colloid solution was excellent.

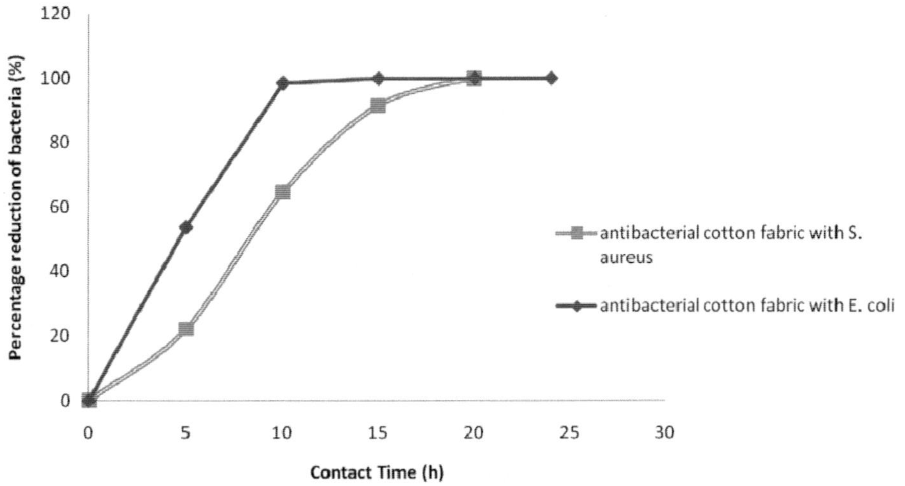

Figure 6. Antibacterial cotton fabrics with against *E. coli* and *S. aureus*.

3.3.3. *Effect to washing times on the reduction of bacteria*

The silver nanoparticles on cotton surface may be washed away when immersed in water. Therefore, the adhesion of silver nanoparticles on cotton fabrics was evaluated by immersing the sample (the specimens were padded with concentration of nano-sized silver colloid 100 ppm) in water and keeping stirred for a given time. Antibacterial activity was tested after the immersion. Table 3 shows the results for the washing 5, 10, 15 times. It can be seen that the antibacterial percentage of the sample washed for 15 times was lower than that of one's washed for 5 times and that of the content of silver nanoparticles on cotton fabrics decreased with the increase of washing times. This phenomenon is associated with the weak physical bonding between the silver nanoparticles and cotton surface. However, even after washed for 10 times, the content of silver nanoparticles on cotton fabrics about 370 mg/kg gave an antibacterial percentage of around 50%. Therefore, the durability of the antibacterial cotton fabrics was satisfied.

Table 3. The percentage reduction of *E. coli* and *S. aureus* bacteria's with cotton fabrics immersed by nano-sized silver colloid 100 ppm.

Washing times	Content of nano-sized silver on cotton fabric (mg/kg)	Bacterial reduction (%)	
		E. coli	*S. aureus*
5	570	97.42	95,87
10	370	55.66	41.67
15	230	36.85	26.42

4. Conclusions

Silver nanoparticles were synthesized by a reduction method where microwave radiation was used as heating source, the as-prepared silver nanoparticles were nearly spherical particle with the particle size distribution about 7-11 nm. The structure of nano Ag had no change after adhesiving cotton surface.

The silver nanoparticles were grafted and well dispersed on the surface of the cotton fabrics. With the increased concentration of nano-sized silver colloid during immersing, especially for concentration over 80ppm, the particles of nano Ag became agglomerated on the cotton surface.

The antibacterial percentage of cotton fabrics showed excellent antibacterial activity against the tested germs. Content of nano silver on cotton fabric and the antibacterial percentage were adjustable by controlling the concentration of nano-sized silver colloid during immersing. When the concentration was 80 ppm, the antibacterial percentage was 99.91% against *E. coli* and 99.31% against *S. aureus*. In addition, a good laundering durability of the antibacterial cotton fabrics was obtained, after washed for 10 times.

References

[1] Burnision N, Bygott C and Stratton J 2004 *J. Surface Coating International Part A* 179
[2] Fei B, Z Deng, Xin J H, Xhang Y, Pang G 2006 *Nanotechnology J.* **17** 1927
[3] Shi Z L, Neoh K G, Kang E T 2005 *Biomaterials* **26** 501
[4] Choi S H, Zhang Y P, Gopalan A, Lee K P, Kang H D 2005 *Colloids and surface A: Physicochem. Eng. Aspects* 165
[5] Satio M 1993 *J. Coated Fabrics* **23** 150
[6] Lee H J and Jeong S H 2005 *Textile Res. J.* **75** 551
[7] Silver S, Phung L T, Silver G 2006 *J. Ind. Microbiol. Biotechno.* **33** 627
[8] Park H P, Kim S H, Kim H J, Choi S H 2006 *Plant Pathol. J.* **22** 295
[9] Vaseashta A, Malinvoska D D 2005 *Sic. Technol. Adv. Mater.* **6** 312
[10] Jain P, Pradee T 2005 *Biotechnol. Bioeng* **90** 59
[11] Gorensek M and Recel P 2007 *Textile Res. J.* **77** 138
[12] Yeo S Y, Lee H J and Jeong S H 2003 *J. Mater. Sci.* **38** 2199
[13] Heard S M, Grieser F, Barraclough C G and Sanders V J 1983 *Colloid Interface Sci.* **93** 545
[14] Jiang H, Moon K, Zhang Z, Pothukuchi S and Wong C P 2006 *J. Nanoparticles Research* **8** 117
[15] Shrivastava S, T Bera, Roy A, Singh G, Ramachandrarao P and Dash D 2007 *Nanotechnology J.* **18** 225103

Acknowledgements

This work was financially supported by Vietnam National University Ho Chi Minh City through Laboratory for Nanotechnology at Vietnam National University-Ho Chi Minh City.

Magnetoresistance and magnetocaloric properties of $La_{0.7}Sr_{0.3}Co_{0.95}Mn_{0.05}O_3$ compound

Le Viet Bau[1], Nguyen Van Khiem[1], Nguyen Xuan Phuc[2], Le Van Hong[2] and Dao Nguyen Hoai Nam[2]

[1]Hong Duc University, 307 Le Lai Street, Thanh Hoa City, Vietnam
[2]Institute of Materials Science, VAST, 18 Hoang Quoc Viet Road, Hanoi, Vietnam

E-mail: levietbau@yahoo.com

Abstract. The magnetoresistance and magnetocaloric effects were studied via magnetic and electric properties measurements. The ferromagnetic transition temperature and critical exponents, which are determined by analyzing the Arott plots, to be $T_C = 190.665$K, $\beta = 0.456$, $\gamma = 1.0623$ and $\delta = 3.2845$, respectively. Although ferro-paramagnetic transition was observed clearly with T_C of 190 K but the insulating behaviour is remained in a whole in a temperature region of 5 K up to 300 K. Moreover the negative magnetoresistance shows a shoulder at $\sim T_C$, broadens toward lower temperature and reach maximum of 11.5% at 174 K in applied field of 5 T. This result is explained as due to the competing between ferromagnetic and non-ferromagnetic phases in the sample. The magnetization measurements were performed in a narrow temperature range, close to T_C and external magnetic fields up to 4.5T. The adiabatic magnetic entropy changes, ΔS, determined from magnetization data, shows a maximum at T_C.

Keywords: Magnetoresistance, magnetocaloric, cobaltite, perovskite.

1. Introduction

Much attention has been devoted to the hole-doped manganite $Ln_{1-x}A_xMnO_3$ and cobaltite $Ln_{1-x}A_xCoO_3$ (*Ln*: lanthanides, A: alkali elements) systems that are strongly attractive to research and technology due to their rich physics. Colossal magnetoresistance phenomena were observed in the perovskite-type hole-doped manganites and explained by DE mechanism, phase separation and spin-polarized tunneling effect [1-3]. The chemical randomness or the impurity doping may cause major modifications in the electronic phase diagram as well as in the magnetoelectronic. Besides, magnetic materials showing a large magnetocaloric effect (MCE) have attracted considerable attention for their potential application in magnetic refrigeration technology [4, 5]. Refrigeration in the temperature range 250-300 K is of particular interest due to the potential impact on energy savings and environmental concerns perovskite-type hole-doped manganites in which the double-exchange ferromagnetic metal phase and the charge–orbital ordered antiferromagnetic phase compete with each other. Cobaltites and magnanites are double-exchange ferromagnetic compounds.

Doping on the Mn- or Co-site has been widely used as a method to directly probe the character of the interaction between the magnetic ions. Magnetic and transport properties of Co-doped $La_{0.7}Sr_{0.3}MnO_3$ were previously reported by several other authors [6, 7]. Although both $La_{0.7}Sr_{0.3}MnO_3$ and $La_{0.7}Sr_{0.3}CoO_3$ are double-exchange (DE) ferromagnets, the interaction between Mn and Co ions in fact is super-exchange antiferromagnetic (AF), as observed in $La_{0.7}Sr_{0.3}Co_{1-x}Mn_xO_3$ [8, 9].

© 2009 IOP Publishing Ltd

Journal of Physics: Conference Series **187** (2009) 012073 doi:10.1088/1742-6596/187/1/012073

The observation of the magnetoresistance in cobaltites is more difficult than in maganites. Moreover the magnetocaloric in maganites is larger than in cobaltites. In this paper we analyze the magneto-caloric and magneto-resistance properties in the mixed perovskite: $La_{0.7}Sr_{0.3}Co_{0.95}Mn_{0.05}O_3$

2. Experiment

Polycrystalline $La_{0.7}Sr_{0.3}Co_{0.95}Mn_{0.05}O_3$ sample was prepared by the conventional solid-state reaction method. X-ray powder diffraction measurements confirmed the single-phase property of the sample. Temperature dependences of resistivity were measured using a Closed Cycle Helium Refrigerator. Temperature dependent field-cooled, $M_{FC}(T)$, zero-field-cooled, $M_{ZFC}(T)$, and field dependent magnetization $M(H)$ measurements were carried out by a SQUID Quantum Design MPMS5 Magnetometer. Resistance measurements are carried out by a 4-probe method, in an OXFORD 2000 system at the Ångström laboratory, Sweden. The magnetic entropy changes were determined from magnetization isotherms, between zero field and a maximum field (H_0) using the thermodynamic relation:

$$\Delta S_m(T,\Delta H) = \int_0^{H_{max}} \left(\frac{\partial M(T,H)}{\partial T}\right)_H dH \cdot$$

3. Results and discussion

Figure 1 presents the temperature dependence of the magnetization of in both two modes: field-cooled, M_{FC}(T), and zero-field-cooled, M_{ZFC}(T). As seen, there is a cusp in the M_{FC}(T) curves and a large separation between M_{FC}(T) and M_{ZFC}(T) curves with decreasing temperature below T_C. A quite similar behaviour has been observed in $La_{0.7}Sr_{0.3}CoO_3$. $La_{0.5}Sr_{0.5}CoO_3$ compound was classified as "cluster glass" (CG) compounds [10]. The ferromagnetic-paramagnetic transition observed for x=0 is at T_C=220 K and the ferromagnetic property of this sample is explained by DE mechanism [1].

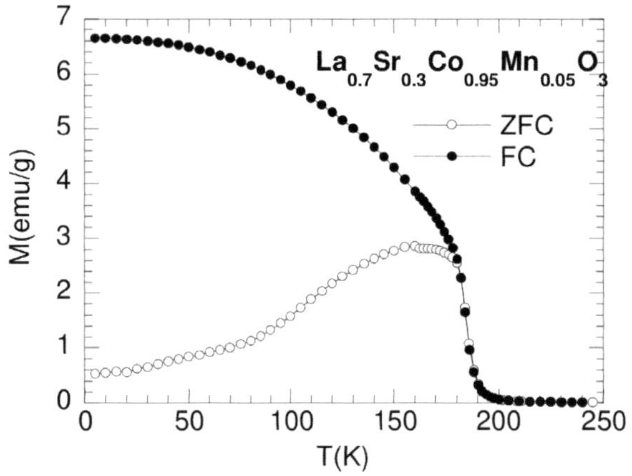

Figure 1. The dependence temperature of M_{FC} and M_{ZFC} of $La_{0.7}Sr_{0.3}Co_{0.95}Mn_{0.05}O_3$.

The doping 5% Mn for Co ions drastically decreases magnetization value and T_C decreases down to ~190 K. Those results, probably due to the replacement of Mn for Co, cause a dilution of the magnetic network formed by Co-O-Co DE interaction and create a new Mn-O-Co super-exchange interaction competing with the remained DE interaction [8]. The substitution of Mn for Co in low concentration region results in an increase of degree of frustration and disorder.

To understand the essence of phase transition in this sample, we have used Arott plots derived from magnetization isotherm curves (not shown here) to determine the Curie temperature and critical exponent parameters in the vicinity of phase transition temperature. The critical exponents to be determined are β, γ, and δ, among which β describes the temperature dependence of the spontaneous magnetization, γ describe the temperature dependence of the zero-field susceptibility, and δ describes the field dependence of the magnetization at the Curie temperature, T_C. The values of T_C obtained in the present investigation match very well those reported in the literature.

In the vicinity of a second-order phase transition with a Curie temperature T_C, the scaling laws for the spontaneous magnetization M_S and susceptibility χ are given by [11-13]

$$M_S(T) = M_0\left(\frac{T_C - T}{T_C}\right)^{\beta} , \qquad (T<T_C) \qquad (1)$$

$$\chi_0^{-1} = \frac{h_0}{M_0}\left(\frac{T - T_C}{T_C}\right)^{\gamma} . \qquad (T>T_C) \qquad (2)$$

These relations define the critical exponents β and γ, where the spontaneous magnetization is determined by $M_S(T) = \lim_{H\to 0} H$ below T_C, and the inverse initial susceptibility is determined by $\chi_0^{-1} = \lim_{M\to 0}\frac{H}{M}$. At the transition temperature, the dependence of the magnetization on the applied field is of the form

$$M = D.H^{\frac{1}{\delta}} \qquad (T=T_C)$$

with the critical exponent $\delta = 1 + \gamma/\beta$. The exponent δ can be obtained from the slope of the log(M) versus log(H) plot at T_C. The critical exponent as well as the critical amplitudes M_0, h_0/M_0 and D, exhibits the universal behavior near the phase transition point.

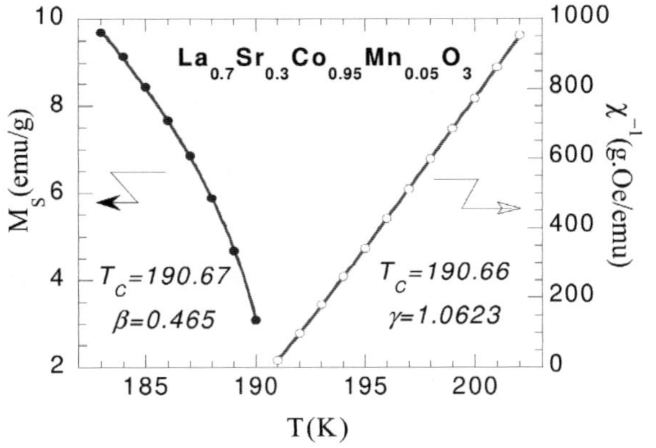

Figure 2. The M_S(T) and χ^{-1}(T) derived by extrapolation the Arott plots M^2 vs. H^2 to H=0 and M=0 respectively. The solid lines are best fits to equation (1) and (2) in the text.

By extrapolating the Arott plots M^2 vs. H^2 curves to H^2=0 for $T<T_C$ and M^2=0 for $T>T_C$, the spontaneous magnetisation M_S(T) and the inverse initial susceptibility values have been recovered. Those results are displayed in the figure 2. By fitting those data to equation (1) and (2) we obtain T_C= 190,67 K, β= 0.465, γ= 1.0623 and, therefore, δ = $1+\gamma/\beta$ =3.2845. Those values of critical exponents are close to the mean field values (β= 0.5, γ= 1 and δ = 3). Hence, we believe that the experimental data agree with the mean-field model.

For ferromagnetic manganites, the conductivity is metallic below T_C but insulating above Tc showing a metallic-insulating (MI) transition and the magnetoresistance is observed at T_C, in the consistence with the DE mechanism. However the ρ(T) curve of $La_{0.7}Sr_{0.3}CoO_3$ merely changes its slope at T_C and still exhibits metallic behaviour in the PM state [8]. This feature is not expected for a typical DE system but can be understood considering the fact that the electronic configuration of the Co ions does not strictly obey Hund's rule, an ingredient of the DE mechanism. That makes the e_g electron of the Co ions quite mobile in the sense that it can hop between Co sites without strict requirement of a parallel localized-t_{2g}-spin configuration or a high applied field. By the way, this is one of the reasons for the magneto-resistance in cobaltites to be usually small compared to that of the manganites.

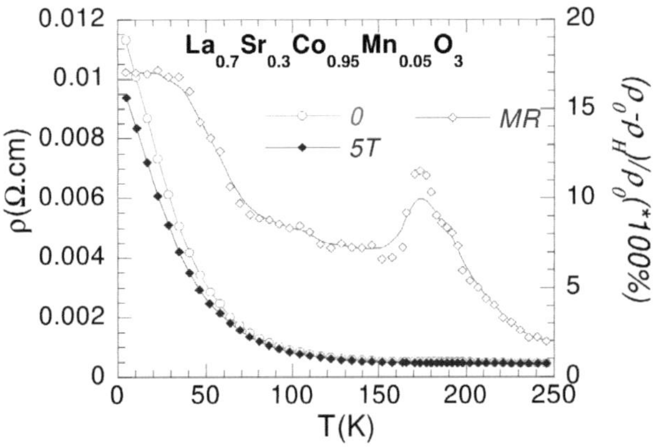

Figure 3. The dependence temperature of resistivity (left axis) and magneto-resistance of sample (right axis).

As shown in the figure 3, by doping 5% Mn for Co ions, the metallic-insulating transition is observed at T_C. In the previous article [8] we suggested that an appreciable amount of the Co atoms is converted to a higher-spin state when an amount of Co ions are replaced by Mn ions in $La_{0.7}Sr_{0.3}CoO_3$. Moreover, the substitution Mn for Co can decreases the concentration of charge carriers [14]. In the results, the spin configurations of Co ions became more stable. The conductivity then becomes consistent with the DE mechanism. In figure 3, the negative magnetoresistance shows a shoulder at ~T_C (can be explained by DE mechanism) and broadens toward lower temperature and reaches maximum of 11.5% at 174 K in applied filed of 5T. This result can be a consequence of the competition between metallic-ferromagnetic and insulator-ferromagnetic phases in the sample. It is interesting that the magnetoresistance at low temperatures far from T_C increases as decreasing temperature. This can be explained by spin-polarized tunnelling effect, similar to maganites [3] and the competition of phases in the sample.

In Ref. [15] the maximum value of entropy change is found to be 1.37 J/kgK. This value is smaller than those evidenced in manganites [16, 17]. This can be explained by the existence of various spin states of Co ions and by the broadening magnetic transition in cobaltites. Substitution of 5% Mn for Co can favour the entropy change.

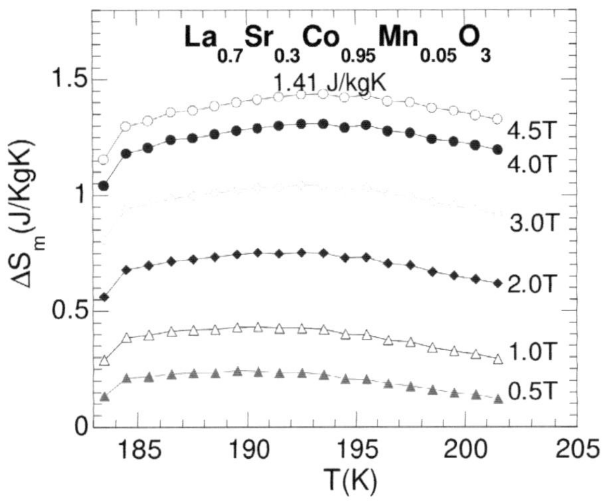

Figure 4. The adiabatic magnetic entropy changes in different applied field.

The temperature dependence of magnetic entropy on external applied fields for $La_{0.7}Sr_{0.3}Co_{0.95}Mn_{0.05}O_3$ compound is plotted in figure 4. The maximum values of entropy change occur almost around the transition temperature. In our case the maximum value is around 1.41 J/(kg K) in a 4.5T magnetic field. The magnetic entropy changes about 0.43 J/kgK in field of 1T. Despite of small value the entropy change is high enough for technical interest and especially in broad temperature range which is beneficial for active magnetic refrigeration.

4. Conclusions

A metallic-insulator transition is observed in the $La_{0.7}Sr_{0.3}Co_{0.95}Mn_{0.05}O_3$. The T_C is fixed at 190.66K and the critical exponents values suggest that the experimental data agree with the mean-field model. The magnetoresistance is observed at T_C and reaches maximum at T~174 K, below T_C, especially increases in low temperatures far from T_C as decreasing temperature. The magnetoresistance property can be explained by DE mechanism, the competition between magnetic phase and the spin-polarized tunnelling in the sample. The magnetocaloric effect is investigated in this compound. The entropy changes in a broad temperature range are beneficial for active magnetic refrigeration.

References

[1] Zener 1951 *Phys.Rev.* **82** 403
[2] Dagotto E 2002 *Nanoscale Phase Separation and Colossal Magnetoresistanc* (Berlin Heidelberg New York: Springer-Verlag)
[3] Hwang H Y, Cheong S-W, Ong N P, and Batlogg B 1996 *Phys. Rev. Lett.* **77** 2041
[4] Anderson PW, Hasegawa H 1955 *Phys. Rev.* **100** 675
[5] Schiffer P, Ramirez A P, Bao W, Cheong S W 1955 *Phys. Rev. Lett.* **75** 3336
[6] Shi J B, Wu F C, Lin C T 1999 *Appl. Phys. A* **68** 577
[7] Fan X J, Zhang J H, Li X G, Wu W B, Wan J Y, Lee T J, Ku H C 1999 *J. Phys.: Condens. Matter* **11** 3141
[8] Phuc N X, Bau L V, Khiem N V, Son L H and Nam D N H 2003 *Physica B* **327** 177
[9] Bau L V, Khiem n V, Hong L V, Phuc N X, Nordblad P and Nam D N H 2004 *J. Magn. Magn. Mater.* **272 - 276** e975
[10] Balcells L, Martınez B, Sandiumenge F, Fontcuberta J 2000 *J. Magn. Magn. Mater.* **211** 193
[11] Kadano L P, Gotze W, Hamblen D, Hecht R, Lewis E A S, Palciauskas V V, Rayl M, Swift J, Aspnes D and Kane J 1967 *Rev. Mod. Phys.* **39** 395
[12] Arrott A and Noakes J E 1967 *Phys. Rev. Lett.* **19** 786
[13] Fisher E M, Ma S K and Nickel B G 1972 *Phys. Rev. Lett.* **29** 917
[14] Wandekar R V, Wani B N, Bharadwaj S R 2008 *Solid State Sciences* **XX** 1
[15] Tetean R, Deac I G, Burzo E and Bezergheanu A 2008 *J. Magn. Magn. Mater.* **320** (14) e179
[16] Dai N V, Son D V, Yu S C, Bau L V, Hong L V, Phuc N X, and Nam D N H 2007 *Phys. Stat. Sol.* (b) **244** (12) 4570
[17] Nam D N H, Dai N V, Hong L V, Phuc N X, Yu S C, Tachibana M, and Takayama-Muromachi E 2008 *J. Appl. Phys.* **103** 043905

Acknowledgments

This work has been sponsored by the Institute of Materials Science (VAST, Vietnam) and Hong Duc University (Vietnam). L V Bau would like to thank Prof. Per. Nordblad for support and performing the experimental measurements.

Synthesis of Y_2O_3:Eu^{3+} micro- and nanophosphors by sol-gel process

Bui Van Hao[1, 2], Pham Thanh Huy[2, 3], Tran Ngoc Khiem[2], Nguyen Thi Thanh Ngan[4] and Pham Hong Duong[4]

[1]Faculty of Engineering Physics, Quy Nhon University, 170 An Duong Vuong, Quy Nhon City, Vietnam
[2]International Training Institute for Materials Science (ITIMS), Hanoi University of Technology (HUT), Hanoi, Vietnam
[3]Hanoi Advanced School of Science and Technology (HAST), HUT, Hanoi, Vietnam
[4]Institute of Materials Science, Vietnam Academy of Science and Technology, 18 Hoang Quoc Viet Road, Cau Giay Dicstrict, Hanoi, Vietnam

E-mail: haobv_itims2006@yahoo.com

Abstract. In this paper, optical properties of Y_2O_3:Eu^{3+} micro- and nanophosphors synthesized by sol-gel process are reported. Citric acid and tartaric acid were employed as chelating agents. Different factors effecting structures and properties of the phosphors, such as concentration of Eu^{3+}, sintered temperature and the ratio of metal ions to tartaric acid, were examined. Crystal structure of the particles was investigated by X-ray diffraction (XRD) technique. The results showed that the phosphor powders completely crystallized at 600°C. Morphology and particle size of the phosphors were studied by scanning electron microscope (SEM) and transmission electron microscope (TEM). The particle size of the obtained phosphors varied from 1 to 6 μm (using citric acid as chelating agent) and from 20 to 30 nm (using tartaric acid as chelating agent). Photoluminescence measurements of the Y_2O_3:Eu^{3+} particles showed red emission peaks at 610 - 612 nm ($^5D_0 \rightarrow {}^7F_2$). The optimal molar concentration to obtain the highest photoluminescent intensity for the phosphors was 7% of Eu^{3+}.

Keywords: Micro- and nanophosphors, sol-gel process, photoluminescent, optical properties.

1. Introduction

The red phosphor Y_2O_3:Eu^{3+} has been studied and widely applied in fluorescent lamps because of its high quantum efficiency, and especially, it can be effectively excited by 254 nm radiation, the strongest emission of mercury vapor [1]. While in many applications, such as fluorescent lamps (FL), cathode ray tubes (CRT), plasma display panels (PDP), particle size of the phosphor is in the range of several micrometers [2, 3], nanoscaled Y_2O_3:Eu^{3+} has been indicated to be one of the most promising red phosphors for field emission displays (FED) [4, 5]. Therefore, the studies on the preparation and luminescent properties of this phosphor attracted extensive interests.

Y_2O_3:Eu^{3+} has been prepared by various methods. Vu *et al.* [6] used combustion reaction to synthesize the phosphor with the particle size estimated in the range of 10 to 20 nm. Nanocrystalline cubic Y_2O_3:Eu^{3+} with an average particle size of only 10 nm was fabricated using chemical vapor technique by Konrad *et al.* [7]. A laser-heated gas condensation technique has also been employed to

© 2009 IOP Publishing Ltd

Journal of Physics: Conference Series **187** (2009) 012074 doi:10.1088/1742-6596/187/1/012074

prepare nanocrystalline Y_2O_3:Eu^{3+} with the monoclinic crystal structure [8]. Zhang et al. [9] prepared the phosphor with the grain size in the range of 0.5 to 2 μm by sol-gel method which employed citric acid as a chelating agent. In these works, however, most of the authors have concentrated on studying synthesis process and photoluminescent properties of either nanoscaled materials or microphosphors. Many researchers have investigated the luminescent properties of bulk and nanocrystalline cubic yttrium oxide prepared by chemical vapor synthesis [7, 10], combustion synthesis [11], and fast thermal decomposition of metal nitrate solution [12].

In this work, we present structure and optical properties of Y_2O_3:Eu^{3+} micro- and nanophosphors synthesized by sol-gel process. Citric acid and tartaric acid were employed as chelating agents in making sol. A general comparison of crystallite structure, morphology and optical properties of two kinds of phosphors was realized.

2. Experimental

2.1. Preparation of Y_2O_3:Eu^{3+}

Synthesis process of the phosphors is shown in figure 1. The initial rare earth oxides Y_2O_3 and Eu_2O_3 (99.99%, Aldrich, USA) with different atomic ratios were dissolved in a definite amount of diluted nitric acid (Merck) and stirred to obtain a transparent liquid. Next, chelating agent was fed on. The solution was constantly stirred for 12 hours at room temperature. Then, the mixture was heated at 80°C and a transparent sol was formed. In the process of heating at 100°C, the condensation reaction between –COOH groups occurred and led to the formation of water. The evaporation of water made the volume of the sol to be reduced quickly. When most of water was removed, the sol turned into a transparent gel and then a viscous resin was formed. Finally, further heating caused the combustion of the gel and the powder with a brownish color was obtained, which is called dry gel. The dry gel was then sintered at different temperatures to obtain the phosphors.

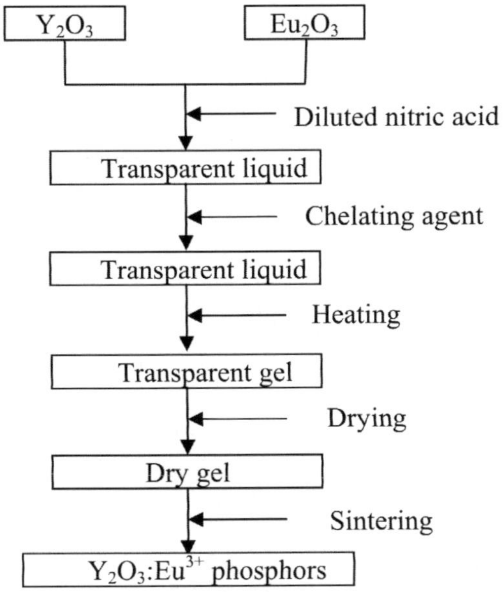

Figure 1. Synthesis process of the phosphors.

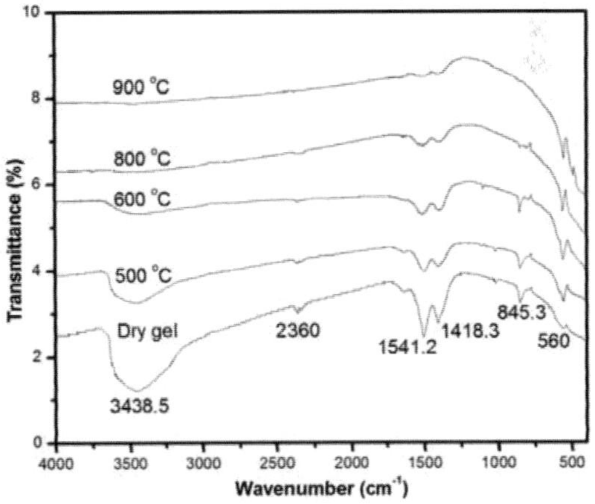

Figure 2. IR spectra of the nanophosphors before sintering and sintered at different temperatures.

2.2. Characterization of the phosphors

The IR spectra of the gels sintered at different temperatures were measured by using Nicolet 6700 FT-IR Spectrometer. X-ray diffraction images were studied with Brucker D8-Advance XRD meter using

Journal of Physics: Conference Series **187** (2009) 012074 doi:10.1088/1742-6596/187/1/012074

Cu K_α radiation. The morphology and particle size of the phosphors were observed by using Hitachi S-4800 Field Emission Scanning Electron Microscope (FESEM) and JEOL JEM-1010 Transmission Electron Microscope (TEM). Photoluminescence measurements were performed using a continuous wave He–Cd laser (325 nm) as the excitation source.

3. Results and Discussion

3.1. Formation of $Y_2O_3:Eu^{3+}$ phosphor in the sintering process

Figure 2 shows the IR spectra of $Y_2O_3:Eu^{3+}$ powders before sintering and sintered at different temperatures. For the dry gel before calcination, the IR spectrum consists mainly of three parts: the first part with a peak at 3438.5 cm^{-1} arises from the absorption of O–H vibration; the second part with the main peaks at 1418.3 and 1541.2 cm^{-1} comes from citrate groups and the third part with a weak peak at 845.3 cm^{-1} due to the absorption of small amount of CO_3^{2-}. After sintering at 600°C, the absorption peaks from the organic components decrease, and a new sharp peak at 560 arises due to the absorption of Y_2O_3. The absorption peak due to Y_2O_3 became stronger indicating the complete crystallization of Y_2O_3. The absorption of organic composition almost disappeared after sintering at 900°C.

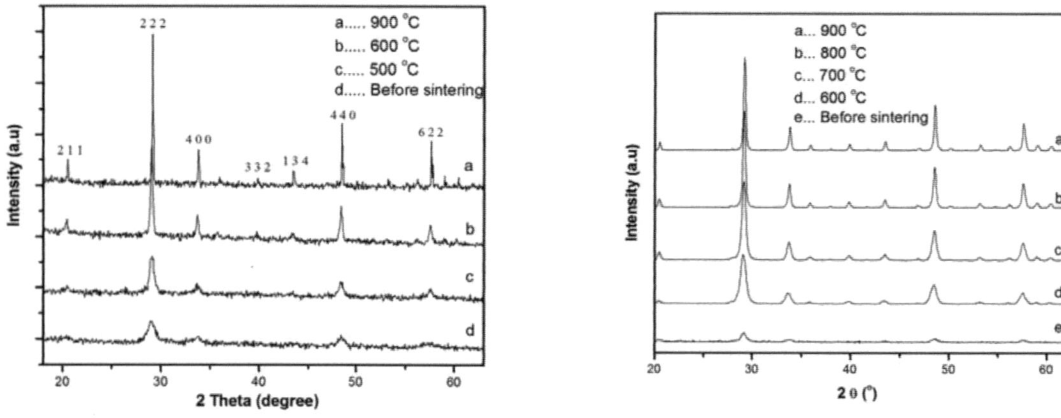

Figure 3. XRD patterns of microphosphors (left) and nanophosphors (right) sintered at different temperatures.

A similar result was obtained for the nanophosphors using tartaric acid as chelating agent. XRD patterns of phosphors sintered at different temperatures are shown in figure 3. The powders completely crystallized when sintered at the temperature of 600°C. The difference in width of peaks indicates the difference in particle size.

3.2. The morphology of the phosphors

Figure 4 (a) shows SEM image of the micro-particles obtained by sintering the dry gel (using citric acid) at 900°C for 3 hours. It can be seen that nearly spherical particles are obtained and the mean particle size is estimated from 1 to 6 µm. On the other hand, nanophosphors were achieved in the case of tartaric acid was used (figure 4 (b) and figure 5). From these images we see that the particles are rather homogeneous and the grain size is in the range of 20 to 30 nm. Crystallite size of the nanophosphors can be calculated according to Scherrer formula [13]: $D = \dfrac{0.89\lambda}{\beta\cos\theta}$, where D is the crystallite diameter (nm), λ the X-ray wavelength (0.15405 nm), θ is the diffraction angle and β is the full-width at half-maximum (FWHM, in radian) of (222) peak. The results are summarized in the table 1. These results are in good agreement with the particle sizes obtained by FE-SEM and TEM. A slight agglomeration of the particles occurs in both of cases. This can be explained as a consequence of high temperature sintering process.

(a) (b)

Figure 4. SEM images of the microphosphors (a) and nanophosphors (b).

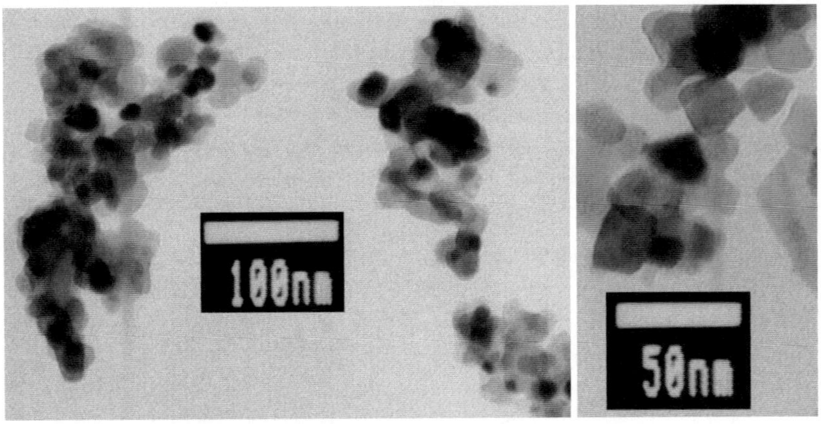

Figure 5. TEM image of the nanophosphor.

Table 1. Crystallite size of the phosphors.

Sample	Sintering temperature	Peak position	FWHM	D (nm)
1	600°C	29.03°	0.883°	9
2	700°C	29.12°	0.543°	15
3	800°C	29.12°	0.505°	16
4	900°C	29.15°	0.410°	20

3.3. The morphology of the phosphors

Emission spectra of the phosphors with Europium atom concentration of 7% sintered at 800°C (nanophosphors) and 900°C (microphosphors) for 3 hours when excited by 325 nm beam are shown in figure 6.

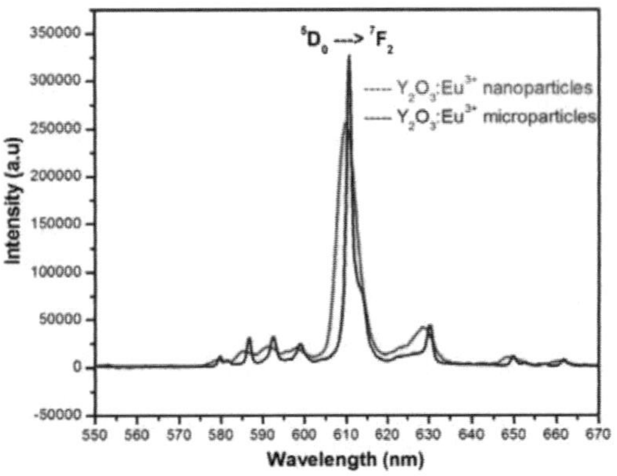

Figure 6. Photoluminescent spectra of the phosphors.

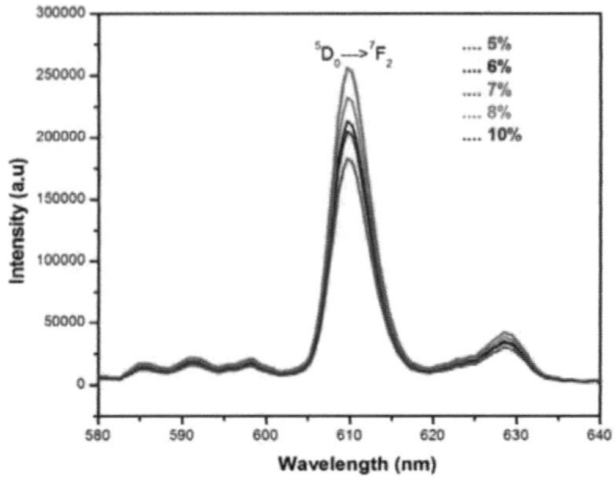

Figure 7. Dependence of photoluminescent intensity on Eu^{3+} concentration.

It can be seen that the phosphors have an intensive fluorescence with sharp peaks at the wavelength of 611.5 nm (micro-particles) and 609.5 nm (nano-particles) corresponding to the transition between $^5D_0$ and $^7F_2$ levels of Eu^{3+} ion. No significant difference in position of the peaks between two spectra, but there is a little difference in shape of the curves. The emission bandwidth of the nanophosphors is larger than that of the microphosphors. The increasing bandwidth results from the increase in the lattice relaxation energy of the excited state with decreasing particle size [10].

The emission intensity of the phosphors depends on the Eu^{3+} concentration and sintering temperature as shown in figure 7. In both cases of the micro- and nanophosphors, the strongest emission intensity corresponds to the concentration of 7%. Higher concentrations caused the decrease in emission intensity. This effect may be attributed to the concentration quenching mechanism [14].

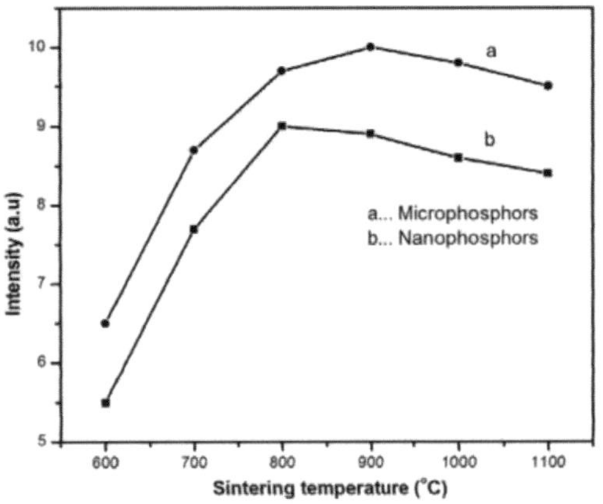

Figure 8. Emission intensity of $Y_{2-x}O_3:xEu^{3+}$ (x = 7 mol%) as a function of sintering temperature.

Figure 9. SEM image of nanophosphor synthesized with the ratio of metal ions to tartaric acid 1:1.

The effect of sintering temperature on photoluminescent intensity was investigated. The results showed that emission intensity increases with the raise of temperature. The optimal sintering temperature to produce high intensity is 800°C in the case of nanophosphors, while that of the microphosphors is 900°C. Higher sintering temperatures caused the decrease of emission intensity.

This can be explained that at higher temperatures, particles tend to move into clusters and, as a result, the emission intensity will be decreased.

From these results, we have a general comparison of emission intensities of two kinds of phosphors. It can be seen that the microphosphors have stronger emission intensity. Nanoscaled phosphors have lower quantum efficiency compared to microscaled particles. This is attributed to the large surface area, which amplifies quenching processes [15].

3.4. The effect of the ratio of metal ions to tartaric acid on particle size and luminescent properties of nanoparticles

In synthesis process of nanophosphors, different molar ratios of metal ions to tartaric acid (1:1, 1:2, 1:3, and 1:4) were studied. At the ratio of 1:1, the particles were in an inhomogeneous form (Fig. 9). Other ratios did not make an obvious change in particle size.

When the ratio of metal ion to tartaric acid was 1:2, the highest emission intensity was obtained. This can be explained that high tartaric acid concentration caused a higher carbon impurity left in the powder which decreased luminescent intensity while the lower tartaric acid amount accelerated the rate of hydrolysis in sol-gel process, as a result, the nanoparticles could not be homogeneous.

4. Conclusions

Y_2O_3:Eu^{3+} micro- and nanophosphors have been synthesized by simple sol-gel process. The luminescent properties of Eu^{3+} doped Y_2O_3 as a function of Eu concentration and sintering temperature have been investigated. The optimal Eu concentration is 7 at.%. The photoluminescent intensity increases with the increase of sintering temperature up to 800°C and 900°C for the nanophosphors and microphosphors, respectively. Higher sintering temperatures reduce the emission intensity. The extension of emission bandwidth of the nanophosphors in comparison with that of the microphosphors was observed. The shape of the phosphor particles is nearly spherical, and particle size distribution is rather narrow.

References

[1] Blasse G, Geiabmaier B C, 1994 *Luminescent Materials* (Berlin: Springer) p 116
[2] Pao R P 1996 *J. Electrochem. Soc.* **143** 189
[3] Mored A, Khiati N EL. 1993 *J. Electrochem. Soc.* **140** 2019
[4] Jing X, Ireland T, Gibbons C, Barber D Silver J J, Vecht A, Fern G, Trowga P, Morton D C, 1999 *J. Electrochem. Soc.* **146** 4654
[5] Lee M H, Oh S G, Yi S C, Seo D S, Yoo J S 2000 *J. Electrochem. Soc.* **147** 3139
[6] Vu N, Anh T K , Yi G C, Strek W 2007 *J. Lumin.* **122–123** 77
[7] Konrad A, Fries T, Gahn A, Kummer F, Herr U, Tidecks R, Samwer K 1999 *J. Appl. Phys.* **86** 3129
[8] Eilers H, Tissue B M 1996 *Chem. Phys. Lett.* **251** 74
[9] Zhang J, Zhang Z, Tang Z, Lin Y, Zheng Z 2002 *J. Mater. Pro. Tech.* **121** 265
[10] Konrad, U Herr, R Tidecks, F Kummer, K Samwer 2001 *J. Appl. Phys.* **90** 3516
[11] Fu Z, Zhou S, Pan T, Zhang S 2007 *J. Lumin.* **124** 213
[12] Jia M et al. 2004 *Chem. Phys. Lett.* **384** 193
[13] Gusev A I and Rempel A A 2003 *Nanocrystalline Materials* (Cambridge International Science Publishing) p 140
[14] Singh L R, Ningthoujam R S, Sudarsan V, Srivastava I, Singh S D, Dey G K and Kulshreshtha S K 2008 *Nanotechnology* **19** 055201
[15] Cees Ronda 2008 *Luminescence: From Theory to Applications* (Wiley-VCH) p 56

Acknowledgments

This work is supported by the National Key Science and Technology Program of Vietnam Grant number KC.02.10/06-10.

Fe nanodot system fabricated by non-lithographic method and its structural properties

Chu Van Chiem[1], Nguyen Thi Thu Ha[1], Ngo Thi Thanh Tam[1], Nguyen Van Chuc[1], Huying Li[2], Seo Jae Muyng[2] and Phan Ngoc Minh[1]

[1]Institute of Materials Science, Vietnam Academy of Science and Technology, 18 Hoang Quoc Viet Road, Cau Giay Distr., Hanoi, Vietnam
[2]Physics department, Chonbuk National University, South Korea

E-mail: chucnv@ims.vast.ac.vn

Abstract. In this work, we study the magnetic structure and morphology of the Fe nanodot system fabricated by the non-lithographic method, using anodic aluminum oxide (AAO) membrane as a template. By the two-steps aluminum anodization, the AAO patterns with the hexagonal pore arrangement have been achieved. Using AAO pattern as a template, under suitable conditions we successfully deposited the iron metal in the pores by the thermal vacuum evaporation. By the exposure of the deposited system from the bottom of the AAO membrane, the hexagonal ordered Fe nanodot system has been obtained. The morphologies of the nanodot system were imaged by the Atomic Force Microscopy (AFM) and Field Emission Scanning Microscopy (FESEM) methods. The magnetic structures were investigated by the Energy Dispersive X-Ray Fluorescence Spectroscopy (EDS) and Magnetic Force Microscopy (MFM) methods. Experimental results confirmed that the MFM image of the fabricated Fe nanodot system is similar to their AFM image.

Keywords: AAO, AFM, MFM, nanodot system, magnetic structures.

1. Introduction

Magnetic nanoparticle monolayer has been proposed as suitable for future ultra-high-density magnetic recording media and other applications [1, 2]. A great deal of brilliant science and engineering involving mesoscopic structures was made possible over the past decade through the use of advanced lithographic technologies. These techniques seem to be economical, but they have technological limits for fabricating one-dimensional (1D) and zero-dimensional (0D) nanostructures. Thereby, creating an alternative technique for fabricating nanostructures fabrication by selective deposition into templates called non-lithographic method is an interesting work.

Non-lithographic fabrications of low-dimensional structures can be performed by either electrochemical or physical methods, using AAO as a patterning template. An attractive approach to get 1D and 0D structures is carried out using anodic aluminum oxide (AAO) template which self-organizes into a hexagonal array of uniform nanopores with pore density up to 10^{12} cm^{-2} and pore diameters down to 4 nm [3]. A variety of nanostructures has been successfully fabricated using AAO templates by electrochemical deposition [4-6], but only a few nanostructures created by physical deposition have been reported [7-9]. The electrochemical deposition entails synthesizing the desired material within the pores of a membrane, resulting in nanorod (nanowire) or nanotube formations. The

© 2009 IOP Publishing Ltd

physical method using AAO as a depositing mask leads to nanodot formation on surface of any substrates. For the electrochemical deposition, the problem of filling the pores is critical and this has been successfully solved by application of some voltage bias with high ac frequency. For the physical deposition, the penetration through the AAO pores of evaporated atoms (particles) is crucial, so it is very difficult to transfer the AAO structure to deposited nanodot system on the substrates. This penetrating process is complicated and still not understood. The process of the selective etching/removal of the deposited product on the template surface is very complicated and in many cases it is impossible.

In this work, we present a unique technique to get symmetric iron (Fe) nanodot system by using AAO templates. We develop a simple technique without selective etching process. The technique involves filling Fe into pores of the AAO. Then the remained aluminum substrate was etched out. Finally, the deposited Fe in the AAO pore was exposed by removing the bottom of the AAO barrier-layer.

2. Experimental

The AAO template was prepared by two-steps anodization of aluminum foil of 3 mm-thick and (5x1.7) cm^2 in size. First, 99.99% pure Al foil was ultrasonically cleaned in acetone and alcohol for 10 min. and 20 min., respectively. Then the back surface and edges of the Al foil were covered by epoxy, so all steps of AAO fabrication can be proceded only from the non covered surface of the Al foil. The flatness of the cleaned aluminum foil was obtained by chemical electro-polishing in the 1:4 %vol. perchloric acid and alcohol under voltage of 20V for 2 minutes at 7°C. The anodization process is carried out in 0.3 M oxalic acid solution at the temperature of 15°C under voltage of 40 V and current of 10 mA. The first anodization was carried out for 10 hours. The aluminum oxide was then etched in 1M chromic phosphoric solution at 65°C for 5 hours. The second anodization was carried out for various times from 2 minutes to 10 minutes. Under these anodization conditions, the pore diameter of the initial AAO template was in the range from 15 nm to 20 nm. Pore-widening process was done in the 0.1 M phosphoric solution at 30°C. After widening process, the pore diameter was about 60 nm. Filling of the Fe in the pores was carried out by vacuum thermal evaporation. Vacuum level, deposition rate and source-substrate distance were 10^{-7} Torr, 0.2 nm/min. and 50 cm, respectively. Aluminum layer was dissolved in the saturated solution of mercury (II) chloride $HgCl_2$. Aluminum barrier oxide was removed by etching in 1 M phosphoric solution for 1 hour at room temperature. Figure 1 presents schematic procedure for fabrication Fe dots. The various kinds of Fe clusters or Fe nanotube in the AAO pores depended on the pore diameter and aspect ratio of the pores. In our work, the Fe nanodot system had a symmetry exactly identified with AAO symmetry. The morphology of the AAO template and prepared Fe nanotube system were analyzed by FESEM and AFM methods. The magnetic structure of the fabricated Fe nanodot system was investigated by MFM method.

3. Results and discussion

The typical surface and cross-section morphology of the prepared two-steps anodization AAO template are presented in figure 2 These images apparently show the hexagonal arrangement of the pores and the pore uniformity both in horizontal (figure 2a) and vertical shapes (figure 2b). As shown in the figure 2, the obtained AAO structure has the pore-pore distance of ~ 95 nm corresponding to the pore density of about $4x10^{11}cm^{-2}$. The pore diameter is about 60 nm that means the thickness of the porous aluminum anodic oxide wall is ~ 35 nm.

Experimental data indicated that, the penetration of the evaporated iron (atoms or group of atoms) into AAO pores was strongly dependent on the pore diameter and aspect ratio of the AAO pore. For the initial AAO pattern with pore diameter about 20 nm, most of the iron evaporated products tend to perform clusters during penetration into the pore. As a result, various small Fe clusters distributed along the pore wall were observed as seen in figure 3a. It is clear that, under these conditions, the AAO structure cannot make uniform Fe systems. To overcome this problem, we have developed a technique for widening the diameter of the pore. When the pore diameter is larger, about 60 nm and pore aspect ratio is less than 10, the evaporated iron metal successfully penetrated along the pore wall. The evaporated iron metal created the single Fe nanotube in each pore, as seen in figure 3b. It is clear that, the bottom of the tubes is closed. As mentioned above, the iron nanodots deposited in the AAO pores were exposed from the bottom side by removing barrier aluminum oxide. The SEM image (figure 4a) clearly shows the typical surface morphology of prepared Fe nanodots. From figure 4a, it is clear that the iron dots have semispherical shape after removing the AAO barrier. We can note that the arrangement of the Fe nanodots was identical with the pore arrangement of the used AAO template.

It is found from the EDS characterizations (figure 4b) that the peaks at 6.398 keV and 7.057 keV corresponding to the Fe element. Experiment results presented in figure 4(a, b) indicated that we had successfully realized all steps for the Fe nanodot system by using AAO template. It is noted that using the developed technique, Fe nanodots with highly uniform arrangements can be formed. This system consists of hexagonal Fe nanodots separated each other by porous anodic aluminum oxide wall.

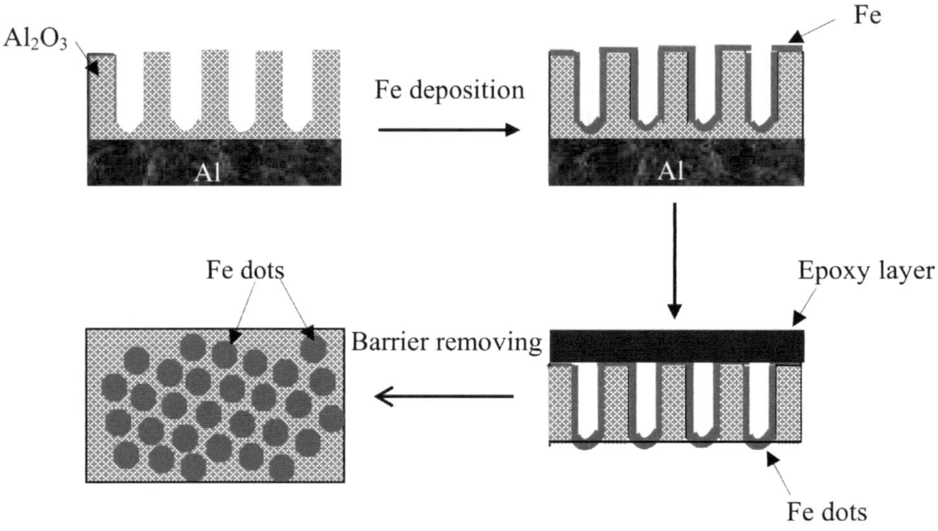

Figure 1. Schematic procedure for fabrication Fe dots.

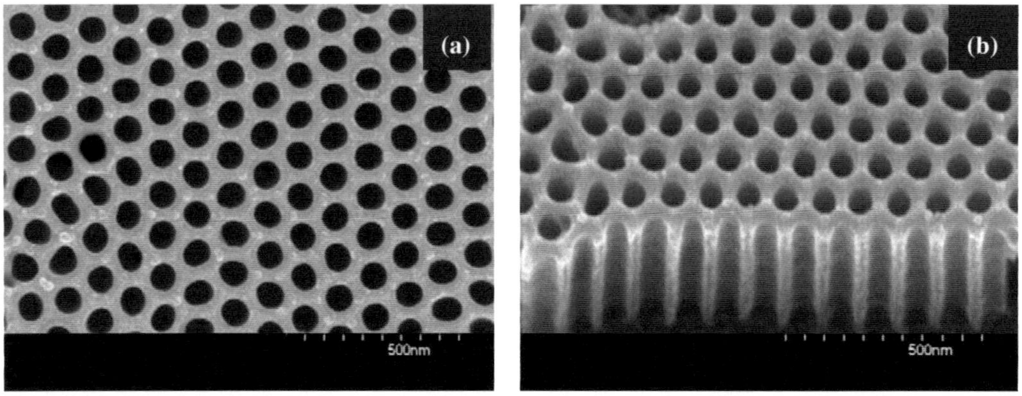

Figure 2. Surface (a) and cross-section (b) morphology of the 60 minutes widened AAO.

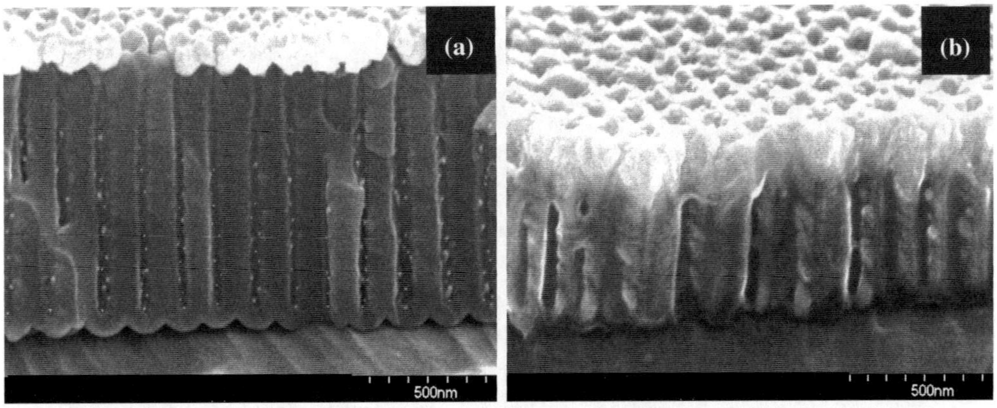

Figure 3. Various Fe clusters formed in the narrow pores of initial AAO (a), single Fe nanotube formed in the widened pores of widening AAO (b).

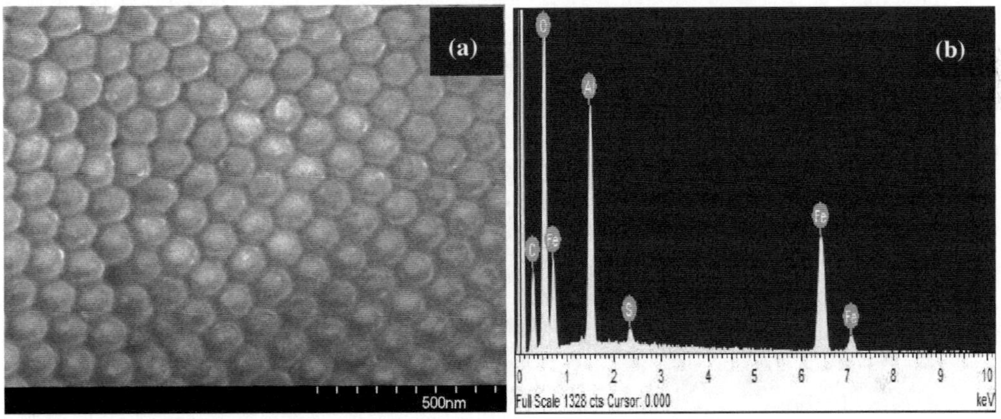

Figure 4. FESEM image (a) and EDS (b) spectrum of the fabricated Fe nanodot system.

It is well-known that MFM is the direct method to observe the magnetic structure of magnetic materials. However, for a system of nanometer size, this method still needs to be improved and developed [10]. In order to get MFM image of a nano system, it is necessary to distinguish the effect caused by the Van der Waals forces and the effect caused by the magnetic forces. In principle, comparing the MFM image of the sample surface with its tapping AFM mode image, we can evaluate the influence of the Van der Waals forces to the MFM tip. But it is impossible to get AFM image of the tapping mode by the magnetic tip. So, it is impossible to simultaneously get the MFM and AFM tapping mode images. To solve this problem, we chose indirect comparison. We compared the AFM image measured by the contact mode with the ones by the tapping mode; then we compared the AFM contact mode image with the MFM image. The result of the comparison between AFM contact mode and AFM tapping mode is shown in figure 5.

It must be noted that these images were simultaneously obtained by scanning line by line in the sequence of one line AFM contact mode and one line AFM tapping mode. It is clear that the AFM contact mode image and AFM tapping mode image show the same structure, except the small difference in the width of the AAO wall. Comparing with that from FESEM image (figure 4a), it is clear that the difference of width also appeared. This difference is due to the resolution of different applied AFM modes and measured methods. But the important conclusion is that all obtained images show the same arrangement of the pores and nanodots. These results confirm that the AFM image of the contact mode (figure 5a) presents the real geometrical distribution of the Fe nanodots. Figure 6 presents the typical result of the comparison between AFM contact mode and MFM mode.

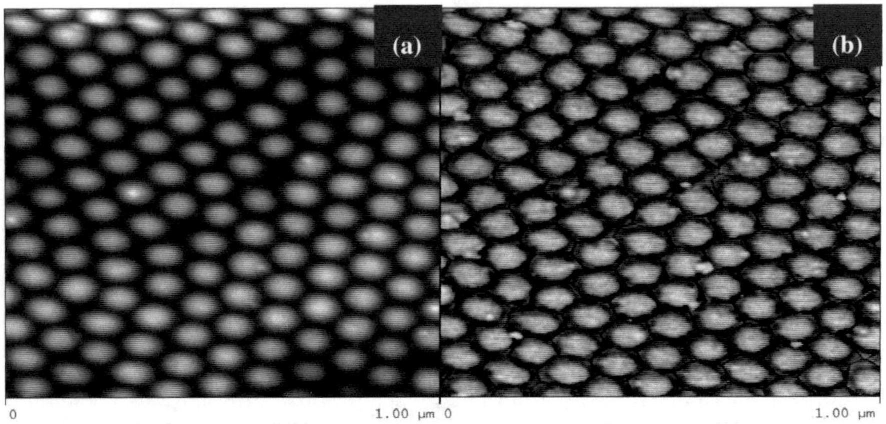

Figure 5. AFM images: contact (a) and tapping modes (b).

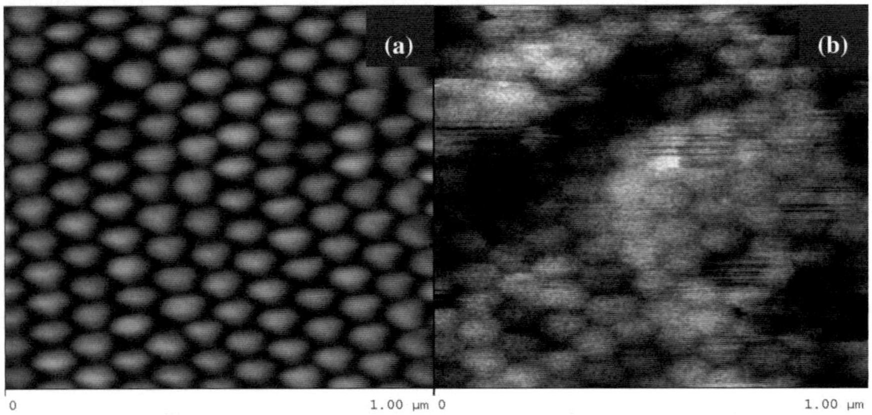

Figure 6. Contact mode AFM image (a) and MFM image (b).

The MFM image was obtained by tip-scanning at a distance of 60 nm from the sample surface. Here, we emphasize that the AFM and MFM images were simultaneously measured, so the AFM and MFM signals had provided images of the same position. Firstly, we note that the dot distribution in the MFM image determined by magnetic forces is identical with that from AFM contact mode measured due to the Van der Waals forces. Secondly, it seems to be the large domain in the MFM image, wich looks like clouds (figure 6b). We propose that the cloud-like signals related with the size difference of Fe dots. A similar result concerned with the size of the magnetic particles was reported in [11] and has been explained by the collective behavior of the Fe dots. Based on the dot distribution and the cloud-like distribution, we can confirm that the obtained MFM image presented in figure 5b is really due to magnetic forces. The dot distribution is due to individual magnetic dipoles of the Fe nanodots, and the cloud-like distribution is due to the dot interaction. Finally, experiment results indicate that by the developed technique we successfully fabricated the hexagonal ordered Fe nanodot system, which has the magnetic structure identified with their topology.

4. Conclusion

By two-step aluminum anodization, the AAO patterns with the hexagonal pore arrangement have been achieved. Using AAO pattern as a template, we successfully deposited iron metal in the pores by the thermal vacuum evaporation. The deposited hexagonal ordered Fe nanodots have been exposed by etching aluminum and alumina oxide layers from the bottom of the AAO template. The morphology of the Fe nanodot system can be clearly identified by SEM and AFM studies. EDS and MFM were employed to confirm the existence of Fe in the nanodot system. The fabricated Fe nanodot system has the magnetic image similar to their topography.

References

[1] Sun S and Murray C B 1999 *J. Appl. Phys.* **85** 4325
[2] Sun S, Murray C B, Weller D, Folks L and Moser A 2000 *Science* **287** 1989
[3] Davydov D N, Haruyama J, Routkevitch D, Statt B W, Ellis D, Moskovits M, Xu J M 1998 *Physical Review B* **57** 13550
[4] Sauer G, Brehm G, Schneider S, Nielsch K, Wehrspohn R B, Choi J, Hofmeister H and Gosele U 2002 *J. Appl. Phys.* **91** 3243
[5] Zeng H, Zheng M, Skomski R, Sellmyer D J, Liu Y, Menon L and Bandyopadhyay S 2000 *J. Appl. Phys.* **87** 4718
[6] Kazadi Mukenga Bantu et al. 2001 *J. Appl. Phys.* **89** 3393
[7] Masuda H, Yasui K, Nishio K 2000 *Adv. Mater.* **12** 1031
[8] Liang J, Chik H, Yin A, Xu J M 2002 *J. Appl. Phys.* **91** 2544
[9] Chik H, Xu J M 2004 *Mater. Sci. Eng. R.* **43** 103
[10] Steffen Porthun, Leon Abelmann, Cock Lodder 1998 *J. Magn. Magn. Mater.* **182** 238
[11] Victor F P, Pau G, Deborah M A, Neus G B 2004 *Nature materials* **3** 263

Acknowledgements

This work is supported by Fundamental Research Projects of the Institute of Materials Science. A part of the work was done with the help of Key laboratory for Electronic Materials & Devices, Institute of Materials Science.

Fabrication of Nd-Fe-B exchange-spring magnets

Nguyen Mau Lam[1], Nguyen Thi Thanh Huyen[2], Do Hung Manh[3], Vu Hong Ky[3], Do Khanh Tung[3] and Nguyen Huy Dan[3]

[1] Department of Physics, Pedagogic University of Hanoi N° 2,
Xuan Hoa, Phuc Yen, Vinh Phuc, Vietnam
[2] University of Industry, Dong Trieu, Quang Ninh, Vietnam
[3] Institute of Materials Science, Vietnam Academy of Science and Technology,
18 Hoang Quoc Viet, Cau Giay, Hanoi, Vietnam

E-mail: dannh@ims.vast.ac.vn

Abstract. In this report we present our recent results on fabrication of Nd-Fe-B exchange-spring magnets using melt-spinning method. Additive elements were used to improve useful parameters of this kind of hard magnetic materials. Hot pressing technique was applied to increase mass density of the magnets. The compositions and technology routes for fabricating the Nd-Fe-B exchange-spring magnets with maximum energy product $(BH)_{max}$ above 10 MGOe were shown.

Keywords: Nanostructure magnetic materials, exchange-spring magnets, hard magnetic materials, rapidly quenched method, rare earth-transition metal alloys.

1. Introduction

The Nd-Fe-B based exchange-spring magnets can be considered as an application of nanotechnology in real life. These magnets consist of hard and soft magnetic phases, which are coupled via exchange interaction in nanometer scale. The hard magnetic phase is $Nd_2Fe_{14}B$ and the soft magnetic phase is mainly α-Fe. The combination of high coercivity of the hard magnetic phase and high remanence of the soft magnetic phase leads to a high maximum energy product, $(BH)_{max}$, for the material. The ideal microstructure of the exchange spring magnets is formed by identical hard magnetic nanocrystallites homogeneously dispersed in a soft magnetic matrix [1]. There are two ways to control the formation of the nanocrystallites in the alloys prepared by rapidly quenching method. The first way is based on the variation of the annealing conditions for the over quenched alloys, which are amorphous in the as-quenched state. The second way is based on the change of the quenching rate of the melted alloys. The problems of the undoped Nd-Fe-B exchange spring magnets are rather low Curie temperature and coercivity. Besides that, the sensitivity of the structure of the materials with the fabrication conditions is also another problem. To enhance Curie temperature of the material, the addition of Co is the most suitable way because Co can substitute for Fe in $Nd_2(Fe,Co)_{14}B$ phase. Moreover, Co can improve saturation magnetization and remanence of the material. In order to control the formation of the crystallites in the Fe-based magnetic materials, Nb is commonly added to the materials because Nb can surround the crystallites and prevent the growth of the crystallites. With an appropriate concentration of additional elements, the crystallites can be formed more identically in nanometer scale leading to better performance of the Nd-Fe-B based exchange spring magnets. Other additive

© 2009 IOP Publishing Ltd

elements such as Dy, Pr, Cu, Cr... are also used to improve the parameters of the hard magnetic behavior and fabrication technology of the materials. Beside of the considerations on quality of the alloys, some aspects in bonding the magnets such as grain size of alloy powder, quality of glue, technique of pressing are also important. In our previous researches [2-10], we only concentrated to study influence of additive elements and fabrication conditions on structure and magnetic properties of the alloys. In this work, we tried to produce some exchange spring magnets with various compositions and different techniques of pressing.

2. Experimental

The pre-alloys with desired compositions were prepared by arc-melting method to ensure their homogeneity. The melt-spinning method was used to fabricate the ribbon samples from the pre-alloys. The quenching rates were high enough to make all the ribbons in amorphous state. The ribbons were annealed in temperatures ranging from 600 to 750°C under Ar atmosphere. The annealed ribbons were then ground into powder having grain size of about 400 μm. Both the cold and hot pressing techniques were applied to bond the magnets. The temperature was raised up to 270°C during hot-pressing process. The structure of the samples was investigated by powder X-ray diffraction (XRD). Magnetic properties of the material were investigated by magnetization measurements on a pulsed field magnetometer. The demagnetization effect, which depends on shape of measuring specimens, was taken into determination of $(BH)_{max}$ of the material.

3. Results and discussion

The first stage of fabricating the Nd-Fe-B based exchange spring magnets is determination of composition. Although this kind of hard magnetic materials can exist in a wide range of composition, the applicable compositions are limited. There have been many compositions reported for producing the exchange spring magnets. However, the results could be found different from one report to another. From our point of view, the best way to choose a composition for the exchange magnet with desired magnetic properties is re-examining it.

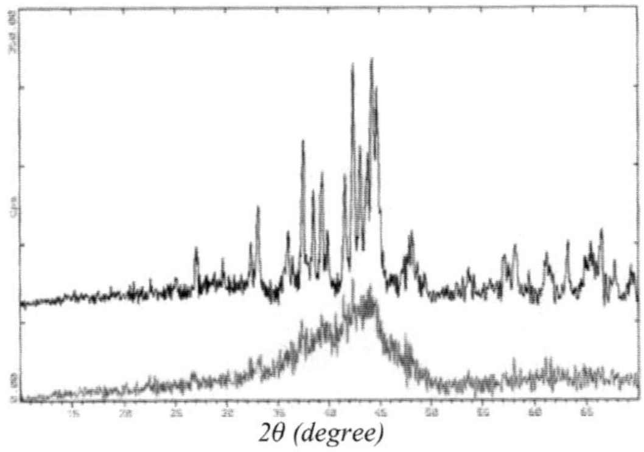

Figure 1. XRD patterns of as-spun (lower) and annealed (upper) $Nd_{10.5}Nb_{1.5}Fe_{82}B_6$ ribbons indicating the formation of expected $Nd_2Fe_{14}B$ and α-Fe phases.

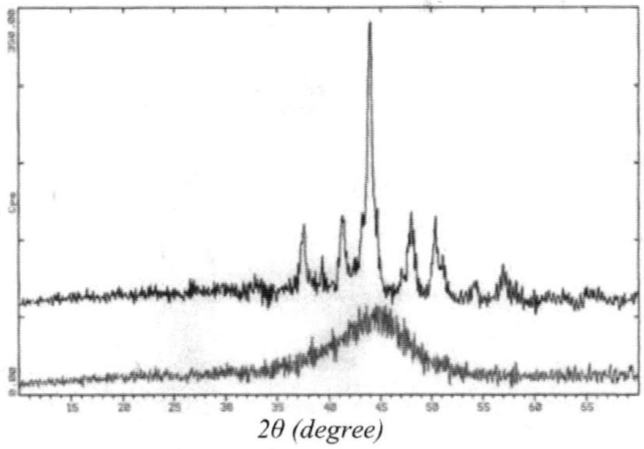

Figure 2. XRD patterns of as-spun (lower) and annealed (upper) $Nd_2Pr_2Nb_{1.5}Fe_{78}B_{16.5}$ ribbons indicating the formation of unexpected $Fe_{23}B_6$ phase.

As mentioned in the introduction section, there are two way to achieve the optimal nanostructure for the exchange spring magnets, by controlling quenching rates or annealing conditions. The former method can produce the magnets directly from melts, while the latter one requires another procedure, annealing process. However, the controlling quenching rate is much more difficult than controlling annealing conditions. Furthermore, the hard magnetic phase cannot be formed directly from melts with a number of compositions, especially with ones containing low content of rare earth components [8].

Journal of Physics: Conference Series **187** (2009) 012076 doi:10.1088/1742-6596/187/1/012076

In this case, the annealing process is required. The influence of additive elements on phase formation might be also different from one composition to another [2-9]. As an example, with the same addition concentration of Nb, the desired crystalline phases such as $Nd_2Fe_{14}B$ and α-Fe are still formed in the $Nd_{10.5}Nb_{1.5}Fe_{82}B_6$ alloy (figure 1), while these phases almost disappear in the $Nd_2Pr_2Nb_{1.5}Fe_{78}B_{16.5}$ alloy (figure 2). The influence of Nb on structure and magnetic properties of some Nd-Fe-B based alloys were investigated quite thoroughly in our previous work [2, 4, 9].

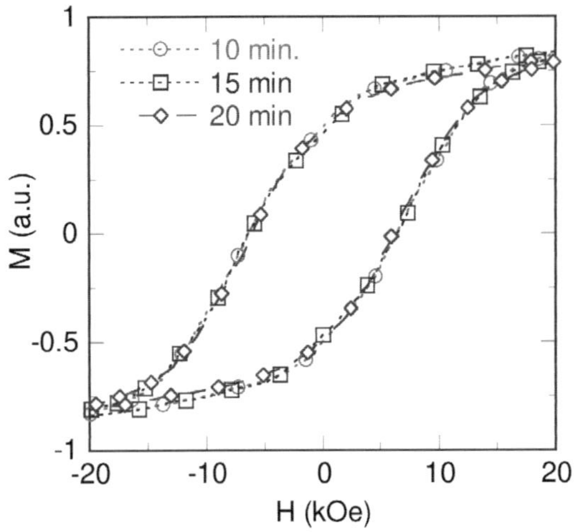

Figure 3. Hysteresis loops of $Nd_{10.5}Nb_{1.5}Fe_{82}B_6$ alloy annealed at 650°C for various time, demonstrating the stability of magnetic properties with variation of fabrication conditions.

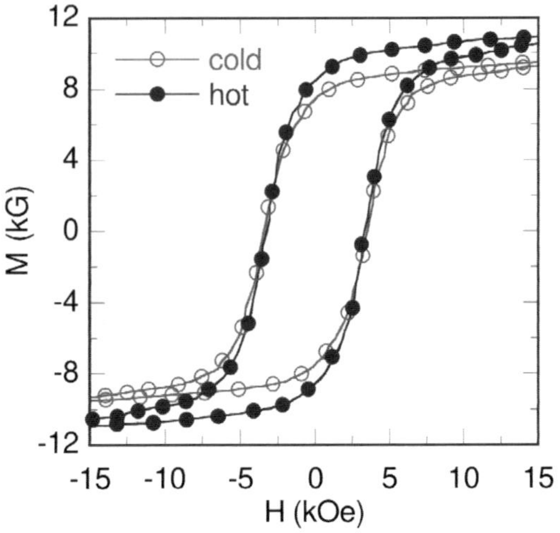

Figure 4. Hysteresis loops of cold- and hot-pressed $Nd_2Pr_2Fe_{78}B_{18}$ exchange-spring magnets showing the enhancement of saturation magnetization and remanence by hot-pressing.

Figure 5. Picture of some produced Nd-Fe-B exchange-spring magnets.

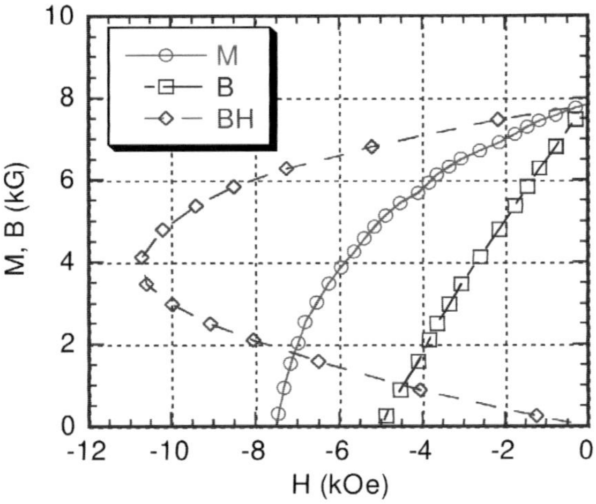

Figure 6. Characteristic curves of a represented Nd-Fe-B exchange-spring magnet.

The stability of magnetic parameters of the material with variation of fabrication conditions is a crucial criterion for making magnets of the same quality. With appropriate compositions and annealing conditions, this requirement can be fulfilled. Figure 3 demonstrates that the magnetic properties of the $Nd_{10.5}Nb_{1.5}Fe_{82}B_6$ alloy are stable with variation of annealing time from 10 to 20 minutes. The additive elements, especially Nb, play an important role for stabilizing quality of the magnets when a fabrication condition varies. As evidence, the materials containing large amount of amorphous phase are more stable with variation of fabrication conditions [6]. The results in this report [6] show that the glass forming ability and performance of Nd-Fe-B based nanocomposite materials can be improved by appropriate concentrations of Co and Nb. The materials still gives a good performance with quite large fractions of amorphous phase. The coercivity of the materials can be improved by the amorphous matrix. Especially, the hard magnetic parameters of this kind of hard magnetic material are quite stable with variation of the fabrication conditions. The optimal range concentrations of Co and Nb for the Nd-Fe-B systems are about 2 ÷ 4 % and 1.5 ÷ 3, respectively.

After annealing in optimal conditions, the alloys in ribbon form were ground into powder with grain size of about 400 μm. Commercial glue was mixed with the powder and then the mixture was made dry. The mixture was pressed into cylindrical tablets with diameter of 16 mm by both the cold and hot techniques. Hot-pressing technique was used to improve mass density of the magnets resulting in enhancement of maximum energy product. The temperature was raised up to 270°C during hot-pressing process. Figure 4 shows hysteresis loops of cold and hot pressed $Nd_2Pr_2Fe_{78}B_{18}$ exchange-spring magnets. One can realize that both the saturation magnetization and remanence of the magnet are enhanced by hot-pressing. Figure 5 is a picture of our produced exchange spring magnets and figure 6 shows some characteristic curves of their representative.

Table 1. Compositions and parameters of the fabricated exchange spring magnets.

Composition	Type of press	Pressure (Ton/cm^2)	Mass density (g/cm^3)	$_MH_c$ (kOe)	$_BH_c$ (kOe)	B_r (kG)	$(BH)_{max}$ (MGOe)
$Nd_4Fe_{78}B_{18}$	cold	7.5	5.7	2.9	2.3	7.5	5.5
	hot	7.5	5.9	2.8	2.3	7.9	6.1
$Nd_2Pr_2Fe_{78}B_{18}$	cold	7.5	5.6	3.4	2.7	7.4	6.4
	hot	7.5	5.8	3.3	2.7	8.4	7.8
$Nd_{10.5}Nb_{1.5}Fe_{82}B_6$	cold	7.5	5.6	7.5	4.8	7.3	9.6
	hot	7.5	5.9	7.5	5.0	7.8	10.7

Based on the obtained results [2-10], we have tried to fabricate some Nd-Fe-B based exchange spring magnets completely. Their compositions and parameters are presented in Table 1. We can see that by hot pressing, the maximum energy product $(BH)_{max}$ of all the exchange spring magnets is increased by an amount larger than 10% in comparison with that of the cold-pressing magnets having the same composition. The $(BH)_{max}$ above 10 MGOe was achieved. We can also realize that $(BH)_{max}$ of these magnets is probably increased further because the mass density of this kind of magnet can reach 6.5 g/cm^3. Perhaps the content of glue and pressure applied for the fabricated magnets are not in optimum. The quality of glue and pressing time might also influence the quality of the magnets.

4. Conclusion

The influence of addition elements and fabrication conditions on quality of the Nd-Fe-B based exchange spring magnets have been investigated. Hot pressing technique was applied to increase mass density of the magnets. The compositions and technology routes for fabricating the Nd-Fe-B exchange-spring magnets with maximum energy product $(BH)_{max}$ above 10 MGOe were achieved.

References

[1] Kneller E F and Hawig R 1991 *IEEE Trans. Magn.* **27** 3588
[2] Thuy D M, Linh N T, Duc V T, Nam N H, Cong P T, An T Qu, Tuan N Anh, Bang D and Dan N H 2006 *Proc. of Inter. Conf. on Engineering Physics (Hanoi, Vietnam)* p 131

[3] Duc V T, Thuy D M, Kien N C, Tuan N A, Bang D and Dan N H 2006 *Proc. of 1st IWOFM-3rd IWONN Conference (Ha Long, Vietnam)* p 298
[4] Thuy D M, Duc V T, Tuan N A, Manh D H, Kien N C and Dan N H 2006 *Proc. of 1st IWOFM-3rd IWONN Conference (Ha Long, Vietnam)* p 320
[5] Dan N H, Thuy D M, Hung L T, Huyen P T T, Huyen N T T, Manh D H and Tuan N A 2007 *Proc. of the 1st Inter. workshop on nanotechnology and application - IWNA (Ho Chi Minh City, Vietnam)* p 253
[6] Huyen N T T and Dan N H 2008 *Proc. of the Eleventh Vietnamese-German Seminar on Physics and Engineering (Nha Trang City, Vietnam)* p 322
[7] Huyen P T T, Huyen N T T and Dan N H 2007 Advances in Natural Sciences **8** 431
[8] Huyen N T T, Thuy D M, Huyen P T T, Duc N H and Dan N H 2007 *Advances in Natural Sciences* **8** 423
[9] Thuy D M and Dan N H 2008 *J. Korean Phys. Society* **52** 1465
[10] Dan N H 2008 *J. Korean Phys. Society* **52** 1443

Acknowledgments
This work was supported by National Basic Research Fund and Quang Ninh University of Industry, Vietnam. A part of the work was done in the National Key Laboratory for Electronic Materials and Devices, Institute of Materials Science, Vietnam.

Preparation of silver nanoparticles by pulse sonoelectrochemical method and studying their characteristics

Le Van Vu, Nguyen Ngoc Long, Sai Cong Doanh and Bui Quoc Trung

Faculty of Physics, Hanoi University of Science, 334 NguyenTrai, Hanoi, Vietnam

E-mail: levanvu@hus.edu.vn

Abstract. Silver nanoparticles with spheres, rods have been prepared by a pulse sonoelectrochemical technique from an aqueous solution of $AgNO_3$ in the presence of sodium dodecyl sulfate $C_{12}H_{25}NaO_4S$ -(SDS). The as-prepared silver nanoparticles are characterized by electron microscopy (TEM, SEM), powder X-ray diffraction (XRD), and UV-vis absorption spectrum. It was found that the concentration of $AgNO_3$ and SDS affects the shape of the nanoparticles. The crystal size could be varied from 5 nm up to 200 nm by controlling the various electrodeposition and sonic parameters.

Keywords: Silver nanoparticles, sonoelectrochemical method, metallic nanoparticles, surface plasmon resonance absorption.

1. Introduction

In recent years, metallic nanoparticles are a strong research focus because of their unique functional properties which lead to varied application in the areas of plasmonics [1], surface enhanced Raman scattering [2], catalysis, medicine and biological sensing [3, 4] It is well-known that catalytic reactivity depends on the size and shape of the metal nanoparticles and that the synthesis of well-controlled shapes and the size of particles could be critical for their applications [5]. The developed methods for nanoparticle fabrication include chemical reduction [6], seed-mediated growth procedure [7, 8], sonochemical reduction [9], microwave-assisted synthesis [11], laser ablation, metal evaporation [12], etc.

Sonochemistry is a very useful method that has been known since the 1930s and has increased in popularity over the past decade. The effects of ultrasonic radiation on chemical reactions are due to the very high temperatures and pressures which develop in and around collapsing bubbles. However, the potential benefits of combining sonochemistry with electrochemistry have become increasingly studied. Some of these beneficial effects include acceleration of mass transport, cleaning and degassing of the electrode surface, and increased reaction rates. Reisse and co-workers have described a novel device for the production of metal powders using pulsed sonoelectrochemical reduction [9, 10]. This device exposes only the flat circular area at the end of the sonic tip to the electrodeposition solution. The exposed area acts as both cathode and ultrasound emitter. The metals were obtained as chemically pure, fine crystalline powders with a high surface area giving an average particle size of 100 nm. The present understanding of the sonoelectrochemical mechanism describes an electrochemical deposition of a thin layer on the sonoelectrode during the electrical pulse. The thin layer is then destroyed by the coming sonic pulse, to form jet-suspended nanoparticles into the

© 2009 IOP Publishing Ltd

solution. Thus, the electrochemical and sonic parameters, including the pulse duration, can be used to control the particles' parameters.

In the present paper, we report the use of the sonoelectrochemical method for the preparation of silver nanopaticles with different shapes including spherical and rodlike nanoparticles. The sonoelectrochemical device employed is different from that described by Reisse and co-workers. A titanium horn acted as the ultrasound emitter only in the experimental setup. This experiment is based on the electroreduction of $AgNO_3$ in aqueous solution in the presence of SDS.

2. Experimental

The schema of the experimental setup assembly for these experiments is shown in figure 1a. It contains an electrochemical cell (cathode, anode, and electrolyte), an ultrasonic pulse generator (sonicator), a direct-current power source, and a synchronizer. The synchronizer is used to drive the durations of ultrasonic and electric pulses. Each sonoelectrochemical cycle lasts 0.9 s, consisting of a 300 ms electric current pulse of 40 mA, a 300 ms sonic pulse immediately following the electric pulse, and a 300 ms rest (figure 1b). The catholic deposition of the silver was done on stainless steel foil. The other stainless steel foil served as the anode. A Vibrate 700 W was used as the sonicator. The power source was limited to 20 V. The experimental cell had dimensions as follows: The effective surface area of the stainless steel foil was 2 cm^2. The distance between the anode and the cathode was set to 2 mm. In all of the experiments, the amount of electrolyte was 50 cm^3. The temperature was stabilized by a water bath and kept in the temperature range 25-28°C. The vessel was cover in by using aluminum paper. The usual value of electric current density was 20 mA/cm^2. The appropriate ultrasonic power density was about 100 W/cm^2.

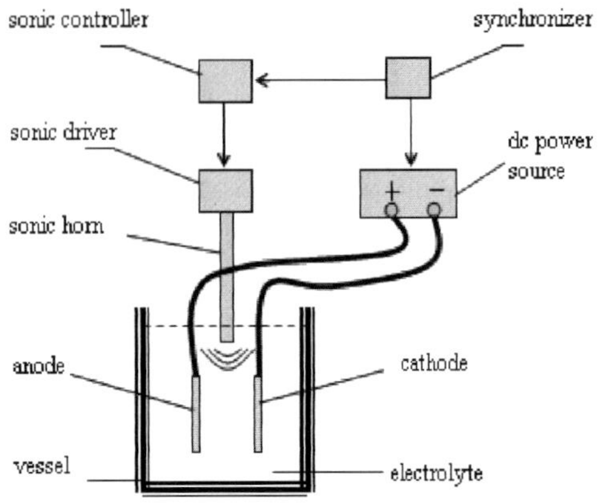

Figure 1a. The schema of the experimental setup.

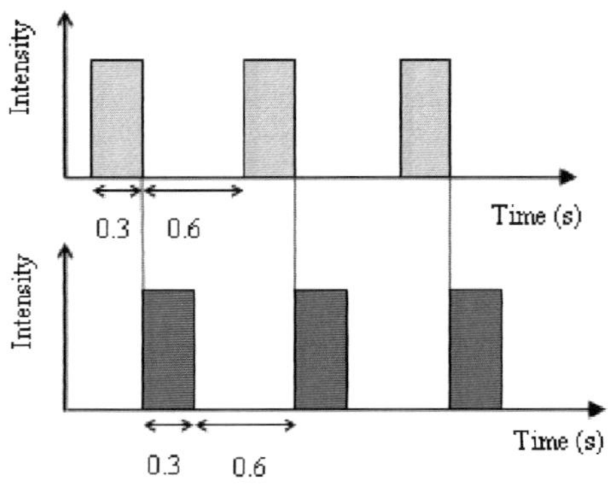

Figure 1b. The sonoelectric cycle: electric current pulse of 0.3 s duration; sonic pulse of 0.3 s duration; the total cycle is 0.9 s long.

Quantities of 1, 5, and 20 g/L $AgNO_3$ with sodium dodecyl sulfate $C_{12}H_{25}NaO_4S$-SDS (Merck) were mixed in an aqueous solution under a N_2 atmosphere. The deposition was carried out by using different reaction times. Oxygen was removed from the water by nitrogen bubbling, and the electrolyte was mixed under a nitrogen atmosphere. A complexing agent, sodium dodecyl sulfate $C_{12}H_{25}NaO_4S$-SDS, was added at the 40 g/L level to avoid aggregation. The products were washed with doubly distilled water and collected by using a centrifuge (Hettich Universal 320, 9000 RPM, 30 minutes). The silver particles were characterized by transmission electron microscopy (TEM) and by scanning electron microscopy (SEM), powder X-ray diffraction (XRD) using Cu-Kα radiation, and UV-vis absorption spectrum. The TEM studies were performed with a JEOL, JEM-1010 microscope. The TEM provides quantitative information for each size distribution calculation using the Image J 1.37 v software.

3. Results and discussion

Figure 2a presents a typical TEM image of the product obtained by the sonoelectrochemical method with 1 g/L $AgNO_3$ and 40 g/L SDS. The deposition was carried out for 5 minutes. It shows that silver nanoparticles are spherical and well-dispersed. The particle size distribution measured directly from TEM micrographs is shown as a histogram in figure 3. The average size of particles is about 8 nm. One can readily see that, at the initial time interval, up to about 15 minutes, single crystallites grow. The mean size of the crystallites increases from 8 nm after 5 minutes of sonoelectrochemical processing to 20 nm after 15 minutes (figure 2) and the aggregation is observed if the reaction is carried out for longer periods of time.

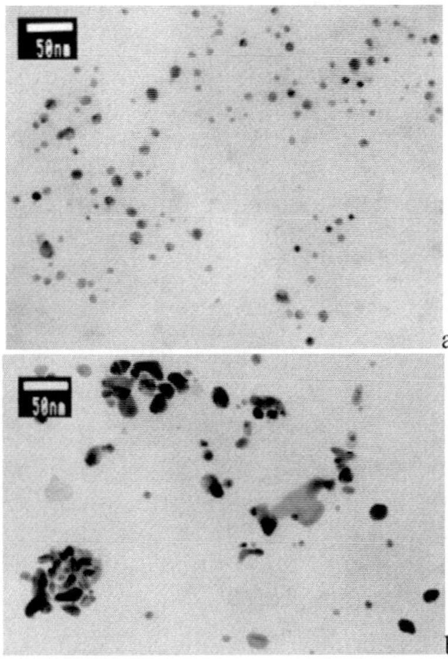

a)

b)

Figure 2. TEM pictures of silver particles obtained by pulsed sonoelectrochemical treatment of different durations for 1 g/L $AgNO_3$ and 40 g/L SDS solution: a) 5 minutes, b) 15 minutes.

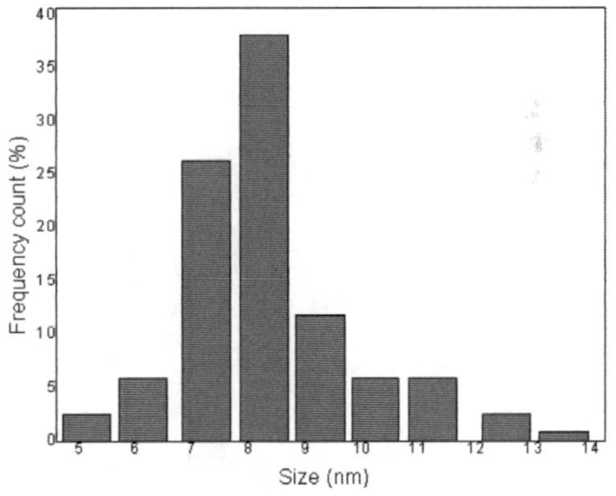

Figure 3. Size distribution measured from TEM micrographs in figure 2a.

Experimental results indicate that the concentration of $AgNO_3$ plays a significant role in the formation and growth of the silver nanoparticles. The TEM image of the products of the 5 g/L $AgNO_3$ electrolysis shows a nanorod structure with a width of 10-20 nm in diameter, the reaction was carried out for 5 minutes (figure 4a). Figure 4b shows TEM image of the product obtained with sonoelectrochemical processing duration of 15 minutes. The nanorods grow longer and have higher density. A detailed observation of the image shows that many convex areas appeared on the surface of the silver nanorods. These convex areas may further develop into dendrites. It is possible that the excess of silver in the solution favors the aggregation and grows into the dendritic structure of Ag clusters.

SDS molecules are used to disperse the silver nanoparticles and also to stabilize the formation of shaped silver nanoparticles. When the concentration of SDS is low, much larger silver particles are obtained. Figure 5 presents a typical TEM image of the product obtained by the sonoelectrochemical method with 10 g/L $AgNO_3$ and 5 g/L SDS. It shows that the product consists of single crystallites with a rod structure having a width of 150-200 nm in diameter. In the current investigation, we have found that the concentration of $AgNO_3$ and SDS affects a key factor in controlling the size and the shape of Ag nanoparticles. During the reaction, the particles are not completely detached from the cathode by the ultrasonic waves, so that the particles formed in the next pulse can further grow on seed left on the electrode. This growth can promote the formation of different shaped silver particles.

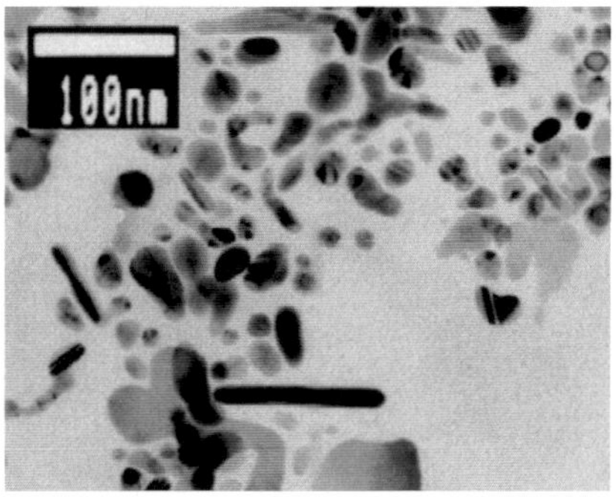

Figure 4a. TEM picture of silver nanorods (5 g/L $AgNO_3$; electrosonication time: 5 minutes).

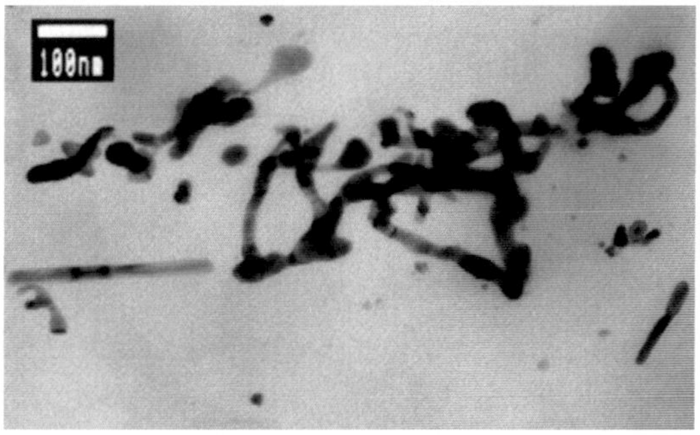

Figure 4b. TEM picture of silver nanorods and dendritic nanoparticles (5 g/L $AgNO_3$; electrosonication time: 15 minutes).

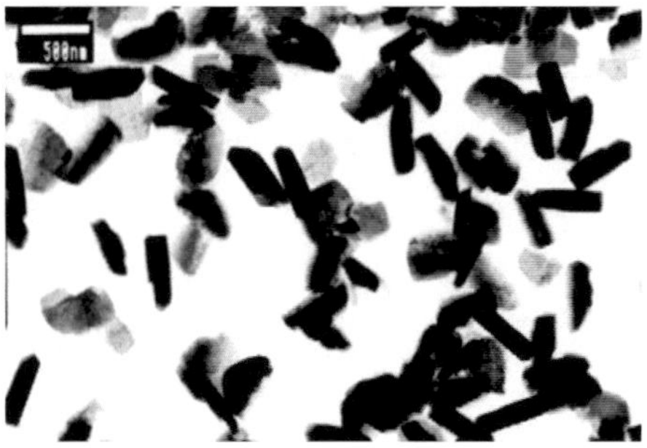

Figure 5. TEM picture of silver rods (10 g/L $AgNO_3$ and 5 g/L SDS; electrosonication time: 15 minutes).

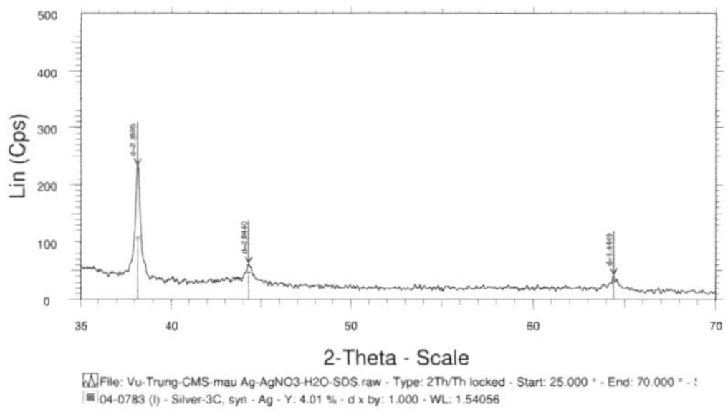

Figure 6. XRD pattern of the silver nanoparticles (5 g/L $AgNO_3$; 40 g/L SDS; electrosonication time: 15 minutes).

A typical XRD pattern of the as-prepared silver nanoparticles is shown in the figure 6. All the peaks in the XRD pattern can be indexed as a face centered cubic (fcc) structure (JCPDS, file no. 4-0783). XRD pattern shows the presence of the diffraction peaks corresponding to the (111), (200), and (220) planes.

The optical absorption spectra for silver nanoparticles prepared with 15 minutes electrosonication time are shown in figure 7. Aqueous solution of the silver spherical nanoparticles developed a surface plasmon resonance band centered at 401 nm that was characteristic for silver nanoparitcles with a dimension less than 20 nm (curve *a*). The curves *b*, *c* and *d* are the optical absorption spectra for aqueous solutions of silver rodlike nanoparticles. For the rodlike nanoparticles the absorption spectra exhibit two bands: one at 418 nm and the other at 811 nm (curve *c*). It is well-known that the size and shape of particles may influence the position and the width of the plasmon resonance absorption band. The red shift of plasmon resonance absorption band of silver particles could be attributed to the increasing size due to the particle aggregation as indicated by the electron microscopy observations. For anisotropic metallic nanoparticles such as nanorods and nanowires, the surface plasmon absorption spectra have two bands: one band at the short wavelength side due to the transverse oscillation of electrons, the other band at the long wavelength side due to the longitudinal oscillation of electrons [7].

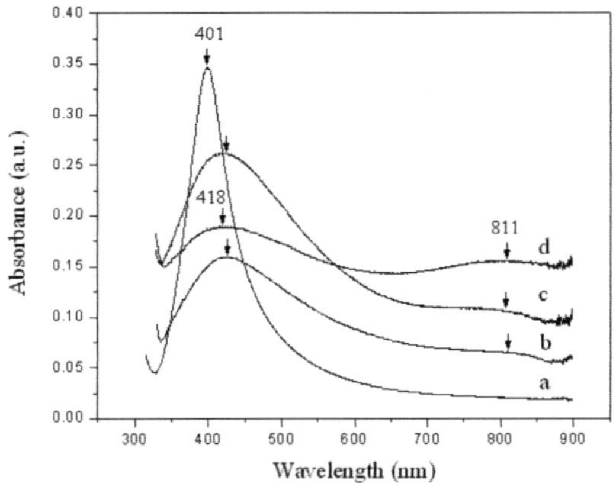

Figure 7. The UV-vis absorption spectra of silver nanoparticles in aqueous solution with 15 minutes electrosonication time: a) 1 g/L $AgNO_3$ and 40g/L SDS; b) 5g/L $AgNO_3$ and 40g/L SDS, c) 5g/L $AgNO_3$ and 20g/L SDS and d) 5g/L $AgNO_3$ and 10g SDS).

3. Conclusion

In summary, the preparation of spherical Ag nanoparticles and Ag nanorods was successfully developed by using a pulse sonoelectrochemical method. By changing the concentration of both $AgNO_3$ and SDS we could control the size distribution and the shape of silver nanoparticles. This present technique can be applied to prepare other metal nanoparticles.

References

[1] Lelly K L, Coronado E, Zhao L L and Schatz G C 2003 *J. Phys. Chem.* B **107** 668
[2] Yu-Chuan Liu, Kuang-Hsuan Yang and Shung-Jim Yang 2006 *Analyt. Chimica Acta* **572** 290
[3] Zharov V P, Mercer K E, Galitovskaya E N and Smeltzery M S 2006 *Biophysical J* **90** 619
[4] Wang J X, Sun X W, Wei A, Lei Y, Cai X P, Li C M and Dong Z L 2006 *Appl. Phys. Lett.* **88** 233106
[5] Yu Chuan Liu , Kuang Hsuan Yang and Shung Jim Yang 2006 *Analyt. Chimica Acta* **572** 290
[6] Suzuki M, Niidome Y, Kuwahara Y, Terasaki N, Inoue K and Yamada S 2004 *Nano Lett. J. Phys. Chem. B* **108** 11660
[7] Tapan K.Sau and Catherine J. Murphy 2004 *Langmuir* **20** 6414
[8] Murphy C J, Sau T K, Gole A M, Orendorff C J, Jinxin Gao, Linfeng Gou, Hunyadi S E and Tan Li 2005 *J. Phys. Chem. B* **109** 13857
[9] Reisse J, Francois H, Vandercammen J, Fabre O, Mesmaeker K D, Maerschalk C and Delphlancke J L 1994 *Electrochim. Acta* **39** 37
[10] Yehoshua Socol, Oleg Abramson, Aharon Gedanken, Yitzhak Meshorer, Larisa Berenstein and Arie Zaban 2002 *Langmuir* **18** 4736
[11] Navaladian S, Viswanathan B, Varadarajan T K and Viswanath R P 2008 *Nanotech.* **19** 045603
[12] Biswas A, Marton Z, Kanzow J, Kruse J, Zaporojtchenko V, Faupel F and Strunskus T 2003 *NanoLett.* **3** 69

Acknowledgements

This work is supported by the Project QG.07.03, Key Project QGTD.08.05 of Vietnam National University and National Natural Science Foundation of Vietnam.

Green synthesis of silver nanoparticles and silver colloidal solutions

Nguyen Thi Phuong Phong, Ngo Hoang Minh, Ngo Vo Ke Thanh and Dang Mau Chien

Laboratory for Nanotechnology, Vietnam National University-HCM City, Linh Trung Ward, Thu Duc District, HCM City, Vietnam

E-mail: ngohoangminh412@yahoo.com.vn

Abstract. In this paper, silver colloidal solutions have been synthesized rapidly in green conditions by using microwave irradiation and non-toxic chemistry substances (acid oxalic, silver nitrate, polyvinyl pyrolidone (PVP; Mw = 55 000)). The particle size and morphology of these solutions can be controlled by altering several factors like the time, the power of microwave exposure, and the ratio of silver oxalate and PVP etc. The silver nanoparticles were fabricated by thermal decomposition of silver oxalate. The synthesized silver colloidal solutions and silver nanoparticles were characterized by several analytical techniques like UV–VIS, XRD, TEM, FESEM/EDS and ICP-AAS studies. Finally, we used the synthesized silver colloidal solutions for antibacterial purpose. The obtained results showed that the synthesized silver colloidal solutions, even at very low concentrations, have highly efficient anti-bacterial property.

Keywords: Silver nanoparticles, silver colloidal solutions, green synthesis.

1. Introduction

Silver nanoparticles have potential applications in various fields such as the environment, biomedicine, catalyst, optics and electronics [1-4]. Silver nanoparticles have also drawn much attention because they show high bactericidal effects [5-9]. Various approaches to synthesis metal nanoparticles have been reported such as chemical reduction, UV photolysis, gamma irradiation, metal vapor deposition, laser ablation, sputtering… Among these, chemical reduction has been the most extensively investigated [1, 2]. In this paper, thermal decomposition of silver oxalate complexes to fabricate high purity silver nanoparticles and reduction of silver oxalate complexes by ethylenglycol using polyvinyl pyrolidone as the capping agent has been explored. These reactions have been assisted by microwave irradiation because of other conventional heating methods. The main advantages include (i) the initial heating process is rapid, (ii) uniform heat transfer to the solution and (iii) microwave induces the generation of localized high temperatures at reaction sites, which enhances the reaction rate [1, 2].

© 2009 IOP Publishing Ltd

2. Experimental

2.1. Preparation of silver oxalate precusor
Silver oxalate was prepared by mixing 50 ml of 0.5 M $AgNO_3$ (Merck, 99%) solution with 30 ml of 0.5 M oxalic acid (Merck, 99%). The white formed precipitate was filtered, washed with distilled water, dried in an air oven for 1 h and stored in a dark bottle [1, 2].

Fabrication of silver nanoparticles: Silver nanoparticles were produced by thermal decomposition of silver oxalate.

Fabrication of silver colloidal solutions: 40 ml of ethylene glycol (China, 99%), 0.25 g of polyvinyl pyrolidone (PVP, Mw ≈ 55 000, BASF-Germany, 99%) was added and stirred. Then, the complete dissolution of PVP, $Ag_2C_2O_4$ was added, stirred for 10 mins and purged with N_2 gas for 10 mins. Immediately, the reaction vessel was irradiated in microwave cavity for different time intervals. All the reactions were carried out in a microwave oven operating in a cycling mode (on for 1 minute, off for 1 minute) to prevent the intense boiling of solvents as well as the agglomeration of metals.

2.2. Characterization
The synthesized silver colloidal solutions and silver nanoparticles were characterized by advanced analytical techniques such as UV–VIS (Cary 100 Conc, Varian), XRD (D8 Advance, Bruker - Germany), TEM (JEM-1400, Japan), FE-SEM/EDS (JSM 7401F, Japan) and ICP-AAS studies. Anti-bacterial (E. Coli) efficiency of silver colloidal solutions was determined by Colony Count Test Method.

3. Results and discussion

3.1. Preparation and characterization of silver oxalate
Silver oxalate was prepared with high yield and was confirmed by powder XRD (figure 1). The interplanar d-spacings of the corresponding lines presented in the powder XRD pattern match those of standard sample, which corresponds to the primitive monoclinic system. Moreover, it is clear from XRD that no other phase of silver oxalate is presented in the as-synthesized silver oxalate. The FE-SEM image of silver oxalate shows particles in the size range of under 2 □m (figure 2) The EDS results of silver oxalate showed that the weight percent of atoms (Ag = 71.22%; C = 8.54%; O = 20.24%), which agrees with the theorical calculations from formula of $Ag_2C_2O_4$ (Ag = 71.05%; C = 7.9%; O = 21.05%) (figure 3).

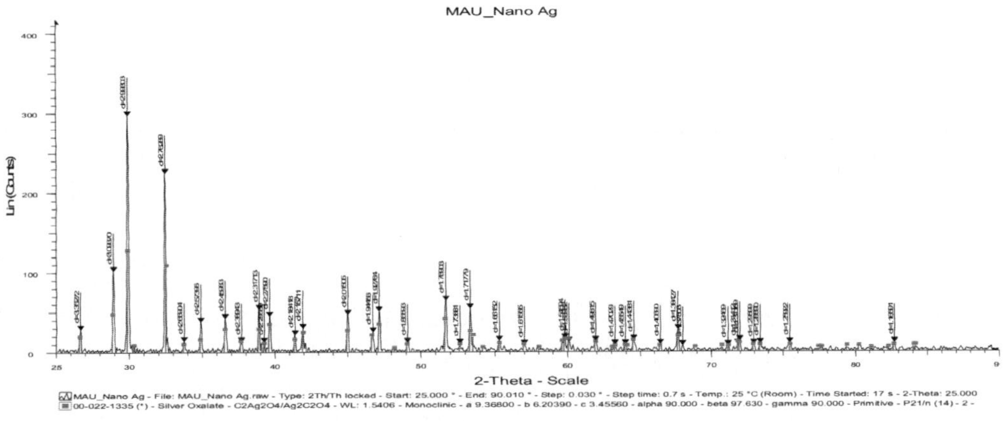

Figure 1. XRD pattern of silver oxalate.

Figure 2. The FE-SEM image of silver oxalate.

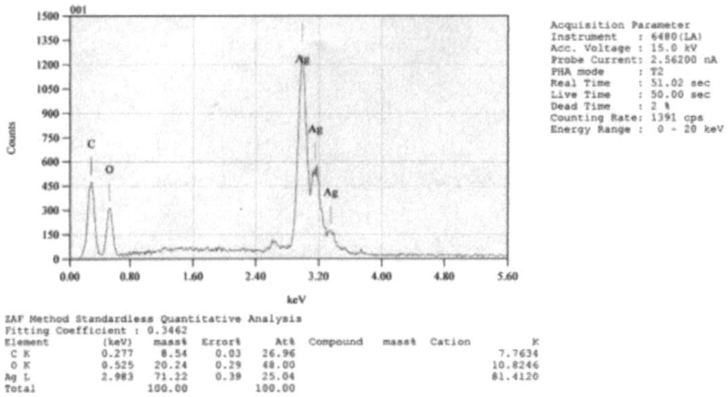

Figure 3. The EDS result of silver oxalate.

3.2. Synthesis of silver nanoparticles

A method for fabrication of highly pure silver nanoparticles without requiring the addition the capping or reducing agents is thermal decomposition of silver oxalates. This high pure silver can be used in the field of electronics requiring high conductivity or in the field of bacteria resistance requiring high purity [1-4].

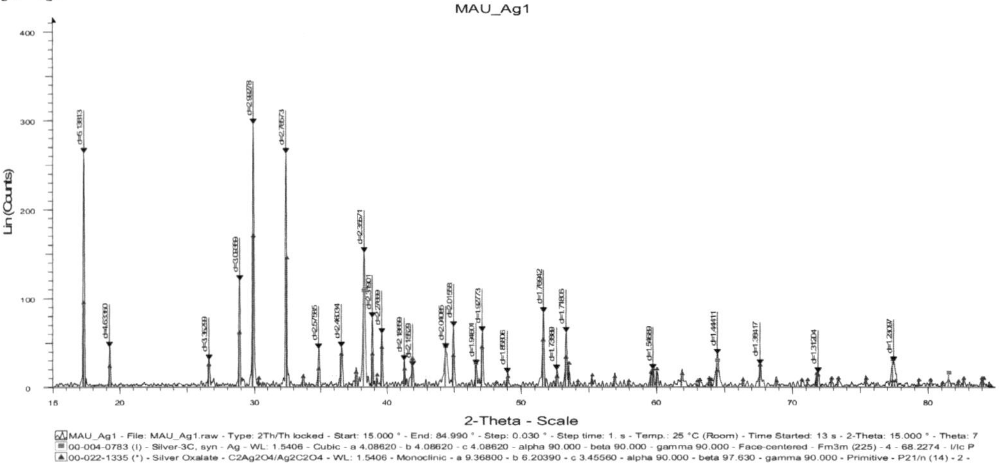

Figure 4. XRD pattern of M1 (140°C in 1 hour).

The XRD pattern of the M1 sample (calcinated at 140°C in 1 h) (figure 4) showed that when calcinating at 140°C in 1 h, silver oxalate was decomposed partially to generate silver (the red peak), the main remaining were also silver oxalates (the blue peak). Thus, we decided to increase the temperature of calcination to 200°C in 1 h and it is interesting that silver oxalate was completely decomposition to generation silver nanoparticles. The XRD powder pattern of the Ag is given in figure 5. Interplanar distance (*d*) corresponding to facets (111), (200), (220), (311) and (222) is d=2.35568 ($2\theta=38^{o}$); 2.04057 ($2\theta=44^{o}$); 1.44352 ($2\theta=64^{o}$); 1.23130 ($2\theta=77^{o}$); 1.17926 ($2\theta=82^{o}$), respectively. These data match Ag standard sample. The crystalline size of Ag nanoparticles was calculated from Sherrer formula is 70 nm.

3.3. Synthesis of silver colloidal solutions

The Ag colloids were synthesized by altering several factors such as concentration of silver nitrate, exposure time and power of microwave. The results were presented in table 1.

Table 1. Silver colloidal solutions.

Samples	Ethylene glycol(ml)	PVP(g)	$Ag_2C_2O_4$(g)	$Ag_2C_2O_4$:PVP	Power(W)	Time
1a	40	0.25	0.05	1:5	160	4 min
1b	40	0.25	0.05	1:5	160	4 min 10 sec
1c	40	0.25	0.05	1:5	160	4 min 20 sec
1d	40	0.25	0.05	1:5	160	4 min 30 sec
1e	40	0.25	0.05	1:5	160	4 min 40 sec
2a	40	0.25	0.025	1:10	160	4 min
2b	40	0.25	0.025	1:10	160	4 min 10 sec
2c	40	0.25	0.025	1:10	160	4 min 20 sec
2d	40	0.25	0.025	1:10	160	4 min 30 sec
2e	40	0.25	0.025	1:10	160	4 min 40 sec
3a	40	0.25	0.01	1:25	160	4 min
3b	40	0.25	0.01	1:25	160	4 min 10 sec
3c	40	0.25	0.01	1:25	160	4 min 20 sec
3d	40	0.25	0.01	1:25	160	4 min 30 sec
3e	40	0.25	0.01	1:25	160	4 min 40 sec
4a	40	0.25	0.025	1:10	800	30 sec
4b	40	0.25	0.025	1:10	800	35 sec
4c	40	0.25	0.025	1:10	800	40 sec
5a	40	0.25	0.01	1:25	800	40 sec
5b	40	0.25	0.01	1:25	800	50 sec

(a) UV-VIS results

The results from UV-VIS of the samples are illustrated in figures 6, 7, 8, 9, and 10. Figure 6 shows the UV-VIS absorbance spectra of the solutions in groups 1a, 1b, 1c, 1d and 1e. These samples are made at high concentration of silver nitrate. The peak intensity increases as the samples are treated in microwave for longer time and it is seen that the position of the peaks are red shifted (from 405 to 410, 411, 416, 420 nm) corresponding with the increasing of MW exposure time (from 4 min to 4 min 10s, 4 min 20s, 4 min 30s). Thus, with the same of concentration of $AgNO_3$, power of MW irradiation, when increasing time of MW exposure, the size of nanoparticles is larger via UV-Vis results.

Figure 7 shows the UV-VIS absorbance spectra of the solutions in groups 2a, 2b, 2c, 2d and 2e. These samples are made at low concentration of silver oxalate. The peak intensity increases as the samples are treated in microwave for longer time and it is seen that the position of the peaks are red shifted (from 404 to 405, 412, 414, 415nm) corresponding with the increasing of MW exposure time (from 4min to 4min 10s, 4min 20s, 4min 30s). However, the peak position changed very slightly.

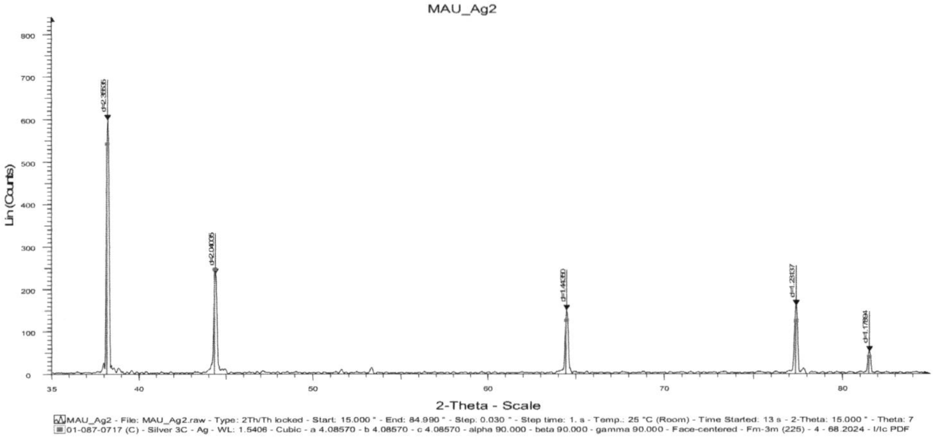

Figure 5. XRD pattern of M2 (200^0C in 1 hour).

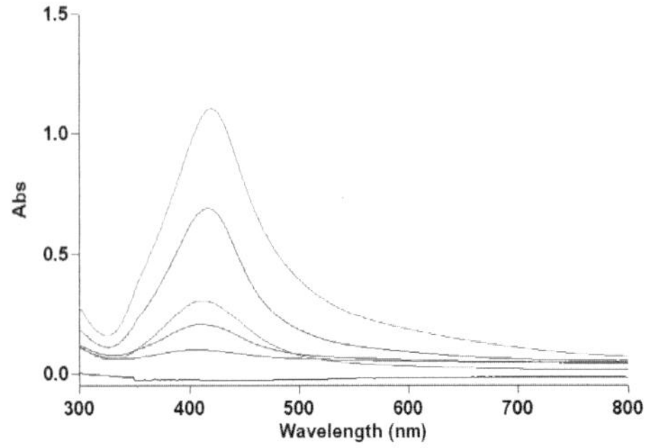

Figure 6. UV-VIS Absorbance spectra of the solutions in groups 1 shows the peak (nm): 1a (405); 1b (410); 1c(411); 1d(416); 1e(420).

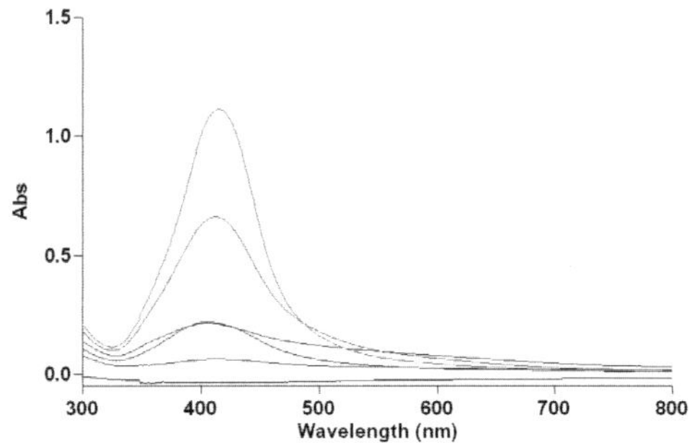

Figure 7. UV-VIS absorbance spectra of the solutions in groups 2 shows the peak (nm): 2a (404); 2b (405); 2c (412); 2d (414); 2e (415).

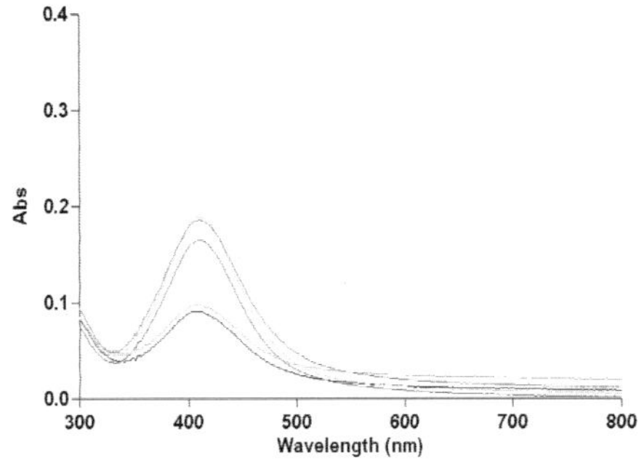

Figure 8. UV-VIS absorbance spectra of the solutions in groups 3 shows the peak (nm): 3a (410); 3b (409); 3c (409); 3d (410); 3e (409).

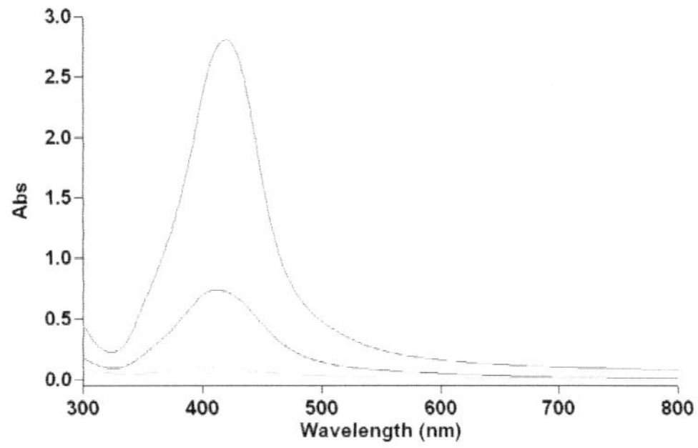

Figure 9. UV-VIS absorbance spectra of the solutions in groups 4 shows the peak (nm): 4a (402); 4b (411); 4c (420).

Journal of Physics: Conference Series **187** (2009) 012078 doi:10.1088/1742-6596/187/1/012078

Figure 8 shows the UV-VIS absorbance spectra of the solutions in groups 3a, 3b, 3c, 3d and 3e. These samples are made at very low concentration of silver oxalate. There is no significant difference on the absorption peak positions at different times of exposure.

Figures 9 and 10 show the UV-VIS absorbance spectra of the solutions in groups 4a, 4b, 5a, 5b. These samples are made at high power of MW. This indicated that the higher the power is, the shorter the time is

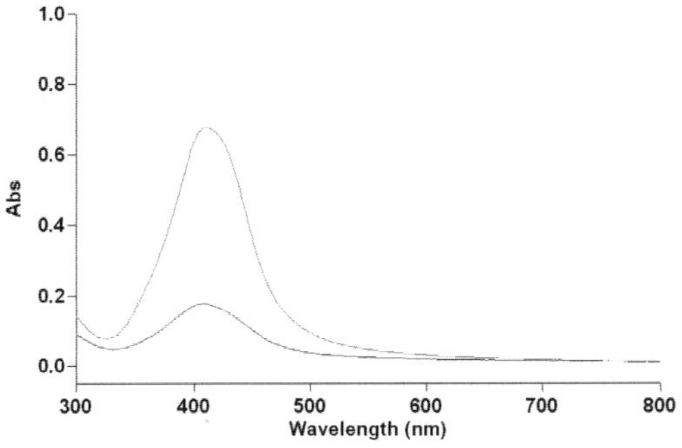

Figure 10. UV-VIS absorbance spectra of the solutions in groups 5 shows the peak (nm): 5a (407); 5b (411).

(b) Characterized by TEM

The TEM image of the 3c sample showed that the average size of silver nanoparticles is about 20nm.

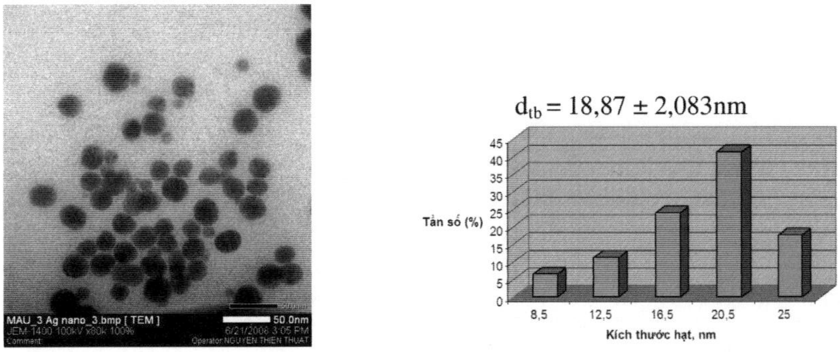

Figure 11. TEM of 3c sample.

(c) Microbiological test

Antibacterial effects of the Ag nanoparticles were studied using Colony Count Test Method. The effect of *E. Coli* (Gram-negative) bacterial strains is determined by the relation

$$\square = (N_1-N_2)/ N_1 * 100\% ,$$

where □ is the percentage of bacterial reduction, N_1 is the number of surviving bacterial colonies from the control sample, and N_2 is the number of surviving colonies from test samples.

All the samples (2c, 2d, 2e, 3c, 3d, 3e) have been diluted 1:12 before testing. The results of *E. Coli* antibacterial property of Ag solutions are shown in figure 12. No bacterium was detected in all the diluted samples (2C, 2D, 2E, 3C, 3D, 3E) after 5 minutes, the antibacteria efficiency is □= 100%.

(d) ICP-AAS results

ICP-AAS results on 3D and 2D solution showed silver nanoparticles content in 3D solution is 17.8mg/l (i.e. concentration of Ag in 3D solution is 17.8 ppm) and in 2D is 12.7mg/l (concentration of Ag in 3D solution is 12.7 ppm). Thus, *E.coli* was killed completely at very low concentration (12.7 ppm). From ICP-AAS results, we calculated the yield of reactions (2D solution: 48.11% and 3D solution: 85.81%).

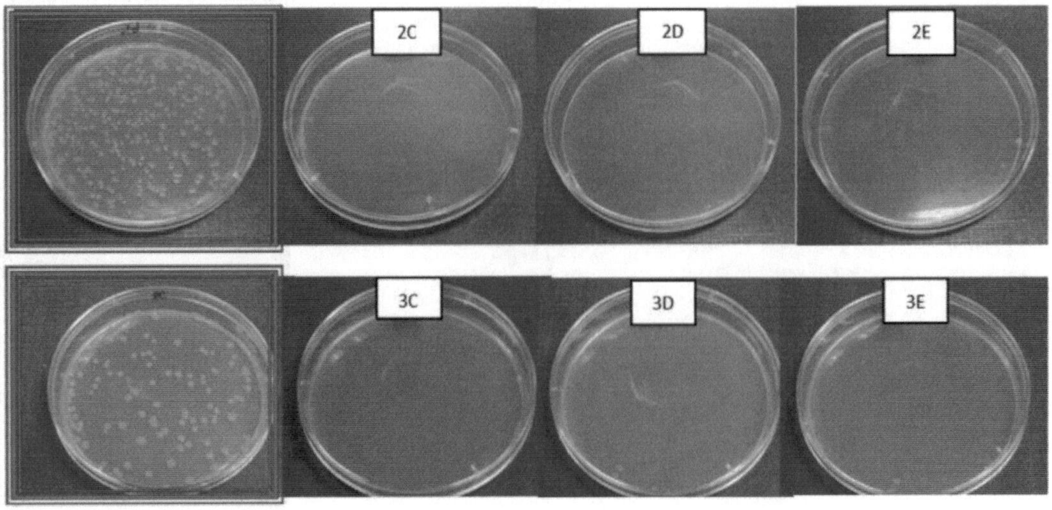

Figure 12. E.coli Antibacterial efficiency of the diluted silver colloidal solutions.

4. Conclusions

- Fabricated successfully silver oxalate with high yield. The average particle sizes are under 2 μm. The results of XRD and FESEM with EDS showed as-fabricated silver oxalate are highly pure.
- The thermal decomposition of silver oxalate at 200°C to receive silver particles. The results of XRD demonstrated these silver particles have high purity with the average size is about 70 nm.
- Silver colloidial solutions have been synthesized rapidly in chemical green condition. The UV-VIS spectra showed that these as-synthesis samples have absorbance peak from 402 to 420 nm. The average size of silver nanoparticles is about 20 nm via TEM images. The yields of reactions are from 48.11% to 85.81%. These colloidal solutions have highly efficient anti-*E. Coli* property at very low concentration (12.7 ppm). Our further research is finding suitable conditions to increase the yield of reaction and to decrease the size of silver nanoparticles.

References

[1] Navaladian S, Viswanathan B, Varadarajan T K, Viswanath R P 2008 *Nanotechnology* **19** 045603
[2] Navaladian S, Viswanathan B, Varadarajan T K and Viswanath R P 2007 *Nanoscale Research Letters* **2** 44
[3] Kvítek L, Prucek R 2005 *J. Materials Science* (10pp.)
[4] Boris V L'vov 2000 *Thermochimica Act.* **364** 99
[5] Popa M, Pradell T, Crespo D 2007 *Colloids and Surfaces, A: Physicochem. Eng. Aspects* **303** 184
[6] Pal S, Tak Y K, and Song J M 2007 *Applied And Environmental Microbiology* **73** 1712
[7] Shrivastava S, Tanmay B, Arnab R 2007 *Nanotechnology* **18** 225103
[8] Baker C, Pradhan A, Pakstis L, Pochan D J, and Shah S I 2005 *J. Nanoscience and Nanotechnology* **5** 244
[9] Sondi I and Sondi S 2004 *J. Colloid and Interface Science* **275** 177

Acknowledgments

This research was supported by the Laboratory for Nanotechnology (LNT) - Vietnam National University-HCM City (VNU-HCMC).

Fabrication of antibacterial water filter by coating silver nanoparticles on flexible polyurethane foams

Nguyen Thi Phuong Phong[1], Ngo Vo Ke Thanh[1] and Phan Hue Phuong[1]

[1] Laboratory for Nanotechnology, Vietnam National University-Ho Chi Minh City Community 6, Linh Trung Ward, Thu Duc District, Ho Chi Minh City-Vietnam

E-mail: ntpphong@vnuhcm.edu.vn

Abstract: In this paper, we fabricated silver-coated polyurethane foams and used it as a bacterial filter for contaminated drinking water. Flexible PU foams were soaked in silver colloidal solutions for 10 h, then washed and air-dried at room temperature. The prepared silver colloidal solutions and silver-coated PU materials were characterized by several techniques including TEM, FESEM/EDS, UV-VIS, ICP-AAS, and Raman spectroscopy. The TEM images showed that the size of silver nanoparticles in colloidal solutions varies from 6 to 12nm. The Raman, FE-SEM/EDS and ICP-AAS data illustrated that silver nanoparticles were stable on the PU foam and were not washed away by water. Furthermore, the microbiological tests (tube tests and flow test) were carried out on silver-coated PU materials with the Coliforms, *E. coli*, and *B. subtilis*. The obtained results showed that the bacteria was killed completely with antibacterial efficiency of 100% being observed. Our research suggests that silver-coated polyurethane foams can be used as excellent antibacterial water filters and would have several applications in other sectors.

Keywords: Silver nanoparticles, polyurethane, *E. coli*, water filter

1. Introduction

Water is the common breeding ground for many pathogens because it contains several bacteria, viruses, etc. The removal and inactivation of pathogenic microorganisms are the last step in the treatment of drinking water [1]. Currently available detection methods do not allow for the routine analysis of all microorganisms that could be present in inadequately treated drinking water. Instead, microbiology quality is determined by testing drinking water for total Coliforms and *Escherichia coli*, a bacterium that is always present in the intestines of humans and other animals. The presence of *E. coli* in drinking water would indicate faecal contamination of the water [1]. According to WHO, the maximum acceptable concentration of *E. coli* in drinking water is none detectable per 100mL [1, 2]. There are many treatment methods such as chemical (chlorine, iodine, etc), physical (ultraviolet light, ozone, radiation) or polymer films or synthetic and natural zeolites, etc [1-3]. During the past few years, advances in nanoscale science and engineering suggest that many of the current problems involving water quality could be resolved or greatly ameliorated using nanosorbents, nanocatalysis, etc and nanoparticles enhanced filtration resulting from the development of nanotechnology [4]. Flexible polyurethane (PU) foams have been extensively used in various applications because of their excellent biocompatibility and mechanical properties [6-8]. In 2004, Prashant Jain *et al.* [2] synthesized the silver colloidal solution by reduction $AgNO_3$ using $NaBH_4$, which is expensive and a toxic chemical,

© 2009 IOP Publishing Ltd

Journal of Physics: Conference Series **187** (2009) 012079 doi:10.1088/1742-6596/187/1/012079

and their microbiological test (flow test) on Ag-coated polyurethane were carried out with false water (purified water loaded *E. coli*). In this paper, we synthesized Ag colloids by reduction of silver nitrate using polyvinyl pyrolidone as the capping agent. Microwave irradiation has been used as it offers several advantages over conventional heating. The main advantages of the method include (i) rapid initial heating process, (ii) uniform heat transfer to the solution, and (iii) significant reaction rate enhancement.

2. Experimental

2.1. Synthesis of silver colloidal solutions

Polyvinyl pyrrolidone (PVP, $Mw \approx 1{,}000{,}000$, China, 99%) and ethylene glycol (Merk, 99.9%) was poured into a 100mL pyrex becher. After the complete dissolution of PVP, silver nitrate $AgNO_3$ was added into the solution and stirred until the $AgNO_3$ was dissolved completely. The solutions were heated in a microwave oven (800W) for various concentrations of $AgNO_3$. Solutions at various $AgNO_3$ concentrations were heated in a domestic microwave oven at 800W.

2.2. Fabrication of silver-coated polyurethane foams

Polyurethane (PU) foams 50x50x6mm was washed by deioned water, dried and then soaked in silver colloidal solutions within 10 h. The sheets were washed repeatedly with water to remove any adsorbed chemical like PVP or ethylenglycol.

2.3. Characterization

The synthesized silver colloidal solutions and silver-coated polyurethane were characterized by analytical techniques such as UV-VIS Spectroscopy (Cary 100 Conc, Varian), TEM (JEM-1400), FESEM with EDS analyzer (JSM 7401F); Raman Scattering Spectroscopy (Horiba Jobin Yvon, Excition laser of 632.81 nm).

2.4. Microbiological tests

Tube Test: *E. coli* (Gram-negative bacteria) and *B. subtilis* (Gram-positive bacteria) were supplied by Faculty of Biology, University of Natural Sciences, VNU-HCMC. Antibacterial effects of the Ag colloidal solutions and Ag-coated polyurethane were studied by Colony Count Test Method [1]. The effect of bacterial strains was calculated, following the relation:

$$\eta = \frac{N1 - N2}{N1} \times 100\%$$

where η is the percentage of bacterial reduction, N_1 is the number of surviving bacterial colonies from the control sample, and N_2 is the number of surviving colonies from test samples.

Flow test: Water (from a well at Thanh Long Company, Binh Chanh Distrist, Ho Chi Minh City), flow rate of 50mL/min. Total Coliforms and *E. coli* were determinate by Most Probable Number (MPN) Method [1]:

Total Coliforms

* For each sample to be tested add the following amounts:

5x10ml to10ml double strength lauryl sulphate broth (LSB)

5x1ml to 5 bottles single strength LSB

5x0.1ml to 5 bottles single strength LSB

* Ensure Durham tubes were completely filled. Incubate at 37°C.

* The tube that produces gas in LSB media was selected and was grown it into Brillant Green Lastose Bile Salt (BGBL) media. Incubate at 37°C

* Examine for: Growth – turbidity; Acid production - a yellow coloration; Gas - trapped in Durham tube. Tubes that produce acid and gas were positive scored.

* Derive the McCarty index and calculate the MPN/100mL. *E. coli*

* The tube that produced gas in LSB media was selected and was grown into EC media. Incubate at

45^0C in 24 h.
* The tube that produce gas in EC media was selected and was grown into EMB media agar Petri dish.
* Purple colonies with gas → presumptive *E. coli* →IMViC test (Indol, Methyl Red, Vogues-Prokaure, Citrate) →If Indol (+), Methyl red (+), Vogues- Prokaure (-), Citrate (-) →*E. coli* (+)
* Derive the McCarty index and calculate the MPN/100mL.

3. Results and discussion

3.1. Synthesis and characterization of Ag colloidal solutions

Four Ag colloidal solutions were synthesized by microwave irradiation (table 1). The results of UV-VIS mearsurements in table 1 showed that silver nanoparticles were presented in solutions. The size and the shape of nanoparticles were illustrated clearly via TEM images (figure 1, figure 2, figure 3 and figure 4). It was observed that the nanoparticles had spherical shape and their major size were essential from 7 – 11 nm (P1); 7 – 14 nm (P2); 6 – 11 nm (P3); 6 – 12 nm (P4). These shapes and sizes are suitable for killing bacteria, virus, and fungus. Indeed, these colloidal solutions exhibit highly efficient antibacterial properties (antibacterial efficiency of 100%) at very low concentration (table 2 and 3). The differences between gram-positive (*B. subtilis*) and gram-negative (*E. coli*) anti-bacteria efficiency were shown clearly (at concentration of 50ppm, antibacterial efficiency of gram-positive is 99.79% and of gram-negative is 100%). These differences could be explained based on the structure of their cell walls. The gram-negative bacteria have a layer of lipopolysaccharide at the exterior, followed underneath by a thin (about 7–8 nm) layer of peptidoglycan. Although the lipopolysaccharides are composed of covalently linked lipids and polysaccharides, they lack strength and rigidity. Negative charges on the lipopolysaccharides are attracted towards weak positive charges available on silver nanoparticles. On the other hand, the cell wall in gram-positive bacteria is principally composed of a thick layer (about 20–80 nm) of peptidoglycan, consisting of linear polysaccharide chains cross-linked by short peptides to form a three dimensional rigid structure. The rigidity and extended cross-linking not only endow the cell walls with fewer anchoring sites for the silver nanoparticles but also make them difficult to penetrate [9].

Table 1. The UV-VIS absorbance peak of nano Ag colloidal solutions.

Samples	Ethylene glycol (ml)	PVP (g)	$AgNO_3$ (g)	Absorbance peak of UV-VIS (nm)
P0	50	0.0752	0	
P1	50	0.0752	0.015	406
P2	50	0.0752	0.051	407
P3	50	0.0752	0.071	408
P4	50	0.0752	0.121	410

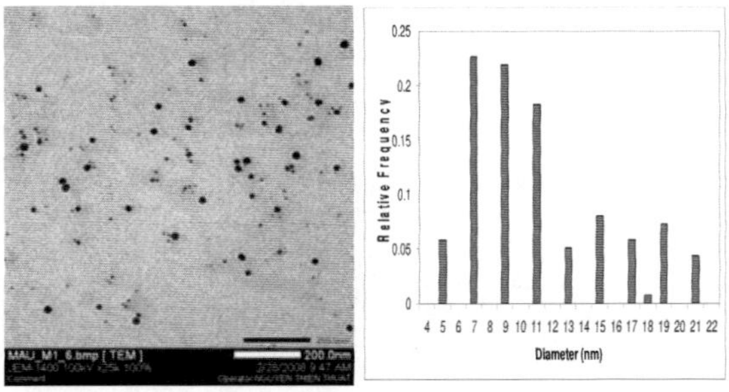

Figure 1. TEM image (magnification of 200 nm) and size distribution of Ag nanoparticles in P1 solution.

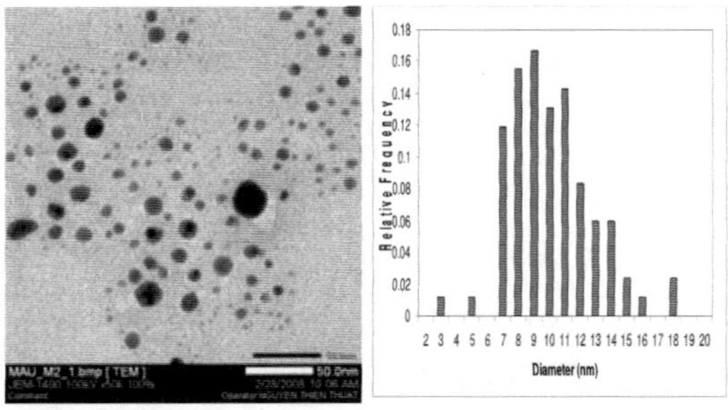

Figure 2. TEM image (magnification of 50 nm) and size distribution of Ag nanoparticles in P2 solution

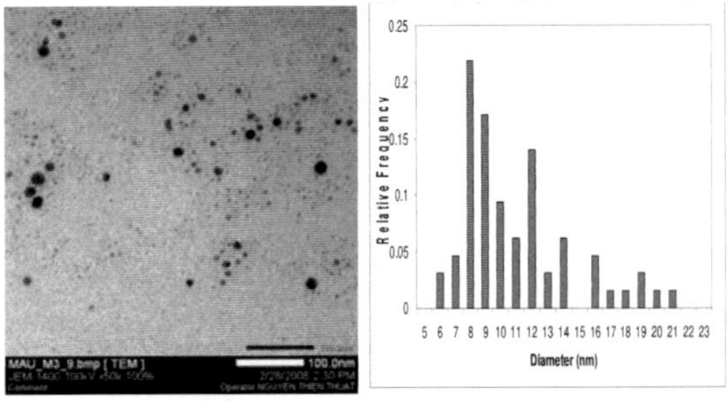

Figure 3. TEM image (magnification of 100 nm) and size distribution of Ag nanoparticles in P3 solution.

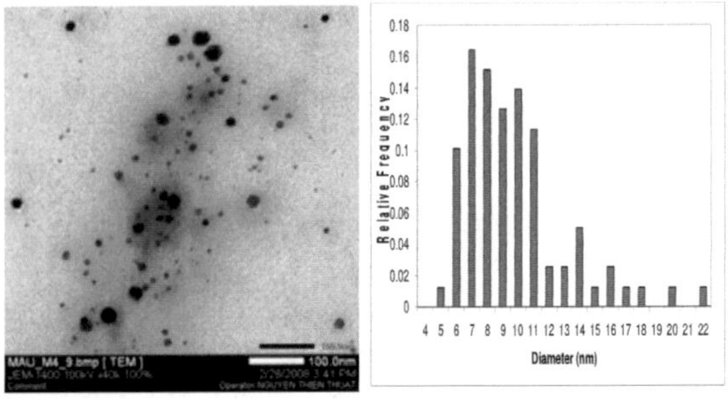

Figure 4. TEM image (magnification of 100 nm) and size distribution of Ag nanoparticles in P4 solution.

Table 2. Colony density and Gram-negative (*E. coli*) antibacterial efficiency of nano Ag colloidal solutions.

Samples	M_i (CFU/ml)			η (%)		
	15 min.	5 h	24 h	15 min.	5 h	24 h
Control sample	$2.25.10^5$	$2.29.10^7$	$2.31.10^9$			
10ppm	$2.90.10^4$	$6.20.10^2$	0	87.11	99.99	100
20ppm	$4.65.10^3$	0	0	97.93	100	100
50ppm	0	0	0	100	100	100

3.2. Fabrication and characterization of Ag-coated polyurethane

Four Ag-coated polyurethane foams were fabricated by soaking in four different Ag colloidal solutions. The results of FESEM/EDS and ICP-AAS proved clearly the presence of Ag nanoparticles in polyurethane foam sheets (figure 5 and table 4). Moreover, Raman spectroscopy of polyurethane (M_0) and Ag-coated polyurethane (M_1, M_2, M_3) showed that linkages of Ag-coated polyurethane samples changed very strongly in comparison with uncoated sample (figure 6, figure 7, figure 8, figure 9). The Raman spectroscopy of uncoated polyurethane (M_0) matched that of polyurethane in the research works of J. W. Cho [5], and C.C Wei [6, 7]. The Raman peaks of M_0 are 1450, 2750, 2900 cm^{-1} (O=C=N-); 1100, 1650 cm^{-1} (C=O); 1185 cm^{-1} (N-H). In Raman spectroscopy of M_1, M_2, M_3, these Raman peaks disappear. This could explain the interaction between the nitrogen (of the N-H bond), oxygen (of the C=O or N=C=O bond) and silver nanoparticles.

Table 3. Colony density and Gram-positive (*B. subtilis*) antibacterial efficiency of nano Ag colloidal solutions.

Samples	M_i (CFU/ml)			η (%)		
	15 min.	5 h	24 h	15 min.	5 h	24 h
Control samle	$1.95.10^5$	$2.41.10^7$	$1.42.10^9$			
N_1 (10 ppm)	$1.96.10^4$	$1.29.10^2$	0	89.95	99.99	100
N_2 (20 ppm)	$9.0.10^3$	$7.7.10^1$	0	95.85	99.99	100
N_3 (50 ppm)	$4{,}45 * 10^2$	0	0	99.79	100	100

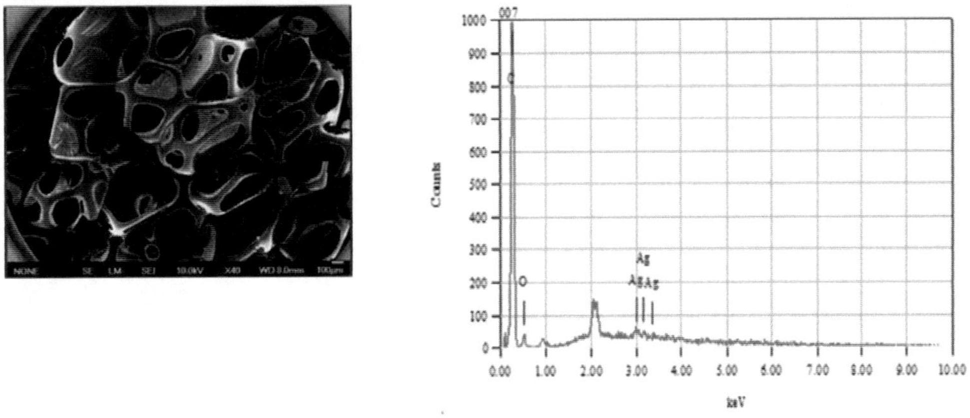

Figure 5. FESEM image and EDS analysis of Ag-coated Polyurethane foam C%:92.32; O%:1.39; Ag%:6.29.

Table 4. The ICP-AAS results of Ag nanoparticles content of polyurethane sheets.

Samples	Ag nanoparticles content (g) of polyurethane sheets (kg)
M1	0.25
M2	0.72
M3	0.96
M4	2.80

3.3. Microbiological test on Ag-coated polyurethane

3.3.1. Tube test

The results of microbiological test in tube test conditions were presented in table 5. The *E. coli* was killed after 15 min. with the antibacterial efficiency of over 95% being observed, and were killed completely after 5 h for all Ag-coated polyurethane sheets. Contrary to Ag-coated polyurethane sheets, it was found that colony density of uncoated polyurethane sample (M_0) increase significantly to soaking time (from $2.45.10^5$ after 15 min. to $1.43.10^6$ after 5 h and to $2.43.10^9$ after 24 h).

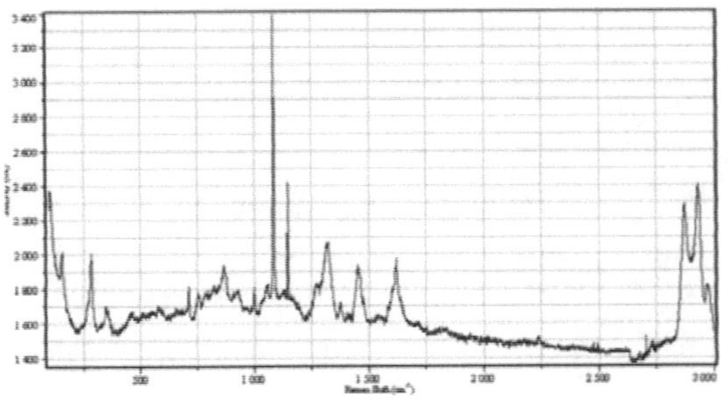

Figure 6. Raman spestrocopy of polyurethane foam.

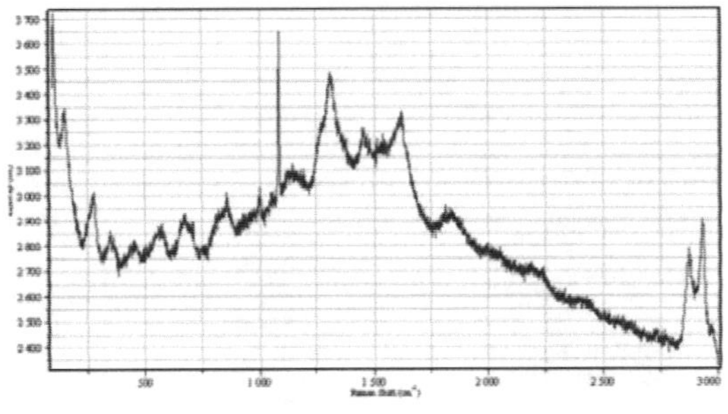

Figure 7. Raman spectroscopy of Ag-coated polyurethane foam (M1).

Journal of Physics: Conference Series **187** (2009) 012079 doi:10.1088/1742-6596/187/1/012079

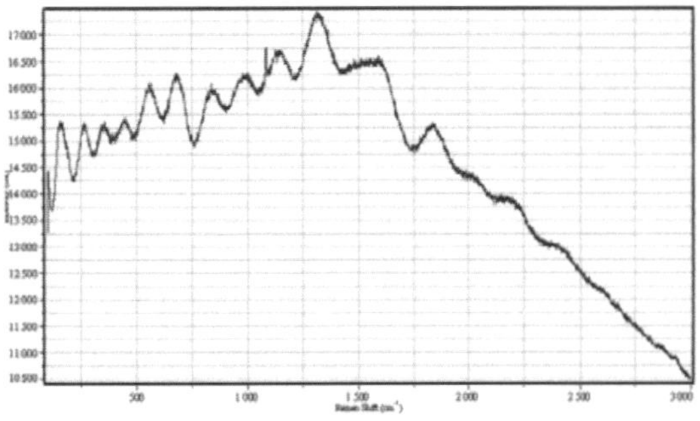

Figure 8. Raman spectroscopy of Ag-coated polyurethane foam (M2).

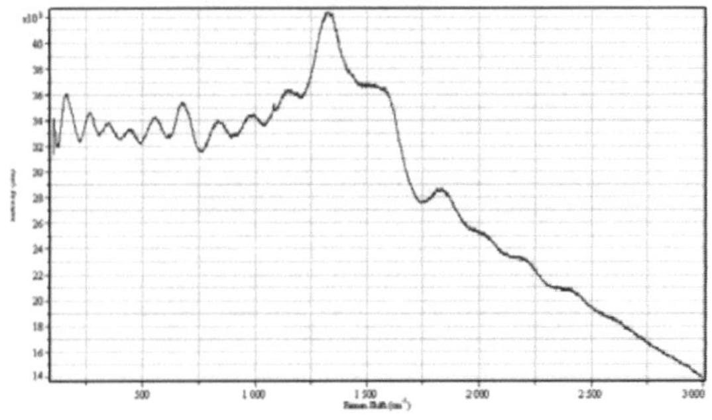

Figure 9. Raman spectroscopy of Ag-coated polyurethane foam (M3).

Table 5. Colony density and Gram-negative (*E. coli*) antibacterial efficiency of Ag-coated polyurethane foam.

Samples	M_i (CFU/ml)			η (%)		
	15 min.	5 h	24 h	15 min.	5 h	24 h
M_0	$2.45.10^5$	$1.43.10^6$	$2.43.10^9$			
M_1	$1.20.10^3$	0	0	95.10	100	100
M_2	$7.65.10^2$	0	0	96.88	100	100
M_3	$5.95.10^2$	0	0	97.57	100	100
M_4	<10	0	0	99.99	100	100

3.3.2. Flow test

The results of the microbiological test on Ag-coated polyurethane in flow test condition confirmed the antibacterial property of these materials. Input water samples, which were received from a well at Binh Chanh District, had total coliforms of 45 MPN/100 ml, *E. coli* density of 30 MPN/100 ml (sample 1) and coliforms density of 110 MNP/100 ml, *E. coli* density of 70 MPN/100 ml (sample 2) (figure 10, figure 11, table 6). After testing two samples, it was found that no bacteria were detected.

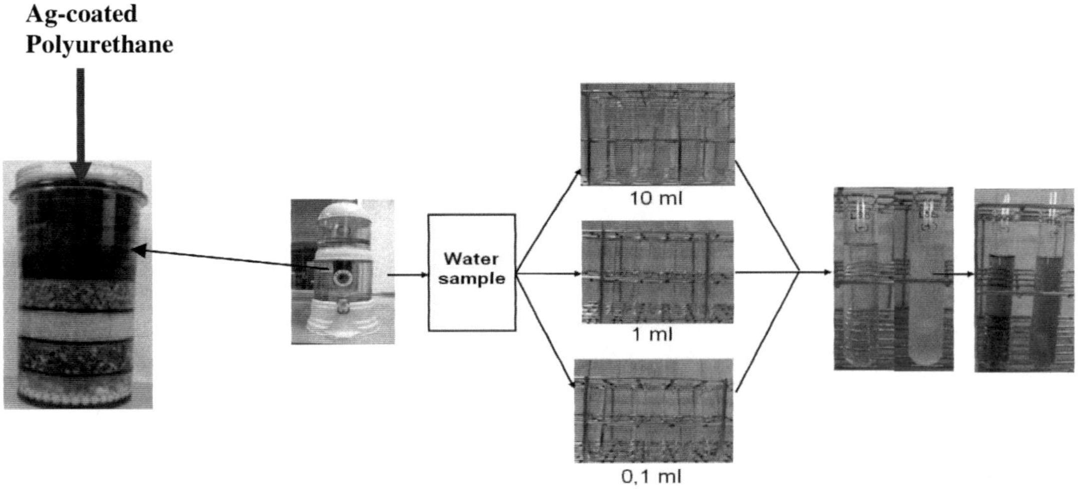

Figure 10. Anti-Coliforms ability of water filter containing Ag-coated polyurethane.

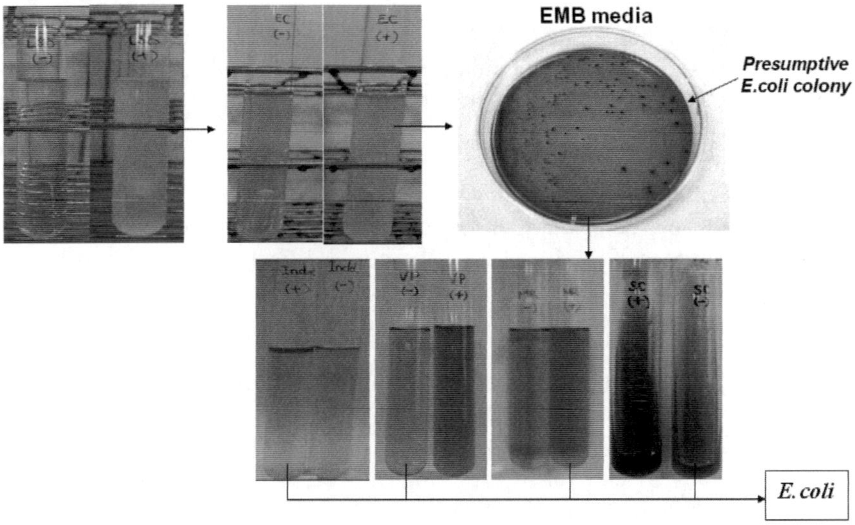

Figure 11. Anti-E. coli ability of water filter containing Ag-coated polyurethane

Table 6. The Coliforms and *E. coli* density results of water treated with Ag-coated and uncoated polyurethane water filter.

Samples	Coliforms Density (MPN/100 ml)		*E. coli* Density (MPN/100 ml)	
	Uncoated PU	Ag-coated PU	Uncoated PU	Ag-coated PU
Number 1	45	0	30	0
Number 2	110	0	70	0

Figure 12. Bacteria contaminated drinking water treatment system by Ag-coated polyurethane foams.

4. Conclusions

Silver colloidal solutions were synthesized rapidly by using microwave irradiation. The UV-VIS measurements showed that these as-synthesized samples have absorbance peak at 406-410 nm. The major size of silver nanoparticles was about 6-12 nm via TEM images. These colloidal solutions had highly efficient gram-negative (*E. coli)* and gram-positive (*B. subtilis*) antibacterial property at very low concentration. The antibacterial efficiency increased when increasing the concentration of silver solutions and the exposure time.

The Ag-coated polyurethane foams were fabricated successfully. The results of Raman, FE-SEM/EDS and ICP-AAS demonstrated that silver nanoparticles strongly immobilized to the polyurethane foams sheet by the chemical linkage. The results of the microbiological test on Ag-coated polyurethane in tube test conditions showed that these materials exhibited highly efficient gram-negative antibacterial property.

The results of the microbiological test on Ag-coated polyurethane in flow test condition confirmed the antibacterial property of these materials. Using input water received from a well at Binh Chanh, it was found that no bacteria were detected. Our research suggests that silver-coated polyurethane foams can be used as excellent antibacterial water filters and could offer several applications in other sectors with simple treatment systems (figure 12).

References

[1] Greenberg A E, Clesceri L S, Andrew D E 1992 *Standard Methods 18th Edition, for examination of water and wastewater*
[2] Jain P, Pradeep T 2005 *Biotechnology and bioengineering* **90** No.1
[3] Chen Y, Wang L, Jiang S 2003 *J. Polymer Materials* **20** 279
[4] Savage N, Diallo M S 2005 *J. Nanoparticle Resarch* **7** 331
[5] Cho J W, Jung H S 2003 *Materials Letter* **60** 2653
[6] Chih-Wei C, Shan-hui et al H 2006 Journal of Antimicrobial Agents 23S1, S75-S78
[7] Shan-Hui H, Chei-Wei C 2006 *Polymer Degr ation and Stability* 675
[8] Siddhartha S, Tanmay B 2007 *Nanotechnology* **18** 225103
[9] Shane P, Min K, Cakmak M 2003 *Polymer* **44** 5137

Acknowledgments

The authors would like to thank the Faculty of Biology, University of Natural Sciences (VNU-HCMC) for providing the bacteria and technical assistance; Institute of Chemical Technology for conducting the flow test. Especially, Department of Science and Technology (VNU-HCMC) is acknowledged for financial support.

A simple method for estimating of spontaneous polarization in perovskite-like nanocrystallites

Pham The Tan[1], Nguyen Kien Cuong[2], Fedor Valach[3] and Hoang Nam Nhat[1]

[1] Faculty of Physics, College of Sciences, Vietnam National University, 334 Nguyen Trai, Thanh Xuan, Hanoi, Vietnam
[2] Institute of Materials Science, Vietnam Academy of Science and Technology, 18 Hoang Quoc Viet Road, Cau Giay District, Hanoi, Vietnam
[3] Department of Chemical Physics, Faculty of Chemical and Food Technology, Slovak Technical University, Radlinského 9, 812 37 Bratislava 1, Slovak Republic

E-mail: namnhat@gmail.com

Abstract. The bond-valence sum approach to spontaneous valence polarization in perovskite-like nanocrystallites ABO_3 is presented. As the bond-valence method allows estimating the amount of valence charge distributed within the given bond according to bond length, the microscopic polarization in the ABO_3 structure may be directly derived from the valence deviation in two conjunctive bonding spheres $A-O_{12}$ and $B-O_6$. The concept was demonstrated on examples and a comparable agreement with data taken either from experiments or first principle calculation was seen. The bond-valence polarization appeared as one of the possible sources contributing to the final macroscopic polarization of the bulk materials.

Keywords: Perovskite, polarization, bond-valence, nanocrystallite.

1. Introduction

In the classical view [1, 2] the polarization of crystal is defined as the dipole moment per unit cell volume:

$$P = \frac{p}{V} = \sum_i \frac{q_i r_i}{V}, \tag{1}$$

where q_i stands for charge and r_i position vector of the atom i. The polarization P may also be decomposed into the product of atomic dipole moment p_i and concentration N_i. The atomic dipole moment is again the product of atomic polarizability α_i and local electrical field E_i:

$$P = \sum N_i p_i = \sum N_i \alpha_i E_i . \tag{2}$$

In this model the displacement of atoms from cubic positions may result in the separation of centers of positive and negative charges and accounts for the spontaneous polarization in absence of outer field. As an empirical rule, if the positive charge ions (e.g. Ba^{2+} and Ti^{3+} in $BaTiO_3$) were moved away from the negative O^{2-} ions by δ Å then the dipole of a cell would be $p \sim 6e\delta$. Therefore, a displacement of 0.1Å would imply a local polarization $p = 1.2\times10^{-18}$ Cm.

© 2009 IOP Publishing Ltd

In contrast to this theory, recent developments consider the macroscopic polarization of crystal as a totally independent observable from the unit cell polarization. In quantum-mechanical approach, the center of negative charge is evaluated according to the value of position operator for the Berry's phase of the Bloch wave function of the valence-band many-electron system [3,4,5]. So the polarization is described in terms of wave functions instead of real charge. Taken from [10] the polarization of a single electron is defined as:

$$P_e = \lim_{L\to\infty} e\langle X\rangle / L \,, \tag{3}$$

where e is the electron charge, L the segment length along which the polarization is defined and $<X>$ the value of position operator for single-electron system. This position operator is defined as trivial

$$\hat{X} = \sum x_i \,,$$

with the ground-state expectation value

$$\langle X\rangle = \langle \Phi_0 | \hat{X} | \Phi_0\rangle = \int dx\, x n(x) \,.$$

In many-electron system the position operator takes form

$$\hat{\aleph} = \exp\left(i2\pi\hat{X} / L\right) \,,$$

with ground-state value defined as

$$< \aleph > = \frac{L}{2\pi} \operatorname{Im} Log < \Psi_0 | \exp\left(i\frac{2\pi}{L}\hat{X} \right) | \Psi_0 > \,.$$

With this position operator the many-electron electronic polarization reads:

$$P = \lim_{L\to\infty} \frac{e}{2\pi} \operatorname{Im} Log < \Psi_0 | e^{i\frac{2\pi}{L}\hat{X}} | \Psi_0 > \,. \tag{4}$$

The macroscopic polarization based on this electronic contribution will have unit of [*dipole per unit length*]. This short outline of the modern development was followed mainly Resta [3, 4, 10, 11] whereas the approaches from the other authors, e.g. King-Smith & Vanderbilt [5], Selloni [6] and Ortiz & Martin [9] were different. They all, however, originate in a common concept of the Berry's phase of the Bloch wave function [7-8]. To unify both classical and modern approaches we re-formulate the macroscopic polarization as follows:

$$P = \hat{P} P_{el} \,, \tag{5}$$

where $\hat{P}$ is some operator acting on elementary contribution P_{el} and producing the final macroscopic polarization P. Depending on either the classical or modern approach, this operator will take different forms and meanings. For classical approach $\hat{P}$ is identity operator $\hat{I}$ and P_{el} is the unit cell dipole moment

$$P = p/V = \sum q_i r_i / V \,.$$

For a modern, "*à la Resta*" approach, $\hat{P}$ takes the form

$$P_e = \lim_{L\to\infty} e\langle X\rangle / L \,.$$

In this paper we explain the alternative approach to the elementary polarization P_{el} and show its direct association with the structure. This approach has its origin in the statistical evaluation of valence distribution in metal-to-oxygen bonding spheres [13-16]. It was empirically so accurate that it was frequently used in the structural science to predict the positions of atoms. The resulting formulae, which will be clear in the next sections, can be given as follows:

$$P_{el} = \left(\Delta v_A - \Delta v_B\right) r \tag{6}$$

$$P = \sum_{i=1}^{N} p_i \cdot \left(P_{el}\right)_i \tag{7}$$

where Δv_A and Δv_B stand for the valence excess/deficit in the coordination spheres of positive ions A and B respectively, r the position vector to the center of A...B and p_i the probability of elementary contribution P_{el}. The next section describes how can P_{el} be obtained from the unit cell parameters.

2. Nominal length and nominal valence

The bond-valence is originally defined as the atomic valence per bond [13, 14], i.e. it is a certain portion of valence charge distributed within a given bond. Evidently, the bond-valence distribution differs in each bond type.

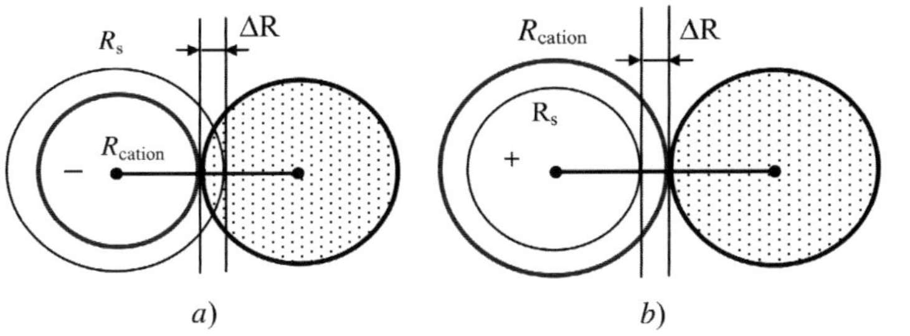

Figure 1. Nominal length R_s shows the radius where the valence of cation is exactly equal its nominal value (oxidation state). Comparison of R_s and ionic radii: cation with excessive negative charge (*a*); cation with excessive positive charge (*b*).

Table 1. Nominal length R_s (Å) for some cations frequently encountered in perovskites.

A-position cations			B-position cations		
Cation	$A\text{–}O_{12}$	$A\text{–}O_6$	Cation	$B\text{–}O_{12}$	$B\text{–}O_6$
Ba^{2+}	2.948	2.691	Cu^{+}	2.519	2.263
Sr^{2+}	2.781	2.524	Cu^{3+}	2.252	1.995
Pb^{2+}	2.775	2.518	Mn^{2+}	2.453	2.196
Ag^{+}	2.761	2.505	Mn^{3+}	2.273	2.016
Na^{+}	2.722	2.466	Mn^{4+}	2.159	1.903
La^{3+}	2.685	2.428	Co^{3+}	2.183	1.926
Pr^{3+}	2.651	2.394	Co^{4+}	2.046	1.790
Hg^{2+}	2.635	2.378	Pb^{4+}	2.448	2.192
Ca^{2+}	2.630	2.373	Ti^{4+}	2.221	1.965
Nd^{3+}	2.618	2.361	Fe^{2+}	2.397	2.140
K^{+}	3.051	2.795	Cr^{6+}	2.050	1.794
Er^{3+}	2.501	2.244	Fe^{3+}	2.272	2.015
Eu^{3+}	2.587	2.330	Fe^{4+}	2.186	1.930

Various studies [15, 16] showed that the bond-valence depends on bond length according to the simple relation

$$v_i = e^{(R_0 - R_i)/B} \quad . \tag{8}$$

This relation holds well for the ionic compounds with erroneous level less than 5% (the R_0 was experimentally determined and B was preset to 0.37). Mathematically read, v_i is the Laplace transformation of R_0/B. In perovskites the only bond types of interest is the metal-to-oxygen bond in two bonding spheres $A\text{–}O_{12}$ and $B\text{–}O_6$. The total atomic valence of metal in each sphere is the sum over all bond-valences in the sphere. The detailed vector calculation for the perovskite-like structure

ABO_3 can be found in [17]. One aspect that comes straightforward from the relation (8) is that in symmetric coordination sphere of n bonds, the average bond-valence v_s is exactly equal to the oxidation state v dividing by n. This average bond-valence v_s may be used to classify the coordination sphere since it refers to a *nominal value* of bond-length R_s for which the atomic valence is equal to the oxidation state. This R_s sets a balance level, for which when the bond contracts or stretches the coordination sphere becomes overcharged or discharged with respect to the cation oxidation state.

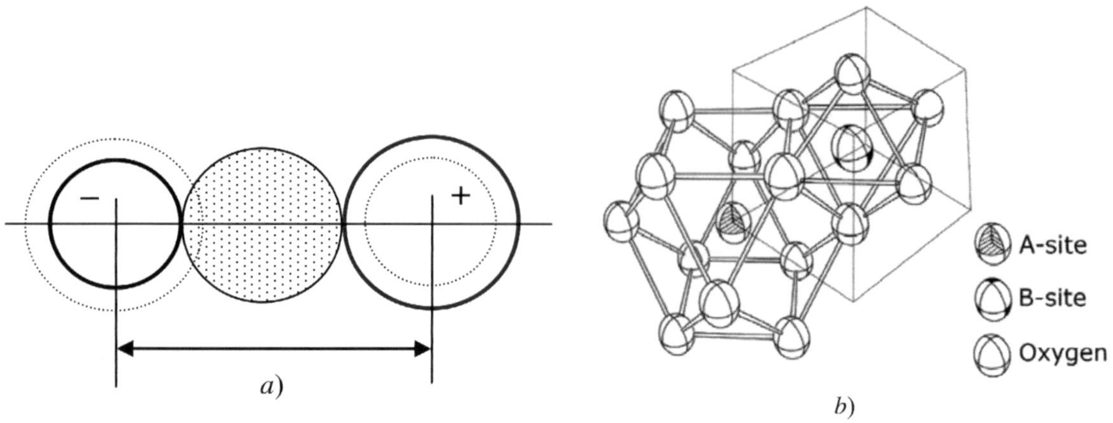

Figure 2. *a*) The typical formulation of dipole in [A–O–B] chain; *b*) Coordination spheres of cations in the perovskite-type arrangement.

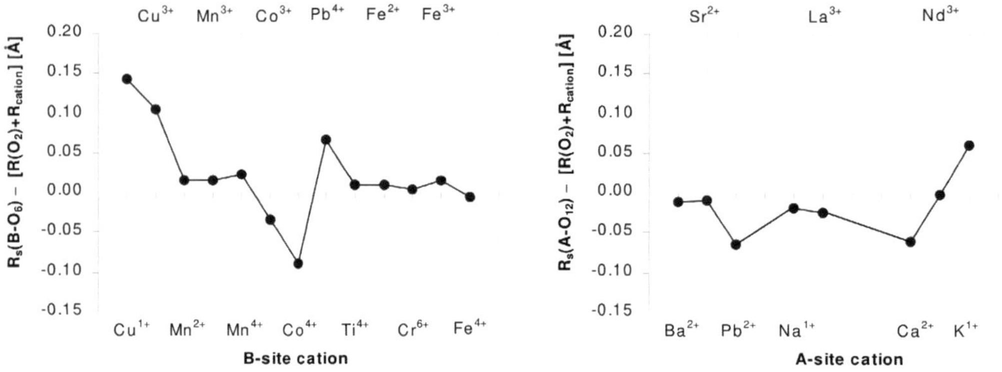

Figure 3. Values of R_s– [R_{cation}+$R(O^{2-})$] for B-site and A-site cations.

In perovskites the calculated atomic valence for the cations usually differs from its oxidation state. For example, in $LaCoO_3$ the total valence over La^{3+}–O_{12} was found a little below +3 whereas the same over Co^{3+}–O_6 was above. This situation can be explained using figure1. In rigid bonding to oxygen, the metal ionic radius allows the closest contact but since the ionic radius is not equal to R_s for the same atom, the metal will have different amounts of bonding electrons in comparison with its nominal oxidation number.

Table 1 lists some R_s for the cations frequently encountered in perovskites. These values were calculated using the empirical valence-to-length values given in [15]. It shows that the right-side cations (e.g. Cu^{3+}, Ti^{4+}, Fe^{2+}, Cr^{6+}, Mn^{3+}) can hardly fall into the A-position (forcing B–O_{12} coordination) since the average radius of the B–O_{12} coordination spheres for these cations is around

2.7Å which is too larger than their listed nominal lengths R_s. So these cations will always be discharged if closely packed. Similarly, the left-side cations (e.g. Eu^{3+} , K^+, Ca^{2+}, Hg^{2+}, Nd^{3+}) can also hardly fall into the B-position (forcing A–O_6 coordination) because of a much smaller average radius of the A–O_6 coordination sphere (~1.9Å) in comparison with the nominal lengths. So these cations will be heavily overcharged if closely packed. In case such unlikely occupation happens, the structure is predicatively distorted with high local polarization being foreseen.

Table 2. The yield of 0.1Å of R_s–[R_{cation}+$R(O^{2-})$] in [b.v.] unit.

Cation	Bonding domain	Bond-valence [b.v]
Cu^+	B–O_6	0.17
Cu^{3+}	B–O_6	0.34
Co^{4+}	B–O_6	−1.06
Pb^{4+}	B–O_6	1.10
Pb^{2+}	A–O_{12}	−0.46
Ca^{2+}	A–O_{12}	−0.49
K^+	A–O_{12}	0.24

3. Polarization of coordination domains

The incompatibility between ionic radii and nominal length R_s seems to have a strong impact on the microscopic polarization in perovskites. Figure 2 shows the possible formulation of dipole in the chain of two cation coordination spheres [A–O_{12}]...[B-O_6].

For the cations listed in table 1, the figure 3 and table 2 demonstrate how possible is this polarization effect and its expected value. For the listed data, the $R(O^{2-})$ was preset to 1.21Å. In most of B-site cations, R_s(A–O_{12}) > [R_{cation}+$R(O^{2-})$] so the difference between the two, $\Delta R > 0$, except for Co^{3+} and Co^{4+}. Therefore, Co^{3+} and Co^{4+} can never reach their nominal oxidation states in perovskite-packed structures. The matching of oxidation state is adequate for Mn^{2+}, Mn^{3+}, Mn^{4+}, Fe^{2+}, Fe^{3+}, Fe^{3+}, Ti^{4+}, Cr^{6+} (ΔR < 0.05Å), but Cu^+, Cu^{3+}, Pb^{4+} (ΔR > 0.05Å) show the excess in valence.

The A-site cations, on the other hand, show the valence deficit for most of cases except for K^+. The positive charge may be excessive as for Pb^{2+}, Ca^{2+} (ΔR < −0.05Å), or adequate as for Ba^{2+}, Sr^{2+}, Na^+, La^{3+} ($\Delta R \leq$ −0.025Å), or equilibrated as for Nd^{3+} (ΔR = 0Å). Generally speaking, the cations Mn^{2+}, Mn^{3+}, Mn^{4+}, Fe^{2+}, Fe^{3+}, Fe^{3+}, Ti^{4+}, Cr^{6+} (B–O_6) and Ba^{2+}, Sr^{2+}, Na^+, La^{3+}, Nd^{3+} (A–O_{12}) are the neutral cations with respect to the matching of their oxidation state, whereas the Cu^+, Cu^{3+}, Pb^{4+}, Co^{4+} (B–O_6) and Pb^{2+}, Ca^{2+}, K^+ (A–O_{12}) can be thought as the highly polarized cations.

According to this analysis the compounds like $Pb^{2+}Co^{4+}O_3$ and $Ca^{2+}Co^{4+}O_3$, which contain strongly polarized cations, are weakly polarized. This is because of the cancelation of the polarization effect by both cations. Similarly, the combination like $K^+Pb^{4+}O_3$ can bring weak polarization effect. A weak effect is also expected for the compounds containing only neutral cations such as (Ba^{2+}, Sr^{2+}, Nd^{3+})(Ti^{4+}, Fe^{2+}, Fe^{3+}, Cr^{6+})O_3. For example, the polarization in $BaTiO_3$ is not large, the calculation for closely-packed structure showed the dipole near p = 0.18e×3.3Å = 0.59eÅ = 1.0×10^{-29} Cm. The compounds that could lead to a strong polarization should contain the Pb^{2+}, Pb^{4+}, Co^{4+} or Ca^{2+} cations. This claim agrees well with experiments. For the numeric calculation of valence polarization we adopt the following definition.

Definition. The coordination domain ABO_3 is referred to as polarized if the decline of the cation bond-valence from its nominal oxidation state differs in two bonding spheres A–O_{12} and B–O_6, i.e. the differences in matching oxidation states for both cations cause the atomic valence imbalance:

$$\Delta v = (v - v_S)_A - (v - v_S)_B \neq 0 \ , \tag{9}$$

where $(v - v_S)_A$ and $(v - v_S)_B$ are the declines of the cation bond-valences from their proper nominal values. The unit of Δv is evidently [valence/domain]. The dipole of domain is understood as common as for the general case:

$$P_{el} = \Delta v \times r \ , \tag{10}$$

where r is the distance from cation A (or B) to the A...B center, the direction of r is chosen so that Δv is positive. In the next section we compare this dipole with some other concepts that were obtained from the experiments or first principle calculation.

Table 3. Born effective charges (1st row) versus bond-valence charge (2nd row) for $BaTiO_3$.

a(Å)	Z(Ba)	Z(Ti)	Z(O)	Δv	p(eÅ)
3.64	2.95	7.23	-3.95		1.02
	5.19	5.68	-3.80	0.52	1.64
3.94	2.77	7.25	-3.93		1.00
	3.10	3.95	-2.35	0.45	1.54
4.00	2.74	7.29	-3.94		1.00
	2.76	3.64	-2.13	0.44	1.53
4.40	2.60	7.78	-4.17		1.04
	1.29	2.12	-1.14	0.42	1.57

4. Test examples and comparison

Table 3 lists the results from the first principle calculations of dynamic charge and associated dipole for $BaTiO_3$ [12] and compares them with our values (we used $p = \Delta v(a\sqrt{3})/2$ where a is the cubic lattice constant; the authors in [12] did not list dipoles, so we recalculated according to the given charges). The samples were under isotropic compression and in cubic phase. The bond valence seems to be more sensitive. Note that the charge associated with bond valence is the static charge, whereas the Born effective charge (BEC) is the dynamic charge, i.e. the charge observed by dipole and contains certain amount of electrons from the inner core. The BEC is not the valence charge and it is evident that this charge depends only a little on pressure, reflecting a small change on final dipole.

How well do the bond-valences reflect the reality of valence charge distribution within the $BaTiO_3$ structure, this can be estimated by comparing them with other concepts of static charge. For Ba^{2+}, the empirical models gave 2.00 and 1.86; the first principle calculation gave 2.00, 2.12 and 1.39, whereas the bond-valence gave its best estimation near 2.56 for the cell constant approaching 4.041Å. For the larger cell, the bond valence might drop to 1.29 as seen in table 3.

For Ti^{4+} the situation was similar, the empirical models gave 1.88 and 3.18; the first principle calculation gave 2.89, 2.43 and 2.79, whereas the bond-valence developed from 2.12 to 5.68 and gave the best estimation near 3.44 for the same cell around 4.041Å. The situation for oxygen was different since none of the calculations either based on empirical or first principle models gave values below −2.0. The bond-valence model, on the other hand, developed from -1.14 to -3.62 reflecting the core theoretical presumption about its elasticity. This means, for the $BaTiO_3$, that under certain circumstances the oxygen might accept more than 2 valence electrons.

In Table 4 we summarized the measured spontaneous polarizations of various samples taken from the classical literature [2] and compared them with our results. The listed P_s is equal p/V, where p is dipole and V is unit cell volume. For the cited compounds the structural data were retrieved from the Powder Diffraction Database [18]. When many samples existed then the one with lowest profile-weighted R_p factor has been chosen. For all compounds the cations were assumed to occur only in one oxidation state, e.g. Ti^{4+}, Ba^{2+}, Sr^{2+}, Pb^{2+}, Nb^{5+}, Ta^{5+} and Li^{+}. The mix valence state was omitted. The structural models were simple for these cases. As seen, the observed and calculated values are quite close to each other, although the absolute agreement is not expected, since the listed polarizations belong to two different concepts. However we might certainly conclude that the bond valence polarization should be considered as one of the local sources that contribute to the final macroscopic polarization.

Table 4. The spontaneous polarization P_s at room temperature for several classical perovskites.

Compound	P_s (C/m^2)	
	Reference [2]	This work
$BaTiO_3$	0.26	0.22
$SrTiO_3$	0.20	0.17
$PbTiO_3$	0.50	0.65
$LiNbO_3$	0.71	0.87
$LiTaO_3$	0.50	0.53

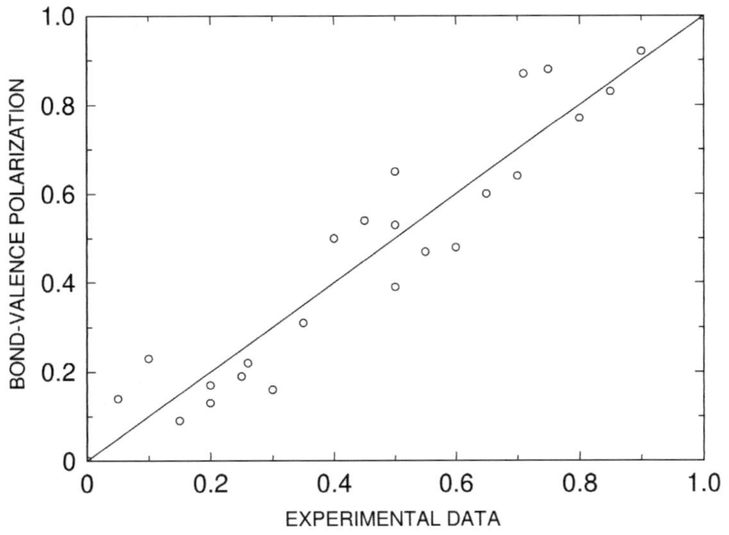

Figure 4. The bond-valence polarization versus experimental data for the perovskite structures retrieved from Powder Diffraction Database [18].

Figure 4 compares the calculated bond-valence polarization with the experimental data. As seen, the agreement is good for many cases and the erroneous level holds below 15% for all cases. At the first guess, this is an excellent result. It demonstrates the applicability of the bond-valence method to the study of polarization effect in perovskite-like materials. The method may be used as guidance for content selection when preparing the new piezoelectric materials.

5. Conclusion

The polarization effect was the main aspect in many studies concerning the perovskites and perovskite-like compounds. However, we found that there was no attempt to discuss the polarization in terms of the classical bond-valences sum concept, although it appeared simple and straightforward to derive the microscopic polarization from the bond-valence model. This investigation, though given in the primary stage, revealed that the bond-valence approach provided the results that are comparable to those of the experiments or quantum-mechanical models. Furthermore, it gave a clear classification of cations according to their tendency to form the polarized structures. This method, however, might

Journal of Physics: Conference Series **187** (2009) 012080 doi:10.1088/1742-6596/187/1/012080

produce mismatched valence configuration if the crystal structure was not modelled correctly. The main difficulty was that with the substitution of foreign elements, the correct location for each of the substituted elements could not be exactly determined. So many structural models must be taken into account to avoid the possible misconstruction. Despite this difficulty the ease of using this method in many common cases persuades its usefulness and applicability in the study of perovskites and perovskite-like compounds.

References

[1] Landau L D and Lifshitz E M 1984 *Electrodynamics of Continuous Media* (Oxford: Pergamon Press)
[2] Kittel C 1996 *Introduction to Solid State Physics 7th edition* (New York: Wiley)
[3] Resta R 1993 *Europhys. Lett.* **22** 133
[4] Resta R 1994 *Rev. Mod. Phys.* **66** 899
[5] King-Smith R D and Vanderbilt D 1993 *Phys. Rev.* B **47** 1651
[6] Selloni A, Carnevali P, Car R and Parrinello M 1987 *Phys. Rev. Lett.* **59** 823
[7] Berry M V 1984 *Proc. Roy. Soc. London A* **45** 392
[8] Shapere A and Wilczek F (ed.) 1989 *Geometric Phases in Physics* (Singapore: World Scientific)
[9] Ortiz G and Martin R M 1994 *Phys. Rev.* B **43** 14202
[10] Resta R 1997 *Phys. Rev. Lett* **26** 919
[11] Resta R 2000 *Condensed Matter arXiv*: cond-mat/0030809
[12] Ghosez Ph, Michenaud J-P and Gonze X 1998 *Phys. Rev.* B **58**(10) 6224
[13] Pauling L 1929 *J. Amer. Chem. Soc.* **51** 1010
[14] Pauling L *Nature of Chemical Bonds 3rd ed.* 1960 (New York: Cornell Univ. Press)
[15] Altermatt D and Brown I D 1985 *Acta Cryst. B* **41** 244-247
[16] Brown I D 2002 The Chemical Bond in Inorganic Chemistry-The Bond-Valence Model (New York, Oxford: University Press)
[17] Nhat H N and Minh D L 2002 *Proc. of the 5th Vietnam-Germany Summit on Advanced Physics and Engineering* p. 110
[18] Powder Diffraction Database PDF-2 2002 *International Centre for Diffraction Data*

Acknowledgments

This work is supported in part by the Asia Research Center and the Grant-in-Aid for Scientific Research from Vietnam National University QGTD 09-02 “Optomagnetic nanoparticles in coating technology”.

Preparation of ZnS nanoparticles by hydrothermal method

Tran Thi Quynh Hoa, Le Van Vu, Ta Dinh Canh and Nguyen Ngoc Long

Faculty of Physics, College of Science, Vietnam National University,
334 Nguyen Trai, Thanh Xuan, Hanoi, Vietnam

E-mail: tranthiquynhhoa@gmail.com

Abstract. ZnS nanoparticles were fabricated by hydrothermal method without using any capping agent. The hydrothermal temperature was 220°C. The product was received in powder form. The structure, composition and optical property of the resultant product were characterized by means of X-ray diffraction (XRD), transmission electron microscopy (TEM), UV-visible optical spectroscopy and photoluminescence (PL). The XRD pattern showed that the prepared ZnS nanoparticles have a zinc blende structure. The TEM image demonstrates clearly that the ZnS nanoparticles are in a spherical form with an average diameter of 25 nm. The UV-VIS spectra exhibited a blue-shift with respect to that of the bulk samples. The shift was attributed to the quantum size effect. The band gap of those samples have been calculated from absorption spectra and is found to be about 3.7 eV.

Keywords: ZnS nanoparticles, hydrothermal method, structural property.

1. Introduction

Zinc sulfide is an important II-IV semiconductor material with a wide band gap (3.72 eV for the cubic zinc blende phase and 3.77 eV for the hexagonal wurtzite phase at 300 K) and a large exciton binding energy (40 meV) [1]. It has a wide range of applications such as in optical sensor, solid state solar window layers, photoconductors, phosphors and catalysts. It has been extensively studied with the aim of controlling the size, morphology and crystallinity of ZnS nanocrystals in order to obtain desired physical properties.

In recent years, zero-dimensional (0D) ZnS nanostructures has attracted much attention because it can be used for biological detection and tagging [2]. Many methods have been used to synthesize ZnS nanoparticles such as sol - gel, electrochemical deposition, sonochemical, hydrothermal technique…

In this paper, we report the synthesis and characteristics of zinc sulphide nanoparticles obtained by using hydrothermal method with different reaction conditions. This is a simple method to fabricate nanocrystals from aqueous solutions at high temperatures and high vapour pressures. The advantage of this method over others includes the ability to create crystalline phases which are not stable at the melting point and create materials which have a high vapour pressure near their melting points. In the process, the decomposition of the precursors in a particular solvent depends on the temperature and pressure which was related to the filling fraction of the solvent within the reaction vessel and was kept constant for all the experiments. The important aim of studies is finding the effect of the synthesis conditions on the size, shape and crystal structure of the ZnS products.

© 2009 IOP Publishing Ltd

Journal of Physics: Conference Series **187** (2009) 012081 doi:10.1088/1742-6596/187/1/012081

2. Experimental

An autoclave, which is a thick-walled steel cylinder hermetically sealed in order to withstand high temperatures and pressures for prolonged periods of time, is used for hydrothermal method. In the internal cavity of the autoclave, a Teflon vessel used to prevent corrosion is inserted. The vessel has an inner diameter of 30 mm and a capacity of about 120 ml.

All the chemicals were of analytical grade and used without any further purification. We have performed experiments with different molar ratios of the compositions of Zn^{2+} and S^{2-} at the temperature of 220°C. Experiment details were as follows: $ZnSO_4.5H_2O$ powder and $Na_2S.7H_2O$ powder used as starting materials were dissolved separately in de-ionized water and stirred at room temperature for 30 min. Then they were mixed by dripping the Na_2S solution in drops into $ZnSO_4$ solution while stirring. Zinc and sulfur sources taken in 1:0.7, 1:1 and 1:1.3 molar ratios were used for synthesis of sample 1, sample 2 and sample 3, respectively. The chemical reaction may be described as follows [3]:

$$ZnSO_4 \leftrightarrow Zn^{2+} + SO_4^{2-} \quad , \qquad (1)$$

$$Zn^{2+} + Na_2S \leftrightarrow ZnS + 2Na^{+} \quad . \qquad (2)$$

The white gels were charged into the Teflon-lined stainless steel autoclave, which was then filled with the mixed solution up to 50% of its volume. The closed chamber was placed inside a box furnace (Mermert - 500) at the temperature of 220°C for 12 h and then cooled down to room temperature naturally. The solution containing products were then centrifuged, had the supernatant layer poured off, washed with double distilled water. The wash procedure was repeated ten times to remove impurities in the sample. The final white products were dried at 60°C in air for 12 h.

The structure of the ZnS samples was analyzed by X-ray diffraction using a Brucker D5005 diffractometer. The microstructures and structures of the samples were characterized by transmission electron microscopy (JEOL JEM 1010) and high - resolution transmission electron microscopy (FEI Tecnai TF20 TEM/STEM). The photoluminescence (PL) and photoluminescence excitation (PLE) spectra were recorded by using a FL3-22 Jobin Yvon Spex spectrofluorometer. The absorption spectra were obtained by a Shimadzu UV 2450 PC spectrometer. The diffuse reflectance was measured by a UV - VIS NIR Cary 5G spectroscopy.

3. Results and discussion

XRD patterns of ZnS products are given in figure 1. The spectra show various diffraction peaks at 2θ values of 28.6°, 33.1°, 47.6° and 56.4° corresponding to the diffraction planes (111), (200), (220) and (311), respectively. The peaks were perfectly indexed to the cubic zinc blende phase of ZnS.

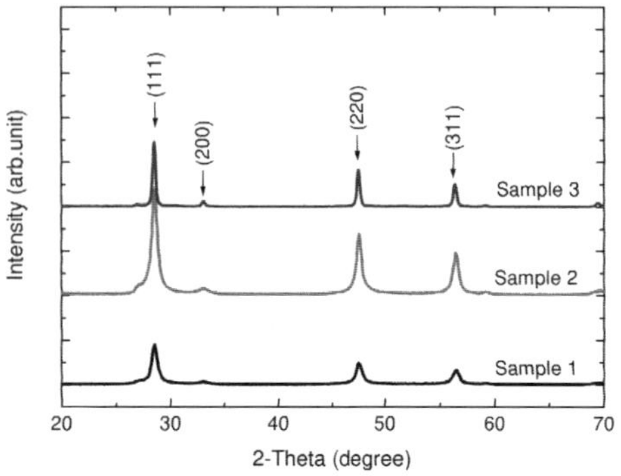

Figure 1. XRD patterns of ZnS nanostructures with 1:0.7, 1:1 and 1:1.3 molar ratios of Zn^{2+}:S^{2-}.

The distances d_{hkl} between successive lattice planes are calculated from Wulf - Bragg relation:

$$2d_{hkl} \sin\theta = n\lambda \quad , \qquad (3)$$

where λ is the wavelength of X-ray, θ is the angle of incidence and the integer n is known as the order of the corresponding reflection. The cell constant of the ZnS crystallites was determined from the XRD patterns using the following formula:

$$d_{hkl} = \frac{a}{\sqrt{(h^2+k^2+l^2)}} \quad . \tag{4}$$

The results shown in table 1 are slightly larger than that of the standard ZnS phase (a = 5.398 Å).

The information on the crystalline size has been obtained from the following Debye - Scherrer relations [4]:

$$L = \frac{0.9\lambda}{\beta\cos\theta} \quad , \tag{5}$$

where β is the full width at half maximum (FWHM) of the diffraction peaks. The grain sizes are estimated to be 13.1, 14.6 and 28.1 nm for sample 1, sample 2 and sample 3, respectively.

The TEM and HRTEM analysis gives more detailed information on the microstructures of the ZnS nanoparticles. The TEM images (figure 2) show that the subspherical nanoparticles have an average diameter of about 15, 20 and 35 nm for samples 1, 2 and 3, respectively. It indicated that the higher Zn^{2+}:S^{2-} molar ratios are, the smaller particle sizes become. Thus, by controlling the molar ratio the particle size can be varied.

HRTEM images (figures 3a, b) of sample 1 show that ZnS nanoparticles are of good crystallinity. Fast Fourier transform (FFT) pattern shown in figure 3c indicates the observed on figure 3b lattice plane is (111) plane. Lattice plane spacing was estimated to be around 0.31 nm. This value is in good accordance with the result of the other authors [1, 5, 6].

Table 1. Lattice constants and nanocrystal sizes calculated from XRD patterns for sample 1, sample 2 and sample 3

Sample	Plane	d_{hkl} (Å)	a (Å)	$\bar{a}$ (Å)	L (nm)	$\bar{L}$ (nm)
1	(111)	3.12	5.41	5.41	13.68	13.08
	(220)	1.91	5.40		12.50	
	(311)	1.63	5.41		13.07	
2	(111)	3.12	5.41	5.41	13.70	14.57
	(220)	1.91	5.40		14.93	
	(311)	1.63	5.40		15.09	
3	(111)	3.13	5.41	5.41	29.20	28.06
	(220)	1.91	5.41		27.84	
	(311)	1.63	5.41		27.14	

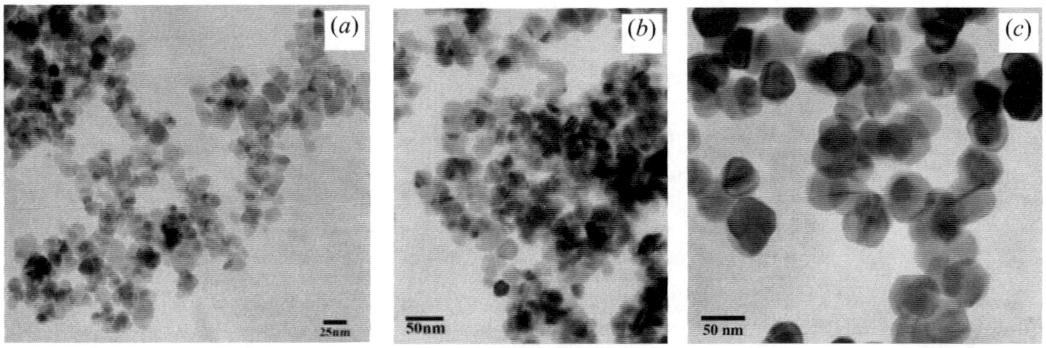

Figure 2. TEM images of ZnS nanostructures with different molar ratios of Zn^{2+}:S^{2-}
a) 1:0.7 b) 1:1 c) 1:1.3

The Selected Area Electron Diffraction (SAED) pattern of ZnS nanoparticles is shown in figure 3d. As expected, the SAED pattern shows a set of rings instead of spots due to the random orientation of the crystallites, which relate to diffraction from different planes of the ZnS nanocrystallites [7, 8]. The

SAED pattern shows many rings, among them the strongest four rings correspond to the (111), (200), (220) and (311) planes of the cubic phase of ZnS.

In order to study optical absorption property of the samples, the ZnS powders were ultrasonically dispersed in double distilled water. The optical absorption spectra of the samples are shown in figure 4. From the curves, we can see the shoulder near 330 nm in each spectrum. In the case of the sample 3, the particle grain size is larger than that of other samples, so the shoulder becomes a weak peak at 340 nm. The shoulder or peak of the spectra corresponds to the fundamental absorption edges in the samples, which can be used to estimate the band gap of material. From the absorption measurements the band gap were obtained as 3.76 eV (330 nm), 3.74 eV (332 nm) and 3.65 eV (339 nm) for the samples synthesized with Zn^{2+}:S^{2-} molar ratios of 1:0.7, 1:1 and 1:1.3, respectively. The band gap values were slightly decreased due to the increase of particle size. The band gap energy blue-shift with respect to that of the bulk samples was attributed to the quantum size effect.

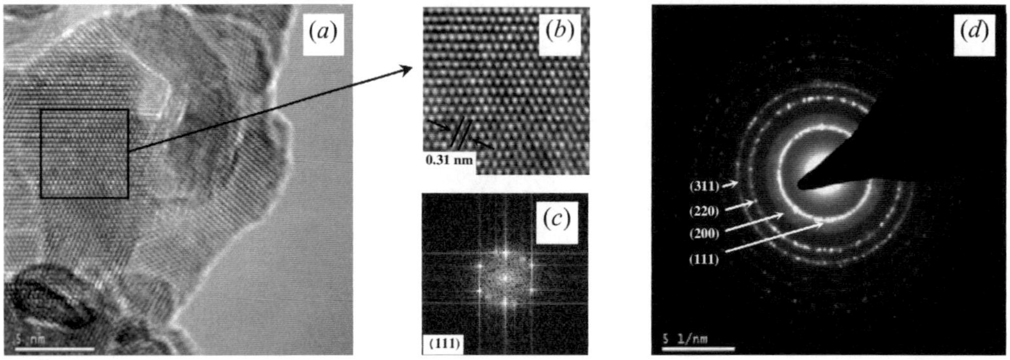

Figure 3. a), b) HRTEM images, c) FFT pattern and d) SAED image of ZnS nanopowder prepared with Zn^{2+}:S^{2-} molar ratio of 1:0.7

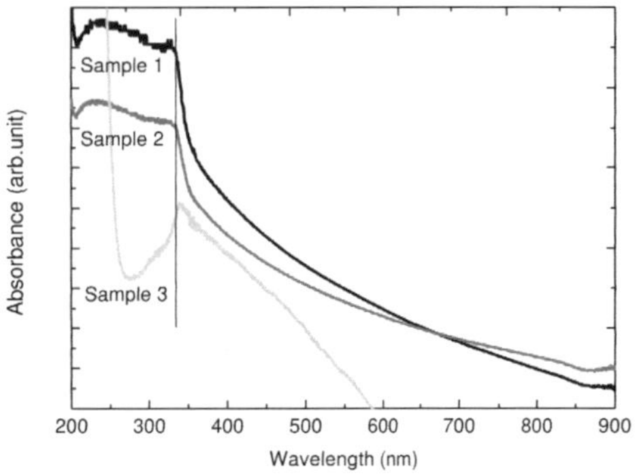

Figure 4. Absorption spectra of ZnS nanostructures with 1:0.7, 1:1 and 1:1.3 molar ratios of Zn^{2+}:S^{2-}

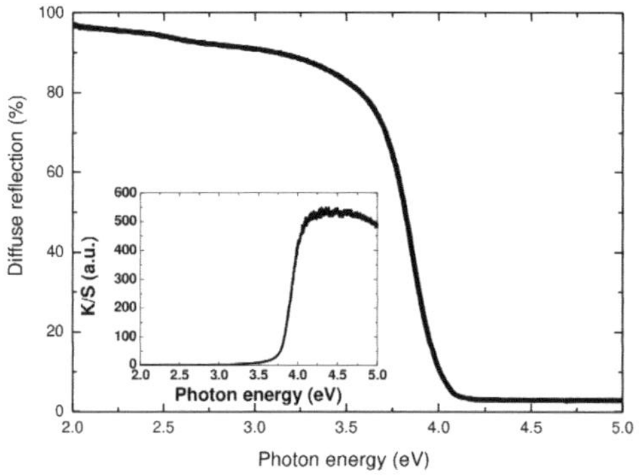

Figure 5. Diffuse reflection spectra of ZnS nanostructure with Zn^{2+}:S^{2-} molar ratio of 1:0.7

Figure 5 shows the diffuse reflection spectrum of sample 1 with Zn^{2+}:S^{2-} molar ratio of 1:0.7. It shows a strong drop in reflection in the UV range approximately 4 eV, corresponding to the valence-to-conduction band transitions of the ZnS nanostructure. In order to verify absorption of the sample, the Kubelka - Munk function was used.

$$F(R) = (1-R)^2\, 2/R = K/S \quad , \tag{5}$$

where *R*, *K* and *S* are the reflection, absorption and scattering coefficient, respectively. The (*K*/*S*) spectrum of the nanostructured ZnS derived from the Kubelka - Munk function is shown in the inset of figure 5. The value of the optical band gap can be calculated by extrapolating the Kubelka - Munk function to $K/S = 0$ [9]. Its optical band gap is calculated to be about 3.78 eV (328 nm). The band gap is slightly larger than the value 3.76 eV which is calculated from the absorption spectra.

Photoluminescence property of the samples was investigated at room temperature. Emission spectra of the samples with the excitation wavelength of 320 nm (3.87 eV) are shown in figure 6*a*. The spectra of sample 1 and sample 2 showed a broad band peaked at 489 nm, 498 nm, respectively, and a shoulder nearby 454 nm, while that of sample 3 showed only a broad band peaking at 530 nm.

Broadening of the emission peak could be attributed to both the size distribution and the increase of the surface states owing to the increase in surface to volume ratio for smaller nanoparticles [1]. The origin of the emission band centered at 490 nm was attributed to the Zn-vacancy [1, 5, 6, 9, 10]. Thus, the blue-green emission could be assigned to some self-activated defect centers related to Zn-vacancies.

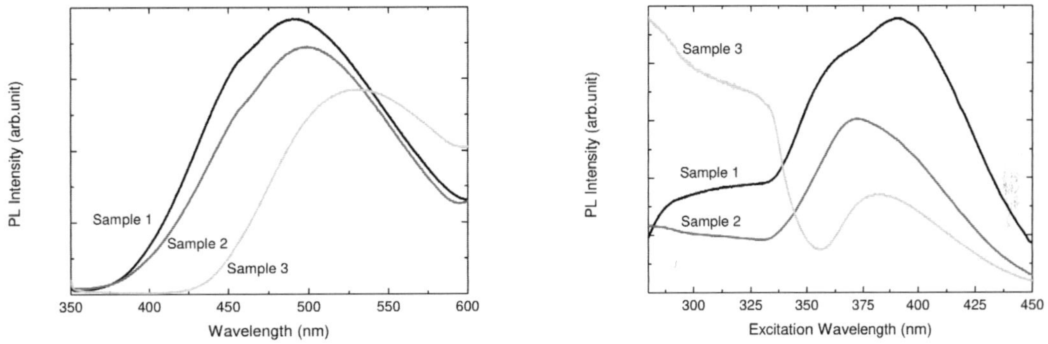

Figure 6. *a*) PL and *b*) PLE spectra of nanostructured ZnS with 1:0.7, 1:1 and 1:1.3 molar ratios of Zn^{2+}:S^{2-}

For investigation of the luminescence mechanism causing the blue-green emission, PLE spectra monitored at wavelength of 500 nm have been recorded. Figure 6b shows the PLE spectra of the samples. The blue-green emission band can be excited both at energies near the band edge of the ZnS (the UV region) and at energies below the band edge (the visible region). In the visible region, for sample 1 two peaks at 359 nm (3.45 eV) and 390 nm (3.18 eV) were observed; but for sample 2 or sample 3 only one peak at 369 nm (3.36 eV) or 378 nm (3.28 eV), respectively, was obtained. These peaks are related to the defect states and the surface states located within the band gap. In the UV region, only for sample 3 shoulder at 332 nm (3.73 eV) related to the near-band-edge absorption was observed. For samples 1 and 2, absorption related to the near band edge was not exhibited. The reason of this fact is that the nanoparticles of samples 1 and 2 have smaller sizes than those of nanoparticles of sample 3; therefore, number of the surface states is increased due to the increase of the surface to volume ratio. In this case, excitation via the surface states is dominated.

4. Conclusions

Nanocrystalline ZnS have been successfully synthesized by a hydrothemal method. X-ray diffraction studies revealed that the synthesized particles have cubic zinc blende structure. The nanoparticle size increases from 13.1 nm to 28.1 nm with decreasing the Zn^{2+}:S^{2-} molar ratio from 1:0.7 to 1:1.3. The band gap values increases from 3.65 eV to 3.76 eV with decreasing the nanoparticle size. The blue-green emission band in photoluminescence spectra is related to the zinc vacancy in the lattice.

References

[1] Biswas S and Kar S 2008 *Nanotechnology* **19** 045710
[2] Tan W B, Huang N and Zhang Y 2007 *Colloid and Interface* **310** 464
[3] Denzier D, Olschewski M and Sattler K 1998 *Applied Physics* **84** 2841
[4] Cullity B D 1978 *Elements of X-ray diffraction* Addision - Wesley Publishing Company, Inc

Journal of Physics: Conference Series **187** (2009) 012081 doi:10.1088/1742-6596/187/1/012081

[5] Biswas S, Kar S and Chaudhuri S 2005 *Journal of Physical Chemistry B* **19** 17526
[6] Liu J Z, Yan P X, Yue G H, Kong L B, Zhuo R F and Qu D M 2006 *Materials Letters* **60** 3471
[7] Lee H J, Lee S I 2007 *Current Applied Physics* **7** 193
[8] Maity R and Chottopadhyay K K 2004 *Nanotechnology* **15** 812
[9] Duan C J, Delsing A C A and Hintzen H T 2009 *Chemistry of Materials* **21** (in press)
[10] Suyver J F, Wuister S F, Kelly J J and Meijerink A 2001 *Nano Letters* **1** 429

Acknowledgements
This work is supported by the Fundamental Research Program, Ministry of Science and Technology of Vietnam (Project No 4 055 06). Authors of this paper would like to express their sincere gratitude to the Center for Materials Science (CMS), Faculty of Physics, College of Science, Vietnam National University for permission to use facilities.

Preparation, characterization and optical properties of $ZrO_2:Er^{3+}$, Yb^{3+} nanomaterials

Tran Thu Huong[1], Nguyen Thanh Huong[1], Lam Thi Kieu Giang[1], Tran Kim Anh[1], Witold Lojkowski[2] and Le Quoc Minh[1, 3]

[1] Institute of Materials Science, Vietnam Academy of Science and Technology, 18 Hoang Quoc Viet Road, Cau Giay Distr., Hanoi, Vietnam
[2] Institute of High Pressure Physics, PAS, Poland
[3] College of Technology, National University Hanoi

E-mail: lequocminh@ims.vast.ac.vn

Abstract. Luminescent nanomaterials from lanthanide are one of most promising materials for applications in medical diagnostics. In this paper, we present for the first time the ZrO_2 nanomaterials with capsule like nanostructure with the sizes of 150 nm in length and 50 nm in width. These nanocapsule like ZrO_2 doped with Er^{3+}, Yb^{3+} ions were prepared by soft template methods. Morphological characterization of nano - $ZrO_2:Er^{3+}$, Yb^{3+} was performed by using FESEM, X-ray diffraction, micro Raman. Up conversion luminescence and photoluminescence have been also studied with IR-laser excitation at 940 nm and UV excitation at 365 nm.

Keywords: Nanomaterials, ZrO_2: Er^{3+},Yb^{3+} photoluminescence, UV excitation.

1. Introduction

Biomedical materials are one of the most attractive subjects in materials science in both sense of basic research and applications [1-3]. For the case of lanthanide luminescence which is applicable in the medical diagnostics, nano-sized up-converting phosphor particles is required [4]. In this paper, we report the fabrication of new nanomaterials $ZrO_2:Er^{3+}$, Yb^{3+} by soft template methods. Morphological characterization of $ZrO_2:Er^{3+}$, Yb^{3+} nanomaterials was performed by using FESEM, X-ray diffraction, Raman scattering spectra. Photoluminescence (PL) spectra were also studied and discussed upon the UV excitation and IR-laser excitation at 940 nm.

2. Experimental

2.1. Materials

Zirconium oxychloride ($ZrOCl_2.8H_2O$, 96%) was purchased from Poole England. Ytterbium chloride ($YbCl_3.6H_2O$, 99.99%) and erbium chloride ($ErCl_3.6H_2O$, 99.99%) were purchased from Aldrich, USA. Sodium hydroxide (NaOH, 99%) and polyethylenglycol (PEG) 4000 were purchased from Merck, Germany.

2.2. Synthesis of $ZrO_2:Yb^{3+}$, Er^{3+} by soft template method

The $ZrO_2:Yb^{3+}$, Er^{3+} nanomaterials were prepared by soft template method. In a typical synthesis, $ZrOCl_2.8H_2O$ powder and PEG powder were completely dissolved in 90 ml H_2O. After that,

© 2009 IOP Publishing Ltd

$YbCl_3.6H_2O$ and $ErCl_3.6H_2O$ were added to the solution in a 100 ml round-bottomed flask and this was followed by a magnetic stirring for 120 min. The pH of the solution after the reaction was in the range of 12-12.5 by 10% NaOH solution. The content ratio of Yb^{3+} and Er^{3+} were 1:1 mole, respectively. After that, the solution was poured into a Teflon lined stainless steel autoclave. The autoclave was sealed tightly and heat 200°C for 20 h, and then cooled down slowly to room temperature. The resulting products were collected, and centrifuged at 5900 rpm, washed several times using water and then dried in air at 60°C for 6 h.

2.3. Characterization

The morphology of the as-synthesized samples was observed using a field emission scanning electron microscope (FESEM, Hitachi). X-ray diffraction (XRD) measurements of the products were performed on an X-ray diffractometer (Siemens D5000 with λ = 1.5406 Å in the range of $15^o \leq 2\theta \leq 75^o$). Raman scattering spectra of samples were also investigated. The PL spectra of the ZrO_2; ZrO_2:Er^{3+}; ZrO_2:Er^{3+}, Yb^{3+} samples were measured at room temperature using a visible-ultraviolet spectrophotometer with a xenon lamp and an IR-laser (940 nm) as the excitation light sources.

3. Results and discussion

The morphology of the obtained ZrO_2 products was measured by field emission scanning electron microscope (FESEM). An FESEM image taken of a large area (figure 1) reveals that the ZrO_2 product consists of nanomaterials with lengths ranging from 100 to 150 nm, widths ranging from 30 to 50 nm. It is clearly seen that these capsule - like shapes are formed.

Figure 1. FESEM images of the ZrO_2 sample at 200°C for 20 h.

Figure 2. FESEM images of the ZrO_2:Er^{3+} (1% mol) sample at 200°C for 20 h.

FESEM images of the ZrO_2:Er^{3+} and ZrO_2: Er^{3+}, Yb^{3+} nanomaterials are presented on figures 2 and 3, respectively. They have capsule like form with lengths ranging from 100 to 150 nm, widths ranging from 30 to 50 nm. The homornology of Er^{3+}, Yb^{3+} doped ZrO_2 nano crystalline is less than that of undoped ZrO_2 ones.

The X-ray diffraction patterns of the ZrO_2:Er^{3+}, Yb^{3+} nanomaterials showed that these crystal structures are isostructural to that of Baddeleyite and the nanomaterials prepared with the soft-template synthesis method were essentially pure (figure 4).

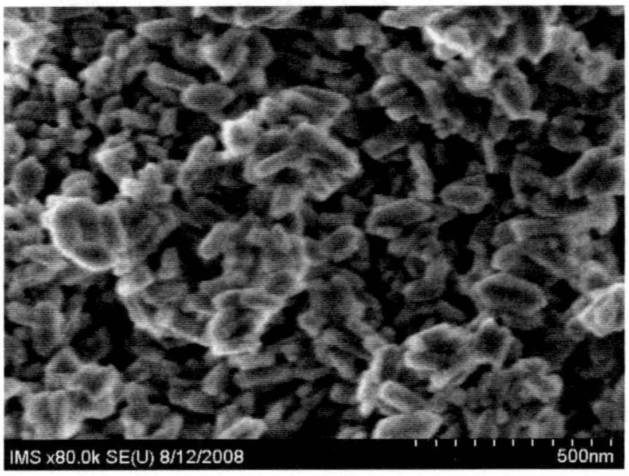

Figure 3. FESEM images of the ZrO_2:Er^{3+}(1% mol), Yb^{3+}(1% mol) sample at 200°C for 20 h.

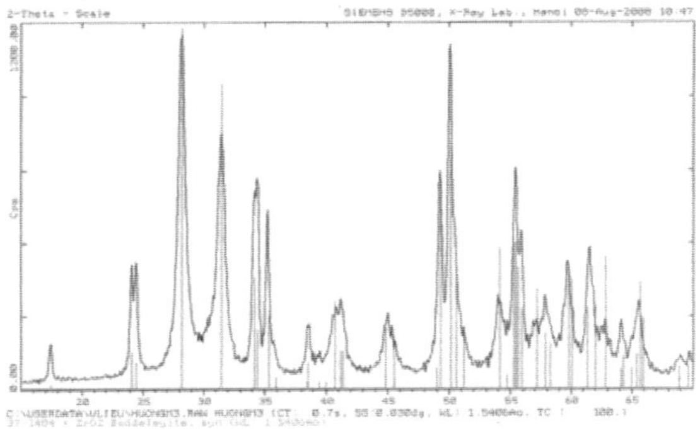

Figure 4. X-ray diffraction pattern of the ZrO_2:Er^{3+}(1% mol), Yb^{3+}(1% mol) nanomaterial at 200°C for 20 h.

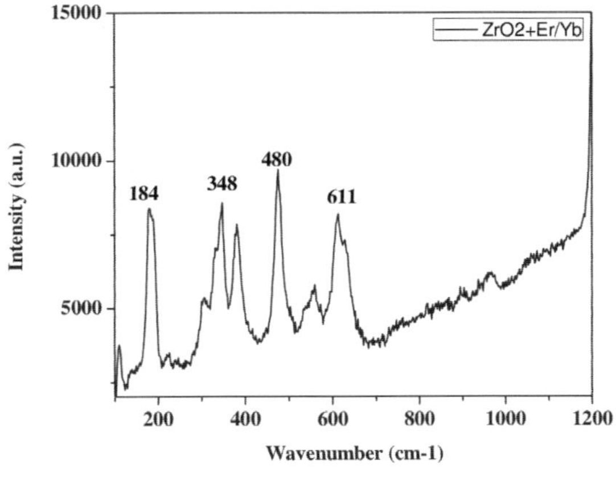

Figure 5. Raman scattering of the ZrO_2:Er^{3+} (1% mol), Yb^{3+} (1% mol) nanomaterial at 200°C for 20 h.

Figure 5 shows Raman scattering of the ZrO_2: Er^{3+} (1% mol), Yb^{3+} (1% mol) nanomaterial at 200°C, 20 h. It is note that there are peaks at 184, 384, 390, 496, and 692 cm^{-1}. That shows vibration modes of Zr–O, O–H bonds. The PL spectrum of the ZrO_2: Er^{3+} (1% mol), Yb^{3+} (1% mol) nanomaterial at 200°C, 20 h under an excitation wavelength of 365 nm at room temperature is displayed in figure 6.

Up conversion luminescence spectrum of the ZrO_2: Er^{3+} (1% mol), Yb^{3+} (1% mol) nanomaterial under an excitation wavelength of 940 nm at room temperature is presented figure 7. The ZrO_2:Yb^{3+}, Er^{3+} nanomaterials showed red (630–710 nm) and green (510 –570 nm) up conversion luminescence due to the $^4F_{9/2} \rightarrow {}^4I_{15/2}$ and $({}^2H_{11/2}, {}^4S_{3/2}) \rightarrow {}^4I_{15/2}$ transitions of Er^{3+}, respectively.

The up conversion luminescence, where the absorption of two or more low energy photons is followed by the emission of a photon of higher energy, has witnessed numerous breakthroughs during the past decades. Up-converting phosphors have a variety of potential applications as lasers, displays

Journal of Physics: Conference Series **187** (2009) 012082 doi:10.1088/1742-6596/187/1/012082

and inks for security printing. In addition, another promising way to overcome the problems with the blood absorption is to use a long-wavelength excitation and benefit the up conversion luminescence. There is practically no absorption by the whole blood in the near IR region from 800 to 1000 nm and it has no capability for up conversion in the excitation wavelength region of the conventional up-converting phosphor based on the Yb^{3+} (sensitizer) and Er^{3+} (activator) combination. It is also possible to find a red emitting ion (Er^{3+}). For coupling to biological compounds, nano-sized up-converting phosphor particles are required. Moreover, nanophosphors with high luminescent efficiency are vital in the development of novel homogeneous label technology for quantitative all-inone whole-blood immunoassay which uses low-cost measurement devices [5].

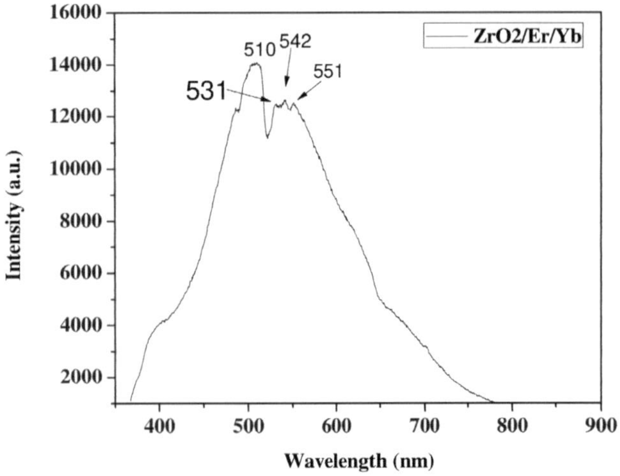

Figure 6. PL spectrum of the ZrO_2: Er^{3+} (1% mol), Yb^{3+} (1% mol) nanomaterial at 200°C for 20 h under an excitation wavelength of 365 nm at room temperature.

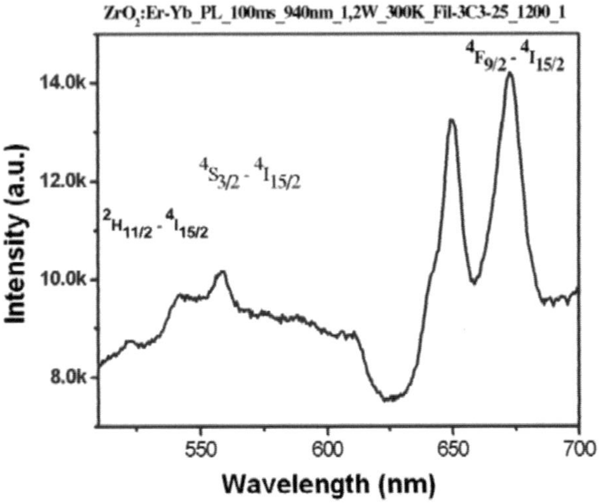

Figure 7. Up conversion luminescence spectra of the ZrO_2: Er^{3+} (1% mol), Yb^{3+} (1% mol) nanomaterial under an excitation wavelength of 940 nm at room temperature.

4. Conclusions

In conclusion, the nanocapsule-like ZrO_2 and Yb^{3+}, Er^{3+} doped ZrO_2 with the sizes of 150 nm in length and 50 nm in width have been prepared for the first time by using PEG 4000 by soft template method. Morphological characterization of the ZrO_2: Er^{3+}, Yb^{3+} nanomaterials was performed by using FESEM. The X-ray diffraction pattern of the ZrO_2: Er^{3+}, Yb^{3+} nanomaterials showed that these crystal structures are isostructural to that of Baddeleyite. Vibration modes of Zr–O, O–H bonds were detected by Raman scattering spectra, PL spectra have been employed to characterize the as-prepared samples.

For efficient use in bioassays, more work is needed to yield nanomaterials with smaller and more uniform crystallite sizes. Surface modifications need to be studied to allow for the dispersion in aqueous solutions. On the other hand, further work must be carried out to optimize the persistent luminescence of the $ZrO_2:Yb^{3+}$, Er^{3+} nanomaterials to allow for their use as efficient immunoassay nanomaterials combining the advantages of both up conversion and persistent luminescence.

References

[1] Salata O V 2004 *J. Nanobiotechnology* **2** 3 doi:10.1186/1477-3155-2-3

[2] Patra C R, Bhattacharya R, Patra S, Basu S, Mukherjee P and Mukhopadhyay D 2006 *J. Nanobiotech.* **4** 11 doi: 10.1186/1477-3155-4-11

[3] Beaurepaire E, Buissette V, Sauviat M P 2004 *Nano Lett.* **4** 2079

[4] Misra S N, Gagnani M A, Devi I, and Shukla R S 2004 *Bioinorganic Chemi. and Appl.* **2** 155

[5] Iko Hyppänen, Jorma HölsäJouko Kankare, Mika Lastusaari and Laura Pihlgren 2007 *J. Nanomaterials* **Vol. 2007** Article ID 16391 doi:10.1155 16391

Acknowledgments

This work was partly supported by Research Programming of Institute of Materials Science, project CS05.08 and National Key Laboratory of Electronic Materials and Devices.

Co-deposition and microstructure of Ni-nano SiC coating on metal

Huynh Thi Ha[1], Cao Tuan Anh[2], Nguyen Thu Ha[1] and Dao Tran Cao[1]

[1] Institute of Materials Science, Vietnam Academy of Science and Technology,
18 Hoang Quoc Viet Road, Cau Giay Distr., Hanoi, Vietnam
[2] Institute of Physics, Vietnam Academy of Science and Technology,
10 Dao Tan Street, Ba Dinh Distr., Hanoi, Vietnam

E-mail: ctanh@iop.vast.ac.vn

Abstract. The study on co-deposition of Ni-nano SiC in sulphamate bath was conducted with different nano SiC concentrations from 1 to 20 g/l. The deposition of light inert SiC nanoparticles in Ni plating was investigated by scanning electron microscopy SEM and X-ray dispersive spectrometer EDX measurements. The microhardness and corrosion resistance were also measured to evaluate mechanical and corrosion properties of Ni-nano SiC coating on metal.

Keywords: Co-deposition, SiC nanoparticles, composite coating, microstructure.

1. Introduction

The co–deposition of particles, such as oxides, carbides with metals has been a goal of composite development to achieve a combination of properties that is not achievable by any of the materials acting alone. Several studies have showed that electrochemically embedded particles bring special properties to produced composite coatings, which can meet industrial requests. For example, diesel and gas turbine engines are subjected to high temperature corrosion and there is a great need to develop beneficial coatings for them. Recently, the ability to produce new composite materials based on reinforcing powders of decreasing particle sizes is very promising. It is proved that uniform dispersion of the co-deposited particles leads to the improvement of the mechanical, tribiological, anti-corrosive and antioxidized properties of the coatings [1, 2].

At present, the electrochemical deposition of nano-size particles in a metallic matrix has led to a new generation of composites due to the advantages of this technique. Ni–nano SiC composite, in particular, has been investigated for its high wear resistance. This property has led to commercial use in protection of friction parts, combustion engines and casting moulds [3, 4]. The main challenges in the co–deposition of nano–size SiC inert particles seem to be the co–deposition of a sufficient number of particles and avoiding the agglomeration of particles suspended in the plating solution. Our work aims to compare the mechanical properties and wear corrosion of pure Ni and Ni–SiC nano-composite coatings.

2. Experimental

Nickel coating was electroplated on brass plates from a 55°C Ni sulphamate bath at current density of

© 2009 IOP Publishing Ltd

4 Adm^{-2}. Prior to electroplating, the brass plates were polished with emery paper up to grade 2000 and rinsed in distilled water. The Ni sulphamate solution contained 90 g l^{-1} Ni ions as Ni sulphamate, 3 g l^{-1} $NiCl_2$, 40 g l^{-1} H_3BO_4 and 2 ml l^{-1} wetting agent (1% sodium dodecyle sulfate). Solution pH was adjusted to 4.5–4.7. To make Ni–SiC composite coatings, SiC powders with different concentrations of 1, 5, 10, 15, and 20 gl^{-1} were gradually added to the Ni sulphamate solution. The selected commercial SiC powder with an average diameter of 45-55 nm (Nanostructured & Amorphous Materials Inc., USA, β-SiC of 98% purity) was used as received without any treatment. The agglomeration was prevented by suspending the SiC particles in electrolyte and subjecting to magnetic stirring for a period of 15 hours. Electroplating was carried out for 60 mins with two stirring speeds (speed 5 and 8), and the thickness of the resulting deposits was in the range of 25-30 μm. Ultrasonic cleaning of composite surface was then executed after the electrolytic deposition to ensure that SiC particles were not only absorbed on the surface but embedded in the Ni matrix.

The microstructure of various deposits was observed using SEM spectrometer (SEM HITACHI 4800). The mass percentage of co-deposited SiC was measured via EDX system. The microhardness of the distinct deposits was measured by indenting across the polished cross sections and reported as an average of three values (Microhardness Tester MHT-10, Anton Paar). The corrosion potential of Ni-nano SiC coating was estimated by porosity and polarisation resistance measurements. An electrolyte was used as 3.5% NaCl solution open cell. Electrochemical experiments were performed using a potentiostat connected with a frequency response analyser AUTOLAB PGS30. A platinum counter electrode and a Calomen SCE reference electrode were also used.

3. Results and discussion

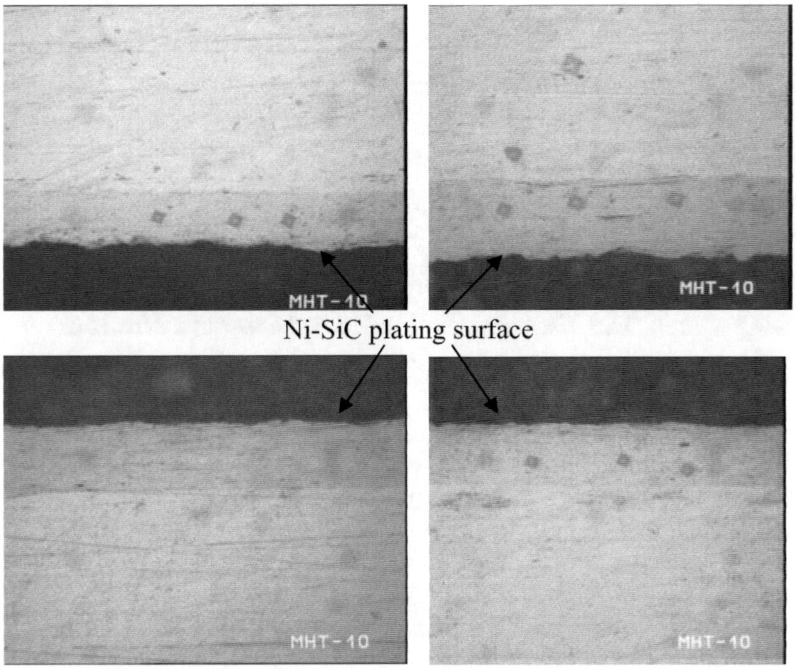

Figure 1. Cross-sections of Ni-SiC plating surface with 1 (a), 5 (b), 15 (c), and 20 (d) gl^{-1} SiC concentrations.

Figure 1 shows the cross-sections of the various deposits observed under an optical microscope (x500). The SiC particles are not clearly visible on the surface because of their small dimensions. It is only noted that the surface of Ni-SiC coating with low SiC concentration 1-5 gl^{-1} is not flat (figures 1a, 1b). As the inert particles' concentration in the solution increases, the flatness of coating surface increases as well (figures 1c, 1d; 15-20 gl^{-1}).

Micrographies presented by SEM measurements allow comparison between a pure Ni coating (figure 2) and Ni with nano-silicone carbide coating (figure 3). From the scanning electron microscope images we can see the differences on surface morphology of pure Ni and composite coatings. The pure Ni deposit has a rather regular surface, whereas the composite coating develops in lumpy (nodular) disturbed surface structure. Smaller grain sizes could be observed in the composite coating (figure 3b) than in pure nickel coating. The surface structures of the two types of coating are different and they are expected to have behavior different with respect to corrosion property.

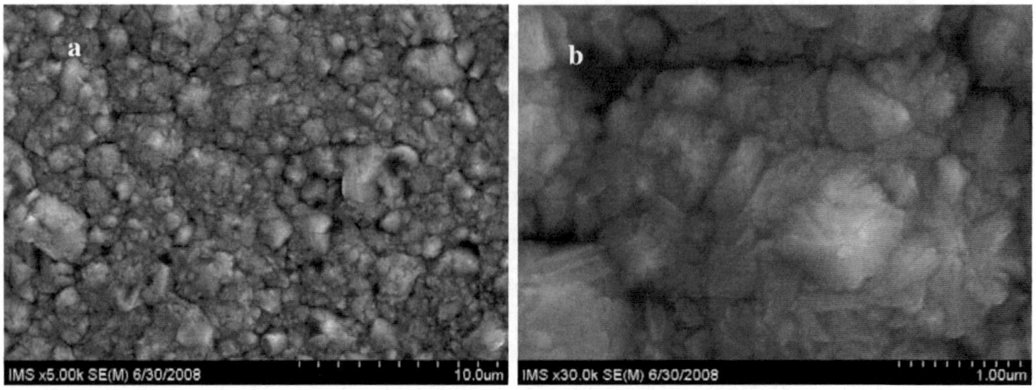

Figure 2. SEM images of pure Ni plating.

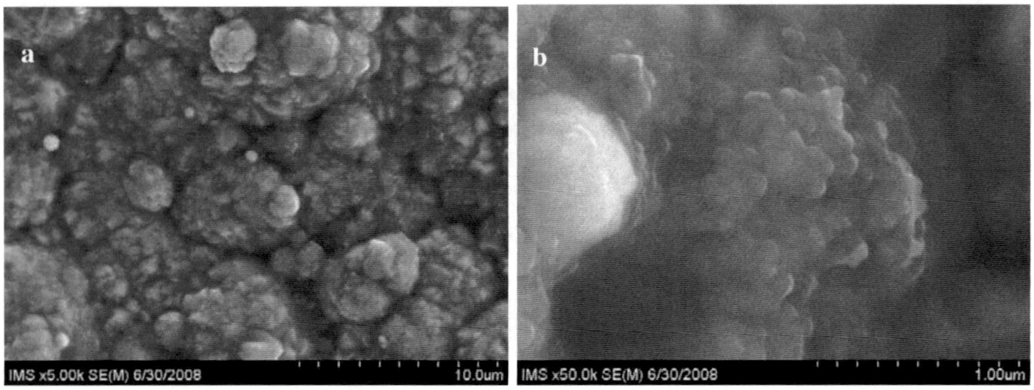

Figure 3. SEM images of Ni- nano SiC plating with 5 gl^{-1} of SiC in the plating bath.

The presence of silicone carbide is determined by EDX spectrum (figure 4). From the general EDX analysis, the total amount of nano SiC particles inside the deposit with concentration of SiC 5 gl^{-1} in the plating bath was calculated at 5.40 wt.% (figure 4a). Figure 4b displays the position of the calculated sample. It is certain to conclude that as we increase the inert particles concentration in the solution, the particles amount in the deposit increases too. It can be proved by the microhardness measurement of Ni-nano SiC coatings

The relationship between inert SiC particles concentration and hardness of coatings is showed in table 1. The microhardness of the plated layers was determined through optic microscopy on cross-sections of the coatings (figure 1). The obtained results indicate the hardness of the various deposits was in the range of 319 to 466 HV, which was higher than that of pure Ni coating (<300 HV). It means that the strength of the coatings was evidently enhanced with deposits of small dimensional SiC particles.

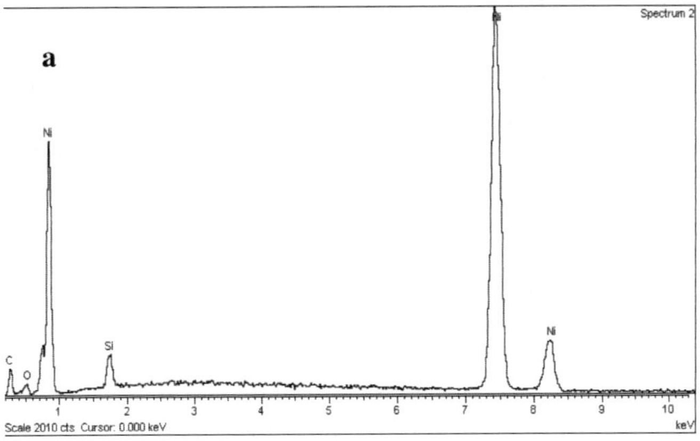

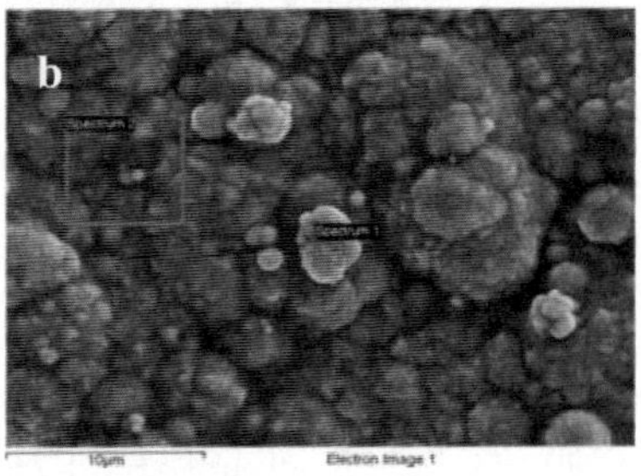

Figure 4. (a) EDX spectrum of Ni-SiC nano coating, (b) Position image of the calculated sample.

Table 1. The relationship between SiC concentrations, hardness and corrosion porosity of Ni-nano SiC coatings.

SiC concentration in solution, gl^{-1}	Hardness HV, kg/mm^2		Corrosion porosity density, cm^{-2}	
	Speed 5	Speed 8	Speed 5	Speed 8
0	< 300		7	
1	319	339	14	10
5	328	328	11	8
10	364	382	7	8
15	378	393	6	7
20	387	466	4	5

Concerned with the different surface morphology of pure nickel and composite coatings, corrosion properties of coatings were tested. By porosity testing method, corrosion pore density was measured. The solution was a mixture of $K_3[Fe(CN)_6]_2$ 20 gl^{-1} and NaCl 10 gl^{-1}. The results shown in table 1 reveal that at low SiC nanoparticles concentration porosity density of the deposit is higher than pure nickel deposits, but it has decreased when the SiC concentration in deposits has increased. All the measurements in table 1 were carried out on coatings of two stirring speeds 5 and 8. It can be seen that at higher stirring speed, the hardness of the coating is higher but the porosity performs higher also when SiC concentration increases.

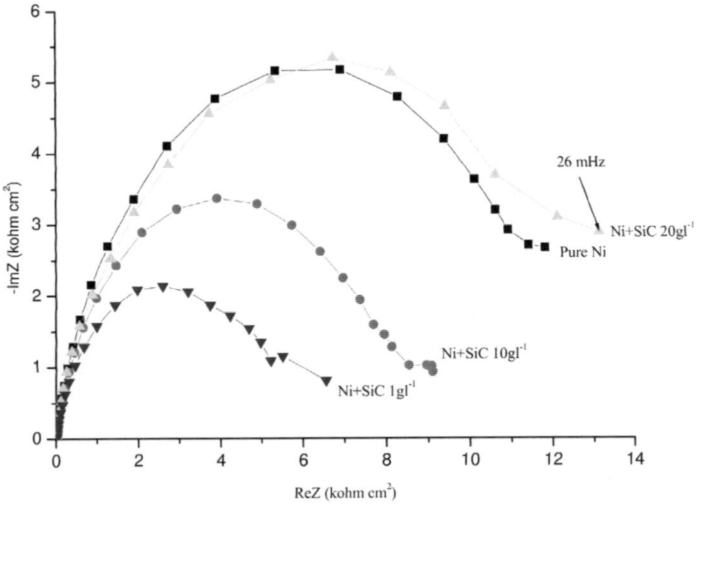

Figure 5. Impedance diagrams performed in NaCl 3,5% at different SiC concentrations of Ni-SiC coatings.

To verify corrosion property of Ni-SiC coatings polarisation resistance was measured also. It is evident that corrosion speed is low when resistance is high. Figure 5 shows the impedance diagram performed in 3.5% NaCl solution. The polarisation resistance of nano-composite coating is lower than that of pure nickel coating (12 $k\Omega cm^2$) when SiC concentration is low (7 $k\Omega cm^2$ at 1gl^{-1} SiC concentration and 9 $k\Omega cm^2$at 10 1gl^{-1} SiC concentration), but it is higher when SiC concentration increases (13 $k\Omega cm^2$ at 20 gl^{-1} SiC concentration). This result coincides with the measurement of porosity density of coatings, where increasing SiC deposit the porosity of coating decreases and hence corrosion resistance improves.

Gyftou et al. [5] in tribiological study of Ni matrix composite coatings containing nano and micro SiC particles considers that there are two factors correlated to improving wear resistance in Ni/SiC deposits: a variation of incorporation percentage of the SiC particles and the modification of crystallographic orientation. In the case of nano-size particles the influence on the adsorption-desorption phenomena during electrocrystallization is mild and all the deposits conserve the preferred orientation, but the observed changes of its quality are sufficient to alter the wear resistance. It is confirmed with our study on microstructure of Ni-nano SiC coatings on metal, where different deposits of SiC nanoparticles in plating clearly affect its corrosion properties.

4. Conclusion

In this work, the co-deposition and microstructure behaviour of a pure Ni and Ni with nano-structured SiC composite coatings were studied.

- An increase of coating hardness was observed with an increase of inert SiC particles amount in the deposit.
- At SiC concentration of 15-20 gl^{-1} with particles size 45-55 nm and magnetic stirring, the hardness of the deposit is between 400 and 450 HV.
- The porosity of coatings decreases with increasing SiC concentration above 15 gl^{-1} , and corrosion resistance of nano-structured SiC-nickel composite coatings performs better compared to Ni coating.
- A proper choice of concentration and plating condition can lead to production of composite deposits that enhances both hardness and wear resistance.

References

[1] Sharma G, Yadava R K and Sharma V K 2006 *Bull. Mater. Sci.* **29** (5) 491
[2] Srivastava M, Grips V K W, Jain A and Rajam K S 2007 *Surface & Coating Tech.* **202** 310
[3] Benea L, Bonora P L, Borello A and Martelli S 2002 *Wear* **249** 995
[4] Lin C S and Huang K C 2004 *Applied Electrochemistry* **34** 1013
[5] Han W, Gyftou P, Stroumbouli M, Pavlatou E A, Asimidis P and Spyrellis N 2005 *Electrochemica Acta* **50** 4544

Acknowledgments

The authors wish to thank Dr. Ph. T. San, Dr. P. D. Long and Mr. D. H. Manh in the Institute of Materials science, VAST for their help in structure measurements. This research is supported by the Vietnamese National Fundamental Research program 2006-2008.

Uniformity's influence of silica xerogel on synthesis efficiency of SiC nanorods by carbothermal reduction

Huynh Thi Ha[1], Cao Tuan Anh[2], Nguyen Thi Thu Ha[1], Phan Viet Phong[1], Dao Tran Cao[1], Le Quang Huy[1] and Nguyen The Quynh[1]

[1] Institute of Materials Science, Vietnam Academy of Science and Technology, 18 Hoang Quoc Viet Road, Cau Giay Distr., Hanoi, Vietnam
[2] Institute of Physics, Vietnam Academy of Science and Technology, 10 Dao Tan Street, Badinh Distr., Hanoi, Vietnam

E-mail: haht@ims.vast.ac.vn

Abstract. Silica xerogels containing nanocarbon were milled for several hours before the carbothermal reduction processing. The formation of SiC nanorods was investigated at different annealing temperatures. It is shown that the formation temperature of SiC nanorods can be decreased if the powder size of milled silica xerogels is lower than 100 nm. The morphology, structure and production efficiency of synthesized SiC were determined by scanning electron microscopy SEM, X-ray diffraction XRD and thermo-gravimetric analysis TGA measurements. It is evident that production efficiency of synthesized SiC nanorods clearly depends on uniformity of precursor materials.

Keywords: Silica xerogel, SiC nanorods, carbothermal reduction, grinding.

1. Introduction

Nanostructured silicon carbide has unique properties in microelectronics, optoelectronics as well as mechanical engineering. Nowadays, nanostructured SiC material manufacturing with high efficiency is an attraction in nanotechnology's application. SiC nanowires have been produced by several methods: carbothermal reduction of silica [1], reaction between silicon halides and CCl_4 [2], etc. An advantage of carbothermal reduction of silica is the relatively low cost in mass production. Some researchers successfully synthesized SiC nanowires by using precursor materials like nano-sized SiO or SiO_2 to react with carbon nanotubes under catalysts such as Fe, Al, etc. [3, 4]. However, under working conditions without any catalyst, the production efficiency of this method depends significantly on the size and uniformity of silica precursors. In current works, silica xerogel containing carbon nanoparticles has been produced from tetraethoxysilane and saccharose. In order to investigate the uniformity's influence of silica xerogel containing carbon particles on synthesis efficiency of SiC nanorods, grinding processes were carried out before and after formation of xerogel.

In this work, the synthesis of SiC nanorods via a two-step reaction in carbothermal reduction has been developed. These studies address the overall structure and growth efficiency of SiC nanorods.

2. Experimental

The first stage in preparation of silica xerogel containing carbon particles is the silica sol making. It was done as follows: 100 ml $Si(OC_2H_5)_4$ (p.a), 100 ml ethanol (p.a) and 100 ml deionized water were

© 2009 IOP Publishing Ltd

firstly mixed together by magnetic stirring for 15 minutes. A few drop of concentrated nitride acid (~ 5 ml, MERKC, p.a) was then added into the mixture under continuous stirring for another 15 minutes. In order to obtain carbon nanoparticles in silica gel, saccharose $C_{12}H_{22}O_{11}$ was next gradually added into the solution under vigorous stirring. The amount of saccharose was adjusted to get molar ratio of C to Si of 4.1:1. The stirring was done for 20 hours in ambient air to achieve homogeneity in gel.

In the second stage, gel was dried at 90°C in a dry box and then was calcined at 700°C for 2 hours in flowing N_2 atmosphere to completely decompose saccharose into carbon and water vapor. Gel was then divided into smaller pieces to undergo different experiments. A part was grinded before calcination. Another part was grinded after xerogel formation. After the grinding, they were then divided again and milled for 5, 10, and 15 hours respectively in ambient air. Another part of xerogel was milled for 5 hour in ethanol. SiO_2 xerogel containing carbon particles was milled by high-energy horizontal vibrating ball mill SPEX 8000D in using Tunsten carbide (WC) grinding vial.

The apparatus for the synthesis of SiC nanorods is a horizontal tube vacuum furnace GLX1600-80X. The milled xerogels were loaded in a graphite boat and placed at the center inside the alumina tube. The samples were heated at different temperatures at 1400, 1450, 1500 and 1550 °C for 3 hours under continuous flow of high-purity argon gas.

Upon completing the reaction, the collected SiC nanorods were characterized by X-ray diffraction (XRD Siemens D5000), transmission electron microscopy (TEM JEM1010 JEOL), scanning electron microscopy (SEM Hitachi S-4800), and thermo-gravimetric analyzer (TG-DTA/DSC Setaram Instrumentation).

3. Results and discussion

In our work, the size of silica xerogel containing carbon nanoparticles grinded by different conditions was carefully tested. Figure 1 displays SEM images of silica xerogel containing carbon nanoparticles.

Figure 1. SEM images of silica xerogel containing carbon nanoparticles under:
a) HEM for 5 hours after xerogel formation,
b) rough grinding before calcination of gel+HEM for 15 hours after xerogel formation,
c) fine grinding before calcination of gel+HEM for 15 hours after xerogel formation,
d) rough grinding before calcination of gel+HEM for 05 hours after xerogel formation in ethanol.

If there was only high-energy mill (HEM) for 5 hours after xerogel formation, the size of xerogel was around several hundred nanometers (figure 1a). In the case where there was a rough grinding before calcination of gel, the size of xerogel appreciably decreased below 100 nm (figure 1b). When there was a fine grinding before calcination of gel, the size of xerogel was smaller than 50 nm though the boundaries were still not clear (figure 1c). In the case where there was a rough grind before calcination of gel and high-energy mill carried out in ethanol, the size of xerogel was smaller and grain boundaries became more evident (figure 1d).

XRD spectrum and diffraction pattern of the product in figure 2 show two constituents. The analysis from figure 2a indicates that the product was mainly β-SiC and contained part of remaining amorphous carbon and silicone oxide. The result from figure 2b reveals that the product was a mixture of predominantly β- SiC with crystalline WC. The appearance of WC particles was due to the WC grinding vial when xerogel was milled in ethanol. The result also shows that at heating temperature 1450 °C the crystal β- SiC could be received.

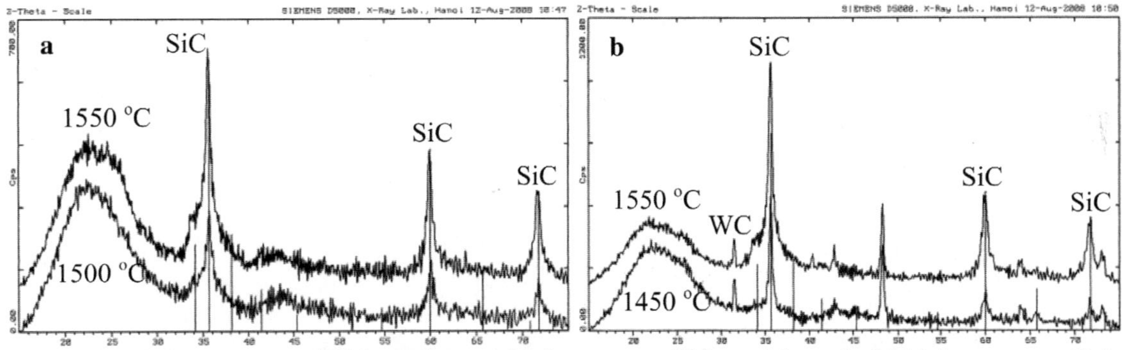

Figure 2. XRD spectrum of SiC product: a) high-energy mill in ambient air; b) high-energy mill in ethanol.

Figure 3 below shows TEM measurements of SiC nanorods heated at 1550°C. The result indicates that outside diameters of SiC nanorods ranged from 10 to 30 nm (figures 3a, b). In the case where high-energy mill was carried out in ethanol, they were smaller (figure 3c, 10 - 20 nm). The morphology of nanorods was straight and solid. Stacking faults were found on SiC nanorods.

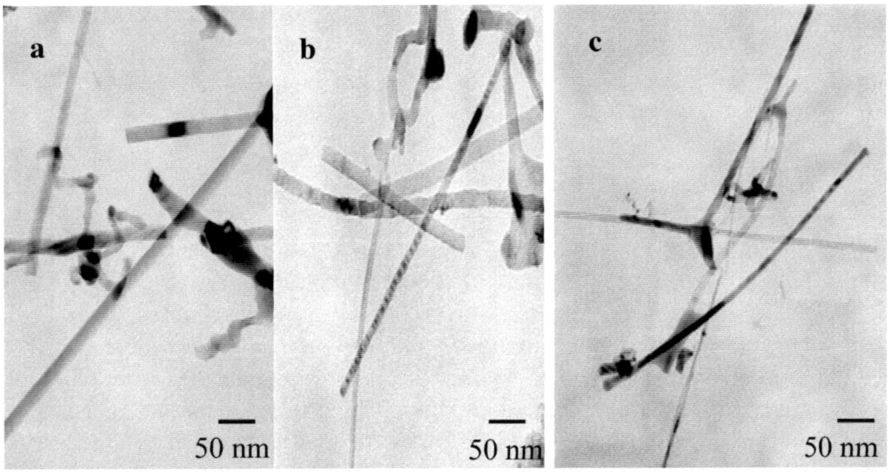

Figure 3. TEM images of SiC nanorods at heating temperature 1550°C with: a,b) high-energy mill in ambient air, c) high-energy mill in ethanol.

Journal of Physics: Conference Series **187** (2009) 012084 doi:10.1088/1742-6596/187/1/012084

SEM images of SiC nanorods are shown in figure 4. At heating temperature 1400°C, rods randomly appeared (figure 4a) and there were no crystalline phases in XRD analysis. At 1450°C, a number of SiC nanorods were obtained with outside diameters of roughly 100 nm and lengths of 5-10 μm (figure 4b). At 1550°C, a large amount of nanorods were formed with thinner outside diameters and shorter lengths (figure 4c).

In the two-step reaction process, the reaction equations are:

$$SiO_2(s) + C(s) \rightarrow SiO(v) + CO(v) \quad , \tag{1}$$

$$SiO(v) + C(s) \rightarrow SiC(s) + CO(v) \quad . \tag{2}$$

According to suggestions by W. Han [5] the generated CO vapor in reaction (1) can react with SiO vapor on the generated SiC nanorods surface as follows:

$$SiO(v) + CO(v) \rightarrow SiC(s) + CO_2(v) \quad . \tag{3}$$

In such a case, the epitaxial growth of SiC on the surface of initially produced SiC nanorods from reaction (2) makes the diameter of some final SiC nanorods larger. We think that at lower heating temperature (1450°C), the lost amount of SiO and CO vapours because of argon flow was smaller than that at a higher temperatures (1550°C); therefore the probability of reaction (3) was larger, which explains why in our results the diameters of SiC nanorods produced at 1450°C were larger.

Figure 4. SEM images of SiC nanorods at heating temperature:
a) 1400°C, b) 1450°C, and c) 1550°C.

Figure 5 is the thermogravimetric analysis (TGA) of SiC product. The measurement was carried out in ambient air in temperature range 20-1350°C with heating rate of 10°C/min. The SiC products under different grinded and heated temperature conditions were tested. The XRD measurements show that a part of amorphous un-reacted carbon still existed in the synthesized product, hence in TGA result it could have been the heat emission process where carbon was burned in a temperature range below 1000°C. The percentage value of lost mass in heat emission implies an indirect relationship between the amount of un-reacted free carbon with Si in solid state and the efficiency of different SiC synthesis processes. Table 1 presents the TGA data of SiC synthesis processes, where TGA measurements were done with different kinds of grinding: non-grinding before xerogel formation sample (NG), rough grinding before xerogel formation (RG), fine grinding before xerogel formation (FG), high-energy mill in ethanol (E) and at different temperatures 1450, 1500, and 1550°C. After xerogel formation the samples were milled by high-energy mill for 5 or 15 hours.

The results show that when the size of xerogel was in the range of several hundred nanometers (with HEM for only 5 hours) the lost mass of free carbon was rather large 33-39%, while with the size in the range of several ten nanometers, the lost mass of free carbon was significantly smaller, even at lower heating temperatures (24.05 %, 1450°C).

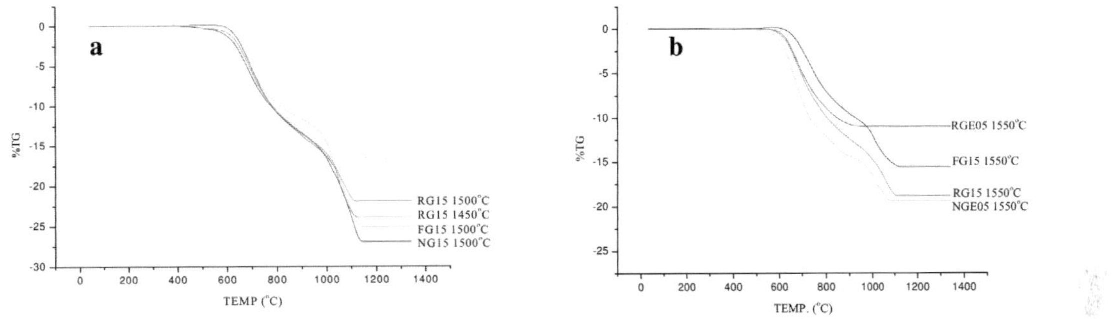

Figure 5. TGA measurements of synthesized production heated at: a) 1450°C and 1500°C, b) 1550°C.

The results also indicate that the products of fine grinding and the rough grinding before xerogel formation gave similar loss on free carbon, while non-grinding before xerogel formation gave larger ones. The loss mass of free carbon in the case of rough grinding before xerogel formation with high-energy mill in ethanol for 5 hours was the least of all. It corresponded to the best uniformity and smallest size of prepared xerogel powder.

Table 1. Relationship of remaing free carbon in SiC product with grinding regimes of SiO_2 xerogel containing carbon nanoparticles.

Heating temperature (°C)	Remaining free carbon, %						
	0.7 MPa Ar pressure[a]	1.4 MPa Ar pressure[a]	NG15[b]	RG15[b]	FG15[b]	NGE05[b]	RGE05[b]
1450	no nanorod detected	no nanorod detected		-24.05			
1500	no nanorod detected	no nanorod detected	-26.94	-21.88	-25.08		
1550	-39	-33		-18.81	-15.25	-19.36	-10.93

[a] *-hundred-nm size xerogel based* [b] *-ten-nm size xerogel based (at 0.7 MPa Ar gas pressure)*

Through all the results obtained above, it can be seen that in carbothermal reduction of SiO_2 without any catalyst the uniform small size of silica xerogel containing carbon nanoparticles plays an important role in production efficiency of SiC nanorods. Of course, high-energy milling produces smaller grain size, possibly higher surface area, which should contribute a reduced reaction on vaporization temperature and a higher chemical reactivity in SiC nucleation, but it is obviously seen that the different grind regimes give different size and uniformity of xerogel, therefore different production efficiencies in result.

4. Conclusion

The morphology of SiC products synthesized in a two-step carbothermal reduction process of SiO_2 without any catalytic particles was investigated.

- The β-SiC nanorods were 10-100 nm in diameter and 1-10 μm in length dependent on the uniformity and the size of silica xerogel containing carbon nanoparticles.
- With fine uniformity and small-size (<100 nm) silica xerogel precursors, the formation temperature of SiC nanorods decreased.
- The production efficiency of synthesized SiC nanorods clearly is affected by uniformity of precursor materials and grinding regimes.

References

[1] Meng G W, Zhang L D, Mo C M, Zhang S Y, Li H J, Qin Y and Feng S P 1999 *Metallurgical and Materials Transactions A* **30A** 213
[2] Lu Q, Hu J, Tang K, Qian Y and Zhang Y 1999 *J. Appl. Phys. Lett* **75** 507
[3] Chen Y, Li C P, Chen H and Chen Y 2006 *Sci. and Tech. of Adv. Mater.* **7** 839
[4] Deng S Z, Wu Z S, Zhou J, Xu N S and Chen J 2002 *Chem. Phys. Lett.* **356** 511
[5] Han W, Fan Sh, Li Qu, Liang W, Gu B and Yu D 1997 *Chem. Phys. Lett.* **265** 374

Acknowledgements

The authors would like to thank their colleagues D. H. Manh and Tr. D. Thanh in the Institute of Materials Science for their help in structure measurements. This work is supported by Research Program and National Key Laboratory for Electronic materials and devices of Institute of Materials science - Vietnam Academy of Science and Technology.

Grain size effect on the permittivity of $La_{1.5}Sr_{0.5}NiO_4$ nanoparticles

Tran Dang Thanh and Le Van Hong

Institute of Materials Science, Vietnam Academy of Science and Technology, 18 Hoang Quoc Viet Road, Cau Giay Distr., Hanoi, Vietnam

E-mail: thanhtd@ims.vast.ac.vn

Abstract. Using the annealing at different temperatures the $La_{1.5}Sr_{0.5}NiO_4$ ceramic samples with different mean grain size were manufactured. Mean grain size (<D>) of the samples was evaluated by Warren-Averbach method and their SEM images. The obtained results of both methods are almost the same, changing from 16.2 to 95 nm in dependence on the annealing temperature. The frequency dependence of dielectric constant in the frequency range of (1-13 MHz) was recorded for all samples. The real (ε') and the imaginary parts (ε'') of the permittivity of $La_{1.5}Sr_{0.5}NiO_4$ samples abnormally depend on the frequency, exhibiting a dielectric resonance around 500 kHz. R-L-C in series equivalent-circuit fitted well for the obtained result. It was supposed that there exists magnetic contribution in material that suggests the material is a multiferroic one. Dependence of the (ε') on the mean grain size supposed that the colossal dielectric property is an intrinsic behaviour of $La_{1.5}Sr_{0.5}NiO_4$ material.

Keywords: Colossal permittivity, multiferroics, dielectric resonance.

1. Introduction

The high permittivity material always is a current interest, because of its potential applications. Various materials of high permittivity have been reported such as the $BaTiO_3$ doped with La that has dielectric constant of 10^4 [1], $CaCu_3Ti_4O_{12}$ with the permittivity of about $8x10^4$ at temperature of 250 K and frequency up to 1 MHz [2]. Kim and co-workers have improved dielectric constant to 10^5 for $Pb_{1-3x/2}La_xTiO_3$, $Pb_{1-x}La_xTi_{1-x/4}O_3$ [3]. There was an interested recent result of Biskup *et al.* [4] that reported results of study on $Pr_{0.6}Ca_{0.4}MnO_3$ perovskite, which informed a high dielectric constant of about 10^4- 10^5. In the last years, La_2NiO_4 compound has attracted much attention not only because of its potential applications in the solid oxide fuel cell cathodes [5], but also due to its very interesting structural, unusual magnetic, conducting properties, especially its permittivity. This material system is of the low dimension structure type which exhibits Charge-Density-Wave (CDW) property. Its permittivity is higher than 10^5 at room temperature in a wide frequency range from 1 KHz up to 1 MHz [6]. So it has been called the colossal dielectric material. The interesting dielectric properties of this material have attracted researchers in both fundamental scientific and technological studies. Colossal dielectric constant behaviour is one of the most technically decisive factors for many applications of the high dielectric constant materials, such as random access memories based on capacitive elements (DRAM). The dependence of dielectric constant on frequency and temperature of $La_{1.5}Sr_{0.5}NiO_4$ material is very similar to that observed in the canonical CDW [7-9], as well as in the

© 2009 IOP Publishing Ltd

left handed material (LHD) made by a composite of both ferromagnetic and ferroelectric constituents [10]. In this paper we present an observation of the negative permittivity of the material at the frequency range around the resonance frequency. The obtained experimental results are fitted well by an R-L-C circuit, R, L and C are equivalent conductance, inductance and capacitance.

2. Experiment

La_2O_3, SrO and NiO with purity of 99.9% were used as starting materials. The starting materials were thoroughly mixed and milled for 6 hours by using the reactive mechanical milling technique on D8000-Spex [11]. The milled powder was pressed into 7 pellets with diameter of 13 mm and height of 1.5 mm, heated in air for 1 hour at 7 different temperatures changing from 700 to 1000°C with the temperature step of 50°C, noted as S1, S2, S3, S4, S5, S6, and S7. Phase structure of all the samples was checked by powder X-ray diffraction by means of a SIEMENS D5000 diffractometer equipped with Cu-kα radiation (λ = 1.5406 Å) at room temperature. By using Warren-Averbach method combined with Win-Crysize program the crystalline grain size was estimated for all samples. Grain size of all the samples was estimated also from their FESEM images done by using the S-4800 FESEM microscopy. Dielectric constant at room temperature of all the samples was estimated via the capacitance and conductance of the disk capacitor in configuration of $Ag/La_{1.5}Sr_{0.5}NiO_4/Ag$ that were measured in a frequency range of (1 KHz-13 MHz) by using a HP 4192A impedance analyzer.

3. Results and Discussion

Figure 1 shows X-ray powder diffraction patterns of the samples recorded at room temperature. Comparing the PDF (Powder Diffraction File) card with number 32-1241 [12] identified that the prepared samples are of $La_{1.5}Sr_{0.5}NiO_4$ single phase, belong to the F_4K_2Ni-type, tetragonal structure with space group of I4/mmm (no. 139). $La_{1.5}Sr_{0.5}NiO_4$ displays a quasi two-dimension structure in which the perovskite blocks are separated by the presence of rock-salt type (La-Sr/O) layers along the c axis. The full width at half maximum of the X-ray diffraction lines decreases with the samples annealed at higher temperature. Probably this is an evidence of growing up crystalline grain. The average crystalline size (<D>) of $La_{1.5}Sr_{0.5}NiO_4$ grain, determined by Warren-Averbach method for all the samples, was presented in table 1.

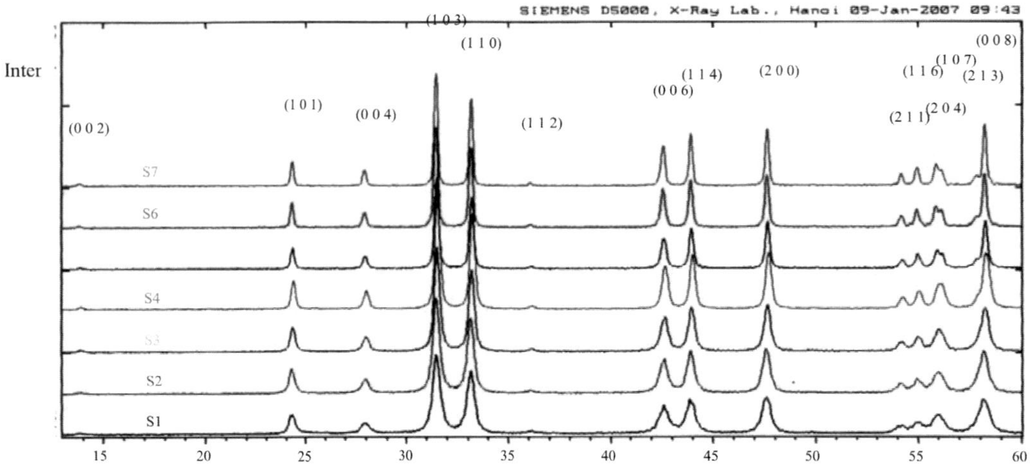

Figure 1. The X-ray powder diffraction patterns of 7 samples.

Table 1. Average grain size (<D>) and the results of fitting the equations (5) and (6) for $\varepsilon'(f)$) and ($\varepsilon''(f)$ to the experimental data for the 7 samples.

Sample (notation)	700°C (S1)	750°C (S2)	800°C (S3)	850°C (S4)	900°C (S5)	950°C (S6)	1000°C (S7)
<D> nm	16.2	17.0	23.5	35.0	41.6	61.3	95.0
f_0 (Hz) x 10^5	5.607	4.886	5.131	5.411	6.807	7.645	7.608
α x 10^6	3.327	3.208	2.758	2.005	1.659	1.235	1.027
R (Ω)	3.741	3.638	3.086	2.251	1.342	0.992	0.852
L (μH)	0.562	0.567	0.559	0.561	0.405	0.402	0.415
τ(ns)	284	326	310	294	234	208	209

Figure 2. FESEM image of two samples S1 and S7.

It is clear that the grain size increases significantly with increasing annealing temperature. The mean grain size of the samples was also estimated by means of FE-SEM image (figure 2). It was shown that the obtained result is comparable to that obtained by Warren-Averbach method.

It was known the frequency dependence of capacitance and dissipation factor of the flat disk capacitor can be expressed as follows:

$$C(\omega) = \frac{\varepsilon'(\omega)\varepsilon_0 S}{d}, \tag{1}$$

$$tg\delta(\omega) = \frac{\varepsilon''(\omega)}{\varepsilon'(\omega)}, \tag{2}$$

where S and d are the area and thickness of capacitor, respectively, ε_o = 8.85 x 10^{-12} F/m is the dielectric constant of vacuum, $\varepsilon'(\omega)$ and $\varepsilon''(\omega)$ are the real and imaginary part of the dielectric constant, respectively.

$$\varepsilon(\omega) = \varepsilon'(\omega) - i\varepsilon''(\omega), \tag{3}$$

In this work, the capacitor with S = 78.5 mm^2 and d = 1.5 mm was used.

Figure 3 presents the frequency dependence of the real part $\varepsilon'(\omega)$ and imaginary part $\varepsilon''(\omega)$ of the permittivity at room temperature of three samples S1 (a), S4 (b), and S7 (c). A dielectric resonance at a frequency range $4 \times 10^5 \div 8 \times 10^5$ Hz was observed and its resonance frequency shifts to higher frequency when increasing grain size.

The dielectric constant at low frequency range is positive and higher than 10^5 for all the samples. But it exhibits a negative dielectric constant at frequency over the resonance frequency f_0 and looks like that of a composite Left Handed Material (LHM) consisting of both ferromagnetic and ferroelectric compounds [10].

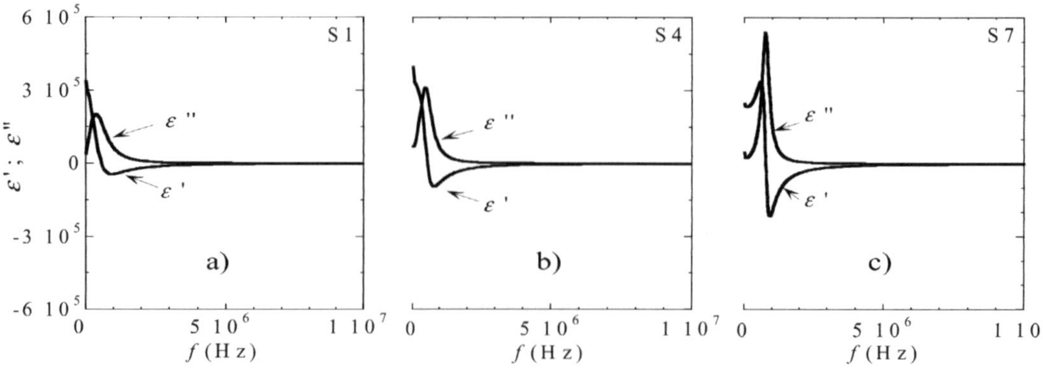

Figure 3. The dependence on frequency of the real part $\varepsilon'(\omega)$ and imaginary part $\varepsilon''(\omega)$ of the permittivity at 300 K for three examples samples S1 (a), S4 (b), and S7 (c).

As known the collective oscillator system with damping is determined by equation [13]

$$\frac{d^2P}{dt^2}+2\alpha\frac{dP}{dt}+(\omega_0^2-\frac{Ne^2}{3m\varepsilon_0})P=\frac{Ne^2}{m}E\,, \tag{4}$$

where P is the polarization of material, α is the damping factor, ω_0 is the resonance frequency ($\omega_0 = 2\pi f_0$), N and m are the number and the induced mass of oscillators, respectively. The equation (4) is similar to the equation for the RLC series circuit, where R, C and L are the resistance, capacitance and inductance of the equivalent circuit, respectively. Therefore the solution of equation (4) could be accepted as a solution of the equation describing the RLC series circuit with the damping factor:

$$\alpha=\frac{\Delta\omega_h}{2}=\frac{R}{2L}\,. \tag{5}$$

By solving equation (4) we have the frequency-dependence expressions for real part ($\varepsilon'(\omega)$) and imaginary part ($\varepsilon''(\omega)$) of the permittivity:

$$\varepsilon'(\omega)=\frac{\left(\omega_0^2-\omega^2\right)}{L\left[\left(\omega_0^2-\omega^2\right)^2+(2\alpha\omega)^2\right]}\times\frac{d}{\varepsilon_0 S}\,, \tag{6}$$

$$\varepsilon''(\omega)=\frac{2\alpha\omega}{L\left[\left(\omega_0^2-\omega^2\right)^2+(2\alpha\omega)^2\right]}\times\frac{d}{\varepsilon_0 S}\,. \tag{7}$$

The dielectric relaxation time (τ) is defined by formula

$$\tau=\frac{1}{\omega_0}\,. \tag{8}$$

Figure 4 presents the experimental data of the real $\varepsilon'(\omega)$ and the imaginary parts $\varepsilon''(\omega)$ (mark points) and fitted results by using equations (6) and (7) in a frequency range around the resonance value (lines). The obtained fitting parameters such as resonance frequency value f_0, damping factor α, R, L, and relaxation time (τ) are presented in table 1.

As seen in table 1 the obtained fitting values L are almost not changed with increasing grain size, but the obtained values of C increases. The relaxation time and dissipation parameter decrease with increasing grain size. It is supposed to be related with an effect of the grain boundary that decreases with increasing grain size. Bai *et al.* [10] observed the same phenomenon in a composite created by mixing a ferromagnetic with a ferroelectric composition. He observed resonance at different frequencies in range of GHz, depending on the ratio of two components. We observed the resonance frequency much lower than that reported by Bai *et al.* It is supposed to be related with the high dielectric constant of the $La_{1.5}Sr_{0.5}NiO_4$ material.

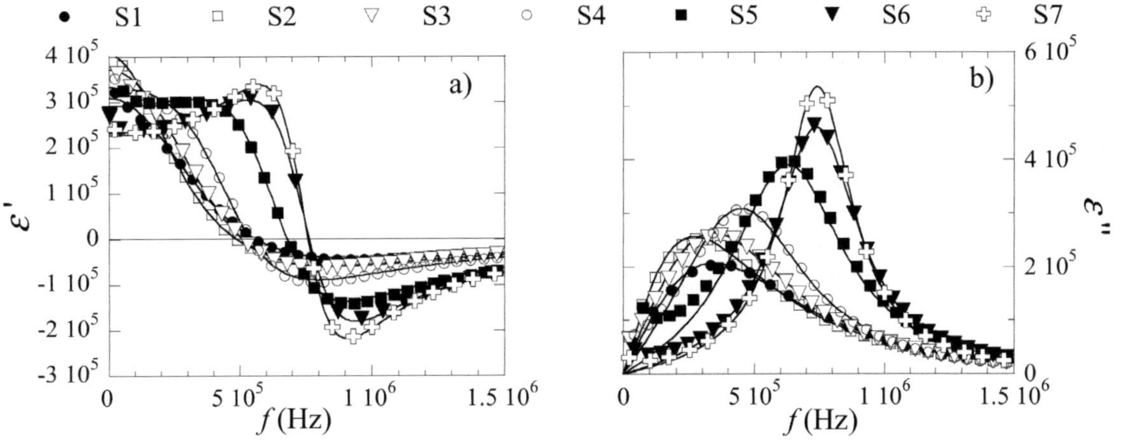

Figure 4. The dependence on frequency of the ε' (a) and ε'' (b) of the permittivity at 300 K for all samples. The solid lines are fitting results by using equations (5) and (6).

The dissipation factor is the most important parameter of the insulator material, suggesting the possibilities of its application. Therefore in the aim of searching the application possibilities we plot curves of dissipation factor at low frequencies ($f < f_0$) at 300 K for all the $La_{1.5}Sr_{0.5}NiO_4$ samples (figure 5). It is clearly seen that all the $La_{1.5}Sr_{0.5}NiO_4$ samples have quite low dielectric loss (~ 10^{-1} – 10^{0}) in a wide frequency range, decreasing in dependence of the grain size of $La_{1.5}Sr_{0.5}NiO_4$. Finally we would like to confirm that the colossal dielectric constant that obtained for $La_{1.5}Sr_{0.5}NiO_4$ samples is an intrinsic behaviour of material as reported by Rivas *et al.* [6]. As shown in figure 6, the dielectric constant of samples is about 10^5, increasing with grain size. This obtained result is opposite to report of Lunkenheimer *et al.* [14], who theoretically suggested that the grain effect is a main reason of the colossal dielectric constant that observed in the related oxides materials. By means of Lunkenheimer [14] the dielectric constant should increase with decreasing grain size, but in our case we have obtained an opposite result. It is supposed to conclude that the observed colossal permittivity is an intrinsic behaviour of $La_{1.5}Sr_{0.5}NiO_4$ material.

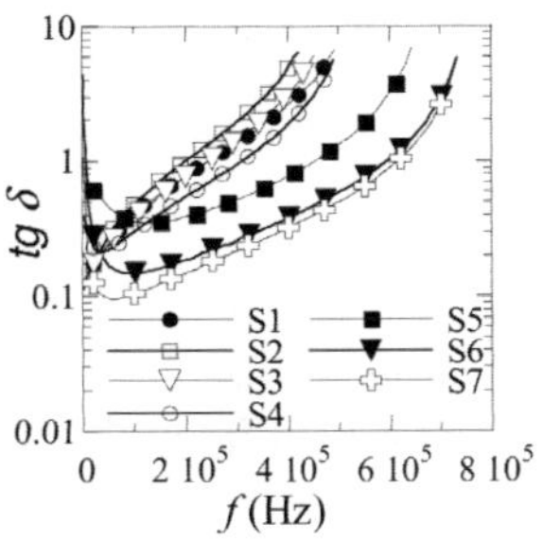

Figure 5. The dissipation factor in the low frequencies ($f < f_0$) at 300 K for all the samples.

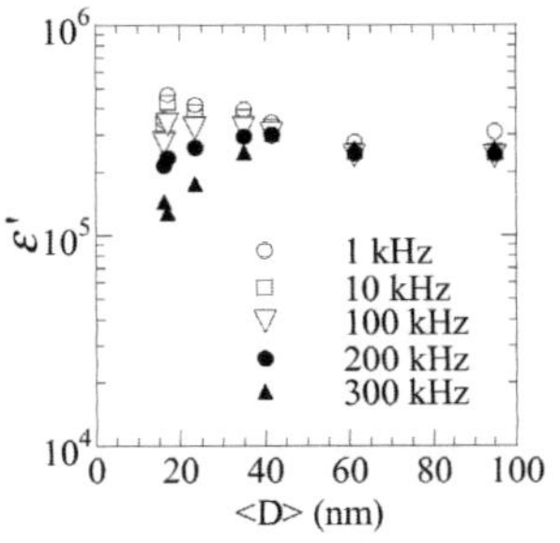

Figure 6. The dependence of the real part permittivity on mean grain size of $La_{1.5}Sr_{0.5}NiO_4$.

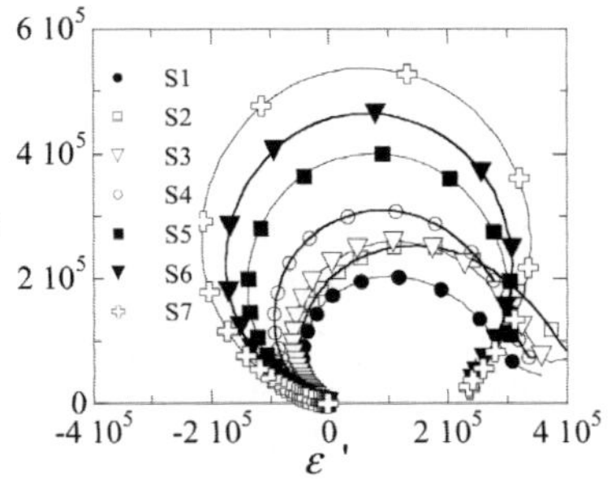

Figure 7. The Cole-Cole curves of $La_{1.5}Sr_{0.5}NiO_4$ samples with 7 different mean grain size.

In aim of searching the influence of grain boundary effect we set up the dependence between $\varepsilon'(\omega)$ and $\varepsilon''(\omega)$ that called Cole-Cole diagram for all the samples, presented in figure 7. It is clear that the Cole-Cole diagram of samples consisting of grains with size larger than 40 nm exhibits circles.

It means that the dielectric relaxation in these samples is a single relaxation process and the influence of the long time relaxation process concerning the surface states at grain boundary is reasonably neglected. The grain boundary effect is clearly observed at the Cole-Cole diagrams of the samples with grain size smaller than 40 nm. Cole-Cole diagrams of the samples deviated essentially at the low frequency region. It is evidence related with the effect induced by the grain boundary states, which normally characterize the low frequency process.

4. Conclusions

The $La_{1.5}Sr_{0.5}NiO_4$ ceramic samples with the average crystals size (<D>) changing from 16.2 to 95 nm were successfully manufactured by using the reactive mechanical milling technique combined with annealing at different temperatures in air. All the samples $La_{1.5}Sr_{0.5}NiO_4$ have perovskite layers structure of K_2NiF_4-type. They exhibit a colossal dielectric constant about 10^5 and low dielectric loss ($\sim 10^{-1} - 10^0$) in a wide frequency range. The colossal dielectric constant is an intrinsic behaviour of the $La_{1.5}Sr_{0.5}NiO_4$ material. A low frequency resonance (about 10^5 Hz) was observed. It behaves like the R-L-C circuit. L and C components probably are contributed by the FM and the FE composition existing in $La_{1.5}Sr_{0.5}NiO_4$, so called the multiferroics one.

References

[1] Olivas J D and Bolin S 1990 *JOM.* **50** 38
[2] Homes C C, Vogt T, Shapiro M, Wakimoto S and Rairez A P 2001 *Science* **293** 673
[3] Kim B G, Cho S M, Kim T Y and Jang H M 2001 *Phys. Rev. Lett.* **86** 3404
[4] Biskup N, Andre's A de, Martinez J L and Perca C 2005 *Phys. Rev.* B **72** 024115
[5] Kharton V V, Viskup A P, Kovalevslky A V, Naumovich E N and Marques F M B 2001 *Solid State Ionics* **143** 337
[6] Rivas J, Rivas-Murias B, Fondado A, Mira J and Senaris-Rodriguez M A 2004 *Appl. Phys. Lett.* **85** 6224
[7] Cava R J, Fleming R M, Littlewood P, Rietmann E A, Schneemeyer L F and Dunn R G 1984 *Phys. Rev.* B **30** 3228
[8] Cava R J, Littlewood P, Fleming R M, and Dunn R G 1985 *Phys. Rev.* B **33** 2439
[9] Ramirez A P, Subramanian M A, Gardel M, Blumberg G, Li D, Vogt T, Shapiro S M 2000 *Solid State Com.* **115** 217
[10] Bai Y, Zhou J, Sun Y, Li B, Yue Z, Gui Z and Li L 2006 *App. Phys. Let.* **89** 112907
[11] Thanh T D, Manh D H, Phuc N X and Hong L V 2005 *Proc. of the eighth German-Vietnamese Seminar on Physics And Engineering (Erlangen, Germany, 03-08, April)* p. 124
[12] 2005 SIEMENS, JCPDS-ICDD, PDF card with number of 32-1241
[13] Hippel A R von 1954 *Dielectrics and waves* (New York:John Wiley & Sons)
[14] Lunkenheimer P, Bobnar V, Pronin A V, Ritus A I, Volkov A A and Loidl A 2002 *Phys. Rev. B* **66** 052105

Acknowledgments

This work is supported in part by the National Fundamental Research Program in Natural Sciences and the scientific project of the IMS. The National Key Laboratory of Electronic Materials and Devices is highly acknowledged.

Synthesis and LPG-sensing properties of TiO_2 nanowires

Dang Thi Thanh Le[1], Dang Duc Vuong[2] and Nguyen Duc Chien[2]

[1]International Training Institute for Materials Science (ITIMS)
[2]Institute of Engineering Physics, Hanoi University of Technology,
1 Dai Co Viet Street, Hanoi, Vietnam

E-mail: thanhle@itims.edu.vn

Abstract. TiO_2 nanowires were synthesized by a hydrothermal process in KOH solution at 190°C using TiO_2 powder as a source material. Characterizations indicate the high degree of crystallinity with diameter of ca. 10-20 nm and length in micrometer scale of the obtained TiO_2 nanowires. Multiple nanowires gas sensors were fabricated by dispersing the TiO_2 nanowires on an interdigit Pt-electrode. The TiO_2 nanowires sensors show a high sensitivity to Liquefied Petroleum Gas (LPG) at working temperature of 400°C. The results demonstrate the possibility of using TiO_2 nanowires for gas-sensing application.

Keywords: TiO_2, nanowires, LPG, gas sensors.

1. Introduction

The synthesis and characterization of one-dimensional (1D) nanostructures (nanowires, nanotubes, nanorods) have received considerable attention due to their unique properties and novel applications. Miniaturization may represent the trend in a range of novel technologies. It is clear that a wealth of interesting and new phenomena are associated with nanoscale structures, with the best established examples including size-dependent excitation or emission [1], quantized conductance [2], Coulomb blockade (or single-electron tunneling, SET) [3], and metal-insulator transition [4]. It is generally accepted that the 1D nanostructures provide a good system to investigate the dependence of electrical and thermal transport or mechanical properties on dimensionality and size reduction (or quantum confinement). They are also expected to play an important role as both interconnects and functional units in fabricating electronic, optoelectronic, electrochemical, and electromechanical devices with nanoscale dimensions. In comparison with quantum dots and wells, the advancement of 1D nanostructures has been slow until very recently and has been hindered by the difficulties associated with the synthesis and fabrication of these nanostructures with well-controlled dimensions, morphology, phase purity, and chemical composition.

Although 1D nanostructures can now be fabricated (in the setting of a research laboratory) using a number of advanced nanolithographic techniques, further development of these techniques into practical routes to large quantities of 1D nanostructures from a diversified range of materials, and at reasonably low cost, still requires great ingenuity. In contrast, unconventional methods based on chemical synthesis might provide an alternative and intriguing strategy for generating 1D nanostructures in terms of material diversity, cost, throughput, and the potential for high-volume production [5].

Many methods have been successfully developed for the fabrication of 1D nanostructures,

© 2009 IOP Publishing Ltd

including VS (vapor-solid), VLS (vapor-liquid-solid), SLS (solution-liquid-solid), template-based synthetic approaches and laser ablation [6]. However, almost all of these methods used either catalyst materials or physical templates, which unavoidably brought some contamination to the products. Therefore, it is very interesting to explore a new approach to synthesize 1D nanomaterials without using preformed templates or catalyst materials. Recently, the reaction between different TiO_2 precursors and a concentrated alkaline solution under moderate hydrothermal method is observed to be an effective approach to prepare 1D nanostructures of titania. Among various technologies available today in advanced materials processing, the hydrothermal technique occupies a unique place owing to its advantages over conventional technologies. A simple and cost-effective hydrothermal method for the large scale production of pure titania nanotubes of small diameters was introduced by Kasuga et al [7, 8].

Much effort has been concentrated on the important metal oxides such as SnO_2, TiO_2, ZnO [9]. TiO_2 is an n-type wide band-gap oxide semiconductor used for variety of applications such as gas sensors, dyesensitized solar cells, environmental purification, nanodevices, and photocatalysts [10-15]. TiO_2 nanostructured materials are particularly interesting because of their high specific surface area.

We adopt a simple wet chemical approach to synthesize single crystalline TiO_2 nanowires. Moreover, this is a simple and cheap approach which is suitable for domestic research conditions.

In this report, we present the synthesis, microstructure, and the Liquefied Petroleum Gas (LPG) sensing characteristics of TiO_2 nanowires.

2. Experimental

2.1. Synthesis of nanowires based on hydrothermal processes

The TiO_2 nanowires were synthesized by a hydrothermal process and using commercial TiO_2 powder (Merck, 98%) as source material. In detail, 1 g of TiO_2 powder and 25 ml of 10 M KOH solution are stirred at room temperature for 24 h. The mixture was put into a teflon-lined stainless autoclave which was heated at 190°C for 24 h. After that, it was naturally cooled down to room temperature. A white mixture was washed by deionized water to reduce KOH concentration. The obtained precipitation was filtered and washed with 1 M HCl solution until a pH value near 7 for eliminating potassium ions. After treating by HCl solution, the mixture was filtered and washed by deionized water, all washing steps were carried out by using a centrifuge. The experimental procedures flow is shown in figure. 1.

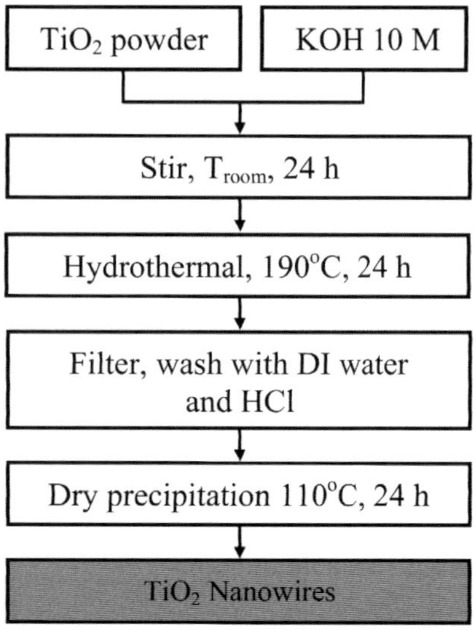

Figure 1. Schematic diagram of experimental procedure.

The obtained titanium oxide nanowires were annealed at 400°C for 30 min with the temperature increasing at the rate of 100°C/h. The collected TiO_2 nanowires were ultrasonically dispersed in polyethylenglycol and then deposited on Pt-interdigitated electrode by drop-coating. The electrode was fabricated using conventional photolithography method as previously reported [16]. Finally, the sensors were heat-treated at 600°C for 30 min.

2.2. Characterizations

The morphologies and crystal structures of the TiO_2 nanowires were characterised using field emission scanning electron microscopy (FE-SEM, HITACHI S4800), transmission electron microscopy (TEM, JEM-100cx instrument with an accelerating voltage of 80 kV) and X-ray diffractometer (PANalytical-Philips X'pert Pro with Cu $K_{\alpha 1}$ radiation $\lambda = 1.54056$ Å).

The LPG sensing properties of TiO_2 nanowires were studied through the resistance change under LPG vapour atmosphere, for LPG concentrations of 500, 1000, 2000, 4000 and 8000 ppm, and at operating temperatures of 400°C. The response and recovery characteristics were monitored and recorded by using a precision semiconductor parameter analyzer (HP 4156A). Herein, the response was defined as the ratio R_{air}/R_{gas}, where R_{air} is the sensor resistance in air and R_{gas} is the resistance in the presence of reducing gas.

3. Results and discussion

3.1. Microstructure characteristics

Figure 2 shows FE-SEM image of TiO_2 nanowires after heat treatment at 400°C for 30 min. The presence of numerous TiO_2 nanowires with diameter of ca. 10-20 nm and length of ca. 700-800 nm was observed (see figure 3). In comparison with Ref. [17], it can be seen that the obtained TiO_2 nanowires are more uniform, longer and have a larger aspect ratio.

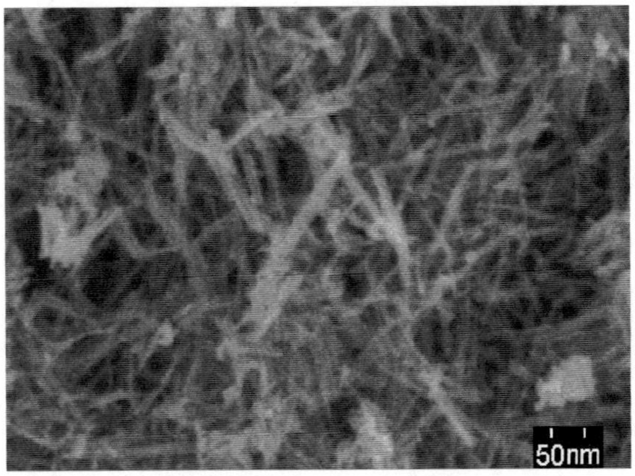

Figure 2. FE-SEM image of TiO_2 nanowires annealed at 400°C.

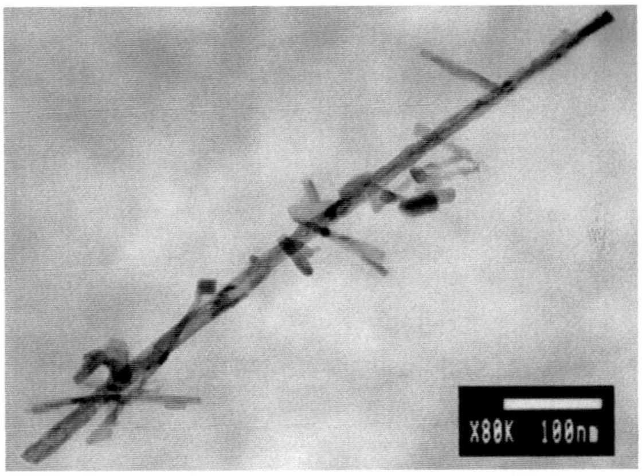

Figure 3. TEM image of TiO_2 nanowires annealed at 400°C.

3.2. The gas sensing characteristics

The gas sensing property of the TiO_2 nanowires were evaluated upon exposure to LPG. The sensors show an obvious response to the target gas at operating temperatures of 400°C. Figure 4 shows the dynamic responses of the TiO_2 nanowires gas sensor towards 500, 1000, 2000, 4000 and 8000 ppm LPG vapour. One can see that the electrical resistance of the material dramatically reduced from 78.6 MΩ to 20.07 MΩ in the presence of 500 ppm LPG with response time of 25.2 s. The corresponding sensitivity of 3.9 was obtained. The response curve shows that the resistance of the sensor varies over time with various cyclic tests for different LPG concentrations (500, 1000, 2000, 4000 and 8000 ppm). It can be seen that the resistance increases and returns to the original value upon exposure to air. Thus, the as-synthesized materials are potential candidate for gas sensor application.

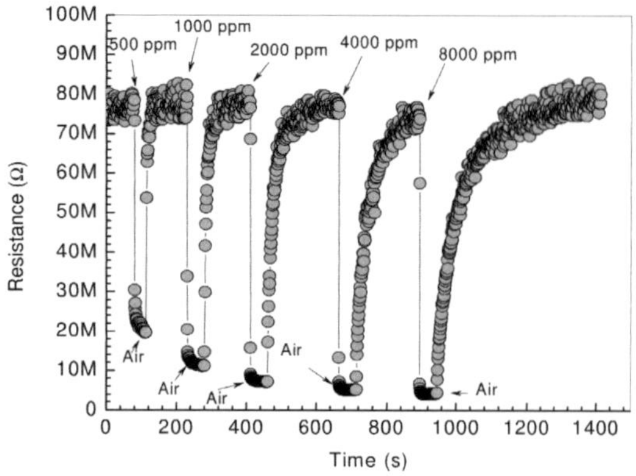

Figure 4. Response to LPG of TiO_2 nanowires at 400°C.

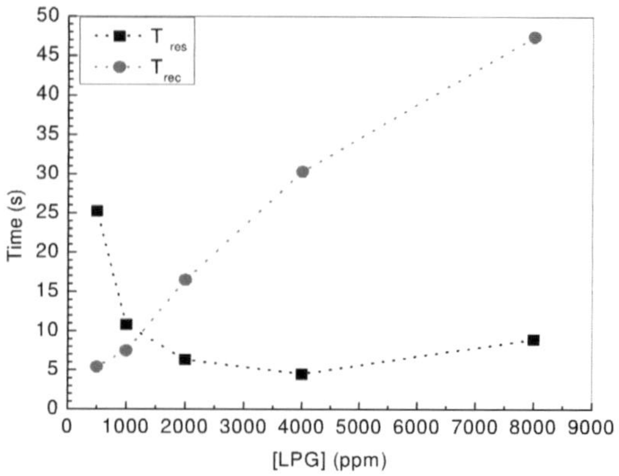

Figure 5. Dependence of τ_{res} and τ_{rec} on LPG concentration.

These measurements highlight also the complete recovery of the signal when the air flux is stored after the gas test. The response time, τ_{res}, is defined here as the time it takes for the conductance of the gas sensor to increase to 90% of the maximum conductance when the target gas is introduced into an environment of air. The recovery time, τ_{rec}, is the time required for 70% reduction in conductance when the target gas is turned off and air is reintroduced into the chamber (i.e., the time to reach 30% of maximum conductance value in the presence of the target gas) [18]. Response time decreases rapidly with increased gas concentration, whereas the recovery time increases with increasing gas concentration (see Figure 5).

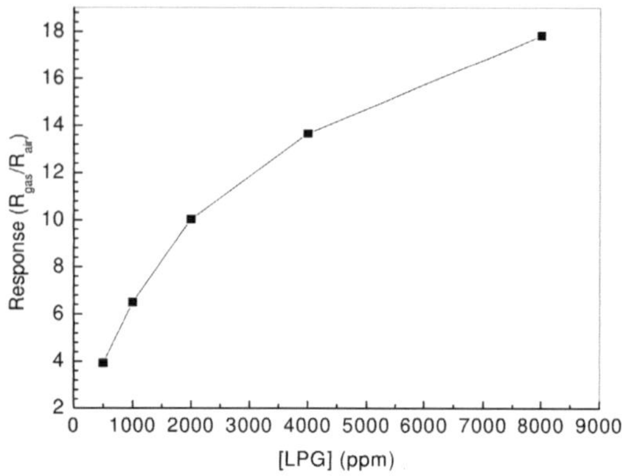

Figure 6. Dependence of response on LPG concentration.

Figure 6 shows the relationship between sensitivity and LPG concentration for the sensor operating at 400°C. It can be seen that the response of the sensor to LPG increases rather steeply with increasing gas concentration, without showing a saturating tendency up to 8000 ppm because there are still sufficient numbers of available surface states to act on LPG vapour. Linearity of response in the low concentration (< 2000 ppm) range is properly suited for the use of these TiO_2 nanowires in checking the concentration of LPG in domestic applications.

4. Conclusion

TiO_2 nanowires have been successfully synthesized by a hydrothermal process at 190°C for 24 h. The obtained nanowires exhibited diameter of ca. 10-20 nm and length of ca. 700-800 nm. The obtained TiO_2 nanowires are uniform, long and have a large aspect ratio. The response of the obtained nanowires to LPG concentrations of 500, 1000, 2000, 4000 and 8000 ppm at operating temperature of 400°C has been characterized.

Journal of Physics: Conference Series **187** (2009) 012086 doi:10.1088/1742-6596/187/1/012086

The obtained results open new perspectives on fabricating devices based on TiO_2 nanowires for detecting LPG using wet chemical approaches.

References

[1] Murray C B, Kagan C R, Bawendi M G 2000 *Annu. Rev. Mater. Sci.* **30** 545. b) Alivisatos A P 1996 *Science* **271** 933. c) Brus L 1994 *J. Phys. Chem.* **98** 3575. d) Bawendi M G, Steigerwald M L, Brus L 1990 *Annu. Rev. Phys. Chem.* **41** 477

[2] Krans J M, Rutenbeek J M van, Fisun V V, Yanson I K, Jogh L J de 1995 *Nature* **375** 767. b) Al'tshuler B L, Lee P A 1988 *Phy. Today* **December** 36

[3] Likharev K K, Cleason T 1992 *Sci. Am.* **June** 80. b) Likharev K K 1988 *IBM J. Res. Dev.* **32** 144

[4] Markovich G, Collier C P, Henrichs S E, Remacle F, Levine R D, Health J R 1999 *Acc. Chem. Res.* **32** 415

[5] Younan Xia, Peidong Yang, Yugang Sun, Yiying Wu, Brian Mayers, Byron Gates, Yadong Yin, Franklin Kim, and Haoquan Yan 2003 *Adv. Mater.* **15** (5) 353

[6] Zhang Y X, Li G H, Jin Y X, Zhang Y, Zhang J, Zhang L D 2002 *Chem. Phys. Lett.* **365** 300

[7] Kasuga T, Hiramatsu M, Hoson A, Sekino T and Nihara K 1998 *Langmuir* **14** 3160

[8] Kasuga T, Hiramatsu M, Hoson A, Sekino T and Nihara K 1999 *Adv. Mater.* **11** 1307

[9] Lu Jia Grace, Chang Paichun, Fan Zhiyong 2006 *Mater. Sci. Eng.* R **52** 49

[10] Chen Q, Zhou W Z, Du G H, Peng L M 2000 *Adv. Mater.* **14** 1208

[11] Du G H, Chen Q, Che R C, Yuan Z Y, Peng L P 2001 *Appl. Phys. Lett.* **79** 3702

[12] Yao B D, Chan Y F, Zhang X Y, Zhang W F, Yang Z Y, Wang N 2003 *Appl. Phys. Lett* **82** 281

[13] Yuan Z Y, Zhou W, Su B L 2002 *Chem. Commun.* 1202.

[14] Yu J G, Yu J C, Ho W K, Wu L, Wang X C 2004 *J. Am. Chem. Soc.* **126** 3422

[15] Zhang D Y, Qi L M 2005 *Chem. Commun.* 2735

[16] Chien N D, Le D T T, Vinh P V, Quy N V, Toan N N, Duy N V 2003 *The conference on Science and Nano Technology* (Hanoi, May)

[17] Le Dang T T, Vuong Dang D, Chien Nguyen Duc 2008 *Adv. Nat. Sci.* **9** (1) 303

[18] Neri G, Bonavita A, Ipsale S, Rizzo G, Baratto C, Faglia G, Sberveglieri G 2007 *Mater. Sci. Eng. B* **139** 41

Acknowledgements

We sincerely thank the financial support of the project B2008-01-218, State Research Programme under project KC.02.05/0610 and the Vietnam-Italy International Cooperation project for the period of 2006-2008.

Journal of Physics: Conference Series **187** (2009) 012087 doi:10.1088/1742-6596/187/1/012087

Modelling molecular field effect transistor using non-equilibrium Green function method

Dinh Sy Hien and Huynh Hoang Trung

HCM University of Natural Sciences,
227 Nguyen Van Cu Street, 5 District, HCM City, Vietnam

E-mail: lhoangtrungedu@yahoo.com

Abstract. In this paper, we introduce a model of molecular field effect transistor (MFET). Because of the small size of these devices (about 1 nm), and the near one-dimensional nature of charge transport within them, MFET modelling demands a rigorous quantum-mechanical basis. This is achieved in this model by using non-equilibrium Green function method to compute transport function and ultimately, the current-voltage (I-V) characteristics. The programme is written by using graphic user guide (GUI) in Matlab. The effects of material, temperature, and bias on I-V characteristics of the MFET have been considered.

Keywords: Molecular transistor, molecular field effect transistor, molecular electronic device.

1. Introduction

Current interest in molecular electronics is largely driven by expectations that molecules can be used as nanoelectronic components able to complement or replace standard silicon CMOS technology on the way down to 10 nm circuit components. The first speculations about molecular electronic devices (diodes, rectifiers) were apparently made in the mid-1970s [1]. That original suggestion of a molecular rectifier has generated a large interest in the field and a flurry of suggestions of various molecular electronic components.

The theoretical description of electron transport through molecular electronics (ME) systems is a complex many-particle problem, where the full solution is not possible in practice. The available approaches use certain approximations, however, the basis for transport calculations is the description of the ME device itself, which implies that approaches to the simulation of transport and structure cannot be chosen independently. A diverse spectrum of approaches to the transport calculations was developed to study the whole spectrum of problems ranging from exact many-particle problem treatment within some simplified models (e.g., the two-barrier potential) for the system, to semi-empirical models for transport through quite realistic structures consisting of a few hundred atoms. Although some of these methods were developed for purposes apparently different from the ME design, like intra-molecular donor–acceptor charge transfer or scanning tunnelling microscopy (STM) image calculation, they are often applicable to ME systems.

Two distinct approaches to the transport calculations are suitable for proper accounting of the full chemical structure of the system: the *Green function* (GF) method and the *density functional* (DF) method. The conceptual difference is that in the GF approach the interactions within the system are treated in the phenomenological way, in the sense that transport calculations and calculations of system properties are separated, thus enabling any level of input to be used. The range of examples

© 2009 IOP Publishing Ltd

Journal of Physics: Conference Series **187** (2009) 012087 doi:10.1088/1742-6596/187/1/012087

varies from semi-empirical models to discrete tight-binding or continuous DF descriptions. Within the DFT framework, the charge carrier density is the unique parameter, hence the current is naturally calculated simultaneously with other system properties that depend on or constitute the density. Consequently, at least in theory, the system could be seamlessly restructured under the bias, but neither the tight-binding nor semi-empirical level of the description is naturally allowed within the DF method. Hence, GF is often used in a semi-empirical way to fit the experimental data. The fitting parameter is usually the shape of the voltage drop over the system.

In this work, we introduce a model of MFET, review the transport calculation using NEGF, and use graphic user interface (GUI) in Matlab to compute current-voltage characteristics of MFET. This is achieved in this model by using non-equilibrium Green function method to compute transport function and ultimately, the current-voltage (I-V) characteristics. As examples, typical simulations of MFET are presented. The effects of material, temperature, and bias on I-V characteristics of the MFET are also considered.

2. Simulation results

The central focus of this work is to introduce a model of three-terminal molecular devices, namely the molecular field effect transistor (MFET). MFET is a promising alternative candidate of traditional MOSFET in future due to its small size, low power and high speed. The structure of the MFET is in shape like traditional MOSFET, but its conductive channel is replaced by a benzene-1,4-dithiolate molecule.

A schematic view of a molecule coupled to gold source (S) and drain (D) contacts is shown in figure 1. As an example we use the benzene-1,4-dithiol molecule which consists of a phenyl ring with thiol (-SH) end groups. A gate terminal may be used to modulate the conductance of the molecule. The coupling between the gate and the molecule is purely capacitive – there is no gate current.

The molecular energy levels consist of a set of occupied levels separated by a gap from a set of unoccupied levels. At equilibrium, the Fermi energy is typically located in the gap between the highest occupied molecular orbital (HOMO) and the lowest unoccupied molecular orbital (LUMO), but when a bias is applied, the Fermi energy in the source contact (μ_1) floats up by qV_D relative to the Fermi level in the drain contact (μ_2). The molecule conducts when the bias is large enough that one or more of the molecular energy levels lie between μ_1 and μ_2.

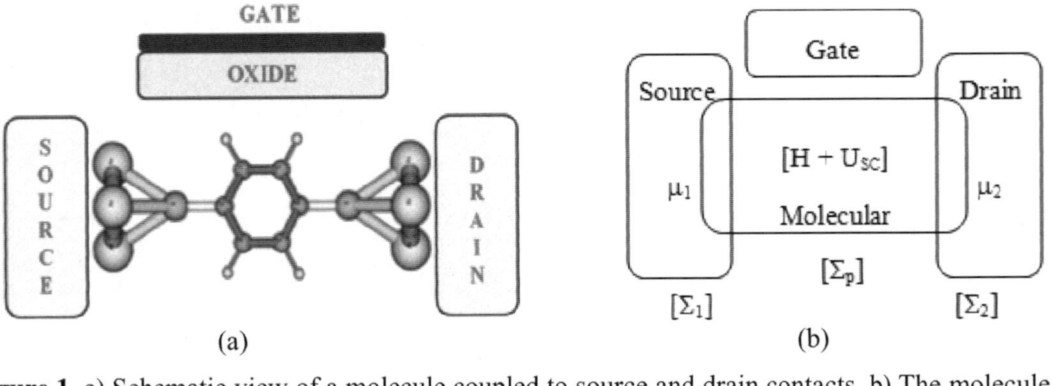

Figure 1. a) Schematic view of a molecule coupled to source and drain contacts. b) The molecule is described by a Hamiltonian H and a self – consistent potential U_{SC}. The effect of the large contacts is described using self – energy matrices $\Sigma_{1,2}$. Scattering processes may be described using another self – energy matrix Σ_p. The source and drain contacts are identified by their respective Fermi levels μ_1 and μ_2.

2.1. The non-equilibrium Green function formalism (NEGF)

The source and drain contacts are identified by their respective Fermi levels μ_1 and μ_2. If a positive voltage V_D is applied externally to the drain contact with respect to the source contact, then the drain has Fermi energy level lower than that of the source by qV_D: $\mu_1 - \mu_2 = qV_D$. The source and drain contacts thus have different Fermi functions and each seeks to bring the molecule into equilibrium with itself. The source keeps pumping electrons in hoping to establish equilibrium. But equilibrium is never achieved as the drain keeps pulling electrons out in its bid to establish equilibrium. The molecule is thus forced into a balancing act between two reservoirs with different agendas which sends it into a non-equilibrium state intermediate between what the source would like to see and what the drain would like to see [2-5]. To describe this balancing process we need a kinetic equation that keeps track of the input and output from each of the reservoirs as shown in figure 2.

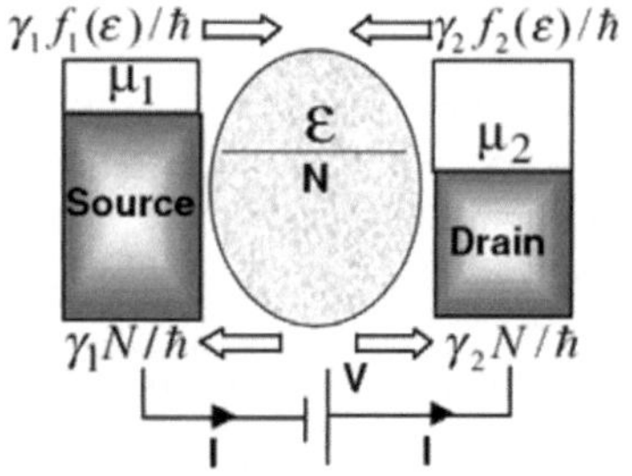

Figure 2. Current flows through a single energy level. One contact keeps filling it up and the other keeps emptying it causing a net current flow from one contact to the other.

2.2. *Current through single energy level*

We treated the energy level ε as discrete, ignoring the broadening $\gamma = \gamma_1 + \gamma_2$ that arises due to the coupling with the source and drain contacts. To take the broadening into account we may replace the discrete level with a Lorentzian density of state (DOS) D(E):

$$D_\varepsilon(E) = \frac{\gamma/2\pi}{(E-\varepsilon)^2 + (\gamma/2)^2} \quad , \gamma = \gamma_1 + \gamma_2 \quad . \tag{1}$$

Let us model a nano-transistor by writing the DOS as:

$$D_\varepsilon(E) = \frac{m^* \; WL}{\pi \; \hbar^2} \theta(E - E_C) \tag{2}$$

where E_C is the energy of the conduction band edge, W is the width of the molecule channel, L is the length of the molecule channel.

The self-consistent potential is:

$$U = U_L + \frac{q^2}{C_E}(N - N_0) \; , \tag{3}$$

where $$U_L = \frac{C_G}{C_E}(-qV_G) + \frac{C_D}{C_E}(-qV_D) \; , \tag{4}$$

is the Laplace potential and

$$C_E = C_S + C_G + C_D \; , \tag{5}$$

is the total capacitance.

The effect of the self-consistent potential U is to raise the density of state (DOS) in energy and can be included in our expressions for the number of electrons N and the current I_D which include the effect of broadening.

The number of electrons N is accounted for the broadened DOS:

$$N = \int_{-\infty}^{+\infty} dE \; D_\varepsilon(E-U) \frac{\gamma_1 f_1(E) + \gamma_2 f_2(E)}{\gamma_1 + \gamma_2} \; , \tag{6}$$

and the current I_D is given as

$$I_D = \frac{2q}{h} \int_{-\infty}^{+\infty} dE \; 2\pi \; D_\varepsilon(E-U) \frac{\gamma_1 \gamma_2}{\gamma_1 + \gamma_2} [f_1(E) - f_2(E)]. \tag{7}$$

We could write it in the form

$$I_D = \frac{2q}{h}\int_{-\infty}^{+\infty} dE\,\bar{T}(E-U)\left[f_1(E)-f_2(E)\right], \tag{8}$$

where the transmission function is defined as follows

$$\bar{T}(E)=2\pi\,D_\varepsilon(E)\frac{\gamma_1\gamma_2}{\gamma_1+\gamma_2}=\frac{\gamma_1\gamma_2}{(E-\varepsilon)^2+(\gamma/2)^2}\,. \tag{9}$$

2.3. NEGF equation as multi – level generalization

We use a simple semi-empirical Hamiltonian H and a self-consistent potential U_{SC} to describe the molecule. The effect of the source and drain contacts are taken into account using self-energy functions Σ_1 and Σ_2. Scattering processes may be described using another self-energy matrix Σ_p. However, in this work we focus on coherent or ballistic transport $\Sigma_p = 0$. The source and drain contacts are identified with their respective Fermi levels μ_1 and μ_2. Our simulation consists of iteratively solving a set of coupled equations and the self-consistent potential U_{SC}. Given H, U_{SC}, Σ_1, Σ_2, μ_1 and μ_2 the NEGF formalism has clear prescriptions to obtain the density matrix ρ from which the electron and current may be calculated. Once the electron density is calculated we solve the Poisson's equation to obtain the self-consistent potential U_{SC} [5, 6].

Self-energy formally arises out of partitioning the molecule-contact system into a molecule subsystem and a contact subsystem. The contact self-energy matrix Σ is calculated knowing the contact surface Green's function G and the coupling between the molecule and contact τ. For a molecule coupled to two contacts (source and drain) the molecular Green's function at an energy E is the written as follows:

$$G(E) = [EI - H - U_{SC} - \Sigma_1 - \Sigma_2]^{-1}, \tag{10}$$

where, I is the identity matrix, H is the molecule Hamiltonian and U_{SC} is the self-consistent potential. The contact self-energy matrices are

$$\Sigma_{1,2}(E)=\tau_{1,2}\,g_{1,2}\,\tau^{+}_{1,2}\,. \tag{11}$$

Given H, $\Sigma_{1,2}$, contact Fermi energies $\mu_{1,2}$ and the self-consistent potential U_{SC}, NEGF has clear prescriptions to obtain the density matrix ρ. The density matrix can be expressed as an energy integral over the correlation function – $iG^{<}(E)$, which can be viewed as an energy-resolved density matrix:

$$\rho = \int dE\left[-iG^{<}(E)/2\pi\right]. \tag{12}$$

The correlation function is obtained from the Green's function G and the broadening function $\Gamma_{1,2}$:

$$-iG^{<}(E) = G(f_1\Gamma_1 + f_2\Gamma_2)G^{+}\,. \tag{13}$$

Unlike the Hamiltonian, the self-energy matrices are non-Hermitian. The anti-Hermitian part of the self-energy is also known as the broadening function

$$\Gamma_{1,2}=i\left(\Sigma_{1,2}-\Sigma^{+}_{1,2}\right), \tag{14}$$

where $f_{1,2}(E)$ are Fermi functions with energy Fermi levels $\mu_{1,2}$:

$$f_{1,2}(E) = \frac{1}{e^{[(E-\mu_{1,2})/k_BT]}+1}\,. \tag{15}$$

The density matrix so obtained can be used to calculate the electron density $n(\vec{r})$ expressed in terms of the eigenvectors $\Psi_\alpha(\vec{r})$ of the Hamiltonian

$$n(\vec{r}) = \sum_{\alpha,\beta} \Psi_\alpha(\vec{r})\Psi^*_\beta(\vec{r})\rho_{\alpha\beta} \quad . \tag{16}$$

The Poisson's equation relates the real space potential distribution $U(\vec{r})$ in a system to the charge density $n(\vec{r})$. The Poisson's equation is then solved for the change in the charge density n – n_0 from the nominal value:

$$\vec{\nabla}.\left(\varepsilon\vec{\nabla}U(\vec{r})\right) = -q^2\left(n(\vec{r}) - n_0(\vec{r})\right) \quad . \tag{17}$$

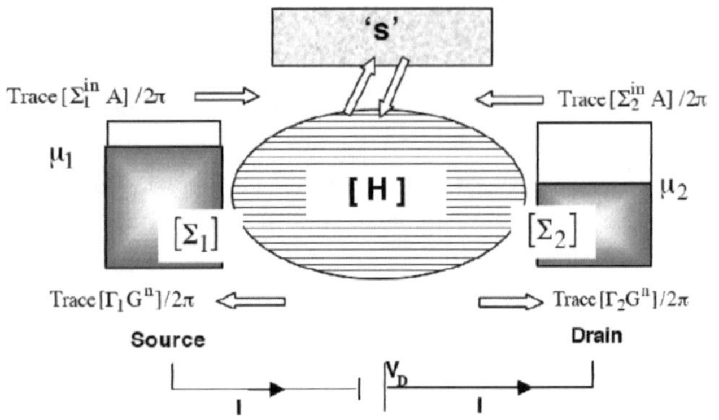

Figure 4. Model for a multi-level conductor with n energy levels is presented. General matrix model is based on the non-equilibrium Green's function (NEGF) formalism. A channel is connected to two contacts (source and drain). The "s-contact" distributed throughout the channel, describes incoherent scattering processes. In general, this contact cannot be described by Fermi function, unlike the real contacts.

The total number of electrons N may be obtained from the density matrix

$$N = Trace\left(\rho\right) . \tag{18}$$

The density matrix may also be used to obtain the terminal current. For coherent transport, we can simplify the calculation of current by using the transmission formalism, where the transmissions function

$$T(E) = Trace\left[\Gamma_1\, G\, \Gamma_2\, G^+\right] \tag{19}$$

is used to calculate the terminal current

$$I_D = \frac{2q}{h}\int_{-\infty}^{+\infty} dE\; T(E)\left[f_1(E) - f_2(E)\right] \quad . \tag{20}$$

In this work the program is written by using graphic user guide (GUI) in Matlab. We have found difference of I-V characteristics between traditional MOSFET and MFET. In addition, effects of type of materials, temperature, and bias on I-V characteristics of the MFET have been considered. Using GUI in Mablab, simulation results are intuitively displayed.

Current–voltage (I-V) characteristics of the molecular field effect transistor (MFET) at different values of bias voltage are shown in figure 5.

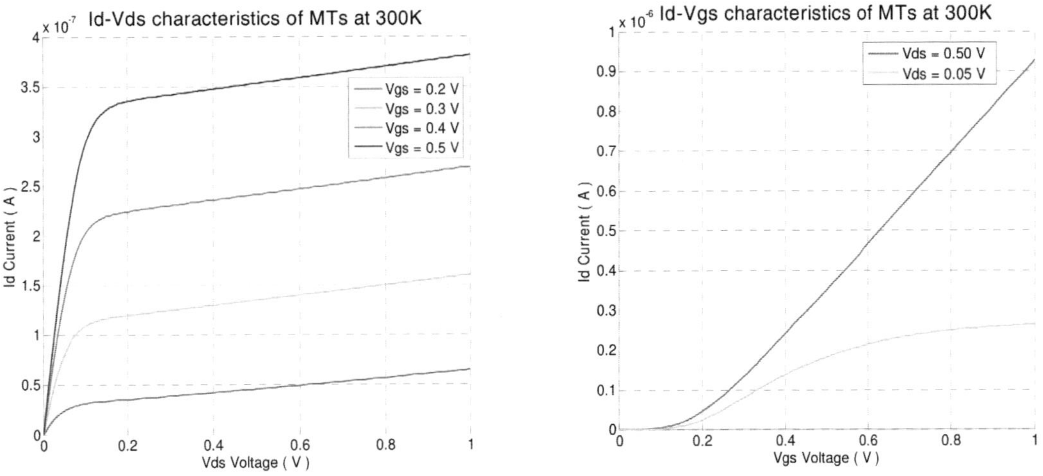

(a) (b)

Figure 5. Current – voltage characteristics of the MFET: (a) $I_D = f(V_{DS})$ for different values of the gate voltage and (b) $I_D = f(V_{GS})$ for different values of the drain voltage.

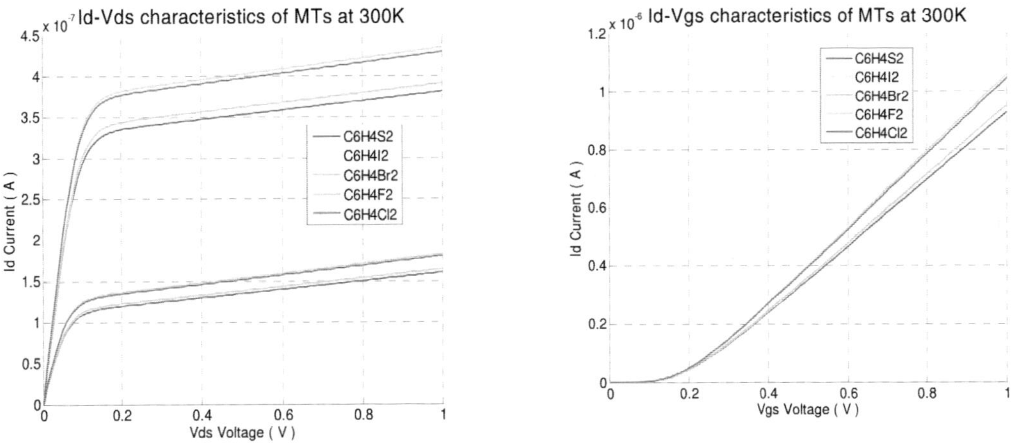

(a) (b)

Figure 6. Voltage – current characteristics of MFET for different materials of conductive channel: a) $I_D = f(V_{DS})$ and b) $I_D = f(V_{GS})$.

Current–voltage (I-V) characteristics of the molecular field effect transistor (MFET) for different values of the length (L) and the width (W) of the conductive molecule channel are shown in figure 6. Benzene-1,4-dithiol molecules have different values of the length (L) and the width (W) (table 1).

Table 1. Benzene-1,4-dithiol molecules have different values of the length (L) and the width (W).

Molecules	$C_6H_4S_2$	$C_6H_4I_2$	$C_6H_4Br_2$	$C_6H_4F_2$	$C_6H_4Cl_2$
Length L (nm)	0,8460	1,0321	1,0025	0,7957	0,9340
Width W (nm)	0,6542	0,6637	0,6714	0,7460	0,7365

The effects of temperature on I-V characteristics of the MFET are shown in figure 7. When temperature increased, the drain current is also increased. The change of drain current in temperature is 10^{-10} A/K.

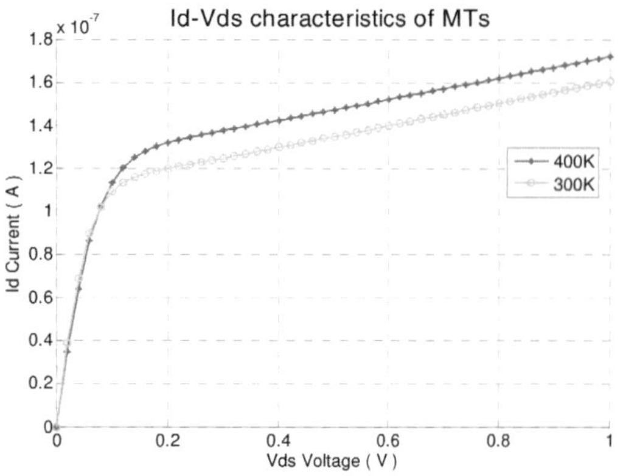

Figure 7. The effects of temperature on drain current-voltage characteristics of MFET with the gate voltage of $V_G = 0.3V$.

2.4. Analytical expression of drain current

From figure 8 we find out that the curves of drain current-voltage characteristics of the MFET can be divided into two regions: linear and saturation areas with evolution point as

$$V_{DS(sat)} = V_{GS} - V_T \ . \tag{21}$$

Drain current in linear area (i.e. $V_{DS} \leq V_{DS(sat)}$) is

$$I_D = \frac{W}{L}\frac{\mu_{Ball}C_G}{2}\left[2(V_{GS}-V_T)V_{DS}-V_{DS}^2\right] . \tag{22}$$

The gate capacitance of the MFET is given by

$$C_G = \frac{\varepsilon_0\,\varepsilon_r\,W\,L}{t_{ox}} , \tag{23}$$

where t_{ox} is the thickness of insulator SiO_2, $\varepsilon_0 = 8{,}854.10^{-12}$ F/m is dielectric constant in a vacuum, $\varepsilon_r = 3{,}9$ is relative dielectric constant of SiO_2. For very small drain-source voltage, V_{DS}, equation (22) becomes

$$I_D = \frac{W}{L}\mu_{Ball}C_G(V_{GS}-V_T)V_{DS} \ . \tag{24}$$

For higher drain – source voltages (i.e. $V_{DS} \geq V_{DS(sat)}$), equation (22) (i.e. saturation drain current) becomes

$$I_{D(sat)} = \frac{W}{L}\frac{\mu_{Ball}\,C_G}{2}(V_{GS}-V_T)^2 + \alpha V_{DS} , \tag{25}$$

where α is the slope: $\alpha = dI_D/dV_{DS}$.

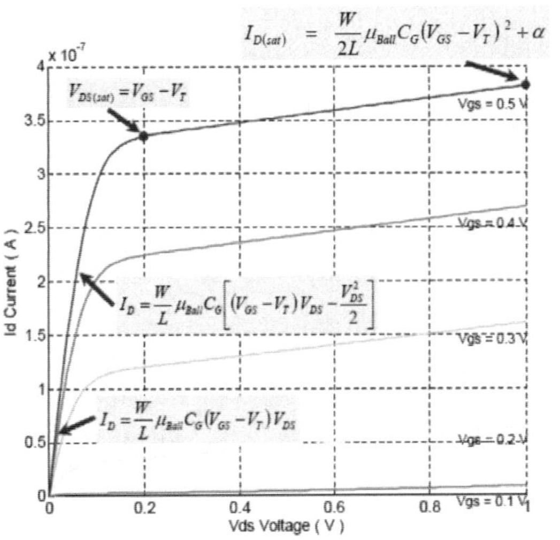

Figure 8. I-V characteristics of the MFET for different values of the gate voltage, and algebraic expressions of these I-V characteristics.

3. Conclusions

We use non-equilibrium Green's function method to compute transport function of charges and ultimately, the current-voltage (I-V) characteristics. The program is written by using graphic user guide (GUI) in Matlab. We have found difference of I-V characteristics between MOSFET and MFET. In addition, effects of type of material, temperature, and bias on I-V characteristics of the MFET have been considered. Using GUI in Mablab, obtained results of simulations are intuitively displayed.

References

[1] Aviram A and Ratner M A 1974 *Chem. Phys. Lett.* **29**, 277
[2] Su W 2006 *First principles study of Molecular electronic devices* (Royal Institute of Technology, Stockholm, Sweden)
[3] Damle P S 2003 *Nanoscale device modeling: from MOSFETs to Molecules* (PhD thesis, Purdue University, USA)
[4] Supriyo Datta 2005 *Quantum transport: Atom to Transistor* (Cambridge University Press)
[5] Lundstrom M and H. Pal 2007 *Nanoscale MOSFET Physics for Compact Models* (Purdue University, USA)
[6] Lundstrom M 2007 *Simple Theory of the Ballistic Nanotransistor* (Purdue University, USA)

Acknowledgments

The authors are grateful to Prof. Dr. Dang Luong Mo of Hosei University, Japan for useful discussions.

Development of quantum device simulator NEMO-VN1

Dinh Sy Hien, Nguyen Thi Luong, Le Hoang Minh, Tran Tien Phuc, Pham Thanh Trung, Bui An Dong, Huynh Lam Thu Thao, Nguyen Van Le Thanh, Thi Tran Anh Tuan, Huynh Hoang Trung, Nguyen Thi Thanh Nhan and Dinh Viet Nga

HCM University of Natural Sciences,
227 Nguyen Van Cu Street, 5 District, Ho Chi Minh City, Vietnam

E-mail: dshien52@yahoo.com

Abstract. We have developed NEMO-VN1 (NanoElectronic MOdelling), a new modelling tool that simulates a wide variety of quantum devices including Quantum Dot (QD), Resonant Tunneling Diode (RTD), Resonant Tunneling Transistor (RTT), Single Electron Transistor (SET), Molecular FET (MFET), Carbon Nanotube FET (CNTFET), Spin FET (SPINFET). It has a collection of models that allow user to trade off between calculation speed and accuracy. NEMO-VN1 also includes a graphic user interface of Matlab that enables parameter entry, calculation control, intuitive display of calculation results, and in-situ data analysis methods.

Keywords: Quantum device, single electron transistor, nanoelectronic modelling.

1. Introduction

The dimensional scaling of CMOS device and process technology will become much more difficult as the industry approaches 10 nm around the year 2020 and will eventually reach asymptotic end according to the International Technology Roadmap for Semiconductor for emerging research devices [1]. Beyond this period of traditional CMOS it may be possible to continue functional scaling by integrating an alternative electronic device on to a silicon platform. These alternative electronic devices in the future include 1D structures (such as CNTs and compound semiconductor nanowires), RTDs, SET, molecular and spin devices, all of which are discussed in this work.

Despite these exciting possibilities, nanoelectronic devices are still in their relative infancy. The expense and difficulty of device fabrication precludes simply building and testing vast arrays of quantum devices. To focus efficiently on the best design, engineers need a tool that predicts electronic characteristics as a function of the device geometry and composition. In the more scientific mode, such a simulator would greatly enhance the understanding of quantum effects that drive the transport process and provide a means to investigate new device concepts.

Even conventional devices require a correction for quantum effects associated with the smaller device features. MOS devices, for example, exhibit electron confinement effects in the inversion layer. This phenomenon is a function of decreasing oxide thickness rather than the overall size of the device. Quantum effects become important as the oxide layer thickness decreases below 2 nm, which will soon be a standard for manufactured integrated circuits. Problems of this nature will become more prevalent as device geometries continue to shrink.

© 2009 IOP Publishing Ltd

Journal of Physics: Conference Series **187** (2009) 012088 doi:10.1088/1742-6596/187/1/012088

Nanoelectronic device modelling requires a fundamental quantum-mechanical approach. Many forms of quantum correction to classical electronic device models have been proposed or implemented. These include MOSFET-specific quantum corrections [2, 3] and generic quantum corrections to the drift-diffusion [4], hydrodynamic [5], and Boltzmann transport equation models [6]. Therefore, the semiconductor industry needs a new fully quantum-mechanically based TCAD (technology computer aided design) tool.

To address this problem, we developed a general purpose quantum device simulator called NEMO-VN1 (NEMO in Vietnam). NEMO-VN1 can simulate a wide variety of nanoelectronic devices, including QD, RTD, RTT, SET, MFET, CNFET and SPINFET. It has a collection of models that allow user to trade off between calculation speed and accuracy. NEMO-VN1 also includes a graphic user interface (GUI) of Matlab that enables parameter entry, calculation control, display of calculation results, and in-situ data analysis methods.

This work reviews the capabilities of NEMO-VN1, summarize the theoretical approach and experimental results, and give examples of typical NEMO-VN1 simulations.

2. Simulation results

2.1. NEMO-VN1 GUI

2.1.1. Overview

Another important goal of the NEMO-VN1 project was to make a user-friendly simulator that provides as much control as possible over every aspect of the simulation. Flexibility and ease of use are difficult to achieve simultaneously, but given the complexity of quantum device simulations it became clear that both criteria were vital to program success. Consequently, GUI development was major part of the NEMO-VN1 program. Here we present an overview of the most important GUI features, including:

- Main Screen
- Resonant Tunnelling Diode (RTD)
- Resonant Tunnelling Transistor (RTT)
- Quantum Dot (QD)
- Quantum Well (QWe)
- Quantum Wire (QWi)
- Single Electron Transistor (SET)
- Carbon Nanotube Field Effect Transistor (CNTFET)
- Molecular Field Effect Transistor (MFET)
- Spin Field Effect Transistor (SPINFET)

2.1.2. Main screen

NEMO-VN1 has a rich variety of simulation models, while this provides the maximum flexibility in terms of applicability to different types of devices and test conditions. The problem is that NEMO-VN1 requires over 100 simulation parameters. Traditional device simulators force the users to familiarize themselves with all available simulation parameters and ensure that they are set correctly. To minimize this burden for the users, NEMO-VN1 uses a hierarchical approach to input and displays simulation parameter values. The top level of this hierarchy specifies the highest level option (type of device). Subsequent levels contain more detailed options such as current-voltage characteristics of devices, types of material, size of devices, size of barriers, temperature, colours, etc.

The main screen shown in figure 1a is the central location where the user controls the NEMO-VN1 simulation. From main screen, the user can choose various types of quantum device simulations by clicking the mouse pointer on component items (left top corner). In this manner, the user can quickly enter the device list with minimum of typing. Double clicking the left mouse pointer on each item in the device list initiates the selection of models which is used to calculate the current voltage characteristics (figure 1b).

(a) (b)

Figure 1. a) The NEMO-VN1 main screen. File of components contains a device list. b) Pressing left mouse pointer on "components" displays a list of simulation quantum devices.

2.2. NEMO-VN1 simulations

In the course of the NEMO-VN1 program, we have simulated QD, QWi, QWe, RTD, RTT, SET, CNTFET, MFET, and SPINFET. As examples, in the following sections, we consider current voltage characteristics of these quantum devices.

2.2.1. RTD and RTT simulations

Resonant tunnelling devices **-** Resonant tunnelling devices for logic applications include RTTs and hybrid devices incorporated with RTDs and one or more FETs (RTD-FET). The RTDs are two terminal devices that have a very high switching speed and exhibit a region of negative differential resistance in their I-V curves. These two characteristics make them potentially attractive as high-speed switching devices.

Adding a control terminal to RTDs extends their usability to a variety of applications. This approach has been used to build RTT. RTTs have a negative transconductance that can be used in several logic circuits, e.g., in XOR gate with only one transistor.

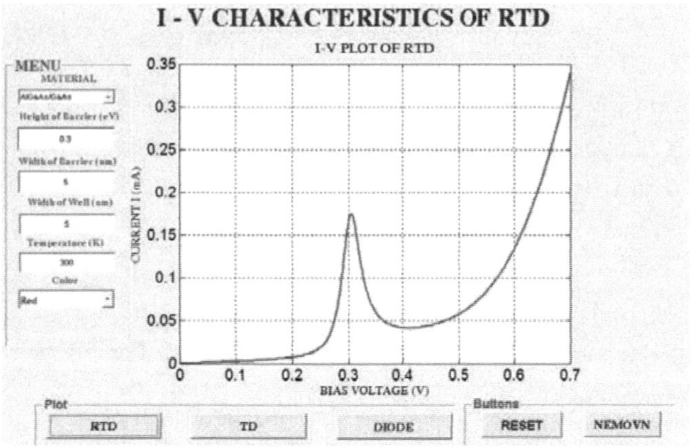

Figure 2. NEMO-VN1 plots current voltage characteristics for 2-barrier RTD device. The DC bias is applied and current is measured as a function of bias voltage at 300 K. For this device, RTD is formed on AlGaAs/GaAs, height of barrier is 0.3 eV, width of barrier is 5 nm, width of well is 5 nm.

Traditionally, RTDs have been fabricated in III-V material systems that has limited widespread their applicability. Recently, several papers have described fabrication of group IV devices with silicon compatible materials. Overall, the resonant tunnelling devices may be used for certain applications requiring high speed and low dynamic range and low peak currents provided the manufacturing issues associated with uniformity of the tunnelling barrier can be resolved. The principle focus of the recent research activity involving RTDs has been in the area of integration on the silicon platform. The current voltage characteristics of RTD are shown in figure 2.

In fact, there are two approaches for RTT model (hybrid RTD-FET) that can be described as follows. RTD is connected with FET in series or in parallel. Here we use the later configuration. The model of RTT is displayed in figure 3a. Current voltage characteristics are shown in figure 3b.

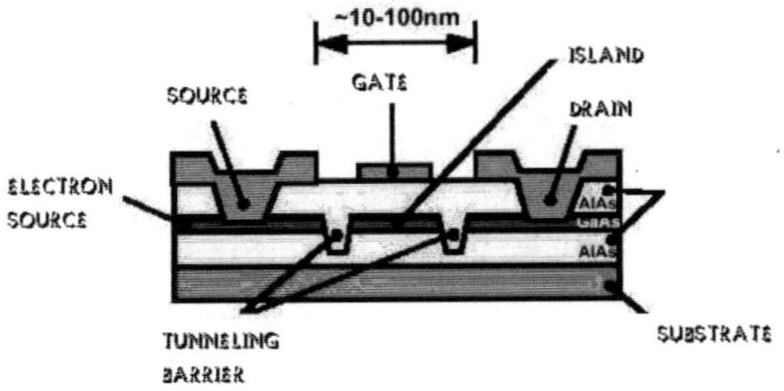

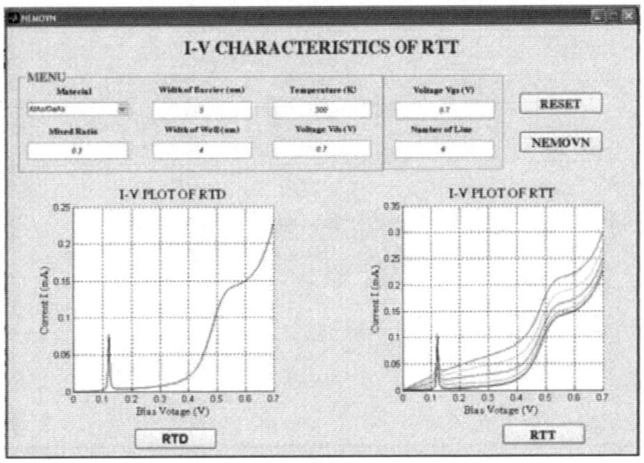

(a) (b)

Figure 3. a) The model of RTT, b) NEMO-VN1 plots current voltage characteristics for RTT device. The drain current is measured as a function of gate voltage (right). The current voltage characteristic of RTD is used for comparison (left).

The algorithm for the calculation of drain current of RTD in the model can be described as follows. Because of the small size of these devices, and the near one-dimensional nature of charge transport within them, RTD modelling demands a rigorous quantum-mechanical basis. This is achieved in this model by using the effective-mass Schrödinger equation to compute transport function of electrons, and by using transport function to compute the source-drain current. Total current can be obtained by summing current of RTD and current of nFET.

2.2.2. SET simulations

Single electron transistors **-** SETs are three-terminal switching devices that can transform electrons from source to drain one by one. Potentially, they can deliver high device density and power efficiency at good speed if the issues of noise immunity and low fan-out can be solved.

The structure of SETs is almost the same as that of FETs. The important difference, however, is that in a SET the channel is separated from source and drain by tunnelling junctions, and the role of the channel is played by a QD. Operational parameters of SETs depend on the size of QD. Single electron devices operating at room temperature were experimentally demonstrated. However, operation of complex and fast SET logic circuits is generally limited to very low temperatures due to low noise immunity.

The algorithm for the calculation of drain current in the model can be described as follows. Because of the small size of these devices, and the near one-dimensional nature of charge transport within them, SET modelling demands a rigorous quantum-mechanical basis. This is achieved in this model by using QD model to compute electro-chemical potential, density of electron states, Fermi-Dirac distribution and to compute the drain current. The model of SET is displayed in figure 4a. Current voltage characteristics are shown in figure 4b.

2.2.3. CNTFET simulations

1D structures (Carbon Nanotubes and Nanowires) **-** Recently, the research activity in this area has matured to the point where one can identify the critical problems associated with 1D structures. The activities have centred around four main poles: 1) understand the basic physical mechanisms at work in quantum confined transport including coherent transport; 2) synthesize nanotubes and nanowire materials with predictable and controllable characteristics; 3) address device fabrication issues like placement, contacting, doping, dielectric and gate material integration issues; and lastly, 4) characterize device issues such as transport efficiency, subthreshold slope, ambipolar conduction, RF response, I_{ON}/I_{OFF} ratios and others.

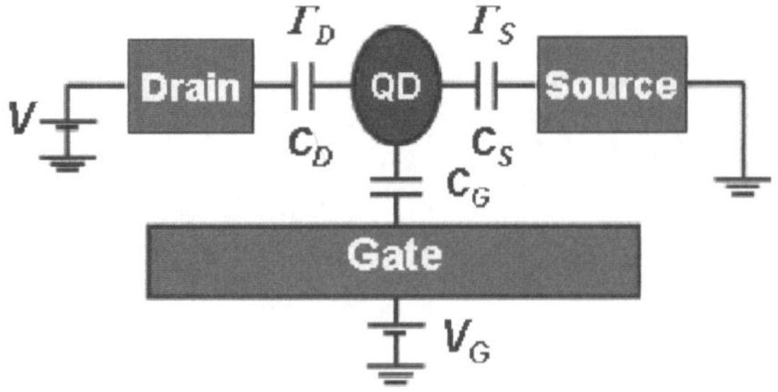

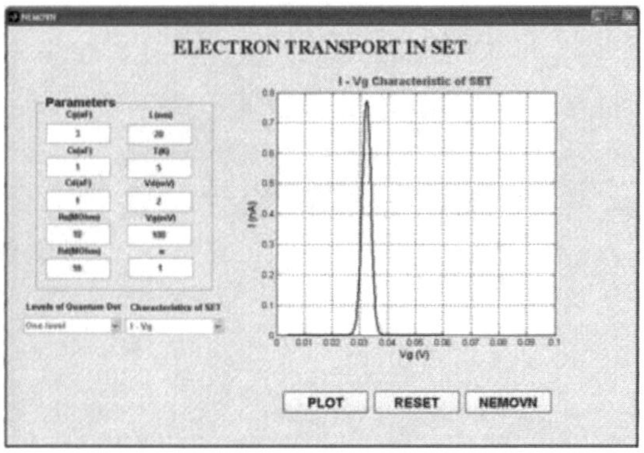

(a) (b)

Figure 4. a) Circuit model of SET. The QD is connected to the source and drain electrodes through small tunnel barriers. The potential in the QD can be modified by gate electrode which is capacitive coupled to the QD. The DC bias (V) is applied and the current is measured as a function of the V and V_g. b) Coulomb blockage oscillations of SET. The current is measured at 5 K as a function of the gate voltage. For this device, $C_G = 3$ aF, $C_S = 1$ aF, $C_D = 1$ aF, V = 2 mV.

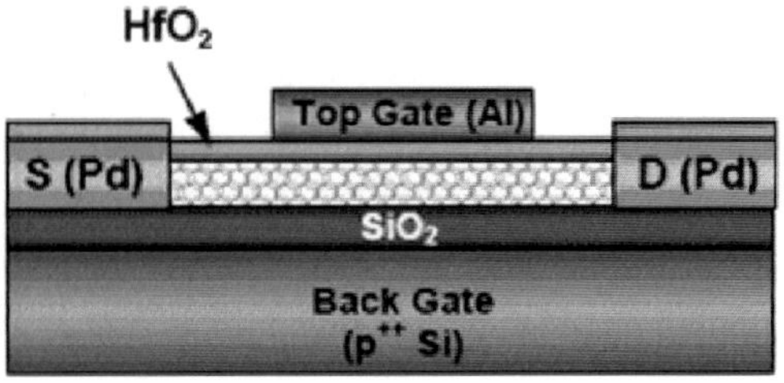

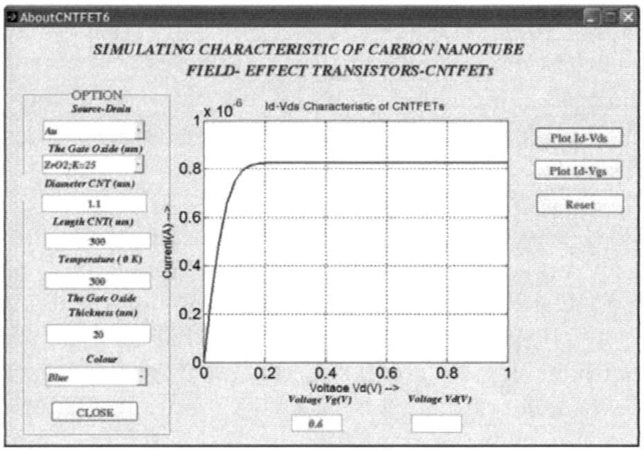

(a) (b)

Figure 5. a) The model of CNTFET, b) NEMO-VN1 plots current voltage characteristics for CNTFET. Current is measured as a function of drain voltage at 300 K. For this device, source-drain of CNTFET is made of Au, diameter of CNT is 1.1 nm, length of CNT is 300 nm, gate thickness is 20 nm.

Significant progress has been made in the synthesis of nanotubes with controllable properties through the use of plasma enhanced chemical vapour deposition. It is now possible to produce nanotubes, over 90% of which have semiconductor properties. While this is a long way from the purity requirements in manufacturing settings, it does represent significant progress.

One of the fabrication challenges associated with CNT-FETs is reliable positioning and growth of nanotubes at lithographically defined sites.

The algorithm for the calculation of drain current in the model can be described as follows. Because of the small size of these devices, and the near one-dimensional nature of charge transport within them, CNFET modelling demands a rigorous quantum-mechanical basis. This is achieved in this model by using non-equilibrium Green's function method to compute transport of the electron and hole charges in nanotubes, and by using the Landauer equation to compute the drain current. The model of CNT-FET is displayed in figure 5a. Current voltage characteristics are shown in figure 5b.

The current voltage curve can be divided into two regions: linear and saturation. Drain current in the linear region of CNTFET can be described as follows:

$$I_d = \frac{W}{L}\mu C_{ox}[(V_{gs} - V_T)V_{ds} - \frac{V_{ds}^2}{2}] \tag{1}$$

or

$$I_d = K_n\left[2(V_{gs} - V_T)V_{ds} - V_{ds}^2\right], \tag{2}$$

where K_n is conductance of CNTFET, W is the width of CNTFET, L is the length of CNTFET, μ is the mobility of carriers, C_{ox} is oxide gate capacitance.

We can also obtain saturation current of CNTFET by replacing $V_{ds(sat)} = V_{gs} - V_T$. Then the expression of saturation current of CNTFET can be written:

$$I_{d(sat)} = K_n(V_{gs} - V_T)^2 . \tag{3}$$

2.2.4. MFET simulations

Molecular devices –The concept of molecular electronic devices is based on electronic properties of individual molecules tailored to perform logic operations and on the assembly of a large number of these functional building blocks into molecular circuits. Logic functions can be provided by electron transport and controlled switching behaviour of the molecules and may involve designated energy levels and charge states. Two-terminal devices such as resistive switches as well as three-terminal devices such as gated, transistor-like molecules are envisioned. The integration of these molecules into circuits requires the fabrication of contacts with atomic level precision and involves organic or inorganic interconnects on nanometer scale.

Many attempts have been made to build and characterize single molecular electronic devices, mainly since the mid-1990. Molecular functions that have been reported include rectifications, controlled conduction through gating as well as switching. Typically, these results have been obtained in mechanical break junction experiments or by scanning probe techniques.

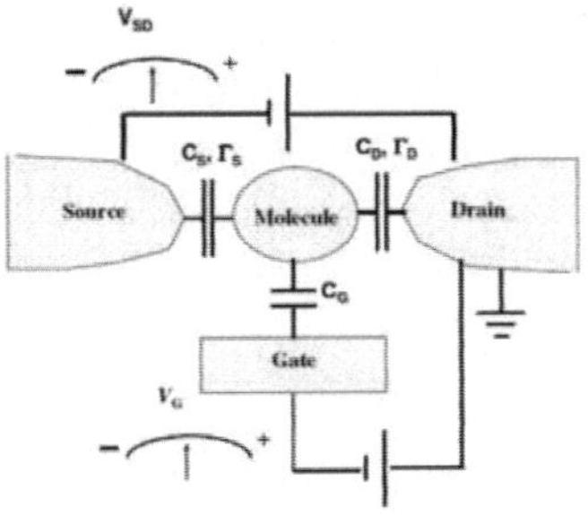

(a)

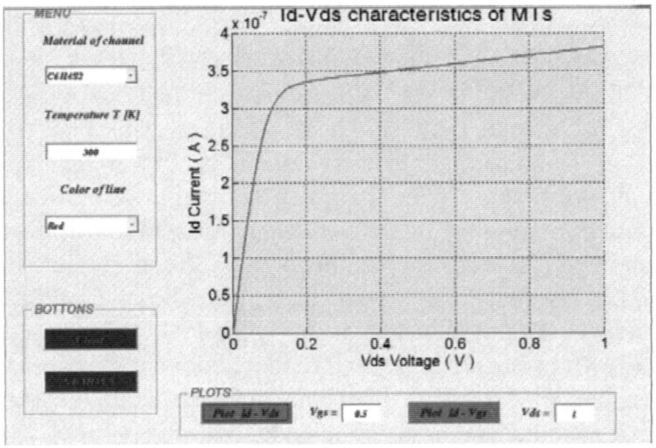

(b)

Figure 6. a) Circuit model of MFET. b) NEMO-VN1 plots current voltage characteristics for MFET. For this device, material is benzene. The drain current is measured as a function of V_{ds} at fixed voltage, V_g.

The algorithm for the calculation of drain current in the model can be described as follows. Because of the small size of these devices, and the near one-dimensional nature of charge transport within them, MFET modelling demands a rigorous quantum-mechanical basis. This is achieved in this model by using non-equilibrium Green's function method to compute transport of the electron and hole charges in the molecular, and by using the transport function to compute the drain current. The model of MFET is displayed in figure 6a. Current voltage characteristics are shown in figure 6b.

The current voltage curve of MFET can be divided into two regions: linear and saturation. Drain current of MFET can be described as follows:

$$I_d = \frac{W}{L}\mu C_{ox}[(V_{gs} - V_T)V_{ds} - \frac{V_{ds}^2}{2}] \tag{4}$$

or

$$I_d = K_n\left[2(V_{gs} - V_T)V_{ds} - V_{ds}^2\right], \tag{5}$$

where K_n is conductance of MFET, W is the width of CNTFET, L is the length of CNTFET, μ is mobility of carriers, C_{ox} is oxide gate capacitance.

We can also obtain saturation current of MFET by replacing $V_{ds(sat)} = V_{gs} - V_T$. Then the expression of saturation current of MFET can be written:

$$I_{d(sat)} = K_n(V_{gs} - V_T)^2 + \alpha V_{ds}, \tag{6}$$

where α is the slope of drain current in saturation region. Here coefficient α equals 75 nA/V.

2.2.5. SPINFET simulations

Spin devices **-** They include a novel MOSFET. This device consists of a MOS gate structure and half-metallic ferromagnetic contacts for the source and drain. Other devices include spin torque transistor, magneto-resistive element (MRE), and spin gain transistor. The spin MOSFET uses a high spin polarized source in conjunction with spin dependent scattering in the drain as spin dependent effects that augment the gate controlled electrostatic transport in heterojunction MOSFET. The resulting hybrid device combines the electrostatic and spin dependent control of the drain current to provide a device concept, which in theory meets several important criteria. Spin MOSFET devices have been simulated to show the above criteria but no experimental demonstration has been published. The principal experimental difficulties will likely centre on the half-metal source-drain materials and problem with injection of spin-polarized electrons into the channel.

The "spin torque transistor" is based on the experimentally verified spin torque effect to modulate a source-drain current flow. This device concept has been reduced to practice and it is quite illustrative of the extended device functionalities created by the fundamental interplay between electric charge and electronic spin. It also illustrates that new materials can play the role in alternative logic technologies beyond CMOS. The "spin gain transistor" is based on experimentally observed and theoretically understood carrier mediated ferromagnetic phase transitions in dilute magnetic semiconductors. This device is similar to a magnetic bipolar transistor, where spin gain is achieved via creation of condition for the ferromagnetic transition. Spin gain in this case is defined to be the ratio of spin polarized (for example, spin up) electrons in collector current relative to the number of spin up electrons in the base. This concept has not been experimentally demonstrated. The MRE devices utilize four or more magneto-resistive junctions, each one of which is composed of a hard magnetic layer used for reference and a soft magnetic layer that can switch polarity dependent on the magnitude and direction of current in two input line. If the magnetic layers are aligned, the resistance of the junction is low and if the magnetic layers are anti-parallel, the resistance will be high.

A few concepts have emerged as indicted above but so far, no viable devices have been demonstrated, but the field of spin transistors is the subject of a great deal of continuing research activity. The algorithm for the calculation of drain current in the model can be described as follows. Because of the small size of these devices, and the near one-dimensional nature of charge transport within them, SPINFET modelling demands a rigorous quantum-mechanical basis. This is achieved in this model by using non-equilibrium Green's function method to compute transport function of electrons, and by using transport function to compute the drain current. The model of SPINFET is displayed in figure 7a. Current voltage characteristics are shown in figure 7b.

The current voltage curve can be divided into two regions: linear and saturation. Drain current in the linear region of SPINFET can be described as follows:

$$I_d = \frac{W}{L}\mu C_{ox}[(V_{gs} - V_T)V_{ds} - \frac{V_{ds}^2}{2}] \tag{7}$$

or

$$I_d = K_n\left[2(V_{gs} - V_T)V_{ds} - V_{ds}^2\right], \tag{8}$$

where K_n is conductance of SPINFET, W is the width of CNTFET, L is the length of CNTFET, μ is the mobility of carriers, C_{ox} is oxide gate capacitance.

We can also obtain saturation current of SPINFET by replacing $V_{ds(sat)} = V_{gs} - V_T$. Then the expression of saturation current of SPINFET can be written:

$$I_d = K_n\left[(V_{gs} - V_T)^2\right]. \tag{9}$$

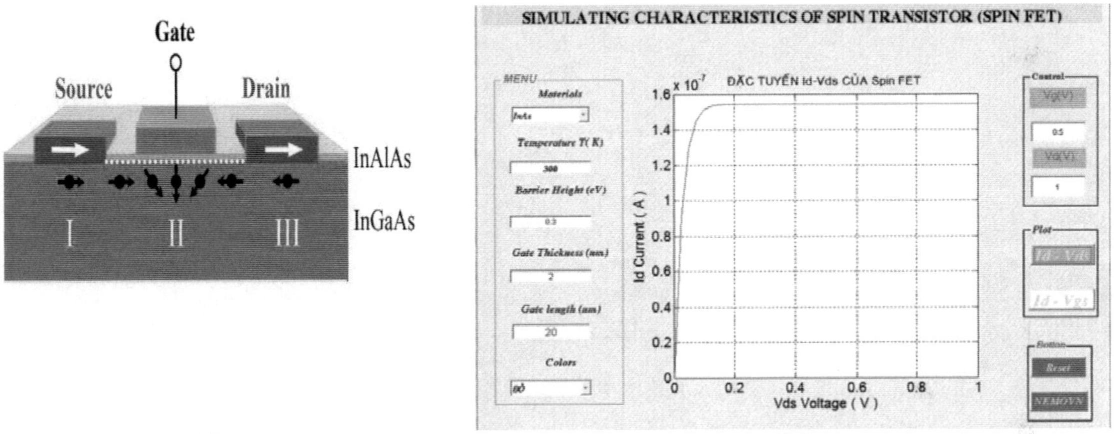

(a) (b)

Figure 7. a) Model of SPINFET. b) NEMO-VN1 plots current voltage characteristics for SPINFET. The drain current of SPINFET is measured at 300 K as a function of V_g and V_{ds}. For this device, material is InAs, height of barrier is 0.3 eV.

3. Conclusions

We have created a comprehensive, versatile and user-friendly quantum device modelling tool. The simulator has predictive capability that we have tested for a number of materials and different devices.

References

[1] The International Technology Roadmap for Semiconductor 2005 Edition
[2] Rios R, Arora N D, Huang C L, Khalil N, Faricelli J and Gruber L 1995 *Int. Electron Devices Meeting Technical Digest (IEEE)* p 937
[3] Voorde A P Vand, Griffin P B, Yu Z, Oh S Y and Dutton R W 1996 *Int. Electron Devices Meeting Technical Digest (IEEE)* p 811
[4] Ancona M G 1989 *Physical Review B* **39**(13) 9536
[5] Kreskovsky J P and Grubin H L 1995 *VLSI Design* **3**(2) 179
[6] Grubin H L and Krekovsky J P 1989 *Solid State Electronics* **32**(12) 1071

Acknowledgments

The authors would like to thank professor Dang Luong Mo of Hosei University, Japan for his very interesting comments and helpful discussions. This work is supported in part by the scientific research project of Vietnam National University Ho Chi Minh City, No. 2005-18-06.

The effect of image force and temperature on electrical characteristics of layer-type MTJs

Nguyen Anh Tuan and Do Phuong Lien

ITIMS, Hanoi University of Technology (HUT), 1 Dai Co Viet Rd.,
Hai Ba Trung Dist., Hanoi, Vietnam
IEP, Hanoi University of Technology (HUT), 1 Dai Co Viet Rd.,
Hai Ba Trung Dist., Hanoi, Vietnam

E-mail: tuanna@itims.edu.vn

Abstract. Tunnel resistivity R (= V/J) as a function of voltage has been calculated for magnetic tunnel junction systems, such as ferromagnet-insulator-ferromagnet three-layer structures. Our early study [1] has been carried out for determining the barrier characteristics in the MTJs based on $Ni_{80}Fe_{20}/Al_2O_3/Co$ systems, by fitting the experimental *I-V* characteristics to Simmons' and Brinkmann's models without image force. In this paper, the image force is taken into account for approaching closely to a real barrier potential. The effect of dielectric constant, and temperature on the tunnel characteristics has also been investigated.

Keywords: Magnetic tunnel junction, three-layer structures, tunnel characteristics.

1. Introduction

Magnetic tunnel junction (MTJ) structures like ferromagnet/insulator/ferromagnet exhibit large tunneling magnetoresistance (TMR) which makes them attractive for magnetic-field sensors, non-volatile magnetic memory and other spintronics devices [1, 2]. It is known that heterogeneous junction surfaces between ferromagnetic metals and very thin isolating layers in the MTJ structures form a potential barrier. Spin-dependent electron tunnelling through the barrier has been studied theoretically and experimentally [3, 4]. Essentially this is an electron scattering process at the surfaces. Important factors in the electronic properties of surfaces are the bound states of the surfaces, so called surface states, localized in the direction normal to the surface [5]. Image potential states are formed from the quantized surface states that exist at metal surfaces with a band gap near the vacuum level.

In the last paper [1], a fitting procedure of our experimental *I-V* characteristic data to theoretical Simmon's equation [6] allowing the determination of the potential barrier's thickness and barrier height in the MTJ $Ni_{80}Fe_{20}/Al_2O_3/Co$ has been presented. The obtained reasonable values are proof of the fact that tunneling dominates the conduction. However in that study we have used the rectangular potential barrier without *image forces* of the Simmon's model. That means the electronic properties of surfaces at the barrier were not concerned. Whereas the practical barrier is a rectangular barrier with the *image potential* included. The image potential states have been considered to provide a unique model system for studying the various scattering processes in detail with the available experimental capabilities [7].

In this paper, a new fitting procedure of the experimental data for the MTJs to Simmon's model has been realized taking into account the image force in the potential formula applied for all voltage

© 2009 IOP Publishing Ltd

Journal of Physics: Conference Series **187** (2009) 012089 doi:10.1088/1742-6596/187/1/012089

ranges. The new mean barrier height in the insulating layer is adjusted to obtain the experimental data. The effect of the dielectric constant of the insulating film is discussed in detail, and, it is shown that this constant affects the temperature dependence of the *J-V* characteristic of the tunnel junction.

2. Theory

2.1. Rectangular barrier

When two metallic electrodes are separated by a thin insulating film, the action of the film is to introduce a potential barrier between the electrodes which impedes the flow of electrons between them. Normally the potential barrier is an ideal rectangular if image forces are excluded, as presented in figure 1a for the case of biased voltage $V = 0$, where ϕ_0 is the height of the barrier at zero bias and s is the wide of the barrier. Under application of a biased voltage $V \neq 0$, the barrier deforms in to trapezium as seen in figures 1b and 1c. The generalized theory developed by Simmons is applied to a rectangular barrier with and without image forces. Theoretical calculated results from the Simmon's model have indicated that, for a given s, ϕ and V, the tunnel resistivity is lower for the practical barrier (with image forces) than for the ideal barrier (without image forces) [6].

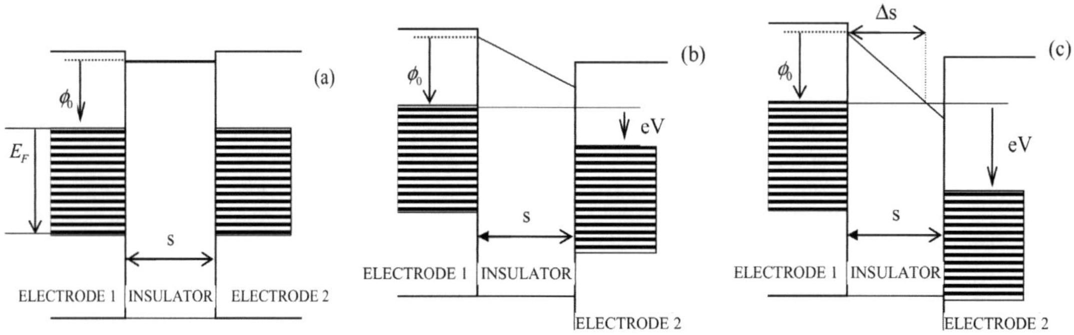

Figure 1. a) Rectangular tunnel barrier at $V = 0$; b) $V < \phi_0/e$; c) $V > \phi_0/e$.

2.2. The image potential

The surface barrier potential can confine electrons in surface states. The Coulomb-like attractive image force, experienced by any charged particles in front of a conductive material, and the repulsive surface then form a potential well for weakly bound electrons. These surface states are quantized to form energy levels described by a Ryberg series similar to electrons in the hydrogen atom, and lead to form a long coulombic tail on this potential of the surface barrier [5]. The effect of the image force is to reduce the area of the potential barrier by rounding off the corners and reducing the thickness of the barrier (figure 2a) and, therefore, increasing the flow of current between the electrodes.

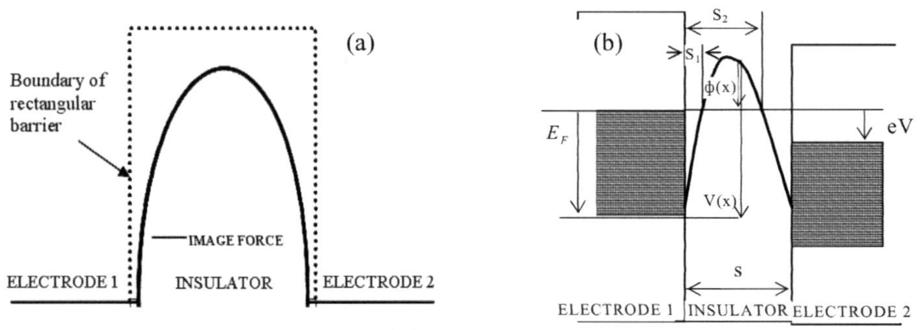

Figure 2. Generalized barrier with image potential for $V = 0$ (a) and biased for $V \neq 0$ (b).

The image potential is readily determined using image force method and is given, in a good approximation, by $V_i = -1.15\lambda d^2/[x(s-x)]$, where $\lambda = e^2 \ln 2/8\pi\varepsilon d$ [6]. In these equations, x is the distance of the electron from negatively biased electrode (electrode 1), s is the insulating film's thickness, e - electron charge, ε - permittivity of the insulating film.

When the image potential expressed by V_i is taken into account, the potential barrier is written as: $\phi(x) = \phi_0 - eVx/s - V_i$, For this case, the mean barrier height above Fermi label of negatively biased electrode is

$$\bar{\phi} = \frac{1}{\bar{d}} \int_{s_1}^{s_2} \phi(x)dx \qquad (1)$$

where s_1 and s_2 are limits of barrier at Fermi level (see figure 2):

$$\bar{d} = s_2 - s_1 \qquad (2)$$

is the width or the thickness of the barrier when image force is included.

The current-voltage characteristic of a tunnel junction in the Simmon's model can be determined as

$$J(A.cm^{-2}) = (6,2.10^{10}/\bar{d}^2)\{\phi_I \exp(-1.025.\bar{d}.\phi_I^{1/2}) - (\phi_I + V)\exp[-1.025.\bar{d}.(\phi_I + V)^{1/2}]\}, \qquad (3)$$

where

$$\phi_I = \phi_0 - (V/2s)(s_1 + s_2) - [5,75/K(s_2 - s_1)]\ln[(s_2(s - s_1)/s_1(s - s_2)] \qquad (4)$$

and

$$\left.\begin{aligned} s_1 &= 6/K\phi_0 \\ s_2 &= s[1 - 46/(3\phi_0 Ks + 20 - 2VKs)] + 6/K\phi_0 \end{aligned}\right\} \quad V < \phi_0$$

$$\left.\begin{aligned} s_1 &= 6/K\phi_0 \\ s_2 &= (\phi_0 Ks - 28)KV \end{aligned}\right\} \quad V > \phi_0 .$$

The first term in the equation (3) can be interpreted as a current density flowing from electrode 1 to electrode 2 and the second – a current density from electrode 2 to electrode 1, resulting in a net current density *J*. It is clearly seen that its value is dependent on the barrier thickness, mean barrier height and the applied voltage. Thus, the fitting procedure of the *I-V* experimental data to the mentioned above Simmon's model should help to determine $\bar{d}$ and $\bar{\phi}$.

3. Experimental

We have fabricated cross-shaped MTJ samples (three $Ni_{80}Fe_{20}/Al_2O_3/Co$ samples) by *rf* sputtering technique and measured their *I-V* characteristics. Figure 3 present the experimental current-voltage curve for the $Ni_{80}Fe_{20}/Al_2O_3/Co$ (1.7 nm) sample. It is in good agreement with another experimental current-voltage presented in Ref. [8].

The barrier parameters were then extracted by fitting the experimental curve to the Simmon [6] and Brinkman [9] models. It is shown that the annealing process (60 minutes at 300^0C for $Ni_{80}Fe_{20}/Al_2O_3/Co$ (1.7 nm) sample, for example) strongly affects the barrier parameters, both the thickness and height, and in this way, on the conduction mechanism in these MTJ samples.

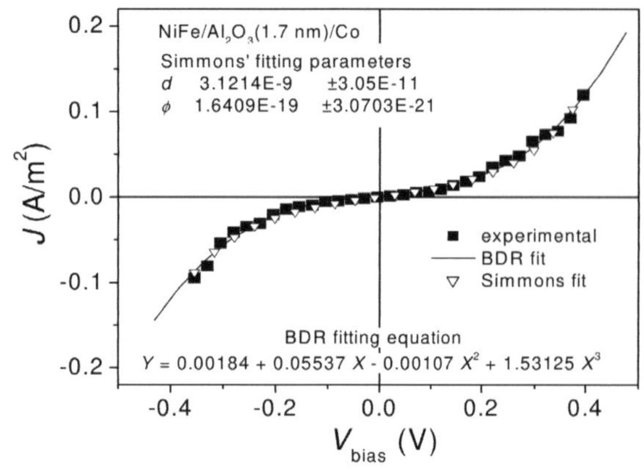

Figure 3. Typical experimental *I-V* curve for $Ni_{80}Fe_{20}/Al_2O_3/Co$ (1.7 nm) sample measured at room temperature.

4. Results and Discussions

4.1. Barrier parameters

By using equations (1) and (3) and coupling them with experimental curve of figure 3, some parameters concerning the barrier in the insulating film have been obtained and summarized in table 1. The reasonable values of these parameters are proof of the fact that tunnelling dominates the conduction in MTJ samples. The thickness takes a value which is comparable with the one calculated from the calibrated curve (not shown here). Usually the thickness found with this approach is different from the structural one (TEM image, for example). One of the reasons for it is caused by a limit of field view which does not allow an accurate determination of the averaged thickness over the sample.

The fitting to the experiment has given a higher value for the height of the rectangular potential when the image force is included, that is a good approach because the effect of the image force is to enhance the current of electrons through the barrier by reducing its width which can be clearly seen from the Table 1 (from 31 to 28,51 Å). From now on, all theoretical calculations take the value of 9.8 for the dielectric constant K of the Al_2O_3 insulating film of the MTJ sample.

Table 1. Barrier parameters extracted by fitting procedure for the rectangular potential without and with image force for $Ni_{80}Fe_{20}/Al_2O_3/Co$ sample.

	Rectangular potential without image force in the Simmon's model (1)	Rectangular potential with image force in the Simmon's model (2)
ϕ_0 (eV)	1	1.26
$\bar{\phi}$ (eV)	$\bar{\phi} = \phi_0$	1.14
s (Å)	31	31
s_1 (Å)	0	0.48
s_2 (Å)	$s_2 = s$	29.75
$\bar{d} = s_2 - s_1$ (Å)	$\bar{d} = s$	28.51

4.2. Current voltage characteristic I-V

The current-voltage of the sample is calculated by substituting to the equation (3) of the Simmon's model the values of the mean thickness and height of the barrier potential obtained when image force is included (data of column (2) in table 1). It is presented in figure 4.

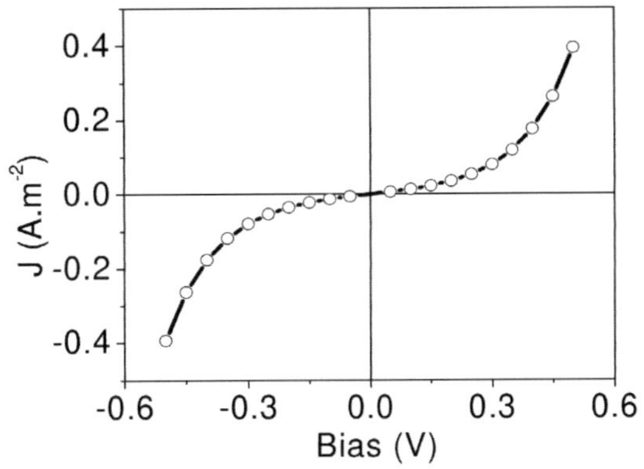

Figure 4. Theoretical *I-V* characteristic of $Ni_{80}Fe_{20}/Al_2O_3/Co$ sample.

4.3. Dependence of tunnel resistivity on applied bias

Tunnel resistivity ($R = V / J\ (A.cm^{-2})$) is calculated from equation (3) including the image force and is compared with the one case without image force. They are depicted in figure 5. It can be observed that, for given s , $\bar{\phi}$ and V , the tunnel resistivity is lower for the practical barrier, as anticipated, although the general shapes of the curves are preserved. Moreover, the difference of the resistivity between two cases becomes more apparent, because of the influence of the image force which is greater for small barriers. If the voltage applied to the junction is great enough, the barrier is lowered below the Fermi level of the negatively biased electrode, and then the current is not impeded in the insulator.

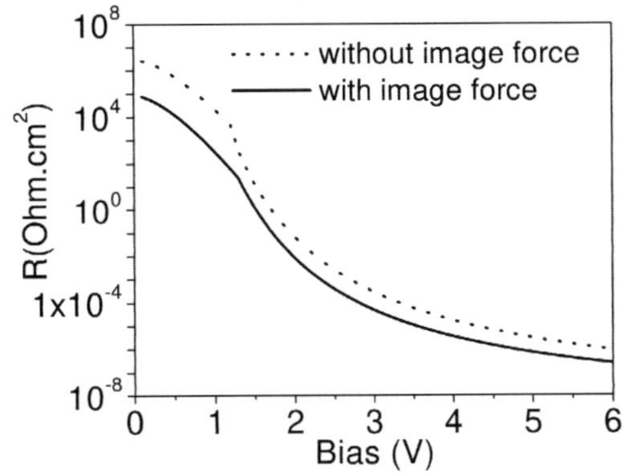

Figure 5. Theoretical *R-V* characteristic of a tunnel junction of the $Ni_{80}Fe_{20}/Al_2O_3/Co$ sample.

4.4. Influence of the dielectric constant on the electrical characteristics of the junction

Inspired by the idea of [8] who studied experimentally the temperature dependence of tunnel resistance and TMR ratio of a MTJ (the ratio TMR is defined ad resistivity relative to the value at saturated magnetic field), we carried out a study of this dependence in our sample through the dependence of the dielectric constant K of the insulating film in equation (3). The theoretical results of the tunnel resistivity R as a function of the film thickness and applied bias are illustrated respectively in figures 6 and 7.

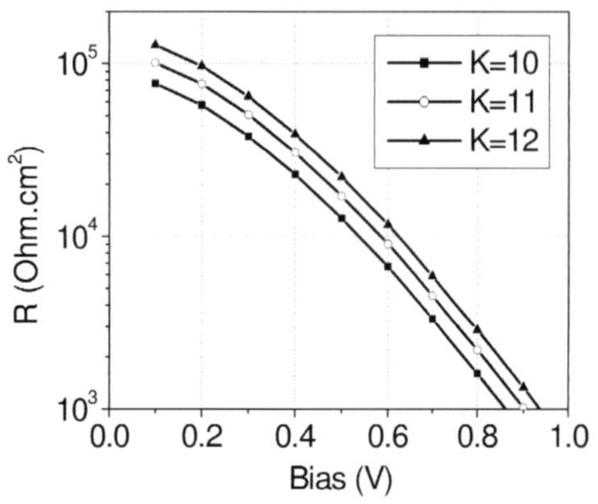

Figure 6. Theoretical *R-V* for different values of K.

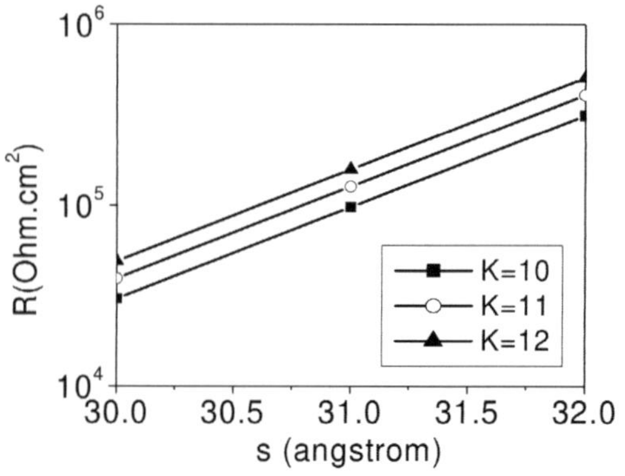

Figure 7. Theoretical *R-s* for different values of K.

It is shown from these figures that the smaller the value of K , the lower the tunnel resistivity. As we have known that the dielectric constant of most materials changes when temperature changes, the electrical characteristics of the junction of MTJs depend therefore on temperature through dielectric constant. However, the law of the tunnel resistance's dependence on temperature must need more study, because not only dielectric constant but also other parameters of the junction, for example film thickness [10], vary with temperature.

5. Conclusion

A Simmon's model based approach, including image force, to practical tunneling image force in the magnetic tunnel junction $Ni_{80}Fe_{20}/Al_2O_3/Co$ has been presented. The fitting procedure of the theory to the experiment allows determining barrier parameters such as effective thickness and mean height. The dependence of electrical characteristics of the tunnel on the dielectric constant and temperature has also been investigated. The derived results for the sample have shown useful to qualify the MTJs.

References

[1] Tuan N A, Minh P L and Dung T T 2006 *Communications in Phys.* **16** 7
[2] Jiang X and Parkin S 2006 *Concepts in Spin Electronics,* ed. Maekawa S. (New York: Oxford Univ. Press) chapter 6 pp 239
[3] Zhang J and White R M 1998 *J. Appl. Phys.* **83** 6512
[4] Tsymbal E V , Mryasov O N and Claire P R Le 2003 *J. Phys.: Condens. Matter* **15** R109
[5] Echenique P M and Pendry J B 1978 *J. Phys. C: Solid State Phys.* **11** 2065
[6] Simmons J G, Mryasov O N and Claire P R Le 1963 *J. Appl. Phys.* **34** 1793
[7] Fauster T, Weinelt M and Höfer U 2007 *Progr. Surf. Sci.* **82** 224
[8] Maekawa S and Shinjo T 2002 *Spin Dependent transport in Magnetic Nanostructures* (CRS Press)
[9] Brinkman W F, Dynes R C, Rowell J M 1970 *J. Appl. Phys.* **41** 1915
[10] Hyun J G, Lee S, Cho S-D and Paik K-K 2005 *Electronic Components and Technology Conference* (IEEE 2005) p 1241

Acknowledgments

This work was supported by the Grant-in-Aid for National Fundamental Research Foundation in Natural Science from the Ministry of Science and Technology of Vietnam.

Effect of strontium doping level on electrical transport and magnetic properties of $La_{1-x}Sr_xMnO_3$ perovskite nanoparticles

Phan Van Cuong[1] **and Do-Hyung Kim**[2]

[1] Department of Physics, Nha Trang University
2 Nguyen Dinh Chieu, Nha Trang, Vietnam
[2] Nano Applied Physics Laboratory (NAPL), Department of Physics
Kyungpook National University, Daegu 702-701, Republic of Korea

E-mail: cuongpv@cb.ntu.edu.vn

Abstract. $La_{1-x}Sr_xMnO_3$ (x = 0.1, 0.2, and 0.3) nanoparticles (LSMO NPs) were successfully synthesized by a simple sonochemical-assisted method. The prepared samples were investigated by X-ray diffraction (XRD), scanning electron microscopy (SEM), and transmission electron microscopy (TEM). The compositions of samples were determined by energy-dispersive spectroscopy (EDS) analysis. The temperature dependence of the resistivity of samples was measured using four-probe technique, and the magnetic measurement was performed in a Magnetic Property Measurement System (MPMS XL 7.0). SEM, TEM, and XRD show that LSMO NPs are 40~80 nm in diameter with a very clean surface, and exhibit single phase perovskite structure. The ferromagnetic transition temperature (T_c) and metal-insulator transition temperature (T_p) of the NPs increase with the doping level x. The maximum values of T_c (377 K) and T_p (264 K) occur at x = 0.3. In this work, the effect of strontium doping level on electrical transport and magnetic properties was systematically studied.

Keywords: Lanthanum compounds, nanoparticle, sonochemical, LSMO.

1. Introduction

The perovskite manganites $La_{1-x}A_xMnO_3$ (A = Sr, Ba, Ca, Pb) have attracted scientific investigations for many decades [1, 2]. These materials have novel physical properties and potential applications, such as magnetic-field sensors [3, 4], hard disk read heads [5], fuel cells [6, 7], infrared devices [8], spintronics, and micro-wave active components, etc. The doped perovskite manganites of $La_{1-x}A_xMnO_3$ have attracted much attention since the discovery of the colossal magnetoresistance effect and their intriguing physics [9]. Furthermore, the nanostructure material has been predicted to provide some new features in magnetic properties in comparison with bulk materials [10, 11]. The nano-sized particles may lead to some interesting magnetic and magnetoresistance properties than the bulk materials [12]. There have been several methods to prepare nano-sized manganites, such as the versatile chemical technique [13], anodized alumina oxide (AAO) template [14], hydrothermal [15-17], homogenous co-precipitation [18], etc. These methods have some disadvantages, for example, complex operating procedure [14], long reaction and treatment time [15-17], etc. The structure, electronic, and magnetic properties of $La_{1-x}A_xMnO_3$ perovskite usually depend on the doping level [19, 20] or the preparation routes [21-25].

© 2009 IOP Publishing Ltd

Journal of Physics: Conference Series **187** (2009) 012090 doi:10.1088/1742-6596/187/1/012090

In this work, using a sonochemical-assisted method, single crystalline phase $La_{1-x}Sr_xMnO_3$ (x = 0.1, 0.2, and 0.3) nanoparticles were prepared. Compared to the above methods, the sonochemical-assisted method is a preferable alternative due to very short reaction time, simple synthetic apparatus and simple operating procedure. This method also allows to synthesis of LSMO NPs in large scale up to several tenth grams.

2. Experimental

The synthetic procedure was as follows. High-purity $KMnO_4$, $MnCl_2.4H_2O$, $La(NO_3)_3.6H_2O$, and $Sr(NO_3)_2$ were used as starting materials and KOH served as mineralizer. All chemicals were purchased from Sigma-Aldrich (USA) without further purification. All chemicals were dissolved in de-ionized water by stirring at room temperature. Finally, KOH was added and stirred vigorously to adjust the alkalinity. The sonication of the mixed solution was carried out for 5 minutes in air using an ultrasonic probe with a 1/2 inch titanium tip (Branson Digital Sonifier, Model 450). The probe tip was kept 1 cm from the top of the solution surface. After 5 minutes of sonicating, the solution was cooled down, centrifuged, and washed several times with de-ionized water. Then the samples were dried in a vacuum oven at 70°C for 12 h and a black powder was finally obtained. Subsequently this powder was pressed in the same pressure to make the pellets with the size of 10 mm × 4 mm × 1 mm. Then the pellets of $La_{1-x}Sr_xMnO_3$ x = 0.1, 0.2, and 0.3 (NPs) were finally annealed at 950°C for 10 h.

The prepared samples were investigated by X-ray diffraction (XRD, Rigaku, Japan), field emission scanning electron microscopy (FESEM, Hitachi S-4300, Japan), high-resolution transmission electron microscopy (HRTEM), and selected-area electron diffraction (SAED, Philips CM 200 STEM). The chemical compositions of samples were determined by energy-dispersive spectroscopy (EDS) analysis. The temperature dependence of the resistivity of samples was measured by using four-probe technique, and the magnetic measurement was performed by using Magnetic Property Measurement System (Quantum Design, MPMS XL 7.0).

3. Results and discussion

LSMO NPs were synthesized by this method should be due to the generation of efficient droplet size induced by ultrasound. Ultrasound, itself, generates sufficient and strong mechanical forces which can easily and evenly disperse the aqueous solution containing all chemical reagents [26]. In sonochemistry, the three main involved events are the creation, growth, and collapse of bubbles that are formed in the liquid. The bubbles will be collapsed when their sizes reach the maximum value. According to the hot-spot mechanism, this implosive collapse can raise the local temperature up to 5000°C, the pressure up to 500 atm. and lifetimes of a few microseconds [27]. In alkaline environment, strong mechanical forces together with high temperature in a short time could be the mains reason for assisting the formation of LSMO NPs. It is worth to mention that micro-sized LSMO particles are usually obtained from the synthesis using solution-annealing process or solid sintering method. In addition, the nano-sized LSMO samples could not be obtained without sonochemical assistance in our previous experiments. Therefore, sonochemical-treatment plays an important role to assist the pre-formation of nano-sized particles. After sonicating and drying, nanoparticles could be obtained. The as-synthesized samples showed paramagnetic properties and the crystallinities were poor. Annealing step is very important one to enhance the crystallinity and to form desired properties of LSMO NPs. This method allows us to synthesize a large amount of LSMO NPs up to several tenth grams with the very simple equipment and process.

The chemical composition of all samples is examined by energy-dispersive spectroscopic. The results indicate that La:Sr:Mn molar ratios of the samples (x = 0.1, 0.2, and 0.3) are 0.89:0.11:1.00, 0.80:0.19:1.00, and 0.72:0.29:1.0, respectively. These values are consistent with nominal compositions within error of 5%.

XRD patterns of the annealed $La_{1-x}Sr_xMnO_3$ samples (x = 0.1, 0.2 and 0.3) and as-synthesis $La_{0.7}Sr_{0.3}MnO_3$ sample are shown in figure 1. Inset shows Rietveld refinement pattern for the annealed $La_{0.7}Sr_{0.3}MnO_3$ sample as an typical sample. The XRD patterns of annealed samples show that all diffraction peaks are narrower in width and strongly higher in intensity than those of the as-synthesized samples. Whole pattern fitting refinements of the XRD patterns of all three samples were performed using the program FULLPROF, based on the Rietveld method. We obtained the best fit for the rhombohedral (space group of $R\bar{3}c$) structure of these three samples. These results are nearly consistent with those reported in reference [18].

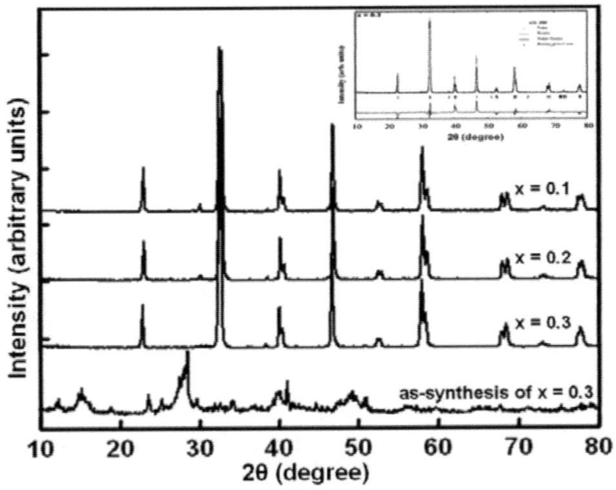

Figure 1. XRD patterns of the sintered $La_{1-x}Sr_xMnO_3$ samples (x = 0.1, 0.2, and 0.3) and as-synthesis $La_{0.7}Sr_{0.3}MnO_3$ sample. Inset: Rietveld refinement pattern of the sintered $La_{0.7}Sr_{0.3}MnO_3$.

Figure 2 presents the SEM images for typical $La_{1-x}Sr_xMnO_3$ NPs with x = 0.3. The LSMO NPs exhibit a sphere-like shape and their sizes are estimated to be within 40 and 80 nm.

Figure 2. SEM image of $La_{0.7}Sr_{0.3}MnO_3$ nanoparticles, the inset is that with the higher magnification.

Figure 3 shows TEM and HRTEM images of $La_{0.8}Sr_{0.2}MnO_3$ NPs. The images reveal that the surface of LSMO NPs is very clean without any sheathed amorphous phase. The inset shows the corresponding SAED pattern with $[1\bar{1}0]$ zone-axis and it indicates that LSMO NPs are single crystalline phase to be. The HRTEM image exhibits a lattice spacing of NPs of about 0.38 nm which corresponds to (100) plane. The ring patterns in the SAED were originated by the copper grid.

Figure 4 shows the dependence of the resistivity ρ on temperature measured in the temperature range of 5 - 300 K at zero field for all three mentioned samples. The inset presents the dependence of T_p, ρ_p on doping level (ρ_p is resistivity at T_p). All samples show the metal-insulator transition. The transition temperature T_p, increases in conjunction with the increasing of the doping level, and the difference of T_p (99 K) between sample x = 0.1 and sample x = 0.3 is clearly observed. The magnitude of ρ decreases with increasing of the doping level. The insulator-metal transition temperatures T_p of our samples are 165, 248, and 264 K for x = 0.1, 0.2, and 0.3, respectively. These T_p values are lower than that of other reports for nano-sized LSMO due to lower annealing temperatures [13, 19], but higher than that of bulk LSMO [28].

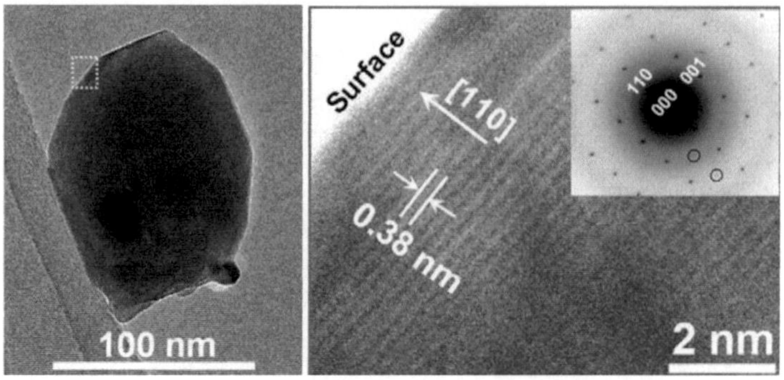

Figure 3. Typical TEM (left) and HRTEM (right) images of $La_{0.8}Sr_{0.2}MnO_3$ nanoparticles. The inset shows the corresponding SAED pattern.

The measurement of magnetization M as a function of temperature carried out with a magnetic field of 0.5 T in the temperature range of 25 - 400 K for $La_{1-x}Sr_xMnO_3$ NPs (figure 5). The inset presents the dependence of T_c on the doping level. Our results clearly show that all samples undergo a ferromagnetic transition. This transition temperature T_c increases with the doping level, thus, the T_c values of samples (x = 0.1, 0.2, and 0.3) are found to be 321, 349, and 377 K, respectively. These values are higher than those of previous reports for single phase LSMO NPs [13, 18, 19]. It is very interesting that with the increasing of the doping level, the ferromagnetic transition temperature (T_c) is almost increased linearly. The values of magnetization M at 25 K of samples (x = 0.1, 0.2, and 0.3) are 57, 69, and 70 emu/g, respectively.

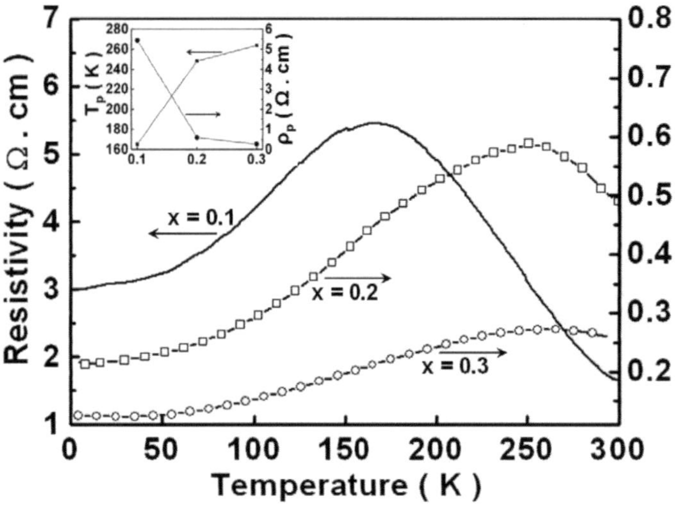

Figure 4. Electrical resistivity vs temperature for samples (x = 0.1, 0.2, and 0.3).
The inset: The dependence of T_p and ρ_p depend on doping level (ρ_p is resistivity at T_p)

The effect of the doping level on the LSMO magnetization was already studied in some literatures [18-20]. They found that the magnetization increases with the doping level. For our result, however, for samples x = 0.2 and 0.3, the magnetization values are almost the same. It can be explained that the chemical reaction may occur incompletely during synthesis process to provide a non-stoichiometric compound even the chemical analysis result for the $La_{0.7}Sr_{0.3}MnO_3$ shows the La:Sr:Mn of 0.72:0.29:1.0.

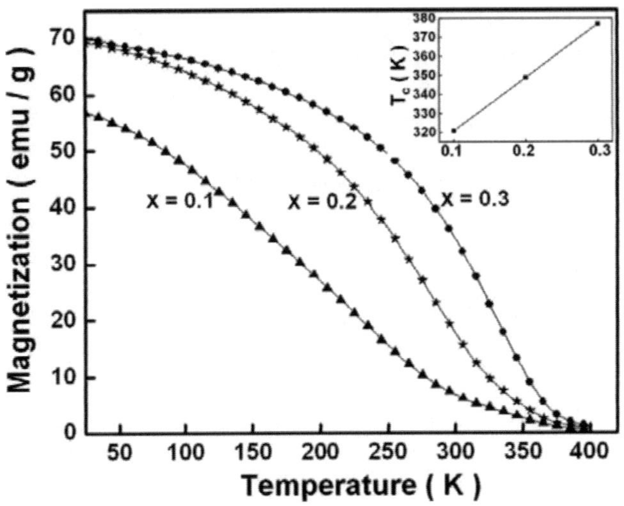

Figure 5. Magnetization M as a function of temperature measured at 0.5 T magnetic field for the three samples (x = 0.1, 0.2, and 0.3). The inset: The dependence of T_c on the doping level.

4. Conclusion

In summary, we have successfully synthesized single crystalline phase $La_{1-x}Sr_xMnO_3$ (x = 0.1, 0.2, and 0.3) nanoparticles by a simple sonochemical-assisted method with a significantly short reaction time. XRD pattern analysis shows that the samples have a perovskite structure of $R\bar{3}c$ symmetry. The results of SEM, TEM, HRTEM, and SAED measurements of studied compounds indicate that the particle size ranging from 40 to 80 nm, and the single crystalline phase was obtained for each with clean surface without any sheathed amorphous phase. The magnetic and transport properties of LSMO NPs are also investigated. The T_p and T_c increases systematically in proportion to the strontium doping level in $La_{1-x}Sr_xMnO_3$ nanoparticles and the maximum values of T_p and T_c occur at $x = 0.3$. In brief, it is important to adjust the doping level and synthesis route to get nano-sized particles and properties for application of these materials.

References

[1] Wollan E O and Koehler W C 1955 *Phys. Rev.* **100** 545
[2] Zener C 1951 *Phys. Rev.* **82** 403
[3] Jin S, McCormack M, Tiefel T H and Ramesh R 1994 *AIP* 6929
[4] Ll B, Enrich R, Mora J, Calleja A, Fontcuberta J and Obradors X 1996 *Appl. Phys. Lett.* **69** 1486
[5] Derbyshire K and Korczynski E 1995 *Solid State Technol.* **38** 57
[6] Lussier A, et al. 2008 *Thin Solid Films* **516** 880
[7] Miyazaki K, Sugimura N, Matsuoka K, Iriyama Y, Abe T, Matsuoka M and Ogumi Z 2008 *J. Power Sources* **178** 683
[8] Alvydas L, Khartsev S I and Alex G 2000 *Appl. Phys. Lett.* **77** 756
[9] Coey J M D, Viret M, Von M and Aacute R, S. 1999 *Advances in Physics* **48** 167
[10] Wang G, Wang Z D and Zhang L D 2005 *Mater. Sci. Eng. B* **116** 183
[11] Yamamoto T A, Tanaka M, Misaka Y, Nakagawa T, Nakayama T, Niihara K and Numazawa T 2002 *Scripta Mater.* **46** 89
[12] Yi T, Gao S, Qi X, Zhu Y, Cheng F, Ma B, Huang Y, Liao C and Yan C 2000 *J. Phys. Chem. Solids* **61** 1407
[13] Sahu D R, Roul B K, Pramanik P and Huang J-L 2005 *Physica B: Condensed Matter* **369** 209
[14] Chen F, Liu H W, Wang K F, Yu H, Dong S, Chen X Y, Jiang X P, Ren Z F and Liu J-M 2005 *J. Phys.: Condens. Matter.* **17** L467
[15] Zhang T, Jin C G, Qian T, Lu X L, Bai J M and Li X G 2004 *J. Mater. Chem.* **14** 2787
[16] Zhu D, Zhu H and Zhang Y H 2002 *J. Phys.: Condens. Matter* **14** L519

[17] Zhua D, Zhu H and Zhang Y 2002 *Appl. Phys. Lett.* **80** 1634
[18] Marques R F C, Jafelicci M, Paiva-Santos C O, Jardim R F, Souza J A, Varanda L C and Godoi R H M 2002 *IEEE T MAGN* **38** 2892
[19] Urushibara A, Moritomo Y, Arima T, Asamitsu A, Kido G and Tokura Y 1995 *Phys. Rev. B* **51** 14103
[20] Zhang N, Yang W, Ding W, Xing D and Du Y 1999 *Solid State Commun.* **109** 537
[21] Dutta A, Gayathri N and Ranganathan R 2003 *Phys. Rev. B* **68** 054432
[22] Fu Y 2000 *Appl. Phys. Lett.* **77** 118
[23] Lei L W, Fu Z Y and Zhang J Y 2006 *Mater. Lett.* **60** 970
[24] Mahesh R, Mahendiran R, Raychaudhuri A K and Rao C N R 1996 *Appl. Phys. Lett.* **68** 2291
[25] Rao C N R, Mahesh R, Raychaudhuri A K and Mahendiran R 1998 *J. Phys. Chem. Solids* **59** 487
[26] Sivakumar M, Towata A, Yasui K, Tuziuti T and Iida Y 2006 *Curr. Appl. Phys.* **6** 591
[27] Suslick K S 1990 *Science* **247** 1439
[28] Huang Y-H, Xu Z-G, Yan C-H, Wang Z-M, Zhu T, Liao C-S, Gao S and Xu G-X 2000 *Solid State Commun.* **114** 43

Acknowledgment
This work was supported by KRF grant (No. KRF-2006-C00536).

AUTHOR INDEX

AUTHOR INDEX

AUTHOR INDEX